FUNDAMENTALS
OF
ENGINEERING MATHEMATICS

VOLUME II

[For the Students of B.E./B.Tech./B. Arch. II Semester Uttarakhand Technical University, Dehradun]

AS PER NEW SYLLABUS 2010-2011

H.K. DASS
M.Sc.
Diploma in Specialist Studies (Maths.)
University of Hull
(England)

Dr. RAMA VERMA
M.Sc. (Gold Medalist), Ph.D.
HOD Mathematics
Mata Sundri College
Delhi University

S Chand And Company Limited
(AN ISO 9001: 2008 COMPANY)
RAM NAGAR, NEW DELHI-110055

S Chand And Company Limited

(An ISO 9001:2008 Company)

Head Office: 7361, RAM NAGAR, NEW DELHI - 110 055
Phone: 23672080-81-82, 9899107446, 9911310888 Fax: 91-11-23677446
www.schandpublishing.com; e-mail: helpdesk@schandpublishing.com

Branches

Ahmedabad	:	Ph: 27541965, 27542369, ahmedabad@schandpublishing.com
Bengaluru	:	Ph: 22268048, 22354008, bangalore@schandpublishing.com
Bhopal	:	Ph: 4274723, 4209587, bhopal@schandpublishing.com
Chandigarh	:	Ph: 2625356, 2625546, chandigarh@schandpublishing.com
Chennai	:	Ph: 28410027, 28410058, chennai@schandpublishing.com
Coimbatore	:	Ph: 2323620, 4217136, coimbatore@schandpublishing.com (Marketing Office)
Cuttack	:	Ph: 2332580, 2332581, cuttack@schandpublishing.com
Dehradun	:	Ph: 2711101, 2710861, dehradun@schandpublishing.com
Guwahati	:	Ph: 2738811, 2735640, guwahati@schandpublishing.com
Hyderabad	:	Ph: 27550194, 27550195, hyderabad@schandpublishing.com
Jaipur	:	Ph: 2219175, 2219176, jaipur@schandpublishing.com
Jalandhar	:	Ph: 2401630, 5000630, jalandhar@schandpublishing.com
Kochi	:	Ph: 2378740, 2378207-08, cochin@schandpublishing.com
Kolkata	:	Ph: 22367459, 22373914, kolkata@schandpublishing.com
Lucknow	:	Ph: 4026791, 4065646, lucknow@schandpublishing.com
Mumbai	:	Ph: 22690881, 22610885, mumbai@schandpublishing.com
Nagpur	:	Ph: 6451311, 2720523, 2777666, nagpur@schandpublishing.com
Patna	:	Ph: 2300489, 2302100, patna@schandpublishing.com
Pune	:	Ph: 64017298, pune@schandpublishing.com
Raipur	:	Ph: 2443142, raipur@schandpublishing.com (Marketing Office)
Ranchi	:	Ph: 2361178, ranchi@schandpublishing.com
Siliguri	:	Ph: 2520750, siliguri@schandpublishing.com (Marketing Office)
Visakhapatnam	:	Ph: 2782609, visakhapatnam@schandpublishing.com (Marketing Office)

First Edition 2008
Subsequent Editions and Reprints 2009, 2010, 2011, 2012, 2013 (Twice), 2014, 2015
Reprint 2017

ISBN : 978-81-219-2942-4 **Code :** 1010D 356

PRINTED IN INDIA

By Vikas Publishing House Pvt. Ltd., Plot 20/4, Site-IV, Industrial Area Sahibabad, Ghaziabad-201010 and published by S Chand And Company Limited, 7361, Ram Nagar, New Delhi -110 055.

PREFACE TO THE SIXTH REVISED EDITION

The textbook is revised thoroughly and the misprints are removed to make the book error free and more user friendly.

Solutions of Question Papers of June 2010, Jan. 2011 and 2012 are added in the respective chapters of the textbook.

I will feel amply rewarded if suggestions and constructive feedback is provided by the students and teachers, which will enable me to make book more user friendly.

We are thankful to the Management Team and the Editorial Department of S. Chand & Company Pvt. Ltd. for all help and support in the publication of this book.

For any clarification, we may be contacted on the following address:

H.K. DASS

D-1/87, Janakpuri, New Delhi-110058
Tel. 011-28525078, 32985078
Mob. 09350055078
hk_dass@yahoo.com

PREFACE TO THE FIRST EDITION 2006-07

This book is primarily written according to the unified syllabus, 2006 of Mathematics of first semester of all Engineering Colleges affiliated to Uttarakhand University, Dehradun and other States of India. This book also covers the B.Tech./B.E./B.Arch. First Year Course of other Indian Engineering Colleges.

The subject matter is presented in a very systematic and logical manner. Every endeavour has been made to make the contents as simple and lucid as possible. Emphasis has been laid on making the concepts clear. Lot of pains and concentration on the part of the author has gone in solving the examples in the best possible way. In providing the solution of the problems, care has been taken not to miss even minor step so that the students can follow the subject even without the guidance of the teacher.

The book contains fairly large number of solved examples from question papers of examinations recently conducted by different universities and Engineering Colleges so that the students may not find any difficulty while answering these problems in their final examination. Latest question paper of First Semester of Uttarakhand Technical University, Dehradun for the year 2006 - 2007 has been fully solved in this book.

The author possesses very long and rich experience of teaching Mathematics to the engineering students of degree classes. and has first hand experience of the problems and difficulties that they generally face.

I take this opportunity to express my deep sense of gratitude to Mrs. Nirmala Gupta, Managing Director, and the other staff of S. Chand & Co. Ltd. for publishing this book in a short time.

I will feel amply rewarded. if suggestions and constructive feedback is provided by the students and teachers, which will enable us to make the book error-free and more user-friendly.

For any clarification, we may be contacted on the following address.

D-1/87, Janakpuri, New Delhi-110058 **H.K. DASS**
Tel. 011-28525078, 32985078, 28521776
Mob. 09350055078
hk dass@yahoo.com

Secular India Award - 98 *for National Integration and Communal Harmony*
By Prime Minister **Sh. Atal Behari Vajpayee 12-06-99**

UTTARAKHAND TECHNICAL UNIVERSITY
DEHRADUN

NEW SYLLABUS FOR 2010-11

MATHEMATICS – II

Unit - I : Differential Equations

Ordinary differential equations of first order, Exact differential equations, Linear differential equations of first order, Linear differential equations of nth order with constant coefficients, Complementary functions and particular integrals, Simultaneous linear differential equations, Solutions of second order differential equations by changing the dependent and independent variables, Method of variation of parameters.

Unit - II : Laplace Transform

Laplace transform, Existence theorem, Laplace transform of derivatives and integrals, Laplace transform, Laplace transform of periodic function, Unit step function, Convolution theorem, Applications to solve simple linear and simultaneous liner differential equations.

Unit - III : Infinite Series

Introduction, Sequences, Series: Convergence, Series of positive terms, Comparison tests, Integral tests, Comparison Test, D' Alembert ratio test, Raabe's test, Cauchy root test, Alternating series: Leibnitz rule, Power series, Uniform convergence, Weierstress's M-test, Properties of uniformly convergent series.

Unit - IV : Fourier Series and Partial Differential Equations

Periodic functions, Trigonometric series, Fourier series of periodic function, Euler's formula, Functions having arbitrary period, Change of intervals, Even and odd functions, Half range sine and cosine series.

Introduction of partial differential equations, Linear partial differential equations with constant coefficients of second order and their classifications, parabolic, elliptic and hyperbolic with illustrative examples.

Unit - V : Applications of Partial Differential Equations

Method of separation of variables for solving partial differential equations, One dimensional Wave equation, Laplace equation in two-dimensions, Heat conduction equations of one dimension and two dimension.

CONTENTS

UNIT – I

UNIT–II

UNIT–III

UNIT–IV

UNIT– V

UNIT–I

1 CHAPTER

ORDINARY DIFFERENTIAL EQUATIONS OF FIRST ORDER

1.1 DEFINITION

An equation which involves differential co-efficient is called a differential equation.

For example,

1. $\frac{dy}{dx} = \frac{1+x^2}{1-y^2}$ **2.** $\frac{d^2y}{dx^2} + 2\frac{dy}{dx} - 8y = 0$ **3.** $\left[1+\left(\frac{dy}{dx}\right)^2\right]^{3/2} = k\frac{d^2y}{dx^2}$

1.2 ORDER AND DEGREE OF A DIFFERENTIAL EQUATION

The *order* of a differential equation is the order of the highest differential co-efficient present in the equation. Consider

1. $L\frac{d^2q}{dt^2} + R\frac{dq}{dt} + \frac{q}{c} = E \sin wt.$ **2.** $\cos x \frac{d^2y}{dx^2} + \sin x \left(\frac{dy}{dx}\right)^2 + 8y = \tan x.$

3. $\left[1+\left(\frac{dy}{dx}\right)^2\right]^3 = \left(\frac{d^2y}{dx^2}\right)^2$

The order of the above equations is 2.

The degree of a differential equation is the degree of the highest derivative after removing the radical sign and fraction.

The degree of the equation (1) and (2) is 1. The degree of the equation (3) is 2.

1.3 FORMATION OF DIFFERENTIAL EQUATIONS

The differential eqautions can be formed by differentiating the ordinary equation and eliminating the arbitray constants.

Example 1. *Form the differential equation by eliminating arbitary constants, in the following cases and also write down the order of the differential equations obtained.*

(*a*) $y = Ax + A^2$ (*b*) $y = A \cos x + B \sin x$ (*c*) $y^2 = Ax^2 + Bx + C.$

Solution. (*a*) $y = Ax + A^2$ ... (1)

On differentiation $\frac{dy}{dx} = A$

Put the value of A in (1), we get $y = x\frac{dy}{dx} + \left(\frac{dy}{dx}\right)^2$ **Ans.**

On eliminating one constant A we get the differential equation of order 1.

(*b*) $y = A \cos x + B \sin x$

On differentiation $\quad \dfrac{dy}{dx} = -A \sin x + B \cos x$

Again differentiating

$$\frac{d^2y}{dx^2} = -A \cos x - B \sin x \quad \Rightarrow \quad \frac{d^2y}{dx^2} = -(A \cos x + B \sin x)$$

$$\Rightarrow \quad \frac{d^2y}{dx^2} = -y \quad \Rightarrow \quad \frac{d^2y}{dx^2} + y = 0 \qquad \textbf{Ans.}$$

This is differential equation of order 2 obtained by eliminating two constants A and B.

(*c*) $y^2 = Ax^2 + Bx + C$.

On differentiation $2y\dfrac{dy}{dx} = 2Ax + B$.

Again differentiating $2y\dfrac{d^2y}{dx^2} + 2\left(\dfrac{dy}{dx}\right)^2 = 2A$

On differentiating again $y\dfrac{d^3y}{dx^3} + \dfrac{dy}{dx}\dfrac{d^2y}{dx^2} + 2\dfrac{dy}{dx}\dfrac{d^2y}{dx^2} = 0 \Rightarrow y\dfrac{d^3y}{dx^3} + 3\dfrac{dy}{dx}\dfrac{d^2y}{dx^2} = 0$ **Ans.**

This is the differential equation of order 3, obtained by eliminating three constants A, B, C.

Example 2. *Determine the differential equation whose set of independent solution is* $\{e^x, xe^x, x^2 e^x\}$. (*U.P., II Semester, Summer* 2002)

Solution. Here, we have

$$y = c_1 e^x + c_2 x e^x + c_3 x^2 e^x \qquad \text{... (1)}$$

Differentiating both sides, we get

$$y' = c_1 e^x + c_2 x e^x + c_2 e^x + c_3 x^2 e^x + c_3 2x e^x$$
$$= (c_1 e^x + c_2 x e^x + c_3 x^2 e^x) + c_2 e^x + c_3 2x e^x$$
$$= y + c_2 e^x + c_3 2x e^x \qquad \text{[Using (1)]}$$
$$\Rightarrow \quad y' - y = c_2 e^x + c_3 2x e^x \qquad \text{... (2)}$$

Again, differentiating both sides, we get

$$\Rightarrow \quad y'' - y' = c_2 e^x + 2c_3 e^x + 2x c_3 e^x$$
$$\Rightarrow \quad y'' - y' = (c_2 e^x + 2x c_3 e^x) + 2c_3 e^x$$
$$\Rightarrow \quad y'' - y' = y' - y + 2c_3 e^x \qquad \text{[Using (2)]}$$
$$\Rightarrow \quad y'' - 2y' + y = 2c_3 e^x \qquad \text{... (3)}$$

Finally, on differentiating both sides, we get

$\Rightarrow \quad y''' - 2y'' + y' = 2c_3 e^x$

$\Rightarrow \quad y''' - 2y'' + y' = y'' - 2y' + y$

$\Rightarrow \quad y''' - 2y'' - y'' + y' + 2y' - y = 0$ [Using (3)]

$\Rightarrow \quad y''' - 3y'' + 3y' - y = 0 \Rightarrow (D-1)^3 y = 0$ **Ans.**

Example 3. *By the elemination of the constants A and B obtain the differential equation of which* $xy = Ae^x + Be^{-x} + x^2$ *is the solution.* (*U.P.B. Pharma* (*C.O.*) 2005)

Solution. We have,

$$xy = A e^x + Be^{-x} + x^2 \qquad \text{... (1)}$$

On differentiating (1), we get

$$x\frac{dy}{dx} + y = Ae^x - Be^{-x} + 2x \qquad \text{...}(2)$$

Again differentiating (2), we get

$$x\frac{d^2y}{dx^2} + \frac{dy}{dx} + \frac{dy}{dx} = A\,e^x + Be^{-x} + 2 \qquad \text{...}(3)$$

From (1), we have $Ae^x + Be^{-x} = xy - x^2$ ... (4)

Putting the value of $Ae^x + Be^{-x}$ from (4) in (3), we have

$$x\frac{d^2y}{dx^2} + 2\frac{dy}{dx} = xy - x^2 + 2$$

$$x\frac{d^2y}{dx^2} + 2\frac{dy}{dx} - xy = 2 - x^2 \qquad \textbf{Ans.}$$

EXERCISE 1.1

1. Write the order and the degree of the following differential equations.

(*i*) $\frac{d^2y}{dx^2} + a^2x = 0$ (*b*) $\left[1 + \left(\frac{dy}{dx}\right)^2\right]^{\frac{3}{2}} = \frac{d^2y}{dx^2}$

(*c*) $x^2\left(\frac{d^2y}{dx^2}\right)^3 + y\left(\frac{dy}{dx}\right)^4 + y^4 = 0$ **Ans.** (*i*) 2, 1 (*b*) 2, 2 (*c*) 2, 3

2. Give an example of each of the following type of differential equations.

(*i*) A linear-differential equation of second order and first degree. **Ans.** Q. 1 (*i*)

(*b*) A non-linear differential equation of second order and second degree **Ans.** Q. 1 (*b*)

(*c*) Second order and third degree. **Ans.** Q. 1 (*c*)

3. Define with examples, order and degree of a differential equation.

4. Obtain the differential equation of which $y^2 = 4a\,(x + a)$ is a solution.

Ans. $y^2\left(\frac{dy}{dx}\right)^2 + 2\,xy\frac{dy}{dx} - y^2 = 0$

5. Obtain the differential equation associated with the primitive $Ax^2 + By^2 = 1$.

Ans. $xy\frac{d^2y}{dx^2} + x\left(\frac{dy}{dx}\right)^2 - y\frac{dy}{dx} = 0$

6. Find the differential equation corresponding to

$y = a\,e^{3x} + be^x$. **Ans.** $\frac{d^2y}{dx^2} - 4\frac{dy}{dx} + 3y = 0$

7. By the elimination of constants A and B, find the differential equation of which $y = e^x\,(A\cos x + B\sin x)$ is a solution. **Ans.** $\frac{d^2y}{dx^2} - 2\frac{dy}{dx} + 2y = 0$

8. Find the differential equation whose solution is $y = a\cos\,(x + 3)$. (*A.M.I.E., Summer* 2000)

Ans. $\frac{dy}{dx} = -\tan\,(x + 3)$

1.4 SOLUTION OF A DIFFERENTIAL EQUATION

In the example 1(*b*), $y = A\cos x + B\sin x$, on eliminating A and B we get the differential equation.

$$\frac{d^2y}{dx^2} + y = 0$$

$y = A\cos x + B\sin x$ is called the solution of the differential equation $\frac{d^2y}{dx^2} + y = 0$.

The order of the differential equation $\frac{d^2y}{dx^2} + y = 0$ is two and the solution $y = A\cos x + B\sin x$ contains two arbitrary constants. The number of arbitrary constants in the solution is equal to the order of the differential equation.

An equation containing dependent variable (y) and independent variable (x) and free from derivative, which satisfies the differential equation, is called the solution (primitive) of the differential equation.

1.5 DIFFERENTIAL EQUATIONS OF THE FIRST ORDER AND FIRST DEGREE

We will discuss the standard methods of solving the differential equations of the following types.

(*i*) Equations solvable by separation of the variables.

(*ii*) Homogeneous equations.

(*iii*) Linear equations of the first order.

(*iv*) Exact differential equations.

1.6 VARIABLE SEPARABLE

If a differential equation can be written in the form

$$f(y)\,dy = \phi(x)\,dx$$

We say that variables are separable, y on left hand side and x on right hand side.

We get the solution by integrating both sides.

Example 5. *Solve* :

(*i*) $(x+1)\frac{dy}{dx} = x(y^2+1)$ (*ii*) $(xy^2 + x)\,dx + (yx^2 + y)\,dy = 0$

Solution. (*i*) $(x+1)\frac{dy}{dx} = x(y^2+1) \quad \Rightarrow \quad (x+1)\,dy = x(y^2+1)\,dx$

$$\Rightarrow \quad \frac{dy}{y^2+1} = \frac{x\,dx}{x+1} \qquad \Rightarrow \quad \frac{1}{1+y^2}\,dy = \left(1 - \frac{1}{x+1}\right)dx$$

Integrating $\int \frac{1}{1+y^2}\,dy = \int dx - \int \frac{1}{x+1}\,dx$

$$\tan^{-1} y = x - \log(1+x) + C$$ **Ans.**

(*ii*) $(xy^2 + x)\,dx + (yx^2 + y)\,dy = 0$

$$\Rightarrow \quad (y^2+1)\,x\,dx = -(x^2+1)\,y\,dy \qquad \Rightarrow \quad \frac{y\,dy}{y^2+1} = -\frac{x}{x^2+1}\,dx$$

$$\Rightarrow \quad \frac{1}{2}\int\frac{2y}{y^2+1}\,dy = -\frac{1}{2}\int\frac{2x}{x^2+1}\,dx \quad \Rightarrow \frac{1}{2}\log(y^2+1) = -\frac{1}{2}\log(x^2+1)+\frac{1}{2}C$$

$$\Rightarrow \quad \log(y^2+1)+\log(x^2+1) = \log C'$$

$$(y^2+1)(x^2+1) = C' \qquad \textbf{Ans.}$$

Example 6. *Solve*: $\dfrac{dy}{dx} = \dfrac{x(2\log x+1)}{\sin y + y\cos y}$ *(U.P., II Sem. June 2008, U.P.B. Pharm* (*C.O.*) 2005)

Solution. We have,

$$\frac{dy}{dx} = \frac{x(2\log x+1)}{\sin y + y\cos y}$$

$\Rightarrow (\sin y + y\cos y)\,dy = \{x(2\log x+1)\}\,dx$

Integrating both the sides, we get

$$\int(\sin y + y\cos y)\,dy = \int\{x(2\log x+1)\}\,dx$$

$$-\cos y + y\sin y - \int(1)\cdot\sin y\,dy = 2\int\log x\cdot x\,dx + \int x\,dx$$

$$\Rightarrow \quad -\cos y + y\sin y + \cos y = 2\left[\log x\cdot\frac{x^2}{2} - \int\frac{1}{x}\cdot\frac{x^2}{2}\,dx\right] + \frac{x^2}{2} + C$$

$$\Rightarrow \quad y\sin y = 2\log x\cdot\frac{x^2}{2} - \int x\,dx + \frac{x^2}{2} + C$$

$$\Rightarrow \quad y\sin y = 2\log x\cdot\frac{x^2}{2} - \frac{x^2}{2} + \frac{x^2}{2} + C$$

$$\Rightarrow \quad y\sin y = x^2\log x + C \qquad \textbf{Ans.}$$

Example 7. *Solve the differential equation:* $\dfrac{x+y-a}{x+y-b}\dfrac{dy}{dx} = \dfrac{x+y+a}{x+y+b}$

Solution. On putting $x + y = z$, so that $1+\dfrac{dy}{dx} = \dfrac{dz}{dx}$

$$\frac{z-a}{z-b}\left(\frac{dz}{dx}-1\right) = \frac{z+a}{z+b} \quad \Rightarrow \quad \frac{dz}{dx}-1 = \frac{(z+a)(z-b)}{(z+b)(z-a)}$$

$$\Rightarrow \quad \frac{dz}{dx} = \frac{z^2+az-bz-ab}{z^2+bz-az-ab}+1 = \frac{z^2+az-bz-ab+z^2+bz-az-ab}{z^2+bz-az-ab}$$

$$\Rightarrow \quad \frac{dz}{dx} = \frac{2z^2-2ab}{z^2-az+bz-ab} \quad \Rightarrow \quad \frac{z^2-az+bz-ab}{z^2-ab}\,dz = 2\,dx$$

$$\left[1-\frac{(a-b)z}{z^2-ab}\right]dz = 2\,dx$$

On integrating, we get

$$z - \frac{a-b}{2}\log(z^2-ab) = 2x+\frac{C}{2} \Rightarrow 2z-(a-b)\log(z^2-ab) = 4x + C$$

$$\Rightarrow \quad 2(x+y)-(a-b)\log[(x+y)^2-ab] = 4x + C$$

$$\Rightarrow \quad 2(x-y)+(a-b)\log(x^2+y^2+2xy-ab)+C = 0 \qquad \textbf{Ans.}$$

Example 8. *Solve* $x\dfrac{dy}{dx} + \cot y = 0,$ *given* $y = \dfrac{\pi}{4}$, *where* $x = \sqrt{2}$

Solution. $x\dfrac{dy}{dx} + \cot y = 0$

$$\Rightarrow \quad x\,dy = -\cot y\,dx \qquad \Rightarrow \quad \tan y\,dy = -\frac{dx}{x}$$

$$\Rightarrow \quad \int \tan y\,dy = -\int \frac{dx}{x} \qquad \Rightarrow \quad \log \sec y = -\log x + \log A$$

$$\Rightarrow \quad \log \sec y + \log x = \log A \qquad \Rightarrow \quad \log \frac{x}{\cos y} = \log A \qquad \Rightarrow x = A \cos y \qquad \dots (1)$$

On putting $y = \dfrac{\pi}{4}$ and $x = \sqrt{2}$ in (1), we get $\sqrt{2} = A \cos \dfrac{\pi}{4} \Rightarrow \sqrt{2} = \dfrac{A}{\sqrt{2}} \Rightarrow A = 2$

Substituting the value of A in (1), we have $x = 2 \cos y$ **Ans.**

Example 9. *Solve the following*

$$(x+y)^2\left(x\frac{dy}{dx} + y\right) = xy\left(1 + \frac{dy}{dx}\right)$$

Solution. $(x+y)^2\left(x\dfrac{dy}{dx} + y\right) = xy\left(1 + \dfrac{dy}{dx}\right) \Rightarrow \dfrac{x\dfrac{dy}{dx} + y}{xy} = \dfrac{1 + \dfrac{dy}{dx}}{(x+y)^2}$

On integrating, we get $\log x\,y = -\dfrac{1}{x+y} + c$ **Ans.**

Example 10. *Solve* : $\cos(x+y)\,dy = dx$

Solution. $\cos(x+y)\,dy = dx \Rightarrow \dfrac{dy}{dx} = \sec(x+y)$

On putting $x + y = z$

so that $$1 + \frac{dy}{dx} = \frac{dz}{dx} \Rightarrow \frac{dy}{dx} = \frac{dz}{dx} - 1$$

$$\frac{dz}{dx} - 1 = \sec z \Rightarrow \frac{dz}{dx} = 1 + \sec z$$

$$\frac{dz}{1 + \sec z} = dx$$

$$\int \frac{\cos z}{\cos z + 1}\,dz = \int dx \quad \Rightarrow \quad \int \left[1 - \frac{1}{\cos z + 1}\right] dz = x + C$$

$$\int \left[1 - \frac{1}{2\cos^2 \frac{z}{2} - 1 + 1}\right] dz = x + C$$

$$\int \left(1 - \frac{1}{2}\sec^2 \frac{z}{2}\right) dz = x + C \quad \Rightarrow \quad z - \tan \frac{z}{2} = x + C$$

$$x + y - \tan \frac{x+y}{2} = x + C$$

$$y - \tan \frac{x+y}{2} = C$$ **Ans.**

Example 11. *Find the solution of the differential equation* $\frac{dy}{dx} - x\tan(y - x) = 1$

Solution. We have, $\frac{dy}{dx} - x\tan(y-x) = 1$ ...(1)

On putting $y - x = t$ so that $\frac{dy}{dx} - 1 = \frac{dt}{dx}$ in (1), we get

$$\left(1 + \frac{dt}{dx}\right) - x\tan t = 1, \quad \frac{dt}{dx} = x\tan t \quad \Rightarrow \quad \cot t\, dt = x\, dx$$

Integrating, we have $\log \sin t = \frac{x^2}{2} + C$

$$\log \sin(y - x) = \frac{x^2}{2} + C$$ **Ans.**

Example 12. *Solve the equation.*

$(2x^2 + 3y^2 - 7)\, x\, dx - (3x^2 + 2y^2 - 8)\, y\, dy = 0$ (*U.P. II Semester Summer* 2005)

Solution. We have,

$(2x^2 + 3y^2 - 7)\, x\, dx - (3x^2 + 2y^2 - 8)\, y\, dy = 0$ ... (1)

Re arranging (1), we get

$$\frac{x}{y}\frac{dx}{dy} = \frac{3x^2 + 2y^2 - 8}{2x^2 + 3y^2 - 7}$$

Applying componendo and dividendo rule, we get

$$\frac{x\,dx + y\,dy}{x\,dx - y\,dy} = \frac{5x^2 + 5y^2 - 15}{x^2 - y^2 - 1}$$

$$\Rightarrow \quad \frac{x\,dx + y\,dy}{x^2 + y^2 - 3} = 5\left(\frac{x\,dx - y\,dy}{x^2 - y^2 - 1}\right)$$

Multiplying by 2 both the sides, we get

$$\Rightarrow \quad \left(\frac{2x\,dx + 2y\,dy}{x^2 + y^2 - 3}\right) = 5\left(\frac{2x\,dx - 2y\,dy}{x^2 - y^2 - 1}\right)$$

Integrating both sides, we get

$$\log(x^2 + y^2 - 3) = 5\log(x^2 - y^2 - 1) + \log c$$

$$\Rightarrow \quad x^2 + y^2 - 3 = c\,(x^2 - y^2 - 1)^5$$

where c is arbitrary constant of integration. **Ans.**

EXERCISE 1.2

Solve the following differential equations :

1. $\frac{dx}{x} = \tan y.\, dy$ **Ans.** $x\cos y = C$

2. $\frac{dy}{dx} = \frac{\sqrt{1-y^2}}{\sqrt{1-x^2}}$ **Ans.** $\sin^{-1} y = \sin^{-1} x + C$

3. $y(1+x^2)^{1/2}\,dy + x\sqrt{1+y^2}\,dx = 0$ **Ans.** $\sqrt{1+y^2}+\sqrt{1+x^2}=C$

4. $\sec^2 x \tan y\, dx + \sec^2 y \tan x\, dy = 0$ **Ans.** $\tan x \tan y = C$

5. $(1+x^2)\,dy - xy\,dx = 0$ **Ans.** $y^2 = C(1+x^2)$

6. $(e^y+1)\cos x\,dx + e^y \sin x\,dy = 0$ **Ans.** $(e^y+1)\sin x = C$

7. $y - x\dfrac{dy}{dx} = a\left(y^2+\dfrac{dy}{dx}\right)$ **Ans.** $(ay-1)(x+a) = Cy$

8. $3e^x \tan y\,dx + (1-e^x)\sec^2 y\,dy = 0$ **Ans.** $(1-e^x)^3 = C\tan y$

9. $y\sec^2 x + (y+7)\tan x\dfrac{dy}{dx} = 0$ **Ans.** $y^7 \tan x = Ce^{-y}$

10. $(e^y+2)\sin x\,dx - e^y\cos x\,dy = 0$ **Ans.** $(e^y+2)\cos x = c$

11. $\dfrac{dy}{dx} = e^{x-y} + x^2 e^{-y}$ **Ans.** $e^y = e^x + \dfrac{x^3}{3} + C$

12. $(xy+x)\,dy - (xy+y)\,dx = 0$ **Ans.** $x = c\,y\,e^{y-x}$

13. $(x+1)\dfrac{dy}{dx} + 1 = 2e^{-y}$ **Ans.** $c(x+1)(2-e^y) = 1$

14. $\dfrac{dy}{dx} = 1 + \tan(y-x)$ [Put $y-x=z$] **Ans.** $\sin(y-x) = e^{x+c}$

15. $(4x+y)^2\dfrac{dx}{dy} = 1$ **Ans.** $\tan^{-1}\dfrac{4x+y}{2} = 2x+c$

16. $\dfrac{dy}{dx} = (4x+y+1)^2$ **Ans.** $\tan^{-1}\dfrac{4x+y+1}{2} = 2x+C$

17. $\dfrac{dy}{dx} = \dfrac{\left(\sin x + \dfrac{\log x}{x}\right)}{(\cos y - \sec^2 y)}$ **Ans.** $\sin y - \tan y = -\cos x + \dfrac{1}{2}(\log x)^2 + C$

1.7 HOMOGENEOUS DIFFERENTIAL EQUATIONS

A differential equation of the form

$$\frac{dy}{dx} = \frac{f(x,y)}{\phi(x,y)}$$

is called a homogeneous equation if each term of $f(x,y)$ and $\phi(x,y)$ is of the same degree *e.g.*,

$$\frac{dy}{dx} = \frac{3xy+y^2}{3x^2+xy}$$

In such cases we put $y = vx$ and $\dfrac{dy}{dx} = v + x\dfrac{dv}{dx}$

The reduced equation involves v and x only. This new differential equation can be solved by ***variables separable*** method.

Example 13. *Solve* : $x^2\,dy + y(x+y)\,dx = 0$

Solution. $x^2\,dy + y(x+y)\,dx = 0$

$$\frac{dy}{dx} = -\frac{xy+y^2}{x^2}$$

Put $y = vx$ then $\dfrac{dy}{dx} = v + x\dfrac{dv}{dx}$

On substitution, the given equation becomes

$$v + x\frac{dv}{dx} = -\frac{v\,x^2 + v^2\,x^2}{x^2} = -\frac{v + v^2}{1} \quad\Rightarrow\quad x\frac{dv}{dx} = -2v - v^2$$

Separating the variables, we get

$$\frac{dv}{v^2 + 2v} = \frac{-dx}{x} \Rightarrow \quad \frac{dv}{2v} - \frac{dv}{2\,(v+2)} = -\frac{dx}{x}$$

$$\Rightarrow \quad \frac{1}{2}\int\frac{dv}{v} - \frac{1}{2}\int\frac{dv}{v+2} = -\int\frac{dx}{x} \Rightarrow \frac{1}{2}\log v - \frac{1}{2}\log(v+2) = -\log x + C$$

$$\Rightarrow \quad \log v - \log(v+2) = -2\log x + 2C$$

$$\Rightarrow \quad \log\frac{v}{v+2} + \log x^2 = 2C$$

$$\Rightarrow \quad \log\frac{v\,x^2}{v+2} = \log a \quad\Rightarrow\quad \frac{v\,x^2}{v+2} = a \qquad \left(\text{Put } v = \frac{y}{x}\right)$$

$$\Rightarrow \quad \frac{\frac{y}{x}\cdot x^2}{\frac{y}{x} + 2} = a \Rightarrow \quad \frac{x^2\,y}{2x + y} = a$$

$$\Rightarrow \quad x^2y = a\,(2x + y)$$ **Ans.**

Example 14. *Solve the equation* :

$$\frac{dy}{dx} = \frac{y}{x} + x\sin\frac{y}{x}$$

Solution. We have, $\quad \dfrac{dy}{dx} = \dfrac{y}{x} + x\sin\dfrac{y}{x}$ ... (1)

Put $y = vx$ in (1) so that $\quad \dfrac{dy}{dx} = v + x\dfrac{dv}{dx}$

$$\Rightarrow \quad v + x\frac{dv}{dx} = x\sin v \quad\Rightarrow\quad x\frac{dv}{dx} = x\sin v \quad\Rightarrow\quad \frac{dv}{dx} = \sin v$$

$$\Rightarrow \quad \frac{dv}{\sin v} = dx \quad\Rightarrow\quad \int \text{cosec}\, v\, dv = \int dx$$

$$\log\tan\frac{v}{2} = x + C \quad\Rightarrow\quad \log\tan\frac{y}{2x} = x + C$$ **Ans.**

EXERCISE 1.3

Solve the following differential equations.

1. $(y^2 - xy)\,dx + x^2\,dy = 0.$ **Ans.** $\dfrac{x}{y} = \log x + c$
2. $(x^2 - y^2)\,dx + 2xy\,dy = 0.$ **Ans.** $x^2 + y^2 = ax$
3. $x\,(y - x)\dfrac{dy}{dx} = y\,(y + x).$ **Ans.** $\dfrac{y}{x} - \log xy = a$
4. $x\,(x - y)\,dy + y^2\,dx = 0$ *(U.P. B. Pharm (C.O.) 2005)* **Ans.** $y = x\log c\,y$
5. $\dfrac{dy}{dx} + \dfrac{x - 2\,y}{2\,x - y} = 0$ **Ans.** $y - x = C(x + y)^3$

6. $\frac{dy}{dx} = \tan\frac{y}{x} + \frac{y}{x}$ **Ans.** $\sin\frac{y}{x} = cx$

7. $\frac{dy}{dx} = \frac{3xy + y^2}{3x^2}$ **Ans.** $3x + y\log x + Cy = 0$

8. $\frac{dy}{dx} = \frac{x^2 - 2y^2}{2xy}$ **Ans.** $4y^2 - x^2 = C/x^2$

9. $(x^2 + y^2)\,dy = xy\,dx$ **Ans.** $-\frac{x^2}{2y^2} + \log y = C$

10. $x^2y\,dx - (x^3 + y^3)\,dy = 0$ **Ans.** $\frac{-x^3}{3y^3} + \log y = C$

11. $(y^2 + 2xy)\,dx + (2x^2 + 3xy)\,dy = 0$ (*AMIETE, Summer* 2004) **Ans.** $xy^2(x + y) = C$

12. $(2xy^2 - x^3)\,dy + (y^3 - 2yx^2)\,dx = 0$ **Ans.** $y^2(y^2 - x^2) = Cx^{-2}$

13. $(x^3 - 3xy^2)\,dx + (y^3 - 3x^2y)\,dy = 0,\ y(0) = 1$ **Ans.** $x^4 - 6x^2y^2 + y^4 = 1$

14. $2xy^2\,dy - (x^3 + 2y^3)\,dx = 0$ **Ans.** $2y^3 = 3x^3\log x + 3x^3 + C$

15. $x\sin\frac{y}{x}\,dy = \left(y\sin\frac{y}{x} - x\right)dx$ **Ans.** $\cos\frac{y}{x} = \log x + C$

16. $\left\{x\cos\frac{y}{x} + y\sin\frac{y}{x}\right\}y - \left\{y\sin\frac{y}{x} - x\cos\frac{y}{x}\right\}x\frac{dy}{dx} = 0$ **Ans.** $xy\cos\frac{y}{x} = a$

17. $\frac{dy}{dx} = \frac{y + \sqrt{x^2 + y^2}}{x}$ (*U.P., II Semester, June* 2008) **Ans.** $y + \sqrt{x^2 + y^2} = cx^2$

18. $x\frac{dy}{dx} = y(\log y - \log x + 1)$ (*AMIETE, Summer* 2004) **Ans.** $\log\frac{y}{x} = cx$

1.8 EQUATIONS REDUCIBLE TO HOMOGENEOUS FORM

The equation of the form

$$\frac{dy}{dx} = \frac{ax + by + c}{Ax + By + C}$$

can be reduced to the homogeneous form by the substitution

$$x = X + h,\quad y = Y + k \qquad (h, k \text{ being constants})$$

$$\therefore \quad \frac{dy}{dx} = \frac{dY}{dX}$$

The given differential equation reduces to

$$\frac{dY}{dX} = \frac{a(X + h) + b(Y + k) + c}{A(X + h) + B(Y + k) + C} = \frac{aX + bY + ah + bk + c}{AX + BY + Ah + Bk + C}$$

Choose h, k so that $ah + bk + c = 0$

$$Ah + Bk + C = 0$$

Then the given equation becomes homogeneous

$$\frac{dY}{dX} = \frac{aX + bY}{AX + BY}$$

Case of failure. If $\frac{a}{A} = \frac{b}{B}$ then the value of h, k will not be finite.

$$\frac{a}{A} = \frac{b}{B} = \frac{1}{m} \qquad \text{(say)}$$

$A = am,\ B = bm$

The given equation becomes $\frac{dy}{dx} = \frac{ax + by + c}{m(ax + by) + c}$

Now put $ax + by = z$ and apply the method of variables separable.

Example 15. *Solve* : $\frac{dy}{dx} = \frac{x + 2y - 3}{2x + y - 3}$

Solution. Put $x = X + h,\ y = Y + k.$

The given equation reduces to

$$\therefore \qquad \frac{dY}{dX} = \frac{(X + h) + 2(Y + k) - 3}{2(X + h) + (Y + k) - 3}$$

$$= \frac{X + 2Y + (h + 2k - 3)}{2X + Y + (2h + k - 3)} \qquad \ldots (1)$$

Now choose h and k so that $h + 2k - 3 = 0,\ 2h + k - 3 = 0$

Solving these equations, we get $h = k = 1$

$$\frac{dY}{dX} = \frac{X + 2Y}{2X + Y} \qquad \ldots (2)$$

Put $\qquad Y = vX$ so that $\frac{dY}{dX} = v + X\frac{dv}{dX}$

The equation (2) is transformed as

$$v + X\frac{dv}{dX} = \frac{X + 2vX}{2X + vX} = \frac{1 + 2v}{2 + v}$$

$$X\frac{dv}{dX} = \frac{1 + 2v}{2 + v} - v = \frac{1 + v^2}{2 + v} \quad \Rightarrow \quad \left(\frac{2 + v}{1 - v^2}\right) dv = \frac{dX}{X}$$

$$\Rightarrow \quad \frac{1}{2}\frac{1}{(1 + v)}\,dv + \frac{3}{2}\frac{1}{(1 - v)}\,dv = \frac{dX}{X} \qquad \text{(Partial fractions)}$$

On integrating, we have

$$\frac{1}{2}\log(1 + v) - \frac{3}{2}\log(1 - v) = \log X + \log C$$

$$\Rightarrow \quad \log\frac{1 + v}{(1 - v)^3} = \log C^2X^2 \quad \Rightarrow \quad \frac{1 + v}{(1 - v)^3} = C^2X^2$$

$$\frac{1 + \frac{Y}{X}}{\left(1 - \frac{Y}{X}\right)^3} = C^2X^2 \quad \Rightarrow \quad \frac{X + Y}{(X - Y)^3} = C^2 \text{ or } X + Y = C^2(X - Y)^3$$

Put $\qquad X = x - 1$ and $Y = y - 1$

$$x + y - 2 = a(x - y)^3 \qquad \textbf{Ans.}$$

Example 16. *Solve* : $(x + 2y)(dx - dy) = dx + dy$

Solution. $(x + 2y)(dx - dy) = dx + dy \Rightarrow (x + 2y - 1)\,dx - (x + 2y + 1)\,dy = 0$

$$\Rightarrow \quad \frac{dy}{dx} = \frac{x+2y-1}{x+2y+1}$$

Hence $\dfrac{a}{A} = \dfrac{b}{B}$ *i.e.,* $\dfrac{1}{1} = \dfrac{2}{2}$ (Case of failure)

Now put $x + 2y = z$ so that

$$1+2\frac{dy}{dx} = \frac{dz}{dx}$$

Equation (1) becomes

$$\frac{1}{2}\frac{dz}{dx} - \frac{1}{2} = \frac{z-1}{z+1} \Rightarrow \frac{dz}{dx} = 2\frac{(z-1)}{z+1} + 1 = \frac{3z-1}{z+1}$$

$$\Rightarrow \quad \frac{z+1}{3z-1}\,dz = dx \Rightarrow \left(\frac{1}{3} + \frac{4}{3}\frac{1}{3z-1}\right)dz = dx$$

On integrating, $\dfrac{z}{3} + \dfrac{4}{9}\log(3z-1) = x + C$

$$3z + 4\log(3z - 1) = 9x + 9C$$

$$\Rightarrow \quad 3(x + 2y) + 4\log(3x + 6y - 1) = 9x + 9C$$

$$3x - 3y + a = 2\log(3x + 6y - 1)$$ **Ans.**

EXERCISE 1.4

Solve the following differential equations :

1. $\dfrac{dy}{dx} = \dfrac{2x+9y-20}{6x+2y-10}$ **Ans.** $(2x - y)^2 = c\,(x + 2y - 5)$
2. $\dfrac{dy}{dx} = \dfrac{y-x+1}{y+x+5}$ **Ans.** $\log[(y + 3)^2 + (x + 2)^2] + 2\tan^{-1}\dfrac{y+3}{x+2} = a.$
3. $\dfrac{dy}{dx} = \dfrac{x-y-2}{x+y+6}$ **Ans.** $(y + 4)^2 + 2(x + 2)(y + 4) - (x + 2)^2 = a^2$
4. $\dfrac{dy}{dx} = \dfrac{y+x-2}{y-x-4}$ **Ans.** $-(y - 3)^2 + 2(x + 1)(y - 3) + (x + 1)^2 = a.$
5. $\dfrac{dy}{dx} = \dfrac{2x-5y+3}{2x+4y-6}$ **Ans.** $(x - 4y + 3)(2x + y - 3) = a$
6. $(x + y + 3)\,dy = (x + y - 3)\,dx$ **Ans.** $-x + y - 3\log(x + y) = a$
7. $(2x - y + 2)\,dx + (4x - 2y - 1)\,dy = 0$ **Ans.** $2(2x - y) - \log(2x - y) = 5x + c$
8. $(2x + y + 1)\,dx + (4x + 2y - 1)\,dy = 0$ **Ans.** $2(2x + y) + \log(2x + y - 1) = 3x + c$
9. $(x - y - 2)\,dx - (2x - 2y - 3)\,dy = 0$ **Ans.** $\log(x - y - 1) = x - 2y + c$
   (U. P. B Pharm (C.O.) 2005)

1.9 LINEAR DIFFERENTIAL EQUATIONS

A differential equation of the form

$$\frac{dy}{dx} + Py = Q \qquad ...(1)$$

is called a linear differential equation, where P and Q, are functions of x (but not of y) or constants.

In such case, multiply both sides of (1) by $e^{\int P dx}$

$$e^{\int P dx}\left(\frac{dy}{dx}+Py\right) = Q\cdot e^{\int P dx} \qquad \text{...(2)}$$

The left hand side of (2) is

$$\frac{d}{dx}\left[y\cdot e^{\int P dx}\right]$$

$$\therefore \quad \frac{d}{dx}\left[y\cdot e^{\int P dx}\right] = Q\cdot e^{\int P dx}$$

Integrating both sides, we get

$$y\cdot e^{\int P dx} = \int Q\cdot e^{\int P dx}\, dx + C$$

This is the required solution.

Note. $e^{\int P dx}$ is called the integrating factor.

Solution is $\boxed{y\times[\text{I.F.}] = \int Q\,[I.F.]\,dx + c}$

Example 17. *Solve:* $(x+1)\dfrac{dy}{dx} - y = e^x\,(x+1)^2$ *(A.M.I.E.T.E., Summer 2002)*

Solution. We have $\dfrac{dy}{dx} - \dfrac{y}{x+1} = e^x\,(x+1)$

Integrating factor $= e^{-\int \frac{dx}{x+1}} = e^{-\log(x+1)} = e^{\log(x+1)^{-1}} = \dfrac{1}{x+1}$

The solution is
$$y\cdot\frac{1}{x+1} = \int e^x\cdot(x+1)\cdot\frac{1}{x+1}\,dx = \int e^x\,dx$$

$$\frac{y}{x+1} = e^x + C \qquad \textbf{Ans.}$$

Example 18. *Solve:* $x(x-1)\dfrac{dy}{dx} - (x-2)\,y = x^2\,(2x-1)$

Solution. Here, $\dfrac{dy}{dx} - \dfrac{x-2}{x(x-1)}\,y = \dfrac{x^2(2x-1)}{x(x-1)}$

Hence, the integrating factor

$$e^{-\int \frac{x-2}{x(x-1)}dx} = e^{\int\left(\frac{1}{x-1}-\frac{2}{x}\right)dx} = e^{\log(x-1)-2\log x} = e^{\log\frac{x-1}{x^2}}$$

$$= \frac{x-1}{x^2}$$

The solution is

$$y\cdot\frac{x-1}{x^2} = \int \frac{x^2(2x-1)}{x(x-1)}\cdot\frac{x-1}{x^2}\,dx = \int\frac{2x-1}{x}\,dx$$

$$= \int\left(2-\frac{1}{x}\right)dx = 2x - \log x + C \qquad \textbf{Ans.}$$

Example 19. *If* $\frac{dy}{dx}+2y\tan x=\sin x$, *and* $y=0$ *for* $x=\frac{\pi}{3}$, *show that maximum value of* y *is* $\frac{1}{8}$.

Solution. We have, $\frac{dy}{dx}+2y\tan x=\sin x$

$$\text{I.F.} = e^{2\int \tan x\,dx} = e^{2\log\sec x} = e^{\log\sec^2 x} = \sec^2 x$$

Hence, solution is

$$y\cdot\sec^2 x = \int \sec^2 x\cdot\sin x\,dx = \int \tan x\sec x\,dx$$

$$= \sec x + c \qquad ...(1)$$

Put $x=\frac{\pi}{3}$, $y=0$ in (1), $0\times\sec^2\frac{\pi}{3} = \sec\frac{\pi}{3}+c$

$$0 = 2 + c \text{ or } c = -2$$

Substituting $c=-2$ in (1), we get

$$y\sec^2 x = \sec x - 2$$

$$y = \cos x - 2\cos^2 x, \Rightarrow \frac{dy}{dx} = -\sin x + 4\cos x\sin x$$

$$0 = -\sin x + 4\sin x\cos x \Rightarrow -\sin x(1-4\cos x) = 0 \Rightarrow \sin x = 0 \text{ or } \cos x = \frac{1}{4}$$

$$\frac{d^2y}{dx^2} = -\cos x - 4\sin^2 x + 4\cos^2 x$$

If $\cos x = \frac{1}{4}$, $\frac{d^2y}{dx^2} = -\frac{1}{4}-4\left(1-\frac{1}{16}\right)+4\left(\frac{1}{16}\right) = -\frac{1}{4}-\frac{15}{4}+\frac{1}{4} = -\frac{15}{4}$

As $\frac{d^2y}{dx^2} = -$ ve, so y is maximum if $\cos x = \frac{1}{4}$, Maximum value of $y = \frac{1}{4}-\frac{2}{16} = \frac{1}{8}$ **Ans.**

EXERCISE 1.5

Solve the following differential equations :

1. $\frac{dy}{dx}+\frac{1}{x}y = x^3-3$ **Ans.** $xy = \frac{x^5}{5}-\frac{3x^2}{2}+C$

2. $(2y-3x)\,dx + x\,dy = 0$ **Ans.** $yx^2 = x^3 + C$

3. $\frac{dy}{dx}+y\cot x = \cos x.$ **Ans.** $y\sin x = \frac{\sin^2 x}{2}+C$

4. $(x+1)\frac{dy}{dx}-2y = (x+1)^4.$ **Ans.** $y = \left(\frac{x^2}{2}+x+C\right)(x+1)^2$

5. $\frac{dy}{dx}+2xy = 2e^{-x^2}$ **Ans.** $y = (2x+C)e^{-x^2}$

6. $\frac{dy}{dx}+3y = e^{2x}$ **Ans.** $y = \frac{1}{5}e^{2x}+Ce^{-3x}$

7. $\frac{dy}{dx}+y\sec x = \tan x$ **Ans.** $y = \frac{C-x}{\sec x+\tan x}+1$

8. $\cos^2 x\dfrac{dy}{dx}+y = \tan x$ **Ans.** $y = \tan x - 1 + Ce^{-\tan x}$

9. $(x+a)\dfrac{dy}{dx}-3y = (x+a)^5$ **Ans.** $2y = (x+a)^5 + 2C(x+a)^3$

10. $x\cos x\dfrac{dy}{dx}+y(x\sin x+\cos x) = 1$ **Ans.** $xy = \sin x + C\cos x$

11. $x\log x\dfrac{dy}{dx}+y = 2\log x$ **Ans.** $y\log x = (\log x)^2 + C$

12. $x\dfrac{dy}{dx}+2y = x^2\log x$ **Ans.** $yx^2 = \dfrac{x^4}{4}\log x-\dfrac{x^4}{16}+C$

13. $dr + (2r\cot\theta + \sin 2\theta)\, d\theta = 0$ **Ans.** $r\sin^2\theta = \dfrac{-\sin^4\theta}{2}+C$

14. $\dfrac{dy}{dx}+y\cos x = \dfrac{1}{2}\sin 2x$ **Ans.** $y = \sin x - 1 + Ce^{-\sin x}$

15. $(1-x^2)\dfrac{dy}{dx}+2xy = x(1-x^2)^{1/2}$ **Ans.** $y = \sqrt{1-x^2} + C(1-x^2)$

1.10 EQUATIONS REDUCIBLE TO THE LINEAR FORM (Bernoulli's Equation)

The equation of the form

$$\frac{dy}{dx} + Py = Qy^n \qquad ...(1)$$

where **P** and **Q** are constants or functions of x can be reduced to the linear form on dividing by y^n and substituting $\dfrac{1}{y^{n-1}} = z$.

On dividing both sides of (1) by y^n, we get

$$\frac{1}{y^n}\frac{dy}{dx} + \frac{1}{y^{n-1}}P = Q \qquad ...(2)$$

Put $\dfrac{1}{y^{n-1}} = z$, so that $\dfrac{(1-n)}{y^n}\dfrac{dy}{dx} = \dfrac{dz}{dx}$

$\therefore$ (2) becomes $\dfrac{1}{1-n}\dfrac{dz}{dx} + Pz = Q$ or $\dfrac{dz}{dx}+P(1-n)z = Q(1-n)$

which is a linear equation and can be solved easily by the previous metnod discussed in article 1.9 on page 12.

Example 20. *Solve* $x^2 dy + y(x+y)\, dx = 0$ *(U.P., II Semester, 2006)*

Solution. We have, $x^2 dy + y(x+y)\, dx = 0$

$$\Rightarrow \quad \frac{dy}{dx}+\frac{y}{x} = -\frac{y^2}{x^2} \Rightarrow \frac{1}{y^2}\frac{dy}{dx}+\frac{1}{xy} = -\frac{1}{x^2}$$

Put $-\dfrac{1}{y} = z$ so that $\dfrac{1}{y^2}\dfrac{dy}{dx} = \dfrac{dz}{dx}$

The given equation reduces to a linear differential equation in z.

$$\frac{dz}{dx} - \frac{z}{x} = -\frac{1}{x^2}$$

$$\text{I.F.} = e^{-\int 1/x\, dx} = e^{-\log x} = e^{\log 1/x} = 1/x.$$

Hence, the solution is

$$z \,.\, 1/x = \int -\frac{1}{x^2}\cdot\frac{1}{x}\, dx \Rightarrow \frac{z}{x} = \int -x^{-3}\, dx$$

$$\Rightarrow \qquad -\frac{1}{xy} = -\frac{x^{-2}}{-2} + C \Rightarrow \frac{1}{xy} = -\frac{1}{2x^2} - C \qquad \textbf{Ans.}$$

Example 21. *Solve* : $x\dfrac{dy}{dx} + y \log y = xye^x$ *(A.M.I.E., Summer 2000)*

Solution. We have $x\dfrac{dy}{dx} + y \log y = xye^x$

Dividing by xy, we get $\dfrac{1}{y}\dfrac{dy}{dx} + \dfrac{1}{x}\log y = e^x$ ...(1)

Put $\log y = z$, so that $\dfrac{1}{y}\dfrac{dy}{dx} = \dfrac{dz}{dx}$

Equation (1) becomes, $\dfrac{dz}{dx} + \dfrac{z}{x} = e^x$

$$\text{I.F.} = e^{\int \frac{1}{x} dx} = e^{\log x} = x$$

Solution is $\quad z\,x = \int x\, e^x\, dx$

$$z\,x = xe^x - e^x + C$$

$$\Rightarrow \qquad x \log y = x\, e^x - e^x + C \qquad \textbf{Ans.}$$

Example 22. *Solve: $(\sec x \tan x \tan y - e^x)\, dx + \sec x \sec^2 y\, dy = 0$*

Solution. We have, $\sec x \tan x \tan y - e^x + \sec x \sec^2 y \dfrac{dy}{dx} = 0$

$$\Rightarrow \qquad \sec^2 y \frac{dy}{dx} + \tan x \tan y = e^x \cos x$$

Put $\tan y = z$, so that $\sec^2 y \dfrac{dy}{dx} = \dfrac{dz}{dx}$

$$\frac{dz}{dx} + z \tan x = e^x \cos x$$

$$\text{I.F.} = e^{\int \tan x\, dx} = e^{\log \sec x} = \sec x$$

Hence, the solution is $z \sec x = \int e^x \cos x \sec x\, dx$

$$z \sec x = \int e^x\, dx$$

$$\tan y \sec x = e^x + C \qquad \textbf{Ans.}$$

Example 23. *Solve* : $\dfrac{dy}{dx} - \dfrac{\tan y}{1+x} = (1+x)\, e^x \sec y.$

Solution. We have, $\dfrac{dy}{dx} - \dfrac{\tan y}{1+x} = (1+x)\, e^x \sec y$

$$\Rightarrow \qquad \cos y \frac{dy}{dx} - \frac{\sin y}{1+x} = (1+x)\, e^x \qquad \text{...(1)}$$

Put $\quad \sin y = z$, so that $\cos y \dfrac{dy}{dx} = \dfrac{dz}{dx}$

(1) becomes $\quad \dfrac{dz}{dx} - \dfrac{z}{1+x} = (1+x)\, e^x$

$$\text{I.F} = e^{-\int \frac{1}{1+x} dx} = e^{-\log(1+x)} = e^{\log \frac{1}{1+x}} = \frac{1}{1+x}$$

Solution is $\quad z \cdot \dfrac{1}{1+x} = \displaystyle\int (1+x) e^x \cdot \frac{1}{1+x}\, dx = \int e^x\, dx$

$$\frac{\sin y}{1+x} = e^x + C \qquad \textbf{Ans.}$$

Example 24. *Solve* $\tan y \dfrac{dy}{dx} + \tan x = \cos y \cos^2 x$

Solution. We have, $\tan y \dfrac{dy}{dx} + \tan x = \cos y \cos^2 x$

$$\sec y \tan y \frac{dy}{dx} + \sec y \tan x = \cos^2 x$$

Writing $\quad z = \sec y$, so that $\dfrac{dz}{dx} = \sec y \tan y \dfrac{dy}{dx}$

The equation becomes $\dfrac{dz}{dx} + z \tan x = \cos^2 x$

$$\text{I.F} = e^{\int \tan x\, dx} = e^{\log \sec x} = \sec x$$

$\therefore$ The solution of the equation is

$$z \sec x = \int \cos^2 x \sec x\, dx$$

$$\sec y \sec x = \int \cos x\, dx = \sin x$$

$$\sec y = (\sin x + C) \cos x \qquad \textbf{Ans.}$$

Example 25. *Solve :* $\sec^2 y \dfrac{dy}{dx} + x \tan y = x^3$

Solution. We have $\sec^2 y \dfrac{dy}{dx} + x \tan y = x^3 \qquad \text{...(1)}$

Putting $\tan y = z$, so that $\sec^2 y \dfrac{dy}{dx} = \dfrac{dz}{dx}$

$\therefore$ (1) becomes $\quad \dfrac{dz}{dx} + xz = x^3$

Here, the integrating factor is $e^{\int x\, dx} = e^{x^2/2}$

$\therefore$ The solution is $\quad z \cdot e^{x^2/2} = \displaystyle\int e^{x^2/2} \cdot x^3 dx = \int x^2 (x e^{x^2/2})\, dx$

Integrating by parts

$$\Rightarrow \quad z\cdot e^{x^2/2} = x^2\cdot e^{x^2/2} - 2\int x\cdot e^{x^2/2}\,dx$$

$$\tan y\cdot e^{x^2/2} = x^2\cdot e^{x^2/2} - 2e^{x^2/2} + C$$

$$\Rightarrow \quad \tan y = x^2 - 2 + Ce^{-x^2/2}$$ **Ans.**

Example 26. *Solve* $(1 + y^2)\,dx = (\tan^{-1} y - x)\,dy.$

(Uttarakhand II Sem. 2012, A.M.I.E., Summer 2004, U.P. (B. Pharm) 2005)

Solution. We have $(1 + y^2)\,dx = (\tan^{-1} y - x)\,dy.$

$$\frac{dx}{dy} = \frac{\tan^{-1} y - x}{1+y^2} \quad \Rightarrow \quad \frac{dx}{dy} + \frac{x}{1+y^2} = \frac{\tan^{-1} y}{1+y^2}$$

This is a linear differential equation.

$$\text{I.F.} = e^{\int \frac{1}{1+y^2}dy} = e^{\tan^{-1} y}$$

Its solution is $\quad x\,.\,e^{\tan^{-1} y} = \int e^{\tan^{-1} y}\,\frac{\tan^{-1} y}{1+y^2}\,dy$

Put $\tan^{-1} y = t$ on R.H.S., so that $\dfrac{1}{1+y^2}\,dy = dt$

$$x\,.\,e^{\tan^{-1} y} = \int e^t\,.t\,dt = t.e^t - e^t + C = e^{\tan^{-1} y}(\tan^{-1} y - 1) + C$$

$$x = (\tan^{-1} y - 1) + Ce^{-\tan^{-1} y}$$ **Ans.**

Example 27. *Solve:* $y\log y\dfrac{dx}{dy} + x - \log y = 0$ *(Uttarakhand, 2007, U.P. II Semester,* 2004)

Solution. $y\log y\dfrac{dx}{dy} + x - \log y = 0$ or $y\log y\dfrac{dx}{dy} + x = \log y$ or $\dfrac{dx}{dy} + \dfrac{x}{y\log y} = \dfrac{1}{y}$

$$\text{I.F.} = e^{\int \frac{dy}{y\log y}} = e^{\log(\log y)} = \log y$$

$\therefore$ The solution is

$$x\,.\,\log y = \int \frac{1}{y}(\log y)\,dy$$

$$x\,.\,\log y = \frac{1}{2}(\log y)^2 + C$$ **Ans.**

EXERCISE 1.6

Solve the following differential equations :

1. $\dfrac{1}{y^2}\dfrac{dy}{dx} - \dfrac{1}{y} = 2\,x\,e^{-x}$ **Ans.** $e^x + x^2\,y + Cy = 0$

2. $3\dfrac{dy}{dx} + 3\dfrac{y}{x} = 2\,x^4\,y^4$ **Ans.** $\dfrac{1}{y^3} = -\,x^5 + C\,x^3$

3. $\dfrac{dy}{dx} = y\tan x - y^2\sec x$ **Ans.** $\sec x = (\tan x + C)\,y$

4. $\dfrac{dy}{dx} = 2y\tan x + y^2\tan^2 x$, if $y = 1$ at $x = 0$. **Ans.** $\dfrac{1}{y}\sec^2 x = -\dfrac{\tan^3 x}{3} + 1$

5. $\frac{dy}{dx} + \tan x \tan y = \cos x \sec y$ **Ans.** $\sin y \sec x = x + C$

6. $dy + y \tan x \,.\, dx = y^2 \sec x \,.\, dx$ **Ans.** $y(x + C) + \cos x = 0$

7. $(x^2 y^2 + xy)\, y\, dx + (x^2 y^2 - 1)\, x\, dy = 0$ **Ans.** $xy = \log Cy$

8. $(x^2 + y^2 + x)\, dx + xy\, dy = 0$ **Ans.** $x^2 y^2 = -\frac{x^4}{2} - \frac{2x^3}{3} + C$

9. $\frac{dy}{dx} + y = 3 e^x y^3$ **Ans.** $\frac{1}{y^2} = 6e^x + C e^{2x}$

10. $(x - y^2)\, dx + 2xy\, dy = 0$ **Ans.** $\frac{y^2}{x} + \log y = C$

11. $e^y\left(\frac{dy}{dx} + 1\right) = e^x$ **Ans.** $e^{x+y} = \frac{e^{2x}}{2} + C$

12. $x^2 y - x^3 \frac{dy}{dx} = y^4 \cos x$ **Ans.** $x^3 = y^3 (3 \sin x - C)$

13. $3\frac{dy}{dx} + \frac{2}{x+1} \cdot y = \frac{x^2}{y^2}$ **Ans.** $y^3 (x + 1)^2 = \frac{x^5}{5} + \frac{x^4}{2} + \frac{x^3}{3} + C$

14. $\cos x \frac{dy}{dx} + 4y \sin x = 4\sqrt{y} \sec x$ **Ans.** $\sqrt{y} \sec^2 x = 2\left[\tan x + \frac{\tan^3 x}{3}\right] + C$

15. $\frac{dy}{dx} + x \sin 2y = x^3 \cos^2 y$ **Ans.** $\tan y = \frac{1}{2}(x^2 - 1) + Ce^{-x^2}$

16. $\frac{1}{1+y^2}\frac{dy}{dx} + 2x \tan^{-1} y = x^3$ **Ans.** $\tan^{-1} y = \frac{1}{2}(x^2 - 1) + Ce^{-x^2}$

17. $(2x - 10y^3)\frac{dy}{dx} + y = 0$ **Ans.** $x = 2y^3 + Cy^{-2}$

18. $(x + 2y^3)\, dy = y \,.\, dx$ **Ans.** $x = y^3 + Cy$

19. $e^{-y} \sec^2 y\, dy = dx + x\, dy$ **Ans.** $x e^y = \tan y + C$

20. $(x + y + 1)\frac{dy}{dx} = 1$ **Ans.** $x + y + 2 = Ce^y$

21. $\frac{dy}{dx} = \frac{y^3}{e^{2x} + y^2}$ **Ans.** $e^{-2x} y^2 + 2 \log y + C = 0$

22. $dx - xy(1 + xy^2)\, dy = 0$ **Ans.** $-\frac{1}{x} = y^2 - 2 + Ce^{-y^2/2}$

23. $\frac{dy}{dx} + \frac{y}{x}\log y = \frac{y}{x^2}(\log y)^2$ **Ans.** $\frac{1}{x \log y} = \frac{1}{2x^2} + C$

(*A.M.I.E.T.E., Summer* 2004, 2003, *Winter* 2003, 2001)

24. $x(3x + 2y^2)\, dx + 2y(1 + x^2)\, dy = 0$ **Ans.** $y^2(1 + x^2) = -x^3 + C$

25. $\sin y \frac{dy}{dx} = \cos x (2 \cos y - \sin^2 x)$ **Ans.** $\cos y = \frac{1}{4}(2\sin^2 x + 2 \sin x - 1 + 4C\, e^{-2\sin x})$

1.11 EXACT DIFFERENTIAL EQUATION

An exact differential equation is formed by directly differentiating its primitive (solution) without any other process.

$$Mdx + Ndy = 0$$

is said to be an exact differential equation if it satisfies the following condition

$$\frac{\partial M}{\partial y} = \frac{\partial N}{\partial x}$$

where $\frac{\partial M}{\partial y}$ denotes the differential co-efficient of M with respect to y keeping x constant and $\frac{\partial N}{\partial x}$, the differential coefficient of N with respect to x, keeping y constant.

Method for Solving Exact Differential Equations

Step 1. Integrate M w.r.t. x keeping y constant.

Step 2. Integrate w.r.t. y, only those terms of N which do not contain x.

Step 3. Result of I + Result of II = Constant.

Example 28. *Find the value of λ for which the differential equation :*

$$(x y^2 + \lambda x^3y^2)\, dx + (x^3y + y x)\, x\, dy = 0$$ *(Uttarakhand, IInd Sem., Summer 2010)*

is exact. Hence, solve.

Solution Here we have

$$(x y^2 + \lambda x^3 y^2)\, dx + (x^3 y + y x)\, x\, dy = 0 \qquad ...(1)$$

$$M = x y^2 + \lambda x^3 y^2 \qquad N = (x^3 y + y x) x$$

$$\frac{\partial M}{\partial y} = 2xy + 2\lambda x^3 y \qquad \frac{\partial N}{\partial x} = 4x^3 y + 2x y$$

If (1) in an exact differential equation, then $\frac{\partial M}{\partial y} = \frac{\partial N}{\partial x}$

$$2x y + 2\lambda x^3 y = 4x^3 y + 2xy$$

$$2\lambda x^3 y = 4x^3 y \Rightarrow 2\lambda = 4 \Rightarrow \lambda = 2$$ **Ans.**

Example 29. *Solve*

$$(5 x^4 + 3 x^2 y^2 - 2 x y^3)\, dx + (2 x^3 y - 3 x^2 y^2 - 5 y^4)\, dy = 0$$

Solution. Here, $M = 5 x^4 + 3 x^2 y^2 - 2 x y^3$, $N = 2x^3y - 3x^2y^2 - 5y^4$

$$\frac{\partial M}{\partial y} = 6 x^2y - 6 x y^2, \qquad \frac{\partial N}{\partial x} = 6 x^2y - 6xy^2$$

Since, $\frac{\partial M}{\partial y} = \frac{\partial N}{\partial x}$, the given equation is exact.

Now $\int M\, dx + \int$ (terms of N not containing x) $dy = C$
(y constant)

$$\int (5x^4 + 3x^2 y^2 - 2xy^3)\, dx + \int -5y^4\, dy = C$$

$$\Rightarrow \qquad x^5 + x^3 y^2 - x^2 y^3 - y^5 = C$$ **Ans.**

Example 30. *Solve :* $(y^2 e^{xy^2} + 4x^3)\, dx + (2xy\, e^{xy^2} - 3y^2)\, dy = 0$

(Uttarakhand II Sem. 2012)

Solution. Here $M = y^2 e^{xy^2} + 4x^3$ $\qquad N = 2 x y e^{xy^2} - 3y^2$

$$\frac{\partial M}{\partial y} = y^2 (e^{xy^2}\, 2xy) + (2y)\, e^{xy^2} \qquad \frac{\partial N}{\partial x} = (2y)\left[x (e^{xy^2}\, y^2) + e^{xy^2}\right]$$

$$= 2 xy^3 \cdot e^{xy^2} + 2y \cdot e^{xy^2} \qquad = 2 xy^3 \cdot e^{xy^2} + 2y \cdot e^{xy^2}$$

Since $\dfrac{\partial M}{\partial y} = \dfrac{\partial N}{\partial x}$, the given equation is exact.

Now $\int_{(y\ \text{constant})} M dx + \int$ (terms of N not containing x) $dy = C$

$$\int (y^2 \cdot e^{xy^2} + 4x^3)\, dx + \int (-3y^2) dy = C$$

$$e^{xy^2} + x^4 - y^3 = C$$ **Ans.**

Example 31. *Solve:* $x\,dx + y\,dy = \dfrac{a^2\,(x\,dy - y\,dx)}{x^2 + y^2}$ (*U.P. Second semester Summer* 2005)

Solution. We have, $x\,dx + y\,dy = \dfrac{a^2\,(x\,dy - y\,dx)}{x^2 + y^2}$

$$\Rightarrow \quad \left(x + \frac{a^2 y}{x^2 + y^2}\right) dx + \left(y - \frac{a^2 x}{x^2 + y^2}\right) dy = 0$$

Here, $$M = x + \frac{a^2 y}{x^2 + y^2},\ N = y - \frac{a^2 x}{x^2 + y^2}$$

Now, $$\frac{\partial M}{\partial y} = \frac{a^2 (x^2 - y^2)}{(x^2 + y^2)^2},\ \frac{\partial N}{\partial x} = \frac{a^2 (x^2 - y^2)}{(x^2 + y^2)^2}$$

Since, $$\frac{\partial M}{\partial y} = \frac{\partial N}{\partial x}$$

Therefore equation is exact. Hence,

$$\int \left(x + \frac{a^2 y}{x^2 + y^2}\right) dx + \int y\,dy = C$$

$$\Rightarrow \quad \frac{x^2}{2} + a^2 y \cdot \frac{1}{y} \tan^{-1}\left(\frac{x}{y}\right) + \frac{y^2}{2} = C$$

$$\Rightarrow \quad \left(\frac{x^2 + y^2}{2}\right) + a^2 \tan^{-1}\left(\frac{x}{y}\right) = C$$ **Ans.**

Example 32. *Test for exactness*

$$(e^y + 1) \cos x\, dx + e^y \sin x\, dy = 0$$

Solution. Here, $M = (e^y + 1) \cos x$, $N = e^y \sin x$

$$\frac{\partial M}{\partial y} = e^y \cos x, \qquad \frac{\partial N}{\partial x} = e^y \cos x$$

Since $$\frac{\partial M}{\partial y} = \frac{\partial N}{\partial x}$$

Hence, the given equation is exact.

The required solution is exact.

$$\int_{(y\ \text{constant})} M\,dx + \int (\text{terms in } N \text{ not containing } x)\, dy = C$$

$$\int (e^y + 1) \cos x\, dx = C$$

$$(e^y + 1) \sin x = C$$ **Ans.**

Example 33. *Solve the equation :*

$$\left[(\cos x)\log_e(2y-8)+\frac{1}{x}\right]dx+\frac{\sin x}{y-4}\,dy=0,$$

Solution. Here, $M=(\cos x)\log_e(2y-8)+\frac{1}{x}$, $N=\frac{\sin x}{y-4}$

$$\frac{\partial M}{\partial y}=\frac{2\cos x}{2y-8}=\frac{\cos x}{y-4},\qquad \frac{\partial N}{\partial x}=\frac{\cos x}{y-4}$$

Since $\frac{\partial M}{\partial y}=\frac{\partial N}{\partial x}$ the given differential equation is exact.

The required solution is

$$\int M dx+\int(\text{terms in } N \text{ not containing } x)\,dy=C$$

(y constant)

$$\int\left[(\cos x)\log_e(2y-8)+\frac{1}{x}\right]dx=C$$

$$(\sin x)\log_e(2y-8)+\log x=C \qquad ...(1)$$

On putting $x=1$ and $y=\frac{9}{2}$ in (1), we get

$$\sin(1)\log(9-8)+\log 1=C \text{ or } C=0$$

Putting the value of C in (1), we have

$$(\sin x)\log(2y-8)+\log x=0$$ **Ans.**

EXERCISE 1.7

Solve the following differential Equations:

1. $(x+y-10)\,dx+(x-y-2)\,dy=0$ **Ans.** $\frac{x^2}{2}+xy-10x-\frac{y^2}{2}-2y=C$
2. $(y^2-x^2)\,dx+2xy\,dy=0$ **Ans.** $\frac{x^3}{3}=xy^2+C$
3. $(1+3e^{x/y})\,dx+3e^{x/y}\left(1-\frac{x}{y}\right)dy=0$ **Ans.** $x+3y\,e^{x/y}=C$
4. $(2x-y)\,dx=(x-y)\,dy$ **Ans.** $xy=x^2+\frac{y^2}{2}+C$
5. $(y\sec^2 x+\sec x\tan x)\,dx+(\tan x+2y)\,dy=0$ **Ans.** $y\tan x+\sec x+y^2=C$
6. $(ax+hy+g)\,dx+(hx+by+f)\,dy=0$ **Ans.** $ax^2+2hxy+by^2+2gx+2fy+C=0$
7. $(x^4-2xy^2+y^4)\,dx-(2x^2y-4xy^3+\sin y)\,dy=0$ **Ans.** $\frac{x^5}{5}-x^2y^2+xy^4+\cos y=C$
8. $(2xy+e^y)\,dx+(x^2+xe^y)\,dy=0$ **Ans.** $x^2y+xe^y=C$
9. $(x^2+2ye^{2x})\,dy+(2xy+2y^2e^{2x})\,dx=0$ **Ans.** $x^2y+y^2e^{2x}=C$
10. $[y(1+\frac{1}{x})+\cos y]\,dx+(x+\log x-x\sin y)\,dy=0$ **Ans.** $y(x+\log x)+x\cos y=C$

11. $(x^3 - 3xy^2)\,dx + (y^3 - 3x^2y)\,dy = 0,\ y(0) = 1$ **Ans.** $x^4 - 6x^2y^2 + y^4 = 1$

12. $(e^y + 2)\sin x\,dx - e^y \cos x\,dy = 0$ **Ans.** $(e^y + 2)\cos x + C = 0$

13. Find the value of λ, for which the differential equation

$$(x\,y^2 + \lambda\,x^2\,y)\,dx + (x + y)\,x^2\,dy = 0$$

is exact. Solve the equation for this value of λ. **Ans.** $\lambda = 3,\ x^2\,y^2 + 2\,x^3\,y = C$

1.12 EQUATIONS REDUCIBLES TO THE EXACT EQUATIONS

Sometimes a differential equation which is not exact may become so, on multiplication by a suitable function known as the integrating factor.

RULE I. If $\dfrac{\dfrac{\partial M}{\partial y} - \dfrac{\partial N}{\partial x}}{N}$ is a function of x alone, say $f(x)$, then I.F. $= e^{\int f(x)dx}$

Example 34. *Solve* $(2x \log x - xy)\,dy + 2y\,dx = 0$ ... (1)

Soution. $M = 2y,\ N = 2x\log x - xy$

$$\therefore \quad \frac{\partial M}{\partial y} = 2, \quad \frac{\partial N}{\partial x} = 2\,(1 + \log x) - y$$

Here
$$\frac{\dfrac{\partial M}{\partial y} - \dfrac{\partial N}{\partial x}}{N} = \frac{-2\log x + y}{2x\log x - xy} = -\frac{1}{x} = f(x)$$

$$\text{I.F.} = e^{\int f(x)dx} = e^{\int -\frac{1}{x}dx} = e^{-\log x} = e^{\log x^{-1}} = x^{-1} = \frac{1}{x}$$

On multiplying the given differential equation (1) by $\dfrac{1}{x}$, we get

$$\frac{2y}{x}\,dx + (2\log x - y)\,dy = 0 \quad \Rightarrow \quad \int \frac{2y}{x}\,dx + \int -y\,dy = C$$

$$\Rightarrow \quad 2y\log x - \frac{1}{2}y^2 = C$$ **Ans.**

EXERCISE 1.8

Solve the following differential equations :

1. $(y\log y)\,dx + (x - \log y)\,dy = 0$ **Ans.** $2x\log y = C + (\log y)^2$

2. $\left(y + \frac{1}{3}y^3 + \frac{1}{2}x^2\right)dx + \frac{1}{4}(1 + y^2)\,x\,dy = 0$ **Ans.** $\dfrac{yx^4}{4} + \dfrac{y^3x^4}{12} + \dfrac{x^6}{12} = C$

3. $2y\,dx + x\,(2\log x - y)\,dy = 0$ **Ans.** $2y\log x - \dfrac{y^2}{2} = C$

4. $(y - 2x^3)\,dx - x(1 - xy)\,dy = 0$ **Ans.** $-\dfrac{y}{x} - x^2 + \dfrac{y^2}{2} = C$

5. $(x\sec^2 y - x^2\cos y)dy = (\tan y - 3x^4)\,dx$ **Ans.** $-\dfrac{1}{x}\tan y - x^3 + \sin y = C$

6. $(x - y^2)\,dx + 2xy\,dy = 0$ **Ans.** $y^2 = Cx - x\log x$

Rule II. If $\dfrac{\dfrac{\partial N}{\partial x}-\dfrac{\partial M}{\partial y}}{M}$ is a function of y alone, say $f(y)$, then

$$\text{I.F.} = e^{\int f(y)\,dy}$$

Example 35. *Solve* $(y^4+2y)\,dx+(xy^3+2y^4-4x)\,dy=0$

Solution. Here, $M=y^4+2y;\ N=xy^3+2y^4-4x$ ...(1)

$$\therefore \quad \frac{\partial M}{\partial y}=4y^3+2; \qquad \frac{\partial N}{\partial x}=y^3-4$$

$$\therefore \quad \frac{\dfrac{\partial N}{\partial x}-\dfrac{\partial M}{\partial y}}{M}=\frac{(y^3-4)-(4y^3+2)}{y^4+2y}=\frac{-3\,(y^3+2)}{y\,(y^3+2)}=-\frac{3}{y}=f(y)$$

$$\text{I.F.}=e^{\int f(y)\,dy}=e^{\int -\frac{3}{y}dy}=e^{-3\log y}=e^{\log y^{-3}}=y^{-3}=\frac{1}{y^3}$$

On multiplying the given equation (1) by $\dfrac{1}{y^3}$ we get the exact differential equation.

$$\left(y+\frac{2}{y^2}\right)dx+\left(x+2y-\frac{4x}{y^3}\right)dy=0$$

$$\int\left(y+\frac{2}{y^2}\right)dx+\int 2y\,dy=C \quad\Rightarrow\quad x\left(y+\frac{2}{y^2}\right)+y^2=C \qquad \textbf{Ans.}$$

EXERCISE 1.9

Solve the following differential equations:

1. $(3x^2y^4+2xy)\,dx+(2x^3y^3-x^2)dy=0$ **Ans.** $x^3y^2+\dfrac{x^2}{y}=C$

2. $(xy^3+y)dx+2(x^2y^2+x+y^4)\,dy=0$ **Ans.** $\dfrac{x^2y^4}{2}+xy^2+\dfrac{y^6}{3}=C$

3. $y\,(x^2y+e^x)\,dx-e^x\,dy=0$ **Ans.** $\dfrac{x^3}{3}+\dfrac{e^x}{y}=C$

4. $(2x^4y^4e^y+2xy^3+y)dx+(x^2y^4e^y-x^2y^2-3x)dy=0$ **Ans.** $x^2e^y+\dfrac{x^2}{y}+\dfrac{x}{y^3}=C$

5. $(x\sec^2 y-x^2\cos y)\,dy=(\tan y-3x^4)\,dx$ **Ans.** $-\dfrac{1}{x}\tan y-x^3+\sin y=C$

Rule III. If M is of the form $M=y\,f_1(xy)$ and N is of the form $N=x\,f_2(xy)$

then $\text{I.F.}=\dfrac{1}{M.x-N.y}$;

Example 36. Solve $y(xy+2x^2y^2)\,dx+x\,(xy-x^2y^2)\,dy=0$

Solution. $y\,(x\,y+2\,x^2\,y^2)\,dx+x\,(xy-x^2y^2)\,dy=0$ ...(1)

Dividing (1) by xy, we get

$y(1+2xy)\,dx+x\,(1-xy)\,dy=0$ ...(2)

$$\text{M}=y\,f_1(xy),\ N=x\,f_2(xy)$$

$$\text{I.F.} = \frac{1}{Mx - Ny} = \frac{1}{xy(1+2xy) - xy(1-xy)} = \frac{1}{3x^2 y^2}$$

Multiplying (2) by $\frac{1}{3x^2 y^2}$, we have an exact differential equation.

$$\left(\frac{1}{3x^2 y} + \frac{2}{3x}\right) dx + \left(\frac{1}{3xy^2} - \frac{1}{3y}\right) dy = 0$$

$$\int \left(\frac{1}{3x^2 y} + \frac{2}{3x}\right) dx + \int -\frac{1}{3y}\, dy = C$$

$$\Rightarrow \quad -\frac{1}{3xy} + \frac{2}{3} \log x - \frac{1}{3} \log y = C$$

$$\Rightarrow \quad -\frac{1}{xy} + 2 \log x - \log y = b$$ **Ans.**

EXERCISE 1.10

Solve the following differential equations:

1. $(y - xy^2)\, dx - (x + x^2 y)\, dy = 0$ **Ans.** $\log\left(\frac{x}{y}\right) - xy = A$

2. $y(1 + xy)\, dx + x(1 - xy)\, dy = 0$ **Ans.** $xy \log\left(\frac{y}{x}\right) = Cxy - 1$

3. $y(1 + xy)\, dx + x(1 + xy + x^2y^2)\, dy = 0$ **Ans.** $\frac{1}{2x^2y^2} + \frac{1}{xy} - \log y = C$

4. $(xy \sin xy + \cos xy)\, y\, dx + (xy \sin xy - \cos xy)\, x\, dy = 0$ **Ans.** $y \cos xy = Cx$

Rule IV.

For of this type of $x^m y^n (ay\, dx + bx\, dy) + x^{m'} y^{n'} (a' y\, dx + b' x\, dy) = 0$, the integrating factor is $x^h y^k$.

where $\frac{m+h+1}{a} = \frac{n+k+1}{b}$, and $\frac{m'+h+1}{a'} = \frac{n'+k+1}{b'}$

Example 37. *Solve* $(y^3 - 2x^2 y)\, dx + (2xy^2 - x^3)\, dy = 0$

Solution. $(y^3 - 2x^2 y)\, dx + (2xy^2 - x^3)\, dy = 0$

$$y^2 (ydx + 2xdy) + x^2(-2ydx - xdy) = 0$$

Here $m = 0$, $n = 2$, $a = 1$, $b = 2$, $m' = 2$, $n' = 0$, $a' = -2$, $b' = -1$

$$\frac{0+h+1}{1} = \frac{2+k+1}{2} \text{ and } \frac{2+h+1}{-2} = \frac{0+k+1}{-1}$$

$$\Rightarrow \quad 2h + 2 = 2 + k + 1 \text{ and } h + 3 = 2k + 2$$

$$\Rightarrow \quad 2h - k = 1 \text{ and } h - 2k = -1$$

On solving $h = k = 1$.Integrating Factor $= xy$

Multiplying the given equation by xy, we get

$$(xy^4 - 2x^3 y^2)\, dx + (2x^2 y^3 - x^4 y)\, dy = 0$$

which is an exact differential equation.

$$\int (xy^4 - 2x^3 y^2)\, dx = C$$

$\Rightarrow \quad \dfrac{x^2 y^4}{2} - \dfrac{2x^4 y^2}{4} = C \qquad \Rightarrow x^2 y^4 - x^4 y^2 = C'$

$\Rightarrow \quad x^2 y^2 (y^2 - x^2) = C'$ **Ans.**

Example 38. *Solve* $(3y - 2xy^3)\, dx + (4x - 3x^2y^2)\, dy = 0$ (*U.P., II Semester, June* 2007)

Solution. $(3y - 2xy^3)\, dx + (4x - 3x^2y^2)\, dy = 0$

$\Rightarrow \quad (3y\, dx + 4x\, dy) + xy^2 (-2y\, dx - 3x dy) = 0$ ...(1)

Comparing the coefficients of (1) with

$$x^m y^n (a\, y\, dx + b\, x\, dy) + x^{m'} y^{n'} (a'\, y\, dx + b'\, x\, dy) = 0, \text{ we get}$$

$$m = 0,\ n = 0,\ a = 3,\ b = 4$$

$$m' = 1,\ n' = 2,\ a' = -2,\ b' = -3$$

To find the integrating factor $x^h y^k$

$$\frac{m+h+1}{a} = \frac{n+k+1}{b} \text{ and } \frac{m'+h+1}{a'} = \frac{n'+k+1}{b'}$$

$$\frac{0+h+1}{3} = \frac{0+k+1}{4} \text{ and } \frac{1+h+1}{-2} = \frac{2+k+1}{-3}$$

$$\Rightarrow \quad \frac{h+1}{3} = \frac{k+1}{4} \text{ and } \frac{h+2}{2} = \frac{k+3}{3}$$

$$\Rightarrow \quad 4h - 3k + 1 = 0 \qquad \text{...(2)}$$

and $$3h - 2k = 0 \Rightarrow h = \frac{2k}{3} \qquad \text{...(3)}$$

Putting the value of h from (3) in (2), we get

$$\frac{8k}{3} - 3k + 1 = 0$$

$$\Rightarrow \quad -\frac{k}{3} + 1 = 0 \quad \Rightarrow \quad k = 3$$

Putting $k = 3$ in (2), we get

$$h = \frac{2k}{3} = \frac{2 \times 3}{3} = 2$$

$$\text{I.F.} = x^h y^k = x^2 y^3$$

On multiplying the given differential equation by x^2y^3, we get

$$x^2y^3 (3y - 2xy^3)\, dx + x^2y^3 (4x - 3x^2y^2)\, dy = 0$$

$$(3x^2y^4 - 2x^3y^6)\, dx + (4x^3y^3 - 3x^4y^5)\, dy = 0$$

This is the exact differential equation.

Its solution is

$$\int (3x^2 y^4 - 2x^3 y^6)\, dx = 0 \Rightarrow x^3 y^4 - \frac{x^4}{2} y^6 = C$$ **Ans.**

EXERCISE 1.11

Solve the following differential equations:

1. $(2\, y\, dx + 3x\, dv) + 2xy\, (3y\, dx + 4x\, dy) = 0$ **Ans.** $x^2y^3(1 + 2xy) = C$
2. $(y^2 + 2yx^2)\, dx + (2x^3 - xy)\, dy = 0$ **Ans.** $4\,(xy)^{1/2} - \dfrac{2}{3}\left(\dfrac{y}{x}\right)^{3/2} = C$
3. $(3x + 2y^2)y\, dx + 2x\, (2x + 3y^2)\, dy = 0$ **Ans.** $x^2y^4\, (x + y^2) = C$

4. $(2x^2y^2 + y)\, dx - (x^3 y - 3x)\, dy = 0$ **Ans.** $\frac{7}{5}x^{10/7}\, y^{-\frac{5}{7}} - \frac{7}{4}x^{-4/7}\, y^{-12/7} = C$

5. $x\,(3y\, dx + 2x\, dy) + 8y^4\,(y\, dx + 3x\, dy) = 0$ **Ans.** $x^3y^2 + 4x^2y^6 = C$

Rule V.

If the given equation $Mdx + Ndy = 0$ is homogeneous equation and $Mx + Ny \neq 0$, then $\frac{1}{Mx + Ny}$ is an integrating factor.

Example 39. *Solve* $\frac{dy}{dx} = \frac{x^3 + y^3}{x\,y^2}$

Solution. $(x^3 + y^3)\, dx - (xy^2)\, dy = 0$ ...(1)

Here, $M = x^3 + y^3$, $N = -xy^2$

$$\text{I.F.} = \frac{1}{M\,x + N\,y} = \frac{1}{x(x^3 + y^3) - x\,y^2\,(y)} = \frac{1}{x^4}$$

Multiplying (1) by $\frac{1}{x^4}$, we get $\frac{1}{x^4}(x^3 + y^3)\, dx + \frac{1}{x^4}(-xy^2)\, dy = 0$

$\Rightarrow \left(\frac{1}{x} + \frac{y^3}{x^4}\right) dx - \frac{y^2}{x^3}\, dy = 0$, which is an exact differential equation.

$$\int\left(\frac{1}{x} + \frac{y^3}{x^4}\right) dx = C \Rightarrow \log x - \frac{y^3}{3x^3} = C$$ **Ans.**

EXERCISE 1.12

Solve the following differential equations:

1. $x^2y\, dx - (x^3 + y^3)dy = 0$ **Ans.** $-\frac{x^3}{3\,y^3} + \log y = C$

2. $(y^3 - 2x^2)dx + (2x^2y - xy^2)\, dy = 0$ **Ans.** $\frac{y}{x} + 3\log x - 2\log y = c$

3. $(x^2y - 2xy^2)\, dx - (x^3 - 3x^2y)\, dy = 0$ **Ans.** $\frac{x}{y} - 2\log x + 3\log y = c$

4. $(y^3 - 2y\,x^2)\, dx + (2\,xy^2 - x^3)\, dy = 0$ **Ans.** $x^2y^4 - x^4y^2 = c$

1.13 DIFFERENTIAL EQUATIONS REDUCIBLE TO EXACT FORM (BY INSPECTION)

The following differentials, which commonly occur, help in selecting the suitable integrating factor.

(*i*) $y\, dx + x\, dy = d\,[xy]$ (*b*) $\frac{x\, dy - y\, dx}{x^2} = d\left[\frac{y}{x}\right]$

(*c*) $\frac{y\, dx - x\, dy}{y^2} = d\left[\frac{x}{y}\right]$ (*d*) $\frac{x\, dy - y\, dx}{x^2 + y^2} = d\left[\tan^{-1}\left(\frac{y}{x}\right)\right]$

(*v*) $\frac{x\, dy - y\, dx}{xy} = d\left[\log\left(\frac{y}{x}\right)\right]$ (*vi*) $\frac{x\, dx + y\, dy}{x^2 + y^2} = d\left[\frac{1}{2}\log\,(x^2 + y^2)\right]$

(*vii*) $\frac{x\, dy - y\, dx}{x^2 - y^2} = d\left[\frac{1}{2}\log\frac{x + y}{x - y}\right]$ (*viii*) $\frac{x\, dy + y\, dx}{x^2y^2} = d\left[-\frac{1}{xy}\right]$

1.14 EQUATIONS OF FIRST ORDER AND HIGHER DEGREE

The differential equations will involve $\frac{dy}{dx}$ in higher degree and $\frac{dy}{dx}$ will be denoted by p.

The differential equation will be of the form $f(x, y, p) = 0$.

Case I. Equations solvable for p.

Example 40. *Solve*: $x^2 = 1 + p^2$

Solution. $x^2 = 1 + p^2 \quad \Rightarrow p^2 = x^2 - 1$

$$\Rightarrow \quad p = \pm\sqrt{x^2 - 1} \quad \Rightarrow \quad \frac{dy}{dx} = \pm\sqrt{x^2 - 1} \quad \Rightarrow \quad dy = \pm\sqrt{x^2 - 1}\, dx$$

which gives on integration $y = \pm\frac{x}{2}\sqrt{x^2 - 1} \mp \frac{1}{2}\log(x + \sqrt{x^2 - 1}) \quad C$ **Ans.**

Case II. Equations solvable for y.

(*i*) Differentiate the given equation w.r.t. "x".

(*ii*) Eliminate p from the given equation and the equation obtained as above.

(*iii*) The eliminant is the required solution.

Example 41. Solve: $y = (x - a)\, p - p^2$

Solution. We have, $\quad y = (x - a)\, p - p^2 \quad$ ...(1)

Differentiating (1) w.r.t. "x", we obtain

$$\frac{dy}{dx} = p + (x - a)\frac{dp}{dx} - 2p\frac{dp}{dx}$$

$$p = p + (x - a)\frac{dp}{dx} - 2p\frac{dp}{dx}$$

$$\Rightarrow \quad 0 = (x - a)\frac{dp}{dx} - 2p\frac{dp}{dx}$$

$$\Rightarrow \quad 0 = \frac{dp}{dx}[x - a - 2p] \quad \Rightarrow \quad \frac{dp}{dx} = 0$$

On integration, we get $p = C$.

Putting the value of p in (1), we get

$$y = (x - a)\, C - C^2$$ **Ans.**

Case III. Equations solvable for x

(*i*) Differentiate the given equation w.r.t. "y".

(*ii*) Solve the equation obtained as in (1) for p.

(*iii*) Eliminate p, by putting the value of p in the given equation.

(*iv*) The eliminant is the required solution.

Example 42. *Solve*: $y = 2\,p\,x + y\,p^2$

Solution. We have, $\quad y = 2\,px + y\,p^2 \quad$ ...(1)

$$\Rightarrow \quad 2\,p\,x = y - y\,p^2 \Rightarrow 2x = \frac{y}{p} - y\,p \quad \text{...(2)}$$

Differentiating (2) w.r.t. "y", we get

$$2\frac{dx}{dy} = \frac{1}{p} - \frac{y}{p^2}\frac{dp}{dy} - p - y\frac{dp}{dy}$$

$$\Rightarrow \quad \frac{2}{p} = \frac{1}{p} - \frac{y}{p^2}\frac{dp}{dy} - p - y\frac{dp}{dy} \qquad \Rightarrow \frac{1}{p} + p = -\frac{y}{p^2}\frac{dp}{dy} - y\frac{dp}{dy}$$

$$\Rightarrow \quad \frac{1}{p} + p = -y\left(\frac{1}{p^2}+1\right)\frac{dp}{dy} \qquad \Rightarrow \frac{1+p^2}{p} = -y\frac{1+p^2}{p^2}\frac{dp}{dy}$$

$$\Rightarrow \quad 1 = -\frac{y}{p}\frac{dp}{dy} \qquad \Rightarrow -\frac{dy}{y} = \frac{dp}{p} \quad \Rightarrow -\log y = \log p + \log C'$$

$$\Rightarrow \quad \log p\, y = \log C \qquad \Rightarrow \quad p\, y = C \quad \Rightarrow \quad p = \frac{C}{y},$$

Putting the value of p in (1), we get

$$y = 2\left(\frac{C}{y}\right)x + y\left(\frac{C^2}{y^2}\right) \quad \Rightarrow \quad y^2 = 2Cx + C^2$$

$$\Rightarrow \quad y^2 = C\,(2x + C)$$ **Ans.**

Case IV. Clairaut's Equation.

The equation $y = px + f\,(p)$ is known as Clairaut's equation. ...(1)

Differentiating (1) w.r.t. "x", we get

$$\frac{dy}{dx} = p + x\frac{dp}{dx} + f'\,(p)\frac{dp}{dx}$$

$$\Rightarrow \quad p = p + x\frac{dp}{dx} + f'\,(p)\frac{dp}{dx} \quad \Rightarrow \quad 0 = x\frac{dp}{dx} + f'\,(p)\frac{dp}{dx}$$

$$\Rightarrow \quad [x + f'\,(p)]\frac{dp}{dx} = 0 \quad \Rightarrow \quad \frac{dp}{dx} = 0 \quad \Rightarrow \quad p = a \quad \text{(constant)}$$

Putting the value of p in (1), we have

$$y = ax + f\,(a)$$

which is the required solution.

Method. In the Clairaut's equation, on replacing p by a (constant), we get the solution of the equation.

Example 43. *Solve*: $p = \log\,(p\,x - y)$

Solution. We have, $p = \log\,(p\,x - y) \Rightarrow e^p = p\,x - y \Rightarrow y = p\,x - e^p$

which is Clairaut's equation.

Hence, its solution is $y = a\,x - e^a$ **Ans.**

EXERCISE 1.13

Solve the following differential equations:

1. $x\,p^2 + x = 2y\,p$ — **Ans.** $2\,C\,y = C^2x^2 + 1$

2. $x\,(1 + p^2) = 1$ — **Ans.** $y - C = \sqrt{(x - x^2)} - \tan^{-1}\sqrt{\frac{1-x}{x}}$

3. $x^2\,p^2 + x\,y\,p - 6y^2 = 0$ — **Ans.** $y = \frac{C}{x^3},\ y = C_1x^2$

4. $\frac{dy}{dx} - \frac{dx}{dy} = \frac{x}{y} - \frac{y}{x}$ — **Ans.** $xy - C,\ x^2 - y^2 - C$

5. $y = px + p^3$ — **Ans.** $y = a\,x + a^3$

6. $x^2\,(y - px) = yp^2$ — **Ans.** $y^2 = C\,x^2 + C^2$

OBJECTIVE TYPE QUESTIONS

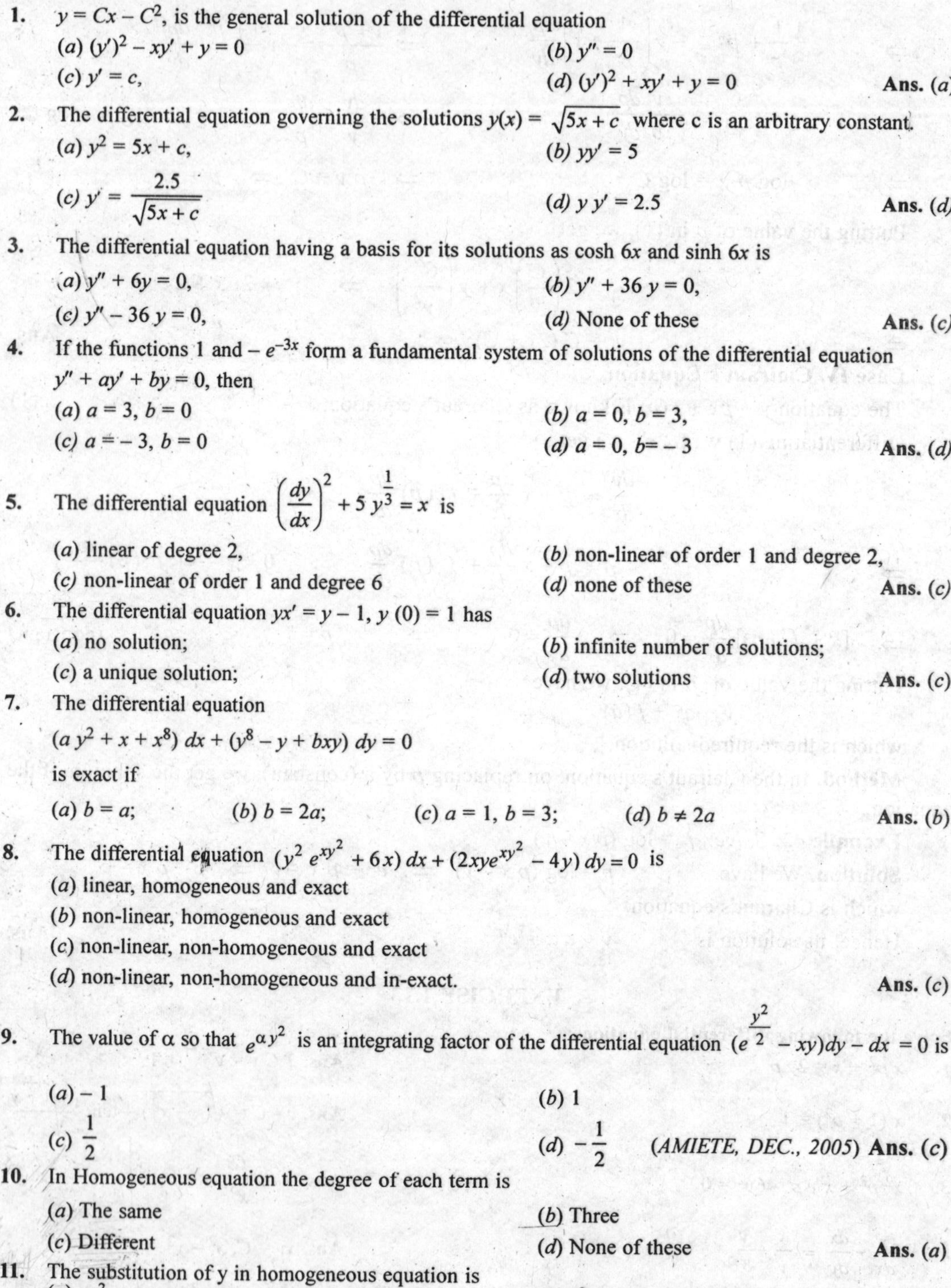

Choose the correct answer.

1. $y = Cx - C^2$, is the general solution of the differential equation
(a) $(y')^2 - xy' + y = 0$ (b) $y'' = 0$
(c) $y' = c$, (d) $(y')^2 + xy' + y = 0$ **Ans.** (a)

2. The differential equation governing the solutions $y(x) = \sqrt{5x + c}$ where c is an arbitrary constant.
(a) $y^2 = 5x + c$, (b) $yy' = 5$
(c) $y' = \dfrac{2.5}{\sqrt{5x+c}}$ (d) $y\,y' = 2.5$ **Ans.** (d)

3. The differential equation having a basis for its solutions as cosh $6x$ and sinh $6x$ is
(a) $y'' + 6y = 0$, (b) $y'' + 36\,y = 0$,
(c) $y'' - 36\,y = 0$, (d) None of these **Ans.** (c)

4. If the functions 1 and $-\,e^{-3x}$ form a fundamental system of solutions of the differential equation $y'' + ay' + by = 0$, then
(a) $a = 3, b = 0$ (b) $a = 0, b = 3$,
(c) $a = -3, b = 0$ (d) $a = 0, b = -3$ **Ans.** (d)

5. The differential equation $\left(\dfrac{dy}{dx}\right)^2 + 5\,y^{\frac{1}{3}} = x$ is
(a) linear of degree 2, (b) non-linear of order 1 and degree 2,
(c) non-linear of order 1 and degree 6 (d) none of these **Ans.** (c)

6. The differential equation $yx' = y - 1$, $y(0) = 1$ has
(a) no solution; (b) infinite number of solutions;
(c) a unique solution; (d) two solutions **Ans.** (c)

7. The differential equation
$(a\,y^2 + x + x^8)\,dx + (y^8 - y + bxy)\,dy = 0$
is exact if
(a) $b = a$; (b) $b = 2a$; (c) $a = 1, b = 3$; (d) $b \neq 2a$ **Ans.** (b)

8. The differential equation $(y^2\,e^{xy^2} + 6x)\,dx + (2xye^{xy^2} - 4y)\,dy = 0$ is
(a) linear, homogeneous and exact
(b) non-linear, homogeneous and exact
(c) non-linear, non-homogeneous and exact
(d) non-linear, non-homogeneous and in-exact. **Ans.** (c)

9. The value of α so that $e^{\alpha y^2}$ is an integrating factor of the differential equation $(e^{\frac{y^2}{2}} - xy)dy - dx = 0$ is
(a) -1 (b) 1
(c) $\dfrac{1}{2}$ (d) $-\dfrac{1}{2}$ *(AMIETE, DEC., 2005)* **Ans.** (c)

10. In Homogeneous equation the degree of each term is
(a) The same (b) Three
(c) Different (d) None of these **Ans.** (a)

11. The substitution of y in homogeneous equation is
(a) vx^2 (b) v^2x
(c) vx (d) $\dfrac{v}{x}$ **Ans.** (c)

12. If the substitution for $y = vx$ then $\frac{dv}{dx}$ is

(a) $v - x\frac{dv}{dx}$ (b) $v + x\frac{dv}{dx}$

(c) $x - v\frac{dv}{dx}$ (d) $x + v\frac{dv}{dx}$ **Ans.** (b)

13. Which is the homogeneous equation

(a) $\frac{dy}{dx} = \frac{x^2 + 2y^2}{3x^2 + 2y^3}$ (b) $\frac{dy}{dx} = \frac{x^3 + 2y^3}{3x^2 + 2y^2}$

(c) $\frac{dy}{dx} = \frac{x^2 + 2y^2}{3x^2 + 4y^2}$ (d) $\frac{dy}{dx} = \frac{x^2 + 2y^2}{3x^2 + 2y}$ **Ans.** (c)

14. The solution of the differential equation$(1 - x^2)\, y' = 2y$, $y(0) = 1$ is

(a) $\left(\frac{1+x}{1-x}\right)^2$ (b) $\left(\frac{1-x}{1+x}\right)^2$

(c) $\left(\frac{1-x}{1+x}\right)$ (d) $\left(\frac{1+x}{1-x}\right)$ **Ans.** (d)

15. One of the solution of the differential equation $\frac{dy}{dx} = \sin\left(y - x\frac{dy}{dx}\right)$ is

(a) $y = cx + \sin^{-1} C$ (b) $y = cx^2 + \sin^{-1}C$

(c) $y = x + C$ (d) None of these **Ans.** (c)

16. Integrating factor of $(x + 1)\frac{dy}{dx} - y = e^{3x}(1 + x)^2$ is :

(a) $x + 1$ (b) $\log(x + 1)$

(c) $\frac{1}{x+1}$ (d) $e^{(x + 1)}$

(R.G.P.V Bhopal, I Semester June 2007) **Ans.** (c)

17. The basis of solutions for the following differential equation $x^2\frac{d^2y}{dx^2} - x\frac{dy}{dx} + y = 0$ are

(a) $x, x \ln x$, (b) $\ln x, e^x$

(c) $\frac{1}{x}, \frac{1}{x^2}$ (d) $\left(\frac{1}{x^2}\right)e^x, x \ln x$. *(A.M.I.E.T.E., Winter., 2001)*

Ans. (a)

18. What is the degree of the differential equation $\frac{d^2y}{dx^2} + \sqrt{1 + \left(\frac{dy}{dx}\right)} = 0$

(a) 0 (b) 1

(c) 2 (d) 4 *(R.G.P.V Bhopal, I Semester June 2006)*

Ans. (c)

19. The differential equation whose solution is of the form $y = A\cos x + B\sin x$ is

(a) $\frac{d^2y}{dx^2} + y = 0$ (b) $\frac{d^2y}{dx^2} - y = 0$

(c) $\frac{dy}{dx} + y = A\cos x$ (d) $\frac{dy}{dx} + y = B\sin x$ **Ans.** (a)

20. The differential equation $y''(x) - [y'(x)]^2 - y(x)\, y'(x) + y^2(x) = 0$, is non linear, because
(*a*) The dependent variable is raised to the power two
(*b*) The dependent variable and its first derivative are multiplied
(*c*) The first derivative is raised to the power two
(*d*) The second order derivative is not raised to any power **Ans.** (*c*)

21. The differential equation, a set of whose linearly independent solution is $\{1, x, e^{-x}\}$ is
(*a*) $(D^3 + D^2)\, y = 0$ (*b*) $(D^3 + D + 1)\, y = 0$
(*c*) $(D^2 + D)y = 0$ (*d*) $(D^3 - D)y = 0$ **Ans.** (*a*)

22. The Integrating factor in solving $\frac{dy}{dx} + Py = Q$ is
(*a*) I.F. = $e^{\int Q dy}$ (*b*) I.F. = $e^{\int P dy}$
(*c*) I.F. = $e^{\int P dx}$ (*d*) I.F = $e^{\int Q dx}$ **Ans.** (*c*)

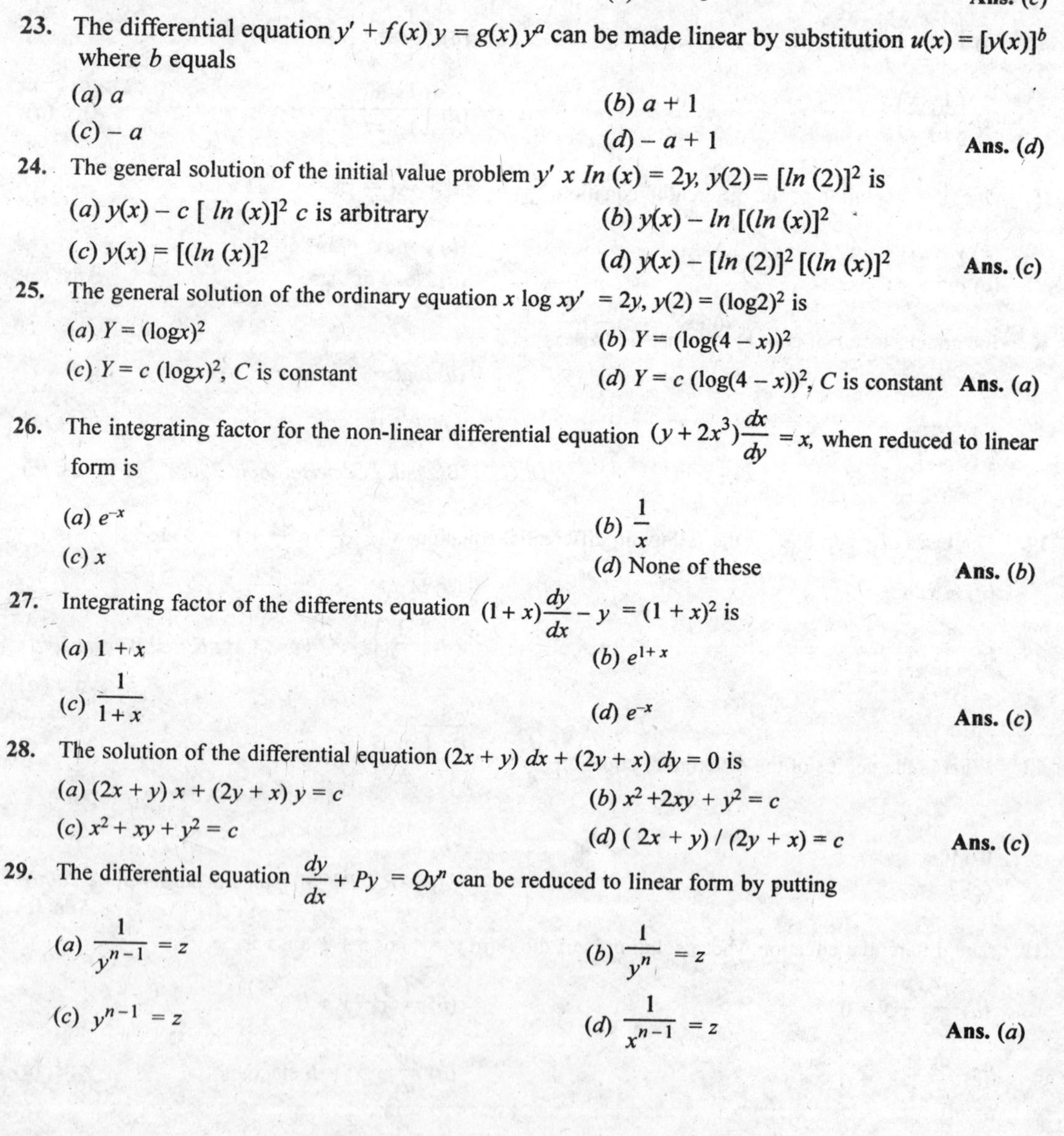

23. The differential equation $y' + f(x)\, y = g(x)\, y^a$ can be made linear by substitution $u(x) = [y(x)]^b$ where b equals
(*a*) a (*b*) $a + 1$
(*c*) $-a$ (*d*) $-a + 1$ **Ans.** (*d*)

24. The general solution of the initial value problem $y'\, x\, ln\,(x) = 2y$, $y(2) = [ln\,(2)]^2$ is
(*a*) $y(x) - c\,[\,ln\,(x)]^2$ c is arbitrary (*b*) $y(x) - ln\,[(ln\,(x)]^2$
(*c*) $y(x) = [(ln\,(x)]^2$ (*d*) $y(x) - [ln\,(2)]^2\,[(ln\,(x)]^2$ **Ans.** (*c*)

25. The general solution of the ordinary equation $x \log xy' = 2y$, $y(2) = (\log 2)^2$ is
(*a*) $Y = (\log x)^2$ (*b*) $Y = (\log(4 - x))^2$
(*c*) $Y = c\,(\log x)^2$, C is constant (*d*) $Y = c\,(\log(4 - x))^2$, C is constant **Ans.** (*a*)

26. The integrating factor for the non-linear differential equation $(y + 2x^3)\frac{dx}{dy} = x$, when reduced to linear form is
(*a*) e^{-x} (*b*) $\frac{1}{x}$
(*c*) x (*d*) None of these **Ans.** (*b*)

27. Integrating factor of the differents equation $(1 + x)\frac{dy}{dx} - y = (1 + x)^2$ is
(*a*) $1 + x$ (*b*) e^{1+x}
(*c*) $\frac{1}{1+x}$ (*d*) e^{-x} **Ans.** (*c*)

28. The solution of the differential equation $(2x + y)\, dx + (2y + x)\, dy = 0$ is
(*a*) $(2x + y)\, x + (2y + x)\, y = c$ (*b*) $x^2 + 2xy + y^2 = c$
(*c*) $x^2 + xy + y^2 = c$ (*d*) $(2x + y) / (2y + x) = c$ **Ans.** (*c*)

29. The differential equation $\frac{dy}{dx} + Py = Qy^n$ can be reduced to linear form by putting
(*a*) $\frac{1}{y^{n-1}} = z$ (*b*) $\frac{1}{y^n} = z$
(*c*) $y^{n-1} = z$ (*d*) $\frac{1}{x^{n-1}} = z$ **Ans.** (*a*)

30. The substitution, to reduce $\frac{dy}{dx} + \frac{y}{x} = \frac{y^2}{x^2}$ to linear form is

(*a*) $y = z$ (*b*) $\frac{1}{y} = z$

(*c*) $x = z$ (*d*) $\frac{1}{x} = z$ **Ans.** (*b*)

31. In solving $\frac{dy}{dx} = \frac{2x+3y+1}{x+y+2}$ the substitutions are

(*a*) $x = X + h,\ y = Y - k$ (*b*) $x = X + h,\ y = Y + k$

(*c*) $x = X^2 + h,\ y = Y^2 + k$ (*d*) $x = x + 1,\ y = Y + 1$ **Ans.** (*b*)

32. In solving $(2x - y + 2)\, dx + (6x - 3y + 4)\, dy = 0$ the substitution is

(*a*) $2x - y = z$ (*b*) $2x + y = z$

(*c*) $x - y = z$ (*d*) $x + y = z$ **Ans.** (*a*)

33. If the substitutions are $x = X + h,\ Y = Y + k$ then $\frac{dy}{dx}$ is

(*a*) $\frac{dX}{dY}$ (*b*) $\frac{dY}{dX}$

(*c*) $\frac{dY}{dx} = h + k$ (*d*) $\frac{dX}{dY} = h + k$ **Ans.** (*b*)

34. If the substitutions in sovling $\frac{dy}{dx} = \frac{ax+by+c}{Ax+By+d}$, $x = X + h,\ y = Y + k$ then the condition is

(*a*) $\frac{a}{A} = \frac{b}{B}$ (*b*) $\frac{a}{A} \neq \frac{C}{d}$

(*c*) $\frac{a}{A} = \frac{c}{d}$ (*d*) $\frac{a}{A} \neq \frac{b}{B}$ **Ans.** (*d*)

35. The differential equation $Mdx + Ndy = 0$ is exact if

(*a*) $\frac{\partial M}{\partial N} = \frac{\partial N}{\partial y}$ (*b*) $\frac{\partial M}{\partial y} = \frac{\partial N}{\partial x}$

(*c*) $\frac{\partial M}{\partial y} + \frac{\partial N}{\partial x} = 0$ (*d*) $\frac{\partial M}{\partial x} + \frac{\partial N}{\partial y} = 0$ **Ans.** (*b*)

36. The solution of the exact differential equation $Mdx + Ndy = 0$ is

(*a*) $\int Mdx = \int Ndy = C$, where *y* in *M* treated as constant and the terms containing function of *x* in *N* are to be ignored.

(*b*) $\int Mdx + \int Ndy = C$, where *x* in *M* treated as constant and the terms containing function of *y* in *N* are to be ignored.

(*c*) $\int Mdx + \int Ndy = C$, where *y* in *M* treated as constant and the terms containing function of *y* in *N* are to be ignored.

(*d*) $\int Mdx + \int Ndy = C$, where *x* in *M* treated as constant and the terms containing function of *x* in *N* are to be ignored. **Ans.** (*a*)

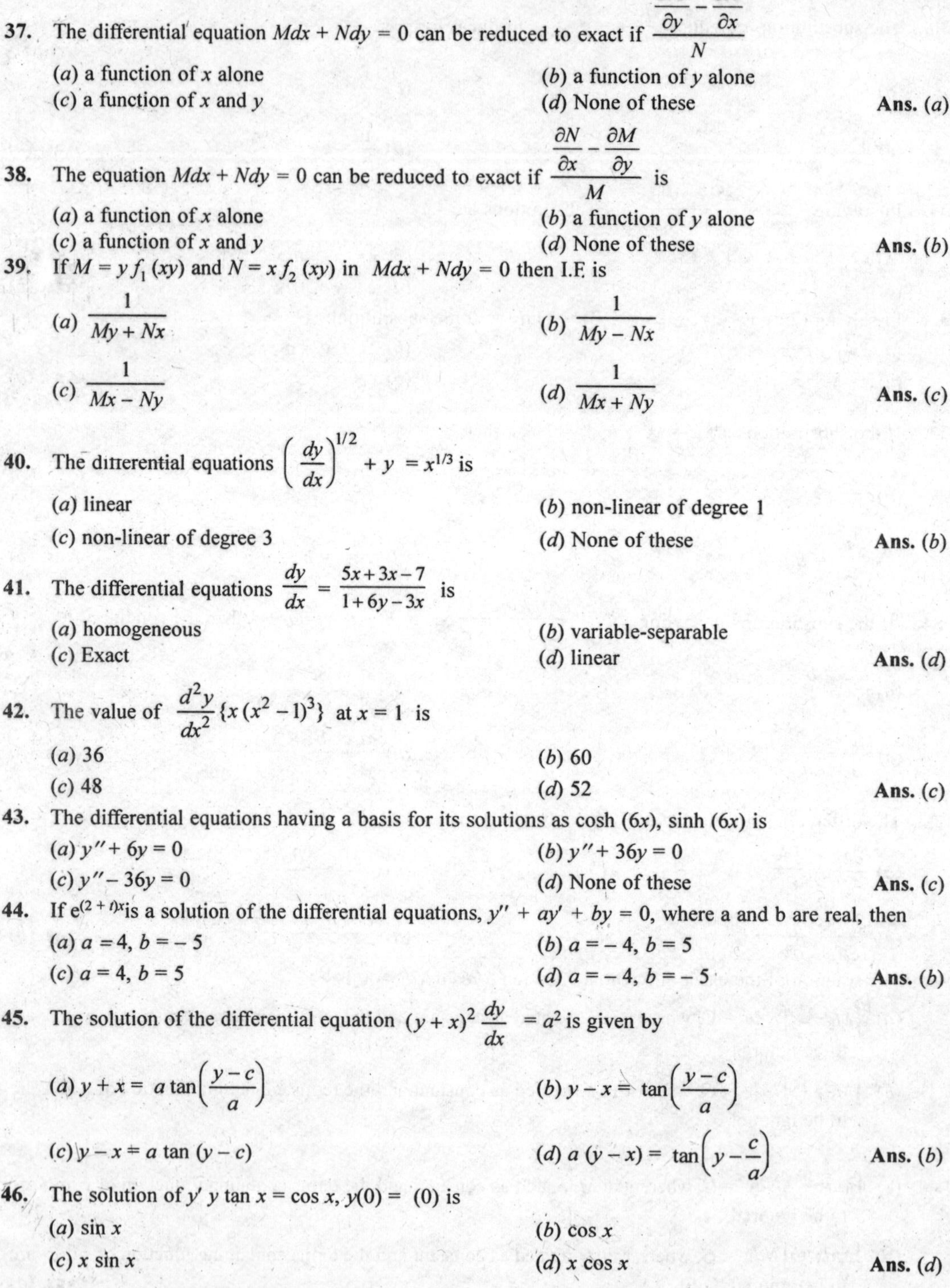

37. The differential equation $Mdx + Ndy = 0$ can be reduced to exact if $\dfrac{\dfrac{\partial M}{\partial y} - \dfrac{\partial N}{\partial x}}{N}$

(*a*) a function of *x* alone (*b*) a function of *y* alone
(*c*) a function of *x* and *y* (*d*) None of these **Ans.** (*a*)

38. The equation $Mdx + Ndy = 0$ can be reduced to exact if $\dfrac{\dfrac{\partial N}{\partial x} - \dfrac{\partial M}{\partial y}}{M}$ is

(*a*) a function of *x* alone (*b*) a function of *y* alone
(*c*) a function of *x* and *y* (*d*) None of these **Ans.** (*b*)

39. If $M = y f_1(xy)$ and $N = x f_2(xy)$ in $Mdx + Ndy = 0$ then I.F. is

(*a*) $\dfrac{1}{My + Nx}$ (*b*) $\dfrac{1}{My - Nx}$

(*c*) $\dfrac{1}{Mx - Ny}$ (*d*) $\dfrac{1}{Mx + Ny}$ **Ans.** (*c*)

40. The diтferential equations $\left(\dfrac{dy}{dx}\right)^{1/2} + y = x^{1/3}$ is

(*a*) linear (*b*) non-linear of degree 1

(*c*) non-linear of degree 3 (*d*) None of these **Ans.** (*b*)

41. The differential equations $\dfrac{dy}{dx} = \dfrac{5x + 3x - 7}{1 + 6y - 3x}$ is

(*a*) homogeneous (*b*) variable-separable
(*c*) Exact (*d*) linear **Ans.** (*d*)

42. The value of $\dfrac{d^2y}{dx^2}\{x(x^2 - 1)^3\}$ at $x = 1$ is

(*a*) 36 (*b*) 60
(*c*) 48 (*d*) 52 **Ans.** (*c*)

43. The differential equations having a basis for its solutions as cosh (6*x*), sinh (6*x*) is

(*a*) $y'' + 6y = 0$ (*b*) $y'' + 36y = 0$
(*c*) $y'' - 36y = 0$ (*d*) None of these **Ans.** (*c*)

44. If $e^{(2+i)x}$ is a solution of the differential equations, $y'' + ay' + by = 0$, where a and b are real, then

(*a*) $a = 4, b = -5$ (*b*) $a = -4, b = 5$
(*c*) $a = 4, b = 5$ (*d*) $a = -4, b = -5$ **Ans.** (*b*)

45. The solution of the differential equation $(y + x)^2 \dfrac{dy}{dx} = a^2$ is given by

(*a*) $y + x = a\tan\left(\dfrac{y - c}{a}\right)$ (*b*) $y - x = \tan\left(\dfrac{y - c}{a}\right)$

(*c*) $y - x = a\tan(y - c)$ (*d*) $a(y - x) = \tan\left(y - \dfrac{c}{a}\right)$ **Ans.** (*b*)

46. The solution of $y'\ y\tan x = \cos x$, $y(0) =$ (0) is

(*a*) sin *x* (*b*) cos *x*

(*c*) *x* sin *x* (*d*) *x* cos *x* **Ans.** (*d*)

(a) $y^2\left(1+(\frac{dy}{dx})^2\right) = r^2$
(b) $x^2\left(1+(\frac{dy}{dx})^2\right) = r^2$

(c) $r^2\left(1+(\frac{dy}{dx})^2\right) = x^2$
(d) $(x^2+y^2)\left(1+(\frac{dy}{dx})^2\right) = x^2$ **Ans.** (a)

48. $e^{-x}\left(c_1 \cos\sqrt{3x} + c_2 \sin\sqrt{3x}\right) + c_3e^{2x}$ is the solution of

(a) $\frac{d^3y}{dx^3} + 4y = 0$
(b) $\frac{d^3y}{dx^3} - 8y = 0$

(c) $\frac{d^3y}{dx^3} + 8y = 0$
(d) $\frac{d^3y}{dx^3} - 2\frac{d^2y}{dx^2} + \frac{dy}{dx} - 2y = 0$ **Ans.** (d)

49. The solution of the differential equation $\frac{dy}{dx} + \frac{y}{x} = x^2$ under the condition $y(1) = 1$ is given by

(a) $4xy = x^3 + 3$
(b) $4xy = x^4 + 3$
(c) $4xy = y^4 + 3$
(d) $4xy = y^3 + 3$ **Ans.** (c)

Fill in the blanks in each of the following:

50. The integrating factor of linear differential equation $\frac{dx}{dy} + Px = Q$, where P and Q are functions of y only, ... **Ans.** $I.F = e^{\int P dy}$

51. The integrating factor of $\frac{dy}{dx} + \frac{y}{x+2} = 0$ is ...... **Ans.** $y = \frac{c}{x+2}$

52. Classify the following differential equations:

(i) $\frac{dy}{dx} + 2y = \sin x$

(a) Homogeneous
(b) Not homogeneous
(c) linear
(d) Exact differential equation **Ans.** (c)

(ii) $\frac{dy}{dx} = \frac{x^2+y^2}{x^2-y^2}$

(a) Homogeneous
(b) Not homogeneous
(c) linear
(d) Exact differential equation **Ans.** (a)

(iii) $\frac{dy}{dx} = \frac{1+x+y}{2-x+y}$

(a) Homogeneous
(b) Not homogeneous
(c) linear
(d) Exact differential equation **Ans.** (b)

(iv) $Mdx + Ndy = 0$ and $\frac{\partial M}{\partial y} = \frac{\partial N}{\partial x}$

(a) Homogeneous
(b) Not homogeneous
(c) linear
(d) Exact differential equation **Ans.** (d)

Indicate True or false for the following:

53. The $I.F.$ of $\dfrac{dy}{dx} + y = x^3$ is e. **Ans.** False

54. We substitute to transform a differential equation $\dfrac{dy}{dx} = \dfrac{2x+3y+4}{x+y+1}$ into homogeneous equation, $x = X + h,\ y = Y + k$. **Ans.** True

55. If the variables are not separated, then we can apply the variable separable method to solve the equation. **Ans.** False

56. The solution of exact differential equation is $\underset{(y\text{-constant})}{\int f(x)dx} + \int (\text{Terms not containing } x)\, dy$ **Ans.** True

UNIT–I

2 CHAPTER — LINEAR DIFFERENTIAL EQUATIONS OF *n*TH ORDER WITH CONSTANT COEFFICIENTS

2.1 LINEAR DIFFERENTIAL EQUATIONS

If the degree of the dependent variable and all derivatives is one, such differential equations are called *linear differential equations e.g.*

(1) $\frac{d^2y}{dx^2}+5\frac{dy}{dx}+6y=x^2+x+1$ (2) $2\frac{d^2x}{dt^2}-\frac{dx}{dt}-3x=f(t)$

2.2 NON LINEAR DIFFERENTIAL EQUATIONS

If the degree of the dependent variable and/or its derivatives are of not greater than 1 such differential equations are called one-linear differential equations.

(1) $\frac{d^2y}{dx^2}+\frac{dy}{dx}+y^2=\sin x$ (2) $\frac{d^2y}{dx^2}+2\left(\frac{dy}{dx}\right)^2+y^2=e^x$

(3) $\left(\frac{d^2x}{dt^2}\right)^2+\frac{dx}{dt}+x=f(t)$

The order of a differential equation is the highest order of the derivative involved. All the above differential equations are of second order.

Fourier and Laplace transforms are mathematical tools to solve the differential equations.

2.3 LINEAR DIFFERENTIAL EQUATIONS OF SECOND ORDER WITH CONSTANT COEFFICIENTS

The general form of the linear differential equation of second order is

$$\frac{d^2y}{dx^2}+P\frac{dy}{dx}+Qy=R$$

where P and Q are constants and R is a function of x or constant.

Differential operator. Symbol D stands for the operation of differential *i.e.*,

$$Dy=\frac{dy}{dx},\quad D^2y=\frac{d^2y}{dx^2}$$

$\frac{1}{D}$ stands for the operation of integration.

$\frac{1}{D^2}$ stands for the operation of integration twice.

$$\frac{d^2y}{dx^2}+P\frac{dy}{dx}+Qy=R \text{ can be written in the operator form.}$$

$$D^2y+P\,Dy+Qy=R \quad\Rightarrow\quad (D^2+PD+Q)\,y=R$$

2.4 COMPLETE SOLUTION = COMPLEMENTARY FUNCTION + PARTICULAR INTEGRAL

Let us consider a linear differential equation of the first order

$$\frac{dy}{dx}+Py=Q \qquad ...(1)$$

Its solution is $\quad ye^{\int Pdx}=\int(Q\,e^{\int Pdx})\,dx+C$

$\Rightarrow \quad y=Ce^{-\int Pdx}+e^{-\int Pdx}\int(Qe^{\int Pdx})\,dx$

$\Rightarrow \quad y=cu+v\text{(say)} \qquad ...(2)$

where $u=e^{-\int Pdx}$ and $v=e^{-\int Pdx}\int Q\,e^{\int Pdx}\,dx$

(*i*) Now differentiating $u=e^{-\int Pdx}$ w.r.t. x, we get

$$\frac{du}{dx}=-Pe^{-\int Pdx}=-Pu$$

$\Rightarrow \quad \frac{du}{dx}+Pu=0 \quad\Rightarrow\quad \frac{d(cu)}{dx}+P(cu)=0$

which shows that $y=c.u$ is the solution of

$$\frac{dy}{dx}+Py=0$$

(*ii*) Differentiating $v=e^{-\int Pdx}\int(Qe^{\int Pdx}\,dx)$ with respect to x, we get

$$\frac{dv}{dx}=-Pe^{-\int Pdx}\int(Qe^{\int Pdx})\,dx+e^{-\int Pdx}Qe^{\int Pdx}$$

$\Rightarrow \quad \frac{dv}{dx}=-Pv+Q$

$\Rightarrow \quad \frac{dv}{dx}+Pv=Q$, which shows that $y=v$ is the solution of

$$\boxed{\frac{dy}{dx}+Py=Q}$$

Solution of the differential equation (1) is (2) consisting of two parts *i.e.* *cu* and *v*. *cu* is the solution of the differential equation whose R.H.S. is zero. *cu* is known as *complementary function*. Second part of (2) is *v* free from any arbitrary constant and is known as *particular integral*.

Complete Solution = Complementary Function + Particular Integral.

$\Rightarrow \quad \boxed{y=C.F.+P.I.}$

2.5 METHOD FOR FINDING THE COMPLEMENTARY FUNCTION

(1) In finding the complementary function, R.H.S. of the given equation is replaced by zero.

(2) Let $y=C_1\,e^{mx}$ be the C.F. of

$$\frac{d^2y}{dx^2}+P\frac{dy}{dx}+Qy=0 \qquad ...(1)$$

Putting the values of y, $\frac{dy}{dx}$ and $\frac{d^2y}{dx^2}$ in (1) then $C_1e^{mx}(m^2+Pm+Q)=0$

$$\Rightarrow \qquad m^2+Pm+Q=0.$$

It is called **Auxiliary equation.**

(3) Solve the auxiliary equation :

Case I : Roots, Real and Different. If m_1 and m_2 are the roots, then the C.F. is

$$y=C_1e^{m_1x}+C_2e^{m_2x}$$

Case II : Roots, Real and Equal. If both the roots are m_1, m_1 then the C.F. is

$$y=(C_1+C_2x)\,e^{m_1x}$$

Proof. Equation (1) can be written as

$$(D-m_1)(D-m_1)\,y=0 \qquad ...(2)$$

Replacing $(D-m_1)\,y=v$ in (2), we get

$$(D-m_1)\,v=0 \qquad ...(3)$$

$$\frac{dv}{dx}-m_1v=0 \quad\Rightarrow\quad \frac{dv}{v}=m_1\,dx$$

$$\Rightarrow \log v=m_1x+\log C_2 \Rightarrow \log v-\log C_2=m_1x \Rightarrow \log\frac{v}{C_2}=m_1x \Rightarrow \frac{v}{C_2}=e^{m_1x}$$

$$v=C_2e^{m_1x}$$

From (3) $\qquad (D-m_1)\,y=C_2\,e^{m_1x}$

This is the linear differential equation.

$$\text{I.F.}=e^{-m_1\int dx}=e^{-m_1x}$$

Solution is

$$y.e^{-m_1x}=\int(C_2e^{m_1x})(e^{-m_1x})\,dx+C_1$$

$$=\int C_2\,dx+C_1$$

$$=C_2x+C_1$$

$$y=(C_2x+C_1)\,e^{m_1x}$$

$$\text{C.F.}=(C_1+C_2x)\,e^{m_1x}$$

Example 1. *Solve :* $\frac{d^2y}{dx^2}-6\frac{dy}{dx}+9y=0$

Solution. Here, we have

$$\frac{d^2y}{dx^2}-6\frac{dy}{dx}+9y=0 \quad\Rightarrow\quad (D^2-6D+9)\,y=0$$

A.E. is $m^2-6m+9=0 \;\Rightarrow\; (m-3)^2=0 \;\Rightarrow\; m=3, 3.$

$$C.F.=(C_1+C_2x)\,e^{3x}$$ **Ans.**

Example 2. *Find the general solution of the differential equation*

$$\frac{d^5y}{dx^5}-\frac{d^3y}{dx^3}=0.$$ (*U.P., II Semester, 2009*)

Solution. Here, we have

$$\frac{d^5y}{dx^5}-\frac{d^3y}{dx^3}=0$$

$\Rightarrow \quad D^5y - D^3y = 0 \quad \Rightarrow \quad (D^5 - D^3)\,y = 0 \quad \Rightarrow \quad D^3(D^2 - 1)y = 0$

A.E. is $m^3(m^2 - 1) = 0 \quad \Rightarrow \quad m = 0, 0, 0, 1, -1$

Hence, its solution is

$y=(C_1+C_2x+C_3x)+C_4e^x+C_5e^{-x}$ **Ans.**

Case III: Roots, Imaginary. If the roots are $\alpha \pm i\,\beta$, then the solution will be

$$y=C_1e^{(\alpha+i\beta)x}+C_2e^{(\alpha-i\beta)x}=e^{\alpha x}.[C_1e^{i\beta x}+C_2e^{-i\beta x}]$$

$$= e^{\alpha x}[C_1(\cos\beta x+i\sin\beta x)+C_2(\cos\beta x-i\sin\beta x)]$$

$$= e^{\alpha x}[(C_1+C_2)\cos\beta x+i(C_1-C_2)\sin\beta x]$$

$$= \boldsymbol{e^{\alpha x}[A\cos\beta x+B\sin\beta x]}$$

Example 3. *Solve:* $\dfrac{d^2y}{dx^2}-8\dfrac{dy}{dx}+15y=0.$

Solution. Given equation can be written as

$$(D^2 - 8D + 15)\,y = 0$$

Here auxiliary equation is $m^2 - 8m + 15 = 0$

$\Rightarrow \quad (m - 3)(m - 5) = 0 \quad \therefore\ m = 3, 5$

Hence, the required solution is

$$y = C_1\,e^{3x} + C_2e^{5x}$$ **Ans.**

Example 4. *Solve:* $\dfrac{d^2y}{dx^2}+4\dfrac{dy}{dx}+5y=0,$

$$y = 2 \text{ and } \frac{dy}{dx}=\frac{d^2y}{dx^2} \text{ when } x = 0.$$

Solution. Here the auxiliary equation is

$$m^2 + 4m + 5 = 0 \quad \Rightarrow \quad m=\frac{-4\pm\sqrt{16-20}}{2}=-2\pm i$$

Its roots are $-2\pm i$

The complementary function is

$$y = e^{-2x}\,(A\cos x + B\sin x) \quad ...(1)$$

On putting $y = 2$ and $x = 0$ in (1), we get

$$2 = A$$

On putting $A = 2$ in (1), we have

$$y = e^{-2x}[2\cos x + B\sin x] \quad ...(2)$$

On differentiating (2), we get

$$\frac{dy}{dx} = e^{-2x}[-2\sin x + B\cos x] - 2e^{-2x}[2\cos x + B\sin x]$$

$$= e^{-2x}[(-2B-2)\sin x + (B-4)\cos x]$$

$$\frac{d^2y}{dx^2} = e^{-2x}[(-2B-2)\cos x - (B-4)\sin x]$$

$$-2e^{-2x}[(-2B-2)\sin x + (B-4)\cos x]$$

$$= e^{-2x}[(-4B+6)\cos x + (3B+8)\sin x]$$

But $\dfrac{dy}{dx} = \dfrac{d^2y}{dx^2}$ (given)

$$e^{-2x}[(-2B-2)\sin x + (B-4)\cos x] = e^{-2x}[(-4B+6)\cos x + (3B+8)\sin x]$$

On putting $x = 0$, we get

$$B - 4 = -4B + 6 \quad \Rightarrow \quad B = 2$$

(2) becomes, $y = e^{-2x}[2\cos x + 2\sin x]$

$$y = 2e^{-2x}[\sin x + \cos x]$$ **Ans.**

Example 5. *Solve the differential equation:*

$$(D^2 + 1)^3 (D^2 + D + 1)^2 y = 0, \text{ where } D \equiv \frac{d}{dx}.$$

Solution. Here, we have

$$(D^2 + 1)^3 (D^2 + D + 1)^2 y = 0$$

Auxiliary equation is

$$(m^2 + 1)^3 (m^2 + m + 1)^2 = 0$$

$\Rightarrow$ $(m^2 + 1)^3 = 0$ gives $m = \pm i, \pm i, \pm i$

and $(m^2 + m + 1)^2 = 0$ gives

$$m = \frac{-1 \pm \sqrt{1-4}}{2}, \frac{1 \pm \sqrt{1-4}}{2}$$

$$m = \frac{-1 \pm \sqrt{3}i}{2}, \frac{-1 \pm \sqrt{3}i}{2}$$

Thus, $C.F. = e^{0x}[(C_1 + C_2x + C_3x^2)(C_4\cos x + C_5\sin x)]$

$$+ e^{-\frac{x}{2}}\left[(C_6 + C_7x)\left(C_8\cos\frac{\sqrt{3}}{2}x + C_9\sin\frac{\sqrt{3}}{2}x\right)\right]$$

Hence, $y = e^{0x}[(C_1 + C_2x + C_3x^2)(C_4\cos x + C_5\sin x)]$

$$+ e^{\frac{x}{2}}\left[(C_6 + C_7x)\left(C_8\cos\frac{\sqrt{3}}{2}x + C_9\sin\frac{\sqrt{3}}{2}x\right)\right]$$ **Ans.**

EXERCISE 2.1

Solve the following equations :

1. $\frac{d^2y}{dx^2} - 3\frac{dy}{dx} + 2y = 0$ **Ans.** $y = C_1 e^x + C_2 e^{2x}$

2. $\frac{d^2y}{dx^2} + \frac{dy}{dx} - 30y = 0$ **Ans.** $y = C_1 e^{5x} + C_2 e^{-6x}$

3. $\frac{d^2y}{dx^2} - 8\frac{dy}{dx} + 16y = 0$ **Ans.** $y = (C_1 + C_2 x)\, e^{4x}$

4. $\frac{d^2y}{dx^2} + \mu^2 y = 0$ **Ans.** $y = C_1 \cos \mu x + C_2 \sin \mu x$

5. $(D^2 + 2D + 2)y = 0,\ y(0) = 0,\ y'(0) = 1$ *(A.M.I.E.T.E., June 2006)*

6. $\frac{d^3y}{dx^3} - 2\frac{d^2y}{dx^2} + 4\frac{dy}{dx} - 8y = 0$ **Ans.** $y = C_1 e^{2x} + C_2 \cos 2x + C_3 \sin 2x$

7. $\frac{d^4y}{dx^4} + 32\frac{d^2y}{dx^2} + 256 = 0$ *(A.M.I.E.T.E., Dec. 2004)* **Ans.** $y = (C_1 + C_2 x) \cos 4x + (C_3 + C_4 x) \sin 4x$

8. $\frac{d^4y}{dx^4} - 4\frac{d^3y}{dx^3} + 8\frac{d^2y}{dx^2} - 8\frac{dy}{dx} + 4y = 0$ **Ans.** $y = e^x\, [(C_1 + C_2 x) \cos x + (C_3 + C_4 x) \sin x]$

9. $\frac{d^4y}{dx^4} + \frac{d^2y}{dx^2} = 0,\ \ y(0) = y'(0) = y''(0) = 0,\ y'''(0) = 1$ **Ans.** $y = x - \sin x$

10. The equation for the bending of a strut is EI $\frac{d^2y}{dx^2} + Py = 0$

If $y = 0$ when $x = 0$, and $y = a$ when $x = \frac{1}{2}$, find y. **Ans.** $y = \dfrac{a \sin \sqrt{\frac{P}{EI}}\, x}{\sin \sqrt{\frac{P}{EI}}\, \frac{1}{2}}$

11. $\frac{d^3y}{dx^3} + 6\frac{d^2y}{dx^2} + 12\frac{dy}{dx} + 8y = 0,\ y(0) = 0$, and $y'(0) = 0$ and $y''(0) = 2$

(A.M.I.E.T.E. Dec. 2008) **Ans.** $y = x^2 e^{-2x}$

12. $\frac{d^3y}{dx^3} + \frac{d^2y}{dx^2} + \frac{4dy}{dx} + 4y = 0,\ y(0) = 0,\ y'(0) = 0,\ y''(0) = -5,$ **Ans.** $y = -e^{-x} + \cos 2x - \frac{1}{2} \sin 2x$

13. $(D^8 + 6D^6 - 32D^2)y = 0$ *(A.M.I.E.T.E., Summer 2005)*

Ans. $G + c_2 x + c_3 e^{\sqrt{2}x} + cue^{-\sqrt{2}x} + (c_5 + c_6 x)(c_7 \cos 2x + c_8 \sin 2x)$

14. Show that non-trivial solutions of the boundary value problem $y^{(iv)} - w^4 y = 0,\ y(0) = 0 = y''(0)$, $y(L) = 0,\ y''(L) = 0$ are $y(x) = \sum_{n=1}^{\infty} D_n \sin\left(\frac{n\pi x}{L}\right)$ where D_n are constants.

(A.M.I.E.T.E. Dec. 2005)

15. Solve the initial value problem $y''' + 6y'' + 11y' + 6y = 0,\ y(0) = 0,\ y'(0) = 1,\ y''(0) = -1.$

(A.M.I.E.T.E., Dec. 2006) **Ans.** $y = 2e^{-x} - 3e^{-2x} + e^{-3x}$.

2.6 RULES TO FIND PARTICULAR INTEGRAL

(i) $\frac{1}{f(D)} e^{ax} = \frac{1}{f(a)} e^{ax}$

If $f(a) = 0$ then $\frac{1}{f(D)} \cdot e^{ax} = x \cdot \frac{1}{f'(a)} \cdot e^{ax}$

If $f'(a)=0$ then $\frac{1}{f(D)}\cdot e^{ax}=x^2\frac{1}{f''(a)}\cdot e^{ax}$

(*ii*) $\frac{1}{f(D)}x^n=[f(D)]^{-1}x^n$ Expand $[f(D)]^{-1}$ and then operate.

(*iii*) $\frac{1}{f(D^2)}\sin ax=\frac{1}{f(-a^2)}\sin ax$ and $\frac{1}{f(D^2)}\cos ax=\frac{1}{f(-a^2)}\cos ax$

If $f(-a^2)=0$ then $\frac{1}{f(D^2)}\sin ax=x\cdot\frac{1}{f'(-a^2)}\cdot\sin ax$

and $\frac{1}{f(D^2)}\cos ax=x\frac{1}{f'(-a)}\cos ax$

(*iv*) $\frac{1}{f(D)}e^{ax}\cdot\phi(x)=e^{ax}\cdot\frac{1}{f(D+a)}\phi(x)$

(*v*) $\frac{1}{D+a}\phi(x)=e^{-ax}\int e^{ax}\cdot\phi(x)\,dx$

2.7 $\frac{1}{f(D)}e^{ax}=\frac{1}{f(a)}e^{ax}$.

Proof. We know that, $D.e^{ax}=a.e^{ax}$, $D^2e^{ax}=a^2.e^{ax}$,.............., $D^n e^{ax}=a^n e^{ax}$

Let $f(D)\,e^{ax}=(D^n+K_1D^{n-1}+\ldots+K_n)\,e^{ax}=(a^n+K_1a^{n-1}+\ldots+K_n)\,e^{ax}=f(a)\,e^{ax}$.

Operating both sides by $\frac{1}{f(D)}$, we get

$$\frac{1}{f(D)}\cdot f(D)e^{ax}=\frac{1}{f(D)}\cdot f(a)e^{ax}$$

$$\Rightarrow\qquad e^{ax}=f(a)\frac{1}{f(D)}\cdot e^{ax}\quad\Rightarrow\quad\frac{1}{f(D)}e^{ax}=\frac{1}{f(a)}e^{ax}$$

If $f(a)=0$, then the above rule fails.

Then $\frac{1}{f(D)}e^{ax}=x\cdot\frac{1}{f'(D)}e^{ax}=x\frac{1}{f'(a)}e^{ax}$ $\Rightarrow$ $\boxed{\frac{1}{f(D)}e^{ax}=\frac{1}{f(a)}e^{ax}}$

If $f'(a)=0$ then $\boxed{\frac{1}{f(D)}e^{ax}=x^2\frac{1}{f''(a)}e^{ax}}$

Example 6. *Solve the differential equation*

$$\frac{d^2x}{dt^2}+\frac{g}{l}x=\frac{g}{l}L$$

where g, l, L are constants subject to the conditions,

$$x=a,\quad\frac{dx}{dt}=0\quad at\ t=0.$$

Solution. We have, $\frac{d^2x}{dt^2}+\frac{g}{l}x=\frac{g}{l}L\quad\Rightarrow\quad\left(D^2+\frac{g}{l}\right)x=\frac{g}{l}L$

A.E. is $m^2 + \frac{g}{l} = 0 \Rightarrow m = \pm\, i\sqrt{\frac{g}{l}}$

$$\text{C.F.} = C_1 \cos\sqrt{\frac{g}{l}}\,t + C_2 \sin\sqrt{\frac{g}{l}}\,t$$

$$\text{P.I.} = \frac{1}{D^2 + \frac{g}{l}} \cdot \frac{g}{l} L = \frac{g}{l} L \frac{1}{D^2 + \frac{g}{l}} e^{0t} = \frac{g}{l} L \frac{1}{0 + \frac{g}{l}} = L \qquad [D = 0]$$

$\therefore$ General solution is = C.F. + P.I.

$$x = C_1 \cos\left(\sqrt{\frac{g}{l}}\right)t + C_2 \sin\left(\sqrt{\frac{g}{l}}\right)t + L \qquad ...(1)$$

$$\frac{dx}{dt} = -C_1\sqrt{\frac{g}{l}} \sin\left(\sqrt{\frac{g}{l}}\right)t + C_2\sqrt{\frac{g}{l}} \cos\left(\sqrt{\frac{g}{l}}\right)t$$

Put $t = 0$ and $\frac{dx}{dt} = 0$

$$0 = C_2\sqrt{\frac{g}{l}} \Rightarrow C_2 = 0$$

(1) becomes $x = C_1 \cos\sqrt{\frac{g}{l}}\,t + L \qquad ...(2)$

Put $x = a$ and $t = 0$ in (2), we get

$$a = C_1 + L \quad \text{or} \quad C_1 = a - L$$

On putting the value of C_1 in (2), we get

$$x = (a - L)\cos\left(\sqrt{\frac{g}{l}}\right)t + L$$ **Ans.**

Example 7. *Solve* : $\frac{d^2y}{dx^2} + 6\frac{dy}{dx} + 9y = 5e^{3x}$

Solution. Here, we have $(D^2 + 6D + 9)y = 5e^{3x}$

Auxiliary equation is $m^2 + 6m + 9 = 0 \Rightarrow (m + 3)^2 = 0 \Rightarrow m = -3, -3,$

$$\text{C.F.} = (C_1 + C_2x)\, e^{-3x}$$

$$\text{P.I.} = \frac{1}{D^2 + 6D + 9}.5.e^{3x} = 5\frac{e^{3x}}{(3)^2 + 6(3) + 9} = \frac{5e^{3x}}{36}$$

The complete solution is $y = (C_1 + C_2x)e^{-3x} + \frac{5e^{3x}}{36}$ **Ans.**

Example 8. *Solve* $\frac{d^2y}{dx^2} - (a + b)\frac{dy}{dx} + aby = e^{ax} + e^{bx}$ (*M.U. II Semester, 2009*)

Solution. Here, we have

$$\frac{d^2y}{dx^2} - (a + b)\frac{dy}{dx} + aby = e^{ax} + e^{bx}$$

$$\Rightarrow [D^2 - (a + b)\, D + ab]\, y = e^{ax} + e^{bx}$$

A.E. is $m^2 - (a + b)m + ab = 0$

$$\Rightarrow \quad (m-a)(m-b) = 0 \quad \Rightarrow \quad m = a, m = b$$

$$\text{C.F.} = C_1 e^{ax} + C_2 e^{bx}$$

$$\text{P.I.} = \frac{1}{D^2-(a+b)D+ab}(e^{ax}+e^{bx})$$

$$\text{P.I.} = \frac{1}{D^2-(a+b)D+ab}e^{ax} + \frac{1}{D^2-(a+b)D+ab}e^{bx}$$

[Here $f(a) = 0$ and $f(b) = 0$, Rule fails]

$$= x\frac{1}{2D-(a+b)}e^{ax} + x\frac{1}{2D-(a+b)}e^{bx}$$

$$= x\frac{1}{2a-(a+b)}e^{ax} + x\frac{1}{2b-(a+b)}e^{bx}$$

$$= x\frac{1}{a-b}e^{ax} + x\frac{1}{b-a}e^{bx} = \frac{x}{a-b}[e^{ax}-e^{bx}]$$

Complete solution is $\quad y = \text{C.F.} + \text{P.I.}$

$$\Rightarrow \quad y = C_1 e^{ax} + C_2 e^{bx} + \frac{x}{(a-b)}[e^{ax}-e^{bx}]$$ **Ans.**

Example 9. *Solve* : $\quad \frac{d^2y}{dx^2} - 6\frac{dy}{dx} + 9y = 6e^{3x} + 7e^{-2x} - \log 2$

Solution. Here, $\quad (D^2 - 6D + 9)y = 6e^{3x} + 7e^{-2x} - \log 2$

A.E. is $(m^2 - 6m + 9) = 0 \quad \Rightarrow \quad (m-3)^2 = 0 \quad \Rightarrow \quad m = 3, 3$

$$\text{C.F.} = (C_1 + C_2 x)e^{3x}$$

$$\text{P.I.} = \frac{1}{D^2-6D+9}6e^{3x} + \frac{1}{D^2-6D+9}7e^{-2x} + \frac{1}{D^2-6D+9}(-\log 2)$$

$$= x\frac{1}{2D-6}6e^{3x} + \frac{1}{4+12+9}7e^{-2x} - \log 2\frac{1}{D^2-6D+9}e^{0x}$$

$$= x^2\frac{1}{2}\cdot 6\cdot e^{3x} + \frac{7}{25}e^{-2x} - \log 2\left(\frac{1}{9}\right) = 3x^2e^{3x} + \frac{7}{25}e^{-2x} - \frac{1}{9}\log 2$$

Complete solution is $y = (C_1 + C_2\,x)e^{3x} + 3x^2e^{3x} + \frac{7}{25}e^{-2x} - \frac{1}{9}\log 2$ **Ans.**

EXERCISE 2.2

Solve the following differential equations:

1. $\frac{d^2y}{dx^2} - 3\frac{dy}{dx} + 2y = e^{3x}$ *(A.M.I.E.T.E. 2007)* **Ans.** $C_1 e^x + C_2 e^{2x} + \frac{e^{3x}}{2}$

2. $(D^3 + 2D^2 - D - 2)\,y = e^x$ **Ans.** $C_1 e^x + C_2 e^{-x} + C_3 e^{-2x} + \frac{x}{6}e^x$

3. $\frac{d^2y}{dx^2} + 2\frac{dy}{dx} + 2y = \sinh x$ **Ans.** $e^{-x}[C_1\cos x + C_2\sin x] + \frac{e^x}{10} - \frac{e^{-x}}{2}$

4. $\frac{d^2y}{dx^2} + 4\frac{dy}{dx} + 5y = -2\cosh x$ **Ans.** $e^{-2x}(C_1\cos x + C_2\sin x) - \frac{1}{10}e^x - \frac{e^{-x}}{2}$

4. $\dfrac{d^2y}{dx^2}+4\dfrac{dy}{dx}+5y=-2\cosh x$ **Ans.** $e^{-2x}(C_1\cos x+C_2\sin x)-\dfrac{1}{10}e^x-\dfrac{e^{-x}}{2}$

5. $(D^3-2D^2-5D+6)\,y=e^{3x}$ **Ans.** $C_1\,e^x+C_2\,e^{-2x}+C_3\,e^{3x}+\dfrac{x.e^{3x}}{10}$

6. $\dfrac{d^3y}{dx^3}-\dfrac{d^2y}{dx^2}+4\dfrac{dy}{dx}-4y=e^x$ **Ans.** $C_1\,e^x+C_2\cos 2x+C_3\sin 2x+\dfrac{xe^x}{5}$

7. $\dfrac{d^2y}{dx^2}-6\dfrac{dy}{dx}+9y=e^{3x}$ **Ans.** $(C_1+C_2x)\,e^{3x}+\dfrac{x^2}{2}e^{3x}$

8. $\dfrac{d^3y}{dx^3}+3\dfrac{d^2y}{dx^2}+3\dfrac{dy}{dx}+y=e^{-x}$ **Ans.** $(C_1+C_2\,x+C_3\,x^2)\,e^{-x}+\dfrac{x^3}{6}e^{-x}$

9. $\dfrac{d^2y}{dx^2}-\dfrac{dy}{dx}-6y=e^x\cosh 2x$ **Ans.** $C_1\,e^{3x}+C_2\,e^{-2x}+\dfrac{1}{10}x\,e^{3x}-\dfrac{1}{8}e^{-x}$

10. $(D-2)\,(D+1)^2\,y=e^{2x}+e^x$ **Ans.** $C_1\,e^{2x}+(C_2+C_3\,x)\,e^{-x}+\dfrac{x}{9}e^{2x}-\dfrac{e^x}{4}$

11. $(D-1)^3\,y=16\,e^{3x}$ **Ans.** $(C_1+C_2\,x+C_3\,x^2)\,2\,e^{3x}$

2.8 $\dfrac{1}{f(D)}x^n=[f(D)]^{-1}x^n.$

Expand $[f(D)]^{-1}$ by the Binomial theorem in ascending powers of D as far as the result of operation on x^n is zero.

Example 10 *Solve the differential equation*

$$\frac{d^2y}{dx^2}+a^2y=\frac{a^2R}{p}(l-x)$$

where a, R, p and l are constants subject to the conditions

$$y=0,\qquad \frac{dy}{dx}=0\qquad at\ x=0.$$

Solution. $\dfrac{d^2y}{dx^2}+a^2y=\dfrac{a^2}{p}R(l-x)\quad\Rightarrow\quad (D^2+a^2)y=\dfrac{a^2}{p}R(l-x)$

A.E. is $m^2+a^2=0 \quad\Rightarrow\quad m=\pm ia$

C.F. $=C_1\cos ax+C_2\sin ax$

$$\text{P.I.}=\frac{1}{D^2+a^2}\frac{a^2}{p}R(l-x)=\frac{a^2R}{p}\frac{1}{a^2}\left[\frac{1}{1+\frac{D^2}{a^2}}\right](l-x)=\frac{R}{p}\left[1+\frac{D^2}{a^2}\right]^{-1}(l-x)$$

$$=\frac{R}{p}\left[1-\frac{D^2}{a^2}\right](l-x)=\frac{R}{p}(l-x)$$

$$y=C_1\cos ax+C_2\sin ax+\frac{R}{p}(l-x) \qquad ...(1)$$

On putting $y=0$, and $x=0$ in (1), we get

$$0=C_1+\frac{R}{p}l\quad\Rightarrow\quad C_1=-\frac{Rl}{p}$$

On differentiating (1), we get

$$\frac{dy}{dx} = -a\,C_1 \sin ax + a\,C_2 \cos ax - \frac{R}{p} \qquad ...(2)$$

On putting $\frac{dy}{dx} = 0$ and $x = 0$ in (2), we get

$$0 = a\,C_2 - \frac{R}{p} \quad \Rightarrow \quad C_2 = \frac{R}{a.p}$$

On putting the values of C_1 and C_2 in (1), we get

$$y = -\frac{R}{p} l \cos ax + \frac{R}{a.p}\sin ax + \frac{R}{p}(l - x)$$

$$y = \frac{R}{p}\left[\frac{\sin ax}{a} - l\cos ax + l - x\right] \qquad \textbf{Ans.}$$

EXERCISE 2.3

Solve the following equations :

1. $(D^2 + 5D + 4)\,y = 3 - 2x$ **Ans.** $C_1 e^{-x} + C_2 e^{-4x} + \frac{1}{8}(11 - 4x)$

2. $\frac{d^2y}{dx^2} + 2\frac{dy}{dx} + y = x$ **Ans.** $(C_1 + C_2 x)\,e^{-x} + x - 2$

3. $(2D^2 + 3D + 4)\,y = x^2 - 2x$ **Ans.** $e^{-\frac{3}{4}x}\left[A\cos\frac{\sqrt{23}}{4}x + B\sin\frac{\sqrt{23}}{4}x\right] + \frac{1}{32}[8x^2 - 28x + 13]$

4. $(D^2 - 4D + 3)\,y = x^3$ **Ans.** $C_1 e^{x} + C_2 e^{3x} + \frac{1}{27}(9x^3 + 36x^2 + 78x + 80).$

5. $5\frac{d^3y}{dx^3} - \frac{d^2y}{dx^2} - 6\frac{dy}{dx} = 1 + x^2.$ **Ans.** $A + Be^{-2x} + Ce^{3x} - \frac{1}{36}\left(2x^3 - x^2 + \frac{25}{3}x\right)$

6. $\frac{d^4y}{dx^4} + 4y = x^4$ **Ans.** $e^{x}(C_1 \cos x + C_2 \sin x) + e^{-x}(C_3 \cos x + C_4 \sin x) + \frac{1}{4}(x^4 - 6)$

7. $\frac{d^2y}{dx^2} + 2P\frac{dy}{dx} + (p^2 + q^2)y = e^{cx} + p.q\,x^2$

Ans. $e^{-px}[C_1 \cos qx + C_2 \sin qx] + \frac{e^{Cx}}{(p+C)^2 + q^2} + \frac{pq}{p^2 + q^2}\left[x^2 - \frac{4px}{p^2 + q^2} + \frac{6p^2 - 2q^2}{(p^2 + q^2)^2}\right]$

8. $D^2(D^2 + 4)\,y = 96\,x^2$ **Ans.** $C_1 + C_2 x + C_3 \cos 2x + C_4 \sin 2x + 2x^2(x^2 - 3)$

2.9 $\boxed{\frac{1}{f(D^2)}\sin ax = \frac{\sin ax}{f(-a^2)}}$ $\boxed{\frac{1}{f(D^2)}\cdot\cos ax = \frac{\cos ax}{f(-a^2)}}$

$D(\sin ax) = a.\cos ax$, $D^2(\sin ax) = D(a\cos ax) = -a^2.\sin ax$

$D^4(\sin ax) = D^2.D^2(\sin ax) = D^2(-a^2 \sin ax) = (-a^2)^2 \sin ax$

$(D^2)^n \sin ax = (-a^2)^n \sin ax$

Hence, $f(D^2)\sin ax = f(-a^2)\sin ax$

$$\frac{1}{f(D^2)} \cdot f(D^2) \sin ax = \frac{1}{f(D^2)} . f(-a^2) . \sin ax$$

$$\sin ax = f(-a^2) \frac{1}{f(D^2)} \sin ax$$

$$\frac{1}{f(D^2)} \cdot \sin ax = \frac{\sin ax}{f(-a^2)}$$

Similarly, $$\frac{1}{f(D^2)} \cos ax = \frac{\cos ax}{f(-a^2)}$$

If $f(-a^2) = 0$ then above rule fails.

then $$\frac{1}{f(D^2)} \sin ax = x \frac{\sin ax}{f'(-a^2)}$$

If $f'(-a^2) = 0$ then, $\dfrac{1}{f(D^2)} \sin ax = x^2 \dfrac{\sin ax}{f''(-a^2)}$

Example 11. *Solve* : $(D^2 + 4)\, y = \sin 3x$

Solution. We have, $(D^2 + 4)\, y = \sin 3x$

Auxiliary equation is $m^2 + 4 = 0 \Rightarrow m = \pm 2i$,

$$\text{C.F.} = A \cos 2x + B \sin 2x$$

$$\text{P.I.} = \frac{1}{D^2 + 4} . \sin 3x = \frac{\sin 3x}{(-3^2) + 4} = -\frac{1}{5} \sin 3x$$

Complete solution is $y = A \cos 2x + B \sin 2x - \dfrac{1}{5} \sin 3x$ **Ans.**

Example 12. *Solve* : $\dfrac{d^2 y}{dx^2} + \dfrac{dy}{dx} + y = \cos 2x$

Solution. We have, $(D^2 + D + 1)\, y = \cos 2x$

Auxiliary equation is $m^2 + m + 1 = 0$

$$m = \frac{-1 \pm \sqrt{-3}}{2}, \qquad \text{C.F.} = e^{-\frac{x}{2}} [A \cos \frac{\sqrt{3}}{2} x + B \sin \frac{\sqrt{3} x}{2}]$$

$$\text{P.I.} = \frac{1}{D^2 + D + 1} . \cos 2x = \frac{1}{(-2^2) + D + 1} . \cos 2x = \frac{1}{D - 3} . \cos 2x$$

$$= \frac{D + 3}{D^2 - 9} . \cos 2x = \frac{D + 3}{(-2^2) - 9} \cos 2x$$

$$= -\frac{1}{13} (D + 3) \cos 2x = -\frac{1}{13} (-2 \sin 2x + 3 \cos 2x)$$

Complete solution is

$$y = e^{-\frac{x}{2}} \left[A \cos \frac{\sqrt{3}}{2} x + B \sin \frac{\sqrt{3}}{2} x \right] + \frac{1}{13} [2 \sin 2x - 3 \cos 2x]$$ **Ans.**

Example 13. *Solve* : $(D^2 + 4)\, y = \cos 2x$ *(A.M.I.E.T.E. June 2009, Dec 2008)*

Solution. Here, we have $(D^2 + 4)\, y = \cos 2x$

Auxiliary equation is $m^2 + 4 = 0 \Rightarrow m = \pm 2i$

$$\text{C.F.} = A\cos 2x + B\sin 2x$$

$$\text{P.I.} = \frac{1}{D^2+4}\cos 2x$$

$$= x.\frac{1}{2D}\cos 2x = \frac{x}{2}\left(\frac{1}{2}\sin 2x\right) = \frac{x}{4}\sin 2x$$

Complete solution is $y = A\cos 2x + B\sin 2x + \frac{x}{4}\sin 2x$ **Ans.**

Example 14. *Solve* $(D^3 + D^2 + D + 1)\, y = \sin^2 x$ (*M.U. II Semester, 2008*)

Solution. Here, we have

$$(D^3 + D^2 + D + 1)\, y = \sin^2 x$$

A.E. is $m^3 + m^2 + m + 1 = 0 \Rightarrow m^2(m+1) + 1(m+1) = 0$

$(m+1)(m^2+1) = 0 \Rightarrow m = -1,\ m = \pm i$

$$\text{C.F.} = C_1 e^{-x} + (C_2\cos x + C_3\sin x)$$

$$\text{P.I.} = \frac{1}{D^3+D^2+D+1}\sin^2 x,$$

$$\text{P.I.} = \frac{1}{D^3+D^2+D+1}\,\frac{1}{2}(1-\cos 2x)$$

$$\text{P.I.} = \frac{1}{2}\,\frac{1}{D^3+D^2+D+1}(1-\cos 2x) = \frac{1}{2}\left[\frac{1}{D^3+D^2+D+1}e^{0x} - \frac{1}{D^3+D^2+D+1}\cos 2x\right]_{z=1}$$

$$\text{P.I.} = \frac{1}{2}\left[1 - \frac{1}{D(-4)-4+D+1}\cos 2x\right] = \frac{1}{2}\left[1 - \frac{1}{-3D-3}\cos 2x\right]$$

$$\text{P.I.} = \frac{1}{2}\left[1 + \frac{1}{3}\,\frac{1}{D+1}\cos 2x\right] = \frac{1}{2}\left[1 + \frac{1}{3}\,\frac{D-1}{D^2-1}\cos 2x\right]$$

$$= \frac{1}{2}\left[1 + \frac{1}{3}\,\frac{D-1}{-4-1}\cos 2x\right] = \frac{1}{2}\left[1 - \frac{1}{15}(D-1)\cos 2x\right] = \frac{1}{2}\left[1 - \frac{1}{15}(-2\sin 2x - \cos 2x)\right]$$

$$= \frac{1}{30}\,[15 + 2\sin 2x + \cos 2x]$$

Complete solution is y = C.F. + P.I.

$\Rightarrow\ y = C_1 e^{-x} + C_2\cos x + C_3\sin x + \frac{1}{30}\,[15 + 2\sin 2x + \cos 2x]$ **Ans.**

Example 15. *Solve* : $\frac{d^3y}{dx^3} - 3\frac{d^2y}{dx^2} + 4\frac{dy}{dx} - 2y = e^x + \cos x$

(*U.P., II Semester, Summer* 2006, 2001)

Solution. We have, $(D^3 - 3D^2 + 4D - 2)\, y = e^x + \cos x$

A.E. is $m^3 - 3m^2 + 4m - 2 = 0$

$\Rightarrow (m-1)(m^2 - 2m + 2) = 0$, *i.e.*, $m = 1,\ 1 \pm i$

$\therefore$ C.F. $= C_1 e^x + (C_2\cos x + C_3\sin x)\, e^x$

$$\text{P.I.} = \frac{1}{(D-1)(D^2-2D+2)}e^x + \frac{1}{D^3-3D^2+4D-2}\cos x$$

$$= \frac{1}{(D-1)(1-2+2)}e^x + \frac{1}{(-1)D-3(-1)+4D-2}\cos x$$

$$= \frac{1}{(D-1)}e^x + \frac{1}{3D+1}\cos x$$

$$= e^x \frac{1}{D+1-1}.1 + \frac{3D-1}{9D^2-1}\cos x$$

$$= e^x \frac{1}{D}.1 + \frac{(-3\sin x - \cos x)}{-9-1}$$

$$= e^x.x + \frac{1}{10}(3\sin x + \cos x)$$

Hence, complete solution is

$$y = C_1 e^x + (C_2\cos x + C_3 \sin x)\, e^x + x\, e^x + \frac{1}{10}(3\sin x + \cos x)$$ **Ans.**

Example 16. *Solve* : $(D^4 - 3D^2 - 4)y = 5 \sin 2x - e^{-2x}$ (*Nagpur University, Summer* 2001)

Solution. Auxiliary equation is

$$m^4 - 3m^2 - 4 = 0$$

$$(m^2 + 1)(m^2 - 4) = 0$$

$$\Rightarrow \quad m^2 + 1 = 0 \text{ and } m^2 - 4 = 0$$

$$\Rightarrow \quad m = \pm 1 \text{ and } m = \pm 2$$

$$\therefore \quad \text{C.F.} = C_1 e^{2x} + C_2 e^{-2x} + C_3\cos x + C_4 \sin x$$

$$\text{P.I.} = \frac{1}{D^4 - 3D^2 - 4}(5\sin 2x - e^{-2x})$$

$$= 5.\frac{1}{D^4 - 3D^2 - 4}\sin 2x - \frac{1}{D^4 - 3D^2 - 4}e^{-2x}$$

$$= 5\frac{1}{(-2^2)^2 - 3(-2^2) - 4}\sin 2x - \frac{1}{16-12-4}e^{-2x}$$ (Rule fails)

$$= \frac{5}{24}\sin 2x - x\frac{1}{4D^3 - 6D}e^{-2x} = \frac{5}{24}\sin 2x - x\frac{1}{4(-2)^3 - 6(-2)}e^{-2x}$$

$$= \frac{5}{24}\sin 2x + \frac{x e^{-2x}}{20}$$

Hence, the complete solution is

$$y = c_1 e^{2x} + c_2 e^{-2x} + c_3 \cos x + c_4 \sin x + \frac{5}{24}\sin 2x + \frac{x e^{-2x}}{20}$$ **Ans.**

Example 17. *Solve* : $(D^3 + 1)y = \cos^2\left(\frac{x}{2}\right) + e^{-x}$ (*Nagpur University, Summer* 2004)

Solution. We have, $(D^3 + 1)y = \cos^2\left(\frac{x}{2}\right) + e^{-x}$

A.E. is $m^3 + 1 = 0$

$(m + 1)(m^2 - m + 1) = 0 \quad \Rightarrow \quad m = -1$

and $m = \dfrac{-(-1) \pm \sqrt{1-4}}{2} = \dfrac{1 \pm i\sqrt{3}}{2} \quad \Rightarrow \quad m = \dfrac{1}{2} \pm i\dfrac{\sqrt{3}}{2}$

$$\therefore \quad \text{C.F.} = C_1 e^{-x} + e^{\frac{x}{2}}\left[C_2 \cos\frac{\sqrt{3}}{2}x + C_3 \sin\frac{\sqrt{3}}{2}x\right]$$

$$\text{P.I.} = \frac{1}{D^3+1}\left[\cos^2\left(\frac{x}{2}\right) + e^{-x}\right] = \frac{1}{D^3+1}\cos^2\left(\frac{x}{2}\right) + \frac{1}{D^3+1}e^{-x} \qquad [\text{Put } D = -1]$$

$$= \frac{1}{D^3+1}\left(\frac{1+\cos x}{2}\right) + \frac{x}{3D^2}e^{-x}$$

$$= \frac{1}{2}\frac{1}{D^3+1}e^{0x} + \frac{1}{2}\frac{1}{D^3+1}\cos x + \frac{x}{3(-1)^2}e^{-x}$$

$$= \frac{1}{2} + \left(\frac{1}{2}\right)\frac{1}{-D+1}\cos x + \frac{x}{3}e^{-x} = \frac{1}{2} - \frac{1}{2}\frac{(D+1)\cos x}{(D-1)(D+1)} + \frac{x}{3}e^{-x}$$

$$= \frac{1}{2} - \left(\frac{1}{2}\right)\frac{(-\sin x + \cos x)}{(D^2-1)} + \frac{x}{3}e^{-x} = \frac{1}{2} + \frac{1}{2}\frac{\sin x}{(D^2-1)} - \frac{1}{2}\frac{1}{(D^2-1)}\cos x + \frac{x}{3}e^{-x}$$

Put $\quad D^2 = -1 = \frac{1}{2} + \frac{1}{2}\frac{\sin x}{(-1-1)} - \frac{1}{2}\frac{1}{(-1-1)}\cos x + \frac{x}{3}e^{-x} = \frac{1}{2} - \frac{\sin x}{4} + \frac{\cos x}{4} + \frac{x}{3}e^{-x}$

$$\text{P.I.} = \frac{1}{2} + \frac{1}{4}(\cos x - \sin x) + \frac{x}{3}e^{-x}$$

Hence, the complete solution is

$$y = C_1 e^{-x} + e^{\frac{x}{2}}\left[C_2 \cos\frac{\sqrt{3}}{2}x + C_3 \sin\frac{\sqrt{3}}{2}x\right] + \frac{1}{2} + \frac{1}{4}(\cos x - \sin x) + \frac{x}{3}e^{-x} \qquad \textbf{Ans.}$$

Example 18. *Solve the differential equation*

$$\frac{d^2y}{dx^2} - 2\frac{dy}{dx} + 2y = \sinh x + \sin\sqrt{2}\,x. \qquad \textit{(Nagpur University, Winter 2001)}$$

Solution. A.E. is $m^2 - 2m + 2 = 0$

$$\therefore \quad m = \frac{2 \pm \sqrt{4-8}}{2} = \frac{2 \pm 2i}{2} = 1 \pm i$$

$$\text{C.F.} = e^x (C_1 \cos x + C_2 \sin x) \qquad \left(\sinh x = \frac{e^x - e^{-x}}{2}\right)$$

$$\text{P.I.} = \frac{1}{D^2 - 2D + 2}\sinh x + \frac{1}{D^2 - 2D + 2}\sin\sqrt{2}x$$

$$= \frac{1}{D^2 - 2D + 2}\left(\frac{e^x - e^{-x}}{2}\right) + \frac{1}{D^2 - 2D + 2}\sin\sqrt{2}x$$

$$= \frac{1}{2}\left[\frac{1}{1-2+2}e^x - \frac{1}{1+2+2}e^{-x}\right] + \frac{1}{-2-2D+2}\sin\sqrt{2}x$$

$$= \frac{1}{2}e^x - \frac{1}{10}e^{-x} + \frac{1}{2\sqrt{2}}\cos\sqrt{2}x \qquad \left(\frac{1}{D}\sin\sqrt{2}x = \int \sin\sqrt{2}\,x\,dx\right)$$

Hence the solution is,

$$y = e^x (C_1 \cos x + C_2 \sin x) + \frac{1}{2}e^x - \frac{1}{10}e^{-x} + \frac{1}{2\sqrt{2}}\cos\sqrt{2}x. \qquad \textbf{Ans.}$$

Example 19. *Solve* : $\frac{d^2y}{dx^2} + y = \sin x \sin 2x.$ (*M.U. II Semester, 2009*)

Solution. We have,

$$\frac{d^2y}{dx^2} + y = \sin x \sin 2x$$

$$(D^2 + 1)\, y = \sin x \sin 2x$$

A.E. is $m^2 + 1 = 0 \Rightarrow m = \pm i$

$$\text{C.F.} = C_1 \cos x + C_2 \sin x$$

$$\text{P.I.} = \frac{1}{D^2+1} \sin x \sin 2x = \frac{1}{D^2+1}\frac{1}{2}[\cos x - \cos 3x]$$

$$= \frac{1}{2}\left[\frac{1}{D^2+1}\cos x - \frac{1}{D^2+1}\cos 3x\right]$$

$$= \frac{1}{2}\left[x\frac{1}{2D}\cos x - \frac{1}{-9+1}\cos 3x\right] = \frac{1}{2}\left[\frac{x}{2}\sin x + \frac{1}{8}\cos 3x\right]$$

$$= \frac{1}{16}[4x \sin x + \cos 3x]$$

Complete solution is $y = \text{C.F.} + \text{P.I.}$

$$\Rightarrow \quad y = C_1 \cos x + C_2 \sin x + \frac{1}{16}(4x \sin x + \cos 3x)$$ **Ans.**

Example 20. *Integrate the differential equation*

$$\frac{d^2x}{dt^2} + 2k\frac{dx}{dt} + w^2x = a \cos pt$$

and give the physical interpretation of the complete solution. Also deduce that the solution takes the form

$$x = \frac{a\cos(pt - \theta)}{\sqrt{(w^2-p^2)^2 + 4k^2p^2}} \qquad \left[\tan\theta = \frac{2kp}{w^2-p^2}\right]$$

as $t \to \infty$.

Solution. Here, we have

$$\frac{d^2x}{dt^2} + 2k\frac{dx}{dt} + w^2x = a \cos pt$$

$$(D^2 + 2k\,D + w^2)\, x = a \cos pt$$

Auxiliary equation is

$m^2 + 2\,k\,m + w^2 = 0$

$$\Rightarrow \quad m = \frac{-2k \pm \sqrt{4k^2 - 4w^2}}{2} = -k \pm \sqrt{k^2 - w^2} = -k \pm i\sqrt{w^2 - p^2}$$

$$\text{C.F.} = e^{-kt}\left(A\cos\sqrt{w^2-p^2}\,t + B\sin\sqrt{w^2-p^2}\,t\right)$$

$$\text{P.I.} = \frac{1}{D^2 + 2k\,D + w^2}\,a\cos pt = a\frac{1}{-p^2 + 2k\,D + w^2}\cos pt$$

$$= a\frac{2aD - (w^2-p^2)}{2k\,D + (w^2-p^2)}\,\frac{1}{2k\,D - (w^2-p^2)}\cos pt = a\frac{2k\,D - (w^2-p^2)}{4k^2D^2 - (w^2-p^2)^2}\cos pt$$

$$= a\frac{2kD-(w^2-p^2)}{4k^2(-p^2)-(w^2-p^2)^2}\cos pt = a\frac{-2kp\sin pt-(w^2-p^2)\cos pt}{-4k^2p^2-(w^2-p^2)^2}$$

$$= a\frac{2kp\sin pt+(w^2-p^2)\cos pt}{4k^2p^2+(w^2-p^2)^2}$$

$$= a\left[\frac{2kp\sin pt}{\sqrt{4k^2p^2+(w^2-p^2)^2}}+\frac{(w^2-p^2)\cos pt}{\sqrt{4k^2p^2+(w^2-p^2)^2}}\right]\frac{1}{\sqrt{4k^2p^2+(w^2-p^2)^2}}$$

$$= a\left[\sin\theta\sin pt+\cos\theta\cos pt\right]\frac{1}{\sqrt{4a^2p^2-(w^2-p^2)^2}}$$

$$= a\frac{\sin\theta\sin pt+\cos\theta\cos pt}{\sqrt{4k^2p^2+(w^2-p^2)^2}} = \frac{a\cos(pt-\theta)}{\sqrt{4k^2p^2+(w^2-p^2)^2}}$$

$\sqrt{4a^2p^2+(w^2-p^2)^2}$, 2ap, θ, w^2-p^2

$$\left[\tan\theta=\frac{2ap}{(w^2-p^2)}\right]$$

Complete solution = C.F + P.I.

$$\Rightarrow \quad x = e^{-kt}\left[A\cos\sqrt{(w^2-p^2)}\,t+B\sin\sqrt{(w^2-p^2)}\,t\right]+\frac{k\cos(pt-\theta)}{\sqrt{4k^2p^2+(w^2-p^2)^2}}$$

when $t\to\infty$

$$x = \frac{a\cos(pt-\theta)}{\sqrt{4k^2p^2+(w^2-p^2)^2}}$$

Proved.

EXERCISE 2.4

Solve the following differential equations :

1. $\dfrac{d^2y}{dx^2}+6y=\sin 4x$ **Ans.** $C_1\cos\sqrt{6}x+C_2\sin\sqrt{6}x-\dfrac{1}{10}\sin 4x$

2. $\dfrac{d^2x}{dt^2}+2\dfrac{dx}{dt}+3x=\sin t$ **Ans.** $e^{-t}[A\cos\sqrt{2}t+B\sin\sqrt{2}t]-\dfrac{1}{4}(\cos t-\sin t)$

3. $\dfrac{d^2x}{dt^2}+2\dfrac{dx}{dt}+5x=\sin 2t$, given that when $t=0$, $x=3$ and $\dfrac{dx}{dt}=0$

   Ans. $e^{-t}\left[\dfrac{55}{17}\cos 2t+\dfrac{53}{34}\sin 2t\right]-\dfrac{1}{17}(4\cos 2t-\sin 2t)$

4. $\dfrac{d^2y}{dx^2}-7\dfrac{dy}{dx}+6y=2\sin 3x$, given that $y=1$, $\dfrac{dy}{dx}=0$ when $x=0$.

   Ans. $-\dfrac{13}{75}e^{6x}+\dfrac{27}{25}e^{x}+\dfrac{1}{75}(7\cos 3x-\sin 3x)$

5. $(D^3+1)\,y=2\cos^2 x$ **Ans.** $C_1e^{-x}+e^{\frac{1}{2}x}\left(C_2\cos\dfrac{\sqrt{3}}{2}x+C_3\sin\dfrac{\sqrt{3}}{2}x\right)+1+\dfrac{1}{65}(-8\sin 2x+\cos 2x)$

6. $(D^2+a^2)\,y=\sin ax$ (*A.M.I.E.T.E., June 2009*) **Ans.** $C_1\cos ax+C_2\sin ax-\dfrac{x}{2a}\cos ax$

7. $(D^4+2a^2D^2+a^4)\,y=8\cos ax$ **Ans.** $(C_1+C_2x)(C_3\cos ax+C_4\sin ax)-\dfrac{x^2}{a^2}\cos ax$

8. $\dfrac{d^2y}{dx^2}+3\dfrac{dy}{dx}+2y=\sin 2x$ (*AMIETE.,Summer* 2002) **Ans.** $C_1e^{-x}+C_2e^{-2x}-\dfrac{1}{20}(3\cos 2x+\sin 2x)$

9. $\frac{d^2y}{dx^2}+y=\sin 3x\cos 2x$ **Ans.** $C_1\cos x+C_2\sin x+\frac{1}{48}[-\sin 5x-12x\cos x]$

10. $\frac{d^2y}{dx^2}+4y=x^2+\sin 2x$ **Ans.** $C_1\cos 2x+C_2\sin 2x+\frac{1}{4}\left(x^2-x\cos 2x-\frac{1}{2}\right)$

11. $\frac{d^2y}{dx^2}+4y=e^x+\sin 2x$ **Ans.** $C_1\cos 2x+C_2\sin 2x+\frac{1}{5}e^x-\frac{x}{4}\cos 2x$

12. $\frac{d^2y}{dx^2}-2\frac{dy}{dx}-3y=2e^{2x}+10\sin 3x$ given that $y(0)=2$ and $y'(0)=4$

Ans. $\frac{29}{12}e^{3x}-\frac{1}{12}e^{-x}-\frac{2}{3}e^{2x}+\frac{1}{3}[\cos 3x-2\sin 3x]$

13. $\frac{d^2y}{dx^2}-4\frac{dy}{dx}+3y=e^{2x}+\cos x$ **Ans.** $C_1e^x+C_2e^{3x}-e^{2x}+\frac{1}{10}(-2\sin x+\cos x)$

14. $\frac{d^2y}{dx^2}-2\frac{dy}{dx}+3y=\cos x+x^2$

Ans. $e^x[C_1\cos\sqrt{2}\,x+C_2\sin\sqrt{2}\,x]+\frac{1}{4}(\cos x-\sin x)+\frac{1}{3}(x^2+\frac{4}{3}x+\frac{2}{9})$

15. $(D^3-3D^2+4D-2)y=e^x+\cos x$

Ans. $(C_1+C_2\cos x+C_3\sin x)e^x+\frac{1}{10}(3\sin x+\cos x)$

16. $(D^3-4D^2+13D)\,y=1+\cos 2x$

Ans. $C_1+e^{2x}(C_2\cos 3x+C_3\sin 3x)+\frac{1}{290}(9\sin 2x+8\cos 2x)+\frac{x}{13}$

17. $(D^2-4D+4)\,y=e^{2x}+x^3+\cos 2x$

Ans. $(C_1+C_2x)e^{2x}+\frac{1}{2}x^2e^{2x}+\frac{1}{8}(2x^3+6x^2+9x+6)-\frac{1}{8}\sin 2x$

18. $\frac{d^2y}{dx^2}+n^2y=h\sin Px$ $(P\neq n)$

where h, p and n are constants satisfying the conditions

$y=a,\ \frac{dy}{dx}=b$ for $x=0$ **Ans.** $a\cos nx+\left(\frac{b}{n}-\frac{Ph}{n(n^2-p^2)}\right)\sin nx+\frac{h\sin px}{(n^2-p^2)}$

19. $y''+y'-2y=-6\sin 2x-18\cos 2x,\ y(0)=2,\ y'(0)=2$ **Ans.** $-e^{-2x}+3\cos 2x$

2.10 $\frac{1}{f(D)}.e^{ax}\cdot\phi(x)=e^{ax}.\frac{1}{f(D+a)}.\phi(x)$

$$D[e^{ax}\phi(x)]=e^{ax}D\phi(x)+ae^{ax}\phi(x)=e^{ax}(D+a)\phi(x)$$

$$D^2[e^{ax}\phi(x)]=D[e^{ax}(D+a)\phi(x)]$$

$$=e^{ax}(D^2+aD)\phi(x)+ae^{ax}(D+a)\phi(x)$$

$$=e^{ax}(D^2+2aD+a^2)\phi(x)=e^{ax}(D+a)^2\phi(x)$$

Similarly, $$D^n[e^{ax}\phi(x)]=e^{ax}(D+a)^n\phi(x)$$

$$f(D)[e^{ax}\phi(x)]=e^{ax}f(D+a)\phi(x)$$

$$e^{ax}\phi(x) = \frac{1}{f(D)}.[e^{ax} f(D+a)\phi(x)] \quad ...(1)$$

Put $f(D+a)\phi(x) = X$, so that $\phi(x) = \frac{1}{f(D+a)}.X$

Substituting these values in (1), we get

$$e^{ax}\frac{1}{f(D+a)}X = \frac{1}{f(D)}[e^{ax}.X] \Rightarrow \frac{1}{f(D)}[e^{ax}.\phi(x)] = e^{ax}\frac{1}{f(D+a)}\phi(x)$$

Example 21. *Obtain the general solution of the differential equation*

$y'' - 2y' + 2y = x + e^x \cos x.$ (*U.P. II Semester Summer*, 2002)

Solution. We have, $y'' - 2y' + 2y = x + e^x \cos x$

A.E. is $m^2 - 2m + 2 = 0 \quad \Rightarrow m = 1 \pm i$

C.F. $= e^x (A \cos x + B \sin x)$

$$\text{P.I.} = \frac{1}{D^2 - 2D + 2}x + \frac{1}{D^2 - 2D + 2}e^x \cos x$$

where
$$I_1 = \frac{1}{D^2 - 2D + 2}x = \frac{1}{2\left[1 - D + \frac{D^2}{2}\right]}x = \frac{1}{2\left[1 - \left(D - \frac{D^2}{2}\right)\right]}x$$

$$= \frac{1}{2}\left[1 - \left(D - \frac{D^2}{2}\right)\right]^{-1} x = \frac{1}{2}\left[1 + \left(D - \frac{D^2}{2}\right) + \ldots\right]x$$

$$= \frac{1}{2}\left[x + Dx - \frac{D^2}{2}x + \ldots\right] = \frac{1}{2}[x+1]$$

and,
$$I_2 = \frac{1}{D^2 - 2D + 2}e^x \cos x$$

$$= e^x \frac{1}{(D+1)^2 - 2(D+1) + 2}\cos x = e^x \frac{1}{D^2 + 1}\cos x = e^x .x\frac{1}{2D}\cos x$$

$$= \frac{1}{2}xe^x \sin x \quad \left[\text{If } f(-a^2) = 0, \text{ then } \frac{1}{f(D^2)}\phi(x) = x\frac{1}{f'(D)}\phi(x)\right]$$

y = C.F. + P.I.

$$= e^x(A\cos x + B\sin x) + \frac{1}{2}(x+1) + \frac{1}{2}xe^x \sin x.$$ **Ans.**

Example 22. *Solve* : $(D^2 - 4D + 4)\, y = x^3 e^{2x}$

Solution. We have, $(D^2 - 4D + 4)\, y = x^3 e^{2x}$

A.E. is $m^2 - 4m + 4 = 0 \Rightarrow (m-2)^2 = 0 \Rightarrow m = 2, 2$

C.F. $= (C_1 + C_2 x)\, e^{2x}$

$$\text{P.I.} = \frac{1}{D^2 - 4D + 4}x^3 \cdot e^{2x} = e^{2x}\frac{1}{(D+2)^2 - 4(D+2) + 4}x^3$$

$$= e^{2x}\frac{1}{D^2}x^3 = e^{2x}.\frac{1}{D}\left(\frac{x^4}{4}\right) = e^{2x}\cdot\frac{x^5}{20}$$

The complete solution is $y = (C_1 + C_2 x)e^{2x} + e^{2x}.\frac{x^5}{20}$ **Ans.**

Example 23. *Solve* $\frac{d^4 y}{dx^4} - y = \cos x . \cosh x$ *(Nagpur University, Summer* 2003)

Solution. We have, $(D^4 - 1)\, y = \cos x \cosh x$

A.E. is $m^4 - 1 = 0 \Rightarrow (m^2 - 1)(m^2 + 1) = 0 \Rightarrow m = \pm 1, \pm i$

C.F. $= C_1 e^x + C_2 e^{-x} + C_3 \cos x + C_4 \sin x$

$$\text{P.I.} = \frac{1}{D^4 - 1} \cos x \cosh x = \frac{1}{D^4 - 1} \cos x \left(\frac{e^x + e^{-x}}{2}\right)$$

$$= \frac{1}{2}\left[\frac{1}{D^4 - 1} e^x \cos x + \frac{1}{D^4 - 1} e^{-x} \cos x\right]$$

$$= \frac{1}{2}\left[e^x \frac{1}{(D+1)^4 - 1} \cos x + e^{-x} \frac{1}{(D-1)^4 - 1} \cos x\right]$$

$$= \frac{1}{2}\left[e^x \frac{1}{D^4 + 4D^3 + 6D^2 + 4D} \cos x + e^{-x} \frac{1}{D^4 - 4D^3 + 6D^2 - 4D} \cos x\right]$$

$$= \frac{1}{2}\left[e^x \frac{1}{(-1)^2 + 4D(-1) + 6(-1) + 4D} \cos x + e^{-x} \frac{1}{(-1)^2 - 4D(-1) + 6(-1) - 4D} \cos x\right]$$

$$= \frac{1}{2}\left[e^x \frac{1}{-5} \cos x + e^{-x} \frac{1}{-5} \cos x\right]$$

$$= -\frac{1}{5}\left(\frac{e^x + e^{-x}}{2}\right) \cos x = -\frac{1}{5} \cosh x \cos x$$

Hence, the complete solution is

$$y = C_1 e^x + C_2 e^{-x} + C_3 \cos x + C_4 \sin x - \frac{1}{5} \cos x \cosh x$$ **Ans.**

Example 24. *Solve* $(D^4 - 1)\, y = e^x \cos x$ *(Uttrakhand, IInd Semeserer, Summer 2010)*

Solution Here we have

$$(D^4 - 1) y = e^x \cos x$$

A.E. is $(m^4 - 1) = 0 \Rightarrow (m + 1)(m - 1)(m^2 + 1) = 0$

$\Rightarrow$ $m = -1, 1, + i, - i$

C.F. $= c_1 e^{-x} + c_2 e^x + (c_3 \cos x + c_4 \sin x)$

$$\text{P.I.} = \frac{1}{D^4 - 1} e^x \cos x$$

$$= e^x \frac{1}{(D+1)^4 - 1} \cos x \Rightarrow P.I. = e^x \frac{1}{D^4 + 6D^3 + 4D^2 + 6D} \cos x$$

$$= e^x \frac{1}{(-1)^2 + 6(-1)D + 4(-1) + 6D} \cos x$$

$$= e^x \frac{1}{1 - 6D - 4 + 6D} \cos x \Rightarrow P.I. = -\frac{e^x \cos x}{3}$$

Complete solution = C.F. + P.I.

$$y = c_1 e^{-x} + c_2 e^x + (c_3 \cos x + c_4 \sin x) - \frac{e^x \cos x}{3}$$ **Ans.**

Example 25. *Solve* :

$(D^2 - 5D + 6)\, y = e^x \cos 2x$

Solution. We have, $(D^2 - 5D + 6)\, y = e^x \cos 2x$

A.E. is $\quad m^2 - 5m + 6 = 0$

$\Rightarrow \quad (m-2)(m-3) = 0, \quad \Rightarrow \quad m = 2, 3$

$$\text{C.F.} = C_1 e^{2x} + C_2 e^{3x}$$

$$\text{P.I.} = \frac{1}{D^2 - 5D + 6} e^x \cos 2x = e^x \cdot \frac{1}{(D+1)^2 - 5(D+1) + 6} \cos 2x$$

$$= e^x . \frac{1}{D^2 - 3D + 2} \cos 2x = e^x . \frac{1}{-4 - 3D + 2} \cos 2x$$

$$= -e^x \frac{1}{3D+2} \cos 2x = -e^x \frac{3D-2}{9D^2 - 4} \cos 2x$$

$$= -e^x \frac{3D-2}{9(-4)-4} \cos 2x = \frac{e^x}{40}(3D-2) \cos 2x$$

$$= \frac{e^x}{40}(-6 \sin 2x - 2 \cos 2x) = -\frac{e^x}{20}(3 \sin 2x + \cos 2x)$$

The complete solution is $y = C_1 e^{2x} + C_2 e^{3x} - \frac{e^x}{20}(3 \sin 2x + \cos 2x)$ **Ans.**

Example 26. *Solve the differential equation* :

$$\frac{d^3y}{dx^3} - 7\frac{d^2y}{dx^2} + 10\frac{dy}{dx} = e^{2x} \sin x$$ (*Nagpur University, Summer* 2005)

Solution. We have, $\frac{d^3y}{dx^3} - 7\frac{d^2y}{dx^2} + 10\frac{dy}{dx} = e^{2x} \sin x$

$\Rightarrow \quad D^3 y - 7D^2 y + 10\, Dy = e^{2x} \sin x \quad \Rightarrow \quad (D^3 - 7D^2 + 10D)\, y = e^{2x} \sin x$

A.E. is $\quad m^3 - 7m^2 + 10\, m = 0 \quad \Rightarrow \quad (m-2)(m^2 - 5m) = 0$

$\Rightarrow \quad m\,(m-2)(m-5) = 0 \quad \Rightarrow \quad m = 0, 2, 5$

$$\text{C.F} = C_1 e^{0x} + C_2 e^{2x} + C_3 e^{5x}$$

$$\text{P.I.} = \frac{1}{D^3 - 7D^2 + 10D} e^{2x} \sin x = e^{2x} \frac{1}{(D+2)^3 - 7(D+2)^2 + 10(D+2)} . \sin x$$

$$= e^{2x} \frac{1}{D^3 + 6D^2 + 12D + 8 - 7D^2 - 28D - 28 + 10D + 20} . \sin x$$

$$= e^{2x} \frac{1}{D^3 - D^2 - 6D} \sin x = e^{2x} \frac{1}{(-1^2)\, D - (-1^2) - 6D} \sin x$$

$$= e^{2x} \frac{1}{-D + 1 - 6D} \sin x = e^{2x} \frac{1}{1 - 7D} \sin x$$

$$= e^{2x} \frac{1 + 7D}{1 - 49D^2} \sin x = e^{2x} \frac{1 + 7D}{1 - 49(-1^2)} \sin x$$

$$= e^{2x}\frac{1+7D}{50}\sin x = \frac{e^{2x}}{50}(\sin x + 7\cos x)$$

Complete solution is

$$y = \text{C.F.} + \text{P.I.}$$

$$\Rightarrow \qquad y = C_1 + C_2 e^{2x} + C_3 e^{5x} + \frac{e^{2x}}{50}(\sin x + 7\cos x) \qquad \textbf{Ans.}$$

Example 27. *Solve* $(D^2 + 2)\, y = e^x \cos x + x^2 e^{2x}$ *(M.U. II Semester, 2008)*

Solution. Here, we have

$$(D^2 + 2)y = e^x \cos x + x^2 e^{2x}$$

A.E. is $m^2 + 2 = 0 \quad \Rightarrow \quad m = \pm i\sqrt{2}$

C.F. $= C_1 \cos \sqrt{2}\, x + C_2 \sin \sqrt{2}\, x$

$$\text{P.I.} = \frac{1}{D^2+2}(e^x \cos x + x^2 e^{2x}) = \frac{1}{D^2+2} e^x \cos x + \frac{1}{D^2+2} x^2 e^{2x}$$

$$= e^x \frac{1}{(D+1)^2+2}\cos x + e^{2x}\frac{1}{(D+2)^2+2}x^2$$

$$= e^x \frac{1}{D^2+2D+3}\cos x + e^{2x}\frac{1}{D^2+4D+6}x^2$$

$$= e^x \frac{1}{-1+2D+3}\cos x + \frac{e^{2x}}{6}\frac{1}{1+\frac{4D}{6}+\frac{D^2}{6}}x^2$$

$$= e^x \frac{1}{2D+2}\cos x + \frac{e^{2x}}{6}\left[1+\frac{4D}{6}+\frac{D^2}{6}\right]^{-1}x^2$$

$$= \frac{e^x}{2}\frac{D-1}{D^2-1}\cos x + \frac{e^{2x}}{6}\left[1-\frac{4D}{6}-\frac{D^2}{6}+\frac{4}{9}D^2+\ldots\right]x^2$$

$$= \frac{e^x}{2}\frac{D-1}{-1-1}\cos x + \frac{e^{2x}}{6}\left[x^2-\frac{2}{3}(2x)-\frac{1}{6}(2)+\frac{4}{9}(2)\right]$$

$$= -\frac{e^x}{4}(D-1)\cos x + \frac{e^{2x}}{6}\left[x^2-\frac{4x}{3}-\frac{1}{3}+\frac{8}{9}\right]$$

$$= -\frac{e^x}{4}(D\cos x - \cos x) + \frac{e^{2x}}{6}\left[x^2-\frac{4x}{3}+\frac{5}{9}\right]$$

$$= -\frac{e^x}{4}(-\sin x - \cos x) + \frac{e^{2x}}{6}\left[x^2-\frac{4x}{3}+\frac{5}{9}\right] = \frac{e^x}{4}(\sin x + \cos x) + \frac{e^{2x}}{6}\left[x^2-\frac{4x}{3}+\frac{5}{9}\right]$$

Complete solution is $y = \text{C.F.} + \text{P.I.}$

$$y = C_1\cos\sqrt{2}\,x + C_2\sin\sqrt{2}x + \frac{e^x}{4}(\sin x + \cos x) + \frac{e^{2x}}{6}\left(x^2-\frac{4x}{3}+\frac{5}{9}\right) \qquad \textbf{Ans.}$$

Example 28. *A body executes damped forced vibrations given by the equation*

$$\frac{d^2x}{dt^2} + 2k\frac{dx}{dt} + b^2x = e^{-kt}\sin \omega t.$$

Solve the differential equation for both the cases when $\omega^2 \neq b^2 - k^2$ *and when* $\omega^2 = b^2 - k^2$.

(U.P., II Semester, Summer 2002)

Solution. Here, we have $(D^2 + 2kD + b^2)\,x = e^{-kt} \sin \omega t$, ...(1)
which is a linear differential equation with constant coefficients.

A.E. is $m^2 + 2km + b^2 = 0 \Rightarrow m = \dfrac{-2k \pm \sqrt{(4k^2 - 4b^2)}}{2} = -k \pm \sqrt{(k^2 - b^2)}$

As the given problem is on vibration, we must have $k^2 < b^2$

$$m = -k \pm \sqrt{-(b^2 - k^2)} = -k \pm i\sqrt{(b^2 - k^2)}$$

$$\text{C.F.} = e^{-kt}\left\{C_1 \cos\sqrt{(b^2 - k^2)}\,t + C_2 \sin\sqrt{(b^2 - k^2)}\,t\right\}$$

$$\text{P.I.} = \frac{1}{D^2 + 2kD + b^2} e^{-kt} \sin \omega t = e^{-kt} \frac{1}{(D-k)^2 + 2k(D-k) + b^2} \sin \omega t$$

$$= e^{-kt} \frac{1}{D^2 + (b^2 - k^2)} \sin \omega t = e^{-kt} \frac{1}{-\omega^2 + (b^2 - k^2)} \sin \omega t, \text{ if } \omega^2 \neq b^2 - k^2 \quad \text{...(2)}$$

If $\omega^2 = b^2 - k^2$, then $\quad \text{P.I.} = e^{-kt}.t\dfrac{1}{2D}\sin \omega t = e^{-kt}\left(-\dfrac{t}{2\omega}\cos \omega t\right)$, ...(3)

Case I. If $\omega^2 \neq b^2 - k^2$, the complete solution of (1) is

$$x = e^{kt}\left\{C_1 \cos\sqrt{(b^2 - k^2)}\,t + C_2 \sin\sqrt{(b^2 - k^2)}\,t\right\} + \frac{e^{-kt}}{(b^2 - k^2) - \omega^2} \sin \omega t \quad \text{[From (2)}$$

Case II. If $\omega^2 = b^2 - k^2$, the complete solution of (1) is

$$x = e^{-kt}\left\{C_1 \cos \omega t + C_2 \sin \omega t\right\} - \frac{e^{-kt}\,t \cos \omega t}{2w} \quad \text{[From (3)]} \quad \textbf{Ans.}$$

Example 29. *Solve the differential equation*

$$\frac{d^2y}{dx^2} + 2\frac{dy}{dx} + y = \frac{e^{-x}}{x+2}$$

Solution Here, we have

$$\frac{d^2y}{dx^2} + 2\frac{dy}{dx} + y = \frac{e^{-x}}{x+2}$$

Auxiliary equation is

$m^2 + 2m + 1 = 0$

$\Rightarrow (m + 1)^2 = 0 \Rightarrow m = -1, -1$

$\therefore$ C.F. $= (C_1 + C_2x)e^{-x}$

$$\text{P.I.} = \frac{1}{(D+1)^2}\left(\frac{e^{-x}}{x+2}\right) = e^{-x}.\frac{1}{(D-1+1)^2}\left(\frac{1}{x+2}\right)$$

$$= e^{-x}\frac{1}{D^2}\left(\frac{1}{x+2}\right) = e^{-x}.\frac{1}{D}\log(x+2)$$

$$= e^{-x}\left[\log(x+2).x - \int \frac{1}{x+2}.x\,dx\right] = e^{-x}\left[x\log(x+2) - \int\left(1 - \frac{2}{x+2}\right).dx\right]$$

$$= e^{-x}\,[x \log(x+2) - x + 2\log(x+2)] = e^{-x}\,[(x+2)\log(x+2) - x]$$

Hence, the complete solution is

$$y = \text{C.F.} + \text{P.I.} = (C_1 + C_2x)\,e^{-x} + e^{-x}\{(x+2)\log(x+2) - x\} \quad \textbf{Ans.}$$

Example 30. *Solve* $(D^2+6D+9)y=\dfrac{e^{-3x}}{x^3}$.

(*Nagpur University, Summer 2002, A.M.I.E.T.E., June 2009*)

Solution A.E. is $m^2 + 6m + 9 = 0$

$$\Rightarrow \quad (m+3)^2 = 0 \quad \Rightarrow \quad m = -3, -3$$

$$\text{C.F.} = (C_1 + C_2 x)\, e^{-3x}$$

$$\text{P.I.} = \frac{1}{D^2+6D+9}\frac{e^{-3x}}{x^3} = e^{-3x}\frac{1}{(D-3)^2+6(D-3)+9}\,\frac{1}{x^3}$$

$$= e^{-3x}\frac{1}{D^2-6D+9+6D-18+9}\,\frac{1}{x^3} = e^{-3x}\frac{1}{D^2}(x^{-3})$$

$$= e^{-3x}\frac{1}{D}\left(\frac{x^{-2}}{-2}\right) = e^{-3x}\frac{x^{-1}}{(-2)\,(-1)} = \frac{e^{-3x}x^{-1}}{2} = \frac{e^{-3x}}{2x}$$

Hence, the solution is $y = (C_1 + C_2 x)e^{-3x} + \dfrac{e^{-3x}}{2x}$ **Ans.**

Example 31. *Solve* $(D^2 - 4D + 3)\, y = 2x\, e^{3x} + 3e^x \cos 2x$

Solution. The auxiliary equation is

$m^2 - 4\,m + 3 = 0$ which gives

$m = 1, 3$

$\text{C.F.} = C_1 e^x + C_2 e^{3x}$

$$\text{P.I.} = \frac{1}{D^2-4D+3}2\,x\,e^{3x} + \frac{1}{D^2-4D+3}3\,e^x\cos 2x$$

$$= 2\,e^{3x}.\frac{1}{(D+3)^2-4(D+3)+3}x + 3\,e^x.\frac{1}{(D+1)^2-4(D+1)+3}\cos 2x$$

$$= 2\,e^{3x}.\frac{1}{D^2+2D}x + 3\,e^x.\frac{1}{D^2-2D}\cos 2x$$

$$= 2\,e^{3x}.\frac{1}{2D(1+D/2)}x + 3\,e^x.\frac{1}{-4-2D}\cos 2x$$

$$= e^{3x}.\frac{1}{D}\left(1+\frac{D}{2}\right)^{-1}x - \frac{3\,e^x}{2}.\frac{1}{2+D}\cos 2x$$

$$= e^{3x}.\frac{1}{D}\left(1-\frac{D}{2}+\frac{D^2}{4}\ldots\right)x - \frac{3\,e^x}{2}.\frac{2-D}{4-D^2}\cos 2x$$

$$= e^{3x}.\left(\frac{1}{D}-\frac{1}{2}+\frac{D}{4}\ldots\right)x - \frac{3\,e^x}{2}.\frac{2-D}{4+4}\cos 2x$$

$$= e^{3x}\left(\frac{x^2}{2}-\frac{x}{2}+\frac{1}{4}\right) - \frac{3\,e^x}{16}(2\cos 2x + 2\sin 2x)$$

The complete solution is

$$y = C_1\,e^x + C_2\,e^{3x} + e^{3x}\left(\frac{x^2}{2}-\frac{x}{2}+\frac{1}{4}\right) - \frac{3\,e^x}{8}(\cos 2x + \sin 2x)$$ **Ans.**

The term $\frac{e^{3x}}{4}$ has been omitted from the P.I., since C_2e^{3x} is present in the C.F.

Example 32. *Find the complete solution of* $\frac{d^2y}{dx^2}-3\frac{dy}{dx}+2y = xe^{3x}+\sin 2x$

(*U.P. II Semester* 2003)

Solution. The auxiliary equation is

$$m^2 - 3m + 2 = 0 \quad \Rightarrow \quad m^2 - 2m - m + 2 = 0$$

$$\Rightarrow \quad (m-2)(m-1) = 0 \Rightarrow \quad m = 1, 2$$

$$\text{C.F.} = C_1e^x + C_2e^{2x}$$

$$\text{P.I.} = \frac{1}{D^2-3D+2}(xe^{3x}+\sin 2x)$$

$$\text{P.I.} = \frac{1}{D^2-3D+2}xe^{3x}+\frac{1}{D^2-3D+2}\sin 2x$$

$$= e^{3x}\frac{1}{(D+3)^2-3(D+3)+2}x+\frac{1}{-4-3D+2}\sin 2x$$

$$= e^{3x}\frac{1}{D^2+6D+9-3D-9+2}x+\frac{1}{-3D-2}\sin 2x$$

$$= e^{3x}\frac{1}{D^2+3D+2}x-\frac{1}{3D+2}\sin 2x = \frac{e^{3x}}{2}\left[1+\left(\frac{3D+D^2}{2}\right)\right]^{-1}x-\frac{(3D-2)}{9D^2-4}\sin 2x$$

$$= \frac{e^{3x}}{2}\left[1-\left(\frac{3D+D^2}{2}\right)+\ldots\right]x-\frac{(3D-2)}{9(-4)-4}\sin 2x$$

$$= \frac{e^{3x}}{2}\left[x-\left(\frac{3D+D^2}{2}\right)x+\ldots\right]-\frac{(3D-2)}{-36-4}\sin 2x = \frac{e^{3x}}{2}\left[x-\frac{3}{2}\right]+\frac{3D-2}{40}\sin 2x$$

$$\Rightarrow \text{P.I.} = \frac{e^{3x}}{4}(2x-3)+\frac{1}{40}(6\cos 2x-2\sin 2x) = \frac{e^{3x}}{4}(2x-3)+\frac{3}{20}\cos 2x-\frac{1}{20}\sin 2x$$

The complete solution is

$$y = \text{C.F.} + \text{P.I.}$$

$$\Rightarrow \quad y = C_1e^x+C_2e^{2x}+\frac{e^{3x}}{4}(2x-3)+\frac{3}{20}\cos 2x-\frac{1}{20}\sin 2x$$ **Ans.**

Example 33. *Solve* : $\frac{d^2y}{dx^2}-2\frac{dy}{dx}+y = xe^x\cos x$

(*U.P. II Semester 2009, M.U. II Semester, 2009*)

Solution. Here, we have $(D^2 - 2D + 1)\,y = x\,e^x \cos x$

A.E. is $m^2 - 2m + 1 = 0 \quad \Rightarrow \quad (m-1)^2 = 0 \quad \Rightarrow \quad m = 1, 1$

$$\text{C.F.} = (C_1 + C_2x)\,e^x$$

$$\text{P.I.} = \frac{1}{D^2-2D+1}xe^x\cos x = \frac{1}{(D-1)^2}xe^x\cos x$$

$$= e^x\frac{1}{(D-1+1)^2}x\cos x = e^x\frac{1}{D^2}x\cos x$$

$$= e^x \frac{1}{D}[x \sin x - (1)(-\cos x)] = e^x . \frac{1}{D}(x \sin x + \cos x)$$

$$= e^x [x(-\cos x) - (1)(-\sin x) + \sin x] = e^x [-x \cos x + \sin x + \sin x]$$

Complete solution is $y = \text{C.F.} + \text{P.I.}$

$$\Rightarrow \quad y = (C_1 + C_2 x) e^x + e^x [2 \sin x - x \cos x]$$ **Ans.**

Example 34. *Solve* $(D^2 + 5D + 6) y = e^{-2x} \sec^2 x (1 + 2 \tan x)$ *(A.M.I.E.T.E., Summer 2003)*

Solution. Here, we have $(D^2 + 5D + 6) y = e^{-2x} \sec^2 x (1 + 2 \tan x)$

Auxiliary Equation is $m^2 + 5m + 6 = 0$

$\Rightarrow \quad (m + 2)(m + 3) = 0$

$\Rightarrow \quad m = -2, \text{ and } m = -3$

Hence, complementary function (C.F.) $= C_1 e^{-2x} + C_2 e^{-3x}$

$$\text{P.I.} = \frac{1}{D^2 + 5D + 6} e^{-2x} \sec^2 x (1 + 2 \tan x) = e^{-2x} \frac{1}{(D-2)^2 + 5(D-2) + 6} \sec^2 x (1 + 2 \tan x)$$

$$= e^{-2x} \frac{1}{D^2 - 4D + 4 + 5D - 10 + 6} \sec^2 x (1 + 2 \tan x)$$

$$= e^{-2x} \frac{1}{D^2 + D} \sec^2 x (1 + 2 \tan x) = e^{-2x} \left[\frac{\sec^2 x}{D^2 + D} + \frac{2 \tan x \sec^2 x}{D^2 + D} \right]$$

$$= e^{-2x} \left[\frac{1}{D(D+1)} \sec^2 x + \frac{1}{D(D+1)} 2 \tan x \sec^2 x \right]$$

$$= e^{-2x} \left[\left(\frac{1}{D} - \frac{1}{D+1} \right) \sec^2 x + \left(\frac{1}{D} - \frac{1}{D+1} \right) 2 \tan x \sec^2 x \right]$$

$$= e^{-2x} \left[\frac{1}{D} \sec^2 x - \frac{1}{D+1} \sec^2 x + \frac{1}{D} 2 \tan x \sec^2 x - \frac{1}{D+1} 2 \tan x \sec^2 x \right]$$

$$= e^{-2x} \left[\tan x - e^{-x} \int e^x . \sec^2 x \, dx + \tan^2 x - e^{-x} \int 2e^x \tan x \sec^2 x \, dx \right]$$

Now, $$\int e^x \sec^2 x \, dx = e^x \sec^2 x - \int e^x . 2 \sec x \sec x \tan x \, dx$$

$$= e^x \sec^2 x - 2 \int e^x \sec^2 x . \tan x \, dx$$

$$\therefore \text{P.I.} = e^{-2x} \left[\tan x - e^{-x} . e^x \sec^2 x + 2e^{-x} \int e^x \sec^2 x \tan x \, dx + \tan^2 x - 2e^{-x} \int e^x \sec^2 x \tan x \, dx \right]$$

$$= e^{-2x} [\tan x - \sec^2 x + \tan^2 x] = e^{-2x} [\tan x - (\sec^2 x - \tan^2 x)] = e^{-2x} (\tan x - 1)$$

$\therefore$ Complete solution is

$$\Rightarrow y = C.F. + P.I. = C_1 e^{-2x} + C_2 e^{-3x} + e^{-2x} (\tan x - 1)$$ **Ans.**

Example 35. *Solve the differential equation*

$$(D^2 - 4D + 4) y = 8x^2 e^{2x} \sin 2x$$

(U.P. II Semester, Summer 2008, Uttarakhand 2007, 2005, 2004; Nagpur University, June 2008)

Solution. Here, we have $(D^2 - 4D + 4) y = 8x^2 e^{2x} \sin 2x$

A.E. is $(m^2 - 4m + 4) = 0$

$\Rightarrow \quad (m-2)^2 = 0 \quad \Rightarrow \quad m = 2, 2$

C.F. $= (C_1 + C_2 x)\, e^{2x}$

$$\text{P.I.} = \frac{1}{D^2 - 4D + 4} 8\, x^2 e^{2x} \sin 2x = 8 \frac{1}{(D-2)^2} x^2 e^{2x} \sin 2x$$

$$= 8\, e^{2x} \frac{1}{(D-2+2)^2} x^2 \sin 2x = 8e^{2x} \frac{1}{D^2} x^2 \sin 2x$$

$$= 8\, e^{2x} \frac{1}{D}\left[x^2 \frac{(-\cos 2x)}{2} - 2x\left(-\frac{\sin 2x}{4}\right) + 2\left(\frac{\cos 2x}{8}\right)\right]$$

$$= 8\, e^{2x} \frac{1}{D}\left[-\frac{x^2}{2}\cos 2x + \frac{x \sin 2x}{2} + \frac{\cos 2x}{4}\right]$$

$$= 8e^{2x}\left[\begin{array}{l} -\frac{x^2}{2}\left(\frac{\sin 2x}{2}\right) - \left(\frac{-2x}{2}\right)\left(-\frac{\cos 2x}{4}\right) + (-1)\left(-\frac{\sin 2x}{8}\right) \\ \quad + \frac{x}{2}\left(-\frac{\cos 2x}{2}\right) - \left(\frac{1}{2}\right)\left(-\frac{\sin 2x}{4}\right) + \frac{\sin 2x}{8} \end{array}\right]$$

$$= e^{2x}\,[-2x^2 \sin 2x - 2x \cos 2x + \sin 2x - 2x \cos 2x + \sin 2x + \sin 2x]$$

$$= e^{2x}\,[-2x^2 \sin 2x - 4x \cos 2x + 3 \sin 2x]$$

$$= -e^{2x}\,[4x \cos 2x + (2x^2 - 3) \sin 2x]$$

Complete solution is, $y =$ C.F. + P.I.

$$y = (C_1 + C_2 x)\, e^{2x} - e^{2x}\,[4x \cos 2x + (2x^2 - 3) \sin 2x]$$ **Ans.**

EXERCISE 2.5

Solve the following equations:

1. $(D^2 - 5D + 6)\, y = e^x \sin x$ **Ans.** $y = C_1 e^{2x} + C_2 e^{3x} + \frac{e^x}{10}(3\cos x + \sin x)$

2. $\frac{d^3y}{dx^3} - \frac{d^2y}{dx^2} + 3\frac{dy}{dx} + 5y = e^x \cos 2x$

   Ans. $y = ae^{-x} + (b\cos 2x + c\sin 2x)e^x - \frac{1}{16} xe^x(\cos 2x - \sin 2x)$

3. $\frac{d^2y}{dx^2} - 7\frac{dy}{dx} + 10y = e^{2x}\sin x$ **Ans.** $y = C_1 e^{2x} + C_2 e^{5x} + \frac{e^{2x}}{10}(3\cos x - \sin x)$

4. $\frac{d^3y}{dx^3} - 2\frac{dy}{dx} + 4y = e^x \cos x$ **Ans.** $y = C_1 e^{-2x} + e^x(C_2 \cos x + C_3 \sin x) + \frac{xe^x}{20}(3\sin x - \cos x)$

5. $(D^2 - 4D + 13)y = e^{2x} \cos 3x + (x^2 + x + 9)$

   Ans. $y = e^{2x}(C_1 \cos 3x + C_2 \sin 3x) + \frac{xe^{2x} \sin 3x}{6} + \frac{1}{13}\left(x^2 + 21\frac{x}{13} + \frac{1579}{169}\right)$

6. $(D^2 - 4D + 3)\, y = 2xe^{3x} + 3e^{3x} \cos 2x$

   Ans. $y = C_1 e^x + C_2 e^{3x} + \frac{1}{2} e^{3x}(x^2 - x) + \frac{3}{8} e^{3x}(\sin 2x - \cos 2x)$

7. $\frac{d^2y}{dx^2} + 2\frac{dy}{dx} + y = \frac{e^{-x}}{x^2}$ **Ans.** $y = (C_1 + C_2\, x)\, e^{-x} - e^{-x} \log x$

8. $(D^2 - 4)\, y = x^2 e^{3x}$ **Ans.** $y = C_1 e^{2x} + C_2 e^{-2x} + \frac{e^{3x}}{5}\left[x^2 - \frac{12x}{5} + \frac{62}{25}\right]$

9. $(D^2 - 3D + 2)\, y = 2x^2 e^{4x} + 5e^{3x}$ **Ans.** $y = C_1 e^{x} + C_2 e^{2x} + \frac{e^{4x}}{54}[18x^2 - 30x + 19] + \frac{5}{2}e^{3x}$

10. $\frac{d^2y}{dx^2} - 4y = x \sinh x$ **Ans.** $y = C_1 e^{2x} + C_2 e^{-2x} - \frac{x}{3}\sinh x - \frac{2}{9}\cosh x$

11. $\frac{d^2y}{dt^2} + 2h\frac{dy}{dt} + (h^2 + p^2)y = ke^{-ht}\cos pt$ **Ans.** $y = e^{-ht}[A\cos pt + B\sin pt] + \frac{k}{2p}te^{-ht}\sin pt$

2.11 TO FIND THE VALUE OF $\frac{1}{f(D)} x^n \sin ax$.

Now, $$\frac{1}{f(D)}x^n(\cos ax + i\sin ax) = \frac{1}{f(D)}x^n e^{iax} = e^{iax}\frac{1}{f(D+ia)}x^n$$

$$\frac{1}{f(D)}\cdot x^n \sin ax = \text{Imaginary part of } e^{iax}\frac{1}{f(D+ia)}\cdot x^n$$

$$\frac{1}{f(D)}\cdot x^n \cos ax = \text{Real part of } e^{iax}\frac{1}{f(D+ia)}\cdot x^n$$

Example 36. *Solve the differential equation :*

$(D^2 + 2D + 1)\, y = x \cos x$ *(Uttarakhand, II Semester, June 2009)*

Solution. Here, we have

$$(D^2 + 2D + 1)\, y = x \cos x$$

Auxiliary equation is

$$m^2 + 2m + 1 = 0 \quad \Rightarrow \quad (m+1)^2 = 0 \quad \Rightarrow \quad m = -1, -1$$

$$\text{C.F.} = (C_1 + C_2 x)\, e^{-x}$$

$$\text{P.I.} = \frac{1}{(D+1)^2}x\cos x = \text{Real part of } \frac{1}{(D+1)^2}x[\cos x + i\sin x]$$

$$= \text{Real part of } \frac{1}{(D+1)^2}x e^{ix} = \text{Real part of } e^{ix}\frac{1}{(D+i+1)^2}x$$

$$= \text{Real part of } e^{ix}\frac{1}{D^2 + 2(i+1)D + (i+1)^2}x = \text{Real part of } e^{ix}\frac{1}{D^2 + 2(1+i)D + 2i}x$$

$$= \text{Real part of } \frac{e^{ix}}{2i}\frac{1}{1 + \frac{1+i}{i}D + \frac{D^2}{2i}}x = \text{Real part of } \frac{e^{ix}}{2i}\left[1 + \frac{1+i}{i}D + \frac{D^2}{2i}\right]^{-1}x$$

$$= \text{Real part of } \frac{e^{ix}}{2i}\left[1 - \left(\frac{1+i}{i}\right)D + \ldots\right]x = \text{Real part of } \frac{1}{2i}(\cos x + i\sin x)\left[x - \frac{1+i}{i}\right]$$

$$= \text{Real part of } -\frac{i}{2}(\cos x + i\sin x)(x + i - 1)$$

$$= \text{Real part of } -\frac{1}{2}(i\cos x - \sin x)(x + i - 1) = \frac{1}{2}\sin x\,(x-1) + \frac{1}{2}\cos x$$

Complete solution is y = C.F. + P.I.

$$\Rightarrow \quad y = (C_1 + C_2 x)\, e^{-x} + \frac{1}{2}(x-1)\sin x + \frac{1}{2}\cos x$$ **Ans.**

Example 37. *Solve* $\frac{d^2y}{dx^2} - 2\frac{dy}{dx} + y = x\sin x$ *(Uttarakhand, IInd Semester, 2012, June 2010)*

Solution. Auxiliary equation is $m^2 - 2m + 1 = 0 \Rightarrow m = 1, 1$

C.F. $= (C_1 + C_2 x)\, e^x$

$$\text{P.I.} = \frac{1}{D^2 - 2D + 1} x \cdot \sin x \qquad (e^{ix} = \cos x + i \sin x)$$

$$= \text{Imaginary part of } \frac{1}{D^2 - 2D + 1} x(\cos x + i \sin x) = \text{Imaginary part of } \frac{1}{D^2 - 2D + 1} x \cdot e^{ix}$$

$$= \text{Imaginary part of } e^{ix} \frac{1}{(D+i)^2 - 2(D+i) + 1} \cdot x$$

$$= \text{Imaginary part of } e^{ix} \frac{1}{D^2 - 2(1-i)D - 2i} \cdot x$$

$$= \text{Imaginary part of } e^{ix} \frac{1}{-2i}\left[1 - (1+i)D - \frac{1}{2i}D^2\right]^{-1} \cdot x$$

$$= \text{Imaginary part of } (\cos x + i \sin x)\left(\frac{i}{2}\right)[1 + (1+i)D]x$$

$$= \text{Imaginary part of } \frac{1}{2}(i\cos x - \sin x)\,[x + 1 + i]$$

$$\text{P.I.} = \frac{1}{2} x\cos x + \frac{1}{2}\cos x - \frac{1}{2}\sin x$$

Complete solution is y = $(C_1 + C_2 x)e^x + \frac{1}{2}(x\cos x + \cos x - \sin x)$ **Ans.**

Example 38. *Solve* $\frac{d^2y}{dx^2} - 2\frac{dy}{dx} + y = xe^x \sin x$ *(U.P. II Semester, 2009, 2005)*

Solution. $(D^2 - 2D + 1)\, y = x\, e^x \sin x$

Auxiliary equation is $m^2 - 2m + 1 = 0 \Rightarrow \quad (m-1)^2 = 0 \quad \Rightarrow \quad m = 1, 1$

$$\text{C.F.} = (C_1 + C_2 x)\, e^x$$

$$\text{P.I.} = \frac{1}{(D-1)^2} xe^x \sin x = e^x \frac{1}{(D+1-1)^2} \cdot x \sin x$$

$$= e^x \frac{1}{D^2} x \sin x = e^x \frac{1}{D} \int x \sin x\, dx \qquad \left[\frac{1}{D} \to \int\right]$$

$$= e^x \frac{1}{D}(-x\cos x + \sin x) = e^x \int (-x\cos x + \sin x)\, dx$$

$$= e^x\,[-x \sin x - \cos x - \cos x] = -e^x\,[x \sin x + 2\cos x]$$

Complete solution is $\quad y = (C_1 + C_2 x)\, e^x - e^x\,[x \sin x + 2\cos x]$

Ans.

EXERCISE 2.6

Solve the following differential equations :

1. $(D^2 + 4)y = 3x \sin x$ **Ans.** $C_1 \cos 2x + C_2 \sin 2x + x \sin x - \frac{2}{3}\cos x$
2. $\frac{d^2y}{dx^2} - y = x \sin 3x + \cos x$ **Ans.** $C_1 e^x + C_2 e^{-x} - \frac{1}{10}\left[\frac{3}{5}\cos 3x + x \sin 3x + 5\cos x\right]$

3. $\dfrac{d^2y}{dx^2} - y = x\sin x + e^x + x^2e^x$ **Ans.** $C_1e^x + C_2e^{-x} - \dfrac{1}{2}[x\sin x + \cos x] + \dfrac{x}{12}e^x(2x^2 - 3x + 9)$

4. $(D^4 + 2D^2 + 1)\,y = x^2 \cos x$

Ans. $(C_1 + C_2x)\cos x + (C_3 + C_4x)\sin x + \dfrac{1}{12}x^3 \sin x - \dfrac{1}{48}(x^4 - 9x^2)\cos x$

2.12 GENERAL METHOD OF FINDING THE PARTICULAR INTEGRAL OF ANY FUNCTION $\phi(x)$

$$\text{P.I.} = \frac{1}{D-a}\phi(x) = y \qquad \text{...(1)}$$

$$\Rightarrow \quad (D-a)\frac{1}{D-a}\cdot\phi(x) = (D-a)\cdot y$$

$$\phi(x) = (D-a)y \qquad \Rightarrow \qquad \phi(x) = Dy - ay$$

$$\frac{dy}{dx} - ay = \phi(x) \text{ which is the linear differential equation.}$$

Its solution is $\quad ye^{-\int a\,dx} = \int e^{-\int a\,dx}\cdot\phi(x)\,dx \qquad \Rightarrow \qquad ye^{-ax} = \int e^{-ax}\cdot\phi(x)\,dx$

$$y = e^{ax}\int e^{-ax}\cdot\phi(x)\,dx$$

$$\boxed{\frac{1}{D-a}\cdot\phi(x) = e^{ax}\int e^{-ax}\cdot\phi(x)\,dx}$$

Example 39. *Solve* $\dfrac{d^2y}{dx^2} + 9y = \sec 3x.$

Solution. Auxiliary equation is $m^2 + 9 = 0$ or $m = \pm 3i$,

C.F. = $C_1 \cos 3x + C_2 \sin 3x$

$$\text{P.I.} = \frac{1}{D^2+9}\cdot\sec 3x = \frac{1}{(D+3i)(D-3i)}\cdot\sec 3x$$

$$= \frac{1}{6i}\left[\frac{1}{D-3i} - \frac{1}{D+3i}\right]\cdot\sec 3x$$

$$= \frac{1}{6i}\cdot\frac{1}{D-3i}\cdot\sec 3x - \frac{1}{6i}\frac{1}{D+3i}\cdot\sec 3x \qquad \text{...(1)}$$

Now, $\dfrac{1}{D-3i}\sec 3x = e^{3ix}\displaystyle\int e^{-3ix}\sec 3x\,dx$

$$= e^{3ix}\int\frac{\cos 3x - i\sin 3x}{\cos 3x}dx = e^{3ix}\int(1 - i\tan 3x)\,dx = e^{3ix}(x + \frac{i}{3}\log\cos 3x)$$

Changing i to $-i$, we have

$$\frac{1}{D+3i}\sec 3x = e^{-3ix}(x - \frac{i}{3}\log\cos 3x)$$

Putting these values in (1), we get

$$\text{P.I.} = \frac{1}{6i}\left[e^{3ix}\left(x + \frac{i}{3}\log\cos 3x\right) - e^{-3ix}\left(x - \frac{i}{3}\log\cos 3x\right)\right]$$

$$= \frac{x}{6i}e^{3ix} + \frac{e^{3ix}\log\cos 3x}{18} - \frac{xe^{-3ix}}{6i} + \frac{e^{-3ix}}{18}\log\cos 3x$$

$$= \frac{x}{3}\frac{e^{3ix} - e^{-3ix}}{2i} + \frac{1}{9}\cdot\frac{e^{3ix} + e^{-3ix}}{2}\log\cos 3x$$

$$= \frac{x}{3}\sin 3x + \frac{1}{9}\cos 3x \cdot \log\cos 3x$$

Hence, complete solution is $y = C_1 \cos 3x + C_2 \sin 3x + \frac{x}{3}\sin 3x + \frac{1}{9}\cos 3x \cdot \log\cos 3x$ **Ans.**

EXERCISE 2.7

Solve the following differential equations :

1. $\frac{d^2y}{dx^2} + a^2 y = \sec ax$ **Ans.** $C_1 \cos ax + C_2 \sin ax + \frac{x}{a}\sin ax + \frac{1}{a^2}\cos ax \cdot \log\cos ax$

2. $\frac{d^2y}{dx^2} + y = \operatorname{cosec} x$ **Ans.** $C_1 \cos x + C_2 \sin x - x\cos x + \sin x \log \sin x$

3. $\frac{d^2y}{dx^2} + 2\frac{dy}{dx} + 2y = e^{-x}\sec^3 x$ *(Nagpur, Winter 2000)*

 Ans. $e^{-x}\left[C_1 \cos x + C_2 \sin x + \frac{\sin x \tan x}{2}\right]$

4. $(D^2 + 4)\, y = \tan 2x$ **Ans.** $C_1 \cos 2x + C_2 \sin 2x - \frac{1}{4}\cos 2x \log(\sec 2x + \tan 2x)$

5. $\frac{d^2y}{dx^2} - 2\frac{dy}{dx} + 2y = e^x \tan x.$ **Ans.** $e^x (C_1 \cos x + C_2 \sin x) - e^x \cos x \log(\sec x + \tan x)$

6. $\frac{d^2y}{dx^2} + y = (x - \cot x)$ *(A.M.I.E. Winter 2002)*

 Ans. $C_1 \cos x + C_2 \sin x - x\cos^2 x - \sin x \log(\operatorname{cosec} x - \cot x)$

2.13 CAUCHY EULER HOMOGENEOUS LINEAR DIFFERENTIAL EQUATIONS

$$a_n\, x^n \frac{d^n y}{dx^n} + a_{n-1}\, x^{n-1}\frac{d^{n-1} y}{dx^{n-1}} + \dots + a_0\, y = \phi(x) \qquad \dots(1)$$

where $a_0, a_1, a_2, \dots$ are constants, is called a homogeneous equation.

Put $\qquad x = e^z, \qquad z = \log_e x, \qquad \frac{d}{dz} = D$

$$\frac{dy}{dx} = \frac{dy}{dz}\cdot\frac{dz}{dx} = \frac{1}{x}\frac{dy}{dz} \Rightarrow x\frac{dy}{dx} = \frac{dy}{dz} \Rightarrow x\frac{dy}{dx} = Dy$$

Again, $$\frac{d^2y}{dx^2} = \frac{d}{dx}\left(\frac{dy}{dx}\right) = \frac{d}{dx}\left(\frac{1}{x}\frac{dy}{dz}\right) = -\frac{1}{x^2}\frac{dy}{dz} + \frac{1}{x}\frac{d^2y}{dz^2}\frac{dz}{dx}$$

$$= -\frac{1}{x^2}\frac{dy}{dz} + \frac{1}{x}\frac{d^2y}{dz^2}\frac{1}{x} = \frac{1}{x^2}\left(\frac{d^2y}{dz^2} - \frac{dy}{dz}\right) = \frac{1}{x^2}(D^2 - D)\, y$$

$$x^2\frac{d^2y}{dx^2} = (D^2 - D)\, y$$

$$\Rightarrow \qquad \boxed{x^2\frac{d^2y}{dx^2} = D(D-1)\, y}$$

Similarly. $$\boxed{x^3 \frac{d^3 y}{dx^3} = D(D-1)(D-2)\,y}$$

The substitution of these values in (1) reduces the given homogeneous equation to a differential equation with constant coefficients.

Example 40. *Solve*: $x^2 \dfrac{d^2 y}{d x^2} - 2x \dfrac{d y}{d x} - 4y = x^4$ (*A.M.I.E. Summer* 2000)

Solution. We have, $$x^2 \frac{d^2 y}{d x^2} - 2x \frac{d y}{d x} - 4y = x^4 \qquad \ldots (1)$$

Putting $x = e^z$, $D = \dfrac{d}{dz}$, $x\dfrac{dy}{dx} = Dy$, $x^2 \dfrac{d^2 y}{d x^2} = D(D-1)\,y$ in (1), we get

$$D(D-1)\,y - 2\,Dy - 4y = e^{4z} \quad \text{or} \quad (D^2 - 3D - 4)\,y = e^{4z}$$

A.E. is $m^2 - 3m - 4 = 0 \quad \Rightarrow \quad (m-4)(m+1) = 0 \quad \Rightarrow \quad m = -1, 4$

$$\text{C.F.} = C_1 e^{-z} + C_2 e^{4z}$$

$$\text{P.I.} = \frac{1}{D^2 - 3D - 4} e^{4z} \qquad \text{[Rule Fails]}$$

$$= z \frac{1}{2D-3} e^{4z} = z \frac{1}{2(4)-3} e^{4z} = \frac{z e^{4z}}{5}$$

Thus, the complete solution is given by

$$y = C_1 e^{-z} + C_2 e^{4z} + \frac{z e^{4z}}{5}$$

$$y = \frac{C_1}{x} + C_2 x^4 + \frac{1}{5} x^4 \log x \qquad \textbf{Ans.}$$

Example 41. *Solve*: $x^2 \dfrac{d^2 y}{d x^2} + 4x \dfrac{d y}{d x} + 2y = e^x$

(*U.P., II Semester 2005; Nagpur University, Summer* 2001)

Solution. Given equation is $$x^2 \frac{d^2 y}{d x^2} + 4x \frac{d y}{d x} + 2y = e^x \qquad \ldots (1)$$

To solve (1), we put $x = e^z$ or $z = \log x$ and $D = \dfrac{d}{dz}$

$$x \frac{d y}{d x} = Dy \quad \text{and} \quad x^2 \frac{d^2 y}{dx^2} = D(D-1)\,y$$

Substituting these values in (1), it becomes

$$[D(D-1) + 4D + 2]\,y = e^{e^z} \quad \Rightarrow \quad (D^2 + 3D + 2)\,y = e^{e^z}$$

$\therefore$ It's *A.E.* is $m^2 + 3m + 2 = 0$

$\therefore$ $(m+1)(m+2) = 0 \quad \therefore \quad m = -1, -2$

$\therefore$ $$\text{C.F.} = C_1 e^{-z} + C_2 e^{-2z} = \frac{C_1}{x} + \frac{C_2}{x^2}$$

$$\text{P.I.} = \frac{1}{D^2 + 3D + 2} e^{e^z} = \frac{1}{(D+1)(D+2)} e^{e^z}$$

$$= \left(\frac{1}{D+1} - \frac{1}{D+2}\right) e^{e^z} = \frac{1}{D+1} e^{e^z} - \frac{1}{D+2} e^{e^z}$$

$$= e^{-z} \int e^{e^z} . e^z \, dz - e^{-2z} \int e^{e^z} . e^{2z} \, dz$$

$$\left[\text{Since } \frac{1}{D-a} X = e^{ax} \int X e^{-ax} \, dx\right].$$

Put $\quad e^z = t$ so that $e^z dz = dt$

$$\text{P.I.} = e^{-z} \int e^t \, dt - e^{-2z} \int e^t . t \, dt = e^{-z} e^t - e^{-2z} (te^t - e^t)$$

$$= e^{-z} e^{e^z} - e^{-2z} (e^z e^{e^z} - e^{e^z})$$

$$= e^{-z} e^{e^z} - e^{-z} e^{e^z} + e^{-2z} e^{e^z} = e^{-2z} e^{e^z} = x^{-2} e^x = \frac{e^x}{x^2} \quad (\because x = e^z)$$

Hence, the complete solution of (1) is $\; y = \text{C.F.} + \text{P.I.}$

$$\Rightarrow \qquad y = \frac{C_1}{x} + \frac{C_2}{x^2} + \frac{e^x}{x^2} \qquad \textbf{Ans.}$$

Example 42. *Solve* $(x^3 D^3 + x^2 D^2 - 2) y = x - \dfrac{1}{x^3}$ *(Nagpur University, Summer* 2000)

Solution. Put $x = e^z$, $z = \log x$

Let $\quad D_1 = \dfrac{d}{dz}$

Then $\quad x^2 \dfrac{d^2 y}{dx^2} = D_1 (D_1 - 1) y \quad \Rightarrow \quad x^3 \dfrac{d^3 y}{dx^3} = D_1 (D_1 - 1)(D_1 - 2) y$

Substituting these values in the given equation, we get

$$[D_1 (D_1 - 1)(D_1 - 2) + D_1 (D_1 - 1) - 2] y = e^z + e^{-3z}$$

$$\Rightarrow \quad [D_1^3 - 3D_1^2 + 2D_1 + D_1^2 - D_1 - 2] y = e^z + e^{-3z}$$

$$\Rightarrow \quad (D_1^3 - 2D_1^2 + D_1 - 2) y = e^z + e^{-3z}$$

A.E. is $m^3 - 2m^2 + m - 2 = 0$

i.e. $(m - 2)(m^2 + 1) = 0$ *i.e.* $m = 2, \; \pm i.$

$$\text{C.F.} = C_1 e^{2z} + C_2 \cos z + C_3 \sin z$$

$$\text{P.I.} = \frac{1}{D_1^3 - 2D_1^2 + D_1 - 2} e^z + \frac{1}{D_1^3 - 2D_1^2 + D_1 - 2} e^{-3z}$$

$$= \frac{1}{1 - 2 + 1 - 2} e^z + \frac{1}{-27 - 18 - 3 - 2} e^{-3z} \qquad \begin{bmatrix} D_1 = 1 \\ D_1 = -3 \end{bmatrix}$$

$$= -\frac{e^z}{2} - \frac{1}{50} e^{-3z}$$

$$\therefore \quad y = C_1 e^{2z} + C_2 \cos z + C_3 \sin z - \frac{1}{2} e^z - \frac{1}{50} e^{-3z}$$

$$= C_1 x^2 + C_2 \cos(\log x) + C_3 \sin(\log x) - \frac{1}{2} x - \frac{1}{50} . \frac{1}{x^3} \qquad \textbf{Ans.}$$

Example 43. *Solve* $x^2 \dfrac{d^2y}{dx^2} + x\dfrac{dy}{dx} + y = \sin(\log x^2)$ *(Nagpur University, Summer* 2005)

Solution. We have, $x^2 \dfrac{d^2y}{dx^2} + x\dfrac{dy}{dx} + y = \sin(\log x^2)$ ... (1)

Let $x = e^z$, so that $z = \log x$, $D = \dfrac{d}{dz}$

(1) becomes

$$D(D-1)y + Dy + y = \sin 2z \Rightarrow (D^2+1)y = \sin 2z$$

$$\text{A.E. is } m^2 + 1 = 0 \Rightarrow m = \pm i$$

$$C.F. = C_1 \cos z + C_2 \sin z$$

$$\text{P.I.} = \frac{1}{D^2+1}\sin 2z = \frac{1}{-4+1}\sin 2z = -\frac{1}{3}\sin 2z$$

Hence complete solution is $y = \text{C.F.} + \text{P.I.} = C_1\cos z + C_2 \sin z - \dfrac{1}{3}\sin 2z$

$\Rightarrow$ $y = C_1\cos(\log x) + C_2\sin(\log x) - \dfrac{1}{3}\sin(\log x^2)$ **Ans.**

Example 44. *Solve*

$$x^2\frac{d^2y}{dx^2} - x\frac{dy}{dx} + y = \log x$$ *(Uttrakhand, IInd Semester, Summer 2010)*

Solution. Here we have

$$x^2\frac{d^2y}{dx^2} - x\frac{dy}{dx} + y = \log x \quad ...(1)$$

Putting $x = e^z$, $x\dfrac{dy}{dx} = Dy$, and $x^2\dfrac{d^2y}{dx^2} = D(D-1)y$ in (1), we get

where $D \equiv \dfrac{d}{dz}$

$D(D-1)y - Dy + y = z$

$\Rightarrow$ $D^2 - Dy - Dy + y = z$ $\Rightarrow (D^2 - 2D + 1)y = z$

$\Rightarrow$ A.E. is $m^2 - 2m + 1 = 0$ $\Rightarrow (m-1)^2 = 0 \Rightarrow m = 1, 1$

C.F. $= (C_1 + C_2 z)e^z$

$$\text{P.I.} = \frac{1}{(D-1)^2}z \Rightarrow P.I. = \frac{1}{(1-D)^2}z$$

$\Rightarrow$ P.I. $= (1-D)^{-2}z \Rightarrow$ P.I. $= (1+2D)z$

$\Rightarrow$ P.I. $= (z+2)$

Complete solution = C.F. + P.I.

$$y = (C_1 + C_2 z)e^z + (z+2)$$

$$= (C_1 + C_2 \log x)x + \log x + 2$$ **Ans.**

Example 45. *Solve* $x^3\dfrac{d^3y}{dx^3} + 3x^2.\dfrac{d^2y}{dx^2} + x\dfrac{dy}{dx} + y = x + \log x$

(U.P. II Semester, Summer, 2001)

Solution. Putting $x = e^z$ or $z = \log x$ and denoting $\frac{d}{dz}$ by D the equation becomes

$$[D(D-1)(D-2) + 3D(D-1) + D + 1]\,y = e^z + z$$

$$\Rightarrow \qquad [D^3 + 1]\,y = e^z + z$$

$\therefore$ A.E. is $\quad m^3 + 1 = 0$

$$\Rightarrow \quad (m+1)(m^2 - m + 1) = 0 \quad \Rightarrow \quad m = -1, \frac{1 \pm i\sqrt{3}}{2}$$

$$\text{C.F.} = C_1 e^{-z} + e^{\frac{1}{2}z}\left\{C_2 \cos\frac{\sqrt{3}}{2}z + C_3 \sin\frac{\sqrt{3}}{2}z\right\}$$

$$\text{P.I.} = \frac{1}{D^3+1}\{e^z + z\}$$

$$= \frac{1}{D^3+1}e^z + \frac{1}{D^3+1}z = \frac{e^z}{1+1} + (1+D^3)^{-1}z$$

$$= \frac{1}{2}e^z + (1 - D^3 + \ldots)\,z = \frac{1}{2}e^z + z.$$

$\therefore$ Complete solution is

$$y = C_1 e^{-z} + e^{\frac{z}{2}}\left\{C_2 \cos\frac{\sqrt{3}}{2}z + C_3 \sin\frac{\sqrt{3}}{2}z\right\} + \frac{1}{2}e^z + z$$

$$\Rightarrow \quad y = C_1 x^{-1} + \sqrt{x}\left\{C_2 \cos\left(\frac{\sqrt{3}}{2}\log x\right) + C_3 \sin\left(\frac{\sqrt{3}}{2}\log x\right)\right\} + \frac{1}{2}x + \log x \qquad \textbf{Ans.}$$

Example 46. *Solve:* $x^2\frac{d^3y}{dx^3} + 3x\frac{d^2y}{dx^2} + \frac{dy}{dx} = x^2 \log x$ (*Nagpur University, Summer* 2003)

Solution. We have, $x^3\frac{d^3y}{dx^3} + 3x^2\frac{d^2y}{dx^2} + x\frac{dy}{dx} = x^3 \log x$

Let $x = e^z$ so that $z = \log x$, $D = \frac{d}{dz}$

The equation becomes after substitution

$$[D(D-1)(D-2) + 3D(D-1) + D]\,y = z\,e^{3z} \qquad \Rightarrow \qquad D^3 y = ze^{3z}$$

Auxiliary equation is $m^3 = 0 \quad \Rightarrow \quad m = 0, 0, 0.$

C.F. $= C_1 + C_2 z + C_3 z^2 = C_1 + C_2 \log x + C_3 (\log x)^2$

$$\text{P.I.} = \frac{1}{D^3}\cdot z e^{3z} = e^{3z}\cdot\frac{1}{(D+3)^3}\cdot z$$

$$= e^{3z}\frac{1}{27}\left(1 + \frac{D}{3}\right)^{-3} z = \frac{e^{3z}}{27}(1 - D)\,z = \frac{e^{3z}}{27}(z-1) = \frac{x^3}{27}(\log x - 1)$$

Complete solution is $\qquad y = C_1 + C_2 \log x + C_3 (\log x)^2 + \frac{x^3}{27}(\log x - 1) \qquad$ **Ans.**

Example 47. *Solve:* $x^2 \dfrac{d^2y}{dx^2} - x\dfrac{dy}{dx} + 4y = \cos(\log x) + x\sin(\log x)$.

Solution. We have, $x^2 \dfrac{d^2y}{dx^2} - x\dfrac{dy}{dx} + 4y = \cos(\log x) + x\sin(\log x)$. ... (1)

Putting $x = e^z \Rightarrow z = \log x,\ D = \dfrac{d}{dz}$ and $x^2\dfrac{d^2y}{dx^2} = D(D-1)y,\ x\dfrac{dy}{dx} = Dy$ in (1), we get

$$[\,D(D-1) - D + 4\,]\,y = \cos z + e^z \sin z$$

i.e. $$(D^2 - 2D + 4)y = \cos z + e^z \sin z$$

A.E. is $m^2 - 2m + 4 = 0 \Rightarrow \quad m = \dfrac{-(-2) \pm \sqrt{4-16}}{2}$

$\Rightarrow$ $$m = 1 \pm \sqrt{3}\, i$$

$\therefore$ $$\text{C.F.} = e^z\,[\,C_1 \cos\sqrt{3}\,z + C_2 \sin\sqrt{3}\,z\,] \qquad \text{... (2)}$$

$$\text{P.I.} = \frac{1}{D^2 - 2D + 4}(\cos z + e^z \sin z)$$

$$= \frac{1}{D^2 - 2D + 4}\cos z + \frac{1}{D^2 - 2D + 4} e^z \sin z$$

$$= \frac{1}{-1 - 2D + 4}\cos z + e^z \frac{1}{(D+1)^2 - 2(D+1) + 4}\sin z$$

$$= \frac{1}{3 - 2D}\cos z + e^z \frac{1}{D^2 + 2D + 1 - 2D - 2 + 4}\sin z$$

$$= \frac{3 + 2D}{9 - 4D^2}\cos z + e^z \frac{1}{D^2 + 3}\sin z = \frac{3 + 2D}{9 + 4}\cos z + e^z \frac{1}{-1 + 3}\sin z$$

$$= \frac{3 + 2D}{13}\cos z + e^z \frac{1}{2}\sin z = \frac{1}{13}(3\cos z - 2\sin z) + \frac{1}{2}e^z \sin z \qquad \text{... (3)}$$

Complete solution is

$$y = \text{C.F.} + \text{P.I.}$$

$\Rightarrow$ $$y = e^z[C_1 \cos\sqrt{3}\,z + C_2 \sin\sqrt{3}\,z\,] + \frac{1}{13}(3\cos z - 2\sin z) + \frac{1}{2}e^z \sin z \qquad \text{... (4)}$$

Replacing $z = \log x$ and $e^z = x$ in (4), we get

$$y = x\,[\,C_1 \cos\sqrt{3}\,(\log x) + C_2 \sin\sqrt{3}\,(\log x)\,]$$

$$+ \frac{3}{13}\cos(\log x) - \frac{2}{13}\sin(\log x) + \frac{1}{2}x\sin(\log x)$$

Ans.

Example 48. *Solve the differential equation*

$$x^2 \frac{d^2y}{dx^2} - 3x\frac{dy}{dx} + 5y = x\log x.$$

(*Nagpur University, Winter 2002, Summer* 2000)

Solution. Putting $x = e^z$ or $z = \log x$ and $D = \dfrac{d}{dz}$

On putting $x\dfrac{dy}{dx} = Dy$ and $x^2\dfrac{d^2y}{dx^2} = D(D-1)y$ in the given equation, we get

$$[D(D-1) - 3D + 5]\, y = ze^z$$

i.e.
$$(D^2 - 4D + 5)\, y = ze^z$$

It's A.E. is
$$m^2 - 4m + 5 = 0$$

$$\therefore \quad m = \frac{4 \pm \sqrt{16-20}}{2} = \frac{4 \pm 2i}{2} = 2 \pm i$$

$$\text{C.F.} = e^{2z}(C_1 \cos z + C_2 \sin z)$$

$$\text{P.I.} = \frac{1}{D^2 - 4D + 5}\, z e^z$$

$$= e^z \cdot \frac{1}{(D+1)^2 - 4(D+1) + 5}\, z \qquad \text{(On replacing } D \text{ by } D+1)$$

$$= e^z \cdot \frac{1}{D^2 - 2D + 2}\, z = \frac{e^z}{2} \cdot \frac{1}{1 + \left(\dfrac{D^2 - 2D}{2}\right)}\, z$$

$$= \frac{e^z}{2}\left[1 + \left(\frac{D^2 - 2D}{2}\right)\right]^{-1} z = \frac{e^z}{2}\left[1 - \left(\frac{D^2 - 2D}{2}\right) + \ldots\right] z$$

$$= \frac{e^z}{2}[z + Dz] = \frac{e^z}{2}(z+1).$$

Hence, the complete solution is

$$y = e^{2z}[C_1 \cos z + C_2 \sin z] + \frac{e^z}{2}(z+1)$$

$$y = x^2[C_1 \cos(\log x) + C_2 \sin(\log x)] + \frac{x}{2}(1 + \log x) \qquad \textbf{Ans.}$$

Example 49. *Solve the homogeneous differential equation :*

$$x^2\frac{d^2y}{dx^2} - x\frac{dy}{dx} - 3y = x^2 \log x$$ *(Uttarakhand, II Semester, 2009)*

Solution. Here, we have

$$x^2\frac{d^2y}{dx^2} - x\frac{dy}{dx} - 3y = x^2 \log x \qquad \ldots(1)$$

Putting $x = e^z \Rightarrow z = \log x$ and $D = \dfrac{d}{dz}$, $x\dfrac{dy}{dx} = Dy$, $x^2\dfrac{d^2y}{dx^2} = D(D-1)y$ in (1), we get

$$[D(D-1)y - Dy - 3y] = e^{2z}.\, z$$

$$[D^2 - 2D - 3]\, y = z\, e^{2z}$$

A.E. is $\quad m^2 - 2m - 3 = 0 \Rightarrow (m-3)(m+1) = 0 \Rightarrow m = -1, m = 3.$

$$\text{C.F.} = C_1 e^{-z} + C_2 e^{3z}$$

$$\text{P.I.} = \frac{1}{D^2 - 2D - 3} z e^{2z}$$

$$= e^{2z} \frac{1}{(D+2)^2 - 2(D+2) - 3} z = e^{2z} \frac{1}{D^2 + 2D - 3} z$$

$$= \frac{e^{2z}}{-3} \frac{1}{1 - \frac{2D}{3} - \frac{D^2}{3}} z = \frac{e^{2z}}{-3}\left[1 - \frac{2D}{3} - \frac{D^2}{3}\right]^{-1} z$$

$$= \frac{e^{2z}}{-3}\left[1 + \frac{2D}{3} + \frac{D^2}{3} + \ldots\right] z = \frac{e^{2z}}{-3}\left[z + \frac{2}{3}\right]$$

Complete solution is, $y = \text{C.F.} + \text{P.I.}$

$$y = C_1 e^{-z} + C_2 . e^{3z} - \frac{e^{2z}}{3}\left(z + \frac{2}{3}\right)$$

$$y = \frac{C_1}{x} + C_2 x^3 - \frac{x^2}{3}\left(\log x + \frac{2}{3}\right)$$ **Ans.**

Example 50. *Solve the homogeneous linear differential equation.*

$$x^2 \frac{d^2 y}{d x^2} + x \frac{d y}{d x} + y = (\log x) \sin (\log x)$$

(*Nagpur University, Winter* 2002, *U.P. II Semester, Summer* 2002)

Solution. Since given equation is homogeneous,

Put $x = e^z \Rightarrow \log x = z$

Also, $x \frac{dy}{dx} = Dy, \quad x^2 \frac{d^2 y}{dx^2} = D(D-1) y, \quad D \equiv \frac{d}{dz}$

The transformed equation is

$$D(D-1) y + D y + y = z \sin z$$

$$(D^2 - D + D + 1) y = z \sin z$$

$$(D^2 + 1) y = z \sin z$$

A.E. is $m^2 + 1 = 0 \Rightarrow m = \pm i$

$$\text{C.F.} = C_1 \cos z + C_2 \sin z$$

$$P.I. = \frac{1}{D^2 + 1} z \sin z$$

$$= \text{Imaginary part of } \frac{1}{D^2+1} z (\cos z + i \sin z) = \text{Imaginary part of } \frac{1}{D^2+1} z e^{iz}$$

$$= \text{Imaginary part of } e^{iz} \frac{1}{(D+i)^2 + 1} z = \text{Imaginary part of } e^{iz} \frac{1}{D^2 + 2iD - 1 + 1} z$$

$$= \text{Imaginary part of } e^{iz} \frac{1}{D^2 + 2iD} z = \text{Imaginary part of } e^{iz} \frac{1}{2iD} \frac{1}{\left(1 + \frac{D}{2i}\right)} z$$

$$= \text{Imaginary part of } e^{iz} \frac{1}{2iD}\left(1 - \frac{D}{2i}\right) z = \text{Imaginary part of } e^{iz} \frac{1}{2iD}\left(z - \frac{1}{2i}\right)$$

$$= \text{Imaginary part of } e^{iz}\,\frac{1}{2i}\left(\frac{z^2}{2}-\frac{z}{2i}\right)$$

$$= \text{Imaginary part of } \frac{1}{2i}\,(\cos z + i\sin z)\left(\frac{z^2}{2}-\frac{z}{2i}\right)$$

$$= \text{Imaginary part of } (\cos z + i\sin z)\left(\frac{z^2}{4i}+\frac{z}{4}\right)$$

$$= \text{Imaginary part of } (\cos z + i\sin z)\left(-i\frac{z^2}{4}+\frac{z}{4}\right) = -\frac{z^2}{4}\cos z + \frac{z}{4}\sin z$$

Complete solution is, y = C.F. + P.I.

$$y = C_1\cos z + C_2\sin z - \frac{z^2}{4}\cos z + \frac{z}{4}\sin z$$

$$y = C_1\cos(\log x) + C_2\sin(\log x) - \frac{1}{4}(\log x)^2\cos(\log x) + \frac{1}{4}(\log x)\sin(\log x) \quad \textbf{Ans.}$$

Example 51. *The radial displacement in a rotating disc at a distance r from the axis is given by* $r^2\frac{d^2u}{dr^2}+r\frac{du}{dr}-u+kr^3=0$, *where k is a constant. Solve the equation under the conditions u = 0 when r = 0, u = 0 when r = a.* (*Nagpur University, Summer* 2008)

Solution. Here, we have

$$r^2\frac{d^2u}{dr^2}+r\frac{du}{dr}-u+kr^3=0, \qquad \ldots (1)$$

On putting $r = e^z$, $\quad r\frac{du}{dr} = Dz$, $\quad r^2\frac{d^2u}{dr^2} = D(D-1)z \quad$ in (1), we get

$$D(D-1)\,u + D\,u - u = -k\,e^{3z} \qquad \left[D \equiv \frac{d}{dz}\right]$$

$$\Rightarrow \quad (D^2 - D + D - 1)\,u = -k\,e^{3z} \Rightarrow (D^2-1)u = -k\,e^{3z}$$

A.E. is $\quad m^2 - 1 = 0 \quad \Rightarrow \quad m = \pm 1$

$$\text{C.F.} = C_1e^z + C_2e^{-z} = C_1 r + \frac{C_2}{r}$$

$$\text{P.I.} = \frac{1}{D^2-1}(-k\,e^{3z}) = -k\,\frac{1}{(3)^2-1}\,e^{3z} = -\frac{k}{8}e^{3z} = \frac{-k}{8}r^3$$

C.S. = C.F. + P.I.

$$u = C_1 r + \frac{C_2}{r} - \frac{k}{8}r^3 \qquad \ldots (2)$$

Putting $\quad u = 0, \quad r = 0 \quad$ in (2), we get

$$0 = C_2$$

Putting $\quad C_2 = 0$ in (2), we get

$$u = C_1 r - \frac{k}{8}r^3 \qquad \ldots (3)$$

Putting $u = 0$, $r = a$ in (3), we get

$$0 = C_1 a - \frac{k}{8} a^3 \quad \Rightarrow \quad C_1 = \frac{k}{8} a^2$$

Putting $C_1 = \frac{k}{8} a^2$ in (3), we get

$$u = \frac{k}{8} a^2 r - \frac{k}{8} r^3 \quad \Rightarrow \quad u = \frac{k r}{8} (a^2 - r^2)$$ **Ans.**

2.14 LEGENDRE'S HOMOGENEOUS DIFFERENTIAL EQUATIONS

A linear differential equation of the form

$$(a + bx)^n \frac{d^n y}{dx^n} + a_1 (a + bx)^{n-1} \frac{d^{n-1} y}{dx^{n-1}} + ... + a_n y = X \qquad ... (1)$$

where $a, b, a_1, a_2, ... a_n$ are constants and X is a function of x, is called Legendre's linear equation.

Equation (1) can be reduced to linear differential equation with constant coefficients by the substitution.

$$a + b x = e^z \quad \Rightarrow \quad z = \log (a + bx)$$

so that $$\frac{dy}{dx} = \frac{dy}{dz} \cdot \frac{dz}{dx} = \frac{b}{a + bx} \cdot \frac{dy}{dz}$$

$$\Rightarrow \quad (a + bx) \frac{dy}{dx} = b \frac{dy}{dz} = b\, Dy, \; D = \frac{d}{dz} \quad \Rightarrow \quad (a + bx) \frac{dy}{dx} = b\, Dy$$

where

Again $$\frac{d^2 y}{dx^2} = \frac{d}{dx}\left(\frac{dy}{dx}\right) = \frac{d}{dx}\left(\frac{b}{a + bx} \cdot \frac{dy}{dz}\right)$$

$$= -\frac{b^2}{(a + bx)^2} \frac{dy}{dz} + \frac{b}{(a + bx)} \cdot \frac{d^2 y}{dz^2} \cdot \frac{dz}{dx}$$

$$= -\frac{b^2}{(a + bx)^2} \frac{dy}{dz} + \frac{b}{(a + bx)} \cdot \frac{d^2 y}{dz^2} \cdot \frac{b}{(a + bx)}$$

$$\Rightarrow \quad (a + bx)^2 \frac{d^2 y}{dx^2} = -b^2 \frac{dy}{dz} + b^2 \frac{d^2 y}{dz^2}$$

$$= b^2 \left(\frac{d^2 y}{dz^2} - \frac{dy}{dz}\right) = b^2 (D^2 y - D y) = b^2 D (D - 1) y$$

$$\Rightarrow \quad (a + bx)^2 \frac{d^2 y}{dx^2} = b^2 D (D - 1)$$

Similarly, $(a + bx)^3 \frac{d^3 y}{dx^3} = b^3 D (D - 1)(D - 2) y$

....

$$(a + bx)^n \frac{d^n y}{dx^n} = b^n D (D - 1)(D - 2) ..... (D - n + 1) y$$

Substituting these values in equation (1), we get a linear differential equation with constant coefficients, which can be solved by the method given in the previous section.

Example 52. *Solve* $(3x+2)^2 \dfrac{d^2y}{dx^2} + 5(3x+2)\dfrac{dy}{dx} - 3y = x^2 + x + 1$

Solution. We have,
$$(3x+2)^2 \frac{d^2y}{dx^2} + 5(3x+2)\frac{dy}{dx} - 3y = x^2 + x + 1 \qquad \dots (1)$$

Let $3x + 2 = e^z \quad \Rightarrow \quad z = \log(3x+2), \qquad D = \dfrac{d}{dz}$

So that $(3x+2)\dfrac{dy}{dx} = 3Dy$ and $(3x+2)^2 \dfrac{d^2y}{dx^2} = 3^2 D(D-1)y$

On putting these values in (1), we get

$$[3^2 D(D-1) + 5.3D - 3]y = \left(\frac{e^z-2}{3}\right)^2 + \left(\frac{e^z-2}{3}\right) + 1$$

$$\Rightarrow \qquad (9D^2 + 6D - 3)y = \frac{1}{9}e^{2z} - \frac{1}{9}e^z + \frac{7}{9}$$

$\therefore$ Its A.E. is $9m^2 + 6m - 3 = 0 \quad \Rightarrow \quad (9m-3)(m+1) = 0 \quad \Rightarrow m = \dfrac{1}{3}, -1$

$$\therefore \qquad \text{C.F.} = C_1 e^{\frac{z}{3}} + C_2 e^{-z}$$

$$\text{P.I.} = \frac{1}{9D^2+6D-3}\left(\frac{1}{9}e^{2z} - \frac{1}{9}e^z + \frac{7}{9}\right)$$

$$= \frac{1}{9}\cdot\frac{1}{9D^2+6D-3}e^{2z} - \frac{1}{9}\cdot\frac{1}{9D^2+6D-3}e^z + \frac{7}{9}\cdot\frac{1}{9D^2+6D-3}e^{0z}$$

$$= \frac{1}{9}\cdot\frac{1}{9(4)+6(2)-3}e^{2z} - \frac{1}{9}\cdot\frac{1}{9(1)+6(1)-3}e^z + \frac{7}{9}\cdot\frac{1}{9(0)+6(0)-3}e^{0z}$$

$$= \frac{1}{405}e^{2z} - \frac{1}{108}e^z - \frac{7}{27}$$

Hence, the complete solution of (1) is

$$y = C.F. + P.I.$$

$$\Rightarrow \qquad y = C_1 e^{\frac{z}{3}} + C_2 e^{-z} + \frac{1}{405}e^{2z} - \frac{1}{108}e^z - \frac{7}{27}$$

$$\Rightarrow \qquad y = C_1(3x+2)^{\frac{1}{3}} + C_2(3x+2)^{-1} + \frac{1}{405}(3x+2)^2 - \frac{1}{108}(3x+2) - \frac{7}{27} \qquad \textbf{Ans.}$$

Example 53. *Solve:* $(3x+2)^2 \dfrac{d^2y}{dx^2} + 3(3x+2)\dfrac{dy}{dx} - 36y = 3x^2 + 4x + 1.$

(Uttarakhand II, Semester, June 2007)

Solution. Let $3x + 2 = e^z \quad \Rightarrow z = \log(3x+2) \quad \Rightarrow \quad \left[x = \dfrac{e^z - 2}{3}\right]$

So that $(3x+2)\dfrac{dy}{dx} = 3Dy$ and $(3x+2)^2 \dfrac{d^2y}{dx^2} = 9D(D-1)y$ where $D = \dfrac{d}{dz}$

Putting these values in the given differential equation, we get

$$9D(D-1)y + 9Dy - 36y = 3\left(\frac{e^z-2}{3}\right)^2 + 4\left(\frac{e^z-2}{3}\right) + 1$$

$\therefore \quad (9D^2 - 36)\, y = \frac{1}{3}(e^{2z} - 4e^z + 4) + \frac{4}{3}e^z - \frac{8}{3} + 1 = \frac{e^{2z}}{3} - \frac{1}{3}$

A.E. is $\quad 9m^2 - 36 = 0$

$\Rightarrow \quad m^2 - 4 = 0 \qquad \therefore \quad m = \pm 2$

$$\text{C.F.} = C_1 e^{2z} + C_2 e^{-2z}$$

$$\text{P.I.} = \frac{1}{9D^2 - 36}\left[\frac{e^{2z}}{3} - \frac{1}{3}\right] = \frac{1}{27}\frac{1}{D^2 - 4}e^{2z} - \frac{1}{3}\left[\frac{1}{9D^2 - 36}\right]e^{0z}$$

$$= \frac{1}{27}z\frac{1}{2D}e^{2z} - \frac{1}{3}\left[\frac{1}{0 - 36}\right] = \frac{1}{27}z\left(\frac{e^{2z}}{4}\right) + \frac{1}{108}$$

$$= \frac{1}{108}[z e^{2z} + 1]$$

Complete Solution is

$$y = C.F. + P.I. = C_1 e^{2z} + C_2 e^{-2z} + \frac{1}{108}(z e^{2z} + 1)$$

$$y = C_1 (3x + 2)^2 + \frac{C_2}{(3x + 2)^2} + \frac{1}{108}[(3x + 2)^2 \log(3x + 2) + 1]$$ **Ans.**

Example 54. *Solve* $(2x + 1)^2 \frac{d^2y}{dx^2} - 2(2x + 1)\frac{dy}{dx} - 12y = 6x.$ (*M.U., II Semester, 2009*)

Solution. Here, we have

$$(2x + 1)^2 \frac{d^2y}{dx^2} - 2(2x + 1)\frac{dy}{dx} - 12y = 6x \qquad \text{...(1)}$$

Put $2x + 1 = e^z \quad \Rightarrow \quad z = \log(2x + 1)$

$(2x + 1)\frac{dy}{dx} = 2Dy$ and $(2x + 1)^2 \frac{d^2y}{dx^2} = 2^2 D(D - 1)y$ in (1), we get

$$4D(D - 1)y - 2 \times 2Dy - 12y = 6\left[\frac{1}{2}(e^z - 1)\right]$$

$$4D^2 y - 4Dy - 4Dy - 12y = 3(e^z - 1)$$

A.E. is $\quad 4m^2 - 8m - 12 = 0 \quad \Rightarrow \quad m^2 - 2m - 3 = 0$

$\Rightarrow \quad (m - 3)(m + 1) = 0 \quad \Rightarrow \quad m = 3, m = -1$

$$\text{C.F.} = C_1 e^{3z} + C_2 e^{-z}$$

$$\text{P.I.} = \frac{1}{4D^2 - 8D - 12}(3e^z - 3)$$

$$= \frac{1}{4D^2 - 8D - 12}3e^z - 3\frac{1}{4D^2 - 8D - 12}e^{(0)z}$$

$$= 3\frac{1}{4(1)^2 - 8(1) - 12}e^z - 3\frac{1}{0 - 0 - 12}(1)$$

$$= \frac{3}{-16}e^z + \frac{1}{4}$$

Complete solution is $y =$ C.F. + P.I.

$$\Rightarrow \quad y = C_1 e^{3z} + C_2 e^{-z} - \frac{3}{16} e^z + \frac{1}{4}$$

$$\Rightarrow \quad y = C_1 (2x+1)^3 + C_2(2x+1)^{-1} - \frac{3}{16}(2x+1) + \frac{1}{4} \quad \textbf{Ans.}$$

Example 55. *Solve:* $(1+x)^2 \frac{d^2y}{dx^2} + (1+x)\frac{dy}{dx} + y = 4 \cos[\log(1+x)]$

(*M.U. II Semester, 2008*)

Solution. Here, we have

$$(1+x)^2 \frac{d^2y}{dx^2} + (1+x)\frac{dy}{dx} + y = 4 \cos[\log(1+x)] \quad ...(1)$$

Put $1 + x = e^z \quad \Rightarrow \quad \log(1+x) = z$

$(1+x)\frac{dy}{dx} = Dy, \quad (1+x)^2 \frac{d^2y}{dx^2} = D(D-1)y$ in (1), we get

$$\left[D = \frac{d}{dz}\right]$$

$$D(D-1)y + Dy + y = 4\cos z$$

$$\Rightarrow \quad (D^2 - D + D + 1)y = 4\cos z$$

$$\Rightarrow \quad (D^2 + 1)y = 4\cos z$$

A.E. is $\quad m^2 + 1 = 0$

$$\Rightarrow \quad m^2 = -1 \quad \Rightarrow \quad m = \pm i$$

$$\text{C.F.} = [C_1 \cos z + C_2 \sin z]$$

$$\text{P.I.} = \frac{1}{D^2+1} 4\cos z$$

$$= 4z \frac{1}{2D} \cos z = \frac{4}{2} z \int \cos z \, dz$$

$$= 2z \sin z$$

Complete solution $=$ C.F. + P.I.

$$= [C_1 \cos z + C_2 \sin z] + 2z \sin z$$

$$= C_1 \cos \log(1+x) + C_2 \sin \log(1+x)$$

$$+ 2\log(1+x) \sin \log(1+x) \quad \textbf{Ans.}$$

Example 56. *Solve* $(1+x)^2 \frac{d^2y}{dx^2} + (1+x)\frac{dy}{dx} + y = \sin 2\{\log(1+x)\}$

Solution. We have, $(1+x)^2 \frac{d^2y}{dx^2} + (1+x)\frac{dy}{dx} + y = \sin 2\{\log(1+x)\}$

Put $1 + x = e^z$ or $\log(1+x) = z$

$(1+x)\frac{dy}{dx} = Dy$ and $(1+x)^2 \frac{d^2y}{dx^2} = D(D-1)y$, where $D = \frac{d}{dz}$

Putting these values in the given differential equation, we get

$$D(D-1)y + Dy + y = \sin 2z \quad \text{or} \quad (D^2 - D + D + 1)y = \sin 2z$$

$$(D^2 + 1)y = \sin 2z$$

A.E. is $\quad m^2 + 1 = 0 \quad \Rightarrow \quad m = \pm i$

$$C.F. = A\cos z + B\sin z$$

$$\text{P.I.} = \frac{1}{D^2+1}\sin 2z = \frac{1}{-4+1}\sin 2z = -\frac{1}{3}\sin 2z$$

Now, complete solution is y = C.F. + P.I.

$$\Rightarrow \qquad y = A\cos z + B\sin z - \frac{1}{3}\sin 2z$$

$$\Rightarrow \qquad y = A\cos\{\log(1+x)\} + B\sin\{\log(1+x)\} - \frac{1}{3}\sin 2\{\log(1+x)\} \qquad \textbf{Ans.}$$

EXERCISE 2.8

Solve the following differential equations.

1. $x^2\dfrac{d^2y}{dx^2} - x\dfrac{dy}{dx} - 3y = 0$ **Ans.** $\dfrac{C_1}{x} + C_2x^3$

2. $x^2y'' + 9xy' + 17y = 0$ **Ans.** $\dfrac{1}{x^4}[C_1\cos(\log x) + C_2\sin(\log x)]$

3. $x^2\dfrac{d^2y}{dx^2} - 4x\dfrac{dy}{dx} + 6y = \dfrac{42}{x^4}$ **Ans.** $C_1x^2 + C_2x^3 + \dfrac{1}{x^4}$

4. $(x^2D^2 - 3xD + 4)\,y = 2x^2$ **Ans.** $(C_1 + C_2\log x)\,x^2 + x^2(\log x)^2$

5. $(x^2D^2 + 1)\,y = 3x^2$ **Ans.** $\sqrt{x}\left[C_1\cos\left(\dfrac{\sqrt{3}}{2}\log x\right) + C_2\sin\left(\dfrac{\sqrt{3}}{2}\log x\right)\right] + x^2$

6. $x^3\dfrac{d^3y}{dx^3} + 2x^2\dfrac{d^2y}{dx^2} + 2y = 10\left(\dfrac{x+1}{x}\right)$ **Ans.** $\dfrac{C_1}{x} + x\,(C_2\cos(\log x) + C_3\sin(\log x) + 5 + \dfrac{2\log x}{x}$

7. $(x^2D^2 + 2xD)\,y = \log x$ **Ans.** $C_1 + \dfrac{C_2}{x} + \dfrac{1}{2}(\log x)^2 - \log x$

8. $\dfrac{d^2y}{dx^2} + \dfrac{1}{x}\dfrac{dy}{dx} = \dfrac{12\log x}{x^2}$ **Ans.** $C_1 + C_2\log x + 2(\log x)^3$

9. $x^2\dfrac{d^2y}{dx^2} + 6x\dfrac{dy}{dx} + 6y = x\log x$ **Ans:** $y = \dfrac{A}{x} + \dfrac{B}{x^3} + \dfrac{x}{12}\left(\log x - \dfrac{7}{12}\right)$

10. $x^3\dfrac{d^3y}{dx^3} + 2x^2\dfrac{d^2y}{dx^2} + 3x\dfrac{dy}{dx} - 3y = x^2$ **Ans.** $C_1x + C_2\cos\left(\sqrt{3}\log x\right) + C_3\sin\left(\sqrt{3}\log x\right) + \dfrac{x^2}{7}$

11. $x^2\dfrac{d^2y}{dx^2} - 2x\dfrac{dy}{dx} + 2y = x^2 + \sin(5\log x)$

 Ans. $C_1x + C_2x^2 + x^2\log x + \dfrac{1}{754}[15\cos(5\log x) - 23\sin(5\log x)]$

12. $x^3\dfrac{d^2y}{dx^2} + 3x^2\dfrac{dy}{dx} + xy = \sin(\log x)$ **Ans.** $\dfrac{1}{x}[C_1 + C_2\log x - \sin(\log x)]$

13. $x^3y'' - x^2y' + xy = 1,\ y(1) = \dfrac{1}{4},\quad y(0) = e + \dfrac{1}{4e}$ *(A.M.I.E.T.E., Dec. 2004)*

14. $(1+x)^2\dfrac{d^2y}{dx^2} + (1+x)\dfrac{dy}{dx} + y = 2\sin\log(1+x)$

 Ans. $y = C_1\cos\log(1+x) + C_2\sin\log(1+x) - \log(1+x)\cos\log(1+x)$

OBJECTIVE TYPE QUESTIONS

Choose the correct alternative :

1. The particular intergral of $\frac{d^2y}{dx^2}+\frac{6dy}{dx}+9y=5e^{3x}$ is

(a) $\frac{5e^{3x}}{18}$ (b) $\frac{e^{3x}}{36}$

(c) $5e^{3x}$ (d) $\frac{5e^{3x}}{36}$ **Ans.** (d)

2. The particular integral of $\frac{d^2y}{dx^2}+\frac{dy}{dx}+y=e^x$ is

(a) e^x (b) $e^{\frac{x}{3}}$

(c) e^{x^2} (d) $e^{\frac{x}{2}}$ **Ans.** (b)

3. The particular integral of $2\frac{d^2y}{dx^2}+\frac{dy}{dx}+3y=e^{2x}$ is

(a) $\frac{e^x}{13}$ (b) $\frac{e^{2x}}{13}$

(c) $e^{\frac{2x}{27}}$ (d) None of these **Ans.** (b)

4. The particular integral of $4\frac{d^2y}{dx^2}+\frac{3dy}{dx}+2y=e^{3x}$ is

(a) $\frac{e^{3x}}{7}$ (b) $\frac{e^{2x}}{47}$

(c) $\frac{e^{3x}}{27}$ (d) $\frac{e^{3x}}{47}$ **Ans.** (d)

5. The particular integral of $2\frac{d^2y}{dx^2}+3\frac{dy}{dx}+y=e^{-x}$ is

(a) $-x\,e^{-x}$ (b) $x\,e^{-x}$

(c) $-x\,e^x$ (d) xe^{x^2} **Ans.** (a)

6. The particular integral of $\frac{d^2y}{dx^2}-4y=e^{2x}$ is

(a) $x\,e^{2x}$ (b) $2\,x\,e^{2x}$

(c) $\frac{xe^{2x}}{2}$ (d) $\frac{xe^{2x}}{4}$ **Ans.** (d)

7. Particular integral of $\frac{d^2y}{dx^2}+y=x^2$ is

(a) x^2 (b) x^2+2

(c) x^2-2 (d) None of these **Ans.** (c)

8. Particular integral of $\frac{d^2y}{dx^2}+\frac{dy}{dx}+y=x^2$ is

(a) x^2-2x (b) x^2+2x

(c) x^2+2 (d) x^2-2 **Ans.** (a)

9. Particular integral of $2\dfrac{d^2y}{dx^2} - y = x^3$ is

(a) $x^3 + 12x$ (b) $-(x^3 + 12x)$

(c) $x^3 + 12$ (d) $-x^3 - 12$ **Ans.** (b)

10. Particular integral of $\dfrac{d^2y}{dx^2} + y = 1$ is

(a) 0 (b) x

(c) 1 (d) x^2 **Ans.** (c)

11. The particular integral of $(D^2 + 4)\, y = \cos 2x$ is

(a) $\dfrac{x}{4}\cos 2x$ (b) $\dfrac{x}{4}\sin 2x$

(c) $x \cos 2x$ (d) $\dfrac{1}{4}\sin 2x$ **Ans.** (b)

12. The particular integral of $\dfrac{d^2y}{dx^2} + y = \sin x$ is

(a) $-\dfrac{x}{2}\cos x$ (b) $\dfrac{x}{2}\cos x$

(c) $-\dfrac{x}{2}\sin x$ (d) None of these **Ans.** (a)

13. The particular integral of $\dfrac{d^2y}{dx^2} - y = \cos x$ is

(a) $-\dfrac{1}{2}\cos x$ (b) $\dfrac{1}{2}\cos x$

(c) $-\dfrac{1}{2}\sin x$ (d) None of these **Ans.** (a)

14. The particular integral of $\dfrac{d^2y}{dx^2} + \dfrac{dy}{dx} + y = \sin 2x$ is

(a) $3 \sin 2x + 2 \cos 2x$ (b) $\dfrac{1}{13}(3 \sin 2x + 2 \cos 2x)$

(c) $13\,(3 \sin 2x + 2 \cos 2x)$ (d) $-\dfrac{1}{13}(3 \sin 2x + 2 \cos 2x)$ **Ans.** (d)

15. The particular integral of $\dfrac{d^2y}{dx^2} + \dfrac{dy}{dx} + y = e^x \sin x$ is

(a) $-\dfrac{e^x}{13}(3 \cos x - 2 \sin x)$ (b) $\dfrac{e^x}{13}(3 \cos x - 2 \sin x)$

(c) $e^x\,(3 \cos x - 2 \sin x)$ (d) None of these **Ans.** (a)

16. The particular integral of $\dfrac{d^2y}{dx^2} + y = e^{2x} \cos x$ is

(a) $\dfrac{e^x}{4}(\sin x + \cos x)$ (b) $e^{2x}\,(\sin x + \cos x)$

(c) $\dfrac{e^{2x}}{8}(\sin x + \cos x)$ (d) None of these **Ans.**(c)

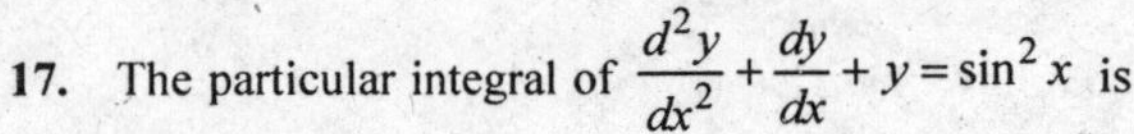

17. The particular integral of $\dfrac{d^2y}{dx^2}+\dfrac{dy}{dx}+y=\sin^2 x$ is

(*a*) $\dfrac{1}{2}-\dfrac{2}{13}(2\sin 2x-2\cos 2x)$ (*b*) $\dfrac{1}{2}+\dfrac{2}{13}(2\sin 2x-3\cos 2x)$

(*c*) $\dfrac{2}{13}(2\sin 2x-3\cos 2x)$ (*d*) None of them **Ans.**(*a*)

18. Particular integral of $(D^2+a^2)\,y=\cos ax$ is :

(*a*) $\dfrac{-x}{2a}\cos a\,x$ (*b*) $\dfrac{1}{2a}\cos a\,x$

(*c*) $\dfrac{1}{2a}\sin a\,x$ (*d*) $\dfrac{x}{2a}\sin a\,x$ **Ans.**(*d*)

(R.G.P.V, Bhopal, 1Semester June,2007)

19. e^{iax} is equal to

(*a*) $\sin a\,x+i\cos ax$ (b) $\cos ax-i\sin ax$

(*c*) $\sin ax-i\cos ax$ (*d*) $\cos ax+i\sin ax$ **Ans.**(*d*)

20. $\dfrac{1}{f(D)}x^n\sin ax$ is equal to

(*a*) $e^{iax}\dfrac{1}{f(D+i\,a)}x^n$ (*b*) Imaginary part of $e^{iax}\dfrac{1}{f(D+i\,a)}x^n$

(*c*) Real part of $e^{iax}\dfrac{1}{f(D+i\,a)}x^n$ (*d*) None of these **Ans.**(*b*)

21. $\dfrac{1}{f(D)}x^n\cos ax$ is equal to

(*a*) Real part of $e^{iax}\dfrac{1}{f(D+ia)}x^n$ (*b*) Imaginary part of $e^{iax}\dfrac{1}{f(D+ia)}x^n$

(*c*) $e^{iax}\dfrac{1}{f(D+ia)}x^n$ (*d*) None of these **Ans.**(*a*)

22. $\dfrac{1}{D-a}\phi(x)$ is equal to

(*a*) $e^{ax}\int\phi(x)\,e^{x}\,dx$ (*b*) $\int e^{-ax}\phi(x)\,dx$

(*c*) $e^{ax}\int e^{-ax}\phi(x)\,dx$ (*d*) $e^{ax}\int\phi(x)\,dx$ **Ans.**(*c*)

23. $\dfrac{1}{D-3i}\sec 3x$ is equal to

(*a*) $e^{3ix}\int e^{-3ix}\sec 3x\,dx$ (*b*) $\int e^{-3ix}\sec 3x\,dx$

(*c*) $e^{3ix}\int e^{-3ix}dx$ (*d*) $e^{-3ix}\int e^{3ix}\sec 3x\,dx$ **Ans.**(*a*)

24. The complementary function for the solution of the differential equation $2x^2y''+3xy'-3y=x^3$ is obtaned as

(*a*) $Ax+Bx^{-3/2}$ (*b*) $Ax+Bx^{3/2}$

(*c*) Ax^2+Bx (*d*) $Ax^{-3/2}+Bx^{3/2}$ **Ans.**(*a*)

25. Solution of $\frac{d^3y}{dx^3}+2\frac{d^2y}{dx^2}+\frac{dy}{dx}=0$ is

(a) $y = C_1e^x + C_2e^{2x}$ (b) $y = C_1 + (C_2 + C_3x)\, e^{-x}$

(c) $y = (C_1 + C_2x + C_3x^2)e^{-x}$ (d) $y = C_1 + C_2\, e^{-x}$ **Ans.**(b)

(A.M.I.E.T.E. June 2009)

26. The basis of solutions for the differential equation $x^2\frac{d^2y}{dx^2}-x\frac{dy}{dx}+y=0$ are

(a) $x, x\, l_n\, x,$ (b) l_nx, e^x

(c) $\frac{1}{x}, \frac{1}{x^2}$ (d) $\frac{1}{x^2}e^x$, xl_nx (R.G.P.V.,Bhopal ,June ,2008)

Ans.(a)

27. The complementary function for the solution of the differential equation $2x^2y'' + 3xy' - 3y = x^3$ is obtained as

(a) $Ax + Bx^{-3/2}$ (b) $Ax + Bx^{3/2}$

(c) $Ax^2 + Bx$ (d) $Ax^{-3/2} + Bx^{3/2}$ **Ans.**(a)

28. The substitutions in solving the homogeneous linear equations are

(a) $x = e^z,\ x\frac{dy}{dx} = Dy,\ x^2\frac{d^2y}{dx^2} = D(D-1)\,y$, where $D = \frac{d}{dz}$

(b) $z = e^x,\ x\frac{dy}{dx} = Dy,\ x^2\frac{d^2y}{dx^2} = D(D-1)\,y$, where $D = \frac{d}{dz}$

(c) $x = e^{-z},\ \frac{dy}{dx} = Dy,\ x^2\frac{d^2y}{dx^2} = D(D-1)\,y$, where $D = \frac{d}{dz}$

(d) None of these **Ans.**(a)

29. The complementary function for the solution of the differential equation.

$x^2y'' + x\,y' + y = \log x \sin(\log x)$ is

(a) $C_1 + C_2 \sin(\log x)$ (b) $C_1 \cos(\log x) + C_2 \sin(\log x)$

(c) $C_1 \cos(\log x) + C_2$ (d) $(C_1 + C_2x)\sin(\log x)$ **Ans.**(b)

30. The integrating factor of $\cos^2 x\frac{dy}{dx} + y = \tan x$ is

(a) $e^{\sec x}$ (b) $e^{-\tan x}$

(c) $e^{\sin x}$ (d) $e^{\tan x}$ **Ans.**(d)

31. The integrating factor of $x(x-1)\frac{dy}{dx} - (x-2)\,y = x^2(2x-1)$ is

(a) $\frac{x^3-1}{x}$ (b) $\frac{x^3+1}{x}$

(c) $\frac{x-1}{x^3}$ (d) $\frac{x+1}{x^3}$ **Ans.**(c)

32. Solution of $(x-y-1)\frac{dy}{dx} = 1$ is

(a) $x - y - 2 = c\, e^y$ (b) $x + y = c\, e^y$

(c) $x - y - 2 = e^y$ (d) $x - y - 2 = c\, e^y$ **Ans.**(a)

33. *P.I.* of $(D^2 + 4)\, y = \sin 3x$ is

(a) $-\frac{1}{10}\sin 3x$ (b) $\frac{1}{2}\sin 3x$

(c) $\frac{1}{3}\sin 2x$ (d) $-\frac{1}{5}\sin 3x$ **Ans.(d)**

34. *P.I.* of $(D^2 - 2D + 1)\, y = e^x$ is

(a) $\frac{x^2}{2}e^x$ (b) $\frac{x}{2}e^x$

(c) $\frac{x}{4}e^{x^2}$ (d) $\frac{x^2}{5}e^{2x}$ **Ans.(a)**

35. On putting $x = e^z$, the transformed differential equation of $x^2\frac{d^2y}{dx^2} + x\frac{dy}{dx} + y = x$ is

(a) $\frac{d^2y}{dz^2} - y = e^x$ (b) $\frac{d^2y}{dz^2} + y = e^x$

(c) $\frac{dy}{dz} + y = e^z$ (d) $\frac{dy}{dz} - y = e^{z^2}$ **Ans. (b)**

36. The linear differential equation in standard form is

(a) $\frac{dy}{dx} + Py = Q$ where P and Q are the functions of x only

(b) $\frac{dy}{dx} + Py = Q$, where P and Q are the functions of y only

(c) $\frac{dy}{dx} + Py = Q$, where P is function of x and Q is function of y

(d) $\frac{dy}{dx} + Py = Q$, where P is funtion y and Q is a function of x. **Ans. (a)**

37. Let y_1, y_2 be two linearly independent solutions of the differential equations $yy'' - (y')^2 = 0$. Then $c_1y_1 + c_2y_2$, where c_1, c_2 are contants is a solution of this differential equation for

(a) $c_1 = c_2 = 0$ only (b) $c_1 - 0$ or $c_2 = 0$

(b) no value of c_1, c_2 (d) all real c_1, c_2 **Ans. (d)**

38. The solution of the differential equation $y'' + 2y' + y = 0$, $y(0) = 1$, $y'(0) = -1$ is

(a) xe^{-x} (b) $-xe^{-x}$ (c) $-e^{-x}$ (d) e^{-x} **Ans. (d)**

39. The solution of the differential equation $y'' + a^2y = 0$; $y(0) = 0$, $y'(0) = a$ is

(a) $\cos x$ (b) $\sin x$ (c) $\cos ax$ (d) $\sin ax$ **Ans. (d)**

40. Which of the following is not a solution of the equation $\frac{d^2y}{dx^2} + \frac{dy}{dx} - 6y = 0$

(a) e^{2x} (b) e^{-3x} (c) $3e^{2x} + 5e^{-3x}$ (d) $c_1e^{2x} + c_2e^{-3x} + 1$ **Ans. (d)**

41. The complementary function of $\frac{d^2y}{dx^2} - 6y\frac{dy}{dx} + 9y = 0$ is

(a) $y = (c_1 + c_2x)e^{3x}$ (b) $y = (c_1 + c_2)e^{3x}$

(c) $y = (c_1 + c_2x)e^{-3x}$ (d) $y = (c_1 + c_2)e^{-3x}$ **Ans. (a)**

42. The equation $\frac{d^2y}{dx^2} - 2\frac{dy}{dx} + 5y = 0$ has the solution.

(a) $c_1e^x + c_2e^{2x}$ (b) $c_1e^{-x} + c_2e^{2x}$

(c) $c_1e^x \cos(2x + c_2)$ (d) $c_1e^{-x} \sin(2x + c_2)$ **Ans.** (c)

43. Complementary function of $(D^4 + 2D^3 - 3D^2)y = x^2$ is:

(a) $c_1x + c_2x^2 + c_3e^x + c_4e^{-3x}$ (b) $(c_1 + c_2x) + c_3e^x + c_4e^{-x}$

(c) $(c_1 + c_2x) + c_3e^x + c_4e^{-3x}$ (d) $(c_1 + c_2x)e^x + c_3e^{2x} + c_4e^{-3x}$ **Ans.** (c)

(R.G.R.V., Bhopal, I Semester June 2007)

44. Complementary function of $(D^2 - 4)y = 0$ is:

(a) $c_1e^{2x} + c_2e^{2x}$ (b) $(c_1 + c_2x)e^{2x}$

(c) $c_1e^{2x} + c_2e^{-2x}$ (d) $c_1 \sin 2x + C_2 \cos 2x$ **Ans.** (c)

(R.G.R.V., Bhopal, I Semester June 2006)

45. The solution of the differntial equation $\frac{d^2y}{dx^2} + y = 0$ satisfying the initial conditions $y(0) = 1, y(\pi/2) = 2$ is

(a) $y = 2\cos x + \sin x$ (b) $y = \cos x + 2\sin x$

(c) $y = \cos x + \sin x$ (d) $y = 2\cos x + 2\sin x$ **Ans.** (b)

46. The value of α so that $e^{\alpha y^2}$ is an integrating factor of the differential equation $\left(e^{-\frac{y^2}{2}} - xy\right)dy - dx = 0$ is

(a) -1 (b) 1 (c) $\frac{1}{2}$ (d) $-\frac{1}{2}$ **Ans.**

47. The solution of $(D^2 + 2D + 2)y = 0, y(0) = 0, y'(0) = 1$ is

(a) $e^x \sin x$ (b) $e^{-x} \cos x$ (c) $e^{-x} \sin x$ (d) $e^x \cos x$ **Ans.** (c)

48. The solution of the differential equation $(D^4 - 6D^3 + 12D^3 - 8D)y = 0$ is given by

(a) $y = a + (b + cx + dx^2)e^{2x}$ (b) $y = (b + cx + dx^2)e^{2x}$

(c) $y = a + bx + cx^2 + dx^3$ (d) $y = a + bx + cx^2 + de^{2x}$

49. The particular integral of the differential equation $\frac{d^2y}{dx^2} + a^2y = \sin ax$ is

(a) $-\frac{x}{2a}\cos ax$ (b) $\frac{x}{2a}\cos ax$

(c) $-\frac{ax}{2}\cos ax$ (d) $\frac{ax}{2}\cos ax$

50. The solution of the differential equation $\frac{d^2y}{dx^2} - 3\frac{dy}{dx} + 2y = e^{3x}$ is

(a) $y = ae^x + be^{2x} + \frac{1}{2}e^{3x}$ (b) $y = ae^{-x} + be^{-2x} + \frac{1}{2}e^{3x}$

(c) $y = ae^x + be^{-2x} + \frac{1}{2}e^{3x}$ (d) $y = ae^{-x} + be^{2x} + \frac{1}{2}e^{3x}$

Fill in the blanks in each of the following:

51. The solution of $\frac{d^3y}{dx^3} - 3\frac{d^2y}{dx^2} + 4y = 0$ is ...... **Ans.** $y = c_1e^{-x} + (c_2 + c_3x)e^{2x}$

52. The particular integral of $(D^2 + 9)\,y = \sin 3x$ is ...... **Ans.** $P.I. = -\frac{x}{6}\cos 3x$

53. The particular integral of $(D^2 + 6D + 5)\,y = 4e^x$ is ...... **Ans.** $P.I. = \frac{1}{3}e^x$

54. The solution of $x^2\frac{d^2y}{dx^2} - x\frac{dy}{dx} = 0$ is ...... **Ans.** $c_1 + c_2x^2$

55. The particular intagral of $\frac{d^2y}{dx^2} + y = 2\cosh 3x$ is .... **Ans.** $-\frac{1}{4}\cosh 3x$

56. The differential equation whose auxiliary equation has the roots 0, – 1, – 1 is ......
Ans. $(D^3 + 2D^2 + D)\,y = 0$

57. The complementary function of the differential equation $x^2y'' + xy' + y = \log x^2$ is ......
Ans. $(c_1 \cos e^z + c_2 \sin e^z)$

58. To transform $x\frac{d^2y}{dx^2} + \frac{dy}{dx} = \frac{2y}{x}$ into the linear differential equation with constant coefficients, we put $x =$ ...... **Ans.** e^z

59. The no. of arbitrary constanats in the general solutions of a differential equation is equal to ...... of the differential equation. **Ans.** Order

Match the following

60. (*i*) $\frac{1}{f(D)}e^{ax}$ (*a*) $\frac{1}{f(ax)}\cos ax$

(*ii*) $\frac{1}{f(D^2)}\cos ax$ (*b*) $[f(d)]^{-1}x^n$

(*iii*) $\frac{1}{f(D)}x^m$ (*c*) $e^{ax}\frac{1}{f(D+a)}\phi(x)$

(*iv*) $\frac{1}{f(D)}e^{ax}\phi(x)$ (*d*) $\frac{1}{f(a)}e^{ax}$

Ans. (*i*) → (*d*)
(*ii*) → (*a*)
(*iii*) → (*b*)
(*iv*) → (*c*)

Indicate true or false for the following :

61. The *P.I* of $\frac{d^2y}{dx^2} + y = \sin 3x$ is $-\frac{1}{8}\sin 3x$ **Ans.** True

62. The P.I of $\frac{d^2y}{dx^2} + 2\frac{dy}{dx} + 2y = e^{3x}$ is $\frac{1}{21}e^{3x}$ **Ans.** False

63. The P.I of $\frac{1}{f(D)}e^{2x}\phi(x) = e^{2x}\frac{1}{f(D+2)}\phi(x)$. **Ans.** True

64. The *P.I.* of $\frac{d^2y}{dx^2} + y = x^2$ is x^2. **Ans.** True

UNIT–I

CHAPTER 3

SIMULTANEOUS LINEAR DIFFERENTIAL EQUATIONS

3.1 SIMULTANEOUS DIFFERENTIAL EQUATIONS

If two or more dependent variables are functions of a single independent variable, the equations involving their derivatives are called simultaneous equations, *e.g.*

$$\frac{dx}{dt} + 4y = t$$

$$\frac{dy}{dt} + 2x = e^t$$

The method of solving these equations is based on the process of elimination, as we solve algebraic simultaneous equations.

Example 1. *The equations of motions of a particle are given by*

$$\frac{dx}{dt} + wy = 0$$

$$\frac{dy}{dt} - wx = 0$$

Find the path of the particle and show that it is a circle. (*U.P. II Semester summer* 2009)

Solution. On putting $\frac{d}{dt} \equiv D$ in the equations, we have

$$Dx + wy = 0 \qquad ...(1)$$

$$-wx + Dy = 0 \qquad ...(2)$$

On multiplying (1) by w and (2) by D, we get

$$wDx + w^2y = 0 \qquad ...(3)$$

$$-wDx + D^2y = 0 \qquad ...(4)$$

On adding (3) and (4), we obtain

$$w^2 y + D^2 y = 0 \quad \Rightarrow \quad (D^2 + w^2)\, y = 0 \qquad ...(5)$$

Now, we have to solve (5) to get the value of y.

A.E. is $m^2 + w^2 = 0 \quad \Rightarrow \quad m^2 = -w^2 \quad \Rightarrow \quad m = \pm iw$

$$\therefore \quad y = A \cos wt + B \sin wt \qquad ...(6)$$

$\Rightarrow Dy = -A\,w \sin wt + Bw \cos wt$

On putting the value of Dy in (2), we get

$-wx - Aw \sin wt + Bw \cos wt = 0$

$$\Rightarrow \quad wx = -Aw \sin wt + Bw \cos wt$$

$\Rightarrow \quad x = -A \sin wt + B \cos wt \quad ...(7)$

On squaring (6) and (7) and adding , we get

$$x^2 + y^2 = A^2(\cos^2 wt + \sin^2 wt) + B^2(\cos^2 wt + \sin^2 wt)$$

$$\Rightarrow \quad x^2 + y^2 = A^2 + B^2$$

This is the equation of circle. **Proved.**

Example 2. *Solve the following differential equation* [*U.P. II Semester, 2009*]

$$\frac{dx}{dt} = y + 1, \quad \frac{dy}{dt} = x + 1$$

Solution. Here, we have

$$Dx - y = 1 \quad ...(1)$$

$$-x + Dy = 1 \quad ...(2)$$

Multiplying (1) by D, we get

$$D^2x - Dy = D.1 \quad ...(3)$$

Adding (2) and (3), we get

$$(D^2 - 1)\, x = 1 + D.1$$

$$\Rightarrow \quad (D^2 - 1)\, x = 1 \text{ or } (D^2 - 1)x = e^0 \quad [D.(1) = 0]$$

A.E. is $\quad m^2 - 1 = 0 \Rightarrow m = \pm 1$

$$\therefore \quad \text{C.F.} = c_1 e^t + c_2 e^{-t}$$

$$\text{P.I.} = \frac{1}{D^2 - 1} . e^0 = \frac{1}{0 - 1} e^0 = -1$$

$$\therefore \quad x = \text{C.F.} + \text{P.I.} = c_1 e^t + c_2 e^{-t} - 1$$

From (1), $y = \dfrac{dx}{dt} - 1$

$$\Rightarrow \quad y = \frac{d}{dt}(c_1 e^t + c_2 e^{-t} - 1) - 1$$

$$\Rightarrow \quad \left.\begin{aligned} y &= c_1 e^t - c_2 e^{-t} - 1 \\ \text{and} \quad x &= c_1 e^t + c_2 e^{-t} - 1 \end{aligned}\right\}$$

Ans.

Example 3. *Solve the following simultaneous differential equations* [*U.P. II Semester, 2008*]

$$\frac{dx}{dt} = 3x + 2y$$

$$\frac{dy}{dt} = 5x + 3y$$

Solution. Here, we have

$$(D - 3)\, x - 2y = 0 \quad ...(1)$$

$$-5x + (D - 3)y = 0 \quad ...(2) \quad \left[D \equiv \frac{d}{dt}\right]$$

Operating (1) by (D – 3) and multiplying equation (2) by 2 then adding, we get

$$(D - 3)^2 x - 2\,(D - 3)y - 10x + 2\,(D - 3)\, y = 0$$

$$\Rightarrow \quad [(D - 3)^2 - 10]\, x = 0$$

$$(D^2 - 6D - 1)\, x = 0$$

Auxiliary equation is

$$m^2 - 6m - 1 = 0 \Rightarrow m = \frac{6 \pm \sqrt{36 + 4}}{2} = 3 \pm \sqrt{10}$$

$$\text{C.F.} = e^{3t}\,(c_1 \cosh\sqrt{10}\,t + c_2 \sinh\sqrt{10}\,t)$$

$$\therefore \quad x = e^{3t}\,(c_1 \cosh\sqrt{10}\,t + c_2 \sinh\sqrt{10}\,t) \quad \text{...(3)}$$

From (1), $2y = \dfrac{dx}{dt} - 3x$

$$= \frac{d}{dt}\left[e^{3t}\,(c_1 \cosh\sqrt{10}\,t + c_2 \sinh\sqrt{10}\,t)\right] - 3e^{3t}\,(c_1 \cosh\sqrt{10}\,t + \sinh\sqrt{10}\,t$$

$$= e^{3t}\sqrt{10}\left(c_1 \sinh\sqrt{10}\,t + c_2 \cosh\sqrt{10}\,t\right) + 3e^{3t}\,(c_1 \cosh\sqrt{10}\,t + c_2 \sinh\sqrt{10}\,t)$$

$$- 3e^{3t}\,(c_1 \cosh\sqrt{10}\,t + c_2 \sinh\sqrt{10}\,t)$$

$$\Rightarrow \quad \left.\begin{aligned} y &= \frac{\sqrt{10}}{2} e^{3t}\left(c_1 \sinh\sqrt{10}\,t + c_2 \cosh\sqrt{10}\,t\right) \\ x &= e^{3t}\,(c_1 \cosh\sqrt{10}\,t + c_2 \sinh\sqrt{10}\,t) \end{aligned}\right\} \quad \textbf{Ans.}$$

Example 4. *Solve the following simultaneous differential equation:*

$$\frac{dx}{dt} + 5x - 2y = t$$

$$\frac{dy}{dt} + 2x + y = 0$$

given that $x = y = 0$ when $t = 0$. [*U.P. II Semester, 2008*]

Solution. Here, we have

$$(D + 5)\,x - 2y = t \quad \text{...(1)}$$

$$2x + (D + 1)\,y = 0 \quad \text{...(2)} \quad \left(D \equiv \frac{d}{dt}\right)$$

Operating (1) by $(D + 1)$, we get

$$(D + 1)\,(D + 5)\,x - 2\,(D + 1)\,y = (D + 1)\,t$$

$$\Rightarrow \quad (D^2 + 6D + 5)\,x - 2\,(D + 1)\,y = 1 + t \quad \text{...(3)}$$

Multiplying (2) by 2, we get

$$4x + 2\,(D + 1)\,y = 0 \quad \text{...(4)}$$

Adding (3) and (4), we get

$$(D^2 + 6D + 9)\,x = 1 + t$$

Auxiliary equation is

$$m^2 + 6m + 9 = 0 \Rightarrow m = -3, -3$$

$$\therefore \quad \text{C.F.} = (c_1 + c_2\,t)\,e^{-3t}$$

$$\text{P.I.} = \frac{1}{(D+3)^2}(1+t) = \frac{1}{9}\left(1 + \frac{D}{3}\right)^{-2}(1+t)$$

$$= \frac{1}{9}\left(1 - \frac{2D}{3}\right)(1+t) = \frac{1}{9}\left(1 + t - \frac{2}{3}\right) = \frac{1}{9}\left(t + \frac{1}{3}\right)$$

$$x = \text{C.F.} + \text{P.I.} = (c_1 + c_2\,t)\,e^{-3t} + \frac{1}{9}\left(t + \frac{1}{3}\right) \quad \text{...(5)}$$

From (1), $2y = \frac{dx}{dt} + 5x - t$

$$= \frac{d}{dt}\left[(c_1 + c_2 t)e^{-3t} + \frac{1}{9}\left(t + \frac{1}{3}\right)\right] + 5\left[(c_1 + c_2 t)\, e^{-3t} + \frac{1}{9}\left(t + \frac{1}{3}\right)\right] - t$$

$$= -3\,(c_1 + c_2 t)e^{-3t} + c_2 e^{-3t} + \frac{1}{9} + 5\,(c_1 + c_2 t)e^{-3t} + \frac{5}{9}\left(t + \frac{1}{3}\right) - t$$

$$= 2(c_1 + c_2 t)e^{-3t} + c_2 e^{-3t} - \frac{4}{9}t + \frac{8}{27}$$

$$y = (c_1 + c_2 t)\, e^{-3t} + \frac{c_2}{2} e^{-3t} - \frac{2}{9}t + \frac{4}{27} \qquad \text{...(6)}$$

Applying condition $x\,(0) = 0$ in (5), we get

$$0 = c_1 + \frac{1}{27} \Rightarrow c_1 = -\frac{1}{27}$$

Applying condition $y\,(0) = 0$ in (6), we get

$$0 = c_1 + \frac{c_2}{2} + \frac{4}{27} = -\frac{1}{27} + \frac{c_2}{2} + \frac{4}{27} = \frac{c_2}{2} + \frac{1}{9} \Rightarrow c_2 = -\frac{2}{9}$$

On putting the values of c_1 and c_2 in (5) and (6), we get

$$x = -\frac{1}{27}(1 + 6t)e^{-3t} + \frac{1}{9}\left(t + \frac{1}{3}\right)$$

$$y = -\frac{2}{27}(2 + 3t)e^{-3t} - \frac{2}{9}t + \frac{4}{27} \qquad \textbf{Ans.}$$

Example 5. *Solve:* $\frac{dx}{dt} + 4x + 3y = t$

$\frac{dy}{dt} + 2x + 5y = e^t$ *(Uttarakhand, II Semester, Jun 2009, U.P. II Semester, 2006)*

Solution. Here, we have

$$(D + 4)x + 3y = t \qquad \text{...(1)}$$

$$2x + (D + 5)y = e^t \qquad \text{...(2)} \quad \left(D = \frac{d}{dt}\right)$$

To eliminate y, operating (1) by $(D + 5)$ and multiplying (2) by 3 then subtracting, we get

$$(D + 5)\,(D + 4)\,x + 3\,(D + 5)\,y - 3\,(2x) - 3\,(D + 5)\,y = (D + 5)t - 3e^t$$

$$[(D + 4)\,(D + 5) - 6]x = (D + 5)t - 3e^t$$

$$(D^2 + 9D + 14)x = 1 + 5t - 3e^t$$

Auxiliary equation is

$$m^2 + 9m + 14 = 0 \Rightarrow m = -2, -7$$

$$\therefore \quad \text{C.F.} = c_1 e^{-2t} + c_2 e^{-7t}$$

$$\text{P.I.} = \frac{1}{D^2 + 9D + 14}(1 + 5t - 3e^t)$$

$$= \frac{1}{D^2 + 9D + 14} e^{0t} + 5\frac{1}{D^2 + 9D + 14} t - 3\frac{1}{D^2 + 9D + 14} e^t$$

$$= \frac{1}{0^2+9(0)+14} e^{0t} + 5 . \frac{1}{14\left(1+\frac{9D}{14}+\frac{D^2}{14}\right)} t - 3\frac{1}{1^2+9(1)+14} e^t$$

$$= \frac{1}{14} + \frac{5}{14}\left[1+\left(\frac{9D}{14}+\frac{D^2}{14}\right)\right]^{-1} t - \frac{1}{8}e^t = \frac{1}{14} + \frac{5}{14}\left[1-\left(\frac{9D}{14}+\frac{D^2}{14}\right)+\cdots\right] t - \frac{1}{8}e^t$$

$$= \frac{1}{14} \cdot \frac{5}{14}\left(t-\frac{9}{14}\right) - \frac{1}{8}e^t = \frac{5}{14}t - \frac{31}{196} - \frac{1}{8}e^t$$

$$x = c_1e^{-2t} + c_2e^{-7t} + \frac{5}{14}t - \frac{31}{196} - \frac{1}{8}e^t$$

$$3y = t - \frac{dx}{dt} - 4x \qquad \text{[From (1)]}$$

$$= t - \frac{d}{dt}\left[c_1e^{-2t} + c_2e^{-7t} + \frac{5}{14}t - \frac{31}{196} - \frac{1}{8}e^t\right] - 4\left[c_1e^{-2t} + c_2e^{-7t} + \frac{5}{14}t - \frac{31}{196} - \frac{e^t}{8}\right]$$

$$3y = t + 2c_1e^{-2t} + 7c_2e^{-7t} - \frac{5}{14} + \frac{1}{8}e^t - 4c_1e^{-2t} - 4c_2e^{-7t} - \frac{10}{7}t + \frac{31}{49} + \frac{1}{2}e^t$$

$$\therefore \quad y = \frac{1}{3}\left[-2c_1e^{-2t} + 3c_2e^{-7t} - \frac{3}{7}t + \frac{27}{98} + \frac{5}{8}e^t\right]$$

Hence,

$$x = c_1e^{-2t} + c_2e^{-7t} + \frac{5}{14}t - \frac{31}{196} - \frac{1}{8}e^t$$

$$y = -\frac{2}{3}c_1e^{-2t} + c_2e^{-7t} - \frac{1}{7}t + \frac{9}{98} + \frac{5}{24}e^t \qquad \textbf{Ans.}$$

Example 6. *Solve:* $\frac{dx}{dt} + 5x + y = e^t, \; \frac{dy}{dt} + x + 5y = e^{5t}.$

Solution. Here, we have

$$(D+5)\,x + y = e^t \qquad ...(1)$$

$$x + (D+5)y = e^{5t} \qquad ...(2)\left(D \equiv \frac{d}{dt}\right)$$

Multiplying (1) by $(D + 5)$, we get

$$(D+5)^2x + (D+5)y = (D+5)\,e^t \qquad ...(3)$$

Subtracting (3) form (2), we get

$$\left\{1-(D+5)^2\right\}x = e^{5t} - (D+5)e^t$$

$$\Rightarrow \quad [\,1 - D^2 - 10D - 25]x = e^{5t} - e^t - 5e^t$$

$$\Rightarrow \quad (D^2 + 10D + 24)x = 6e^t - e^{5t}$$

Auxiliary equation is

$$m^2 + 10m + 24 = 0 \quad \Rightarrow \quad (m+4)(m+6) = 0$$

$$\Rightarrow \quad m = -4, -6$$

$$\therefore \qquad \text{C.F.} = c_1 e^{-4t} + c_2 e^{-6t}$$

$$\text{P.I.} = \frac{1}{D^2 + 10D + 24}(6e^t - e^{5t})$$

$$= \frac{6}{1^2 + 10(1) + 24} e^t - \frac{1}{(5)^2 + 10(5) + 24} e^{5t} = \frac{6e^t}{35} - \frac{e^{5t}}{99}$$

Thus $$x = \text{C.F.} + \text{P.I.} = c_1 e^{-4t} + c_2 e^{-6t} + \frac{6e^t}{35} - \frac{e^{5t}}{99}$$

From (1),

$$y = e^t - (D+5)x$$

$$= e^t - \frac{dx}{dt} - 5x$$

$$= e^t - \frac{d}{dt}\left(c_1 e^{-4t} + c_2 e^{-6t} + \frac{6e^t}{35} - \frac{e^{5t}}{99}\right) - 5\left(c_1 e^{-4t} + c_2 e^{-6t} + \frac{6e^t}{35} - \frac{e^{5t}}{99}\right)$$

$$= e^t + 4c_1 e^{-4t} + 6c_2 e^{-6t} - \frac{6e^t}{35} + \frac{5e^{5t}}{99} - 5c_1 e^{-4t} - 5c_2 e^{-6t} - \frac{30e^t}{35} + \frac{5e^{5t}}{99}$$

$$y = -\frac{1}{35} e^t - c_1 e^{-4t} + \frac{10}{99} e^{5t} + c_2 e^{-6t}$$

$$x = c_1 e^{-6t} + c_2 e^{-4t} + \frac{6e^t}{35} - \frac{e^{5t}}{99}$$ **Ans.**

Example 7. *Solve the following simultaneous equations:*

$$\frac{dx}{dt} + \frac{dy}{dt} + 2x + y = 0$$

$$\frac{dy}{dt} + 5x + 3y = 0$$

Solution. We have, $$\frac{dx}{dt} + \frac{dy}{dt} + 2x + y = 0$$

$$\frac{dy}{dt} + 5x + 3y = 0$$

On putting $\frac{d}{dt} \equiv D$, we have

$$(D+2)\,x + (D+1)\,y = 0 \qquad \text{...(1)}$$

$$5x + (D+3)\,y = 0 \qquad \text{...(2)}$$

Multiplying (1) by 5 and (2) by $(D+2)$, we get

$$5\,(D+2)\,x + 5\,(D+1)\,y = 0 \qquad \text{...(3)}$$

$$5\,(D+2)\,x + (D+2)\,(D+3)\,y = 0 \qquad \text{...(4)}$$

Subtracting (3) from (4), we get

$$(D+2)\,(D+3)\,y - 5\,(D+1)\,y = 0$$

$$(D^2 + 5D + 6 - 5D - 5)\,y = 0 \quad \Rightarrow \quad (D^2 + 1)\,y = 0 \;\Big|\; \text{A.E. is } m^2 + 1 = 0,\ m = \pm i$$

$$y = c_1 \cos t + c_2 \sin t$$

$$\frac{dy}{dt} = -c_1 \sin t + c_2 \cos t$$

Putting the values of Dy and y in (2), we get

$$5x - c_1 \sin t + c_2 \cos t + 3\, c_1 \cos t + 3\, c_2 \sin t = 0$$

$$\Rightarrow \qquad 5x = c_1 \sin t - 3\, c_2 \sin t - 3\, c_1 \cos t - c_2 \cos t$$

$$x = \frac{1}{5}(c_1 - 3c_2)\sin t - \frac{1}{5}(3c_1 + c_2)\cos t$$

$$y = c_1 \cos t + c_2 \sin t \qquad \textbf{Ans.}$$

Example 8. *Solve the following system of differential equations*

$$Dx + Dy + 3x = \sin t$$

$$Dx + y - x = \cos t \qquad (\textit{U.P. II Semester, Summer } 2003)$$

Solution. We have, $(D + 3)\, x + Dy = \sin t$ ...(1)

$$(D - 1)\, x + y = \cos t \qquad ...(2)$$

Operating (2) by D, we get

$$D\,(D - 1)\, x + Dy = -\sin t \qquad ...(3)$$

Subtracting (1) from (3), we get

$$\{D\,(D - 1) - (D + 3)\}\, x = -2 \sin t$$

$$\Rightarrow \quad \{D^2 - D - D - 3\}\, x = -2 \sin t$$

$$\Rightarrow \quad (D^2 - 2D - 3)\, x = -2 \sin t$$

A.E. is $m^2 - 2m - 3 = 0$

$$\Rightarrow \quad (m - 3)(m + 1) = 0 \qquad \Rightarrow m = 3, -1$$

$$\therefore \quad C.F. = c_1 e^{3t} + c_2 e^{-t}$$

$$P.I. = \frac{1}{D^2 - 2D - 3}(-2 \sin t) = -2\frac{1}{(-1) - 2D - 3}\sin t$$

$$= 2 \cdot \frac{1}{2\,(D+2)}\sin t = \frac{(D-2)}{D^2 - 4}\sin t$$

$$= \frac{D-2}{-1-4}\sin t = \frac{\cos t - 2 \sin t}{-5} = \frac{1}{5}(2 \sin t - \cos t)$$

$$x = \text{C.F.} + \text{P.I.} = c_1 e^{3t} + c_2 e^{-t} + \frac{1}{5}(2 \sin t - \cos t) \qquad ...(4)$$

From (2), we get

$$(D - 1)\, x + y = \cos t$$

$$\Rightarrow (D-1)\left\{c_1 e^{3t} + c_2 e^{-t} + \frac{1}{5}(2 \sin t - \cos t)\right\} + y = \cos t$$

$$\Rightarrow y = \cos t - D\left\{c_1 e^{3t} + c_2 e^{-t} + \frac{1}{5}(2 \sin t - \cos t)\right\} + \left\{c_1 e^{3t} + c_2 e^{-t} + \frac{1}{5}(2 \sin t - \cos t)\right\}$$

$$= \cos t - 3c_1 e^{3t} + c_2 e^{-t} - \frac{1}{5}(2 \cos t + \sin t) + c_1 e^{3t} + c_2 e^{-t} + \frac{1}{5}(2 \sin t - \cos t)$$

$$= \cos t - 2c_1 e^{3t} + 2c_2 e^{-t} - \frac{1}{5}[3\cos t - \sin t]$$

$$= \frac{2}{5}\cos t + \frac{1}{5}\sin t + 2c_2 e^{-t} - 2c_1 e^{3t}$$

Here, $y = \frac{1}{5}(2\cos t + \sin t) - 2c_1 e^{3t} + 2c_2 e^{-t}$ ...(5)

(4) & (5) are the required solutions **Ans.**

Example 9. *Solve*

$$\frac{dx}{dt} - \frac{dy}{dt} + 2y = \cos 2t$$

$$\frac{dx}{dt} + \frac{dy}{dt} - 2x = \sin 2t$$

Solution. $Dx - Dy + 2y = \cos 2t$

$Dx + Dy - 2x = \sin 2t$

$\Rightarrow \quad Dx - (D-2)y = \cos 2t$ ...(1)

$(D-2)x + Dy = \sin 2t$ ...(2)

Operating (1) by D and (2) by $(D-2)$, we get

$D^2x - (D^2 - 2D)y = -2\sin 2t$ ...(3)

$(D^2 - 4D + 4)x + (D^2 - 2D)y = 2\cos 2t - 2\sin 2t$ ...(4)

On adding (3) and (4), we get

$$(2D^2 - 4D + 4)x = 2\cos 2t - 4\sin 2t$$

$$(D^2 - 2D + 2)x = \cos 2t - 2\sin 2t$$

A.E. is $\quad m^2 - 2m + 2 = 0 \quad \Rightarrow \quad m = \frac{2 \pm \sqrt{4-8}}{2} = 1 \pm i$

$$\text{C.F.} = e^t[c_1 \cos t + c_2 \sin t]$$

$$\text{P.I.} = \frac{1}{D^2 - 2D + 2}\cos 2t - \frac{2}{D^2 - 2D + 2}\sin 2t$$

$$= \frac{1}{-4 - 2D + 2}\cos 2t - \frac{2}{-4 - 2D + 2}\sin 2t$$

$$= -\frac{1}{2D + 2}\cos 2t + \frac{1}{D+1}\sin 2t$$

$$= -\frac{2D - 2}{4D^2 - 4}\cos 2t + \frac{D-1}{D^2 - 1}\sin 2t$$

$$= -\frac{2D - 2}{-16 - 4}\cos 2t + \frac{D-1}{-4-1}\sin 2t$$

$$= \frac{1}{10}(D-1)\cos 2t - \frac{1}{5}(D-1)\sin 2t$$

$$= \frac{1}{10}(-2\sin 2t - \cos 2t) - \frac{1}{5}(2\cos 2t - \sin 2t) = -\frac{1}{2}\cos 2t$$

Gen. Sol. = C.F. + P.I.

$$x = e^t (c_1 \cos t + c_2 \sin t) - \frac{1}{2} \cos 2t \qquad ...(5)$$

$$\frac{dx}{dt} = e^t (-c_1 \sin t + c_2 \cos t) + e^t (c_1 \cos t + c_2 \sin t) + \sin 2t \qquad ...(6)$$

Adding equations (1) and (2), we get

$$2\frac{dx}{dt} + 2y - 2x = \cos 2t + \sin 2t \qquad ...(7)$$

Putting the values of x and $\frac{dx}{dt}$ from (5) and (6) in (7), we have

$$2 [e^t (-c_1 \sin t + c_2 \cos t) + e^t (c_1 \cos t + c_2 \sin t) + \sin 2t] + 2y$$

$$- 2 [e^t (c_1 \cos t + c_2 \sin t) - \frac{1}{2} \cos 2t] = \cos 2t + \sin 2t$$

Simplifying, we get

$$y = e^t (c_1 \sin t - c_2 \cos t) - \frac{1}{2} \sin 2t \qquad ...(8)$$

Relations (5) and (8) give the required solution. **Ans.**

Example 10. *Solve the simultaneous equations*

$$\frac{dx}{dt} + \frac{dy}{dt} - 2y = 2 \cos t - 7 \sin t \qquad ...(1)$$

$$\frac{dx}{dt} - \frac{dy}{dt} + 2x = 4 \cos t - 3 \sin t \qquad ...(2)$$

(*U.P., B. Pharm* 2005, *II Semester, Summer* 2001)

Solution. $Dx + (D - 2) y = 2 \cos t - 7 \sin t \qquad ...(3)$

$(D + 2) x - Dy = 4 \cos t - 3 \sin t \qquad ...(4)$

Operating (3) by D and (4) by $(D - 2)$, we get

$\Rightarrow$ $D^2x + D (D - 2) y = - 2 \sin t - 7 \cos t \qquad ...(5)$

$(D^2 - 4) x - D (D - 2) y = (D - 2) 4 \cos t - (D - 2) 3 \sin t \qquad ...(6)$

On adding (5) and (6), we get

$$(D^2 + D^2 - 4) x = - 2 \sin t - 7 \cos t - 4 \sin t - 8 \cos t - 3 \cos t + 6 \sin t$$

$$\Rightarrow \quad (2D^2 - 4) x = - 18 \cos t$$

$$\Rightarrow \quad (D^2 - 2) x = - 9 \cos t$$

A.E. is $m^2 - 2 = 0 \Rightarrow m = \pm\sqrt{2}$, C.F. $= c_1 e^{\sqrt{2}t} + c_2 e^{-\sqrt{2}t}$

$$P.I. = \frac{1}{D^2 - 2} (-9 \cos t) = \frac{-9}{-1 - 2} \cos t = 3 \cos t$$

$$x = c_1 e^{\sqrt{2}t} + c_2 e^{-\sqrt{2}t} + 3 \cos t \qquad ...(7)$$

Putting the value of x in (2), we get

$$\frac{dy}{dt} = \sqrt{2}\, c_1 e^{\sqrt{2}t} - \sqrt{2}\, c_2 e^{-\sqrt{2}t} - 3 \sin t + 2c_1 e^{\sqrt{2}t} + 2c_2 e^{-\sqrt{2}t} + 6 \cos t - 4 \cos t + 3 \sin t$$

$$\frac{dy}{dt} = (2+\sqrt{2})\, c_1 e^{\sqrt{2}t} + (2-\sqrt{2})\, c_2 e^{-\sqrt{2}t} + 2\cos t$$

On integrating, we get

$$\Rightarrow y = (\sqrt{2}+1)\, c_1 e^{\sqrt{2}t} - (\sqrt{2}-1)\, c_2 e^{-\sqrt{2}t} + 2\sin t + c_3 \qquad ...(8)$$

Relations (7) and (8) are the required solutions **Ans.**

Example 11. *Solve* $\frac{dx}{dt} = 2y,\ \frac{dy}{dt} = 2z,\ \frac{dz}{dt} = 2x$ *(Uttarakhand, II Semester, June 2007)*

Solution. Here, we have

$$\frac{dx}{dt} = 2y \quad \Rightarrow \quad Dx = 2y \qquad ...(1)$$

$$\frac{dy}{dt} = 2z \quad \Rightarrow \quad Dy = 2z \qquad ...(2)$$

$$\frac{dz}{dt} = 2x \quad \Rightarrow \quad Dz = 2x \qquad ...(3)$$

From (1), we have $\frac{dx}{dt} = 2y$

$$\Rightarrow \quad \frac{d^2x}{dt^2} = \frac{2\,dy}{dt} = 2\,(2z) = 4z \qquad \left[\text{Using (2), } \frac{dy}{dt} = 2z\right]$$

$$\frac{d^3x}{dt^3} = 4\frac{dz}{dt} = 4\,(2x) = 8x \qquad \left[\text{Using (3), } \frac{dz}{dt} = 2x\right]$$

$$\Rightarrow \quad \frac{d^3x}{dt^3} - 8x = 0 \quad \Rightarrow \quad (D^3 - 8)\,x = 0$$

A.E. is $m^3 - 8 = 0 \quad \Rightarrow \quad (m-2)(m^2 + 2m + 4) = 0$

$\Rightarrow \quad m - 2 = 0 \quad \Rightarrow \quad m = 2$

or $\quad m^2 + 2m + 4 = 0 \quad \Rightarrow \quad m = \frac{-2 \pm \sqrt{4-16}}{2} = \frac{-2 \pm i\sqrt{12}}{2} = -1 \pm i\sqrt{3}$

So the C.F. of x is

$$x = C_1 e^{2t} + e^{-t}\,(A\cos\sqrt{3}t + B\sin\sqrt{3}t) \qquad ...(4)$$

$$[A = C_2\cos\alpha,\ B = C_2\sin\alpha] \qquad \left[\tan\alpha = \frac{B}{A},\quad \alpha = \tan^{-1}\left(\frac{B}{A}\right)\right]$$

$$x = C_1 e^{2t} + e^{-t}\,[C_2\cos\alpha\cos\sqrt{3}t + C_2\sin\alpha\sin\sqrt{3}t]$$

$$x = C_1 e^{2t} + e^{-t}\,C_2\cos(\sqrt{3}t - \alpha) = C_1 e^{2t} + C_2 e^{-t}\cos(\sqrt{3}t - \alpha)$$

From (3), we have $\frac{dz}{dt} = 2x$

$$\Rightarrow \quad \frac{dz}{dt} = 2C_1 e^{2t} + 2C_2 e^{-t}\cos(\sqrt{3}t - \alpha) \qquad \text{[On putting the value of } x\text{]}$$

$$z = C_1 e^{2t} + 2C_2 \frac{e^{-t}}{\sqrt{1+3}} \cos(\sqrt{3}t - \alpha - \beta)$$

$$\Rightarrow \quad z = C_1 e^{2t} + 2C_2 \frac{e^{-t}}{\sqrt{1+3}} \cos\left[\sqrt{3}t - \alpha - \frac{2\pi}{3}\right]$$

$$\left[\int e^{ax} \cos bx\, dx = \frac{e^{ax}}{\sqrt{a^2+b^2}} \cos(bx - \beta)\right] \qquad \left[\beta = \tan^{-1}\frac{\sqrt{3}}{-1} = \frac{2\pi}{3}\right]$$

$$\Rightarrow \quad z = C_1 e^{2t} + C_2 e^{-t} \cos\left(\sqrt{3}t - \alpha + \frac{4\pi}{3}\right) \qquad ...(5) \qquad \left[-\frac{2\pi}{3} = \frac{4\pi}{3}\right]$$

From (2), we have $\frac{dy}{dx} = 2z$

$$\Rightarrow \quad \frac{dy}{dt} = 2C_1 e^{2t} + 2C_2 e^{-t} \cos\left(\sqrt{3}t - \alpha + \frac{4\pi}{3}\right)$$ [On putting the value of z]

$$\Rightarrow \quad y = \int 2C_1 e^{2t}\, dt + 2C_2 \int e^{-t} \cos\left(\sqrt{3}t - \alpha + \frac{4\pi}{3}\right) dt \qquad \left(\gamma = \tan^{-1}\frac{\sqrt{3}}{-1} = \frac{2\pi}{3}\right)$$

$$\Rightarrow \quad y = C_1 e^{2t} + 2C_2 \frac{e^{-t}}{\sqrt{1+3}} \cos\left(\sqrt{3}t - \alpha + \frac{4\pi}{3} - \gamma\right)$$

$$\Rightarrow \quad y = C_1 e^{2t} + 2C_2 \frac{e^{-t}}{\sqrt{1+3}} \cos\left(\sqrt{3}t - \alpha + \frac{4\pi}{3} - \frac{2\pi}{3}\right)$$

$$\Rightarrow \quad y = C_1 e^{2t} + C_2 e^{-t} \cos\left(\sqrt{3}t - \alpha + \frac{2\pi}{3}\right) \qquad ...(6)$$

Relations (4), (5) and (6) are the required solutions. **Ans.**

Example 12. *Solve:* $\frac{d^2x}{dt^2} + y = \sin t, \qquad \frac{d^2y}{dt^2} + x = \cos t$

Solution. Here, we have

$$D^2x + y = \sin t \qquad ...(1)$$

$$x + D^2y = \cos t \qquad ...(2) \left(D = \frac{d}{dt}\right)$$

Operating equation (1) by D^2, we get

$$D^4x + D^2y = -\sin t \qquad ...(3)$$

Subtracting (2) from (3), we get

$$(D^4 - 1)x = -\sin t - \cos t$$

Auxiliary equation is

$$m^4 - 1 = 0 \Rightarrow (m^2 - 1)(m^2 + 1) = 0$$

$$\Rightarrow \quad m = 1, -1, \pm i$$

$$\text{C.F.} = c_1 e^t + c_2 e^{-t} + c_3 \cos t + c_4 \sin t$$

$$\text{P.I.} = \frac{1}{D^4 - 1}(-\sin t - \cos t)$$

$$= -t.\frac{1}{4D^3}(\sin t + \cos t)$$

$$= \frac{t}{4}(-\cos t + \sin t) \qquad \left(\frac{1}{D} \equiv \int\right)$$

$$\therefore \quad x = c_1 e^t + c_2 e^{-t} + c_3 \cos t + c_4 \sin t + \frac{t}{4}(\sin t - \cos t) \qquad ...(4)$$

From (1), $y = \sin t - \frac{d^2x}{dt^2}$

$$= \sin t - \frac{d^2}{dt^2}\left[(c_1 e^t + c_2 e^{-t}) + (c_3 \cos t + c_4 \sin t) + \frac{t}{4}(\sin t - \cos t)\right]$$

$$= \sin t - \frac{d}{dt}\left[c_1 e^t - c_2 e^{-t} - c_3 \sin t + c_4 \cos t + \frac{t}{4}(\cos t + \sin t) + \frac{1}{4}(\sin t - \cos t)\right]$$

$$= \sin t - \left[c_1 e^t + c_2 e^{-t} - c_3 \cos t - c_4 \sin t + \frac{t}{4}(-\sin t + \cos t) + \frac{1}{4}(\cos t + \sin t) + \frac{1}{4}(\cos t + \sin t)\right]$$

$$y = -c_1 e^t - c_2 e^{-t} + c_3 \cos t + c_4 \sin t + \frac{t}{4}(\sin t - \cos t) - \frac{1}{2}(\sin t + \cos t)$$

and $\quad x = c_1 e^t + c_2 e^{-t} + c_3 \cos t + c_4 \sin t + \frac{t}{4}(\sin t - \cos t) \qquad$ [From (4)] **Ans.**

Example 13. *Solve:* $\frac{d^2x}{dt^2} + 4x + 5y = t^2$

$$\frac{d^2y}{dt^2} + 5x + 4y = t + 1$$

Solution. Here, we have

$$(D^2 + 4)x + 5y = t^2 \qquad ...(1)$$

and $\quad 5x + (D^2 + 4)y = t + 1 \qquad ...(2) \quad \left(D = \frac{d}{dt}\right)$

To eliminate y, operating (1) by $(D^2 + 4)$ and multiplying (2) by 5 then subtracting, we get

$$(D^2 + 4)(D^2 + 4)x + 5(D^2 + 4)y - 5[5x + (D^2 + 4)y] = (D^2 + 4)t^2 - 5t - 5$$

$$\Rightarrow \quad [(D^2 + 4)^2 - 25]x = (D^2 + 4)t^2 - 5(t + 1)$$

$$\Rightarrow \quad (D^4 + 8D^2 - 9)x = 2 + 4t^2 - 5t - 5$$

$$\Rightarrow \quad (D^4 + 8D^2 - 9)x = 4t^2 - 5t - 3$$

Auxiliary equation is

$$m^4 + 8m^2 - 9 = 0$$

$\Rightarrow (m^2 + 9)(m^2 - 1) = 0 \quad \therefore m = \pm 1, \pm 3i$

$$\therefore \quad \text{C.F.} = c_1 e^t + c_2 e^{-t} + c_3 \cos 3t + c_4 \sin 3t$$

$$\text{P.I.} = \frac{1}{D^4 + 8D^2 - 9}(4t^2 - 5t - 3) = \frac{1}{-9\left(1 - \frac{8D^2}{9} - \frac{D^4}{9}\right)}(4t^2 - 5t - 3)$$

$$= -\frac{1}{9}\left[1-\left(\frac{8D^2}{9}+\frac{D^4}{9}\right)\right]^{-1}(4t^2-5t-3)$$

$$= -\frac{1}{9}\left[1+\left(\frac{8D^2}{9}+\frac{D^4}{9}\right)+\cdots\right](4t^2-5t-3)$$

$$= -\frac{1}{9}\left[4t^2-5t-3+\frac{8}{9}(8)\right]=-\frac{1}{9}\left(4t^2-5t+\frac{37}{9}\right)$$

$$x = \text{C.F.} + \text{P.I.}$$

$$x = c_1e^t + c_2e^{-t} + c_3\cos 3t + c_4\sin 3t - \frac{4}{9}t^2 + \frac{5}{9}t - \frac{37}{81}$$

$$5y = t^2 - 4x - \frac{d^2x}{dt^2} \qquad \text{[From (2)]}$$

$$= t^2 - 4\left[c_1e^t + c_2e^{-t} + c_3\cos 3t + c_4\sin 3t - \frac{4}{9}t^2 + \frac{5}{9}t - \frac{37}{81}\right]$$

$$-\frac{d^2}{dt^2}\left[c_1e^t + c_2e^{-t} + c_3\cos 3t + c_4\sin 3t - \frac{4}{9}t^2 + \frac{5}{9}t - \frac{37}{81}\right]$$

$$= t^2 - 4c_1e^t - 4c_2e^{-t} - 4c_3\cos 3t - 4c_4\sin 3t - \frac{16}{9}t^2 + \frac{20}{9}t - \frac{148}{81}$$

$$-\frac{d}{dt}\left[c_1e^t - c_2e^{-t} - 3c_3\sin 3t + 3c_4\cos 3t - \frac{8}{9}t + \frac{5}{9}\right]$$

$$= t^2 - 4c_1e^t - 4c_2e^{-t} - 4c_3\cos 3t - 4c_4\sin 3t - \frac{16}{9}t^2 + \frac{20}{9}t - \frac{148}{81}$$

$$-\left[c_1e^t + c_2e^{-t} - 9c_3\cos 3t - 9c_4\sin 3t - \frac{8}{9}\right]$$

$$y = \frac{1}{5}\left[-5c_1e^t - 5c_2e^{-t} - 5c_3\cos 3t + 5c_4\sin 3t + \frac{25}{9}t^2 - \frac{20}{9}t + \frac{220}{81}\right]$$

$$y = -c_1e^t - c_2e^{-t} + c_3\cos 3t + c_4\sin 3t + \frac{1}{9}\left(5t^2 - 4t + \frac{44}{9}\right)$$

$$x = c_1e^t + c_2e^{-t} + c_3\cos 3t + c_4\sin 3t - \frac{4}{9}t^2 + \frac{5}{9}t - \frac{37}{81} \qquad \textbf{Ans.}$$

Example 14. *Solve the following simultaneous equations :*

$$\frac{d^2x}{dt^2} - 3x - 4y = 0, \quad \frac{d^2y}{dt^2} + x + y = 0$$

(*Uttarakhand, II Semester, Summer 2010, U.P. II Semester Summer* 2005)

Solution. We have, $\frac{d^2x}{dt^2} - 3x - 4y = 0$

$$\frac{d^2y}{dt^2} + x + y = 0$$

$$(D^2 - 3)x - 4y = 0 \qquad ...(1)$$

$$x + (D^2 + 1)\,y = 0 \qquad ...(2)$$

Operating equation (2) by $(D^2 - 3)$, we get

$$(D^2 - 3)x + (D^2 - 3)(D^2 + 1)\,y = 0 \qquad ...(3)$$

Subtracting (3) from (1), we get

$$-4y - (D^2 - 3)(D^2 + 1)\,y = 0$$

$$\Rightarrow \quad -4y - (D^4 - 2D^2 - 3)\,y = 0$$

$$\Rightarrow \quad (D^4 - 2D^2 - 3 + 4)\,y = 0$$

$$\Rightarrow \quad (D^4 - 2D^2 + 1)\,y = 0$$

$$\Rightarrow \quad (D^2 - 1)^2\,y = 0$$

A.E. is $(m^2 - 1)^2 = 0$

$$\Rightarrow \quad (m^2 - 1) = 0 \quad \Rightarrow \quad m = \pm 1$$

$$y = (c_1 + c_2 t)e^t + (c_3 + c_4 t)e^{-t} \qquad ...(4)$$

From (2), we have

$$\begin{aligned}
x &= -(D^2 + 1)y \\
&= -D^2 y - y \\
&= D^2\,[(c_1 + c_2 t)\,e^t + (c_3 + c_4 t)e^{-t}] - [(c_1 + c_2 t)e^t + (c_3 + c_4 t)e^{-t}] \\
&= -D\,[\{(c_1 + c_2 t)e^t + c_2 e^t\} + \{(c_3 + c_4 t)(-e^{-t}) + c_4 e^{-t}\}] - [(c_1 + c_2 t)e^t + (c_3 + c_4 t)e^{-t}] \\
&= -[(c_1 + c_2 t)\,e^t + c_2 e^t + c_2 e^t + (c_3 + c_4 t)(e^{-t}) - c_4 e^{-t} - c_4 e^{-t}] \\
&\qquad - [(c_1 + c_2 t)\,e^t + (c_3 + c_4 t)e^{-t}] \\
&= -[(c_1 + c_2 t + 2c_2 + c_1 + c_2 t)\,e^t + (c_3 + c_4 t - 2c_4 + c_3 + c_4\,t)\,e^{-t}] \\
&= -[(2c_1 + 2c_2 + 2c_2 t)e^t + (2c_3 - 2c_4 + 2c_4 t)e^{-t}] \qquad ...(5)
\end{aligned}$$

Relations (4) and (5) are the required solution. **Ans.**

Example 15. *Solve :* $\dfrac{d^2x}{dt^2} + \dfrac{dy}{dt} + 3x = e^{-t}$,

$$\frac{d^2y}{dt^2} - 4\frac{dx}{dt} + 3y = \sin 2t$$

(*U.P., II Semester, June* 2007)

Solution. We have, $D^2x + Dy + 3x = e^{-t} \quad \Rightarrow \quad (D^2 + 3)\,x + D\,y = e^{-t} \qquad ...(1)$

$$D^2y - 4Dx + 3y = \sin 2t \quad \Rightarrow \quad -4Dx + (D^2 + 3)\,y = \sin 2t \qquad ...(2)$$

To eliminate y operating (1) by $(D^2 + 3)$ and (2) by D, we get

$$\left[\begin{array}{l} (D^2 + 3)^2\,x + D\,(D^2 + 3)\,y = (D^2 + 3)e^{-t} \\ -4D^2\,x + D(D^2 + 3)\,y = D\sin 2t \end{array}\right.$$

$$\left[\begin{array}{ll} (D^4 + 6D^2 + 9)x + D\,(D^2 + 3)\,y = e^{-t} + 3e^{-t} & ...(3) \\ -4D^2x + D(D^2 + 3)\,y = 2\cos 2t & ...(4) \end{array}\right.$$

Subtracting (4) from (3), we get

$$(D^4 + 10\,D^2 + 9)x = 4e^{-t} - 2\cos 2t$$

A.E. is $m^4 + 10m^2 + 9 = 0 \quad \Rightarrow (m^2 + 1)(m^2 + 9) = 0 \quad \Rightarrow m = \pm\, i,\ m = \pm 3i$

C.F. $= C_1 \cos t + C_2 \sin t + C_3 \cos 3t + C_4 \sin 3t$

$$\text{P.I.} = \frac{1}{D^4 + 10D^2 + 9}\,4e^{-t} - \frac{1}{D^4 + 10D^2 + 9}\,(2\cos 2t)$$

$$= \frac{1}{1+10+9}e^{-t} - \frac{1}{(-4)^2+10(-4)+9}(2\cos 2t)$$

$$= \frac{e^{-t}}{5} + \frac{2}{15}\cos 2t$$

$$x = C_1\cos t + C_2\sin t + C_3\cos 3t + C_4\sin 3t + \frac{e^{-t}}{5} + \frac{2}{15}\cos 2t$$

Putting the value of x in (2), we get

$$-4D\left[C_1\cos t + C_2\sin t + C_3\cos 3t + C_4\sin 3t + \frac{e^{-t}}{5} + \frac{2}{15}\cos 2t\right] + (D^2+3)\,y = \sin 2t$$

$$\Rightarrow \quad -4\left(-C_1\sin t + C_2\cos t - 3C_3\sin 3t + 3C_4\cos 3t - \frac{e^{-t}}{5} - \frac{4}{15}\sin 2t\right) + (D^2+3)\,y = \sin 2t$$

$$\Rightarrow \quad (D^2+3)\,y = \sin 2t - 4C_1\sin t + 4C_2\cos t - 12\,C_3\sin 3t + 12C_4\cos 3t - \frac{4}{5}e^{-t} - \frac{16}{15}\sin 2t$$

A.E. is $m^2 + 3 = 0 \Rightarrow m = \pm i\sqrt{3}$

$$\text{C.F.} = C_1\cos\sqrt{3}\,t + C_2\sin\sqrt{3}\,t$$

$$\text{P.I.} = \frac{1}{D^2+3}\left[\sin 2t - 4C_1\sin t + 4C_2\cos t - 12C_3\sin 3t + 12C_4\cos 3t - \frac{4}{5}e^{-t} - \frac{16}{15}\sin 2t\right]$$

$$= \frac{1}{D^2+3}\left(-\frac{1}{15}\sin 2t\right) + \frac{1}{D^2+3}(-4C_1\sin t) + \frac{1}{D^2+3}4C_2\cos t$$

$$+ \frac{1}{D^2+3}(-12\,C_3\sin 3t) + \frac{1}{D^2+3}(12C_4\cos 3t) + \frac{1}{D^2+3}\left(\frac{-4}{5}e^{-t}\right)$$

$$= \frac{1}{-4+3}\left(-\frac{1}{15}\sin 2t\right) + \frac{1}{-1+3}(-4C_1\sin t) + \frac{1}{-1+3}4C_2\cos t + \frac{1}{-9+3}(-12C_3\sin 3t)$$

$$+ \frac{1}{-9+3}(12\,C_4\cos 3t) + \frac{1}{1+3}\left(\frac{-4}{5}e^{-t}\right)$$

$$= \frac{1}{15}\sin 2t - 2C_1\sin t + 2C_2\cos t + 2C_3\sin 3t - 2C_4\cos 3t - \frac{1}{5}e^{-t}$$

$y = C.F. + P.I.$

$$y = C_1\cos\sqrt{3}t + C_2\sin\sqrt{3}t + \frac{1}{15}\sin 2t - 2C_1\sin t + 2C_2\cos t + 2C_3\sin 3t$$

$$- 2C_4\cos 3t - \frac{1}{5}e^{-t}$$

$$x = C_1\cos t + C_2\sin t + C_3\cos 3t + C_4\sin 3t + \frac{e^{-t}}{5} + \frac{2}{15}\cos 2t \qquad \textbf{Ans.}$$

Example 16. *Solve the simultaneous differential equations*

$$\frac{d^2x}{dt^2} - 4\frac{dx}{dt} + 4x = y,$$

$$\frac{d^2y}{dt^2} + 4\frac{dy}{dt} + 4y = 25x + 16e^t$$

(*U.P. II Semester, 2001*)

Solution. Here, we have

$$(D^2 - 4D + 4)x - y = 0 \qquad ...(1)$$

$$-25x + (D^2 + 4D + 4)y = 16e^t \qquad ...(2) \quad \left(D = \frac{d}{dt}\right)$$

Operating (1) by $D^2 + 4D + 4$ and adding to (2), we get

$\Rightarrow \quad (D^2 + 4D + 4)(D^2 - 4D + 4)x - (D^2 + 4D + 4)y - 25x + (D^2 + 4D + 4)y = 16e^t$

$(D^2 - 4D + 4)(D^2 + 4D + 4)x - 25x = 16e^t$

$\Rightarrow \quad (D^4 - 8D^2 - 9)x = 16e^t$

Auxiliary equation is

$$m^4 - 8m^2 - 9 = 0$$

$\Rightarrow (m^2 - 9)(m^2 + 1) = 0 \Rightarrow m = \pm i, \pm 3$

C.F. $= c_1e^{3t} + c_2e^{-3t} + c_3 \cos t + c_4 \sin t$

P.I. $= \dfrac{1}{D^4 - 8D^2 - 9}(16e^t) = -e^t$

$x =$ C.F. + P.I.

$$x = c_1e^{3t} + c_2e^{-3t} + c_3 \cos t + c_4 \sin t - e^t \qquad ...(3)$$

From (1), $y = \dfrac{d^2x}{dt^2} - 4\dfrac{dx}{dt} + 4x$

$$= \frac{d^2}{dt^2}\left(c_1e^{3t} + c_2e^{-3t} + c_3 \cos t + c_4 \sin t - e^t\right)$$
$$- 4\frac{d}{dt}\left(c_1e^{3t} + c_2e^{-3t} + c_3 \cos t + c_4 \sin t - e^t\right)$$
$$+ 4\left(c_1e^{3t} + c_2e^{-3t} + c_3 \cos t + c_4 \sin t - e^t\right)$$

$$= \frac{d}{dt}\left[3c_1e^{3t} - 3c_2e^{-3t} + c_3(-\sin t) + c_4 \cos t - e^t\right]$$
$$- 4\left[3c_1e^{3t} - 3c_2e^{-3t} + c_3(-\sin t) + c_4 \cos t - e^t\right]$$
$$+ 4\left[c_1e^{3t} + c_2e^{-3t} + c_3 \cos t + c_4 \sin t - e^t\right]$$

$$= 9c_1e^{3t} + 9c_2e^{-3t} - c_3 \cos t - c_4 \sin t - e^t +$$
$$\left[-12c_1e^{3t} + 12c_2e^{-3t} + 4c_3 \sin t - 4c_4 \cos t + 4e^t\right]$$
$$+\left[4c_1e^{3t} + 4c_2e^{-3t} + 4c_3 \cos t + 4c_4 \sin t - 4e^t\right]$$

$$y = c_1e^{3t} + 25c_2e^{-3t} + (3c_3 - 4c_4)\cos t + (4c_3 + 3c_4)\sin t - e^t$$

$$x = c_1e^{3t} + c_2e^{-3t} + c_3 \cos t + c_4 \sin t - e^t \quad \text{[From (3)]}$$ **Ans.**

EXERCISE 3.1

Solve the following simultaneous equations:

1. $\dfrac{dx}{dt}+2x-3y=0$

$\dfrac{dy}{dt}-3x+2y=0$ **Ans.** $x=c_1\,e^t-c_2\,e^{-5t}$, $y=c_1\,e^t+c_2\,e^{-5t}$

2. $\dfrac{d^2y}{dt^2}=x,\quad \dfrac{d^2x}{dt^2}=y$ **Ans.** $x=c_1\,e^t+c_2\,e^{-t}+(c_3\cos t+c_4\sin t)$

$y=c_1\,e^t+c_2\,e^{-t}-(c_3\cos t+c_4\sin t)$

3. $\dfrac{dx}{dt}-y=t$

$\dfrac{dy}{dt}=t^2-x$ **Ans.** $x=c_1\cos t+c_2\sin t+t^2-1$; $y=-c_1\sin t+c_2\cos t+t$

4. $\dfrac{dx}{dt}+2y+\sin t=0$

$\dfrac{dy}{dt}-2x-\cos t=0$ **Ans.** $x=c_1\cos 2t+c_2\sin 2t-\cos t$; $y=c_1\sin 2t-c_2\cos 2t-\sin t$

5. $4\dfrac{dx}{dt}-\dfrac{dy}{dt}+3x=\sin t$

$\dfrac{dx}{dt}+y=\cos t$ **Ans.** $x=c_1\,e^{-t}+c_2\,e^{-3t}$, $y=c_1\,e^{-t}+3\,c_2\,e^{-3t}+\cos t$

6. $\dfrac{dy}{dx}=x$ and $\dfrac{dx}{dt}=y+e^{2t}$ **Ans.** $x=C_1e^t+C_2e^{-t}+\dfrac{2}{3}e^{2t}$, $y=C_1e^t-C_2e^{-t}+\dfrac{1}{3}e^{2t}$

7. $\dfrac{dx}{dt}=y+t,\quad \dfrac{dy}{dx}=-2x+3y+1$ **Ans.** $x=c_1\,e^t+\dfrac{1}{2}c_2\,e^{2t}-\dfrac{3}{2}t-\dfrac{5}{4}$, $y=c_1\,e^t+c_2\,e^{2t}-t-\dfrac{3}{2}$

8. $t\dfrac{dx}{dt}+y=0,\ t\dfrac{dy}{dx}+x=0$ **Ans.** $x=c_1\,t+c_2\,t^{-1}$, $y=c_2\,t^{-1}-c_1\,t$

given $x(1)=1$ and $y(-1)=0$

9. $\dfrac{dx}{dt}+y=\sin t,\ \dfrac{dy}{dt}+x=\cos t,$ given that $x=2$, $y=0$ when $t=0$ *(U.P., II Semester 2004)*

Ans. $x=e^t+e^{-t}$, $y=\sin t-e^t+e^{-t}$

10. $(D-1)\,x+Dy=2t+1$

$(2D+1)\,x+2\,Dy=t$ **Ans.** $x=-t-\dfrac{2}{3}$, $y=\dfrac{t^2}{2}+\dfrac{4}{3}t+C$

11. $\dfrac{dx}{dt}+\dfrac{2}{t}(x-y)=1,$ *(U.P., II Semester, Summer (C.O.) 2005)*

$\dfrac{dy}{dt}+\dfrac{1}{t}(x+5y)=t$ **Ans.** $x=At^{-4}+Bt^{-3}+\dfrac{t^2}{15}+\dfrac{3t}{10}$, $y=-At^{-4}-\dfrac{1}{2}Bt^{-3}+\dfrac{2t^2}{15}-\dfrac{t}{20}$

12. $(D^2-1)x+8Dy=16e^t$ and $Dx+3(D^2+1)y=0$ *(Q. Bank U.P.T.U. 2001)*

Ans. $y=c_1\cos\dfrac{t}{\sqrt{3}}+c_2\sin\dfrac{t}{\sqrt{3}}+c_3\cosh\sqrt{3}\,t+c_4\sin h\sqrt{3}\,t+2e^t$

$x=\sqrt{3}\,c_1\sin\dfrac{t}{\sqrt{3}}-\sqrt{3}\,c_2\cos\dfrac{t}{\sqrt{3}}-3\sqrt{3}\,c_3\sinh\sqrt{3}\,t-3\sqrt{3}\,c_4\cosh\sqrt{3}\,t-6e^t-3t.$

13. $\dfrac{dx}{dt}+\dfrac{2}{t}(x-y)=1,$ (U.P. 2005)

$\dfrac{dy}{dt}+\dfrac{1}{t}(x+5y)=t.$ **Ans.** $x = At^{-4}+Bt^{-3}+\dfrac{t^2}{15}+\dfrac{3t}{10},\ y=-At^{-4}-\dfrac{1}{2}Bt^{-3}+\dfrac{2t^2}{15}-\dfrac{t}{20}$

OBJECTIVE TYPE QUESTIONS

Choose the correct answer:

1. If $\dfrac{dx}{dt}+4y=0$ and $\dfrac{dy}{dt}-4x=0$, then $x=$

(i) $-A\sin 4t+B\cos 4t$ (ii) $A\sin 4t-B\cos 4t$

(iii) $A\sin 4t+B\cos 4t$ (iv) None of these **Ans.** (i)

2. If $\dfrac{dx}{dt}+5y=0$ and $\dfrac{dy}{dt}-5x=0$, then $y=$

(i) $A\cos 5t-B\sin 5t$ (ii) $A\cos 5t+B\sin 5t$

(iii) $-A\cos 5t+B\sin 5t$ (iv) None of these **Ans.** (ii)

3. If $\dfrac{dx}{dt}-y=t$ and $\dfrac{dy}{dt}=t^2-5$, then $x=$

(i) $c_1\cot t+c_2\sin t+t^2-1$ (ii) $c_1\cos t+c_2\sin t-t^2+1$

(iii) $c_1\cos t+c_2\sin t+t^2-1$ (iv) $c_1\cos t-c_2\sin t-t^2+1$ **Ans.** (iii)

4. If $\dfrac{dx}{dt}=3x+2y,\ \dfrac{dy}{dt}=5x+3y$, then $x=$

(i) $e^{3t}(c_1\cos h\sqrt{10}\,t+c_2\sin h\sqrt{10}\,t)$ (ii) $(c_1\cos h\sqrt{10}\,t+c_2\sin h\sqrt{10}\,t)$

(iii) $e^{-3t}(c_1\cos h\sqrt{10}\,t+c_2\sin h\sqrt{10}\,t)$ (iv) $e^{3t}(c_1\cos h\sqrt{10}\,t-c_2\sin h\sqrt{10}\,t)$ **Ans.** (i)

UNIT–I

4 CHAPTER

SOLUTIONS OF SECOND ORDER DIFFERENTIAL EQUATION

(By Changing Dependent and Independent Variables and Method of variation of parameters)

4.1 INTRODUCTION

No general procedure exists for solving second and higher order linear differential equation with variable coefficient. However, by transformation we can solve the second order differential equation with variable coefficients by the methods given in the next Article.

The method of variation of parameters to solve such equation is given in the next chapter.

4.2 SECOND ORDER DIFFERENTIAL EQUATIONS

A differential equation which involves the maximum second order derivative is known as second order differential equation.

e.g.
$$\frac{d^2y}{dx^2} + P\frac{dy}{dx} + Qy = R.$$

Where P, Q and R are the functions of x only. The coefficient of $\frac{d^2y}{dx^2}$ is always one.

We will discuss the following types of differential equations:

1. Differential equations which do not contain y, of the type $\frac{d^2y}{dx^2} = f(x)$.
2. Differential equations which do not contain x, of the type $\frac{d^2y}{dx^2} = f(y)$.
3. Differential equations which do not contain y directly, of the type $f\left(\frac{d^2y}{dx^2}, \frac{dy}{dx}, x\right) = 0$.
4. Differential equations that do not contain x directly, of the type $f\left(\frac{d^2y}{dx^2}, \frac{dy}{dx}, y\right) = 0$.
5. Equations whose one solution of complementary function is known (Method of reduction).
6. By removing the first derivative (Normal form).
7. By changing the independent variable by putting Either $P_1 = 0$ or $Q_1 =$ a constant

4.3 TYPE I

Now we solve the second order differential equations which do not contain y are of the type

$$\boxed{\frac{d^2y}{dx^2} = f(x)}$$

This type of exact differential equations are solved by successive integration.

Example 1. *Solve* $\frac{d^2y}{dx^2} = x^2 \sin x.$

Solution. We have, $\frac{d^2y}{dx^2} = x^2 \sin x$ ...(1)

Integrating the differential equation (1), we get

$$\frac{dy}{dx} = x^2(-\cos x) - (2x)(-\sin x) + (2)(\cos x) + c_1$$

$$\frac{dy}{dx} = -x^2 \cos x + 2x \sin x + 2\cos x + c_1$$ **Ans.**

Example 2. *Solve* $\frac{d^2y}{dx^2} = xe^x$

Solution. Here, we have $\frac{d^2y}{dx^2} = xe^x$ ...(1)

Integrating (1), we get

$$\frac{dy}{dx} = xe^x - e^x + c_1$$ ...(2)

Again integrating (2), we get

$$y = xe^x - e^x - e^x + c_1x + c_2$$

$$\Rightarrow \quad y = (x-2)e^x + c_1x + c_2$$ **Ans.**

4.4 TYPE II

Now we will solve the differential equations of the type $\boxed{\frac{d^2y}{dx^2} = f(y)}$.

Working Rule:

Step 1. Multiplying by $2\frac{dy}{dx}$, we get $2\frac{dy}{dx}\frac{d^2y}{dx^2} = 2f(y)\frac{dy}{dx}$ ...(1)

Step 2. Integrating (1), we have $\left(\frac{dy}{dx}\right)^2 = 2\int f(y)\,dy + c = \phi(y)$ (say)

Step 3. $\frac{dy}{dx} = \sqrt{\phi(y)} \quad \Rightarrow \quad \frac{dy}{\sqrt{\phi(y)}} = dx$

Step 4. $\int \frac{dy}{\sqrt{\phi(y)}} = x + c$

Example 3. *Solve* $\frac{d^2y}{dx^2} = \sqrt{y}$, *under the condition* $y = 1$, $\frac{dy}{dx} = \frac{2}{\sqrt{3}}$ *at* $x = 0$.

Solution. We have, $\frac{d^2y}{dx^2} = \sqrt{y}$ ...(1)

Multiplying (1) by $2\frac{dy}{dx}$, we get

$$2\frac{dy}{dx}\frac{d^2y}{dx^2} = 2\sqrt{y}\frac{dy}{dx}$$ ...(2)

Integrating (2), we get $\left(\dfrac{dy}{dx}\right)^2 = \dfrac{4}{3} y^{3/2} + c_1$ ...(3)

On putting $y = 1$ and $\dfrac{dy}{dx} = \dfrac{2}{\sqrt{3}}$, we have $c_1 = 0$

Equation (3) becomes $\left(\dfrac{dy}{dx}\right)^2 = \dfrac{4}{3} y^{3/2} \Rightarrow \dfrac{dy}{dx} = \dfrac{2}{\sqrt{3}} y^{3/4} \Rightarrow y^{-3/4}\, dy = \dfrac{2}{\sqrt{3}} dx$

Again integrating, we get $\dfrac{y^{1/4}}{\frac{1}{4}} = \dfrac{2}{\sqrt{3}} x + c_2 \Rightarrow 4y^{\frac{1}{4}} = \dfrac{2}{\sqrt{3}} x + c_2$ ...(4)

On putting $x = 0, y = 1$, we get $c_2 = 4$

(4) becomes $4y^{1/4} = \dfrac{2}{\sqrt{3}} x + 4$ **Ans.**

Example 4. *Solve* $\dfrac{d^2y}{dx^2} = \sec^2 y \tan y$ *under the condition* $y = 0$ *and* $\dfrac{dy}{dx} = 1$ *when* $x = 0$.

Solution. We have, $\dfrac{d^2y}{dx^2} = \sec^2 y \tan y$

$$\Rightarrow \quad 2\frac{dy}{dx}\frac{d^2y}{dx^2} = 2\sec^2 y \tan y \frac{dy}{dx}$$

$$\int 2\frac{dy}{dx}\frac{d^2y}{dx^2} = \int 2\sec^2 y \tan y \frac{dy}{dx}$$

$$\left(\frac{dy}{dx}\right)^2 = \tan^2 y + c_1 \Rightarrow \frac{dy}{dx} = \sqrt{\tan^2 y + c_1}$$

On putting $y = 0$ and $\dfrac{dy}{dx} = 1$, we get $c_1 = 1$

Now, $\dfrac{dy}{dx} = \sqrt{\tan^2 y + 1} = \sec y$

$\Rightarrow \quad \cos y\, dy = dx$

On integrating, we get $\sin y = x + c$

On putting $y = 0, x = 0$, we have $c = 0$

$\sin y = x \Rightarrow y = \sin^{-1} x$ **Ans.**

Example 5. *Solve* $\dfrac{d^2y}{dx^2} = 2(y^3 + y)$, *under the condition* $y = 0, \dfrac{dy}{dx} = 1$, *when* $x = 0$.

(*U.P., II Semester, Summer 2003*)

Solution. We have, $\dfrac{d^2y}{dx^2} = 2(y^3 + y) \Rightarrow 2\dfrac{dy}{dx}\dfrac{d^2y}{dx^2} = 4(y^3 + y)\dfrac{dy}{dx}$

Integrating, we get

$$\left(\frac{dy}{dx}\right)^2 = 4\left(\frac{y^4}{4} + \frac{y^2}{2}\right) + c_1 = y^4 + 2y^2 + c_1 \qquad ...(1)$$

On putting $y = 0$ and $\frac{dy}{dx} = 1$ in (1), we get $1 = c_1$

Equation (1) becomes $\left(\frac{dy}{dx}\right)^2 = y^4 + 2y^2 + 1 = (y^2 + 1)^2$

$$\frac{dy}{dx} = y^2 + 1 \Rightarrow \frac{dy}{1+y^2} = dx$$

Again integrating, we get $\tan^{-1} y = x + c_2$ ...(2)

On putting $y = 0$ and $x = 0$ in (2), we have $0 = c_2$

Equation (2) is reduced to $\tan^{-1} y = x \Rightarrow y = \tan x$ **Ans.**

Example 6. *A motion is governed by* $\frac{d^2x}{dt^2} = 36x^{-2}$, *given that at* $t = 0$, $x = 8$ *and* $\frac{dx}{dt} = 0$, *find the displacement at any time t.*

Solution. We have, $\frac{d^2x}{dt^2} = 36x^{-2}$

$$\Rightarrow \quad 2\frac{d^2x}{dt^2}\frac{dx}{dt} = 2 \times 36x^{-2}\frac{dx}{dt} \quad ...(1)$$

Integrating (1), we have $\left(\frac{dx}{dt}\right)^2 = -72x^{-1} + c_1$ ...(2)

Putting $x = 8$ and $\frac{dx}{dt} = 0$ in (2), we get

$$0 = -\frac{72}{8} + c_1 \Rightarrow c_1 = 9$$

(2) becomes $\left(\frac{dx}{dt}\right)^2 = -\frac{72}{x} + 9 \Rightarrow \left(\frac{dx}{dt}\right)^2 = \frac{-72 + 9x}{x}$

$$\Rightarrow \quad \frac{dx}{dt} = 3\sqrt{\frac{(x-8)}{x}}$$

$$\Rightarrow \quad \int \frac{\sqrt{x}\,dx}{\sqrt{x-8}} = 3\int dt + c_2$$

$$\Rightarrow \quad \int \frac{x\,dx}{\sqrt{x^2 - 8x}} = 3t + c_2$$

$$\frac{1}{2}\int \frac{2x - 8 + 8}{\sqrt{x^2 - 8x}}\,dx = 3t + c_2$$

$$\frac{1}{2}\int \frac{2x-8}{\sqrt{x^2-8x}}\,dx + 4\int \frac{1}{\sqrt{(x-4)^2 - (4)^2}}\,dx = 3t + c_2$$

$$\sqrt{x^2 - 8x} + 4\cosh^{-1}\left(\frac{x-4}{4}\right) = 3t + c_2 \quad ...(3)$$

On putting $x = 8$ and $t = 0$ in (3), we get $c_2 = 0$

(3) becomes $\sqrt{x^2 - 8x} + 4\cosh^{-1}\left(\frac{x-4}{4}\right) = 3t$ **Ans.**

EXERCISE 4.1

Solve the following differential equations :

1. $y^3 \dfrac{d^2y}{dx^2} = a$ **Ans.** $c_1 y^2 = (c_1 x + c_2)^2$

2. $e^{2y} \dfrac{d^2y}{dx^2} = 1$ **Ans.** $c_1 e^y = \cosh(c_1 x + c_2)$

3. $\sin^3 y \dfrac{d^2y}{dx^2} = \cos y$ **Ans.** $\sin[(x + c_2)\sqrt{(1 + c_1)}] + \sqrt{\left(\dfrac{1 + c_1}{c_1}\right)} \cos y = 0$

TYPE III.

4.5 EQUATIONS WHICH DO NOT CONTAIN 'y' DIRECTLY

(Change of Dependent variable)

Let $f\left(\dfrac{d^2y}{dx^2}, \dfrac{dy}{dx}, x\right) = 0$ ...(1)

We will solve, the differential equation by changing the dependent variable y by P.

On substituting $\dfrac{dy}{dx} = P$ *i.e.*, $\dfrac{d^2y}{dx^2} = \dfrac{dP}{dx}$ in (1), we get

$$f\left(\frac{dP}{dx}, P, x\right) = 0$$

Example 7. *Solve* $(1 - x^2)\dfrac{d^2y}{dx^2} - x\dfrac{dy}{dx} = 2$

Solution. Here, we have

$$\frac{d^2y}{dx^2} - \frac{x}{(1 - x^2)}\frac{dy}{dx} = 2 \quad \text{...(1)}$$

Put $\dfrac{dy}{dx} = P$ so that $\dfrac{d^2y}{dx^2} = \dfrac{dP}{dx}$ in (1), we get

$$\frac{dP}{dx} - \frac{x}{1 - x^2}P = \frac{2}{1 - x^2}$$

$$\text{I.F.} = e^{-\int\left(\frac{x}{1 - x^2}\right)dx} = e^{\frac{1}{2}\log(1 - x^2)} = e^{\log\sqrt{1 - x^2}} = \sqrt{1 - x^2}$$

Hence, its solution is

$$P\sqrt{1 - x^2} = 2\int\frac{\sqrt{1 - x^2}}{1 - x^2}dx \quad \Rightarrow \quad P\sqrt{1 - x^2} = \int\frac{2}{\sqrt{1 - x^2}}dx$$

$$\Rightarrow \quad P\sqrt{1 - x^2} = 2\sin^{-1}x + c_1 \quad \Rightarrow \quad \frac{dy}{dx} = \frac{2\sin^{-1}x}{\sqrt{1 - x^2}} + \frac{c_1}{\sqrt{1 - x^2}}$$

On integrating, we get

$$y = (\sin^{-1}x)^2 + c_1\sin^{-1}x + c_2$$ **Ans.**

Example 8. *Solve:* $\dfrac{d^2y}{dx^2}+\dfrac{dy}{dx}+\left(\dfrac{dy}{dx}\right)^3=0$ ...(1)

Solution. On putting $\dfrac{dy}{dx}=P$ and $\dfrac{d^2y}{dx^2}=\dfrac{dP}{dx}$, equation (1) becomes

$$\frac{dP}{dx}+P+P^3=0 \quad\Rightarrow\quad \frac{dP}{dx}+P(1+P^2)=0$$

$$\Rightarrow \quad \frac{dP}{dx}=-P(1+P^2) \quad\Rightarrow\quad \frac{dP}{P(1+P^2)}=-dx$$

$$\Rightarrow \quad \left(\frac{1}{P}-\frac{P}{1+P^2}\right)dP=-dx$$

On integrating, we have

$$\log P-\frac{1}{2}\log(1+P^2)=-x+c_1 \quad\Rightarrow\quad \log\frac{P}{\sqrt{1+P^2}}=-x+c_1$$

$$\Rightarrow \quad \frac{P}{\sqrt{1+P^2}}=e^{-x+c_1} \quad\Rightarrow\quad \frac{P^2}{1+P^2}=a^2e^{-2x} \Rightarrow P^2=(1+P^2)\,a^2e^{-2x}$$

$$\Rightarrow \quad P^2(1-a^2e^{-2x})=a^2e^{-2x} \quad\Rightarrow P=\frac{ae^{-x}}{\sqrt{1-a^2e^{-2x}}} \quad\Rightarrow\quad \frac{dy}{dx}=\frac{ae^{-x}}{\sqrt{1-a^2e^{-2x}}}$$

$$dy=\frac{ae^{-x}}{\sqrt{1-a^2e^{-2x}}}\,dx$$

On integration, we get $\quad y=-\sin^{-1}(a\,e^{-x})+b$ **Ans.**

Example 9. *Solve:* $\dfrac{d^2y}{dx^2}=\left[1+\left(\dfrac{dy}{dx}\right)^2\right]^{1/2}$

Solution. Put $\dfrac{dy}{dx}=P$ in the given equation, we get

$$\frac{dP}{dx}=[1+P^2]^{1/2} \Rightarrow \frac{dP}{\sqrt{1+P^2}}=dx$$

On integration, we get $\sinh^{-1}P=x+c_1 \Rightarrow P=\sinh(x+c_1)$

$$\Rightarrow \quad \frac{dy}{dx}=\sinh(x+c_1) \Rightarrow dy=\sinh(x+c_1)\,dx$$

On integrating, we get $\quad y=-\cosh(x+c_1)+c_2$ **Ans.**

Example 10. *Solve:* $\dfrac{d^2y}{dx^2}=\left[1-\left(\dfrac{dy}{dx}\right)^2\right]^{1/2}$ (*U.P. Second Sem., 2002*)

Solution. We have,

$$\frac{d^2y}{dx^2}=\left[1-\left(\frac{dy}{dx}\right)^2\right]^{1/2} \qquad ...(1)$$

Putting $P = \frac{dy}{dx} \quad \Rightarrow \quad \frac{dP}{dx} = \frac{d^2y}{dx^2}$ in (1), we get

$$\frac{dP}{dx} = \sqrt{1-P^2} \quad \Rightarrow \quad \frac{dP}{\sqrt{1-P^2}} = dx$$

On integrating, we have

$$\sin^{-1} P = x + c$$

$$\Rightarrow \quad P = \sin(x+c)$$

$$\Rightarrow \quad \frac{dy}{dx} = \sin(x+c)$$

On integrating, we have

$$y = -\cos(x+c) + c_1$$ **Ans.**

Example 11. *Solve* $x\frac{d^2y}{dx^2} + x\left(\frac{dy}{dx}\right)^2 - \frac{dy}{dx} = 0$ *(U.P. II, Semester 2010)*

Solution. We have, $\frac{d^2y}{dx^2} + \left(\frac{dy}{dx}\right)^2 - \frac{1}{x}\frac{dy}{dx} = 0$

On putting $\frac{dy}{dx} = P$ and $\frac{d^2y}{dx^2} = \frac{dP}{dx}$ in the given equation, we get

$$\frac{dP}{dx} + P^2 - \frac{P}{x} = 0 \Rightarrow \frac{1}{P^2}\frac{dP}{dx} - \frac{1}{P}\frac{1}{x} = -1 \quad ...(1)$$

Again putting $\frac{1}{P} = z$ so that $-\frac{1}{P^2}\frac{dP}{dx} = \frac{dz}{dx}$

Equation (1) becomes $\frac{-dz}{dx} - \frac{z}{x} = -1 \quad \Rightarrow \quad \frac{dz}{dx} + \frac{z}{x} = 1$

$$\text{I.F.} = e^{\int \frac{1}{x} dx} = e^{\log x} = x$$

Hence, solution is $zx = \int x\, dx + c_1 \quad \Rightarrow \quad zx = \frac{x^2}{2} + c_1 \quad \Rightarrow \quad \frac{1}{P}x = \frac{x^2}{2} + c$

$$\Rightarrow \quad \frac{x}{P} = \frac{x^2 + 2c_1}{2} \Rightarrow P = \frac{2x}{x^2 + 2c_1} \Rightarrow \frac{dy}{dx} = \frac{2x}{x^2 + 2c_1} \Rightarrow dy = \frac{2x}{x^2 + 2c_1} dx$$

On integrating, we have $y = \log(x^2 + 2c_1) + c_2$ **Ans.**

EXERCISE 4.2

Solve the following differential equations:

1. $(1+x^2)\frac{d^2y}{dx^2} + x\frac{dy}{dx} + ax = 0$ **Ans.** $y = c_2 - ax + c_1 \log[x + \sqrt{(1+x^2)}]$

2. $(1+x^2)\frac{d^2y}{dx^2} + 1 + \left(\frac{dy}{dx}\right)^2 = 0$ **Ans.** $y = -\frac{x}{k} + \frac{1+k^2}{k^2}\log(1+kx) + a$

3. $e^{x^2/2}\left[x\frac{d^2y}{dx^2} - \frac{dy}{dx}\right] = x^3$ **Ans.** $y = e^{-x^2/2} + c_1\frac{x^2}{2} + c_2$

TYPE IV.

4.6 EQUATIONS THAT DO NOT CONTAIN 'x' DIRECTLY

(Change of dependent variable)

The equations that do not contain x directly are of the form

$$f\left(\frac{d^2y}{dx^2}, \frac{dy}{dx}, y\right) = 0 \qquad \text{...(1)}$$

On substituting $\quad \dfrac{dy}{dx} = P,$

and $\quad \dfrac{d^2y}{dx^2} = \dfrac{dP}{dx} = \dfrac{dP}{dy} \cdot \dfrac{dy}{dx} = \dfrac{dP}{dy} P$

in (1), we get

$$f\left(P\frac{dP}{dy}, P, y\right) = 0 \qquad \text{...(2)}$$

Equation (2) is solved for P. Let

$$P = f_1(y) \Rightarrow \frac{dy}{dx} = f_1(y) \Rightarrow \frac{dy}{f_1(y)} = dx$$

$$\Rightarrow \quad \int \frac{dy}{f_1(y)} = x + c$$

Example 12. *Solve:* $y\dfrac{d^2y}{dx^2} + \left(\dfrac{dy}{dx}\right)^2 = \dfrac{dy}{dx}$ ...(1)

Solution. Here, we have

$$\frac{d^2y}{dx^2} + \frac{1}{y}\left(\frac{dy}{dx}\right)^2 = \frac{1}{y}\frac{dy}{dx}$$

Putting $\dfrac{dy}{dx} = P, \dfrac{d^2y}{dx^2} = \dfrac{dP}{dx} = \dfrac{dP}{dy} \bullet \dfrac{dy}{dx} = \dfrac{dP}{dy} P$ in (1), we get

$$P\frac{dP}{dy} + \frac{P^2}{y} = \frac{P}{y} \Rightarrow P\frac{dP}{dy} = \frac{P}{y}(1-P)$$

$$\Rightarrow \quad \frac{dP}{1-P} = \frac{dy}{y} \Rightarrow -\log(1-P) = \log y + \log c_1$$

$$\Rightarrow \quad \frac{1}{1-P} = c_1 y \Rightarrow P = 1 - \frac{1}{c_1 y} \Rightarrow \frac{dy}{dx} = \frac{c_1 y - 1}{c_1 y}$$

$$\Rightarrow \quad \frac{c_1 y}{c_1 y - 1} dy = dx \Rightarrow \left(1 + \frac{1}{c_1 y - 1}\right) dy = dx$$

$$y + \frac{1}{c_1}\log(c_1 y - 1) = x + c_2 \qquad \textbf{Ans.}$$

Example 13. *Solve:* $y\dfrac{d^2y}{dx^2} + \left(\dfrac{dy}{dx}\right)^2 = y^2$ ...(1)

Solution. Putting $\frac{dy}{dx} = P, \frac{d^2y}{dx^2} = \frac{dP}{dx} = \frac{dP}{dy}\frac{dy}{dx} = P\frac{dP}{dy}$ in (1), we get

$$yP\frac{dP}{dy} + P^2 = y^2 \Rightarrow P\frac{dP}{dy} + \frac{P^2}{y} = y \quad \text{...(2)}$$

Put $P^2 = z \Rightarrow 2P\frac{dP}{dy} = \frac{dz}{dy}$ in (2), $\frac{1}{2}\frac{dz}{dy} + \frac{z}{y} = y \Rightarrow \frac{dz}{dy} + \frac{2z}{y} = 2y$

$$\text{I.F.} = e^{\int \frac{2}{y} dy} = e^{2\log y} = e^{\log y^2} = y^2$$

Hence, the solution is $\quad zy^2 = \int 2y.(y^2)\, dy + c$

$$\Rightarrow \quad P^2 y^2 = \frac{y^4}{2} + c$$

$$\Rightarrow \quad 2P^2 y^2 = y^4 + k \Rightarrow \sqrt{2}\, yP = \sqrt{y^4 + k} \quad [\text{Put } 2c = k]$$

$$\Rightarrow \quad \sqrt{2}\, y\frac{dy}{dx} = \sqrt{y^4 + k} \Rightarrow \sqrt{2}\frac{y\,dy}{\sqrt{y^4 + k}} = dx$$

$$\Rightarrow \quad \frac{1}{\sqrt{2}}\frac{dt}{\sqrt{t^2 + k}} = dx \quad [\text{Put } y^2 = t,\ 2y\,dy = dt]$$

$$\Rightarrow \quad \frac{1}{\sqrt{2}}\sinh^{-1}\frac{t}{\sqrt{k}} = x + C$$

$$\Rightarrow \quad \sinh^{-1}\frac{y^2}{\sqrt{k}} = \sqrt{2}x + C \Rightarrow y^2 = \sqrt{k}\sinh(\sqrt{2}x + C) \quad \textbf{Ans.}$$

Example 14. *Solve:* $y(1 - \log y)\frac{d^2y}{dx^2} + (1 + \log y)\left(\frac{dy}{dx}\right)^2 = 0 \quad \text{... (1)}$

Solution. Put $\frac{dy}{dx} = P, \frac{d^2y}{dx^2} = \frac{dP}{dx} = \frac{dP}{dy}\frac{dy}{dx} = \frac{dP}{dy}P$

(1) becomes $y(1 - \log y)\frac{dP}{dy}P + (1 + \log y)P^2 = 0$

$$\Rightarrow \quad y(1 - \log y)\frac{dP}{dy} + (1 + \log y)P = 0$$

$$\Rightarrow \quad y(1 - \log y)\frac{dP}{dy} = -(1 + \log y)P$$

$$\Rightarrow \quad \frac{dP}{P} = -\frac{(1 + \log y)}{y(1 - \log y)}dy \quad \text{... (2)}$$

Putting $\quad \log y = t \Rightarrow \frac{dy}{y} = dt$ in (2), we get

$$\frac{dP}{P} = -\frac{(1 + t)}{(1 - t)}dt = \left(1 + \frac{2}{t - 1}\right)dt$$

Integrating, we get

$$\log P = t + 2\log(t-1) + \log c$$

$$\Rightarrow \quad \log P = t + \log c(t-1)^2$$

$$\Rightarrow \quad P = ce^t (t-1)^2 = cy(\log y - 1)^2$$

$$\Rightarrow \quad \frac{dy}{dx} = cy(\log y - 1)^2$$

$$\Rightarrow \quad \frac{dy}{y(\log y - 1)^2} = c\,dx$$

On integration, we get $-\dfrac{1}{(\log y - 1)} = cx + d \Rightarrow (1 - \log y) = \dfrac{1}{cx + d}$ **Ans.**

EXERCISE 4.3

Solve the following differential equations:

1. $\dfrac{d^2y}{dx^2} + \left(\dfrac{dy}{dx}\right)^2 = 1$ **Ans.** $y^2 = x^2 + ax + b$

2. $y\dfrac{d^2y}{dx^2} - \left(\dfrac{dy}{dx}\right)^2 + 2\dfrac{dy}{dx} = 0$ **Ans.** $cy + 2 = d\,e^{cx^a}$

3. $2y\dfrac{d^2y}{dx^2} - 3\left(\dfrac{dy}{dx}\right)^2 - 4y^2 = 0$ **Ans.** $y = a\sec^2(x + b)$

4. $y\dfrac{d^2y}{dx^2} + \dfrac{dy}{dx} + \left(\dfrac{dy}{dx}\right)^3 = 0$ **Ans.** $y = a - \sin^{-1}(b\,e^{-x})$

5. $y\dfrac{d^2y}{dx^2} = \left\{\dfrac{dy}{dx}\right\}^2 \left[1 - \dfrac{dy}{dx}\cos y + y\dfrac{dy}{dx}\sin y\right]$ **Ans.** $x = c_1 + c_2 \log y + \sin y$

6. $y\dfrac{d^2y}{dx^2} - \left(\dfrac{dy}{dx}\right)^2 = y^2 \log y$ **Ans.** $\log y = b.\,e^{x} + a\,e^{-x}$

TYPE V

4.7 METHOD OF REDUCTION

(Equations Whose one solution of complementary function is known)

If $y = u$ is given solution belonging to the complementary function of the differential equation. Let the other solution be $y = v$. Then $y = u.\,v$ is complete solution of the differential equation.

Let $\dfrac{d^2y}{dx^2} + P\dfrac{dy}{dx} + Qy = R$ ....(1), be the differential equation and u is the complementary function of (1).

$$\therefore \quad \frac{d^2u}{dx^2} + P\frac{du}{dx} + Qu = 0 \qquad ...(2)$$

$$y = u.\,v \text{ so that } \frac{dy}{dx} = v\frac{du}{dx} + u\frac{dv}{dx}$$

$$\frac{d^2y}{dx^2} = v\frac{d^2u}{dx^2} + 2\frac{dv}{dx}\frac{du}{dx} + u\frac{d^2v}{dx^2}$$

Substituting the values of $y, \frac{dy}{dx}, \frac{d^2y}{dx^2}$ in (1), we get

$$v\frac{d^2u}{dx^2} + 2\frac{dv}{dx}\frac{du}{dx} + u\frac{d^2v}{dx^2} + P\left(v\frac{du}{dx} + u\frac{dv}{dx}\right) + Qu.v = R$$

On arranging

$$\Rightarrow \quad v\left[\frac{d^2u}{dx^2} + P\frac{du}{dx} + Qu\right] + u\left[\frac{d^2v}{dx^2} + P\frac{dv}{dx}\right] + 2\frac{du}{dx}\cdot\frac{dv}{dx} = R$$

The first bracket is zero by virtue of relation (2), and the remaining is divided by u

$$\frac{d^2v}{dx^2} + P\frac{dv}{dx} + \frac{2}{u}\frac{du}{dx}\frac{dv}{dx} = \frac{R}{u}$$

$$\Rightarrow \quad \frac{d^2v}{dx^2} + \left[P + \frac{2}{u}\frac{du}{dx}\right]\frac{dv}{dx} = \frac{R}{u} \qquad ...(3)$$

Let $\frac{dv}{dx} = z$, so that $\frac{d^2v}{dx^2} = \frac{dz}{dx}$

Equation (3) becomes

$$\boxed{\frac{dz}{dx} + \left[P + \frac{2}{u}\frac{du}{dx}\right]z = \frac{R}{u}}$$

This is the linear differential equation of first order and can be solved (z can be found), which will contain one constant.

On integration $z = \frac{dv}{dx}$, we can get v.

Having found v, the solution is $y = uv$.

Example 15. *Solve $y'' - 4xy' + (4x^2 - 2)y = 0$ given that $y = e^{x^2}$ is an integral included in the complementary function.* *(U.P., II Semester, 2004)*

Solution. Here, we have $y'' - 4xy' + (4x^2 - 2)y = 0$ ...(1)

On putting $y = v.e^{x^2}$ in (1), the reduced equations as in the article 4.7.

$$\frac{d^2v}{dx^2} + \left[P + \frac{2}{u}\frac{du}{dx}\right]\frac{dv}{dx} = 0. \qquad [P = -4x,\ Q = 4x^2 - 2,\ R = 0]$$

$$\Rightarrow \frac{d^2v}{dx^2} + \left[-4x + \frac{2}{e^{x^2}}(2x\,e^{x^2})\right]\frac{dv}{dx} = 0$$

$$\Rightarrow \quad \frac{d^2v}{dx^2} + [-4x + 4x]\frac{dv}{dx} = 0 \quad \Rightarrow \quad \frac{d^2v}{dx^2} = 0 \quad \Rightarrow \quad \frac{dv}{dx} = c_1 \quad \Rightarrow \quad v = c_1x + c_2$$

$$\therefore \quad y = uv \qquad [u = e^{x^2}]$$

$$y = e^{x^2}(c_1x + c_2)$$ **Ans.**

Example 16. *Solve* $x^2 y'' - (x^2 + 2x)y' + (x + 2)y = x^3 e^x$ *given that* $y = x$ *is a solution.*

Solution. Here, we have $x^2 y'' - (x^2 + 2x)y' + (x + 2)y = x^3 e^x$

$$\Rightarrow \quad y'' - \frac{x^2+2x}{x^2}y' + \frac{x+2}{x^2}y = xe^x \qquad ...(1)$$

On putting $y = vx$ in (1), the reduced equation as in the article 4.7.

$$\frac{d^2v}{dx^2} + \left\{P + \frac{2}{u}\frac{du}{dx}\right\}\frac{dv}{dx} = \frac{R}{u}$$

$$\frac{d^2v}{dx^2} + \left[-\frac{x^2+2x}{x^2} + \frac{2}{x}(1)\right]\frac{dv}{dx} = \frac{xe^x}{x}$$

$$\Rightarrow \quad \frac{d^2v}{dx^2} - \frac{dv}{dx} = e^x \Rightarrow \frac{dz}{dx} - z = e^x \qquad \left(z = \frac{dv}{dx}\right)$$

which is a linear differential equation

$$I.F. = e^{-\int dx} = e^{-x}$$

Its solution is $\quad z\,e^{-x} = \int e^x . e^{-x}\, dx + c$

$$\Rightarrow \quad z\,e^{-x} = x + c \quad \Rightarrow \quad z = e^x . x + c\,e^x$$

$$\Rightarrow \quad \frac{dv}{dx} = e^x . x + c\,e^x$$

$$\Rightarrow \quad v = x.e^x - e^x + c\,e^x + c_1$$

$$\Rightarrow \quad v = (x - 1 + c)\,e^x + c_1$$

$$y = vx = (x^2 - x + cx)\,e^x + c_1 x$$ **Ans.**

Example 17. *Solve* $x\dfrac{d^2y}{dx^2} - (2x-1)\dfrac{dy}{dx} + (x-1)y = 0$

given that $y = e^x$ *is an integral included in the complementary function.*

Solution. Here, we have $x\dfrac{d^2y}{dx^2} - (2x-1)\dfrac{dy}{dx} + (x-1)\,y = 0$

$$\Rightarrow \quad \frac{d^2y}{dx^2} - \frac{2x-1}{x}\frac{dy}{dx} + \frac{x-1}{x}y = 0 \qquad ...(1)$$

By putting $y = ve^x$ in (1),we get the reduced equation as in the article 4.7.

$$\frac{d^2v}{dx^2} + \left[P + \frac{2}{u}\frac{du}{dx}\right]\frac{dv}{dx} = 0 \qquad ...(2)$$

Putting $u = e^x$ and $\dfrac{dv}{dx} = z$ in (2), we get

$$\frac{dz}{dx} + \left[-\frac{2x-1}{x} + \frac{2}{e^x}e^x\right]z = 0$$

$$\Rightarrow \quad \left[\frac{dz}{dx} + \frac{-2x+1+2x}{x}z\right] = 0 \quad \Rightarrow \quad \frac{dz}{dx} + \frac{z}{x} = 0$$

$$\Rightarrow \quad \frac{dz}{z} = -\frac{dx}{x} \quad \Rightarrow \quad \log z = -\log x + \log c_1$$

$$\Rightarrow \quad z = \frac{c_1}{x} \Rightarrow \frac{dv}{dx} = \frac{c_1}{x} \Rightarrow dv = c_1 \frac{dx}{x} \Rightarrow v = c_1 \log x + c_2$$

$$y = u.\, v = e^x (c_1 \log x + c_2)$$ **Ans.**

4.8 RULE TO FIND OUT PART OF THE COMPLEMENTARY FUNCTION

Rule	*Condition*	Part of Complementary Function = u
1	$1 + P + Q = 0$	e^x
2	$1 - P + Q = 0$	e^{-x}
3	$1 + \frac{P}{a} + \frac{Q}{a^2} = 0$	e^{ax}
4	$P + Qx = 0$	x
5	$2 + 2Px + Qx^2 = 0$	x^2
6	$n(n-1) + Pnx + Qx^2 = 0$	x^n

Example 18. *Solve* $x^2 \frac{d^2y}{dx^2} - 2x[1+x]\frac{dy}{dx} + 2(1+x)y = x^3$

Solution. $x^2 \frac{d^2y}{dx^2} - 2x(1+x)\frac{dy}{dx} + 2(1+x)y = x^3$

$$\Rightarrow \quad \frac{d^2y}{dx^2} - \frac{2x(1+x)}{x^2}\frac{dy}{dx} + \frac{2(1+x)y}{x^2} = x \qquad ...(1)$$

Here, $$P + Qx = -\frac{2x(1+x)}{x^2} + \frac{2(1+x)}{x^2}x = 0$$

Hence $y = x$ is a solution of the C.F. and the other solution is v.
Putting $y = vx$ in (1), we get the reduced equation as in article 4.7

$$\frac{d^2v}{dx^2} + \left\{P + \frac{2}{u}\frac{du}{dx}\right\}\frac{dv}{dx} = \frac{x}{u}$$

$$\frac{d^2v}{dx^2} + \left[\frac{-2x(1+x)}{x^2} + \frac{2}{x}(1)\right]\frac{dv}{dx} = \frac{x}{x}$$

$$\Rightarrow \quad \frac{d^2v}{dx^2} - 2\frac{dv}{dx} = 1 \Rightarrow \frac{dz}{dx} - 2z = 1 \qquad \left[\frac{dv}{dx} = z\right]$$

which is a linear differential equation of first order and $I.F. = e^{\int -2\,dx} = e^{-2x}$

Its solution is $$z\,e^{-2x} = \int e^{-2x}\,dx + c_1$$

$$\Rightarrow \quad z\,e^{-2x} = \frac{e^{-2x}}{-2} + c_1 \quad \Rightarrow \quad z = \frac{-1}{2} + c_1 e^{2x}$$

$$\Rightarrow \quad \frac{dv}{dx} = -\frac{1}{2} + c_1 e^{2x} \Rightarrow dv = \left(-\frac{1}{2} + c_1 e^{2x}\right) dx$$

$$\Rightarrow \quad v = \frac{-x}{2} + \frac{c_1}{2} e^{2x} + c_2$$

$$y = uv = x\left(\frac{-x}{2} + \frac{c_1}{2} e^{2x} + c_2\right) \qquad \textbf{Ans.}$$

Example 19. *Solve* $(x+2)\frac{d^2y}{dx^2} - (2x+5)\frac{dy}{dx} + 2y = (x+1)e^x$

Solution. Here, we have $\frac{d^2y}{dx^2} - \frac{2x+5}{x+2}\frac{dy}{dx} + \frac{2y}{x+2} = \frac{(x+1)\,e^x}{x+2}$ ...(1)

Here $P = \frac{2x+5}{x+2},\ Q = \frac{2}{x+2},\ R = \frac{(x+1)e^x}{x+2}$

$$1 + \frac{P}{a} + \frac{Q}{a^2} = 0, \text{ Choosing } a = 2$$

$$1 + \frac{P}{2} + \frac{Q}{4} = 1 - \frac{2x+5}{2x+4} + \frac{2}{4x+8} = 0$$

Hence $y = e^{2x}$ is a part of C.F.

Putting $y = e^{2x} v$ in (1), the reduced equation as in the article 4.7.

$$\frac{d^2v}{dx^2} + \left[P + \frac{2}{u}\frac{du}{dx}\right]\frac{dv}{dx} = \frac{R}{u}$$

On putting the values of P, Q, and R, we get

$$\Rightarrow \quad \frac{d^2v}{dx^2} + \left[-\frac{2x+5}{x+2} + \frac{2}{e^{2x}} 2e^{2x}\right]\frac{dv}{dx} = \frac{(x+1)\,e^x}{e^{2x}(x+2)}$$

$$\Rightarrow \quad \frac{d^2v}{dx^2} + \left[-\frac{2x+5}{x+2} + 4\right]\frac{dv}{dx} = \frac{x+1}{x+2}e^{-x}$$

$$\frac{d^2v}{dx^2} + \frac{2x+3}{x+2}\frac{dv}{dx} = \frac{x+1}{x+2}e^{-x}$$

$$\Rightarrow \quad \frac{dz}{dx} + \frac{2x+3}{x+2}\,z = \frac{x+1}{x+2}e^{-x} \qquad \left(\frac{dv}{dx} = z\right)$$

which is a linear differential equation,

$$I.F. = e^{\int \frac{2x+3}{x+2}dx} = e^{\int\left(2 - \frac{1}{x+2}\right)dx} = e^{2x - \log(x+2)} = \frac{e^{2x}}{x+2}$$

Its solution is

$$z.\frac{e^{2x}}{x+2} = \int \frac{e^{2x}}{x+2}\left[\frac{x+1}{x+2}\right] e^{-x}\, dx + c$$

$$= \int \frac{e^x\,(x+1)}{(x+2)^2} dx + c = \int e^x \left[\frac{1}{x+2} - \frac{1}{(x+2)^2}\right] dx + c$$

$$= \int \frac{e^x dx}{x+2} - \int \frac{e^x dx}{(x+2)^2} + c = \frac{e^x}{x+2} + \int \frac{e^x dx}{(x+2)^2} - \int \frac{e^x\, dx}{(x+2)^2} + c = \frac{e^x}{x+2} + c$$

$$\Rightarrow \quad z = e^{-x} + c\,(x+2)e^{-2x}$$

$$\Rightarrow \quad \frac{dv}{dx} = e^{-x} + c(x+2)e^{-2x}$$

$$v = \int e^{-x}\,dx + c\int (x+2)e^{-2x}\,dx + c_1$$

$$= -e^{-x} + c\left[\frac{(x+2)\,e^{-2x}}{-2} - \frac{e^{-2x}}{4}\right] + c_1$$

$$= -e^{-x} + \frac{ce^{-2x}}{4}[2x+5] + c_1$$

$$y = u.v$$

$$y = e^{2x}\left[-e^{-x} + \frac{ce^{-2x}}{4}(2x+5) + c_1\right]$$

$$y = -e^{x} - \frac{c}{4}(2x+5) + c_1\,e^{2x}$$ **Ans.**

EXERCISE 4.4

Solve the following differential equations :

1. $(3-x)\frac{d^2y}{dx^2} - (9-4x)\frac{dy}{dx} + (6-3x)\,y = 0$, given $y = e^x$ is a solution.

Ans. $y = \frac{c_1}{8}e^{3x}(4x^3 - 42x^2 + 150x - 183) + c_2e^x$

2. $x\frac{d^2y}{dx^2} - \frac{dy}{dx} + (1-x)y = x^2\,e^{-x}$ given $y = e^x$ is an integral included in C.F.

Ans. $y = c_2e^x + c_1(2x+1)\,e^{-x} - \frac{1}{4}(2x^2+2x+1)e^{-x}$

3. $(1-x^2)\frac{d^2y}{dx^2} + x\frac{dy}{dx} - y = x(1-x^2)^{3/2}$, given $y = x$ is part of C.F.

Ans. $y = -\frac{x}{9}(1-x^2)^{3/2} - c_1\left[\sqrt{(1-x^2)} + x\sin^{-1}x\right] + c_2x.$

4. $\sin^2 x\frac{d^2y}{dx^2} = 2y$, given that $y = \cot x$ is a solution. **Ans.** $cy = 1 + (c_1 - x)\cot x$

5. $\frac{d^2y}{dx^2} - x^2\frac{dy}{dx} + xy = x$, given $y = x$ is a part of C.F. **Ans.** $y = 1 + c_1x\int \frac{1}{x^2}e^{\frac{x^3}{3}}dx + c_2x$

6. $(x\sin x + \cos x)\frac{d^2y}{dx^2} - x\cos x\frac{dy}{dx} + y\cos x = 0$ given $y = x$ is solution.

Ans. $y = c_2x - c_1\cos x$

7. $x^2\frac{d^2y}{dx^2} + x\frac{dy}{dx} - y = 0$, given that $y = x + \frac{1}{x}$ is one integral. **Ans.** $y = c_2\left(x + \frac{1}{x}\right) + \frac{c_1}{x}$

8. $x^2\frac{d^2y}{dx^2} + 2x\frac{dy}{dx} - 12y = x^3\log x$ *(U.P., II Semester 2004)*

[Hint. $n(n-1) + pnx + Qx^2 = 0$, $n = 3$, satisfies this equation. Put $y = vx^3$, $\frac{dv}{dx} = z$]

Ans. $y = \left(c_1x^3 + \frac{c_2}{x^4}\right) + \frac{x^3}{98}\log x\,(7\log x - 2)$

TYPE (VI)

4.9 NORMAL FORM (REMOVAL OF FIRST DERIVATIVE)

When the part of complementary function cannot be found by previous method then we find the complete solution by removing $\frac{dy}{dx}$ from

$$y'' + Py' + Qy = R \quad \text{...(1)}$$

Method:

Let uv be the complete solution of the given differential equation.

$$y = u.v.$$

$$\frac{dy}{dx} = v\frac{du}{dx} + u\frac{dv}{dx}$$

$$\frac{d^2y}{dx^2} = u\frac{d^2v}{dx^2} + 2\frac{du}{dx}\frac{dv}{dx} + v\frac{d^2u}{dx^2}$$

On putting the values of $y, \frac{dy}{dx}, \frac{d^2y}{dx^2}$ in (1), we get

$$\left(u\frac{d^2v}{dx^2} + 2\frac{dv}{dx}\frac{du}{dx} + v\frac{d^2u}{dx^2}\right) + P\left(u\frac{dv}{dx} + v\frac{du}{dx}\right) + Q.uv = R$$

$$\Rightarrow \quad v\frac{d^2u}{dx^2} + \frac{du}{dx}\left(Pv + 2\frac{dv}{dx}\right) + u\left(\frac{d^2v}{dx^2} + P\frac{dv}{dx} + Q.v\right) = R$$

$$\Rightarrow \quad \frac{d^2u}{dx^2} + \frac{du}{dx}\left(P + \frac{2}{v}\frac{dv}{dx}\right) + \frac{u}{v}\left(\frac{d^2v}{dx^2} + P\frac{dv}{dx} + Q.v\right) = \frac{R}{v} \quad \text{..(2)}$$

Here in the last bracket on L.H.S. is not zero as $y = v$ is not a part of C.F.

Here we shall remove the first derivative.

$$P + \frac{2}{v}\frac{dv}{dx} = 0 \Rightarrow \frac{dv}{v} = -\frac{1}{2}P\,dx \Rightarrow \log v = \frac{-1}{2}\int P\,dx$$

$$\Rightarrow \quad v = e^{-\frac{1}{2}\int P\,dx}$$

In (2) we have to find out the value of the last bracket *i.e.*, $\frac{d^2v}{dx^2} + P\frac{dv}{dx} + Qv$

$$\frac{dv}{dx} = -\frac{P}{2}e^{-\frac{1}{2}\int P\,dx} = -\frac{1}{2}Pv \qquad \left[\because v = e^{-1/2\int P dx}\right]$$

$$\frac{d^2v}{dx^2} = -\frac{1}{2}\frac{dP}{dx}v - \frac{P}{2}\frac{dv}{dx} = -\frac{1}{2}\frac{dP}{dx}v - \frac{P}{2}\left(-\frac{1}{2}Pv\right) = -\frac{1}{2}\frac{dP}{dx}v + \frac{1}{4}P^2v$$

$$\therefore \quad \frac{d^2v}{dx^2} + P\frac{dv}{dx} + Qv = -\frac{1}{2}\frac{dP}{dx}v + \frac{1}{4}P^2v + P\left(-\frac{1}{2}Pv\right) + Qv$$

$$= v\left[Q - \frac{1}{2}\frac{dP}{dx} - \frac{1}{4}P^2\right]$$

Equation (1) is transformed as

$$\frac{d^2u}{dx^2} + \frac{u}{v}v\left\{Q - \frac{1}{2}\frac{dP}{dx} - \frac{P^2}{4}\right\} = \frac{R}{v}$$

$$\Rightarrow \quad \frac{d^2u}{dx^2} + u\left\{Q - \frac{1}{2}\frac{dP}{dx} - \frac{P^2}{4}\right\} = R e^{\frac{1}{2}\int P\,dx}$$

$$\frac{d^2u}{dx^2} + Q_1 u = R_1 \qquad \text{...(3), where } Q_1 = \left[Q - \frac{1}{2}\frac{dP}{dx} - \frac{P^2}{4}\right]$$

$$R_1 = R e^{\frac{1}{2}\int P\,dx} \quad \text{or} \quad \frac{R}{v}$$

On putting the values of Q_1 and R_1 in (3), we can find u by solving (3)
Hence the solution of (1) is

$$y = uv$$ **Ans.**

Example 20. *Solve* $\frac{d}{dx}\left[\cos^2 x \frac{dy}{dx}\right] + \cos^2 x . y = 0$ *by removing the first derivative.*

Solution. We have, $\frac{d}{dx}\left(\cos^2 x \frac{dy}{dx}\right) + \cos^2 x . y = 0$

$$\Rightarrow \quad \frac{d^2y}{dx^2}\cos^2 x - 2\cos x \sin x \frac{dy}{dx} + (\cos^2 x) y = 0$$

$$\Rightarrow \quad \frac{d^2y}{dx^2} - 2\tan x . \frac{dy}{dx} + y = 0$$

Here, $P = -2\tan x$, $Q = 1$, $R = 0$

$$Q_1 = Q - \frac{1}{2}\frac{dP}{dx} - \frac{P^2}{4}$$

$$= 1 - \frac{1}{2}(-2\sec^2 x) - \frac{4\tan^2 x}{4}$$

$$= 1 + \sec^2 x - \tan^2 x = 1 + 1 = 2$$

$$R_1 = R\, e^{\frac{1}{2}\int P\,dx} = 0$$

$$v = e^{-\frac{1}{2}\int P dx} = e^{-\frac{1}{2}\int(-2\tan x)dx} = e^{\int \tan x\,dx} = e^{\log \sec x} = \sec x$$

Normal equation is

$$\frac{d^2u}{dx^2} + Q_1 u = R_1$$

$$\frac{d^2u}{dx^2} + 2u = 0 \quad \Rightarrow \quad (D^2 + 2)\,u = 0$$

A.E. is $m^2 + 2 = 0 \quad \Rightarrow \quad m = \pm i\sqrt{2}$

$$u = \text{C.F.} = c_1 \cos\sqrt{2}\,x + c_2 \sin\sqrt{2}\,x$$

$$y = u.v$$

Hence the complete solution is $y = [c_1 \cos\sqrt{2}\,x + c_2 \sin\sqrt{2}x]\sec x$ **Ans.**

Example 21. *Solve* $x^2\frac{d^2y}{dx^2} - 2(x^2 + x)\frac{dy}{dx} + (x^2 + 2x + 2)y = 0$ *by removing the first derivative.*

Solution. We have, $\dfrac{d^2y}{dx^2} - \dfrac{2(x^2+x)}{x^2}\dfrac{dy}{dx} + \left(\dfrac{x^2+2x+2}{x^2}\right)y = 0$ ...(1)

Here, $P = -2\left(1+\dfrac{1}{x}\right),\ Q = \dfrac{x^2+2x+2}{x^2},\ R = 0$

In order to remove the first derivative, we put $y = u.v$ in (1) to get the normal equation

$$\frac{d^2v}{dx^2} + Q_1 v = R_1 \qquad ...(2)$$

where $v = e^{-\frac{1}{2}\int P dx} = e^{-\frac{1}{2}\int -2\left(1+\frac{1}{x}\right)dx} = e^{\int\left(1+\frac{1}{x}\right)dx} = e^x . e^{\log x} = x\, e^x$

$$Q_1 = Q - \frac{1}{2}\frac{dP}{dx} - \frac{P^2}{4} = \frac{x^2+2x+2}{x^2} - \frac{1}{2}\left(\frac{2}{x^2}\right) - \frac{4}{4}\left(1+\frac{1}{x}\right)^2$$

$$= 1 + \frac{2}{x} + \frac{2}{x^2} - \frac{1}{x^2} - 1 - \frac{1}{x^2} - \frac{2}{x} = 0$$

$$R_1 = R e^{\frac{1}{2}\int p dx} = 0$$

On putting the values of Q_1 and R_1 in (2), we get

$$\frac{d^2u}{dx^2} + 0\,(u) = 0 \quad \Rightarrow \quad \frac{d^2u}{dx^2} = 0$$

$$\frac{du}{dx} = c_1 \quad \Rightarrow \quad u = c_1 x + c_2$$

Hence the complete solution is $y = u.v = (c_1 x + c_2)\, x\, e^x$ **Ans.**

Example 22. *Solve:* $\dfrac{d^2y}{dx^2} - 4x\dfrac{dy}{dx} + (4x^2-1)y = -3e^{x^2}\sin 2x$ *by removing the first derivative.* *(U.P. II Semester (C.O.) 2004)*

Solution. We have, $\dfrac{d^2y}{dx^2} - 4x\dfrac{dy}{dx} + (4x^2-1)\,y = -3e^{x^2}\sin 2x$ ...(1)

Here $P = -4x$, $\quad Q = 4x^2 - 1$, $\quad R = -3e^{x^2}\sin 2x$

In order to remove the first derivative, $v = e^{-\frac{1}{2}\int P dx} = e^{-\frac{1}{2}\int -4x\, dx} = e^{2\int x\, dx} = e^{x^2}$

On putting $y = uv$, the normal equation is

$$\frac{d^2u}{dx^2} + Q_1 u = R_1 \qquad ...(2)$$

where $$Q_1 = Q - \frac{1}{2}\frac{dP}{dx} - \frac{P^2}{4} = (4x^2-1) - \frac{1}{2}(-4) - \frac{16x^2}{4}$$

$$= 4x^2 - 1 + 2 - 4x^2 = 1$$

$$R_1 = \frac{R}{v} = \frac{-3e^{x^2}\sin 2x}{e^{x^2}} = -3\sin 2x$$

Equation (2) becomes $\dfrac{d^2u}{dx^2} + u = -3\sin 2x$

$$(D^2+1)u = -3\sin 2x$$

A.E. is $m^2 + 1 = 0 \Rightarrow m = \pm i$

$\Rightarrow$ $C.F. = c_1 \cos x + c_2 \sin x$

$$P.I. = \frac{1}{D^2+1}(-3\sin 2x) = \frac{-3\sin 2x}{-4+1} = \sin 2x$$

Its solution is $u = C.F. + P.I.$

$$u = c_1\cos x + c_2 \sin x + \sin 2x$$

Hence, complete solution is $y = u.\,v = (c_1\cos x + c_2\sin x + \sin 2x)e^{x^2}$ **Ans.**

EXERCISE 4.5

Solve the following differential equations by removing the first derivative:

1. $\dfrac{d^2y}{dx^2} - 2\tan x.\,y - 5y = 0$ **Ans.** $y = (a\,e^{2x} + b\,e^{-3x})\sec x$

2. $\dfrac{d^2y}{dx^2} - 4x\dfrac{dy}{dx} + (4x^2-3)\,y = e^{x^2}$ **Ans.** $y = (c_1e^x + c_2\,e^{-x} - 1)$

3. $\dfrac{d^2y}{dx^2} - 2x\dfrac{dy}{dx} + (x^2+2)y = e^{\frac{1}{2}(x^2+2x)}$ **Ans.** $v = (c_1\cos\sqrt{3}x + c_2\sin\sqrt{3}\,x)\,e^{\frac{x^2}{2}} + \frac{1}{4}e^x.e^{\frac{x^2}{2}}$

4. $\dfrac{d^2y}{dx^2} - \dfrac{2}{x}\dfrac{dy}{dx} + \left(n^2 + \dfrac{2}{x^2}\right)y = 0$ **Ans.** $y = (c_1\cos nx + c_2\sin nx)x$

5. $\dfrac{d^2y}{dx^2} + \dfrac{2}{x}\dfrac{dy}{dx} - n^2y = 0$ **Ans.** $y = (c_1e^{nx} + c_2\,e^{-nx})\dfrac{1}{x}$

6. $\dfrac{d^2y}{dx^2} + \dfrac{1}{x^{\frac{1}{3}}}\dfrac{dy}{dx} + \left(\dfrac{1}{4x^{2/3}} - \dfrac{1}{x^{4/3}} - \dfrac{6}{x^2}\right)y = 0$ **Ans.** $y = (c_1x^3 + c_2\,x^{-2})e^{-\frac{3}{4}x^{\frac{2}{3}}}$

7. $\dfrac{d^2y}{dx^2} - \dfrac{1}{\sqrt{x}}\dfrac{dy}{dx} + \dfrac{y}{4x^2}(-8 + \sqrt{x} + x) = 0$ **Ans.** $y = (c_1x^2 + c_2x^{-1})\,e^{\sqrt{x}}$

TYPE (VII)

4.10 METHOD OF SOLVING LINEAR DIFFERENTIAL EQUATIONS BY CHANGING THE INDEPENDENT VARIABLE

Consider. $$\frac{d^2y}{dx^2} + P\frac{dy}{dx} + Qy = R \qquad ...(1)$$

Let us change the independent variable x to z and $z = f(x)$

$$\frac{dy}{dx} = \frac{dy}{dz}\frac{dz}{dx}$$

$$\frac{d^2y}{dx^2} = \frac{d^2y}{dz^2}\left(\frac{dz}{dx}\right)^2 + \frac{dy}{dz}\frac{d^2z}{dx^2}$$

Putting the values of $\frac{dy}{dx}$ and $\frac{d^2y}{dx^2}$ in (1), we get

$$\left(\frac{d^2y}{dz^2}\left(\frac{dz}{dx}\right)^2+\frac{dy}{dz}\frac{d^2z}{dx^2}\right)+P\left(\frac{dy}{dz}\frac{dz}{dx}\right)+Qy=R$$

$$\Rightarrow \quad \frac{d^2y}{dz^2}\left(\frac{dz}{dx}\right)^2+\left(P\frac{dz}{dx}+\frac{d^2z}{dx^2}\right)\frac{dy}{dz}+Qy=R$$

$$\Rightarrow \quad \frac{d^2y}{dz^2}+\frac{\left(P\frac{dz}{dx}+\frac{d^2z}{dx^2}\right)}{\left(\frac{dz}{dx}\right)^2}\frac{dy}{dz}+\frac{Qy}{\left(\frac{dz}{dx}\right)^2}=\frac{R}{\left(\frac{dz}{dx}\right)^2} \qquad \text{...(2)}$$

On putting $\frac{\left(P\frac{dz}{dx}+\frac{d^2z}{dx^2}\right)}{\left(\frac{dz}{dx}\right)^2}P_1$, $\frac{Q}{\left(\frac{dz}{dx}\right)^2}=Q_1$ and $\frac{R}{\left(\frac{dz}{dx}\right)^2}R_1$ in (2), we get

$$\Rightarrow \quad \frac{d^2y}{dz^2}+P_1\frac{dy}{dz}+Q_1y=R_1 \qquad \text{...(3)}$$

Equation (3) is solved either by taking $P_1 = 0$ (first methrd) or $Q_1 = a$ constant (second method).

Working Rule:

Step 1. Coefficient of $\frac{d^2y}{dx^2}$ should be made as 1 if it is not so.

Step 2. To get P, Q and R, compare the given differential equation with the standard form $y'' + Py' + Qy = R$.

Step 3. Find P_1, Q_1 and R_1 by the following formulae.

$$P_1=\frac{\frac{d^2z}{dx^2}+P\frac{dz}{dx}}{\left(\frac{dz}{dx}\right)^2},\quad Q_1=\frac{Q}{\left(\frac{dz}{dx}\right)^2},\quad R_1=\frac{R}{\left(\frac{dz}{dx}\right)^2}$$

Step 4. Find out the value of z by taking

First Method. $P_1 = 0$ or

Second Method. Q_1 = constant

Step 5. We get a reduced equation

$$\frac{d^2y}{dz^2}+P_1\frac{dy}{dz}+Q_1y=R_1$$

On solving this equation we can find out the value of y in terms of z.
Then write down the solution in terms of ~~of x by~~ replacing the value of z.

Example 23. *Solve* $\frac{d^2y}{dx^2}+\cot x\frac{dy}{dx}+4y\,\text{cosec}^2 x = 0$ *by changing the independent variable.* *(Uttarakhand, II Semester, 2012, RGPV, Bhopal, Dec. 2011)*

Solution. We have, $\dfrac{d^2y}{dx^2} + \cot x \dfrac{dy}{dx} + 4y \operatorname{cosec}^2 x = 0$ ...(1)

Here, $P = \cot x$, $Q = 4 \operatorname{cosec}^2 x$ and $R = 0$

Changing the independent variable from x to z, the equation becomes

$$\frac{d^2y}{dx^2} + P_1 \frac{dy}{dz} + Q_1 y = 0 \quad ...(2)$$

where $P_1 = \dfrac{P\dfrac{dz}{dx} + \dfrac{d^2z}{dx^2}}{\left(\dfrac{dz}{dx}\right)^2}$, $Q_1 = \dfrac{Q}{\left(\dfrac{dz}{dx}\right)^2}$

First Method. Let us take $P_1 = 0$

$$\frac{P\dfrac{dz}{dx} + \dfrac{d^2z}{dx^2}}{\left(\dfrac{dz}{dx}\right)^2} = 0 \Rightarrow P\frac{dz}{dx} + \frac{d^2z}{dx^2} = 0$$

$$\Rightarrow \quad \frac{d^2z}{dx^2} + \cot x \frac{dz}{dx} = 0 \quad ...(3)$$

Put $\dfrac{dz}{dx} = v, \quad \dfrac{d^2z}{dx^2} = \dfrac{dv}{dx}$

(3) becomes $\dfrac{dv}{dx} + (\cot x) v = 0 \Rightarrow \dfrac{dv}{v} = -\cot x.\ dx$

$$\Rightarrow \quad \log v = -\log \sin x + \log c = \log c \operatorname{cosec} x \Rightarrow v = c \operatorname{cosec} x$$

$$\frac{dz}{dx} = c \operatorname{cosec} x \Rightarrow dz = (c \operatorname{cosec} x)\, dx \Rightarrow z = c \log \tan \frac{x}{2}$$

Second Method **OR**

Now let us take Q_1 = Constant.

$$Q_1 = \frac{Q}{\left(\dfrac{dz}{dx}\right)^2} = \frac{4 \operatorname{cosec}^2 x}{c^2 \operatorname{cosec}^2 x} = \frac{4}{c^2} \text{ which is constant}$$

Hence the equation (2) reduces to

$$\frac{d^2y}{dz^2} + 0\frac{dy}{dz} + \frac{4}{c^2}y = 0 \Rightarrow \frac{d^2y}{dz^2} + \frac{4}{c^2}y = 0 \quad \left[\because P_1 = 0, Q_1 = \frac{4}{c^2}\right]$$

$$\Rightarrow \quad \left(D^2 + \frac{4}{c^2}\right)y = 0, \text{ A.E. is } m^2 + \frac{4}{c^2} = 0 \Rightarrow m = \pm i\frac{2}{c}$$

$$\text{C.F.} = c_1 \cos \frac{2z}{c} + c_2 \sin \frac{2z}{c} \quad \left(z = c \log \tan \frac{x}{2}\right)$$

$$\Rightarrow \quad y = c_1 \cos\left(2 \log \tan \frac{x}{2}\right) + c_2 \sin\left(2 \log \tan \frac{x}{2}\right) \quad \textbf{Ans.}$$

Example 24. *Solve* $x^6 \dfrac{d^2y}{dx^2} + 3x^5 \dfrac{dy}{dx} + a^2 y = \dfrac{1}{x^2}$ *by changing the independent variable.*

Solution. We have, $\dfrac{d^2y}{dx^2} + \dfrac{3}{x}\dfrac{dy}{dx} + a^2 \dfrac{y}{x^6} = \dfrac{1}{x^8}$ ...(1)

Here $P = \dfrac{3}{x}$ and $Q = \dfrac{a^2}{x^6}$

On changing the independent variable x to z, the equation (1) is reduced to

$$\frac{d^2y}{dz^2} + P_1 \frac{dy}{dz} + Q_1 y = R_1 \quad ...(2)$$

Using Second Method

Let $Q_1 = a_2$ (constant) $\quad Q_1 = \dfrac{Q}{\left(\dfrac{dz}{dx}\right)^2} = \dfrac{a^2}{x^6\left(\dfrac{dz}{dx}\right)^2}$ = constant = a^2 (say)

$$\therefore \quad x^6\left(\frac{dz}{dx}\right)^2 = 1 \Rightarrow x^3\left(\frac{dz}{dx}\right) = 1 \Rightarrow \frac{dz}{dx} = \frac{1}{x^3}$$

$$\Rightarrow \quad dz = \frac{dx}{x^3} \Rightarrow z = \frac{x^{-2}}{-2} + c$$

On differentiating twice, we have $\dfrac{d^2z}{dx^2} = \dfrac{-3}{x^4}$

$$P_1 = \frac{P\dfrac{dz}{dx} + \dfrac{d^2z}{dx^2}}{\left(\dfrac{dz}{dx}\right)^2} = \frac{\dfrac{3}{x}\cdot\dfrac{1}{x^3} + \left(\dfrac{-3}{x^4}\right)}{\left(\dfrac{1}{x^3}\right)^2} = 0$$

$$R_1 = \frac{R}{\left(\dfrac{dz}{dx}\right)^2} = \frac{\dfrac{1}{x^8}}{\dfrac{1}{x^6}} = \frac{1}{x^2} = -2z$$

On putting the values of P_1, Q_1 and R_1 in (2), we get

$$\frac{d^2y}{dz^2} + a^2 y = -2z \quad \Rightarrow \quad (D^2 + a^2)\, y = -2z$$

A.E. is $m^2 + a^2 = 0$, $m = \pm i a$ $\quad \Rightarrow \quad$ C.F. $= c_1 \cos az + c_2 \sin az$

$$\text{P.I.} = \frac{1}{D^2 + a^2}(-2z) = \frac{1}{a^2}\left[1 + \frac{D^2}{a^2}\right]^{-1}(-2z) = \frac{1}{a^2}\left[1 - \frac{D^2}{a^2}\right](-2z) = \frac{-2z}{a^2} = \frac{1}{a^2x^2}$$

$$y = \text{C.F.} + \text{P.I.}$$

$$y = c_1 \cos\frac{a}{2x^2} - c_2 \sin\frac{a}{2x^2} + \frac{1}{a^2x^2} \quad \textbf{Ans.}$$

Example 25. *Solve* $\frac{d^2y}{dx^2} - \frac{1}{x}\frac{dy}{dx} + 4x^2 y = x^4$ *by changing the independent variable.*

(*U.P., II Semester, Summer 2003, 2002*)

Solution. We have, $\frac{d^2y}{dx^2} - \frac{1}{x}\frac{dy}{dx} + 4x^2 y = x^4$ ...(1)

Here $P = -\frac{1}{x},\ Q = 4x^2,\ R = x^4$

On changing the independent variable x to z, the equation (1) is transformed as

$$\frac{d^2y}{dz^2} + P_1\frac{dy}{dz} + Q_1 y = R_1 \qquad ...(2)$$

Using Second Method

Let $Q_1 = 1$ (constant)

but $$Q_1 = \frac{Q}{\left(\frac{dz}{dx}\right)^2} = \frac{4x^2}{\left(\frac{dz}{dx}\right)^2} = \text{constant} = 1 \quad \text{(say)}$$

$$\left(\frac{dz}{dx}\right)^2 = 4x^2 \Rightarrow \frac{dz}{dx} = 2x$$

$\Rightarrow$ $dz = 2x\, dx \Rightarrow z = x^2 + c$

$\Rightarrow$ $x^2 = z - c$ ...(3)

$$P_1 = \frac{P\left(\frac{dz}{dx}\right) + \frac{d^2z}{dx^2}}{\left(\frac{dz}{dx}\right)^2} = \frac{-\frac{1}{x}(2x) + 2}{(2x)^2} = 0$$

$$R_1 = \frac{R}{\left(\frac{dz}{dx}\right)^2} = \frac{x^4}{4x^2} = \frac{x^2}{4} = \frac{z-c}{4} \qquad \text{[Using (3)]}$$

On putting the values of P_1, Q_1 and R_1 in (2), we get

$$\frac{d^2y}{dz^2} + (0)\frac{dy}{dz} + (1)y = \frac{z-c}{4} \Rightarrow \frac{d^2y}{dz^2} + y = \frac{z-c}{4} \Rightarrow (D^2+1)\,y = \frac{z-c}{4}$$

A.E. is $m^2 + 1 = 0$, $m = \pm i$

$$\text{C.F.} = c_1 \cos z + c_2 \sin z$$

$$\text{P.I.} = \frac{1}{D^2+1}\left(\frac{z-c}{4}\right) = (1 + D^2)^{-1}\,\frac{z-c}{4}$$

$$= (1 - D^2)\frac{z-c}{4} = \frac{z-c}{4}$$

Now complete solution is $y = \text{C.F.} + \text{P.I.}$

$\Rightarrow$ $y = c_1 \cos z + c_2 \sin z + \frac{z-c}{4}$

$\Rightarrow$ $y = c_1 \cos x^2 + c_2 \sin x^2 + \frac{x^2 - c}{4}$ **Ans.**

Example 26. *Solve the following differential equation by changing the independent variable*

$$x\frac{d^2y}{dx^2}+(4x^2-1)\frac{dy}{dx}+4x^3y = 2x^3 \qquad \text{(U.P., II Semester, Summer 2006)}$$

Solution. We have,

$$x\frac{d^2y}{dx^2}+(4x^2-1)\frac{dy}{dx}+4x^3y = 2x^3$$

$$\Rightarrow \qquad \frac{d^2y}{dx^2}+\left(\frac{4x^2-1}{x}\right)\frac{dy}{dx}+4x^2y = 2x^2 \qquad ...(1)$$

Here, $P = \dfrac{4x^2-1}{x}$, $\quad Q = 4x^2$ and $R = 2x^2$

Changing the independent variable from x to z, the equation becomes

$$\frac{d^2y}{dz^2}+P_1\frac{dy}{dz}+Q_1y = R_1 \qquad ...(2)$$

$$\text{where } P_1 = \frac{P\dfrac{dz}{dx}+\dfrac{d^2z}{dx^2}}{\left(\dfrac{dz}{dx}\right)^2},\quad Q_1 = \frac{Q}{\left(\dfrac{dz}{dx}\right)^2},\quad R_1 = \frac{R}{\left(\dfrac{dz}{dx}\right)^2}$$

Using Second Method

Choose $Q_1 = 1$ (constant) $\qquad$ so $1 = \dfrac{Q}{\left(\dfrac{dz}{dx}\right)^2}$

$$\Rightarrow \qquad \left(\frac{dz}{dx}\right)^2 = 4x^2 \qquad [\because Q = 4x^2]$$

$$\Rightarrow \qquad \frac{dz}{dx} = 2x \qquad \Rightarrow \quad dz = 2x\,dx \quad \Rightarrow \quad z = x^2$$

$$\frac{d^2z}{dx^2} = 2$$

and

$$P_1 = \frac{P\left(\dfrac{dz}{dx}\right)+\dfrac{d^2z}{dx^2}}{\left(\dfrac{dz}{dx}\right)^2} = \frac{\left(\dfrac{4x^2-1}{x}\right)(2x)+2}{(2x)^2}$$

$$\Rightarrow \qquad P_1 = \frac{8x^2-2+2}{4x^2} = 2$$

$$R_1 = \frac{R}{\left(\dfrac{dz}{dx}\right)^2} = \frac{2x^2}{4x^2} = \frac{1}{2}$$

The equation (2) is transformed to

$$\frac{d^2y}{dz^2}+2\frac{dy}{dz}+y = \frac{1}{2}$$

A.E. is $m^2 + 2m + 1 = 0$

$\Rightarrow$ $(m + 1)^2 = 0$ $\Rightarrow$ $m = -1, -1$

$$\text{C.F.} = (C_1 + C_2 z)\, e^{-z}$$

$$\text{P.I.} = \frac{1}{D^2 + 2D + 1}\left(\frac{1}{2}\right) = \frac{1}{0 + 0 + 1}\left(\frac{1}{2}\right) = \frac{1}{2}$$

Complete solution is

$$y = C.F. + P.I. = (C_1 + C_2 z) e^{-z} + \frac{1}{2}$$

$$\Rightarrow \qquad y = (C_1 + C_2 x^2)\, e^{-x^2} + \frac{1}{2}$$ **Ans.**

Example 27. *Solve by changing the independent variable*

$$(1+x^2)^2 \frac{d^2y}{dx^2} + 2x(1+x^2)\frac{dy}{dx} + 4y = 0$$ *(Q. Bank U.P., II Semester 2001)*

Solution. Here, we have

$$\frac{d^2y}{dx^2} + \frac{2x}{1+x^2}\frac{dy}{dx} + \frac{4}{(1+x^2)^2} y = 0 \qquad \text{...(1)}$$

$$P = \frac{2x}{1+x^2}, \quad Q = \frac{4}{(1+x^2)^2}, \quad R = 0$$

Choose z such that

$$\left(\frac{dz}{dx}\right)^2 = \frac{4}{(1+x^2)^2}$$

$$\Rightarrow \qquad \frac{dz}{dx} = \frac{2}{1+x^2} \qquad \text{...(2)}$$

Integration yields, $z = 2 \tan^{-1} x$

From (2), $$\frac{d^2z}{dx^2} = \frac{-4x}{(1+x^2)^2}$$

$$\therefore \qquad P_1 = \frac{\dfrac{d^2z}{dx^2} + P\dfrac{dz}{dx}}{\left(\dfrac{dz}{dx}\right)^2} = \frac{\dfrac{-4x}{(1+x^2)^2} + \dfrac{2x}{1+x^2}\cdot\dfrac{2}{1+x^2}}{\dfrac{4x}{(1+x^2)^2}} = 0$$

$$Q_1 = \frac{Q}{\left(\dfrac{dz}{dx}\right)^2} = \frac{\dfrac{4}{(1+x^2)^2}}{\dfrac{4}{(1+x^2)^2}} = 1,$$

$$R_1 = \frac{R}{\left(\dfrac{dz}{dx}\right)^2} = \frac{0}{\left\{\dfrac{4}{(1+x^2)^2}\right\}} = 0$$

Reduced equation is $$\frac{d^2y}{dz^2} + y = 0$$

Auxiliary equation is $m^2 + 1 = 0 \Rightarrow m = \pm i$

$$\text{C.F.} = c_1 \cos z + c_2 \sin z$$

$$\text{P.I.} = 0$$

$\therefore$ $$y = c_1 \cos z + c_2 \sin z$$

Complete solution is $y = c_1 \cos(2\tan^{-1} x) + c_2 \sin(2\tan^{-1} x)$

where c_1 and c_2 are arbitrary constants of integration. **Ans.**

Example 28. *Solve by changing the independent variable*

$$\cos x \frac{d^2y}{dx^2} + \sin x \frac{dy}{dx} - 2y\cos^3 x = 2\cos^5 x.$$ *(Q. Bank U.P. II Semester 2001)*

Solution. Here, we have

$$\frac{d^2y}{dx^2} + \tan x \frac{dy}{dx} - 2y\cos^2 x = 2\cos^4 x$$

Here, $P = \tan x,\quad Q = -2\cos^2 x,\quad R = 2\cos^4 x$

Choose z such that $$\left(\frac{dz}{dx}\right)^2 = \cos^2 x \Rightarrow \frac{dz}{dx} = \cos x \quad ...(1)$$

Integration yields, $$z = \sin x \quad ...(2)$$

From (1), $$\frac{d^2z}{dx^2} = -\sin x$$

$$P_1 = \frac{\frac{d^2z}{dx^2} + P\frac{dz}{dx}}{\left(\frac{dz}{dx}\right)^2} = \frac{-\sin x + \tan x \cos x}{\cos^2 x} = 0$$

$$Q_1 = \frac{Q}{\left(\frac{dz}{dx}\right)^2} = \frac{-2\cos^2 x}{\cos^2 x} = -2$$

$$R_1 = \frac{R}{\left(\frac{dz}{dx}\right)^2} = \frac{2\cos^4 x}{\cos^2 x} = 2\cos^2 x = 2(1 - z^2)$$

Reduced equation is

$$\frac{d^2y}{dz^2} - 2y = 2(1 - z^2)$$

Auxiliary equation is $m^2 - 2 = 0 \Rightarrow m = \pm\sqrt{2}$

$\therefore$ $$\text{C.F.} = c_1 \cos h\sqrt{2}z + c_2 \sin h\sqrt{2}z$$

$$\text{P.I.} = \frac{1}{D^2 - 2}[2(1 - z^2)] = \frac{2}{-2 + D^2}(1 - z^2) = -\left(1 - \frac{D^2}{2}\right)^{-1}(1 - z^2)$$

$$= -\left(1 + \frac{D^2}{2}\right)(1 - z^2) = -\left[1 - z^2 + \frac{1}{2}(-2)\right]$$

$$= -[1 - z^2 - 1] = z^2$$

$\therefore$ $$y = c_1 \cosh\sqrt{2}z + c_2 \sinh\sqrt{2}z + z^2$$

Complete solution is

$$\therefore \qquad y = c_1 \cosh \sqrt{2}(\sin x) + c_2 \sinh \sqrt{2}(\sin x) + \sin^2 x$$

where c_1 and c_2 are arbitrary constants of integration. **Ans.**

Example 29. *Solve:* $(1+x)^2 \dfrac{d^2 y}{dx^2} + (1+x)\dfrac{dy}{dx} + y = 4 \cos \log (1 + x)$ *by changing the independent variable.* *(Q. Bank U.P. II Semester 2001)*

Solution. Here, we have

$$\frac{d^2 y}{dx^2} + \frac{1}{1+x}\frac{dy}{dx} + \frac{y}{(1+x)^2} = \frac{4}{(1+x)^2} \cos \log (1+x) \qquad ...(1)$$

Here,
$$P = \frac{1}{1+x}, \quad Q = \frac{1}{(1+x)^2}, \quad R = \frac{4}{(1+x)^2}$$

Choose z such that,
$$\left(\frac{dz}{dx}\right)^2 = \frac{1}{(1+x)^2}$$

$$\Rightarrow \qquad \frac{dz}{dx} = \frac{1}{1+x} \qquad ...(2)$$

Integration yields,
$$z = \log (1 + x) \qquad ...(3)$$

From (2),
$$\frac{d^2 z}{dx^2} = -\frac{1}{(1+x)^2}$$

$$\therefore \qquad P_1 = \frac{-\dfrac{1}{(1+x)^2} + \dfrac{1}{1+x}.\dfrac{1}{1+x}}{\dfrac{1}{(1+x)^2}} = 0$$

$$Q_1 = \frac{Q}{\left(\dfrac{dz}{dx}\right)^2} = 1$$

From (3)
$$R_1 = \frac{R}{\left(\dfrac{dz}{dx}\right)^2} = 4 \cos \log (1 + x) = 4 \cos z$$

Reduced equation is

$$\frac{d^2 y}{dz^2} + y = 4 \cos z$$

Auxiliary equation is
$$m^2 + 1 = 0 \Rightarrow m = \pm i$$

$$\therefore \qquad \text{C.F.} = c_1 \cos z + c_2 \sin z$$

$$\text{P.I.} = \frac{1}{D^2 + 1}(4 \cos z) = 4.\frac{z}{2} \sin z = 2z \sin z$$

Complete solution is

$$y = c_1 \cos z + c_2 \sin z + 2z \sin z$$

$$y = c_1 \cos \log (1 + x) + c_2 \sin \log (1 + x) + 2 \log (1 + x) \sin \log (1 + x)$$

Ans.

Example 30. *Solve by changing the independent variable*

$$\frac{d^2y}{dx^2}+(3\sin x-\cot x)\frac{dy}{dx}+2y\sin^2 x = e^{-\cos x}\sin^2 x$$

Solution. Here, we have

$$\frac{d^2y}{dx^2}+(3\sin x-\cot x)\frac{dy}{dx}+2y\sin^2 x = e^{-\cos x}\sin^2 x$$

Here, $P = (3\sin x - \cot x)$, $Q = 2\sin^2 x$, $R = e^{-\cos x}\sin^2 x$

Choose z such that

$$\left(\frac{dz}{dx}\right)^2 = \sin^2 x \quad \Rightarrow \quad \frac{dz}{dx} = \sin x \qquad ...(1)$$

Integration yields, $\quad z = -\cos x \qquad ...(2)$

From (1), $\quad \dfrac{d^2z}{dx^2} = \cos x$

$$\therefore \qquad P_1 = \frac{\cos x + (3\sin x - \cot x)\sin x}{\sin^2 x} = 3$$

$$Q_1 = \frac{2\sin^2 x}{\sin^2 x} = 2$$

$$R_1 = \frac{e^{-\cos x}.\sin^2 x}{\sin^2 x} = e^{-\cos x} = e^z$$

Reduced equation is

$$\frac{d^2y}{dz^2}+3\frac{dy}{dz}+2y = e^z$$

Auxiliary equation is $\quad m^2+3m+2=0 \quad \Rightarrow m = -1, -2$

$$\therefore \qquad \text{C.F.} = c_1e^{-z}+c_2e^{-2z}$$

$$\text{P.I.} = \frac{1}{D^2+3D+2}(e^z) = \frac{e^z}{6}$$

Complete solution is, $\quad y = \text{C.F.} + \text{P.I.}$

$$y = c_1e^{-z}+c_2e^{-2z}+\frac{e^z}{6}$$

$$\Rightarrow \qquad y = c_1e^{\cos x}+c_2e^{2\cos x}+\frac{e^{-\cos x}}{6}$$

where c_1 and c_2 are arbitrary constants of integration. **Ans.**

EXERCISE 4.6

Solve the following differential equations by changing the independent variable:

1. $x^4\dfrac{d^2y}{dx^2}+2x^3\dfrac{dy}{dx}+a^2y=0$ **Ans.** $y=c_1\cos\dfrac{a}{x}+c_2\sin\dfrac{a}{x}$

2. $\dfrac{d^2y}{dx^2} + \tan x \dfrac{dy}{dx} + y\cos^2 x = 0$ **Ans.** $y = c_1 \cos(\sin x) + c_2 \sin(\sin x)$

3. $x\dfrac{d^2y}{dx^2} - \dfrac{dy}{dx} - 4x^3y = 8x^3 \sin x^2$ **Ans.** $y = c_1 e^{x^2} + c_2 e^{-x^2} - \sin x^2$

4. $\dfrac{d^2y}{dx^2} + (\tan x - 1)^2 \dfrac{dy}{dx} - n(n-1)y\sec^4 x = 0$ **Ans.** $y = C_1 e^{-n\tan x} + C_2 e^{(n-1)\tan x}$

5. $\dfrac{d^2y}{dx^2} - \cot x \dfrac{dy}{dx} - y\sin^2 x = \cos x - \cos^3 x$ **Ans.** $y = C_1 e^{-\cos x} + C_2 e^{\cos x} - \cos x$

4.11 METHOD [1(a)] OF VARIATION OF PARAMETERS

Let the complementary function be $Ay_1 + By_2$ as calculated in chapter 3 or Chapter 4.

Method I. To find particular integral of

$$\frac{d^2y}{dx^2} + b\frac{dy}{dx} + cy = X \quad \text{... (1)}$$

Let complementary function $= Ay_1 + By_2$, so that y_1 and y_2 satisfy

$$\frac{d^2y}{dx^2} + b\frac{dy}{dx} + cy = 0 \quad \text{... (2)}$$

Let us assume particular integral $y = uy_1 + vy_2$, ... (3)

where u and v are unknown functions of x.

Differentiating (3) *w.r.t.* x, we have $y' = uy_1' + vy_2' + u'y_1 + v'y_2$ assuming that u, v satisfy the equation

$$u'y_1 + v'y_2 = 0 \quad \text{... (4)}$$

then $$y' = uy_1' + vy_2' \quad \text{... (5)}$$

Differentiating (5) *w.r.t.* x, we have $y'' = uy_1'' + u'y_1' + vy_2'' + v'y_2'$

Substituting the values of y, y' and y'' in (1), we get

$$(uy_1'' + u'y_1' + vy_2'' + v'y_2') + b(uy_1' + vy_2') + c(uy_1 + vy_2) = X$$

$$\Rightarrow \quad u(y_1'' + by_1' + cy_1) + v(y_2'' + by_2' + cy_2) + (u'y_1' + v'y_2') = X \quad \text{... (6)}$$

y_1 and y_2 will satisfy equation (1)

$$\therefore \quad y_1'' + by_1' + cy_1 = 0 \quad \text{... (7)}$$

and $$y_2'' + by_2' + cy_2 = 0 \quad \text{... (8)}$$

Putting the values of expressions from (7) and (8) in (6), we get

$$\Rightarrow \quad u'y_1' + v'y_2' = X \quad \text{... (9)}$$

Solving (4) and (9), we get

$$u' = \begin{vmatrix} 0 & y_2 \\ X & y_2' \end{vmatrix} \div \begin{vmatrix} y_1 & y_2 \\ y_1' & y_2' \end{vmatrix} = \frac{-y_2 X}{y_1 y_2' - y_1' y_2}$$

$$v' = \begin{vmatrix} y_1 & 0 \\ y_1' & X \end{vmatrix} \div \begin{vmatrix} y_1 & y_2 \\ y_1' & y_2' \end{vmatrix} = \frac{y_1 X}{y_1 y_2' - y_1' y_2}$$

$$\boxed{u = \int \frac{-y_2 X}{y_1 y_2' - y_1' y_2} dx}$$

$$\boxed{v = \int \frac{y_1 X}{y_1 y_2' - y_1' y_2} dx}$$

General solution = complementary function + particular integral.

Working Rule

Step 1. Find out the *C.F. i.e.*, $A y_1 + B y_2$

Step 2. Particular integral $= u y_1 + v y_2$

Step 3. Find u and v by the formulae

$$u = \int \frac{-y_2 X}{y_1 y_2' - y_1' y_2} dx, \qquad v = \int \frac{y_1 X}{y_1 y_2' - y_1' y_2} dx$$

Example 31. *Solve* $\frac{d^2y}{dx^2} + y = \text{cosec}\, x.$ (*Nagpur University, Summer* 2005)

Solution. Here, we have $(D^2 + 1)\, y = \text{cosec}\, x$

A.E. is $m^2 + 1 = 0 \Rightarrow m = \pm i$

$$\text{C.F.} = A \cos x + B \sin x$$

Here
$$y_1 = \cos x,$$
$$y_2 = \sin x$$
$$\text{P.I.} = y_1\, u + y_2\, v$$

where
$$u = \int \frac{-y_2 \,.\, \text{cosec}\, x\, dx}{y_1 \cdot y'_2 - y'_1 \cdot y_2} = \int \frac{-\sin x \,.\, \text{cosec}\, x\, dx}{\cos x\,(\cos x) - (-\sin x)(\sin x)}$$

$$= \int \frac{-\sin x \,.\, \frac{1}{\sin x}\, dx}{\cos^2 x + \sin^2 x} = -\int dx = -x$$

$$v = \int \frac{y_1 \,.\, X\, dx}{y_1 \cdot y'_2 - y'_1\, y_2} = \int \frac{\cos x \,.\, \text{cosec}\, x\, dx}{\cos x\,(\cos x) - (-\sin x)(\sin x)}$$

$$= \int \frac{\cos x \,.\, \frac{1}{\sin x}}{\cos^2 x + \sin^2 x} dx = \int \frac{\cot x\, dx}{1} = \log \sin x$$

$$P.I. = uy_1 + vy_2 = -x \cos x + \sin x\, (\log \sin x)$$

General solution is
$$y = \text{C.F.} + \text{P.I.}$$
$$y = A \cos x + B \sin x - x \cos x + \sin x\, .(\log \sin x) \quad \textbf{Ans.}$$

Example 32. *Apply the method of variation of parameters to solve*

$$\frac{d^2y}{dx^2} + y = \tan x$$

(*A. M. I. E. T. E., Winter 2001, Summer 2000, U.P. Second Semester, 2009*)

Solution. We have, $\frac{d^2y}{dx^2} + y = \tan x$

$\Rightarrow \quad (D^2 + 1)y = \tan x$

A.E. is $\quad m^2 = -1 \quad$ or $\quad m = \pm i$

C. F. $\quad y = A \cos x + B \sin x$

Here, $\quad y_1 = \cos x,$

$y_2 = \sin x$

$$y_1 \cdot y'_2 - y'_1 \cdot y_2 = \cos x (\cos x) - (-\sin x) \sin x = \cos^2 x + \sin^2 x = 1$$

P.I. $= u. y_1 + v. y_2$ where

$$u = \int \frac{-y_2 \tan x}{y_1 \cdot y'_2 - y'_1 \cdot y_2} dx = -\int \frac{\sin x \tan x}{1} dx$$

$$= -\int \frac{\sin^2 x}{\cos x} dx = -\int \frac{1-\cos^2 x}{\cos x} dx$$

$$= \int (\cos x - \sec x) dx = \sin x - \log(\sec x + \tan x)$$

$$v = \int \frac{y_1 \tan x}{y_1 \cdot y'_2 - y'_1 \cdot y_2} dx$$

$$= \int \frac{\cos x \cdot \tan x}{1} dx = \int \sin x \, dx = -\cos x$$

$$\text{P. I.} = u. y_1 + v. y_2$$

$$= [\sin x - \log(\sec x + \tan x)] \cos x - \cos x \sin x$$

$$= -\cos x \log(\sec x + \tan x)$$

Complete solution is $\quad y = A \cos x + B \sin x - \cos x \log(\sec x + \tan x)$ **Ans.**

Example 33. *Use variation of parameters method to solve* $y'' + y = \sec x$

(*Nagpur University, Winter,* 2002, 2001, *U. P. Second Semester* 2002, *AMIE Summer* 2004)

Solution. We have, $\frac{d^2y}{dx^2} + y = \sec x$ ... (1)

A. E. is $\quad m^2 + 1 = 0 \quad \Rightarrow m^2 = -1 \quad \Rightarrow \quad m = \pm i$

Complementary Function of (1) is

C. F. $= A \cos x + B \sin x$

Here $\quad y_1 = \cos x,\ y_2 = \sin x$

P. I. $= uy_1 + vy_2$ ... (2)

where $u = \int \frac{-y_2 \sec x}{y_1 \cdot y'_2 - y'_1 \cdot y_2} dx$ $\quad \left[\begin{array}{l} y_1 y'_2 - y_1' y_2 = \cos x \cos x - (-\sin x) \sin x \\ = \cos^2 x + \sin^2 x = 1 \end{array} \right]$

On putting the values of y_2 and $y_1 y_2' - y_1' y_2$, we get

$$u = \int \frac{-\sin x \sec x}{1} dx = -\int \tan x \, dx = \log \cos x$$

and $$v = \int \frac{y_1 \sec x}{y_1 y'_2 - y'_1 y_2} dx$$

On putting the values of y_1 and $y_1y_2' - y_1'y_2$, we get

$$v = \int \frac{\cos x \cdot \sec x}{1}\, dx = \int dx = x$$

Putting the values of u and v in (2), we get

$P.\ I. = \cos x \cdot \log \cos x + x \sin x$

Complete solution is

$$y = C.\ F. + P.\ I.$$

$\Rightarrow \quad y = A \cos x + B \sin x + \cos x \log \cos x + x \sin x$ **Ans.**

Example 34. *Obtain general solution of the differential equation* $x^2 y'' + xy' - y = x^3 e^x$.

(*Nagpur University, Summer* 2004, *U. P. II, Semester, Summer* 2002)

Solution. The given differential equation is $x^2 y'' + xy' - y = x^3 e^x$

$$\Rightarrow \quad x^2 \frac{d^2 y}{dx^2} + x \frac{dy}{dx} - y = x^3 e^x \qquad \text{...(1)}$$

Putting $x = e^z \Rightarrow D = \frac{d}{dz},\ x\frac{dy}{dx} = Dy,\ x^2 \frac{d^2 y}{dx^2} = D(D-1)\, y,$ in (1), we get

$$D(D-1)\, y + D\, y - y = e^{3z} e^{e^z}$$

$$\Rightarrow \quad (D^2 - 1) y = e^{3z} e^{e^z}$$

A. E. is $m^2 - 1 = 0 \Rightarrow m = \pm 1$

$\therefore \quad$ C. F. $= C_1 e^z + C_2 e^{-z}$

$\Rightarrow \quad$ C. F. $= C_1 x + C_2 x^{-1} = C_1 x + \dfrac{C_2}{x}$

P.I. $= uy_1 + vy_2$, where $y_1 = e^{-z},\ y_2 = e^{-z} \left[y_1 = x_1,\ y_2 = \frac{1}{x} \right]$

Also $u = -\displaystyle\int \frac{y_2 z}{y_1y_2' - y_1'y_2}\, dz = -\int \frac{e^{-z}.e^{3z}.e^{e^z}}{e^z(-e^{-z}) - e^z(e^{-z})}\, dz = -\int \frac{e^{2z}e^{e^z}}{-1-1}\, dz$

$$= \frac{1}{2}\int e^{2z} e^{e^z}\, dz = \int x^2 e^x \frac{dx}{x} = \frac{1}{2}\int x e^x\, dx \qquad \left[x = e^z,\ dx = e^z dz \quad dz = \frac{dx}{e^z} = \frac{dx}{x} \right.$$

$$= \frac{1}{2}\left[xe^x - (1)\, e^x \right] = \frac{1}{2}(xe^x - e^x)$$

and $v = \displaystyle\int \frac{y_1 z}{y_1y_2' - y_1'y_2}\, dz = \int \frac{e^z.e^{3z}.e^{e^z}}{e^z(-e^{-z}) - e^z(e^{-z})}\, dz = \int \frac{e^{4z}e^{e^z}}{-1-1}\, dz = \int \frac{x^4 e^x}{-2} \frac{dx}{x} = -\frac{1}{2}\int x^3 e^x dx$

$$= -\frac{1}{2}[x^3 e^x - 3x^2 e^x + 6x e^x - 6e^x]$$

P.I. $= uy_1 + vy_2 = \dfrac{1}{2}(xe^x - e^x)\, x - \dfrac{1}{2}(x^3 e^x - 3x^2 e^x + 6xe^x - 6e^x)\dfrac{1}{x}$

$$= \frac{1}{2}\left[x^2 - x - x^2 + 3x - 6 + \frac{6}{x} \right] e^x = \frac{1}{2}\left(2x - 6 + \frac{6}{x} \right) e^x = \left(x - 3 + \frac{3}{x} \right) e^x$$

Complete solution = C.F. + P.I.

$$y=(C_1e^z+C_2e^{-z})+\left(x-3+\frac{3}{x}\right)e^x$$

$$=C_1x+\frac{C_2}{x}+\left(x-3+\frac{3}{x}\right)e^x$$ **Ans.**

Example 35. *Solve by method of variation of parameters:*

$$\frac{d^2y}{dx^2}-y=\frac{2}{1+e^x}$$ (*Uttarakhand, II Semester, June* 2007, *A.M.I.E.T.E., Summer* 2001)

(*U.P. II sem; summer 2001, Nagpur University, Summer* 2001)

Solution. We have $\frac{d^2y}{dx^2}-y=\frac{2}{1+e^x}$

A. E. is $m^2-1=0 \Rightarrow m^2=1 \Rightarrow m=\pm 1$

$$C.F.=C_1e^x+C_2e^{-x}$$

$\therefore$ $P.I.=uy_1+vy_2$

Here, $y_1=e^x, \quad y_2=e^{-x}$

and $y_1 \cdot y'_2-y'_1 \cdot y_2=-e^x \cdot e^{-x}-e^x.e^{-x}=-2$

$$u=\int\frac{-y_2X}{y_1.y'_2-y'_1.y_2}dx=-\int\frac{e^{-x}}{-2}\times\frac{2}{1+e^x}dx$$

$$=\int\frac{e^{-x}}{1+e^x}dx=\int\frac{dx}{e^x(1+e^x)}=\int\left(\frac{1}{e^x}-\frac{1}{1+e^x}\right)dx$$

$$=\int e^{-x}dx-\int\frac{e^{-x}}{e^{-x}+1}dx=-e^{-x}+\log(e^{-x}+1)$$

$$v=\int\frac{y_1X}{y_1.y'_2-y'_1.y_2}dx=\int\frac{e^x}{-2}\left[\frac{2}{1+e^x}\right]dx=-\int\frac{e^x}{1+e^x}dx=-\log(1+e^x)$$

$$\text{P.I.}=u.y_1+v.y_2=[-e^{-x}+\log(e^{-x}+1)]e^x-e^{-x}\log(1+e^x)$$

$$=-1+e^x\log(e^{-x}+1)-e^{-x}\log(e^x+1)$$

Complete solution is $y=C_1e^x+C_2e^{-x}-1+e^x\log(e^{-x}+1)-e^{-x}\log(e^x+1)$ **Ans.**

Example 36. *Apply the method of variation of parameters to solve*

$$\frac{d^2y}{dx^2}-3\frac{dy}{dx}+2y=\frac{e^x}{1+e^x}$$ (*U.P. II Semester Summer* 2005)

Solution. Given equation is $\frac{d^2y}{dx^2}-3\frac{dy}{dx}+2y=\frac{e^x}{1+e^x}$

Auxiliary equation is $m^2-3m+2=0 \Rightarrow (m-1)(m-2)=0 \Rightarrow m=1, 2$

$$\text{C. F.}=C_1e^x+C_2e^{2x}$$

$=C_1y_1+C_2y_2$, Here $y_1=e^x, \quad y_2=e^{2x}$

$$\text{P. I.}=uy_1+vy_2,$$

$$u = \int \frac{-y_2\, X\, dx}{y_1 y_2' - y_2 y_1'}$$

$$= \int \frac{-e^{2x}\, \frac{e^x}{1+e^x}}{e^x\,(2e^{2x}) - e^{2x}\,(e^x)}\, dx = \int \frac{\frac{-e^{3x}}{1+e^x}}{2e^{3x} - e^{3x}}\, dx$$

$$= \int \frac{-e^{3x}}{e^{3x}\,(1+e^x)}\, dx = -\int \frac{1}{1+e^x}\, dx$$

$$= -\int \frac{e^{-x}}{e^{-x}+1}\, dx \qquad \text{[Dividing by } e^x\text{]}$$

$$= \log(e^{-x}+1)$$

Now, $$v = \int \frac{y_1 X\, dx}{y_1 y_2' - y_2 y_1'} = \int \frac{e^x\left(\frac{e^x}{1+e^x}\right) dx}{e^x\,(2e^{2x}) - e^{2x}\,(e^x)}$$

$$= \int \frac{\frac{e^{2x}}{1+e^x}}{2e^{3x} - e^{3x}}\, dx = \int \frac{e^{2x}}{e^{3x}\,(1+e^x)}\, dx$$

$$= \int \frac{1}{e^x\,(1+e^x)}\, dx = \int \left(\frac{1}{e^x} - \frac{1}{1+e^x}\right) dx \qquad \text{[By Partial fraction]}$$

$$= \int \left(e^{-x} - \frac{e^{-x}}{e^{-x}+1}\right) dx = -e^{-x} + \log(e^{-x}+1)$$

P.I. $= uy_1 + vy_2$

P.I. $= e^x \log(e^{-x}+1) + e^{2x}\{-e^{-x} + \log(e^{-x}+1)\}$

P.I. $= e^x \log(e^{-x}+1) - e^x + e^{2x} \log(e^{-x}+1)$

Complete solution is

$y =$ C. F. + P. I.

$= C_1 e^x + C_2 e^{2x} + e^x \log(e^{-x}+1) - e^x + e^{2x} \log(e^{-x}+1)$ **Ans.**

Example 37. *Solve by method of variation of parameters*

$$\frac{d^2y}{dx^2} - y = \left(1 + \frac{1}{e^x}\right)^{-2}$$

(*Nagpur University, Summer* 2000)

Solution. $$\frac{d^2y}{dx^2} - y = \left(\frac{e^x+1}{e^x}\right)^{-2} = \frac{e^{2x}}{(e^x+1)^2}$$

A. E. is $m^2 - 1 = 0 \Rightarrow m = \pm 1$

C. F. $= C_1 e^{-x} + C_2 e^x$

Let P. I. $= uy_1 + vy_2$, where $y_1 = e^{-x}$, $y_2 = e^x$

$y_1 y_2' - y_1' y_2 = e^{-x} . e^x + e^{-x} . e^x = 2$

$$u = \int \frac{-y_2 X}{y_1 y_2' - y_1' y_2} dx = \frac{1}{2}\int -e^x . \frac{e^{2x}}{(1+e^x)^2} dx = -\frac{1}{2}\int \frac{e^{2x}}{(1+e^x)^2} e^x dx$$

Putting $t = 1 + e^x$, $dt = e^x\, dx$, we get

$$u = -\frac{1}{2}\int (t-1)^2 \frac{dt}{t^2}$$

$$u = -\frac{1}{2}\int\left(1 - \frac{2}{t} + t^{-2}\right) dt = -\frac{1}{2}(t - 2\log t - t^{-1})$$

$$u = -\frac{1}{2}(1+e^x) + \log(1+e^x) + \frac{1}{2}(1+e^x)^{-1}$$

Now, $v = \int \frac{y_1 X}{y_1 y_2' - y_1' y_2} dx = \frac{1}{2}\int \frac{e^{-x} . e^{2x}}{(1+e^x)^2} dx = \frac{1}{2}\int \frac{e^x\, dx}{(1+e^x)^2}$

$$= \frac{1}{2}\int \frac{1}{t^2} dt = -\frac{1}{2}\frac{1}{t} = -\frac{1}{2}(1+e^x)^{-1} \qquad \text{where } t = 1 + e^x$$

$$\text{P.I.} = u\, y_1 + v\, y_2$$

$$\text{P.I.} = e^{-x}\left[-\frac{1}{2}(1+e^x) + \log(1+e^x) + \frac{1}{2}(1+e^x)^{-1}\right] - \frac{1}{2}e^x (1+e^x)^{-1}$$

$$= -\frac{1}{2}(1+e^x)^{-1}\{e^x - e^{-x}\} + e^{-x}\left(-\frac{1}{2}\right)(1+e^x) + e^{-x}\log(1+e^x)$$

$$= -(1+e^x)^{-1}\sinh x - \frac{1}{2}e^{-x}(1+e^x) + e^{-x}\log(1+e^x)$$

Hence the solution is

$$y = C_1 e^{-x} + C_2 e^x - (1+e^x)^{-1}\sinh x - \frac{1}{2}e^{-x}(1+e^x) + e^{-x}\log(1+e^x) \qquad \textbf{Ans.}$$

Example 38. *Solve by the method of variation of parameters.*

$$(D^2 - 6D + 9)\, y = \frac{e^{3x}}{x^2} \qquad (\textit{M.U. II Semester 2008})$$

Solution. Here, we have

$$(D^2 - 6\, D + 9)\, y = \frac{e^{3x}}{x^2}$$

A.E. is $m^2 - 6\, m + 9 = 0$

$\Rightarrow \quad (m-3)^2 = 0, \qquad m = 3, 3$

$$\text{C.F.} = (C_1 + C_2 x)\, e^{3x}$$

$$= C_1 e^{3x} + C_2\, x\, e^{3x}$$

$$\text{P.I.} = uy_1 + vy_2 \text{ where } y_1 = e^{3x},\ y_2 = x\, e^{3x},\ X = \frac{e^{3x}}{x^2}$$

$$u = -\int \frac{y_2\, X\, dx}{y_1 y_2' - y_1' y_2} = -\int \frac{x\, e^{3x}\, \frac{e^{3x}}{x^2}}{e^{3x}\,(3x\, e^{3x} + e^{3x}) - 3\, e^{3x}\, .\, x\, e^{3x}} dx$$

$$= -\int \frac{\frac{e^{6x}}{x}}{3x\,e^{6x} + e^{6x} - 3x\,e^{6x}}\,dx = -\int \frac{\frac{e^{6x}}{x}}{e^{6x}}\,dx = -\int \frac{1}{x}\,dx = -\log x$$

$$v = \int \frac{y_1 X}{y_1 y_2' - y_1' y_2}\,dx$$

$$= \int \frac{e^{3x} \cdot \frac{e^{3x}}{x^2}}{e^{3x}\,(3x\,e^{3x} + e^{3x}) - 3\,e^{3x} \cdot x\,e^{3x}}\,dx$$

$$= \int \frac{\frac{e^{6x}}{x^2}}{3x\,e^{6x} + e^{6x} - 3x\,e^{6x}}\,dx = \int \frac{\frac{e^{6x}}{x^2}}{e^{6x}}\,dx = \int \frac{1}{x^2}\,dx = -\frac{1}{x}$$

$$\text{P.I.} = u\,y_1 + v\,y_2$$

$$= (-\log x)\,e^{3x} + \left(-\frac{1}{x}\right) x\,e^{3x} = e^{3x}\,(-\log x - 1) = -e^{3x}\log ex \qquad [\log_e e = 1]$$

Complete solution is y = C.F. + P.I.

$$y = (C_1\,e^{3x} + C_2\,x\,e^{3x}) - e^{3x}\log e\,x$$

$$= e^{3x}\,[C_1 + C_2\,x - \log ex]$$

Ans.

Example 39. *Solve by method of variation of parameters*

$$\frac{d^2y}{dx^2} - 2\frac{dy}{dx} = e^x \sin x$$

(MDU, Dec. 2009, *U.P. II Semester,* 2003)

Solution. We have, $\frac{d^2y}{dx^2} - 2\frac{dy}{dx} = e^x \sin x$

$\Rightarrow$ $(D^2 - 2D)y = e^x \sin x$

A. E. is $m^2 - 2m = 0 \quad \Rightarrow \quad m\,(m - 2) = 0 \quad \Rightarrow \quad m = 0, 2$

$$\text{C. F.} = C_1 + C_2\,e^{2x}$$

P. I. $= u\,y_1 + v\,y_2$ where, $y_1 = 1, \quad y_2 = e^{2x}$

$$\therefore \quad u = \int \frac{-y_2 X\,dx}{y_1 y_2' - y_1' y_2} = \int \frac{-e^{2x} \cdot e^x \sin x}{1\,(2e^{2x}) - 0\,(e^{2x})}\,dx$$

$$= -\frac{1}{2}\int e^x \sin x\,dx \qquad \left[\int e^{ax} \sin b\,x\,dx = \frac{e^{ax}}{a^2 + b^2}(a \sin bx - b \cos b\,x)\right]$$

$$= -\frac{1}{2}\,\frac{e^x}{1+1}\{\sin x - \cos x\} = -\frac{e^x}{4}(\sin x - \cos x)$$

$$v = \int \frac{y_1 X\,dx}{y_1 y_2' - y_1' y_2}$$

$$= \int \frac{1 \cdot e^x \sin x\,dx}{1\,(2e^{2x}) - 0\,(e^{2x})} = \frac{1}{2}\int e^{-x} \sin x\,dx$$

$$= \frac{1}{2}\,\frac{e^{-x}}{(-1)^2 + (1)^2}[-\sin x - \cos x]$$

$$= -\frac{1}{2}\,\frac{e^{-x}}{2}(\sin x + \cos x) = -\frac{e^{-x}}{4}(\sin x + \cos x)$$

$$\text{P. I.} = u\,y_1 + v\,y_2 = -\frac{e^x}{4}(\sin x - \cos x)\,1 - \frac{e^{-x}}{4}(\sin x + \cos x)\,e^{2x}$$

$$= -\frac{e^x}{4}[\sin x - \cos x + \sin x + \cos x] = -\frac{e^x}{2}\sin x$$

The complete solution is

$$y = C.\,F. + P.\,I.$$

$$\Rightarrow \quad y = C_1 + C_2 e^{2x} - \frac{e^x}{2}\sin x$$ **Ans.**

Example 40. *Solve :* $\dfrac{d^2y}{dx^2} + 3\dfrac{dy}{dx} + 2y = \sin(e^x)$ *(Nagpur University, Winter* 2003)

Solution. The given equation can be written as

$$(D^2 + 3D + 2)\,y = \sin e^x$$

A. E. is $(m^2 + 3m + 2) = 0 \Rightarrow (m+1)(m+2) = 0 \Rightarrow m = -1, -2,$

$$\therefore \quad \text{C. F.} = C_1 e^{-x} + C_2 e^{-2x}$$

$$\text{P. I.} = u\,y_1 + v\,y_2$$

where $y_1 = e^{-x}, \quad y_2 = e^{-2x}$

$$y_1' = -e^{-x}, \quad y_2' = -2e^{-2x}$$

$$y_1 y_2' - y_2 y_1' = e^{-x}(-2e^{-2x}) - e^{-2x}(-e^{-x}) = -2e^{-3x} + e^{-3x} = -e^{-3x}$$

$$u = \int \frac{-y_2 X}{y_1 y'_2 - y'_1 y_2}\,dx = \int \frac{-e^{-2x}}{-e^{-3x}}\sin(e^x)\,dx$$

$$= \int e^x \sin(e^x)\,dx = -\cos(e^x)$$

$$v = \int \frac{y_1 X}{y_1 y'_2 - y'_1 y_2}\,dx = \int \frac{e^{-x}\sin(e^x)\,dx}{-e^{-3x}}$$

$$= -\int e^{2x}\sin(e^x)\,dx = -\int t \sin t\,dt \qquad [t = e^x \text{ so that } dt = e^x\,dx]$$

$$= t\cos t - \sin t$$

$$= e^x \cos(e^x) - \sin(e^x)$$

Putting the values of u, v, y_1 and y_2 in (1), we get

$$\text{P.I.} = e^{-x}[-\cos(e^x)] + e^{-2x}[e^x \cos(e^x) - \sin(e^x)] = -e^{-2x}\sin(e^x)$$

The solution is $y = C_1 e^{-x} + C_2 e^{-2x} - e^{-2x}\sin(e^x)$. **Ans.**

Example 41. *Solve by method of variation of parameters:*

$$\frac{d^2y}{dx^2} + y = (x - \cot x)$$ *(Nagpur University, Winter* 2001)

Solution. Here A. E. is $m^2 + 1 = 0 \Rightarrow m = \pm i$

$$C.\,F. = C_1 \cos x + C_2 \sin x = C_1 y_1 + C_2 y_2$$

where

$$y_1 = \cos x, \qquad y_2 = \sin x$$

$$y'_1 = -\sin x, \qquad y'_2 = \cos x$$

$$y_1 y'_2 - y_2 y'_1 = \cos x \,.\cos x + \sin x .\, \sin x = 1$$

Let $$\text{P. I.} = uy_1 + vy_2 \qquad \ldots (1)$$

Where $$u = \int \frac{-y_2 X}{y_1 y_2' - y_2 y_1'} dx = \int \frac{-\sin x(x - \cot x)}{1} dx$$

$$= \int \cos x\, dx - \int x \sin x\, dx$$

$$= \sin x - \{x(-\cos x) + \sin x\} = x \cos x$$

$$v = \int \frac{y_1 X}{y_1 y_2' - y_2 y_1'} dx = \int \cos x\,(x - \cot x)\, dx$$

$$= \int x \cos x\, dx - \int \frac{\cos^2 x}{\sin x} dx = \int x \cos x\, dx - \int \frac{1 - \sin^2 x}{\sin x} dx$$

$$= \int x \cos x\, dx - \int \operatorname{cosec} x\, dx + \int \sin x\, dx$$

$$= x \sin x + \cos x - \log(\operatorname{cosec} x - \cot x) - \cos x$$

$$= x \sin x - \log(\operatorname{cosec} x - \cot x)$$

Putting the values of u, v, y_1, y_2 in (1), we get

$$\text{P. I.} = \cos x . x \cos x + \sin x\,\{x \sin x - \log(\operatorname{cosec} x - \cot x)\}$$

$$= x \cos^2 x + x \sin^2 x - \sin x \log(\operatorname{cosec} x - \cot x)$$

$$= x - \sin x \log(\operatorname{cosec} x - \cot x)$$

Hence, complete solution is

$$y = C_1 \cos x + C_2 \sin x + x - \sin x \log(\operatorname{cosec} x - \cot x) \qquad \textbf{Ans.}$$

Example 42. *Solve* $\dfrac{d^2y}{dx^2} + 2\dfrac{dy}{dx} + 2y = e^{-x} \sec^3 x.$ *(Nagpur University, Winter* 2000)

Solution. A. E. is $m^2 + 2m + 2 = 0$.

$$m = \frac{-2 \pm \sqrt{4-8}}{2} = \frac{-2 \pm 2i}{2} = -1 + i$$

C. F. $= e^{-x}(C_1 \cos x + C_2 \sin x) = C_1 y_1 + C_2 y_2$

where $$y_1 = e^{-x} \cos x \Rightarrow y_1' = -e^{-x} \sin x - e^{-x} \cos x$$

$$y_2 = e^{-x} \sin x \Rightarrow y_2' = -e^{-x} \sin x + e^{-x} \cos x$$

$$y_1 y_2' - y_1' y_2 = e^{-x} \cos x(-e^{-x} \sin x + e^{-x} \cos x) - (-e^{-x} \cos x - e^{-x} \sin x)\, e^{-x} \sin x$$

$$= e^{-2x}(\sin^2 x + \cos^2 x - \sin x \cos x + \sin x \cos x)$$

$$= e^{-2x}(\sin^2 x + \cos^2 x) = e^{-2x}$$

Let $$\text{P.I.} = uy_1 + vy_2$$

where $$u = -\int \frac{y_2 X}{y_1 y_2' - y_1' y_2} dx = -\int \frac{e^{-x} \sin x . e^{-x} \sec^3 x}{e^{-2x}} dx$$

$$= -\int \sin x . \sec^3 x\, dx = -\int \tan x \cdot \sec^2 x\, dx = -\frac{\tan^2 x}{2}$$

$$v = \int \frac{y_1 X}{y_1 y_2' - y_1' y_2} dx = \int \frac{e^{-x} \cos x . e^{-x} \sec^3 x}{e^{-2x}} dx$$

$$= \int \cos x . \sec^3 x \, dx = \int \sec^2 x \, dx = \tan x$$

$$\text{P. I.} = uy_1 + vy_2$$

$$= \frac{-\tan^2 x}{2} e^{-x} \cos x + \tan x . e^{-x} \sin x$$

$$= e^{-x}\left[\frac{-\tan^2 x}{2} \cos x + \tan x . \sin x\right]$$

$$= e^{-x}\left[\frac{-\sin x . \tan x}{2} + \tan x . \sin x\right] = \frac{1}{2} e^{-x} \sin x \tan x$$

Complete solution is

$$y = C.F. + P.I.$$

$$= e^{-x}(C_1 \cos x + C_2 \sin x) + \frac{1}{2} e^{-x} \sin x \tan x \qquad \textbf{Ans.}$$

Example 43. *Solve, by method of variation of parameters*

$$\frac{d^2 y}{d x^2} - y = e^{-2x} \sin (e^{-x})$$ (*Nagpur University, Summer* 2008)

Solution. Here, we have

$$\frac{d^2 y}{d x^2} - y = e^{-2x} \sin (e^{-x})$$

$$\Rightarrow \quad (D^2 - 1) y = e^{-2x} \sin (e^{-x})$$

A. E. is $\quad m^2 - 1 = 0 \Rightarrow m = \pm 1$

$$\text{C. F.} = C_1 e^{x} + C_2 e^{-x}$$

$$\text{P. I.} = uy_1 + vy_2 \quad \text{where} \quad y_1 = e^x, \quad y_2 = e^{-x}$$

$$u = -\int \frac{y_2 X}{y_1 y'_2 - y'_1 y_2} dx = -\int \frac{e^{-x} e^{-2x} \sin (e^{-x})}{e^x(-e^{-x}) - e^x e^{-x}} dx$$

$$= \frac{1}{2}\int e^{-2x} e^{-x} \sin (e^{-x}) \, dx = -\frac{1}{2}\int t^2 \sin t \, dt \qquad [\text{Put } e^{-x} = t]$$

$$= -\frac{1}{2}[t^2(-\cos t) - 2t(-\sin t) + 2\cos t]$$

$$= \frac{1}{2}[t^2 \cos t - 2t \sin t - 2\cos t]$$

$$= \frac{1}{2}[e^{-2x} \cos e^{-x} - 2(e^{-x}) \sin (e^{-x}) - 2\cos(e^{-x})]$$

$$v = \int \frac{y_1 X}{y_1 y'_2 - y'_1 y_2} dx = \int \frac{e^x e^{-2x} \sin (e^{-x})}{e^x(-e^{-x}) - e^x . e^{-x}} dx$$

$$= -\frac{1}{2}\int e^{-x} \sin e^{-x} \, dx = \frac{1}{2}\int \sin t \, dt$$

$$= -\frac{1}{2}\cos t = -\frac{1}{2}\cos e^{-x}$$

$$\text{P. I.} = uy_1 + vy_2$$

$$= \frac{1}{2}[e^{-2x} \cos e^{-x} - 2e^{-x} \sin e^{-x} - 2\cos e^{-x}] e^x + \left(-\frac{1}{2}\cos e^{-x}\right) e^{-x}$$

$$=-\frac{1}{2}[-e^{-x}\cos e^{-x}+2\sin e^{-x}+2e^{x}\cos e^{-x}+e^{-x}\cos e^{-x}]$$

$$=-\frac{1}{2}[2\sin e^{-x}+2e^{x}\cos e^{-x}]=-[\sin e^{-x}+e^{x}\cos e^{-x}]$$

Complete solution is y = C. F. + P. I.

$$y=C_1e^{x}+C_2e^{-x}-[\sin e^{-x}+e^{x}\cos e^{-x}]$$ **Ans.**

Example 44. *Apply the method of variation of parameters to solve*

$$\frac{d^2y}{dx^2}-y=e^{-x}\sin(e^{-x})+\cos(e^{-x})$$

(Nagpur University, Summer 2002; A.M.I.E.T.E., Winter, 2002)

Solution. The auxiliary equation is $m^2-1=0$

$$\therefore \quad m=\pm 1$$

$$\therefore \quad C.\,F.=C_1e^{x}+C_2e^{-x}=C_1y_1+C_2y_2$$

where $y_1=e^{x}$ and $y_2=e^{-x}$

Let P. I. $=uy_1+vy_2$

where $$u=\int\frac{-y_2X}{y_1y'_2-y'_1y_2}dx$$

$$=-\int\frac{e^{-x}\{e^{-x}\sin(e^{-x})+\cos(e^{-x})\}}{e^{x}(-e^{-x})-e^{x}e^{-x}}dx$$

$$=\frac{1}{2}\int[e^{-2x}\sin(e^{-x})+e^{-x}\cos(e^{-x})]\,dx \qquad (\text{Put } e^{-x}=t,\ -e^{-x}dx=dt)$$

$$=-\frac{1}{2}\int(t\sin t+\cos t)\,dt=-\frac{1}{2}\{-t\cos t+\sin t+\sin t\}$$

$$=\frac{1}{2}[t\cos t-2\sin t]$$

$$=\frac{1}{2}[e^{-x}\cos(e^{-x})-2\sin(e^{-x})]$$

$$v=\int\frac{y_1X}{y_1y'_2-y'_1y_2}dx=\int\frac{e^{x}\{e^{-x}\sin(e^{-x})+\cos(e^{-x})\}}{e^{x}(-e^{-x})-e^{x}e^{-x}}dx$$

$$=-\frac{1}{2}\int\{\sin(e^{-x})+e^{x}\cos(e^{-x})\}dx$$

$$\left[e^{-x}=t \text{ and } -e^{-x}dx=dt \text{ i.e. } dx=-\frac{dt}{t}\right]$$

$$=\frac{1}{2}\int\left(\frac{\sin t}{t}+\frac{\cos t}{t^2}\right)dt=\frac{1}{2}\left\{\frac{-\cos t}{t}-\int\frac{\cos t}{t^2}dt+\int\frac{\cos t}{t^2}dt\right\}$$

$$=-\frac{1}{2}\frac{\cos t}{t}=-\frac{1}{2}e^{x}\cos(e^{-x})$$

$$P.I.=e^{x}\left\{\frac{1}{2}e^{-x}\cos(e^{-x})-\sin(e^{-x})\right\}-\frac{1}{2}e^{-x}e^{x}\cos(e^{-x})$$

$$=-e^{x}\sin(e^{-x})$$

$\therefore$ Required solution is y = C. F. + P. I.

i.e. $$y = C_1 e^x + C_2 e^{-x} - e^x \sin(e^{-x})$$ **Ans.**

Example 45. *Use the method of variation of parameters to solve*

$$y'' + 3y' + 2y = e^{e^x}$$ (*M.U. II Semester 2009*)

Solution. Here, we have

$$y'' + 3y' + 2y = e^{e^x}$$

$$\Rightarrow \quad (D^2 + 3D + 2)y = e^{e^x}$$

A.E. is $m^2 + 3m + 2 = 0$

$$\Rightarrow \quad (m+1)(m+2) = 0 \quad \Rightarrow \quad m = -1, m = -2$$

$$\text{C.F.} = C_1 e^{-x} + C_2 e^{-2x}$$

$$\text{P.I.} = u\, y_1 + v\, y_2, \text{ where } y_1 = e^{-x} \text{ and } y_2 = e^{-2x}, X = e^{e^x}$$

$$u = \int \frac{-y_2 X\, dx}{y_1 y_2' - y_1' y_2} = \int \frac{-e^{-2x} e^{e^x} dx}{e^{-x}(-2e^{-2x}) - (-e^{-x})(e^{-2x})}$$

$$= \int \frac{-e^{-2x} e^{e^x} dx}{-2e^{-3x} + e^{-3x}} = \int \frac{-e^{-2x} e^{e^x} dx}{-e^{-3x}}$$

$$= \int e^x e^{e^x} dx \qquad [e^x = t \text{ so that } e^x dx = dt]$$

$$= \int e^t dt = e^t = e^{e^x}$$

$$v = \int \frac{y_1 X\, dx}{y_1 y_2' - y_1' y_2} = \int \frac{e^{-x} e^{e^x} dx}{e^{-x}(-2e^{-2x}) - (-e^{-x})(e^{-2x})}$$

$$= \int \frac{e^{-x} e^{e^x} dx}{-2e^{-3x} + e^{-3x}} = \int \frac{e^{-x} e^{e^x} dx}{-e^{-3x}} = \int -e^{2x} e^{e^x} dx$$

$$= -\int e^x e^x e^{e^x} dx \qquad [\text{Put } e^x = t \Rightarrow e^x dx = dt]$$

$$= -\int t e^t dt = -\left[t e^t - \int 1 . e^t dt\right]$$

$$= -t e^t + e^t$$

$$= -e^x e^{e^x} + e^{e^x} = e^{e^x}(1 - e^x)$$

$$\text{P.I.} = u\, y_1 + v\, y_2 = e^{e^x} e^{-x} + e^{e^x}(1 - e^x) e^{-2x}$$

$$= e^{e^x}[e^{-x} + (1 - e^x) e^{-2x}]$$

$$= e^{e^x}[e^{-x} + e^{-2x} - e^{-x}] = e^{e^x} e^{-2x}$$

Complete solution is y = C.F. + P.I.

$$\Rightarrow \quad y = C_1 e^{-x} + C_2 e^{-2x} + e^{e^x} e^{-2x}$$ **Ans.**

Example 46. *Solve* $(D^2 + 2D + 1)\, y = 4e^{-x} \log x$ *by method of variation of parameters.* *(Nagpur University, Winter* 2004)

Solution. $(D^2 + 2D + 1)\, y = 4e^{-x} \log x$

A. E. is $m^2 + 2m + 1 = 0 \Rightarrow (m+1)^2 = 0 \Rightarrow m = -1, -1$

$$\therefore \quad C.F. = (C_1 + C_2 x)\, e^{-x}$$

$$\Rightarrow \quad C.F. = C_1 e^{-x} + C_2 x e^{-x} = C_1 y_1 + C_2 y_2$$

$$y_1 = e^{-x} \quad \Rightarrow \quad y'_1 = -e^{-x}$$

$$y_2 = xe^{-x} \quad \Rightarrow \quad y'_2 = -xe^{-x} + e^{-x}$$

$$y_1 y'_2 - y'_1 y_2 = e^{-x}(-xe^{-x} + e^{-x}) + e^{-x}(x e^{-x})$$

$$= -xe^{-2x} + e^{-2x} + xe^{-2x} = e^{-2x}$$

Let $\quad$ P. I. $= uy_1 + vy_2 \qquad$ ... (1)

where
$$u = -\int \frac{y_2 X}{y_1 y'_2 - y'_1 y_2}\, dx = -\int \frac{xe^{-x}\, 4e^{-x} \log x}{e^{-2x}}\, dx$$

$$= -4 \int x \log x\, dx = -4\left[\log x \frac{x^2}{2} - \int \frac{1}{x} \frac{x^2}{2}\, dx\right]$$

$$= -4\left[\frac{x^2}{2} \log x - \frac{1}{2}\frac{x^2}{2}\right] = -2x^2 \log x + x^2$$

$$u = x^2 (1 - 2 \log x)$$

and
$$v = \int \frac{y_1 X}{y_1 y'_2 - y'_1 y_2}\, dx = \int \frac{e^{-x} . 4\, e^{-x} \log x}{e^{-2x}}\, dx$$

$$= 4 \int \log x\, dx = 4 \int 1 \cdot \log x\, dx = 4\left[(\log x)\, x - \int \frac{1}{x} . x\, dx\right]$$

$$= 4\,[x \log x - x]$$

$$v = 4x (\log x - 1)$$

Putting the values of u, v, y_1 and y_2 in (1), we get

$$\text{P. I.} = x^2 (1 - 2 \log x)\, e^{-x} + 4x (\log x - 1) x\, e^{-x}$$

$$= x^2 e^{-x} [1 - 2 \log x + 4 \log x - 4]$$

$$\text{P.I.} = x^2 e^{-x} [2 \log x - 3]$$

Complete solution is $\quad y =$ C. F. + P. I.

$$\Rightarrow \quad y = (C_1 + C_2 x)\, e^{-x} + x^2 e^{-x} [2 \log x - 3]$$ **Ans.**

4.12 METHOD [1 (*b*)]

Here, the method to find out C.F. is different from the methods discussed earlier. Now, we will find C.F. by new methods and then will apply the method of variation of parameters as discussed in Article 4.11 on page 134.

Example 47. *By the method of variation of parameters, solve the differential equation.*

$$\frac{d^2 y}{dx^2} + (1 - \cot x)\frac{dy}{dx} - y \cot x = \sin^2 x$$ *(U.P., II Semester, Summer 2002)*

Solution. First we shall find the C.F. of the given equation *i.e.,* the solution of the equation.

$$\frac{d^2y}{dx^2}+(1-\cot x)\frac{dy}{dx}-y\cot x=0 \qquad \left[\frac{d^2y}{dx^2}+P\frac{dy}{dx}+Qy=0\right] \quad ...(1)$$

Here, $P = 1 - \cot x,\ Q = -\cot x$

Now, $1 - P + Q = 1 - (1 - \cot x) - \cot x = 0$, $\therefore\ y = e^{-x}$ is a part of the C.F.

Putting $y = v\,e^{-x}$, $\dfrac{dy}{dx} = -ve^{-x} + e^{-x}\dfrac{dv}{dx}$

$\dfrac{d^2y}{dx^2} = e^{-x}\dfrac{d^2v}{dx^2} - 2e^{-x}\dfrac{dv}{dx} + ve^{-x}$ the equation (1) reduces to

$$\frac{d^2v}{dx^2}-(1+\cot x)\frac{dv}{dx}=0$$

$$\Rightarrow \quad \frac{dP}{dx}-(1+\cot x)P=0 \qquad \text{where } P=\frac{dv}{dx}$$

$$\Rightarrow \quad \frac{dP}{P}=(1+\cot x)\,dx$$

Integrating, we get $\log P = x + \log\sin x + \log C_1$

$\Rightarrow \quad \log P - \log\sin x - \log C_1 = x$

$$\Rightarrow \quad \log\frac{P}{C_1\sin x}=x$$

$$\Rightarrow \quad \frac{P}{C_1\sin x}=e^x \ \Rightarrow\ P=C_1e^x\sin x$$

$$\therefore \quad P=\frac{dv}{dx}=C_1e^x\sin x$$

$$\Rightarrow \quad v=C_1\int e^x\sin x\,dx \qquad \left[\int e^{ax}\sin bx\,dx=\frac{e^{ax}}{a^2+b^2}[a\sin bx-b\cos bx]\right]$$

$\Rightarrow \quad v = C_1.\dfrac{1}{2}e^x(\sin x - \cos x) + C_2 =$ only one part of C.F. of equation (1),

R.H.S. of the given differential equation (1) is zero.

$\therefore$ The solution of (1) *i.e.*, C.F. of the given equation is

$$\text{C.F.}=ve^{-x}=C_1.\frac{1}{2}e^x(\sin x-\cos x)e^{-x}+C_2e^{-x}=\frac{C_1}{2}(\sin x-\cos x)+C_2e^{-x}$$

$$y_1=(\sin x-\cos x),\ y_2=e^{-x}$$

$$\text{P. I.}=Uy_1+Vy_2$$

$$P.I.=U(\sin x-\cos x)+V\,.\,e^x \quad ...(2)$$

where $$U=\int\frac{-y_2X}{y_1y_2'-y_1'y_2}dx,$$

$$=\int\frac{-e^{-x}.\sin^2 x}{(\sin x-\cos x)(-e^{-x})-(\cos x+\sin x)e^{-x}}dx$$

$$= \int \frac{-\sin^2 x}{-\sin x + \cos x - \cos x - \sin x} dx = \int -\frac{\sin^2 x}{-2\sin x} dx$$

$$= \frac{1}{2} \int \sin x \, dx = -\frac{1}{2} \cos x$$

$$V = \int \frac{y_1 X}{y_1 y_2' - y_1' y_2} dx = \int \frac{(\sin x - \cos x).\sin^2 x}{(\sin x - \cos x)(-e^{-x}) - (\cos x + \sin x) e^{-x}} dx$$

$$= \int \frac{(\sin x - \cos x)\sin^2 x. e^x}{-\sin x + \cos x - \cos x - \sin x} dx = \int \frac{(\sin x - \cos x)\sin^2 x. e^x \, dx}{-2\sin x}$$

$$= \frac{1}{2} \int (-\sin x + \cos x)\sin x \, e^x \, dx = \frac{1}{2} \int (-\sin^2 x \, e^x + \sin x \cos x \, e^x) \, dx$$

$$= \frac{1}{2} \int \frac{1}{2} (\cos 2x - 1) e^x \, dx + \frac{1}{4} \int \sin 2x. e^x dx$$

$$= \frac{1}{4} \int \cos 2x. e^x \, dx - \frac{1}{4} \int e^x dx + \frac{1}{4} \int \sin 2x \, e^x dx$$

$$= \frac{1}{4} \frac{e^x}{1+4} [\cos 2x + 2\sin 2x] - \frac{1}{4} e^x + \frac{1}{4} \frac{e^x}{1+4} [\sin 2x - 2\cos 2x]$$

$$= \frac{e^x}{20} [\cos 2x + 2\sin 2x + \sin 2x - 2\cos 2x] - \frac{e^x}{4}$$

$$= \frac{e^x}{20} [-\cos 2x + 3\sin 2x] - \frac{e^x}{4}$$

On putting the values of U and V in (2), we get

$$\text{P.I.} = Uy_1 + Vy_2$$

$$\text{P.I.} = -\frac{1}{2} \cos x (\sin x - \cos x) + \left[\frac{e^x}{20} (-\cos 2x + 3\sin 2x) - \frac{e^x}{4} \right] e^{-x}$$

$$= -\frac{1}{2} \sin x \cos x + \frac{1}{2} \cos^2 x + \frac{1}{20} (-\cos 2x + 3\sin 2x) - \frac{1}{4}$$

$$= -\frac{1}{4} \sin 2x + \frac{1}{2} \cos^2 x - \frac{1}{20} \cos 2x + \frac{3}{20} \sin 2x - \frac{1}{4}$$

$$= -\frac{1}{10} \sin 2x + \frac{1}{4} (\cos 2x + 1) - \frac{1}{20} \cos 2x - \frac{1}{4}$$

$$= -\frac{1}{10} \sin 2x + \frac{1}{5} \cos 2x$$

$$y = \text{C.F.} + \text{P.I.}$$

$$y = C_1' (\sin x - \cos x) + C_2 e^{-x} - \frac{1}{10} \sin 2x + \frac{1}{5} \cos 2x$$

$$y = C_1' (\sin x - \cos x) + C_2 e^{-x} - \frac{1}{10} (\sin 2x - 2\cos 2x)$$ **Ans.**

4.13 SECOND METHOD. The above question can be solved by the following method

Let $y = A(\sin x - \cos x) + Be^{-x}$ be the complete solution of the given equation where A and B are functions of x, so chosen that the given equation will be satisfied.

$$\therefore \quad \frac{dy}{dx} = A(\cos x + \sin x) - Be^{-x} + \frac{dA}{dx}(\sin x - \cos x) + \frac{dB}{dx}e^{-x}. \quad ...(3)$$

Let us choose A and B such that

$$\frac{dA}{dx}(\sin x - \cos x) + \frac{dB}{dx}e^{-x} = 0. \quad ...(4)$$

Now (3) becomes

$$\therefore \quad \frac{dy}{dx} = A(\cos x + \sin x) - Be^{-x}$$

and $$\frac{d^2y}{dx^2} = \frac{dA}{dx}(\cos x + \sin x) - \frac{dB}{dx}e^{-x} + A(-\sin x + \cos x) + Be^{-x}$$

Putting these values of $\frac{d^2y}{dx^2}, \frac{dy}{dx}$ and y in the given equation, we get

$$\frac{dA}{dx}(\cos x + \sin x) - \frac{dB}{dx}e^{-x} + A(-\sin x + \cos x) + Be^{-x}$$
$$+ (1 - \cot x)[A(\cos x + \sin x) - Be^{-x}] - [A(\sin x - \cos x) + Be^{-x}]\cot x = \sin^2 x$$

$$\Rightarrow \quad \frac{dA}{dx}(\cos x + \sin x) - \frac{dB}{dx}e^{-x} = \sin^2 x \quad ...(5)$$

Solving (4) and (5), we get

$$\frac{dA}{dx} = \frac{1}{2}\sin x \quad \text{and} \quad \frac{dB}{dx} = \frac{1}{2}e^{x}(\sin x \cos x - \sin^2 x)$$
$$= \frac{e^x}{4}(\sin 2x + \cos 2x - 1)$$

Integrating these, we get

$$A = -\frac{1}{2}\cos x + C_1$$

and $$B = \frac{1}{4}\int e^x(\sin 2x - 1 + \cos 2x)\,dx$$
$$= \frac{1}{4}\int e^x \sin 2x\,dx - \frac{1}{4}\int e^x\,dx + \frac{1}{4}\int e^x \cos 2x\,dx$$
$$= \frac{1}{4}\cdot\frac{e^x}{5}(\sin 2x - 2\cos 2x) - \frac{e^x}{4} + \frac{1}{4}\cdot\frac{e^x}{5}(\cos 2x + 2\sin 2x) + C_2$$
$$= \frac{e^x}{20}(3\sin 2x - \cos 2x) - \frac{e^x}{4} + C_2$$

Putting the values of A and B in $y = A(\sin x - \cos x) + Be^{-x}$, the general solution of the given equation is

$$y = \left(-\frac{1}{2}\cos x + C_1\right)(\sin x - \cos x) + \left\{\frac{e^x}{20}(3\sin 2x - \cos 2x) - \frac{e^x}{4} + C_2\right\}e^{-x}$$

$$= -\frac{1}{2}\cos x \sin x + \frac{1}{2}\cos^2 x + C_1 \sin x - C_1 \cos x + \frac{3}{20}\sin 2x - \frac{1}{20}\cos 2x - \frac{1}{4} + C_2 e^{-x}$$

$$= -\frac{1}{4}(\sin 2x) + \frac{1}{4}(\cos 2x + 1) + C_1(\sin x - \cos x) + \frac{3}{20}\sin 2x - \frac{1}{20}\cos 2x - \frac{1}{4} + C_2 e^{-x}$$

$$y = C_1(\sin x - \cos x) + C_2 e^{-x} - \frac{1}{10}(\sin 2x - 2\cos 2x)$$ **Ans.**

Example 48. *Solve by the method of variation of parameters:*

$$x^2 \frac{d^2y}{dx^2} - 2x(1+x)\frac{dy}{dx} + 2(1+x)\,y = x^3.$$

Solution. Dividing the given equation by x^2, we get

$$\frac{d^2y}{dx^2} - \frac{2(1+x)}{x}\frac{dy}{dx} + \frac{2(1+x)}{x^2}y = x. \quad ...(1)$$

Firstly, we shall find the C.F. of (1) *i.e.*, the complete solution of the following differential equation

$$\frac{d^2y}{dx^2} - \frac{2(1+x)}{x}\frac{dy}{dx} + \frac{2(1+x)}{x^2}y = 0 \quad ...(2)$$

Here $$P = -\frac{2(1+x)}{x}, Q = \frac{2(1+x)}{x^2}, R = 0$$

Clearly, $$P + Qx = -\frac{2(1+x)}{x} + \frac{2(1+x)}{x} = 0$$

$\therefore$ $y = x$ is part of C.F.

Let $y = vx$ be complete primitive of (2). So putting $y = vx$ in (2), the reduced equation is

$$\frac{d^2v}{dx^2} + \left(\frac{2}{u}\frac{du}{dx} + P\right)\frac{dv}{dx} = \frac{R}{u}$$

i.e., $$\frac{d^2v}{dx^2} + \left\{\frac{2}{x}.1 - \frac{2(1+x)}{x}\right\}\frac{dv}{dx} = 0$$

$$\Rightarrow \quad \frac{d^2v}{dx^2} - \frac{2\,dv}{dx} = 0 \Rightarrow (D^2 - 2D)v = 0 \quad [D \equiv d/dx]$$

$$\Rightarrow \quad D(D-2)\,v = 0, \text{ A.E. is } m(m-2) = 0 \Rightarrow m = 0, 2$$

$\Rightarrow$ C.F. $= C_1 + C_2 e^{2x}$ $\quad v = C_1 + C_2 e^{2x}$

Thus from $y = vx$, the complete solution of (2) *i.e.*, the C.F. of equation (1) is given by

$$y = C_1 x + C_2 xe^{2x} \Rightarrow C.F. = C_1 y_1 + C_2 y_2$$

First Method.

$$\text{P.I.} = u\,y_1 + v\,y_2$$
$$\text{P.I.} = u\,x + v\,x\,e^{2x} \quad ...(3)$$

where $$u = \int -\frac{y_2 X}{y_1 y_2' - y_1' y_2}dx \text{ and } v = \int \frac{y_1 X}{y_1 y_2' - y_1' y_2}dx$$

$$u = \int \frac{-xe^{2x}.x}{x(e^{2x} + 2x\,e^{2x}) - 1.(xe^{2x})}dx$$

$$= \int \frac{-x^2}{x+2x^2-x}\,dx = -\int \frac{x^2}{2x^2}\,dx = -\frac{1}{2}\int dx = -\frac{x}{2}$$

$$v = \int \frac{x.x}{x(e^{2x}+2x\,e^{2x})-1.(x\,e^{2x})}\,dx$$

$$= \int \frac{x}{e^{2x}+2x\,e^{2x}-e^{2x}}\,dx = \frac{1}{2}\int \frac{1}{e^{2x}}\,dx = \frac{1}{2}\int e^{-2x}\,dx = -\frac{e^{-2x}}{4}$$

On putting the values of u and v in (3), we get

$$\text{P.I.} = -\frac{x}{2}x - \frac{e^{-2x}}{4}(x\,e^{2x}) = -\frac{x^2}{2} - \frac{x}{4}$$

$$y = \text{C.F.} + \text{P.I.} = C_1 x + C_2 x\,e^{2x} - \frac{x^2}{2} - \frac{x}{4} \qquad \textbf{Ans.}$$

SECOND METHOD.

Now suppose that the complete solution of (1) is given by

$$y = Ax + B\,x\,e^{2x} \qquad ...(4)$$

where A and B are functions of x.

$$\therefore \quad \frac{dy}{dx} = A + Be^{2x} + 2Bxe^{2x} + \frac{dA}{dx}.x + \frac{dB}{dx}xe^{2x} = A + Be^{2x} + 2Bxe^{2x} \qquad ...(5)$$

$$\text{where,} \quad \frac{dA}{dx} + \frac{dB}{dx}.e^{2x} = 0 \qquad ...(6)$$

Differentiating (5), we have

$$\frac{d^2y}{dx^2} = \frac{dA}{dx} + \left(\frac{dB}{dx}e^{2x} + 2\,Be^{2x}\right) + \left(2\frac{dB}{dx}x\,e^{2x} + 2\,B\,e^{2x} + 4B\,e^{2x}\right)$$

$$= \left(\frac{dA}{dx} + e^{2x}\frac{dB}{dx}\right) + \left(4\,Be^{2x} + 4\,B\,x\,e^{2x} + 2xe^{2x}\frac{dB}{dx}\right)$$

$$= 0 + 4\,B\,e^{2x} + 4\,B\,x\,e^{2x} + 2x\,e^{2x}\frac{dB}{dx} \qquad ...(7) \text{ [Using (6)]}$$

Substituting the values of $\frac{d^2y}{dx^2}, \frac{dy}{dx}$ and y from (7), (5) and (4) respectively in (1), we get

$$\frac{dA}{dx} + e^{2x}(2x+1)\frac{dB}{dx} = x \qquad ...(8)$$

Solving (6) and (8), we get

$$\frac{dA}{dx} = -\frac{1}{2}, \qquad \frac{dB}{dx} = \frac{1}{2}e^{-2x}$$

$$\text{Integrating} \qquad A = -\frac{1}{2}x + C_1, \qquad B = -\frac{1}{4}e^{-2x} + C_2$$

Substituting the values of A and B in (4), the complete solution of (1) is given by

$$y = \left(-\frac{1}{2}x + C_1\right)x + \left(-\frac{1}{4}e^{-2x} + C_2\right)xe^{2x} = C_1 x + C_2 xe^{2x} - \frac{1}{2}x^2 - \frac{1}{4}x. \qquad \textbf{Ans.}$$

EXERCISE 4.7

Solve the following equations by variation of parameters method.

1. $\dfrac{d^2y}{dx^2}-4y=e^{2x}$ **Ans.** $y=C_1e^{2x}+C_2e^{-2x}+\dfrac{x}{4}e^{2x}-\dfrac{e^{2x}}{16}$

2. $\dfrac{d^2y}{dx^2}+y=\sin x$ **Ans.** $y=C_1\cos x+C_2\sin x-\dfrac{x}{2}\cos x+\dfrac{1}{4}\sin x$

3. $\dfrac{d^2y}{dx^2}-3\dfrac{dy}{dx}+2y=\sin x$ **(MDU, 2010)** **Ans.** $y=C_1e^{x}+C_2e^{2x}+\dfrac{1}{10}(3\cos x+\sin x)$

4. $\dfrac{d^2y}{dx^2}+y=\sec x\tan x$ **Ans.** $y=C_1\cos x+C_2\sin x+x\cos x+\sin x\log\sec x-\sin x$

5. Using method of variation of parameters show that A can always be determined so that $y=Ax^2e^{\lambda x}$ is a solution of the differential equation $y''-2\lambda y'+\lambda^2y=2e^{\lambda x}$ *(A.M.I.E.T.E., Summer 2005)*

6. $(1-x)\dfrac{d^2y}{dx^2}+x\dfrac{dy}{dx}-y=(1-x)^2$. **Ans.** $y=C_1e^x+C_2x+x^2+x+1$

7. $(1-x^2)\dfrac{d^2y}{dx^2}-4x\dfrac{dy}{dx}-(1+x^2)y=x$ **Ans.** $y=\dfrac{1}{1-x^2}(C_1\cos x+C_2\sin x+x)$

8. $(1-x^2)\dfrac{d^2y}{dx^2}+x\dfrac{dy}{dx}-y=x(1-x^2)^{\frac{3}{2}}$ **Ans.** $y=C_1(\sqrt{1-x^2}+x\sin^{-1}x)+C_2x-\dfrac{x}{9}(1-x^2)^{3/2}$

9. $x\dfrac{d^2y}{dx^2}-(2x+1)\dfrac{dy}{dx}+(x+1)y=(x^2+x-1)e^{2x}$

   Ans. $y=C_1x+C_2x^{-1}+x\int e^xx^{-3}(x^2-2x+2)\,dx$

10. $x\dfrac{d^2y}{dx^2}-2(x+1)\dfrac{dy}{dx}+(x+2)y=(x-2)e^x$ **Ans.** $y=\dfrac{C_1}{3}x^3e^x+C_2e^x+\left(x-\dfrac{x^2}{2}\right)e^x.$

11. $x^2\dfrac{d^2y}{dx^2}+\dfrac{dy}{dx}-y=x^2e^x$ **Ans.** $y=C_1x+\dfrac{C_2}{x}+e^x-\dfrac{e^x}{x},\ (x>0)$

OBJECTIVE TYPE QUESTIONS

Choose the correct alternative:

1. $y=x$ is a part of complementary function of the equation $\dfrac{d^2y}{dx^2}+P\dfrac{dy}{dx}+Qy=R$

   If: *(i)* $1+P+Q=0$ *(ii)* $1-P+Q=0$
   (iii) $2+2Px+Qx^2=0$ *(iv)* $P+Qx=0$ **Ans.** *(iv)*

   (R.G.P.V., Bhopal, II Semester, June 2007)

2. $y=e^{-x}$ is a part of C.F. of $\dfrac{d^2y}{dx^2}+P\dfrac{dy}{dx}+Qy=R$

   If: *(i)* $1+P+Q=0$ *(ii)* $1-P-Q=0$
   (iii) $1-P+Q=0$ *(iv)* $1+P-Q=0$ **Ans.** *(iii)*

   (R.G.P.V., Bhopal, II Semester, June 2007)

3. P.I. of $\dfrac{d^2y}{dx^2}+9y=\sin 3x$ is

   (i) $-\dfrac{x}{6}\cos 3x$ *(ii)* $\dfrac{x}{6}\cos 3x$

(*iii*) $\frac{x}{6}\sin 3x$ (*iv*) $-\frac{x}{6}\sin 3x$ **Ans.** (*i*)

Fill in the blanks:

4. The integrating factor of $\cos^2 x \frac{dy}{dx} + y = \tan x$ is .... **Ans.** $e^{\tan x}$

5. The integrating factor of $x(x-1)\frac{dy}{dx} - (x-2)y = x^2(2x-1)$ is ....

Ans. $\frac{x-1}{x^3}$

6. Solution of $(x+y+1)\frac{dy}{dx} = 1$ is .... **Ans.** $x + y + 2 = ce^x$

7. $M\,dx + N\,dy$ is an exact differential equation if .... **Ans.** $\frac{\partial M}{\partial y} = \frac{\partial N}{\partial x}$

8. P. I. of $(D^2 + 4)\,y = \sin 3x$ is .... **Ans.** $-\frac{1}{5}\sin 3x$

9. P. I. of $(D^2 - 2D + 1)\,y = e^x$ is .... **Ans.** $\frac{x^2}{2}.e^x$

10. On putting $x = e^z$, the transformed differential equation of $x^2\frac{d^2y}{dx^2} + x\frac{dy}{dx} + y = x$ is......

Ans. $\frac{d^2y}{dx^2} + y = e^x$

11. If the C. F. of $ay'' + by' + cy = X$ is $Ay_1 + By_2$, then P. I. $= uy_1 + vy_2$ where $u =$ .... and $v =$ ....

Ans. $u = \int \frac{-y_2 . X\,dx}{y_1y'_2 - y'_1y_2}$ and $v = \int \frac{y_1 . X\,dx}{y_1y'_2 - y'_1y_2}$

12. In the equation $\frac{d^2y}{dx^2} + P\frac{dy}{dx} + Qy = R$, if we obtain $1 + \frac{P}{2} + \frac{Q}{4} = 0$, then the part of C.F. is...

Ans. e^{2x}

13. Normal form of second order differential equation $\frac{d^2y}{dx^2} + P\frac{dy}{dx} + Qy = R$ is given by

$\frac{d^2v}{dx^2} + Av = B$, where $A =$ .... and $B =$ .... **Ans.** $A = Q - \frac{1}{2}\frac{dP}{dx} - \frac{P^2}{4}$, $B = \frac{R}{u}$

14. If the equation $\frac{d^2y}{dx^2} + P\frac{dy}{dx} + Qy = R$ is transformed by changing the independent variable *i.e.*,

$\frac{d^2y}{dx^2} + P_1\frac{dy}{dx} + Q_1y = R_1$, where $P_1 =$ ...., $Q_1 =$ ...., $R_1 =$ ....

Ans. $P_1 = \frac{\frac{d^2z}{dx^2} + P\frac{dz}{dx}}{\left(\frac{dz}{dx}\right)^2}$, $Q_1 = \frac{Q}{\left(\frac{dz}{dx}\right)^2}$, $R_1 = \frac{R}{\left(\frac{dz}{dx}\right)^2}$

15. If C.F. of $\frac{d^2y}{dx^2} + P\frac{dy}{dx} + Qy = R$ is $c_1y_1 + c_2y_2$ then P.I. $= uy_1 + vy_2$ where $u =$ .... and $v =$ ....

Ans. $u = \int \frac{-y_2.R}{y_1y_2' - y_1'y_2}dx$, $v = \int \frac{y_1.R}{y_1y_2' - y_1'y_2}dx$

UNIT–II

5 CHAPTER LAPLACE TRANSFORM

5.1 INTRODUCTION

Laplace transforms help in solving the differential equations with boundary values without finding the general solution and the values of the arbitrary constants.

5.2 LAPLACE TRANSFORM

Definition. Let $f(t)$ be function defined for all positive values of t, then

$$F(s) = \int_0^{\infty} e^{-st} f(t)\,dt$$

provided the integral exists, is called the **Laplace Transform** of $f(t)$. It is denoted as

$$L[f(t)] = F(s) = \int_0^{\infty} e^{-st} f(t)\,dt$$

5.3 IMPORTANT FORMULAE

1. $L(1) = \frac{1}{s}$

2. $L(t^n) = \frac{n!}{s^{n+1}}$, when $n = 0, 1, 2, 3...$

3. $L(e^{at}) = \frac{1}{s-a}$ $\quad (s > a)$

4. $L(\cosh at) = \frac{s}{s^2 - a^2}$ $\quad (s^2 > a^2)$

5. $L(\sinh at) = \frac{a}{s^2 - a^2}$ $\quad (s^2 > a^2)$

6. $L(\sin at) = \frac{a}{s^2 + a^2}$ $\quad (s > 0)$

7. $L(\cos at) = \frac{s}{s^2 + a^2}$ $\quad (s > 0)$

1. $L(1)=\frac{1}{s}$

Proof. $L(1) = \int_0^{\infty} 1 \cdot e^{-st}\,dt = \left[\frac{e^{-st}}{-s}\right]_0^{\infty} = -\frac{1}{s}\left[\frac{1}{e^{st}}\right]_0^{\infty} = -\frac{1}{s}[0-1] = \frac{1}{s}$

Hence $L(1) = \frac{1}{s}$ **Proved.**

2. $L(t^n)=\frac{n!}{s^{n+1}}$ **where n and s are positive.**

Proof. $L(t^n) = \int_0^{\infty} e^{-st} t^n\,dt$

Putting $st = x$ or $t = \frac{x}{s}$ or $dt = \frac{dx}{s}$

Thus, we have $L(t^n) = \int_0^{\infty} e^{-x}\left(\frac{x}{s}\right)^n \frac{dx}{s} \Rightarrow L(t^n) = \frac{1}{s^{n+1}}\int_0^{\infty} e^{-x}.x^n\,dx$

$\Rightarrow L(t^n) = \frac{\left|\overline{n+1}\right.}{s^{n+1}} \Rightarrow L(t^n) = \frac{n!}{s^{n+1}}$ $\left[\begin{array}{l}\left|\overline{n+1}\right. = \int_0^{\infty} e^{-x}.x^n\,dx \\ \text{and} \quad \left|\overline{n+1}\right. = n!\end{array}\right]$ **Proved.**

3. $L(e^{at})=\frac{1}{s-a}$ **where $s > a$**

Proof. $L(e^{at}) = \int_0^{\infty} e^{-st}.e^{at}\,dt = \int_0^{\infty} e^{-st+at}.dt$

$= \int_0^{\infty} e^{(-s+a)t}.dt = \int_0^{\infty} e^{-(s-a)t}.dt = \left[\frac{e^{-(s-a)t}}{-(s-a)}\right]_0^{\infty} = -\frac{1}{s-a}\left[\frac{1}{e^{(s-a)t}}\right]_0^{\infty}$

$= \frac{-1}{(s-a)}(0-1) = \frac{1}{s-a}$ **Proved.**

4. $L(\cosh at)=\frac{s}{s^2-a^2}$

Proof. $L(\cosh at) = L\left[\frac{e^{at}+e^{-at}}{2}\right]$ $\left(\therefore \cosh at = \frac{e^{at}+e^{-at}}{2}\right)$

$= \frac{1}{2}L(e^{at}) + \frac{1}{2}L(e^{-at})$

$= \frac{1}{2}\left[\frac{1}{s-a}+\frac{1}{s+a}\right]$ $\left[L(e^{at}) = \frac{1}{s-a}\right]$

$= \frac{1}{2}\left[\frac{s+a+s-a}{s^2-a^2}\right] = \frac{s}{s^2-a^2}$ **Proved.**

5. $\boxed{\mathbf{L(\sinh at) = \frac{a}{s^2 - a^2}}}$

Proof. $\mathrm{L}\ (\sinh\ at) \quad = \mathrm{L}\left[\frac{1}{2}\left(e^{at} - e^{-at}\right)\right]$

$$= \frac{1}{2}[L(e^{at}) - L(e^{-at})] = \frac{1}{2}\left[\frac{1}{s-a} - \frac{1}{s+a}\right] = \frac{1}{2}\left[\frac{s+a-s+a}{s^2-a^2}\right]$$

$$= \frac{a}{s^2-a^2}$$ **Proved.**

6. $\boxed{\mathbf{L(\sin at) = \frac{a}{s^2 + a^2}}}$

Proof. $\mathrm{L}\ (\sin\ at) = \mathrm{L}\left[\frac{e^{iat} - e^{-iat}}{2i}\right]$ $\left[\because \sin at = \frac{e^{iat} - e^{-iat}}{2i}\right]$

$$= \frac{1}{2i}\left[L\left(e^{iat} - e^{-iat}\right)\right] = \frac{1}{2i}\left[L\left(e^{iat}\right) - L\left(e^{-iat}\right)\right]$$

$$= \frac{1}{2i}\left[\frac{1}{s-ia} - \frac{1}{s+ia}\right] = \frac{1}{2i}\frac{s+ia-s+ia}{s^2+a^2}$$

$$= \frac{1}{2i}\frac{2ia}{s^2+a^2} = \frac{a}{s^2+a^2}$$ **Proved.**

7. $\boxed{\mathbf{L(\cos at) = \frac{s}{s^2 + a^2}}}$

Proof. $\mathrm{L}(\cos at) = \mathrm{L}\left(\frac{e^{iat} + e^{-iat}}{2}\right)$ $\left[\because \cos at = \frac{e^{iat} + e^{-iat}}{2}\right]$

$$= \frac{1}{2}[\mathrm{L}(e^{iat} + e^{-iat})] = \frac{1}{2}[\mathrm{L}(e^{iat}) + \mathrm{L}(e^{-iat})]$$

$$= \frac{1}{2}\left[\frac{1}{s-ia} + \frac{1}{s+ia}\right] = \frac{1}{2}\frac{s+ia+s-ia}{s^2+a^2}$$

$$= \frac{s}{s^2+a^2}$$ **Proved.**

Example 1. *Find the Laplace transform of f(t) defined as*

$$f(t) = \begin{cases} \dfrac{t}{k}, & when\ \ 0 < t < k \\ 1, & when\ \ t > k \end{cases}$$

Solution. $L\ [f(t)] = \int_0^k \frac{t}{k} e^{-st} dt + \int_k^\infty 1.e^{-st} dt = \frac{1}{k}\left[\left(t\frac{e^{-st}}{-s}\right)_0^k - \int_0^k \frac{e^{-st}}{-s} dt\right] + \left[\frac{e^{-st}}{-s}\right]_k^\infty$

$$= \frac{1}{k}\left[\frac{ke^{-ks}}{-s} - \left(\frac{e^{-st}}{s^2}\right)_0^k\right] + \frac{e^{-ks}}{s} = \frac{1}{k}\left[\frac{ke^{-ks}}{-s} - \frac{e^{-sk}}{s^2} + \frac{1}{s^2}\right] + \frac{e^{-ks}}{s}$$

$$= -\frac{e^{-sk}}{s} - \frac{1}{k}\frac{e^{-ks}}{s^2} + \frac{1}{k}\frac{1}{s^2} + \frac{e^{-ks}}{s} = \frac{1}{ks^2}[-e^{-ks}+1]$$ **Ans.**

Example 2. *Find the Laplace transform of the function* $f(t) = \begin{cases} t-1, & 1<t<2 \\ 3-t, & 2<t<3 \end{cases}$.

(U.P., II Semester, 2009)

Solution. The given function is periodic with period 3.

$$L[f(t)] = \int_1^3 f(t)\,e^{-st}\,dt$$

$$= \left[\int_1^2 f(t)\,e^{-st}\,dt + \int_2^3 f(t)\,e^{-st}\,dt\right]$$

$$= \left[\int_1^2 (t-1)\,e^{-st}\,dt + \int_2^3 (3-t)\,e^{-st}\,dt\right]$$

$$= \left[\left\{(t-1)\frac{e^{-st}}{-s} - \frac{e^{-st}}{(-s)^2}\right\}_1^2 + \left\{(3-t)\frac{e^{-st}}{-s} + \frac{e^{-st}}{(-s)^2}\right\}_2^3\right]$$

$$= \left[\left\{\frac{e^{-2s}}{-s} - \frac{e^{-2s}}{s^2} + \frac{e^{-s}}{s^2}\right\} + \left\{\frac{e^{-3s}}{s^2} - \frac{e^{-2s}}{-s} - \frac{e^{-2s}}{s^2}\right\}\right]$$

$$= \left[-\frac{e^{-2s}}{s} - \frac{e^{-2s}}{s^2} + \frac{e^{-s}}{s^2} + \frac{e^{-3s}}{s^2} + \frac{e^{-2s}}{s} - \frac{e^{-2s}}{s^2}\right]$$

$$= \left[\frac{1}{s^2}(-e^{-2s} + e^{-s} + e^{-3s} - e^{-2s})\right]$$

$$= \frac{1}{s^2}[e^{-s} - 2e^{-2s} + e^{-3s}]$$ **Ans.**

Example 3. *Find the Laplace transform of* $F(t) = \begin{cases} 1, & 0 \le t < 1 \\ t, & 1 \le t < 2 \\ t^2, & 2 \le t < \infty \end{cases}$ *(Q. Bank U.P. 2001)*

Solution. Here, we have $F(t) = \begin{cases} 1, & 0 \le t < 1 \\ t, & 1 \le t < 2 \\ t^2, & 2 \le t < \infty \end{cases}$

$$L[F(t)] = \int_0^\infty e^{-st}\cdot F(t)dt = \int_0^1 e^{-st}dt + \int_1^2 t\,e^{-st}dt + \int_2^\infty t^2 e^{-st}dt$$

$$= \left(\frac{e^{-st}}{-s}\right)_0^1 + \left(t\frac{e^{-st}}{-s} - \frac{e^{-st}}{s^2}\right)_1^2 + \left(t^2\frac{e^{-st}}{-s}\right)_2^\infty - \int_2^\infty 2t\cdot\frac{e^{-st}}{-s}dt$$

$$= \left(\frac{1-e^{-s}}{s}\right) + \left(\frac{-2}{s}e^{-2s} - \frac{e^{-2s}}{s^2}\right) - \left(\frac{e^{-s}}{-s} - \frac{e^{-s}}{s^2}\right) + \frac{4}{s}e^{-2s} + \frac{2}{s}\int_2^\infty t\,e^{-st}\,dt$$

$$= \frac{1}{s} + \frac{2}{s}e^{-2s} + \frac{e^{-s}}{s^2} - \frac{e^{-2s}}{s^2} + \frac{2}{s}\left[\left(t\frac{e^{-st}}{-s}\right)_2^\infty - \int_2^\infty 1\cdot\frac{e^{-st}}{-s}dt\right]$$

$$= \frac{1}{s} + \frac{2}{s}e^{-2s} + \frac{e^{-s}}{s^2} - \frac{e^{-2s}}{s^2} + \frac{2}{s}\left[\frac{2}{s}e^{-2s} + \frac{1}{s}\left(\frac{e^{-st}}{-s}\right)_2^{\infty}\right]$$

$$= \frac{1}{s} + \frac{2}{s}e^{-2s} + \frac{e^{-s}}{s^2} + \frac{3}{s^2}e^{-2s} + \frac{2}{s^3}e^{-2s}. \qquad \textbf{Ans.}$$

Example 4. *Find the Laplace transform of*

$$f(t) = \begin{cases} t^2, & 0 < t < 2 \\ t-1, & 2 < t < 3 \\ 7, & t > 3 \end{cases}$$

(U.P, II Semester, June 2007)

Solution. $L[f(t)] = \int_0^{\infty} e^{-st} f(t)dt = \int_0^2 t^2 e^{-st} dt + \int_2^3 (t-1)e^{-st} dt + \int_3^{\infty} 7e^{-st} dt$

$$\left[\int I\, II = I\, II_1 - I'\, II_{11} + I''\, II_{111} - \ldots\right]$$

$$= \left[t^2\left(\frac{e^{-st}}{(-s)}\right) - 2t\frac{e^{-st}}{(-s)^2} + 2\frac{e^{-st}}{(-s)^3}\right]_0^2 + \left[(t-1)\left(\frac{e^{-st}}{(-s)}\right) - \frac{e^{-st}}{(-s)^2}\right]_2^3 + 7\left[\frac{e^{-st}}{-s}\right]_3^{\infty}$$

$$= \left[-4\left(\frac{e^{-2s}}{s}\right) - 4\left(\frac{e^{-2s}}{s^2}\right) - 2\left(\frac{e^{-2s}}{s^3}\right) + \frac{2}{s^3}\right] + \left[2\left(\frac{e^{-3s}}{-s}\right) - \frac{e^{-3s}}{s^2} + \frac{e^{-2s}}{s} + \frac{e^{-2s}}{s^2}\right] + 7\left(0 + \frac{e^{-3s}}{s}\right)$$

$$= \frac{2}{s^3} + e^{-2s}\left[-\frac{4}{s} - \frac{4}{s^2} - \frac{2}{s^3}\right] + e^{-3s}\left[-\frac{2}{s} - \frac{1}{s^2}\right] + e^{-2s}\left[\frac{1}{s} + \frac{1}{s^2}\right] + e^{-3s}\left[\frac{7}{s}\right]$$

$$= \frac{2}{s^3} + e^{-2s}\left[-\frac{4}{s} - \frac{4}{s^2} - \frac{2}{s^3} + \frac{1}{s} + \frac{1}{s^2}\right] + e^{-3s}\left[-\frac{2}{s} - \frac{1}{s^2} + \frac{7}{s}\right]$$

$$= \frac{2}{s^3} + e^{-2s}\left[-\frac{3}{s} - \frac{3}{s^2} - \frac{2}{s^3}\right] + e^{-3s}\left[\frac{5}{s} - \frac{1}{s^2}\right]$$

$$= \frac{2}{s^3} - \frac{e^{-2s}}{s^3}\left(2 + 3s + 3s^2\right) + \frac{e^{-3s}}{s^2}(5s - 1) \qquad \textbf{Ans.}$$

Example 5. *Find the Laplace transform of (1 + sin 2t).*

Solution. Laplace transform of (1 + sin 2*t*)

$$= \int_0^{\infty} e^{-st}(1 + \sin 2t)\, dt = \int_0^{\infty} e^{-st}\left(1 + \frac{e^{2it} - e^{-2it}}{2i}\right) dt$$

$$= \frac{1}{2i}\int_0^{\infty} [2ie^{-st} + e^{(-s+2i)t} - e^{(-s-2i)t}]dt = \frac{1}{2i}\left[\frac{2ie^{-st}}{-s} + \frac{e^{(-s+2i)t}}{-s+2i} - \frac{e^{(-s-2i)t}}{-s-2i}\right]_0^{\infty}$$

$$= \frac{1}{2i}\left[\left(0 + \frac{2i}{s}\right) + \frac{1}{-s+2i}(0-1) - \frac{1}{-s-2i}(0-1)\right]$$

$$= \frac{1}{2i}\left[\frac{2i}{s} + \frac{1}{s-2i} - \frac{1}{s+2i}\right] = \frac{1}{2}\left[\frac{2}{s} + \frac{4}{s^2+4}\right]$$

$$= \frac{1}{s} + \frac{2}{s^2+4} \qquad \textbf{Ans.}$$

Alternate Method

$$L(1+\sin 2t) = L(1) + \text{L} \sin 2t = \frac{1}{s} + \frac{2}{s^2+4}$$ **Ans.**

5.4 PROPERTIES OF LAPLACE TRANSFORMS

(1) $\text{L}[af_1(t)+bf_2(t)] = a\,\text{L}[f_1(t)] + b\,\text{L}[f_2(t)]$

Proof. $\text{L}[af_1(t)+bf_2(t)] = \int_0^\infty e^{-st}[af_1(t)+bf_2(t)]\,dt$

$$= a\int_0^\infty e^{-st} f_1(t)\,dt + b\int_0^\infty e^{-st} f_2(t)\,dt$$

$$= a\,\text{L}[f_1(t)] + b\,[\text{L}f_2(t)]$$ **Proved.**

5.5 CHANGE OF SCALE PROPERTY

If $\text{L}\{f(t)\} = F(s)$ then $\boxed{\text{L}\{f(at)\} = \frac{1}{a}F\left(\frac{s}{a}\right)}$

Proof. $\text{L}\{f(at)\} = \int_0^\infty e^{-st} f(at)\,dt$

$$= \int_0^\infty e^{-\left(\frac{s}{a}\right)u} f(u)\frac{du}{a} \qquad \left[\text{Put } at = u \Rightarrow dt = \frac{du}{a}\right]$$

$$= \frac{1}{a}\int_0^\infty e^{-\left(\frac{s}{a}\right)u} f(u)\,du$$

$$= \frac{1}{a}\int_0^\infty e^{-\left(\frac{s}{a}\right)t} f(t)\,dt$$

$$= \frac{1}{a}\int_0^\infty e^{-St} f(t)\,dt = \frac{1}{a}\{Lf(t)\} = \frac{1}{a}F(S) \qquad \left[\text{Put } S = \frac{s}{a}\right]$$

$$= \frac{1}{a}F\left(\frac{s}{a}\right)$$ **Proved.**

Example 6. *If* $L\{J_0(\sqrt{t})\} = \frac{e^{-\frac{1}{4s}}}{s}$, *find* $L\{J_0(2\sqrt{t})\}$.

Solution. Here, we have

$$L\{J_0(\sqrt{t})\} = \frac{e^{-\frac{1}{4s}}}{s}$$

By change of scale property,

$$L\{J_0(\sqrt{4t})\} = \frac{1}{4}\cdot\left\{\frac{e^{-\frac{1}{4(s/4)}}}{(s/4)}\right\}$$

$$\Rightarrow \quad L\{J_0(2\sqrt{t})\} = \frac{1}{s}e^{-1/s}.$$ **Ans.**

(2) **First Shifting Theorem.** If $\text{L}\{f(t)\} = \text{F}(s)$, then

$$\boxed{\mathbf{L[e^{at} f(t)] = F(s-a)}}$$

Proof. $L[e^{at} f(t)] = \int_0^\infty e^{-st}.e^{at} f(t)\,dt = \int_0^\infty e^{-(s-a)t} f(t)\,dt$

$= \int_0^\infty e^{-rt} f(t)\,dt$ where $r = s - a$

$= F(r) = F(s-a)$ **Proved.**

With the help of this property, we can have the following important results :

1. $L(e^{at} t^n) = \dfrac{n!}{(s-a)^{n+1}}$

2. $L(e^{at} \cosh bt) = \dfrac{s-a}{(s-a)^2 - b^2}$

3. $L(e^{at} \sinh bt) = \dfrac{b}{(s-a)^2 - b^2}$

4. $L(e^{at} \sin bt) = \dfrac{b}{(s-a)^2 + b^2}$

5. $L(e^{at} \cos bt) = \dfrac{s-a}{(s-a)^2 + b^2}$

5.6 HEAVISIDE'S SHIFTING THEOREM (Second Translation Property)

(*a*) If $L\{f(t)\} = F(s)$ and $g(t) = \begin{cases} f(t-a), & t > a \\ 0, & 0 < t < a \end{cases}$ then prove that

$$\boxed{L\{g(t)\} = e^{-as} F(s)}$$

(U.P. II Semester, Summer 2006)

Proof. $L\{g(t)\} = \int_0^\infty e^{-st} g(t)\,dt$

$= \int_0^a e^{-st} g(t)\,dt + \int_a^\infty e^{-st} g(t)\,dt$ [$g(t) = 0$, when $0 < t < a$]

$= 0 + \int_a^\infty e^{-st}.f(t-a)\,dt$

$= \int_a^\infty e^{-st} f(t-a)\,dt$ [Put $t - a = u \Rightarrow dt = du$]

$= \int_0^\infty e^{-s(u+a)} f(u)\,du = e^{-sa}\int_0^\infty e^{-su} f(u)\,du = e^{-sa}\int_0^\infty e^{-st} f(t)\,dt$

Proved.

$$\boxed{\mathbf{L\{g(t)\} = e^{-as} F(s)}}$$

Example 7. *Find the Laplace transform of* $\cos^2 t$.

Solution. We know that $\cos 2t = 2\cos^2 t - 1$

$$\cos^2 t = \frac{1}{2}[\cos 2t + 1]$$

$$L(\cos^2 t) = L\left[\frac{1}{2}(\cos 2t + 1)\right] = \frac{1}{2}[L(\cos 2t) + L(1)]$$

$$= \frac{1}{2}\left[\frac{s}{s^2 + (2)^2} + \frac{1}{s}\right] = \frac{1}{2}\left[\frac{s}{s^2 + 4} + \frac{1}{s}\right]$$ **Ans.**

Example 8. *If* $L(\cos^2 t) = \dfrac{s^2+2}{s(s^2+4)}$, *find L (cos² at).* *(U.P., II Semester, Summer 2006)*

Solution. We have, $L(\cos^2 t) = \dfrac{s^2+2}{s(s^2+4)}$

By change of scale property, we have

$$L(\cos^2 at) = \frac{1}{a}\cdot\frac{\left(\frac{s}{a}\right)^2+2}{\frac{s}{a}\left[\left(\frac{s}{a}\right)^2+4\right]} = \frac{1}{a}\left[\frac{s^2+2a^2}{\frac{s}{a}(s^2+4a^2)}\right]$$

$$= \frac{s^2+2a^2}{s(s^2+4a^2)}$$ **Ans.**

Example 9. *Find the Laplace transform of* $t^{-\frac{1}{2}}$.

Solution. We know that $L(t^n) = \dfrac{\lceil n+1}{s^{n+1}}$

Put $n = -\dfrac{1}{2}, L\left(t^{-1/2}\right) = \dfrac{\lceil -\frac{1}{2}+1}{s^{-1/2+1}} = \dfrac{\lceil \frac{1}{2}}{\sqrt{s}} = \dfrac{\sqrt{\pi}}{\sqrt{s}}$, where $\lceil \frac{1}{2} = \sqrt{\pi}$ **Ans.**

Example 10. *Find the Laplace Transform of 2 sin 2t cos t.*

Solution. We have

$$f(t) = 2\sin 2t\cos t$$

$$= \sin\frac{2t+4t}{l} + \sin\frac{2t-4t}{l}$$

$$= \sin 6t - \sin 2t$$

$$L f(t) = L(\sin 6t) - L(\sin 4)$$

$$= \frac{6}{s^2+36} - \frac{2}{s^2+4}$$ **Ans.**

Example 11. *Find the Laplace transform of 4 sin³ t.*

Solution. We have

$$f(t) = 4\sin^3 t$$

$$= 3\sin t - \sin 3t \qquad [\sin 3t = 3\sin t - 4\sin^3 t]$$

$$L f(t) = 3 L\sin t - L\sin 3t$$

$$= \frac{3}{s^2+1} - \frac{3}{s^2+9}$$ **Ans.**

Example 12. *Find the Laplace transform of 4 cosh 2t sin 4t*

Solution. We have

$$f(t) = 4\cosh 2t\sin 4t = 4\left(\frac{e^{2t}+e^{-2t}}{2}\right)\left(\frac{e^{4it}-e^{-4it}}{2i}\right)$$

$$= \frac{1}{i}\left[e^{(2+4i)t} - e^{(2-4i)t} + e^{(-2+4i)t} - e^{(-2-4i)t}\right]$$

$$L[f(t)] = -i\left[L(e^{(2+4i)t}) - L(e^{(2-4i)t}) + L(e^{(-2+4i)t}) - L(e^{(-2-4i)t})\right]$$

$$= -i\left[\frac{1}{s-2-4i} - \frac{1}{s-2+4i} + \frac{1}{s+2-4i} - \frac{1}{s+2+4i}\right]$$

$$= -i\left[\left(\frac{1}{s-2-4i} - \frac{1}{s+2+4i}\right) - \left(\frac{1}{s-2+4i} - \frac{1}{s+2-4i}\right)\right]$$

$$= -i\left[\frac{4+8i}{s^2-(2+4i)^2} - \frac{4-8i}{s^2-(2-4i)^2}\right] \quad \textbf{Ans.}$$

EXERCISE 5.1

Find the Laplace transforms of the following:

1. $t + t^2 + t^3$ **Ans.** $\frac{1}{s^2} + \frac{2}{s^3} + \frac{6}{s^4}$

2. $\sin t \cos t$ **Ans.** $\frac{1}{s^2+4}$

3. $t^3 e^{-2t}$ **Ans.** $\frac{6}{(s+2)^4}$

4. $\sin^3 2t$ **Ans.** $\frac{48}{(s^2+4)(s^2+36)}$

5. $e^{-t}\cos^2 t$ **Ans.** $\frac{1}{2s+2} + \frac{s+1}{2s^2+4s+10}$

6. $\sin 2t \cos 3t$ **Ans.** $\frac{2(s^2-5)}{(s^2+1)(s^2+25)}$

7. $\sin 2t \sin 3t$ **Ans.** $\frac{12s}{(s^2+1)(s^2+25)}$

8. $\cos at \sinh at$ **Ans.** $\frac{1}{2}\left[\frac{s-a}{(s-a)^2+a^2} - \frac{s+a}{(s+a)^2+a^2}\right]$

9. $\sinh^3 t$ **Ans.** $\frac{6}{(s^2-1)(s^2-9)}$

10. $\cos t \cos 2t$ **Ans.** $\frac{s(s^2+5)}{(s^2+1)(s^2+9)}$

11. $\cosh at \sin at$ **Ans.** $\frac{a(s^2+2a^2)}{s^4+4a^4}$

12. $f(t) = \begin{cases} \cos\left(t - \frac{2\pi}{3}\right), & t > \frac{2\pi}{3} \\ 0, & t < \frac{2\pi}{3} \end{cases}$ **Ans.** $e^{\frac{-2\pi s}{3}} \cdot \frac{s}{s^2+1}$

5.7 EXISTENCE THEOREM

According to this theorem $\int_0^\infty e^{-st} f(t)\, dt$ exists if $\int_0^\lambda e^{-st} f(t) dt$ can actually be evaluated and its limit as $\lambda \to \infty$ exists.

Otherwise we may use the following theorem:

If $f(t)$ is continuous and $\lim_{t\to\infty} \left[e^{-at} f(t)\right]$ is finite, then Laplace transform of $f(t)$ *i.e.* $\int_0^\infty e^{-st} f(t)\, dt$ exists for $s > a$.

It should however, be kept in mind that the above foresaid conditions are sufficient but not necessary.

For example; $L\left(\frac{1}{\sqrt{t}}\right)$ exists though $\frac{1}{\sqrt{t}}$ is infinite at $t = 0$. Similarly a function $f(t)$ for which $\lim_{t\to\infty}\left[e^{-at} f(t)\right]$ is finite and having a finite discontinuity will have a Laplace transform of $s > a$.

5.8 LAPLACE TRANSFORM OF THE DERIVATIVE OF f (t)

$$L[f'(t)] = sL[f(t)] - f(0) \qquad \text{where} \qquad L[f(t)] = F(s).$$

Proof. $L[f'(t)] = \int_0^\infty e^{-st} f'(t)dt$

Integrating by parts, we get

$$L[f'(t)] = \left[e^{-st} f(t)\right]_0^\infty - \int_0^\infty (-se^{-st}) f(t)\, dt$$

$$= -f(0) + s\int_0^\infty e^{-st} f(t)dt \qquad (e^{-st} f(t) = 0, \text{ when } t = \infty)$$

$$= -f(0) + s\, Lf(t)$$

$$\boxed{L[f'(t)] = s\, Lf(t) - f(0).}$$ **Proved.**

Note. Roughly, Laplace transform of **derivative** of $f(t)$ corresponds to **multiplication** of the Laplace transform of $f(t)$ by s.

5.9 LAPLACE TRANSFORM OF DERIVATIVE OF ORDER n

$$L[f^n(t)] = s^n L[f(t)] - s^{n-1} f(0) - s^{n-2} f'(0) - s^{n-3} f''(0) - \ldots - f^{n-1}(0)$$

Proof. We have already proved in Article **5.8 that**

$$L[f'(t)] = s\, L[f(t)] - f(0) \qquad \ldots(1)$$

Replacing $f(t)$ by $f'(t)$ and $f'(t)$ by $f''(t)$ in (1), we get

$$L[f''(t)] = s\, L[f'(t)] - f'(0) \qquad \ldots(2)$$

Putting the value of $L[f'(t)]$ from (1) in (2), we have

$$L[f''(t)] = s\,[s\, L[f(t)] - f(0)] - f'(0)$$

$\Rightarrow$ $$L[f''(t)] = s^2 L[f(t)] - sf(0) - f'(0)$$

Similarly, $$L[f'''(t)] = s^3 L[f(t)] - s^2 f(0) - s f'(0) - f''(0)$$

$$L[f^{iv}(t)] = s^4 L[f(t)] - s^3 f(0) - s^2 f'(0) - s f''(0) - f'''(0)$$

$$\boxed{L[f^n(t)] = s^n L[f(t)] - s^{n-1} f(0) - s^{n-2} f'(0) - s^{n-3} f''(0) + \ldots\ldots - f^{n-1}(0)}$$

Example 13. *Given* $L\left(2\sqrt{\frac{t}{\pi}}\right) = \frac{1}{s^{3/2}}$, *show that* $L\left(\frac{1}{\sqrt{\pi t}}\right) = \frac{1}{\sqrt{s}}$. *(U.P., II Semester, 2005)*

Solution. Let $F(t) = 2\sqrt{\frac{t}{\pi}} \Rightarrow F'(t) = \frac{1}{\sqrt{\pi t}}$. Also $F(0) = 0$

$\therefore$ $$L\{F'(t)\} = s\, L\{F(t)\} - F(0) = s.\ \frac{1}{s^{3/2}} - 0$$

$\therefore$ $$L\left(\frac{1}{\sqrt{\pi t}}\right) = \frac{1}{\sqrt{s}}.$$ **Proved.**

Example 14. *Find the Laplace transform of sin* $\sqrt{t}$ *; Hence find* $L\left(\frac{\cos\sqrt{t}}{\sqrt{t}}\right)$

Solution. $\sin\sqrt{t} = \sqrt{t} - \frac{(\sqrt{t})^3}{3!} + \frac{(\sqrt{t})^5}{5!} - \ldots\ldots$

$$= t^{1/2} - \frac{t^{3/2}}{3!} + \frac{t^{5/2}}{5!} - \ldots\ldots \qquad \left[\sin x = x - \frac{x^3}{3!} + \frac{x^5}{5!} - \ldots\ldots\right]$$

$$\therefore \quad L(\sin\sqrt{t}) = L\left(t^{1/2} - \frac{t^{3/2}}{3!} + \frac{t^{5/2}}{5!} - \ldots\ldots\right)$$

$$= \frac{\Gamma 3/2}{s^{3/2}} - \frac{\Gamma 5/2}{3!s^{5/2}} + \frac{\Gamma 7/2}{5!s^{7/2}} - \ldots\ldots$$

$$= \frac{\sqrt{\pi}}{2s^{3/2}}\left\{1 - \left(\frac{1}{2^2 s}\right) + \frac{1}{2!}\left(\frac{1}{2^2 s}\right)^2 - \ldots..\right\}$$

$$= \frac{\sqrt{\pi}}{2s^{3/2}}\left\{1 - \left(\frac{1}{2^2 s}\right) + \frac{1}{2!}\left(\frac{1}{2^2 s}\right)^2 - \ldots\ldots\right\} \left[e^x = 1 + \frac{x}{1!} + \frac{x^2}{2!} + \frac{x^3}{3!} + \ldots\ldots\right]$$

$$\Rightarrow \quad L(\sin\sqrt{t}) = \frac{\sqrt{\pi}}{2s^{3/2}} e^{-(1/4s)}$$

Now, $L\left[\frac{d}{dt}(\sin\sqrt{t})\right] = s\,L(\sin\sqrt{t}) - 0 \qquad \left[\because F(0) = 0 \text{ and } L\left[\frac{d}{dt}[F(t)]\right] = sF(s)\right]$

$$L\left(\frac{\cos\sqrt{t}}{2\sqrt{t}}\right) = \frac{\sqrt{\pi}}{2\sqrt{s}} e^{-\left(\frac{1}{4s}\right)} \quad \Rightarrow \quad L\left(\frac{\cos\sqrt{t}}{\sqrt{t}}\right) = \frac{\sqrt{\pi}}{\sqrt{s}} e^{-\frac{1}{4s}} \qquad \textbf{Ans.}$$

5.10 LAPLACE TRANSFORM OF INTEGRAL OF *f(t)*

$$\boxed{L\left[\int_0^t f(t)\,dt\right] = \frac{1}{s}F(s)} \qquad \text{where} \quad L[f(t)] = F(s)$$

Proof. Let $\phi(t) = \int_0^t f(t)\,dt$ and $\phi(0) = 0$ then $\phi'(t) = f(t)$

We know the formula of Laplace transforms of $\phi'(t)$ *i.e.*

$$L[\phi'(t)] = s\,L[\phi(t)] - \phi(0)$$

$$\Rightarrow \quad L[\phi'(t)] = s\,L[\phi(t)] \qquad [\phi(0) = 0]$$

$$\Rightarrow \quad L[\phi(t)] = \frac{1}{s}L[\phi'(t)]$$

Putting the values of $\phi(t)$ and $\phi'(t)$, we get

$$L\left[\int_0^t f(t)\,dt\right] = \frac{1}{s}L[f(t)] \Rightarrow \boxed{\mathbf{L}\left[\int_0^t f(t)\,dt\right] = \frac{1}{s}F(s)} \qquad \textbf{Proved.}$$

Note. (1) Laplace transform of **Integral** of $f(t)$ corresponds to the division of the Laplace transform of $f(t)$ by s.

(2) $$\int_0^t f(t)\,dt = L^{-1}\left[\frac{1}{s}F(s)\right]$$

Example 15. *Evaluate by Laplace transform* $\int_0^{\infty} te^{-3t} \sin t \, dt$

(Uttarakhand IInd Semester, Summer 2010)

Solution. We know that

$$L \sin t = \frac{1}{s^2+1}$$

$$\Rightarrow \quad L\, e^{-3t} \sin t = \frac{1}{(s+3)^2+1} = \frac{1}{s^2+6s+10}$$

$$\Rightarrow \quad L\, t\, e^{-3t} \sin t = -\frac{d}{ds}\left(\frac{1}{s^2+6s+10}\right) = \frac{2s+6}{(s^2+6s+10)^2}$$

$$\Rightarrow \quad \int_0^{\infty} t\, e^{-3t} \sin t = \frac{2s+6}{s\left(s^2+6s+10\right)^2}$$

Ans.

5.11 LAPLACE TRANSFORM OF *t.f(t)* (Multiplication by *t*)

If $L[f(t)] = F(s)$, then

$$L[t^n f(t)] = (-1)^n \frac{d^n}{ds^n}[F(s)].$$

(U.P., II Semester, Summer 2005)

Proof. $L[f(t)] = F(s) = \int_0^{\infty} e^{-st} f(t)\, dt$ ...(1)

Differentiating (1) w.r.t. "*s*", we get

$$\frac{d}{ds}[F(s)] = \frac{d}{ds}\left[\int_0^{\infty} e^{-st} f(t)\, dt\right] = \int_0^{\infty} \frac{\partial}{\partial s}\left(e^{-st}\right) f(t)\, dt$$

$$= \int_0^{\infty} \left(-t e^{-st}\right).f(t)\, dt = \int_0^{\infty} e^{-st}\left[-t.f(t)\right] dt$$

$$= L\left[-t f(t)\right] \quad \text{or} \quad L\left[t f(t)\right] = (-1)^1 \frac{d}{ds}\left[F(s)\right]$$

Similarly, $L\left[t^2 f(t)\right] = (-1)^2 \frac{d^2}{ds^2}\left[F(s)\right]$

$$L\left[t^3 f(t)\right] = (-1)^3 \frac{d^3}{ds^3}\left[F(s)\right]$$

$$\boxed{L\left[t^n f(t)\right] = (-1)^n \frac{d^n}{ds^n}[F(s)]}$$

Proved.

5.12 INITIAL AND FINAL VALUE THEOREMS

(*a*) Initial Value Theorem. $L\{f(t)\} = F(s)$

$$\Rightarrow \quad \lim_{t \to 0} f(t) = \lim_{s \to \infty}\left[sF(s)\right], \text{ provided the limit exists.}$$

Proof. $L\{f'(t)\} = s\, L\{f(t)\} - f(0)$

$$\Rightarrow \quad \int_0^\infty e^{-st} f'(t)\,dt = s\,F(s) - f(0)$$

$$\Rightarrow \quad \lim_{s\to\infty} \int_0^\infty e^{-st} f'(t)\,dt = \lim_{s\to\infty}\left[sF(s) - f(0)\right]$$

$$\Rightarrow \quad \lim_{s\to\infty}\left[s\,F(s)\right] = f(0) + \int_0^\infty \left(\lim_{s\to\infty} e^{-st}\right) f'(t)\,dt$$

$$= f(0) + \int_0^\infty (0)\, f'(t)\,dt \qquad \left(\because \lim_{s\to\infty} e^{-st} = 0\right)$$

$$= f(0) + 0 = f(0) = \lim_{t\to 0} f(t)$$

(*b*) Final Value Theorem. $L\{f(t)\} = F(s)$

$$\Rightarrow \quad \lim_{t\to\infty} f(t) = \lim_{s\to 0}\left[sF(s)\right], \text{ provided the limits exist.}$$

Proof. $L\{f'(t)\} = sL\{f(t)\} - f(0) \quad \Rightarrow \int_0^\infty e^{-st} f'(t)\,dt = sF(s) - f(0)$

$$\Rightarrow \quad \lim_{s\to 0} \int_0^\infty e^{-st} f'(t)\,dt = \lim_{s\to 0}\left[sF(s) - f(0)\right]$$

$$\Rightarrow \quad \lim_{s\to 0}\,[s\,F(s) - f(0)] = \lim_{s\to 0} \int_0^\infty e^{-st} f'(t)\,dt$$

$$\Rightarrow \quad \lim_{s\to 0}\,[s\,F(s)] - f(0) = \lim_{s\to 0} \int_0^\infty e^{-st} f'(t)\,dt$$

$$\Rightarrow \quad \lim_{s\to 0}\left[sF(s)\right] - f(0) = \int_0^\infty \lim_{s\to 0} e^{-st} f'(t)\,dt = \int_0^\infty (1)\, f'(t)\,dt \qquad \left[\because \lim_{s\to 0} e^{-st} = 1\right]$$

$$\Rightarrow \quad \boxed{\lim_{s\to 0}\left[sF(s)\right] = \lim_{t\to\infty} f(t)}$$

Example 16. *If* $L\{F(t)\} = \dfrac{1}{s(s+\beta)}$ *then, find* $\lim\limits_{t\to\infty} F(t)$

Solution. By final-value theorem,

$$\lim_{t\to\infty} F(t) = \lim_{s\to 0} sL\{F(t)\} = \lim_{s\to 0} \frac{s}{s(s+\beta)} = \lim_{s\to 0} \frac{1}{(s+\beta)} = \frac{1}{\beta} \quad \textbf{Ans.}$$

5.13. EXPONENTIAL INTEGRAL FUNCTION $\int_t^\infty \left(\frac{e^{-x}}{x}\right) dx$

Let $f(t) = \int_t^\infty \dfrac{e^{-x}}{x}\,dx$

$$\Rightarrow \quad f'(t) = -\frac{e^{-t}}{t} \quad \Rightarrow \quad tf'(t) = -e^{-t}$$ [Here – ve sign appears due to lower limit]

Taking Laplace Transform of $tf'(t)$, we get

$$L\{t\,f'(t)\} = L\{-e^{-t}\} = -L\{e^{-t}\}$$

$$\Rightarrow \qquad -\frac{d}{ds}\left[sF(s) - f(0)\right] = -\frac{1}{s+1}$$

$$\Rightarrow \qquad \frac{d}{ds}\left[s\,F(s)\right] = \frac{1}{s+1} \qquad \left[\because f(0) = \text{constant} \therefore \frac{d}{ds}f(0) = 0\right]$$

Integrating both the sides, we get

$$s\,F(s) = \log(s+1) + C \qquad \text{...(1)}$$

Now, by final value theorem, we have

$$\lim_{s\to 0} sF(s) = \lim_{t\to\infty} f(t) \qquad \text{...(2)}$$

Hence, $\lim\limits_{s\to 0}\left[sF(s)\right] = \lim\limits_{s\to 0}[\log(s+1) + C] = 0 + C = C$ ...(3)

Also, $\lim\limits_{t\to\infty}\left[f(t)\right] = \lim\limits_{t\to\infty}\int_t^{\infty}\left(\frac{e^{-x}}{x}\right)dx = 0$ ...(4)

Putting the values of $\lim\limits_{s\to 0}\,[s\,F(s)]$ and $\lim\limits_{t\to\infty}\left[f(t)\right]$ from (3) and (4) in (2), we get

$$C = 0.$$

Hence from (1), $sF(s) = \log(s+1) \Rightarrow F(s) = \left\{\frac{\log(s+1)}{s}\right\}$

$$\Rightarrow \qquad \boxed{L\int_t^{\infty}\left(\frac{e^{-x}}{x}\right)dx = \left[\frac{\log(s+1)}{s}\right]}$$

Example 17. *Find the Laplace Transform of t sin at.*

Solution. $L(t\sin at) = L\left(t\,\frac{e^{iat} - e^{-iat}}{2i}\right) = \frac{1}{2i}[L(t.e^{iat}) - L(te^{-iat})]$

$$= \frac{1}{2i}\left[-\frac{d}{ds}\frac{1}{s-ia} + \frac{d}{ds}\frac{1}{s+ia}\right]$$

$$= \frac{1}{2i}\left[\frac{1}{(s-ia)^2} - \frac{1}{(s+ia)^2}\right] = \frac{1}{2i}\left[\frac{(s+ia)^2 - (s-ia)^2}{(s-ia)^2\,(s+ia)^2}\right]$$

$$= \frac{1}{2i}\frac{\left(s^2 + 2ias - a^2\right) - \left(s^2 - 2ias - a^2\right)}{\left(s^2 + a^2\right)^2} = \frac{1}{2i}\frac{4\,ias}{(s^2 + a^2)^2} = \frac{2\,as}{(s^2 + a^2)^2} \qquad \textbf{Ans.}$$

Example 18. *Find the Laplace transform of t sinh at.*

Solution. $L(\sinh at) = \frac{a}{s^2 - a^2}$

$$L\,[t\sinh at] = -\frac{d}{ds}\left(\frac{a}{s^2 - a^2}\right)$$

$$\Rightarrow \qquad = \frac{2as}{\left(s^2 - a^2\right)^2} \qquad \textbf{Ans.}$$

Example 19. *Find the Laplace transform of t^2 cos at*

Solution. $L(\cos at) = \dfrac{s}{s^2+a^2}$

$$L\left(t^2 \cos at\right) = (-1)^2 \frac{d^2}{ds^2}\left[\frac{s}{s^2+a^2}\right] = \frac{d}{ds}\frac{\left(s^2+a^2\right).1 - s(2s)}{\left(s^2+a^2\right)^2} = \frac{d}{ds}\frac{a^2-s^2}{\left(s^2+a^2\right)^2}$$

$$= \frac{\left(s^2+a^2\right)^2(-2s) - \left(a^2-s^2\right).2\left(s^2+a^2\right)(2s)}{\left(s^2+a^2\right)^4} = \frac{(s^2+a^2)(-2s)-(a^2-s^2)4s}{(s^2+a^2)^3}$$

$$= \frac{-2s^3 - 2a^2s - 4a^2s + 4s^3}{\left(s^2+a^2\right)^3}$$

$$= \frac{2s\left(s^2-3a^2\right)}{\left(s^2+a^2\right)^3}$$ **Ans.**

Example 20. *Obtain the Laplace transform of* $t^2e^t.\sin 4t$. *(U.P. II Semester, Summer 2002)*

Solution. L (sin 4*t*) = $\dfrac{4}{s^2+16}$,

$$L\left(e^t \sin 4t\right) = \frac{4}{(s-1)^2+16}$$

$$L\left(t\,e^t \sin 4t\right) = -\frac{d}{ds}\left(\frac{4}{s^2-2s+17}\right) = \frac{4(2s-2)}{\left(s^2-2s+17\right)^2}$$

$$L\left(t^2e^t \sin 4t\right) = -\frac{d}{ds}\left(\frac{4(2s-2)}{\left(s^2-2s+17\right)^2}\right)$$

$$= -4\frac{\left(s^2-2s+17\right)^2 2 - (2s-2)2\left(s^2-2s+17\right)(2s-2)}{\left(s^2-2s+17\right)^4}$$

$$= -4\frac{(s^2-2s+17)\,2 - 2(2s-2)^2}{(s^2-2s+17)^3} = \frac{-4\left(2s^2-4s+34-8s^2+16s-8\right)}{\left(s^2-2s+17\right)^3}$$

$$= \frac{-4\left(-6s^2+12s+26\right)}{\left(s^2-2s+17\right)^3} = \frac{8\left[3s^2-6s-13\right]}{(s^2-2s+17)^3}$$ **Ans.**

Example 21. *Find the Laplace transform of the function*

$$f(t) = te^{-t}\sin 2t$$ *(U.P. II Semester, Summer 2002)*

Solution. $L[\sin 2t] = \dfrac{2}{s^2+4}$

$$L\left[e^{-t}\sin 2t\right] = \frac{2}{(s+1)^2+4} = F(s) \text{ (say)}$$

$$L\left(te^{-t}\sin 2t\right) = -F'(s) = -\frac{d}{ds}\left[\frac{2}{(s+1)^2+4}\right] = \frac{2\cdot 2(s+1)}{\left[(s+1)^2+4\right]^2}$$

$$= \frac{4(s+1)}{\left[(s+1)^2+4\right]^2}$$ **Ans.**

Example 22. *Find*

$$L\left\{t^2 e^{2t}\sin t\right\}.$$ *(Uttarakhand, IInd Semester, Summer 2010)*

Solution. We know that

$$L[\sin t] = \frac{1}{s^2+1}$$

$$L\left[e^{2t}\sin t\right] = \frac{1}{(s-2)^2+1}$$

$$L\left[t\,e^{2t}\sin t\right] = -\frac{d}{ds}\left[\frac{1}{(s-2)^2+1}\right] = -\frac{d}{ds}\left[\frac{1}{s^2-4s+5}\right] = \frac{2s-4}{(s^2-4s+5)^2}$$

$$L\left[t^2 e^{2t}\sin t\right] = -\frac{d}{ds}\left[\frac{2s-4}{(s^2-4s+5)^2}\right]$$

$$= -\frac{(s^2-4s+5)^2(2)-(s-4)2(s^2-4s+5)(2s-4)}{(s^2-4s+5)^4}$$

$$= \frac{-2(s^2-4x+5)+2(2s-4)^2}{(s^2-4s+5)^3} = \frac{6s^2-24s+22}{(s^2-4s+5)^2}$$ **Ans.**

EXERCISE 5.2

Find the Laplace transforms of the following :

1. $t\sin at$ **Ans.** $\dfrac{2as}{\left(s^2+a^2\right)^2}$

2. $t\cosh at$ **Ans.** $\dfrac{s^2+a^2}{\left(s^2-a^2\right)^2}$

3. $t\cos t$ **Ans.** $\dfrac{s^2-1}{\left(s^2+1\right)^2}$

4. $t\cosh t$ **Ans.** $\dfrac{s^2+1}{\left(s^2-1\right)^2}$

5. $t^2\sin t$ **Ans.** $\dfrac{2\left(3s^2-1\right)}{\left(s^2+1\right)^3}$

6. $t^3 e^{-3t}$ **Ans.** $\dfrac{6}{(s+3)^4}$

7. $t\sin^2 3t$ **Ans.** $\dfrac{1}{2}\left[\dfrac{1}{s^2}-\dfrac{s^2-36}{\left(s^2+36\right)^2}\right]$

8. $t\,e^{at}\sin at$ **Ans.** $\dfrac{2a(s-a)}{\left(s^2-2as+2a^2\right)^2}$ *(Uttarakhand II Semester, 2012)*

9. $te^{-t}\cosh t$ **Ans.** $\dfrac{s^2+2s+2}{\left(s^2+2s\right)^2}$

10. $t^2e^{-2t}\cos t$ **Ans.** $\dfrac{2\left(s^3+6s^2+9s+2\right)}{\left(s^2+4s+5\right)^3}$

11. $\int_0^t e^{-2t} t \sin^3 t \, dt$ **Ans.** $\frac{3(s+2)}{2s}\left[\frac{1}{\left[(s+2)^2+9\right]^2}-\frac{1}{\left[(s+2)^2+1\right]^2}\right]$

12. If $f(t)$ is continuous, except for an ordinary discontinuity at $t = a$, $(a > 0)$ as given in the figure, then show that

$$L[f'(t)] = s[f(t)] - f(0) - e^{-as}[f(a+0) - f(a-0)]$$

(U.P. II Semester 2003)

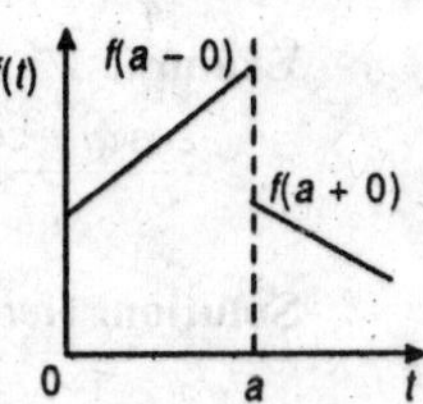

5.14 LAPLACE TRANSFORM OF $\frac{1}{t} f(t)$ (Division by t)

If $L[f(t)] = F(s)$, then $L\left[\frac{1}{t}f(t)\right] = \int_s^\infty f(s)\,ds$ *(U.P. II Semester Summer, 2007, 2005)*

Proof. We know that $L[f(t)] = F(s)$ or $F(s) = \int_0^\infty e^{-st} f(t)\,dt$ ...(1)

Integrating (1) w.r.t. 's', we have

$$\int_s^\infty F(s)\,ds = \int_s^\infty \left[\int_0^\infty e^{-st} f(t)\,dt\right] ds$$

$$= \int_0^\infty f(t)\left[\int_s^\infty e^{-st}\,ds\right] dt = \int_0^\infty f(t)\left[\frac{e^{-st}}{-t}\right]_s^\infty dt$$

$$= \int_0^\infty \frac{-f(t)}{t}\left[e^{-st}\right]_s^\infty dt = \int_0^\infty \frac{-f(t)}{t}\left[0 - e^{-st}\right] dt$$

$$= \int_0^\infty e^{-st}\left\{\frac{1}{t}.f(t)\right\} dt = L\left[\frac{1}{t}f(t)\right]$$

$$\Rightarrow \quad \boxed{L\left[\frac{1}{t}f(t)\right] = \int_0^\infty F(s)\,ds}$$ **Proved.**

Cor. $L^{-1}\int_s^\infty F(s)\,ds = \frac{1}{t} f(t)$

Example 23. *Find the Laplace transform of* $\frac{\sin 2t}{t}$

Solution. $L(\sin 2t) = \frac{2}{s^2+4}$

$$L\left(\frac{\sin 2t}{t}\right) = \int_s^\infty \frac{2}{s^2+4}\,ds = 2.\frac{1}{2}\left[\tan^{-1}\frac{s}{2}\right]_s^\infty$$

$$= \left[\tan^{-1}\infty - \tan^{-1}\frac{s}{2}\right] = \frac{\pi}{2} - \tan^{-1}\frac{s}{2}$$

$$= \cot^{-1}\frac{s}{2}$$ **Ans.**

Example 24. *Find the Laplace transform of* $f(t) = \int_0^t \frac{\sin at}{t}\,dt$

(U.P., II Semester, Summer 2005)

Solution. $L(\sin at) = \frac{a}{s^2+a^2}$

$$L\left(\frac{\sin at}{t}\right)=\int_s^\infty \frac{a}{s^2+a^2}\,ds=\left[\tan^{-1}\frac{s}{a}\right]_s^\infty=\frac{\pi}{2}-\tan^{-1}\frac{s}{a}=\cot^{-1}\frac{s}{a}$$

Hence, $$L\left[\int_0^t \frac{\sin at}{t}\,dt\right]=\frac{1}{s}\cot^{-1}\frac{s}{a}$$ **Ans.**

Example 25. *Find the Laplace transform of :*

$\dfrac{\cos at-\cos bt}{t}$ *(Uttarakhand, II Sem., 2012, June 2010, 2007, U.P., II Semester, 2004)*

Solution. Here, $$f(t)=\frac{\cos at-\cos bt}{t}$$

We know that, $L\,(\cos at-\cos bt)=L\,(\cos at)-L\,(\cos bt)=\dfrac{s}{s^2+a^2}-\dfrac{s}{s^2+b^2}$

$$L\left(\frac{\cos at-\cos bt}{t}\right)=\int_s^\infty\left(\frac{s}{s^2+a^2}-\frac{s}{s^2+b^2}\right)ds$$

$$=\left[\frac{1}{2}\log\left(s^2+a^2\right)-\frac{1}{2}\log\left(s^2+b^2\right)\right]_s^\infty$$

$$=\frac{1}{2}\left[\log\frac{s^2+a^2}{s^2+b^2}\right]_s^\infty=\frac{1}{2}\left[\log\frac{1+\frac{a^2}{s^2}}{1+\frac{b^2}{s^2}}\right]_s^\infty$$

$$=\frac{1}{2}\log 1-\frac{1}{2}\log\frac{1+\frac{a^2}{s^2}}{1+\frac{b^2}{s^2}}=0-\frac{1}{2}\log\frac{s^2+a^2}{s^2+b^2}\qquad [\log 1=0]$$

$$=\frac{1}{2}\log\frac{s^2+b^2}{s^2+a^2}$$ **Ans.**

Example 26. *If* $f(t)=\dfrac{e^{at}-\cos bt}{t}$, *find the Laplace transform of f (t).*

(U.P. II Semester, Summer 2003)

Solution. $$f(t)=\frac{e^{at}-\cos bt}{t}=\frac{e^{at}}{t}-\frac{\cos bt}{t}$$

We know that, $L\left(e^{at}-\cos bt\right)=\left(\dfrac{1}{s-a}-\dfrac{s}{s^2+b^2}\right)$

$$\therefore\quad L\left(\frac{e^{at}-\cos bt}{t}\right)=\int_s^\infty\left(\frac{1}{s-a}-\frac{s}{s^2+b^2}\right)ds=\left[\log(s-a)-\frac{1}{2}\log\left(s^2+b^2\right)\right]_s^\infty$$

$$=\left[\frac{2\log(s-a)-\log\left(s^2+b^2\right)}{2}\right]_s^\infty=\frac{1}{2}\left[\log(s-a)^2-\log\left(s^2+b^2\right)\right]_s^\infty$$

$$= \frac{1}{2}\left[\log\frac{(s-a)^2}{s^2+b^2}\right]_s^\infty = \frac{1}{2}\left[\log\left\{\frac{\left(1-\frac{a}{s}\right)^2}{1+\frac{b^2}{s^2}}\right\}\right]_s^\infty$$

$$= \frac{1}{2}\left[0-\log\frac{\left(1-\frac{a}{s}\right)^2}{\left(1+\frac{b^2}{s^2}\right)}\right] = \frac{1}{2}\left[\log\frac{s^2+b^2}{(s-a)^2}\right]$$ **Ans.**

Example 27. *Find the Laplace transform of* $\frac{1-\cos t}{t^2}$.

Solution. $L(1-\cos t) = L(1) - L(\cos t) = \frac{1}{s} - \frac{s}{s^2+1}$

$$L\left[\frac{1-\cos t}{t}\right] = \int_s^\infty\left(\frac{1}{s}-\frac{s}{s^2+1}\right)ds = \left[\log s - \frac{1}{2}\log\left(s^2+1\right)\right]_s^\infty$$

$$= \frac{1}{2}\left[\log s^2 - \log\left(s^2+1\right)\right]_s^\infty = \frac{1}{2}\left[\log\frac{s^2}{s^2+1}\right]_s^\infty$$

$$= \frac{1}{2}\left[\log\frac{1}{\left(1+\frac{1}{s^2}\right)}\right]_s^\infty = \frac{1}{2}\left[0-\log\frac{s^2}{s^2+1}\right] = -\frac{1}{2}\log\frac{s^2}{s^2+1}$$

Again, $$L\left[\frac{1-\cos t}{t^2}\right] = -\frac{1}{2}\int_s^\infty \log\frac{s^2}{s^2+1}\,ds = -\frac{1}{2}\int_s^\infty\left(\log\frac{s^2}{s^2+1}.1\right)ds$$

Integrating by parts, we have,

$$= -\frac{1}{2}\left[\log\frac{s^2}{s^2+1}.s - \int\frac{s^2+1}{s^2}\frac{\left(s^2+1\right)2s-s^2(2s)}{\left(s^2+1\right)^2}.sds\right]_s^\infty$$

$$= -\frac{1}{2}\left[s\log\frac{s^2}{s^2+1} - 2\int\frac{1}{s^2+1}ds\right]_s^\infty = -\frac{1}{2}\left[s\log\frac{s^2}{s^2+1} - 2\tan^{-1}s\right]_s^\infty$$

$$= -\frac{1}{2}\left[0-2\left(\frac{\pi}{2}\right) - s\log\frac{s^2}{s^2+1} + 2\tan^{-1}s\right] = -\frac{1}{2}\left[-\pi - s\log\frac{s^2}{s^2+1} + 2\tan^{-1}s\right]$$

$$= \frac{\pi}{2} + \frac{s}{2}\log\frac{s^2}{s^2+1} - \tan^{-1}s$$

$$= \left(\frac{\pi}{2} - \tan^{-1}s\right) + \frac{s}{2}\log\frac{s^2}{s^2+1} = \cot^{-1}s + \frac{s}{2}\log\frac{s^2}{s^2+1}$$ **Ans.**

Example 28. *Evaluate* $L\left[e^{-4t}\dfrac{\sin 3t}{t}\right]$.

Solution. $L\,[\sin 3t] = \dfrac{3}{s^2+3^2}$

$$\Rightarrow \quad L\left[\frac{\sin 3t}{t}\right] = \int_s^\infty \frac{3}{s^2+9}\,ds = \left[\frac{3}{3}\tan^{-1}\frac{s}{3}\right]_s^\infty = \left[\tan^{-1}\frac{s}{3}\right]_s^\infty$$

$$= \frac{\pi}{2} - \tan^{-1}\frac{s}{3} = \cot^{-1}\frac{s}{3}$$

$$L\left[e^{-4t}\frac{\sin 3t}{t}\right] = \cot^{-1}\frac{s+4}{3} = \tan^{-1}\frac{3}{s+4}$$ **Ans.**

EXERCISE 5.3

Find Laplace Transform of the following:

1. $\dfrac{1}{t}(1-e^t)$ **Ans.** $\log\dfrac{s-1}{s}$
2. $\dfrac{1}{t}\left(e^{-at}-e^{-bt}\right)$ **Ans.** $\log\dfrac{s+b}{s+a}$
3. $\dfrac{1}{t}(1-\cos at)$ **Ans.** $-\dfrac{1}{2}\log\dfrac{s^2}{s^2+a^2}$
4. $\dfrac{1}{t}\sin^2 t$ **Ans.** $\dfrac{1}{4}\log\dfrac{s^2+4}{s^2}$
5. $\dfrac{1}{t}\sinh t$ **Ans.** $-\dfrac{1}{2}\log\dfrac{s-1}{s+1}$
6. $\dfrac{1}{t}\left(e^{-t}\sin t\right)$ **Ans.** $\cot^{-1}(s+1)$
7. $\dfrac{1}{t}(1-\cos t)$ **Ans.** $\dfrac{1}{2}\left[\log\left(s^2+1\right)-\log s\right]$
8. $\displaystyle\int_0^\infty \frac{1}{t}e^{-2t}\sin t\,dt$ **Ans.** $\dfrac{1}{s}\cot^{-1}(s+2)$
9. $\displaystyle\int_0^\infty \frac{e^{-t}-e^{-3t}}{t}\,dt$ **Ans.** $\log 3$

5.15 LAPLACE TRANSFORM OF ERROR FUNCTION

Example 29. *Find* $L\left\{erf\sqrt{t}\right\}$ *and hence prove that*

$$L\left\{t.erf\,2\sqrt{t}\right\} = \frac{3s+8}{s^2(s+4)^{3/2}}$$ *(U.P. II Semester, Summer 2001)*

Solution. We know that $erf\sqrt{t} = \dfrac{2}{\sqrt{\pi}}\displaystyle\int_0^{\sqrt{t}} e^{-x^2}\,dx$

$$= \frac{2}{\sqrt{\pi}}\int_0^{\sqrt{t}}\left(1-x^2+\frac{x^4}{2!}-\frac{x^6}{3!}+\dots\right)dx = \frac{2}{\sqrt{\pi}}\left[x-\frac{x^3}{3}+\frac{x^5}{10}-\frac{x^7}{42}\dots\right]_0^{\sqrt{t}}$$

$$= \frac{2}{\sqrt{\pi}}\left[\sqrt{t}-\frac{t^{3/2}}{3}+\frac{t^{5/2}}{10}-\frac{t^{7/2}}{42}+\dots\right]$$

$$L\left\{erf\sqrt{t}\right\} = \frac{2}{\sqrt{\pi}}\left[\frac{\left|\frac{3}{2}\right.}{s^{3/2}}-\frac{\left|\frac{5}{2}\right.}{3s^{5/2}}+\frac{\left|\frac{7}{2}\right.}{10s^{7/2}}-\frac{\left|\frac{9}{2}\right.}{42s^{9/2}}+\dots\right]$$

$$= \frac{2}{\sqrt{\pi}}\left[\frac{\frac{1}{2}\left|\frac{1}{2}\right.}{s^{3/2}} - \frac{\frac{3}{2}.\frac{1}{2}\left|\frac{1}{2}\right.}{3s^{5/2}} + \frac{\frac{5}{2}.\frac{3}{2}.\frac{1}{2}\left|\frac{1}{2}\right.}{10s^{7/2}} - \frac{\frac{7}{2}.\frac{5}{2}.\frac{3}{2}.\frac{1}{2}\left|\frac{1}{2}\right.}{42s^{9/2}} + \ldots\right] \qquad \left[\because \left|\frac{1}{2}\right. = \sqrt{\pi}\right]$$

$$= \frac{1}{s^{3/2}} - \frac{1}{2}\frac{1}{s^{5/2}} + \frac{1.3}{2.4}\frac{1}{s^{7/2}} - \frac{1}{2}.\frac{3}{4}.\frac{5}{6}\frac{1}{s^{9/2}} + \ldots$$

$$= \frac{1}{s^{3/2}}\left[1 - \frac{1}{2}.\frac{1}{s} + \frac{1}{2}.\frac{3}{4}.\frac{1}{s^2} - \frac{1}{2}.\frac{3}{4}.\frac{5}{6}\frac{1}{s^3} + \ldots\right]$$

$$= \frac{1}{s^{3/2}}\left[1 - \frac{1}{2}.\frac{1}{s} + \frac{\left(-\frac{1}{2}\right)\left\{-\frac{3}{2}\right\}}{2!}\frac{1}{s^2} + \frac{\left(-\frac{1}{2}\right)\left(-\frac{3}{2}\right)\left(-\frac{5}{2}\right)}{3!}\frac{1}{s^3} + \ldots\right]$$

$$= \frac{1}{s^{3/2}}\left[1 + \frac{1}{s}\right]^{-\frac{1}{2}} = \frac{1}{s^{3/2}}\left[\frac{s}{s+1}\right]^{\frac{1}{2}} = \frac{1}{s\sqrt{(s+1)}} \qquad \textbf{Ans.}$$

Now, $L\left\{erf\left(2\sqrt{t}\right)\right\} = L\left\{erf\sqrt{4t}\right\} = \frac{1}{4}\frac{1}{\frac{s}{4}\sqrt{\frac{s}{4}+1}} = \frac{2}{s\sqrt{s+4}}$

$$L\left\{t.erf\left(2\sqrt{t}\right)\right\} = -\frac{d}{ds}\frac{2}{\sqrt{s^3+4s^2}} = -2\left(-\frac{1}{2}\right)\left[s^3+4s^2\right]^{-\frac{3}{2}}\left(3s^2+8s\right)$$

$$= \frac{3s^2+8s}{\left(s^3+4s^2\right)^{3/2}} = \frac{s(3s+8)}{s^3(s+4)^{3/2}} = \frac{3s+8}{s^2(s+4)^{3/2}} \qquad \textbf{Proved.}$$

5.16. COMPLEMENTARY ERROR FUNCTION

This function is defined by

$$erf_c\left(\sqrt{t}\right) = 1 - erf\left(\sqrt{t}\right) = 1 - \frac{2}{\sqrt{\pi}}\int_0^{\sqrt{t}} e^{-x^2}\,dx$$

Now, $L\left\{erf_c\left(\sqrt{t}\right)\right\} = L\left\{1 - \frac{2}{\sqrt{\pi}}\int_0^{\sqrt{t}} e^{-x^2}\,dx\right\} = L(1) - \frac{2}{\sqrt{\pi}}L\left\{\int_0^{\sqrt{t}} e^{-x^2}\,dx\right\} = \frac{1}{s} - \frac{1}{s\sqrt{s+1}}$

$$= \frac{\sqrt{(s+1)}-1}{s\sqrt{(s+1)}} = \frac{\left\{\sqrt{(s+1)}-1\right\}\left\{\sqrt{(s+1)}+1\right\}}{s\sqrt{(1+s)}\left\{\sqrt{(s+1)}+1\right\}}$$

$$= \frac{\left[\sqrt{(s+1)}-1\right]\left[\sqrt{s+1}+1\right]}{s\sqrt{(s+1)}\,(\sqrt{s+1}+1)} = \frac{s+1-1}{s\sqrt{s+1}\,(\sqrt{s+1}+1)}$$

$$= \frac{1}{\sqrt{(s+1)}\left\{\sqrt{(s+1)}+1\right\}}$$

$$\therefore L\left[erf_c\left(\sqrt{t}\right)\right] = \frac{1}{\sqrt{s+1}\left\{\sqrt{s+1}+1\right\}} \qquad \textbf{Ans.}$$

5.17 UNIT STEP FUNCTION

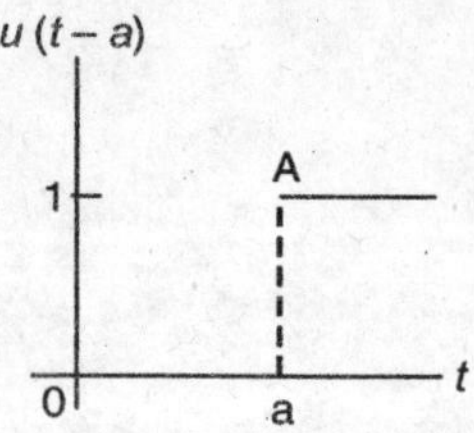

With the help of unit step functions, we can find the inverse transform of functions, which cannot be determined with previous methods.

The unit step functions $u(t-a)$ is defined as follows:

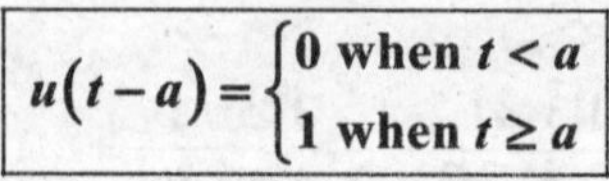

$$u(t-a)=\begin{cases}0 \text{ when } t<a\\ 1 \text{ when } t\geq a\end{cases}$$ where $a\geq 0$

5.18 LAPLACE TRANSFORM OF UNIT FUNCTION

$$L[u(t-a)]=\frac{e^{-as}}{s}$$

Proof. $$L[u(t-a)]=\int_0^\infty e^{-st}u(t-a)\,dt$$

$$=\int_0^a e^{-st}.0dt+\int_a^\infty e^{-st}.1\,dt=0+\left[\frac{e^{-st}}{-s}\right]_a^\infty$$

$$\boxed{L[u(t-a)]=\frac{e^{-as}}{s}}$$ **Proved.**

Example 30. *Express the following function in terms of unit step functions and find its Laplace transform:*

$$f(t)=\begin{cases}8, & t<2\\ 6, & t\geq 2\end{cases}$$

Solution. $$f(t)=\begin{cases}8+0, & t<2\\ 8-2 & t\geq 2\end{cases}=8+\begin{cases}0, & t<2\\ -2, & t\geq 2\end{cases}$$

$$=8+(-2)\begin{cases}0, & t<2\\ 1, & t\geq 2\end{cases}=8-2u(t-2)$$

$$L\{f(t)\}=8L(1)-2Lu(t-2)=\frac{8}{s}-2\frac{e^{-2s}}{s}$$ **Ans.**

Example 31. *Draw the graph of $u(t-a)-u(t-b)$.*

Solution. As in Art 5.17 the graph of *u(t–a)* is a straight line parallel to *t*-axis from *A* to ∞.

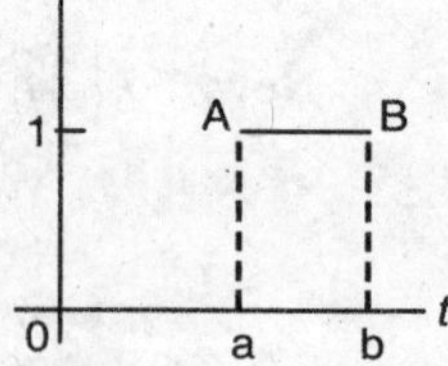

Similarly, the graph of $u(t-b)$ is a straight line parallel to *t*-axis from *B* to ∞.

Hence, the graph of $u(t-a)-u(t-b)$ is *AB*. **Ans.**

Example 32. *Express the following function in terms of unit step function and find its Laplace transform:*

$$f(t)=\begin{cases}E, & a<t<b\\ 0, & t\geq b\end{cases}$$

Solution. $$f(t)=E\begin{cases}1, & a<t<b\\ 0, & t\geq b\end{cases}$$ $$\left[L[f(t-a).\,u(t-a)]=e^{-as}F(s)\right]$$

$$Lf(t)=E\left[\frac{e^{-as}}{s}-\frac{e^{-bs}}{s}\right]$$ **Ans.**

5.19 SECOND SHIFTING THEOREM

If $L[f(t)] = F(s)$, then $\boxed{L[f(t-a).u(t-a)] = e^{-as}F(s)}$

Proof.
$$\begin{aligned} L[f(t-a).u(t-a)] &= \int_0^\infty e^{-st}[f(t-a).u(t-a)]\,dt \\ &= \int_0^a e^{-st} f(t-a).0\,dt + \int_a^\infty e^{-st} f(t-a)(1)\,dt \\ &= \int_a^\infty e^{-st} f(t-a)\,dt \\ &= \int_0^\infty e^{-s(u+a)} f(u)\,du, \qquad \text{where } u = t-a \\ &= e^{-sa}\int_0^\infty e^{-su}.f(u)\,du = e^{-sa}F(s) \end{aligned}$$

Proved.

Example 33. *Express the following function in terms of unit step function :*

$$f(t) = \begin{cases} t-1, & 1<t<2 \\ 3-t, & 2<t<3 \end{cases}$$

and find its Laplace transform. (*U.P.; II Semester, 2009*)

Solution.
$$\begin{aligned} f(t) &= \begin{cases} t-1, & 1<t<2 \\ 3-t, & 2<t<3 \end{cases} \\ &= (t-1)[u(t-1)-u(t-2)]+(3-t)[u(t-2)-u(t-3)] \\ &= (t-1)u(t-1)-(t-1)u(t-2)+(3-t)u(t-2)+(t-3)u(t-3) \\ &= (t-1)u(t-1)-2(t-2)u(t-2)+(t-3)u(t-3) \end{aligned}$$

$$L[f(t)] = e^{-s}L(t) - 2e^{-2s}L(t) - e^{-3s}L(t)$$

$$[L[f(t-a).u(t-a)] = e^{-as}F(s)]$$

$$L[f(t)] = \frac{e^{-s}}{s^2} - 2\frac{e^{-2s}}{s^2} + \frac{e^{-3s}}{s^2}$$ **Ans.**

Example 34. *Find* $L\{F(t)\}$ *if*

$$F(t) = \begin{cases} \sin\left(t-\dfrac{\pi}{3}\right), & t>\dfrac{\pi}{3} \\ 0, & t<\dfrac{\pi}{3} \end{cases}$$

Solution. $L\{F(t)\} = e^{-s\frac{\pi}{3}} L(\sin t)$ $\left[\because a = \dfrac{\pi}{3}\right]$

$$= e^{-s\frac{\pi}{3}} \cdot \frac{1}{s^2+1}$$ **(Using second shifting property) Ans.**

5.20 THEOREM. $Lf(t)u(t-a) = e^{-as}L[f(t+a)]$

Proof.
$$\begin{aligned} Lf(t).u(t-a) &= \int_0^\infty e^{-st}[f(t).u(t-a)]\,dt \\ &= \int_0^a e^{-st}[f(t).u(t-a)]\,dt + \int_a^\infty e^{-st}.[f(t).u(t-a)]\,dt \end{aligned}$$

$$= 0+\int_a^{\infty} e^{-st}.f(t)(1)\,dt = \int_a^{\infty} e^{-s(y+a)}.f(y+a)\,dy$$

$$= e^{-as}\int_a^{\infty} e^{-sy}.f(y+a)\,dy \qquad (t-a=y)$$

$$= e^{-as}\int_a^{\infty} e^{-st} f(t+a)\,dt = e^{-as} Lf(t+a)$$ **Proved.**

Example 35. *Find the Laplace Transform of $t^2\, u\,(t-3)$.*

Solution. $t^2.u(t-3)=\left[(t-3)^2+6(t-3)+9\right]u(t-3)$

$$= (t-3)^2.u(t-3)+6(t-3).u(t-3)+9u(t-3)$$

$$Lt^2.u(t-3) = L(t-3)^2.u(t-3)+6L(t-3)u(t-3)+9Lu(t-3)$$

$$= e^{-3s}\left[\frac{2}{s^3}+\frac{6}{s^2}+\frac{9}{s}\right]$$

Aliter. $Lt^2u(t-3) = e^{-3s}L(t+3)^2 = e^{-3s}L\left[t^2+6t+9\right] = e^{-3s}\left[\frac{2}{s^3}+\frac{6}{s^2}+\frac{9}{s}\right]$ **Ans.**

Example 36. *Find the Laplace transform of $e^{-2t}u_{\pi}(t)$ where*

$$u_{\pi}(t) = \begin{cases} 0; & t<\pi \\ 1; & t>\pi \end{cases}$$

Solution. $u_{\pi}(t) = \begin{cases} 0; & t<\pi \\ 1; & t>\pi \end{cases}$

$$u_{\pi}(t) = u(t-\pi)$$

$$L[u_{\pi}(t)] = L[u(t-\pi)] = \frac{e^{-\pi s}}{s}$$

$$L\left[e^{-2t}u_{\pi}(t)\right] = \frac{e^{-\pi(s+2)}}{s+2}$$ **Ans.**

Example 37. *Express the following function in terms of unit step function and find its Laplace transform* $f(t)=\begin{cases} 0, & 0<t<1 \\ t-1, & 1<t<2 \\ 1, & 2<t \end{cases}$ *(U.P. II Semester, Summer 2002)*

Solution. The above function shown in the figure is expressed in algebraic form

$$f(t)=\begin{cases} 0, & 0<t<1 \\ t-1, & 1<t<2 \\ 1, & 2<t \end{cases} \qquad \ldots (1)$$

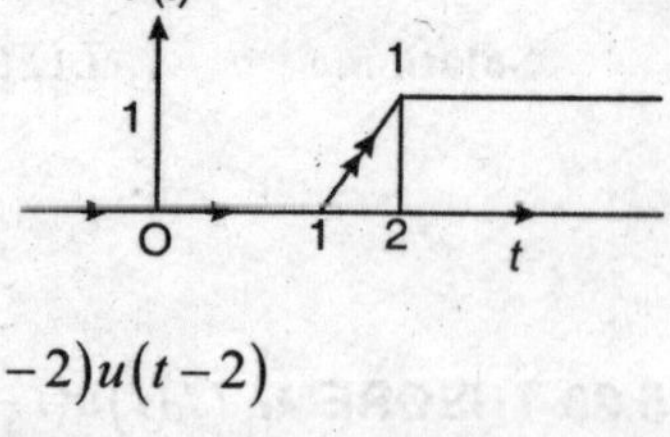

$$f(t)=(t-1)\left[u(t-1)-u(t-2)\right]+u(t-2)$$

$$= (t-1)u(t-1)-u(t-2)\{t-1-1\} = (t-1)u(t-1)-(t-2)u(t-2)$$

$$Lf(t) = L(t-1)u(t-1)-L(t-2)u(t-2)$$

$$= \frac{e^{-s}}{s^2}-\frac{e^{-2s}}{s^2}$$ **Ans.**

Example 38. *Represent f(t) = sin 2t, 2π < t < 4π and f(t) = 0 otherwise, in terms of unit step function and then find its Laplace transform.*

Solution. $f(t) = \begin{cases} \sin 2t, & 2\pi < t < 4\pi \\ 0, & \text{otherwise} \end{cases}$

$$f(t) = \sin 2t\left[u(t-2\pi) - u(t-4\pi)\right]$$

$$Lf(t) = L\sin 2t . u(t-2\pi) - L\sin 2t . u(t-4\pi)$$

$$= e^{-2\pi s} L\sin 2(t+2\pi) - e^{-4\pi s} L\sin 2(t+4\pi)$$

$$= e^{-2\pi s} L\sin 2t - e^{-4\pi s} L\sin(2t)$$

$$= e^{-2\pi s}\frac{2}{s^2+4} - e^{-4\pi s}\frac{2}{s^2+4}$$

$$= \left(e^{-2\pi s} - e^{-4\pi s}\right)\frac{2}{s^2+4}$$ **Ans.**

Example 39. *A function f (t) obeys the equation f (t) + 2* $\int_0^t f(t)\,dt = \cosh 2t$

Find the Laplace transform of f (t). (*U.P. II Semester Summer* 2006)

Solution. We have, $f(t) + 2\int_0^t f(t)\,dt = \cosh 2t$

Taking Laplace transformation of both the sides, we get

$L\{f(t)\} + 2L\int_0^t f(t)\,dt = L(\cosh 2t) \quad \Rightarrow \quad F(s) + 2.\frac{1}{s}F(s) = \frac{s}{s^2-4}$

$\Rightarrow \quad F(s)\left\{1 + \frac{2}{s}\right\} = \frac{s}{s^2-4} \quad \Rightarrow \quad F(s)\left\{\frac{s+2}{s}\right\} = \frac{s}{s^2-4}$

$\Rightarrow \quad F(s) = \left(\frac{s}{s^2-4}\right)\left(\frac{s}{s+2}\right) \quad \Rightarrow \quad F(s) = \frac{s^2}{(s^2-4)(s+2)}$ **Ans.**

EXERCISE 5.4

Find the Laplace transform of the following:

1. $f(t) = \begin{cases} t-1, & 1 < t < 2 \\ 0, & \text{otherwise} \end{cases}$ **Ans.** $\frac{e^{-s} - e^{-2s}}{s^2} - \frac{e^{-2s}}{s}$

2. $e^t u(t-1)$ **Ans.** $\frac{e^{-(s-1)}}{s-1}$

3. $\frac{1-e^{2t}}{s} + tu(t) + \cosh t.\cos t$ **Ans.** $\log\frac{s-2}{s} + \frac{1}{s^2} + \frac{s^3}{s^4+4}$

4. $t^2 u(t-2)$ **Ans.** $\frac{e^{-2s}}{s^3}\left(4s^2+4s+2\right)$

5. $\sin t\, u(t-4)$ **Ans.** $\frac{e^{-4s}}{s^2+1}[\cos 4 + s\sin 4]$

6. $f(t) = K(t-2)[u(t-2) - u(t-3)]$ **Ans.** $\frac{K}{s^2}\left[e^{-2s} - (s+1)e^{-3s}\right]$

7. $f(t) = K\frac{\sin \pi t}{T}[u(t-2T) - u(t-3T)]$ **Ans.** $\frac{K\pi T}{s^2T^2 + \pi^2}\left(e^{-2sT} - e^{-3sT}\right)$

Express the following in terms of unit step functions and obtain Laplace transforms.

8. $f(t) = \begin{cases} t, & 0 < t < 2 \\ 0, & 2 < t \end{cases}$ **Ans.** $u(t) - u(t-2), \frac{1-(2s+1)e^{-2s}}{s^2}$

9. $f(t) = \begin{cases} \sin t, & 0 < t < \pi \\ t, & t > \pi \end{cases}$ **Ans.** $\frac{1+e^{-\pi s}}{s^2+1} + \frac{e^{-\pi s}(\pi s+1)}{s^2}$

10. $f(t) = \begin{cases} 4, & 0 < t < 1 \\ -2, & 0 < t < 3 \\ 5, & t > 3 \end{cases}$ **Ans.** $\frac{4 - 6e^{-s} + 7e^{-3s}}{s}$

5.21. PERIODIC FUNCTIONS

Let $f(t)$ be a periodic function with period T, then

$$\boxed{L[f(t)] = \frac{\int_0^T e^{-st} f(t)\, dt}{1 - e^{-sT}}}$$

Proof. $L[f(t)] = \int_0^\infty e^{-st} f(t)\, dt = \int_0^T e^{-st} f(t)\, dt + \int_T^{2T} e^{-st} f(t)\, dt + \int_{2T}^{3T} e^{-st} f(t)\, dt + \ldots$

Substituting $t = u + T$ in second integral and $t = u + 2T$ in third integral, and so on.

$$L[f(t)] = \int_0^T e^{-st} f(t)\, dt + \int_0^T e^{-s(u+T)} f(u+T)\, du + \int_0^T e^{-s(u+2T)} f(u+2T)\, du + \ldots$$

$$= \int_0^T e^{-st} f(t)\, dt + e^{-sT}\int_0^T e^{-su} f(u)\, du + e^{-2sT}\int_0^T e^{-su} f(u)\, du + \ldots$$

$$[f(u) = f(u+T) = f(u+2T) = f(u+3T) = \ldots]$$

$$= \int_0^T e^{-st} f(t)\, dt + e^{-sT}\int_0^T e^{-st} f(t)\, dt + e^{-2sT}\int_0^T e^{-st} f(t)\, dt + \ldots$$

$$= \left[1 + e^{-sT} + e^{-2sT} + e^{-3sT} + \ldots\right]\int_0^T e^{-st} f(t)\, dt \qquad \left[1 + a + a^2 + a^3 + \ldots = \frac{1}{1-a}\right]$$

$$= \frac{1}{1-e^{-sT}}\int_0^T e^{-st} f(t)\, dt.$$ **Proved.**

Example 40. *Find the Laplace transform of the waveform*

$$f(t) = \left(\frac{2t}{3}\right), 0 \le t \le 3.$$

Solution. $$L[f(t)] = \frac{1}{1-e^{-sT}}\int_0^T e^{-st} f(t)\, dt$$

$$L\left[\frac{2t}{3}\right] = \frac{1}{1-e^{-3s}}\int_0^3 e^{-st}\left(\frac{2}{3}t\right) dt = \frac{1}{1-e^{-3s}}\,\frac{2}{3}\left[\frac{te^{-st}}{-s} - (1)\frac{e^{-st}}{s^2}\right]_0^3$$

$$= \frac{2}{3}\frac{1}{1-e^{-3s}}\left[\frac{3e^{-3s}}{-s} - \frac{e^{-3s}}{s^2} + \frac{1}{s^2}\right] = \frac{2}{3} \cdot \frac{1}{1-e^{-3s}}\left[\frac{3e^{-3s}}{-s} + \frac{1-e^{-3s}}{s^2}\right]$$

$$= \frac{2e^{-3s}}{-s\left(1-e^{-3s}\right)} + \frac{2}{3s^2}$$ **Ans.**

Example 41. *Draw the graph and find the Laplace transform of the triangular wave function of period 2 C given by*

$$f(t) = \begin{cases} t, & 0 < t \le C \\ 2C - t, & C < t < 2C \end{cases}$$ *(Uttarakhand, II Semester, June 2007)*

Solution. Period = 2C = T

Laplace transform of periodic function $f(t)$

f (t)
O a 2a 4a t

$$L\{f(t)\} = \frac{\int_0^T e^{-st} f(t)\,dt}{1-e^{-sT}}$$

$$L\{f(t)\} = \frac{1}{1-e^{-2Cs}} \int_0^{2C} e^{-st} f(t)\,dt \qquad (T = 2c)$$

On putting the values of $f(t)$, we get

$$L[f(t)] = \frac{1}{1-e^{-2Cs}}\left[\int_0^C e^{-st} t\,dt + \int_C^{2C} e^{-st}(2C-t)\,dt\right]$$

$$= \frac{1}{1-e^{-2Cs}}\left[\left\{\frac{te^{-st}}{-s} - 1.\frac{e^{-st}}{(-s)^2}\right\}_0^C + \left\{(2C-t)\frac{e^{-st}}{(-s)} - (-1)\frac{e^{-st}}{(-s)^2}\right\}_C^{2C}\right]$$

$$= \frac{1}{1-e^{-2Cs}}\left[\left\{\frac{C.e^{-Cs}}{-s} - \frac{e^{-Cs}}{(-s)^2} - 0 + \frac{1}{s^2}\right\} + \left\{(2C-2C)\frac{e^{-2Cs}}{(-s)} + \frac{e^{-2Cs}}{s^2} - \left((2C-C)\frac{e^{-Cs}}{-s} + \frac{e^{-Cs}}{s^2}\right)\right\}\right]$$

$$= \frac{1}{1-e^{-2Cs}}\left\{-\frac{Ce^{-Cs}}{s} - \frac{e^{-Cs}}{s^2} + \frac{1}{s^2} + \frac{e^{-2Cs}}{s^2} + \frac{Ce^{-Cs}}{s} - \frac{e^{-Cs}}{s^2}\right\}$$

$$= \frac{1}{1-e^{-2Cs}}\left\{\frac{1}{s^2}(1 - 2e^{-Cs} + e^{-2Cs})\right\} = \frac{\left(1-e^{-Cs}\right)^2}{s^2\left(1+e^{-Cs}\right)\left(1-e^{-Cs}\right)} = \frac{1-e^{-Cs}}{s^2\left(1+e^{-Cs}\right)}$$ **Ans.**

Example 42. *Draw the graph of the periodic function*

$$f(t) = \begin{cases} t, & 0 < t < \pi \\ \pi - t, & \pi < t < 2\pi \end{cases}$$

and find its Laplace transform. *(U.P. Second Semester, 2003)*

Solution. Period $= 2\pi = T$

Laplace transform of Periodic functions

$$L\{f(t)\} = \frac{\int_0^T e^{-st} f(t)\,dt}{1-e^{-sT}}$$

$$=\frac{1}{1-e^{-2\pi s}}\int_0^{2\pi}e^{-st}f(t)\,dt=\frac{1}{1-e^{-2\pi s}}\left[\int_0^{\pi}e^{-st}t\,dt+\int_{\pi}^{2\pi}e^{-st}(\pi-t)\,dt\right]$$

$$=\frac{1}{1-e^{-2\pi s}}\left\{\frac{t.e^{-st}}{-s}-1.\frac{e^{-st}}{(-s)^2}\right\}_0^{\pi}+\left\{(\pi-t)\frac{e^{-st}}{(-s)}-(-1)\frac{e^{-st}}{(-s)^2}\right\}_{\pi}^{2\pi}$$

$$=\frac{1}{1-e^{-2\pi s}}\left[\left\{\frac{\pi.e^{-\pi s}}{-s}-\frac{e^{-\pi s}}{(-s)^2}-0+\frac{1}{s^2}\right\}+\left\{(\pi-2\pi)\frac{e^{-2\pi s}}{-s}+\frac{e^{-2\pi s}}{s^2}-\left((\pi-\pi)\frac{e^{-\pi s}}{-s}+\frac{e^{-\pi s}}{s^2}\right)\right\}\right]$$

$$=\frac{1}{1-e^{-2\pi s}}\left\{-\frac{\pi e^{-\pi s}}{s}-\frac{e^{-\pi s}}{s^2}+\frac{1}{s^2}+\pi\frac{e^{-2\pi s}}{s}+\frac{e^{-2\pi s}}{s^2}-0-\frac{e^{-\pi s}}{s^2}\right\}$$

f(t)
(π, π)
π
O
2π
t
π
–π
(2π, – π)

$$=\frac{1}{1-e^{-2\pi s}}\left\{-\frac{\pi}{s}e^{-\pi s}+\frac{\pi}{s}e^{-2\pi s}+\frac{1}{s^2}-\frac{1}{s^2}.e^{-\pi s}+\frac{1}{s^2}e^{-2\pi s}-\frac{e^{-\pi s}}{s^2}\right\}$$

$$=\frac{1}{1-e^{-2\pi s}}\left[\frac{\pi}{s}\left(e^{-2\pi s}-e^{-\pi s}\right)+\frac{1}{s^2}\left(1+e^{-2\pi s}-2e^{-\pi s}\right)\right]=\frac{-\pi s e^{-\pi s}\left(1-e^{-\pi s}\right)+\left(1-e^{-\pi s}\right)^2}{s^2\left(1+e^{-\pi s}\right)\left(1-e^{-\pi s}\right)}$$

$$=\frac{-\pi s e^{-\pi s}+1-e^{-\pi s}}{s^2\left(1+e^{-\pi s}\right)}$$ **Ans.**

Example 43. *Find the Laplace transform of the function (Half wave rectifier)*

$$f(t)=\begin{cases}\sin\omega t & \text{for } 0<t<\dfrac{\pi}{\omega}\\ 0 & \text{for } \dfrac{\pi}{\omega}<t<\dfrac{2\pi}{\omega}.\end{cases}$$ *(U.P. II Semester, Summer 2002)*

Solution. $L[f(t)]=\frac{1}{1-e^{-sT}}\int_0^T e^{-st}f(t)\,dt$

$$=\frac{1}{1-e^{-\frac{2\pi s}{\omega}}}\int_0^{2\pi/\omega}e^{-st}f(t)\,dt$$ $\left[\begin{array}{l}f(t)\text{ is a periodic function}\\ T=\dfrac{2\pi}{\omega}\end{array}\right.$

$$=\frac{1}{1-e^{-\frac{2\pi s}{\omega}}}\left[\int_0^{\pi/\omega}e^{-st}\sin\omega t\,dt+\int_{\pi/\omega}^{2\pi/\omega}e^{-st}\times 0\times dt\right]$$

$$= \frac{1}{1-e^{-\frac{2\pi s}{\omega}}} \int_0^{\pi/\omega} e^{-st} \sin \omega t \, dt$$

$$\left[\int e^{ax} \sin bx \, dx = e^{ax} \frac{(a \sin bx - b \cos bx)}{a^2 + b^2} \right]$$

$$L[f(t)] = \frac{1}{1-e^{-\frac{2\pi s}{\omega}}} \left[\frac{e^{-st}(-s \sin \omega t - \omega \cos \omega t)}{s^2 + \omega^2} \right]_0^{\pi/\omega}$$

$$= \frac{1}{1-e^{-\frac{2\pi s}{\omega}}} \left[\frac{\omega e^{-\frac{\pi}{\omega}s} + \omega}{s^2 + \omega^2} \right] = \frac{\omega \left[1 + e^{-\frac{\pi s}{\omega}} \right]}{(s^2 + \omega^2)\left[1 - e^{-\frac{2\pi s}{\omega}} \right]} = \frac{\omega \left[1 + e^{-\frac{\pi}{\omega}s} \right]}{(s^2 + \omega^2)\left(1 - e^{-\frac{\pi}{\omega}s} \right)\left(1 + e^{-\frac{\pi}{\omega}s} \right)}$$

$$= \frac{\omega}{(s^2 + \omega^2)\left[1 - e^{-\frac{\pi s}{\omega}} \right]}$$

Ans.

Example 44. *Find the Laplace Transform of the Periodic function (saw tooth wave)*

$$f(t) = \frac{kt}{T} \text{ for } 0 < t < T, \quad f(t+T) = f(t)$$

Solution. $L[f(t)] = \frac{1}{1-e^{-sT}} \int_0^T e^{-st} f(t) \, dt = \frac{1}{1-e^{-sT}} \int_0^T e^{-st} \frac{kt}{T} dt$

$$= \frac{1}{1-e^{-sT}} \frac{k}{T} \int_0^T e^{-st} . t \, dt = \frac{k}{T(1-e^{-sT})} \left[t \frac{e^{-st}}{-s} - \int 1 . \frac{e^{-st}}{-s} dt \right]_0^T$$ Integrating by parts

$$= \frac{k}{T(1-e^{-sT})} \left[\frac{te^{-st}}{-s} - \frac{e^{-st}}{s^2} \right]_0^T = \frac{k}{T(1-e^{-sT})} \left[\frac{Te^{-sT}}{-s} - \frac{e^{-sT}}{s^2} + \frac{1}{s^2} \right]$$

$$= \frac{k}{T(1-e^{-sT})} \left[\frac{Te^{-sT}}{-s} + \frac{1}{s^2}(1-e^{-sT}) \right]$$

$$= -\frac{ke^{-sT}}{s(1-e^{-sT})} + \frac{k}{Ts^2}$$

Ans.

Example 45. *Obtain Laplace transform of rectangular wave given by*

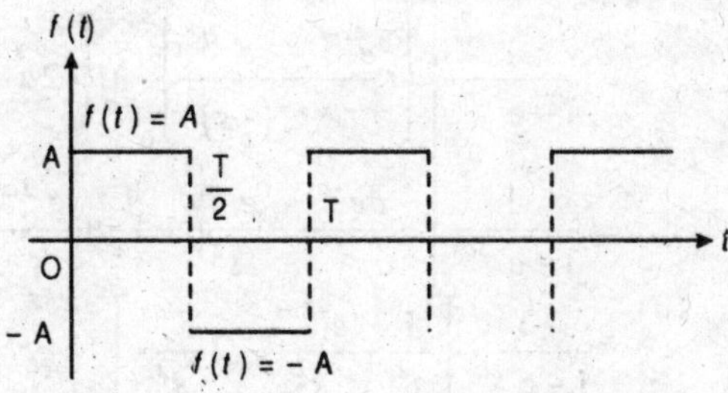

Solution. We know that Laplace transform of a periodic function *i.e.*,

$$L f(t)=\frac{\int_0^T e^{-st} f(t)\,dt}{1-e^{-sT}}=\frac{\int_0^{\frac{T}{2}} e^{-st} A\,dt+\int_{\frac{T}{2}}^{T} e^{-st}(-A)\,dt}{1-e^{-sT}}$$

$$=A\frac{\left[\frac{e^{-st}}{-s}\right]_0^{\frac{T}{2}}-\left[\frac{e^{-st}}{-s}\right]_{\frac{T}{2}}^{T}}{1-e^{-sT}}=\frac{A}{1-e^{-sT}}\left[-\frac{e^{-\frac{sT}{2}}}{s}+\frac{1}{s}+\frac{e^{-sT}}{s}-\frac{e^{-\frac{sT}{2}}}{s}\right]$$

$$=\frac{A}{s\left(1-e^{-sT}\right)}\left[1-2e^{-\frac{sT}{2}}+e^{-sT}\right]=\frac{A}{s\left(1-e^{-sT}\right)}\left[1-e^{-\frac{sT}{2}}\right]^2$$

$$=\frac{A\left[1-e^{-\frac{sT}{2}}\right]^2}{s\left(1+e^{\frac{-sT}{2}}\right)\left(1-e^{-\frac{sT}{2}}\right)}=\frac{A}{s}\frac{\left(1-e^{-\frac{sT}{2}}\right)}{\left(1+e^{-\frac{sT}{2}}\right)}$$

$$=\frac{A\left(e^{\frac{sT}{4}}-e^{-\frac{sT}{4}}\right)}{s\left(e^{\frac{sT}{4}}+e^{-\frac{sT}{4}}\right)}=\frac{A}{s}\tanh\frac{sT}{4}$$ **Ans.**

Example 46. *Draw the graph of the following periodic function and find its Laplace transform:*

$$f(t)=\begin{cases} t & \text{for } 0<t\le a \\ 2a-t & \text{for } a<t<2a \end{cases}$$ *(U.P. II Semester, Summer 2002)*

Solution. The given function is periodic with period $2a$.

f (t)

O a 2a 4a t

$$\therefore \quad Lf(t)=\frac{1}{1-e^{-2as}}\int_0^{2a} f(t)e^{-st}\,dt$$

$$=\frac{1}{1-e^{-2as}}\left[\int_0^{a} f(t)e^{-st}\,dt+\int_a^{2a} f(t)e^{-st}\,dt\right]$$

$$=\frac{1}{1-e^{-2as}}\left[\int_0^{a} t\,e^{-st}\,dt+\int_a^{2a}(2a-t)\,e^{-st}\,dt\right]$$

$$=\frac{1}{1-e^{-2as}}\left[\left\{t\frac{e^{-st}}{-s}-\frac{e^{-st}}{(-s)^2}\right\}_0^{a}+\left\{(2a-t)\frac{e^{-st}}{-s}+\frac{e^{-st}}{(-s)^2}\right\}_a^{2a}\right]$$

$$=\frac{1}{1-e^{-2as}}\left[-\frac{ae^{-as}}{s}-\frac{e^{-as}}{s^2}+\frac{1}{s^2}+\frac{e^{-2as}}{s^2}+\frac{ae^{-as}}{s}-\frac{e^{-as}}{s^2}\right]$$

$$=\frac{1}{1-e^{-2as}}\left[\frac{1}{s^2}+\frac{e^{-2as}}{s^2}-2\frac{e^{-as}}{s^2}\right]$$

$$= \frac{1}{s^2}\frac{1}{\left(1-e^{-2as}\right)}\left(1+e^{-2as}-2e^{-as}\right)$$

$$= \frac{1}{s^2}\frac{\left(1-e^{-as}\right)^2}{\left(1+e^{-as}\right)\left(1-e^{-as}\right)} = \frac{1}{s^2}\left(\frac{1-e^{-as}}{1+e^{-as}}\right)$$

$$= \frac{1}{s^2}\left[\frac{e^{\frac{as}{2}}-e^{-\frac{as}{2}}}{e^{\frac{as}{2}}+e^{-\frac{as}{2}}}\right] = \frac{1}{s^2}\tanh\frac{as}{2} \qquad \textbf{Ans.}$$

Example 47. *A periodic square wave function f(t), in terms of unit step functions, is written as*

$$f(t) = k\left[u_0(t) - 2u_a(t) + 2u_{2a}(t) - 2u_{3a}(t) + \ldots\right]$$

Show that the Laplace transform of f (t) is given by

$$L\left[f(t)\right] = \frac{k}{s}\tanh\left(\frac{as}{2}\right)$$

Solution.

$$f(t) = k\left[u_0(t) - 2u_a(t) + 2u_{2a}(t) - 2u_{3a}(t) + \ldots\right]$$

$$f(t) = k\,[u(t-0) - 2u(t-a) + 2u(t-2a) - 2u(t-3a) + \ldots\ldots]$$

$$L\,f(t) = k\,[L\,u(t-0) - 2\,L\,u(t-a) + 2\,L\,u(t-2a) - 2\,L\,u(t-3a) + \ldots\ldots]$$

$$= k\left[\frac{1}{s} - 2\frac{e^{-as}}{s} + 2\frac{e^{-2as}}{s} - 2\frac{e^{-3as}}{s} + \ldots\right]$$

$$= \frac{k}{s}\left[1 - 2e^{-as} + 2e^{-2as} - 2e^{-3as} + \ldots\right]$$

$$= \frac{k}{s}\left[1 - 2(e^{-as} - e^{-2as} + e^{-3as} - \ldots)\right]$$

$$= \frac{k}{s}\left[1 - 2\frac{e^{-as}}{1+e^{-as}}\right] = \frac{k}{s}\left[\frac{1+e^{-as}-2e^{-as}}{1+e^{-as}}\right] \qquad \left[\text{This is G.P.} \quad s = \frac{a}{1-r}\right]$$

$$= \frac{k}{s}\left[\frac{1-e^{-as}}{1+e^{-as}}\right] = \frac{k}{s}\left[\frac{e^{\frac{as}{2}}-e^{\frac{-as}{2}}}{e^{\frac{as}{2}}+e^{-\frac{as}{2}}}\right] = \frac{k}{s}\tanh\frac{as}{2} \qquad \textbf{Ans.}$$

EXERCISE 5.5

1. Find the Laplace transform of the periodic function

$f(t) = e^t$ for $0 < t < 2\pi$ **Ans.** $\dfrac{e^{2(1-s)\pi}-1}{(1-s)\left(1-e^{-2\pi s}\right)}$

2. Obtain Laplace transform of full wave rectified sine wave given by

$f(t) = \sin \omega t, \quad 0 < t < \dfrac{\pi}{w}$ **Ans.** $\dfrac{\omega}{\left(s^2+\omega^2\right)}\coth\dfrac{\pi s}{2\omega}$

3. Find the Laplace transform of the staircase function

$f(t) = kn,\ np < t < (n+1)p,\ n = 0, 1, 2, 3$ **Ans.** $\dfrac{ke^{ps}}{s\left(1-e^{-ps}\right)}$

Find Laplace transform of the following:

4. $f(t)=t^2,\ 0<t<2,\ f(t+2)=f(t)$ **Ans.** $\dfrac{2-e^{-2s}-4se^{-2s}-4s^2e^{-2s}}{s^3\left(1-e^{-2s}\right)}$

5. $f(t)=\begin{cases}1, & 0\le t\le \dfrac{a}{2}\\ -1, & \dfrac{a}{2}\le t<a\end{cases}$ *(U.P. II Semester, 2004)* **Ans.** $\dfrac{1}{s}\tanh\dfrac{as}{4}$

6. $f(t)=\begin{cases}\cos\omega t, & 0<t<\dfrac{\pi}{\omega}\\ 0, & \dfrac{\pi}{\omega}<t<\dfrac{2\pi}{\omega}\end{cases}$ **Ans.** $\dfrac{s}{\left(s^2+\omega^2\right)\left(1-e^{-\frac{\pi s}{\omega}}\right)}$

7. $f(t)=\begin{cases}t, & 0<t<1\\ 0, & 1<t<2\end{cases}\quad f(t+2)=f(t)$ **Ans.** $\dfrac{1-e^{-s}(s+1)}{s^2\left(1-e^{-2s}\right)}$

8. $f(t)=\begin{cases}\dfrac{2t}{T}, & 0\le t\le\dfrac{T}{2}\\ \dfrac{2}{T}(T-t), & \dfrac{T}{2}\le t\le T\end{cases}\quad f(t+T)=f(t)$ **Ans.** $\dfrac{2}{Ts^2}\tanh\dfrac{sT}{4}-\dfrac{1}{s\left(e^{\frac{sT}{2}}+1\right)}$

5.22 IMPULSE FUNCTION

When a large force acts for a short time, then the product of the force and the time is called impulse in applied mechanics. The unit impulse function is the limiting function.

$$\delta(t-1)=\begin{bmatrix}\dfrac{1}{\varepsilon}, & a<t<a+\varepsilon\\ 0, & \text{otherwise}\end{bmatrix}$$

The value of the function (height of the strip in the figure) becomes infinite as $\varepsilon\to 0$ and the area of the rectangle is unity.

(1) The Unit Impulse function is defined as follows:

$$\delta(t-a)=\begin{cases}\infty & \text{for } t=a\\ 0 & \text{for } t\neq a\end{cases}$$

f(t)
$\frac{1}{\epsilon}$
O
t = a t = a + ∈
t

and $$\int_0^\infty \delta(t-a).dt=1$$ [Area of strip = 1]

(2) Laplace Transform of Unit Impulse function

$$\int_0^\infty f(t)\ \delta(t-a)\,dt=\int_a^{a+\varepsilon} f(t).\frac{1}{\varepsilon}dt$$

$\left\{\begin{array}{l}\text{Mean value Theorem}\\ \displaystyle\int_a^b f(t)\,dt=(b-a)\,f(\eta)\end{array}\right.$

$$=(a+\varepsilon-a)\,f(\eta).\frac{1}{\varepsilon}$$ where $a<\eta<a+\varepsilon$

$$=f(\eta)$$

Property I. $\boxed{\int_0^\infty f(t)\delta(t-a)\,dt=f(a)}$ as $\varepsilon\to 0$

Note. If $f(t)=e^{-st}$ and $L\left[\delta(t-a)\right]=e^{-as}$

Example 48. *Evaluate* $\int_{-\infty}^{\infty} e^{-5t}\delta(t-2)\,dt.$

Solution. $\int_{-\infty}^{\infty} e^{-5t}\delta(t-2)\,dt = e^{-5\times 2} = e^{-10}$ **Ans.**

Property II: $\boxed{\int_{-\infty}^{\infty} f(t)\delta'(t-a)\,dt = -f'(a)}$

Proof. $\int_{-\infty}^{\infty} f(t)\delta'(t-a)\,dt = \left[f(t).\delta(t-a)\right]_{-\infty}^{\infty} - \int_{-\infty}^{\infty} f'(t)\delta(t-a)\,dt$

$$= 0 - 0 - f'(a) = -f'(a)$$

Example 49. *Find the Laplace transform of* $t^3\delta(t-4)$

Solution. $Lt^3\delta(t-4) = \int_0^{\infty} e^{-st}t^3\delta(t-4)\,dt = 4^3 e^{-4s}$ **Ans.**

EXERCISE 5.6

Evaluate the following:

1. $\int_0^{\infty} e^{-3t}\delta(t-4)\,dt$ **Ans.** e^{-12}
2. $\int_{-\infty}^{\infty} \sin 2t\,\delta\left(t-\frac{\pi}{4}\right)dt$ **Ans.** 1
3. $\int_{-\infty}^{\infty} e^{-3t}\delta'(t-2)\,dt$ **Ans.** $3e^{-6}$

Find Laplace transform of

4. $\frac{\delta(t-4)}{t}$ **Ans.** $\frac{e^{-4s}}{4}$
5. $\cos t \log t\, \delta(t-\pi)$ **Ans.** $-e^{-\pi s}\log\pi$
6. $e^{-4t}\delta(t-3)$ **Ans.** $e^{-3(s+4)}$

5.23 CONVOLUTION THEOREM

If $L[f_1(t)] = F_1(s)$ and $L[f_2(t)] = F_2(s)$

then $L\left\{\int_0^t f_1(x) f_2(t-x)\,dx\right\} = F_1(s)\cdot F_2(s)$

or $\boxed{L^{-1}(F_1(s).F_2(s)) = \int_0^t f_1(x) f_2(t-x)\,dx}$

Proof. We have

$$L\left\{\int_0^{\infty} f_1(x) f_2(t-x)\,dx\right\} = \int_0^{\infty} e^{-st}\left[\int_0^t f_1(x) f_2(t-x)\,dx\right]dt \qquad \text{(By Definition)}$$

where the double integral is taken over the infinite region in the first quadrant lying between the lines $x = 0$ and $x = t$.

Here first we are integrating w.r.t. "x", within limits $x = 0$ and $x = t$, and then we will integrate w.r.t. "t" with limits $t = 0$ and $t = \infty$.

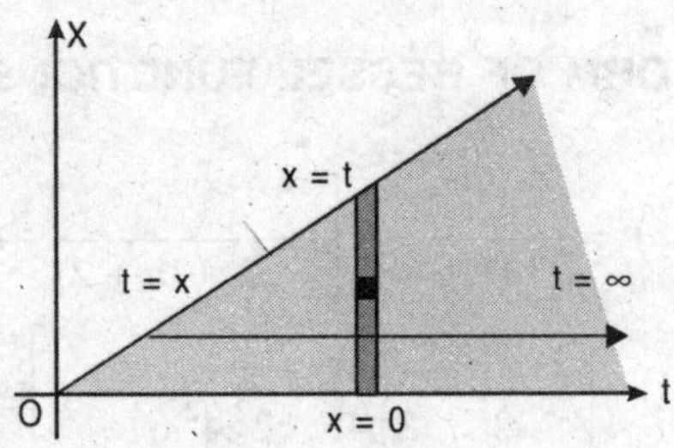

On changing the order of integration first we integrate w.r.t. "t" with limits $t = x$ and $t = \infty$ and then w.r.t. "x" with limits $x = 0$ and $x = \infty$.

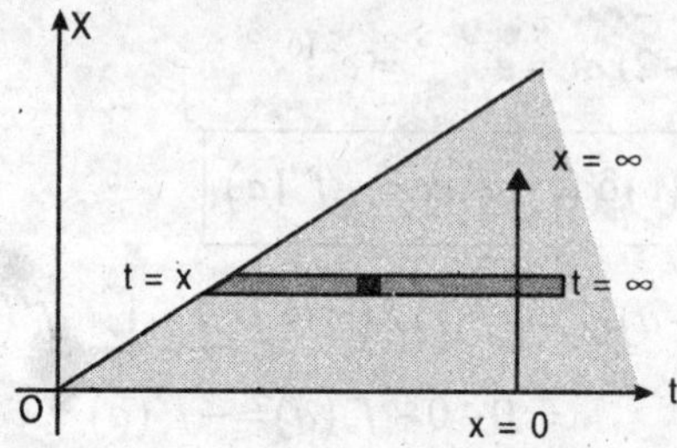

On changing the order of integration, the integral becomes

$$\int_0^\infty dx \left[\int_x^\infty e^{-st} f_1(x).f_2(t-x)\,dt\right]$$

$$= \int_0^\infty dx \left[\int_x^\infty e^{-s(t-x+x)} f_1.(x).f_2(t-x)\,dt\right]$$

$$= \int_0^\infty dx \left[\int_x^\infty e^{-s(t-x)}.e^{-sx} f_1(x).f_2(t-x)\,dt\right]$$

$$= \int_0^\infty e^{-sx} f_1(x)\,dx \left[\int_x^\infty e^{-s(t-x)} f_2(t-x)\,dt\right]$$

$$= \int_0^\infty e^{-sx} f_1(x)\,dx \left[\int_x^\infty e^{-sz} f_2(z)\,dz\right]$$

[Put $t - x = z \Rightarrow dt = dz$]

$$= \int_0^\infty e^{-sx} f_1(x)dx \int_0^\infty e^{-sz} f_2(z)dz,$$

[Lower limit $x - x = z \Rightarrow z = 0$]

$$= \int_0^\infty e^{-sx} f_1(x)F_2(s)dx = \left[\int_0^\infty e^{-sx} f_1(x)dx\right]F_2(s)$$

$$= F_1(s)F_2(s)$$

Proved.

Example 50. *Find the Laplace transform of* $\int_0^t e^x.\sin(t-x)\,dx$

Solution. By Convolution Theorem

$$L\int_0^t f_1(x)\,f_2(t-x)\,dx = F_1(s).\,F_2(s)$$

$$\Rightarrow \quad L\int_0^t e^x.\sin(t-x)\,dx = L(e^x).L\sin t$$

$$= \frac{1}{s-1}\,\frac{1}{s^2+1} = \frac{1}{(s-1)(s^2+1)}$$

Ans.

Note. Convolution Theorem is generally used to find Inverse Laplace transform of the product of two functions, discussed in the next chapter.

5.24 LAPLACE TRANSFORM OF BESSEL FUNCTIONS $J_0(x)$ and $J_1(x)$

We know that

$$J_n(x) = \frac{x^2}{2^n\,\Gamma(n+1)}\left[1 - \frac{x^2}{2.(2n+2)} + \frac{x^4}{2.4\,(2n+2)\,(2n+4)} - .....\right]$$

$$J_0(t) = \left[1 - \frac{t^2}{2^2} + \frac{t^4}{2^2.4^2} - \frac{t^6}{2^2.4^2.6^2} + ...\right]$$

Taking Laplace transforms of both sides, we have

$$LJ_0(t)=\frac{1}{s}-\frac{1}{2^2}\cdot\frac{2!}{s^3}+\frac{1}{2^2.4^2}\frac{4!}{s^5}-\frac{1}{2^2.4^2.6^2}\frac{6!}{s^7}+...$$

$$=\frac{1}{s}\left[1-\frac{1}{2}\left(\frac{1}{s^2}\right)+\frac{1.3}{2.4}\left(\frac{1}{s^4}\right)-\frac{1.3.5}{2.4.6}\left(\frac{1}{s^6}\right)+...\right]$$

$$=\frac{1}{s}\left[1+\left(-\frac{1}{2}\right)\left(\frac{1}{s^2}\right)+\frac{\left(-\frac{1}{2}\right)\left(-\frac{3}{2}\right)}{2!}\left(\frac{1}{s^2}\right)^2+\frac{\left(-\frac{1}{2}\right)\left(-\frac{3}{2}\right)\left(-\frac{5}{2}\right)}{3!}\left(\frac{1}{s^2}\right)^3+...\right]$$

$$=\frac{1}{s}\left[1+\frac{1}{s^2}\right]^{-\frac{1}{2}}$$ (By Binomial theorem)

$$=\frac{1}{s}\left[\frac{s^2+1}{s^2}\right]^{-\frac{1}{2}}=\frac{1}{s}\left[\frac{s^2}{s^2+1}\right]^{\frac{1}{2}}=\frac{1}{\sqrt{s^2+1}}$$ ... (1) **Ans.**

We know that $Lf(at)=\frac{1}{a}F\left(\frac{s}{a}\right)$

$$LJ_0(at)=\frac{1}{a}\frac{1}{\sqrt{\frac{s^2}{a^2}+1}}=\frac{1}{\sqrt{s^2+a^2}}$$ [From (1)]

$$LJ_1(x)=-LJ_0'(x)=-\left[s\,LJ_0(x)-J_0(0)\right]$$

$$=-\left[s.\frac{1}{\sqrt{s^2+1}}-1\right]=1-\frac{s}{\sqrt{s^2+1}}$$ **Ans.**

EXERCISE 5.7

Find the Laplace transform of the following:

1. $e^{ax}J_0(bx)$ **Ans.** $\frac{1}{\sqrt{s^2+2as+a^2+b^2}}$

2. $x\,J_0(ax)$ **Ans.** $\frac{s}{(s^2+a^2)^{3/2}}$

3. $x\,J_1(x)$ **Ans.** $\frac{1}{(s^2+1)^{3/2}}$

5.25 EVALUATION OF INTEGRALS

We can evaluate number of integrals having lower limit 0 and upper limit ∞ by the help of Laplace transform.

Example 51. *Evaluate* $\int_0^\infty te^{-3t}\sin t\,dt$

Solution. $\int_0^\infty te^{-3t}\sin t\,dt=\int_0^\infty te^{-st}\sin t\,dt$ $(s=3)$

$$=L(t\sin t)=-\frac{d}{ds}\left(\frac{1}{s^2+1}\right)=\frac{2s}{\left(s^2+1\right)^2}$$

Putting $s = 3$, we get

$$= \frac{2\times 3}{\left(3^2+1\right)^2} = \frac{6}{100} = \frac{3}{50}$$ **Ans.**

Example 52. *Evaluate* $\int_0^\infty \frac{e^{-t}\sin t}{t}\,dt$ and $\int_0^\infty \frac{\sin t}{t}\,dt$ *(U.P., II Semester, 2009)*

Solution. $\int_0^\infty \frac{e^{-t}\sin t}{t}\,dt = \int_0^\infty e^{-st}\frac{\sin t}{t}\,dt$ $(s = 1)$

$$= L\left[\frac{\sin t}{t}\right] = \int_s^\infty \frac{1}{s^2+1}\,ds = \left[\tan^{-1}s\right]_s^\infty$$

$$= \frac{\pi}{2} - \tan^{-1}s \qquad \text{...(1)}$$

$$= \frac{\pi}{2} - \tan^{-1}(1) \qquad (s = 1)$$

$$= \frac{\pi}{2} - \frac{\pi}{4} = \frac{\pi}{4}$$ **Ans.**

On putting $s = 0$ in (1), we get

$$\int_0^\infty \frac{\sin t}{t}\,dt = \frac{\pi}{2} - \tan^{-1}(0) = \frac{\pi}{2}$$ **Ans.**

Example 53. *Evaluate* $\int_0^\infty \frac{e^{-at} - e^{-bt}}{t}\,dt$

Solution. $\int_0^\infty e^{-st}(e^{-at} - e^{-bt})\,dt = L(e^{-at} - e^{-bt}) = L(e^{-at}) - L(e^{-bt})$

$$= \left(\frac{1}{s+a} - \frac{1}{s+b}\right)$$

$$\int_0^\infty e^{-st}\left(\frac{e^{-at} - e^{-bt}}{t}\right)dt = L\frac{1}{t}(e^{-at} - e^{-bt}) = \int_s^\infty \left(\frac{1}{s+a} - \frac{1}{s+b}\right)ds$$

$$= [\log(s+a) - \log(s+b)]_s^\infty$$

$$= \left[\log\left(\frac{s+a}{s+b}\right)\right]_s^\infty = \left[\log\frac{1+\frac{a}{s}}{1+\frac{b}{s}}\right]_s^\infty = \left[\log 1 - \log\frac{s+a}{s+b}\right]$$

$$= \log\frac{s+b}{s+a}$$

Putting s = 0 in above, we get

$$\int_0^\infty \left(\frac{e^{-at} - e^{-bt}}{t}\right)dt = \log\left(\frac{b}{a}\right)$$ **Ans.**

Example 54. *Show that* $\int_0^\infty t^3 e^{-t}\sin t\,dt = 0$

Solution. $L\{t^3 \sin t\} = (-1)^3 \frac{d^3}{ds^3} L\{\sin t\}$

$$\Rightarrow \quad \int_0^\infty e^{-st}\, t^3 \sin t\,dt = \frac{-d^3}{ds^3}\,\frac{1}{s^2+1} = -\frac{d^2}{ds^2}\left[-\frac{2s}{(s^2+1)^2}\right]$$

$$= \frac{d}{ds}\left[\frac{(s^2+1)^2(2) - 2s\,[2(s^2+1)]\,(2s)}{(s^2+1)^4}\right]$$

$$= \frac{d}{ds}\left[\frac{2(s^2+1) - 8s^2}{(s^2+1)^3}\right] = \frac{d}{ds}\left[\frac{-6s^2+2}{(s^2+1)^3}\right]$$

$$= \frac{(s^2+1)^3(-12s) - (-6s^2+2)3(s^2+1)^2(2s)}{(s^2+1)^6}$$

$$= \frac{(s^2+1)(-12s) - (-6s^2+2)\,6s}{(s^2+1)^4} = \frac{-12s^3 - 12s + 36s^3 - 12s}{(s^2+1)^4}$$

$$\int_0^\infty e^{-st} t^3 \sin t\, dt = \frac{24s^3 - 24s}{(s^2+1)^4} = \frac{24s\,(s^2-1)}{(s^2+1)^4} \quad \text{... (1)}$$

Putting $s = 1$ in (1), we get

$$\int_0^\infty e^{-t}\, t^3 \sin t\, dt = 0 \qquad \textbf{Ans.}$$

Example 55. *Evaluate* $\int_0^\infty t^2 e^{3t} \sin^2 t\, dt$

Solution. We have, $\sin^2 t = \frac{1}{2}(1 - \cos 2t)$

$$\Rightarrow \quad L\{\sin^2 t\} = \frac{1}{2}[L\{1\} - L\{\cos 2t\}] = \frac{1}{2}\left[\frac{1}{s} - \frac{s}{s^2+4}\right]$$

$$\Rightarrow \quad L\left[t^2 \sin^2 t\right] = (-1)^2 \cdot \frac{d^2}{ds^2}\left[\frac{1}{2}\left\{\frac{1}{s} - \frac{s}{s^2+4}\right\}\right]$$

$$\Rightarrow \quad \int_0^\infty e^{-st}\, t^2 \sin^2 t\, dt = \frac{1}{2}\frac{d}{ds}\left[\frac{d}{ds}\left\{\frac{1}{s} - \frac{s}{s^2+4}\right\}\right]$$

$$= \frac{1}{2}\frac{d}{ds}\left[-\frac{1}{s^2} - \frac{(s^2+4)(1) - s(2s)}{(s^2+4)^2}\right]$$

$$= \frac{1}{2}\frac{d}{ds}\left[-\frac{1}{s^2} - \frac{-s^2+4}{(s^2+4)^2}\right]$$

$$= \frac{1}{2}\left[\frac{2}{s^3} - \frac{(s^2+4)^2(-2s) - (-s^2+4)\,2(s^2+4)(2s)}{(s^2+4)^4}\right]$$

$$= \frac{1}{2}\left[\frac{2}{s^3} - \frac{(s^2+4)(-2s) - (-s^2+4)\,4s}{(s^2+4)^3}\right] \quad \text{... (1)}$$

Putting the value of $s = -3$ in (1), we get

$$\int_0^\infty e^{3t}\, t^2 \sin^2 t\, dt = \frac{1}{2}\left[\frac{2}{-27} - \frac{(13)6 - (-5)(-12)}{(9+4)^3}\right]$$

$$= -\frac{1}{27} - \frac{9}{(13)^3} = \frac{-2197 - 243}{59319} = \frac{-2440}{59319} \qquad \textbf{Ans.}$$

EXERCISE 5.8

Evaluate the following by using Laplace Transform:

1. $\int_0^\infty t e^{-4t} \sin t \, dt$ **Ans.** $\frac{8}{289}$

2. $\int_0^\infty \frac{e^{-2t} \sinh t \sin t}{t} dt$ **Ans.** $\frac{1}{2}\tan^{-1}\frac{1}{2}$

3. $\int_0^\infty \frac{\sin^2 t}{t^2} dt$ **Ans.** $i\frac{5}{2}$

4. $\int_0^\infty \frac{e^{-t} - e^{-4t}}{t} dt$ **Ans.** log 4

5.26 FORMULATION OF LAPLACE TRANSFORM

S.No.	f (*t*)	*F*(*s*)
1.	e^{at}	$\frac{1}{s-a}$
2.	t^n	$\frac{\boxed{n+1}}{s^{n+1}}$ or $\frac{n!}{s^{n+1}}$
3.	sin *at*	$\frac{a}{s^2+a^2}$
4.	cos *at*	$\frac{s}{s^2+a^2}$
5.	sinh *at*	$\frac{a}{s^2-a^2}$
6.	cosh *at*	$\frac{s}{s^2-a^2}$
7.	$u(t-a)$	$\frac{e^{-as}}{s}$
8.	$\delta(t-a)$	e^{-as}
9.	$e^{bt} \sin at$	$\frac{a}{(s-b)^2+a^2}$
10.	$e^{bt} \cos at$	$\frac{s-b}{(s-b)^2+a^2}$
11.	$\frac{t}{2a}\sin at$	$\frac{s}{(s^2+a^2)^2}$
12.	$t \cos at$	$\frac{s^2-a^2}{(s^2+a^2)^2}$
13.	$\frac{1}{2a^3}(\sin at - at\cos at)$	$\frac{1}{(s^2+a^2)^2}$
14.	$\frac{1}{2a}(\sin at + at \cos at)$	$\frac{s^2}{(s^2+a^2)^2}$

5.27 PROPERTIES OF LAPLACE TRANSFORM

S.No.	Property	f(t)	F (s)
1.	Scaling	$f(at)$	$\frac{1}{a}F\left(\frac{s}{a}\right)$, $a > 0$
2.	Derivative	$\frac{df(t)}{dt}$	$s\,F(s) - f(0)$, $s > 0$
		$\frac{d^2 f(t)}{dt^2}$	$s^2 F(s) - sf(0) - f'(0)$, $s > 0$
		$\frac{d^3 f(t)}{dt^3}$	$s^3F(s) - s^2f(0) - s\,f'(0) - f''(0)$, $s > 0$
3.	Integral	$\int_0^t f(t)\,dt$	$\frac{1}{s}F(s)$, $s > 0$
4.	Initial Value	$\lim_{t\to 0} f(t)$	$\lim_{s\to\infty} sF(s)$
5.	Final Value	$\lim_{t\to\infty} f(t)$	$\lim_{s\to\infty} sF(s)$
6.	First shifting	$e^{-at} f(t)$	$F(s + a)$
7.	Second shifting	$f(t)\,u(t - a)$	$e^{-a}\,L\,f(t + a)$
8.	Multiplication by t	$t\,f(t)$	$-\frac{d}{ds}F(s)$
9.	Multiplication by t^n	$t^n f(t)$	$(-1)^n \frac{d^n}{ds^n}F(s)$
10.	Division by t	$\frac{1}{t}f(t)$	$\int_s^\infty F(s)\,ds$
11.	Periodic function	$f(t)$	$\frac{\int_0^T e^{-st} f(t)}{1-e^{-st}}$ $f(t + T) = f(t)$
12.	Convolution	$f(t) * g(t)$	$F(s)\,G(s)$

OBJECTIVE TYPE QUESTIONS

Choose the correct alternative :

1. Laplace transform sin $(2t)$ =

(*i*) $\frac{s}{s^2+2}$ (*ii*) $\frac{4}{s^2+2}$ (*iii*) $\frac{2}{s^2+4}$ (*iv*) $\frac{s}{s^2+1}$ **Ans.** (*iii*)

2. L (sinh $3t$) =

(*i*) $\frac{3}{s^2-9}$ (*ii*) $\frac{3}{s^2+9}$ (*iii*) $\frac{1}{s^2+9}$ (*iv*) $\frac{s}{s^2+9}$ **Ans.** (*i*)

3. $L(\cos t) =$

(i) $\dfrac{s}{s^2-1}$ (ii) $\dfrac{s}{s^2+1}$ (iii) $\dfrac{1}{s^2+1}$ (iv) $\dfrac{-s}{s^2+1}$ **Ans.** (ii)

4. $L(t^n) =$

(i) $\dfrac{\lfloor n}{s^n+1}$ (ii) $\dfrac{(n+1)!}{s^2+1}$ (iii) $\dfrac{1}{s^n+1}$ (iv) $\dfrac{\lfloor n+1}{s^n+1}$ **Ans.** (iv)

5. Laplace transform t^3e^{-3t} is :

(i) $\dfrac{7}{(s+4)^3}$ (ii) $\dfrac{s}{(s+3)^3}$ (iii) $\dfrac{6}{(s+3)^4}$ (iv) $\dfrac{2}{(s+6)^3}$ **Ans.** (iii)

(R.G.P.V., Bhopal, II Semester, Fer. 2006)

6. Laplace transform of t^ne^{-at} is

(i) $\dfrac{\lceil n}{(s+a)^n}$ (ii) $\dfrac{(n+1)!}{(s+a)^{n+1}}$ (iii) $\dfrac{n!}{(s+a)^n}$ (iv) $\dfrac{\lceil n+1}{(s+a)^{n+1}}$ **Ans.** (iv)

7. Laplace transform of $f(t) = te^{at}.\sin(at), t > 0$

(i) $\dfrac{2a(s-a)}{\left[(s-a)^2+a^2\right]^2}$ (ii) $\dfrac{a(s-a)}{(s-a)^2+a^2}$ (iii) $\dfrac{s-a}{(s-a)^2+a^2}$ (iv) $\dfrac{(s-a)^2}{(s-a)^2+a^2}$ **Ans.** (i)

8. If $f(x) = x^4 f(x)$, where $f(x)$ has derivatives of all orders, then $L\left[\dfrac{d^4 f(x)}{dx^4}\right]$ is given by

(i) $s^3L[f(x)]$ (ii) $s^4Lf(x)$ (iii) $s^4L[f^3(x)]$ (iv) none of these **Ans.** (ii)

9. The Laplace transform of te^{-t} cosh $2t$ is

(i) $\dfrac{s^2+2s+5}{(s^2+2s-3)^2}$; (ii) $\dfrac{s^2-2s+5}{(s^2+2s-3)^2}$; (iii) $\dfrac{4s+4}{(s^2+2s-3)^2}$; (iv) $\dfrac{4s-4}{(s^2+2s-3)^2}$ **Ans.** (ii)

10. Laplace tranform e^{-2t} sin $4t$ is :

(i) $\dfrac{2}{s^2+4s+20}$ (ii) $\dfrac{s-2}{s^2+4s+20}$ (iii) $\dfrac{s-4}{s^2+4s+20}$ (iv) $\dfrac{4}{s^2+4s+20}$ **Ans.** (iv)

(R.G.P.V., Bhopal, II Semester, June 2007)

11. If $\{F(t)\} = \bar{f}(s)$, then $L\left\{\int_0^t F(x)\,dx\right\}$ is :

(i) $\int_0^s \bar{f}(s)\,ds$ (ii) $\int_0^s \frac{1}{s}\bar{f}(s)\,ds$ (iii) $\frac{1}{s}\bar{f}(s)$ (iv) $s\,\bar{f}(s)$ **Ans.** (iii)

(R.G.P.V., Bhopal, II Semester, June 2006)

12. If $L\{F(t)\} = \bar{f}(s)$, then $L\{t\,F(t)\}$ is :

(i) $\bar{f}'(s)$ (ii) $-\bar{f}'(s)$ (iii) $\bar{f}'(s)+\bar{f}(s)$ (iv) $s\bar{f}'(s)+\bar{f}(s)$ **Ans.** (ii)

(R.G.P.V., Bhopal, II Semester, June 2006)

13. $L\left(\dfrac{1}{t}f(t)\right) =$

(i) $\int_0^s F(s)\,dx$ (ii) $\int_s^\infty F(s)\,dx$ (iii) $\int_0^\infty F(s)\,dx$ (iv) $\int_{-\infty}^\infty F(s)\,dx$ **Ans.** (ii)

14. $L\left(\frac{\sin 2t}{t}\right) =$

(*i*) $\int_0^\infty \frac{2}{s^2+4}\,ds$ (*ii*) $\int_0^\infty \frac{2}{s^2+4}\,ds$ (*iii*) $\int_s^\infty \frac{1}{s^2+4}\,ds$ (*iv*) $\int_{-\infty}^\infty \frac{5}{s^2+1}\,ds$ **Ans.** (*i*)

15. Laplace transform of $\frac{\cos at - \cos bt}{t}$ is

(*i*) $\log \frac{s^2+b^2}{s^2+a^2}$ (*ii*) $\frac{1}{2}\log \frac{s^2+a^2}{s^2+b^2}$ (*iii*) $\frac{1}{2}\log \frac{s^2+b^2}{s^2+a^2}$ (*iv*) $\log \frac{s+b}{s+a}$ **Ans.** (*iii*)

(*R.G.P.V., Bhopal, II Semester, Feb. 2006, 2005*)

16. Laplace transform of $t^m e^{bt}$ is

(*i*) $\frac{\Gamma m}{(s+b)^m}$ (*ii*) $\frac{\Gamma(m+1)}{(s-b)^{m+1}}$ (*iii*) $\frac{(m+1)!}{(s+b)^{m+1}}$ (*iv*) $\frac{m!}{(s-b)^m}$ **Ans.** (*ii*)

17. The Laplace transform of $t\,u_2(t)$ is

(*i*) $\left(\frac{1}{s^2}+\frac{2}{s}\right)e^{-2s}$ (*ii*) $\frac{1}{s^2}e^{-2s}$ (*iii*) $\left(\frac{1}{s^2}-\frac{2}{s}\right)e^{-2s}$ (*iv*) $\frac{e^{-2s}}{s^2}$ **Ans.** (*i*)

18. $L[t^2 u(t-3)]$ is

(*i*) $e^{-3s}\left(\frac{1}{s^3}+\frac{6}{s^2}+\frac{9}{s}\right)$ (*ii*) $e^{-s^3}\left(\frac{1}{s}+\frac{9}{s^2}+\frac{6}{s}\right)$

(*iii*) $e^{-3s}\left(\frac{2}{s^3}+\frac{6}{s^2}+\frac{9}{s}\right)$ (*iv*) None of these **Ans.** (*iii*)

19. Laplace transform of $t\,u\,(t-2)$ is

(*i*) $\frac{1}{s^2}e^{-2s}$ (*ii*) $\left(\frac{1}{s^2}-\frac{2}{s}\right)e^{-2s}$ (*iii*) $\left(\frac{1}{s^2}+\frac{2}{s}\right)e^{-2s}$ (*iv*) $\frac{2e^{-2s}}{s}$ **Ans.** (*iii*)

20. The Laplace transform of the function

$$F(t) = \begin{cases} 1, & 0 \le t < 2 \\ -1, & 2 \le t < 4 \end{cases}, \quad f(t+4) = f(t) \text{ is given as,}$$

(*i*) $\frac{1-e^{-2s}}{s(1+e^{-2s})}$ (*ii*) $\frac{1+e^{-2s}}{s(1+e^{-2s})}$ (*iii*) 0 (*iv*) $\frac{s+1}{s-1}$ **Ans.** (*i*)

(*U.P., II Semester, 2009*)

21. $L\left\{\frac{\delta(t-\pi)}{t}\right\}$ is equal to

(*i*) $\frac{e^{s\pi}}{\pi}$ (*ii*) $\pi e^{-s\pi}$ (*iii*) $e^{-s\pi}$ (*iv*) $\frac{e^{-s\pi}}{\pi}$ **Ans.** (*iv*)

22. $L\left\{\int_0^1 f_1(x) f_2(t-x)\,dx\right\} =$

(*i*) $\int f_1(s)\cdot f_2(s)\,ds$ (*ii*) $F_1(s)\times F_2(s)$

(*iii*) $\frac{d}{dx}F_1(s)\cdot F_2(s)$ (*iv*) $F_1(s)\cdot\frac{d}{dx}F_2(s)$ **Ans.** (*ii*)

23. $L\,f(at) =$

(*i*) $F\left(\frac{s}{a}\right)$ (*ii*) $\frac{2}{a}F\left(\frac{s}{a}\right)$ (*iii*) $\frac{1}{a}F\left(\frac{s}{a}\right)$ (*iv*) $2F\left(\frac{s}{a}\right)$ **Ans.** (*iii*)

24. $L\,\{J_0(t)\} =$

(*i*) $\frac{1}{\sqrt{s^2+1}}$ (*ii*) $\frac{5}{\sqrt{s^2+2}}$ (*iii*) $\frac{s}{\sqrt{s^2+4}}$ (*iv*) $\frac{s^2+1}{s}$ **Ans.** (*i*)

25. $L\,J_1\,(t) =$

(*i*) $\frac{s}{\sqrt{s^2+1}}$ (*ii*) $1-\frac{s}{\sqrt{s^2+1}}$ (*iii*) $\frac{s}{\sqrt{s^2+2}}-1$ (*iv*) $2-\frac{\sqrt{s^2+1}}{s}$ **Ans.** (*ii*)

26. The Laplace transform of a unit step function with step at point a is:

(*i*) e^{-as} (*ii*) e^{-as}/s (*iii*) se^{-as} (*iv*) e^{-as}/s^2 **Ans.** (*ii*)

Fill in the blanks for each of the following questions:

27. $L\,(e^{4t}\sin t) =$ ...... **Ans.** $\frac{1}{s^2-8s+17}$

28. $L(\cosh 5t) =$ ...... **Ans.** $\frac{s}{s^2-25}$

29. $L\left(\frac{1}{\sqrt{t}}\right) =$ ...... **Ans.** $\sqrt{\frac{\pi}{s}}$

30. $L(\sqrt{t^3}) =$ .... **Ans.** $\frac{\sqrt{\frac{5}{2}}}{s^{5/2}}$

31. $\int_0^\infty e^{-2t}\cos 3t\,dt =$ ...... **Ans.** 2/13

32. Laplace transform of ($t\sin 3t$) = ... **Ans.** $\frac{6s}{(s^2+9)^2}$

33. Laplace transform of $F''(t)$ = ... **Ans.** $s^2f(s) - s\,F(0) - F'(0)$

34. The Laplace's transform of

$\int_0^t\int_0^t\int_0^t \cos au\,du\,du\,du$ is given as ...... [*U.P.T.U. (SUM) 2009*] **Ans.** $\frac{1}{s^2(s^2+a^2)}$

35. $L(e^{4t}\cos 3t) =$ ...... **Ans.** $\frac{s-4}{s^2-8s+25}$

36. $L[u\,(t-a)] =$ ...... **Ans.** $\frac{e^{-as}}{s}$

Indicate True or False for the following

37. $L(t^4) = \frac{4!}{s^5}$ **Ans.** True

38. $\int_0^\infty f(t)\,\delta\,(t-a)\,dt = f(a) - f(0)$ **Ans.** False

39. Ordinary and partial differential equations can be solved by Laplace Transformation **Ans.** True

40. $L\left(erf\sqrt{t}\right) = \frac{1}{s\sqrt{s+1}}$ **Ans.** True

Match the items on the right hand side with those on left hand side :

41. (*i*) $L(e^{2t})$	(*a*) $\frac{1}{s-2}$	**Ans.** (*i*) → (*a*)
(*ii*) $L(\sin t)$	(*b*) $\frac{s}{s^2+9}$	**Ans.** (*ii*) → (*d*)
(*iii*) $L(\cos 3t)$	(*c*) $\frac{s}{s^2-25}$	**Ans.** (*iii*) → (*b*)
(*iv*) $L(\cosh 5t)$	(*d*) $\frac{1}{s^2+1}$	**Ans.** (*iv*) → (*c*)
42. (*i*) $L(\sinh t)$	(*a*) $\frac{3!}{s^4}$	**Ans.** (*i*) → (*c*)
(*ii*) $L(t^3)$	(*b*) $\frac{1}{s-4}$	**Ans.** (*ii*) → (*a*)
(*iii*) $L(1)$	(*c*) $\frac{1}{s^2-1}$	**Ans.** (*iii*) → (*d*)
(*iv*) $L(e^{4t})$	(*d*) $\frac{1}{s}$	**Ans.** (*iv*) → (*b*)
43. (*i*) $L(e^{-t}.\ t^3)$	(*a*) $\frac{1}{a}F\left(\frac{s}{a}\right)$	**Ans.** (*i*) → (*d*)
(*ii*) $L\{f(at)\}$	(*b*) $F(s-a)$	**Ans.** (*ii*) → (*a*)
(*iii*) $L\{e^{at}f(t)\}$	(*c*) $\sqrt{\frac{\pi}{s}}$	**Ans.** (*iii*) → (*b*)
(*iv*) $L\left(t^{-\frac{1}{2}}\right)$	(*d*) $\frac{3!}{(s+1)^4}$	**Ans.** (*iv*) → (*c*)
44. (*i*) $L(t^3 e^{-2t})$	(*a*) $\frac{1}{s}F(s)$	**Ans.** (*i*) → (*c*)
(*ii*) $L[f'(t)]$	(*b*) $\lim_{s\to\infty} f(t)$	**Ans.** (*ii*) → (*d*)
(*iii*) $L\left[\int_0^t f(t)\,dt\right]$	(*c*) $\frac{6}{(s+2)^4}$	**Ans.** (*iii*) → (*a*)
(*iv*) $\lim_{s\to 0}[s\,F(s)]$	(*d*) $s\,[L\,f(t)-f(0)]$	**Ans.** (*iv*) → (*b*)
45. (*i*) $L[\sinh 6t]$	(*a*) $\frac{1}{s}\left(1-\frac{1}{\sqrt{s+1}}\right)$	**Ans.** (*i*) → (*b*)
(*ii*) $L\left(erf_c\sqrt{t}\right)$	(*b*) $\frac{s}{s^2-36}$	**Ans.** (*ii*) → (*a*)
(*iii*) $L\left[e^t \sin 4t\right]$	(*c*) $\int_s^\infty f(s)\,ds$	**Ans.** (*iii*) → (*d*)
(*iv*) $L\left[\frac{f(t)}{t}\right]$	(*d*) $\frac{4}{(s-1)^2+16}$	**Ans.** (*iv*) → (*c*)

UNIT–II

6 CHAPTER

INVERSE LAPLACE TRANSFORMS

(SOLUTION OF DIFFERENTIAL EQUATIONS)

6.1 INVERSE LAPLACE TRANSFORMS

If $F(s)$ is the Laplace Transform of a function $f(t)$, then $f(t)$ is known as Inverse Laplace Transform. Now we will discuss how to find $f(t)$ when $F(s)$ is given.

If $L[f(t)] = F(s)$, then $L^{-1}[F(s)] = f(t)$, where L^{-1} is called the Inverse Laplace Transform operator.

From the application point of view, the Inverse Laplace Transform is very useful.

Inverse Laplace Transform is used in solving differential equations without finding the general solution and arbitrary constants.

6.2 IMPORTANT FORMULAE

1. $L^{-1}\left(\frac{1}{s}\right) = 1$

2. $L^{-1}\frac{1}{s^n} = \frac{t^{n-1}}{(n-1)!}$

3. $L^{-1}\frac{1}{s-a} = e^{at}$

4. $L^{-1}\frac{s}{s^2-a^2} = \cosh at$

5. $L^{-1}\frac{1}{s^2-a^2} = \frac{1}{a}\sinh at$

6. $L^{-1}\frac{1}{s^2+a^2} = \frac{1}{a}\sin at$

7. $L^{-1}\frac{s}{s^2+a^2} = \cos at$

8. $L^{-1}F(s-a) = e^{at}f(t)$

9. $L^{-1}\frac{1}{(s-a)^2+b^2} = \frac{1}{b}e^{at}\sin bt$

10. $L^{-1}\frac{s-a}{(s-a)^2+b^2} = e^{at}\cos bt$

11. $L^{-1}\frac{1}{(s-a)^2-b^2} = \frac{1}{b}e^{at}\sinh bt$

12. $L^{-1}\frac{s-a}{(s-a)^2-b^2} = e^{at}\cosh bt$

13. $L^{-1}\frac{1}{(s^2+a^2)^2} = \frac{1}{2a^3}(\sin at - at\cos at)$

14. $L^{-1}\frac{s}{(s^2+a^2)^2} = \frac{1}{2a}t\sin at$

15. $L^{-1}\frac{s^2-a^2}{(s^2+a^2)^2} = t\cos at$

16. $L^{-1}(1) = s(t)$

17. $L^{-1}\frac{s^2}{(s^2+a^2)^2} = \frac{1}{2a}[\sin at + at\cos at]$

18. $L^{-1}\left\{\frac{1}{s}F(s)\right\} = \int_0^t f(t)\,dt$

Example 1. *Show that* $\frac{1}{s^{1/2}} = L\left[\frac{1}{\sqrt{\pi t}}\right]$. *(U.P., II Semester, Summer 2005)*

Solution. We have to show that $\frac{1}{s^{1/2}} = L\left[\frac{1}{\sqrt{\pi t}}\right]$.

Now, $$L^{-1}\left\{\frac{1}{s^n}\right\} = \frac{t^{n-1}}{(n-1)!} = \frac{t^{n-1}}{\lceil n}$$

So $$L^{-1}\left\{\frac{1}{s^{1/2}}\right\} = \frac{t^{\frac{1}{2}-1}}{\lceil 1/2} = \frac{t^{-1/2}}{\lceil 1/2} = \frac{t^{-1/2}}{\sqrt{\pi}}$$

$$\Rightarrow \quad L^{-1}\left\{\frac{1}{s^{1/2}}\right\} = \frac{1}{\sqrt{\pi t}} \quad \Rightarrow \quad \frac{1}{s^{1/2}} = L\left[\frac{1}{\sqrt{\pi t}}\right]$$ **Proved.**

Example 2. *Find the inverse Laplace Transform of the following:*

(i) $\frac{1}{s-2}$ *(ii)* $\frac{1}{s^2-9}$ *(iii)* $\frac{s}{s^2-16}$

(iv) $\frac{1}{s^2+25}$ *(v)* $\frac{s}{s^2+9}$ *(vi)* $\frac{1}{(s-2)^2+1}$

(vii) $\frac{s-1}{(s-1)^2+4}$ *(viii)* $\frac{1}{(s+3)^2-4}$ *(ix)* $\frac{s+2}{(s+2)^2-25}$

(x) $\frac{1}{2s-7}$

Solution.

(i) $L^{-1}\frac{1}{s-2} = e^{2t}$

(ii) $L^{-1}\frac{1}{s^2-9} = L^{-1}\frac{1}{3}\cdot\frac{3}{s^2-(3)^2} = \frac{1}{3}\sinh 3t$

(iii) $L^{-1}\frac{s}{s^2-16} = L^{-1}\frac{s}{s^2-(4)^2} = \cosh 4t$

(iv) $L^{-1}\frac{1}{s^2+25} = \frac{1}{5}\frac{5}{s^2+(5)^2} = \frac{1}{5}\sin 5t$

(v) $L^{-1}\frac{s}{s^2+9} = \frac{s}{s^2+(3)^2} = \cos 3t$

(vi) $L^{-1}\frac{1}{(s-2)^2+1} = e^{2t}\sin t$

(vii) $L^{-1}\frac{s-1}{(s-1)^2+4} = e^t\cos 2t$

(viii) $L^{-1}\frac{1}{(s+3)^2-4} = \frac{1}{2}\frac{2}{(s+3)^2-(2)^2} = \frac{1}{2}e^{-3t}\sinh 2t$

(ix) $L^{-1}\frac{s+2}{(s+2)^2-25} = L^{-1}\frac{(s+2)}{(s+2)^2-(5)^2} = e^{-2t}\cosh 5t$

(x) $L^{-1}\frac{1}{2s-7} = \frac{1}{2}e^{\frac{7}{2}t}$ $\left[L^{-1}F(as) = \frac{1}{a}f\left(\frac{t}{a}\right)\right]$

Example 3. *Find* $L^{-1}\left[\dfrac{s^2+2s+6}{s^3}\right]$ *(M.D.U. 2010)*

Solution. Here, we have

$$L^{-1}\left[\frac{s^2+2s+6}{s^3}\right] = L^{-1}\left[\frac{1}{s}+\frac{2}{s^2}+\frac{6}{s^3}\right] = 1+\frac{2t}{1!}+\frac{6t^2}{2!}$$

$$= 1 + 2t + 3t^2 \qquad \textbf{Ans.}$$

Example 4. *Find Inverse Laplace Transform of*

(*a*) $\left\{\dfrac{6}{2s-3}-\dfrac{3+4s}{9s^2-16}+\dfrac{8-6s}{16s^2+9}\right\}$ (*b*) $\dfrac{2s-5}{9s^2-25}$

(*c*) $\dfrac{s-2}{6s^2+20}$ *(U.P. II. Semester Summer* 2001)

Solution.

$$(a)\ L^{-1}\left\{\frac{6}{2s-3}-\frac{3}{9s^2-16}-\frac{4s}{9s^2-16}+\frac{8}{16s^2+9}-\frac{6s}{16s^2+9}\right\}$$

$$= L^{-1}\left\{\frac{3}{s-\frac{3}{2}}-\frac{\frac{1}{3}}{s^2-\left(\frac{4}{3}\right)^2}-\frac{\frac{4}{9}s}{s^2-\left(\frac{4}{3}\right)^2}+\frac{\frac{1}{2}}{s^2+\left(\frac{3}{4}\right)^2}-\frac{\frac{3}{8}s}{s^2+\left(\frac{3}{4}\right)^2}\right\}$$

$$= L^{-1}\left\{\frac{3}{s-\frac{3}{2}}-\frac{1}{4}\frac{\frac{4}{3}}{s^2-\left(\frac{4}{3}\right)^2}-\frac{4}{9}\frac{s}{s^2-\left(\frac{4}{3}\right)^2}+\frac{2}{3}\frac{\frac{3}{4}}{s^2+\left(\frac{3}{4}\right)^2}-\frac{3}{8}\frac{s}{s^2+\left(\frac{3}{4}\right)^2}\right\}$$

$$= 3\,e^{\frac{3}{2}t}-\frac{1}{4}\sinh\frac{4}{3}t-\frac{4}{9}\cosh\frac{4}{3}t+\frac{2}{3}\sin\frac{3}{4}t-\frac{3}{8}\cos\frac{3}{4}t \qquad \textbf{Ans.}$$

$$(b)\ L^{-1}\frac{2s-5}{9s^2-25} = L^{-1}\left[\frac{2s}{9s^2-25}-\frac{5}{9s^2-25}\right] = L^{-1}\left[\frac{2s}{9\left[s^2-\left(\frac{5}{3}\right)^2\right]}-\frac{5}{9\left[s^2-\left(\frac{5}{3}\right)^2\right]}\right]$$

$$= \frac{2}{9}\cosh\frac{5}{3}t-\frac{1}{3}L^{-1}\left(\frac{\frac{5}{3}}{s^2-\left(\frac{5}{3}\right)^2}\right) = \frac{2}{9}\cosh\frac{5t}{3}-\frac{1}{3}\sin\frac{5t}{3} \qquad \textbf{Ans.}$$

$$(c)\ L^{-1}\frac{s-2}{6s^2+20} = L^{-1}\frac{s}{6s^2+20}-L^{-1}\frac{2}{6s^2+20} = \frac{1}{6}L^{-1}\frac{s}{s^2+\frac{10}{3}}-\frac{1}{3}L^{-1}\frac{1}{s^2+\frac{10}{3}}$$

$$= \frac{1}{6}L^{-1}\frac{s}{s^2+\frac{10}{3}}-\frac{1}{3}\times\sqrt{\frac{3}{10}}\,L^{-1}\frac{\sqrt{\frac{10}{3}}}{s^2+\frac{10}{3}} = \frac{1}{6}\cos\sqrt{\frac{10}{3}}\,t-\frac{1}{\sqrt{30}}\sin\sqrt{\frac{10}{3}}\,t \qquad \textbf{Ans.}$$

Example 5. *Find the inverse Laplace transform of following functions:*

$\dfrac{14s+10}{49s^2+28s+13}$. *[U.P., II Semester, 2007]*

Solution. The given function can be writtan as

$$\frac{14s+10}{49s^2+28s+13}=\frac{14s+10}{(7s+2)^2+9}=\frac{14\left(s+\frac{2}{7}\right)+6}{49\left(s+\frac{2}{7}\right)^2+9}$$

$$\therefore \quad L^{-1}\left(\frac{14\,s+10}{49\,s^2+28\,s+13}\right)=L^{-1}\left[\frac{14\left(s+\frac{2}{7}\right)+6}{49\left(s+\frac{2}{7}\right)^2+9}\right]=e^{-\frac{2t}{7}}L^{-1}\left(\frac{14\,s+6}{49\,s^2+9}\right)$$

$$=e^{-\frac{2}{7}t}L^{-1}\frac{14}{49}\left(\frac{s+\frac{6}{14}}{s^2+\frac{9}{49}}\right)=e^{-\frac{2t}{7}}\left[\frac{14}{49}L^{-1}\left(\frac{s}{s^2+\frac{9}{49}}\right)+\left(\frac{14}{49}\right)\left(\frac{6}{14}\right)L^{-1}\left(\frac{1}{s^2+\frac{9}{49}}\right)\right]$$

$$=e^{-\frac{2t}{7}}\left[\frac{2}{7}\cos\frac{3}{7}t+\frac{6}{49}\cdot\frac{7}{3}\sin\frac{3}{7}t\right]=\frac{2}{7}e^{-\frac{2t}{7}}\left(\cos\frac{3}{7}t+\sin\frac{3}{7}t\right) \quad \textbf{Ans.}$$

EXERCISE 6.1

Find the Inverse Laplace Transform of the following:

1. $\dfrac{3s-8}{4s^2+25}$ **Ans.** $\dfrac{3}{4}\cos\dfrac{5t}{2}-\dfrac{4}{5}\sin\dfrac{5t}{2}$

2. $\dfrac{3\left(s^2-2\right)^2}{2s^5}$ **Ans.** $\dfrac{3}{2}-3t^2+\dfrac{1}{2}t^4$

3. $\dfrac{2s-5}{4s^2+25}+\dfrac{4s-18}{9-s^2}$ **Ans.** $\dfrac{1}{2}\left(\cos\dfrac{5t}{2}-\sin\dfrac{5t}{2}\right)-4\cosh 3t+6\sinh 3t$

4. $\dfrac{5s-10}{9s^2-16}$ **Ans.** $\dfrac{5}{9}\cosh\dfrac{4}{3}t-\dfrac{5}{6}\sinh\dfrac{4}{3}t$

5. $\dfrac{1}{4s}+\dfrac{16}{1-s^2}$ **Ans.** $\dfrac{1}{4}-16\sinh t$

6. $L^{-1}\left\{\dfrac{1}{s^n}\right\}$ exist only when the value of n is:

   (*i*) Negative integer (*ii*) Positive integer

   (*iii*) zero (*iv*) None of these

   (U.P., II Sem. 2010) **Ans.** (*ii*)

6.3 MULTIPLICATION BY *s*

$$L^{-1}\left[s\,F(s)\right]=\frac{d}{dt}f(t)+f(0)\,\delta(t)$$

Example 6*a*. *Find the Inverse Laplace Transform of* (*i*) $\dfrac{s}{s^2+1}$ (*ii*) $\dfrac{s}{4s^2-25}$ (*iii*) $\dfrac{3s}{2s+9}$

Solution.

(*i*) $L^{-1}\dfrac{1}{s^2+1}=\sin t$

$$L^{-1}\frac{s}{s^2+1}=\frac{d}{dt}(\sin t)+\sin(0)\,\delta(t)=\cos t$$ **Ans.**

(*ii*) $L^{-1}\dfrac{1}{4s^2-25}=\dfrac{1}{4}L^{-1}\dfrac{1}{s^2-\frac{25}{4}}=\dfrac{1}{4}\cdot\dfrac{2}{5}L^{-1}\dfrac{\frac{5}{2}}{s^2-\left(\frac{5}{2}\right)^2}=\dfrac{1}{10}\sinh\dfrac{5}{2}t$

$$L^{-1}\frac{s}{4s^2-25}=\frac{1}{10}\frac{d}{dt}\sinh\frac{5}{2}t+\frac{1}{10}\sinh\frac{5}{2}(0)\,\delta(t)$$

$$=\frac{1}{10}\left(\frac{5}{2}\right)\cosh\frac{5}{2}t=\frac{1}{4}\cosh\frac{5}{2}t$$ **Ans.**

(*iii*) $L^{-1}\dfrac{3}{2s+9}=\dfrac{3}{2}L^{-1}\dfrac{1}{s+\frac{9}{2}}=\dfrac{3}{2}e^{-\frac{9}{2}t}$

$$L^{-1}\frac{3s}{2s+9}=\frac{3}{2}\frac{d}{dt}\left(e^{-\frac{9}{2}t}\right)+\frac{3}{2}e^{-\frac{9}{2}(0)\delta(t)}=\frac{3}{2}\left(-\frac{9}{2}\right)e^{-\frac{9}{2}t}+\frac{3}{2}$$

$$=-\frac{27}{4}e^{-\frac{9}{2}t}+\frac{3}{2}$$ **Ans.**

EXERCISE 6.2

Find the Inverse Laplace Transform of the following:

1. $\dfrac{s}{s+5}$ **Ans.** $-5e^{-5t}+1$

2. $\dfrac{2s}{3s+6}$ **Ans.** $\dfrac{2}{3}\left[-2e^{-2t}+1\right]$

3. $\dfrac{s}{2s^2-1}$ **Ans.** $\dfrac{1}{2}\cosh\dfrac{t}{2}$

4. $\dfrac{s^2}{s^2+a^2}$ **Ans.** $-a\sin at+1$

5. $\dfrac{s^2+4}{s^2+9}$ **Ans.** $-\dfrac{5}{3}\sin 3t+1$

6.4 DIVISION BY *s* (MULTIPLICATION BY $\frac{1}{s}$)

$$\boxed{L^{-1}\left[\frac{F(s)}{s}\right]=\int_0^t\left[L^{-1}[F(s)]\right]dt=\int_0^t f(t)\,dt}$$

Example 6*b*. *Find the Inverse Laplace Transform of*

(*i*) $\dfrac{1}{s(s+a)}$ (*ii*) $\dfrac{1}{s(s^2+1)}$ (*iii*) $\dfrac{s^2+3}{s(s^2+9)}$

Solution.

(*i*) $L^{-1}\left(\frac{1}{s+a}\right)=e^{-at}$

$$L^{-1}\left[\frac{1}{s(s+a)}\right]=\int_0^t L^{-1}\left(\frac{1}{s+a}\right)dt=\int_0^t e^{-at}\,dt=\left[\frac{e^{-at}}{-a}\right]_0^t$$

$$=\frac{e^{-at}}{-a}+\frac{1}{a}=\frac{1}{a}\left[1-e^{-at}\right]$$ **Ans.**

(*ii*) $L^{-1}\frac{1}{s^2+1}=\sin t$

$$L^{-1}\frac{1}{s}\left(\frac{1}{s^2+1}\right)=\int_0^t L^{-1}\left(\frac{1}{s^2+1}\right)dt=\int_0^t \sin t\,dt=[-\cos t]_0^t=-\cos t+1$$ **Ans.**

(*iii*) $$L^{-1}\frac{s^2+3}{s(s^2+9)}=L^{-1}\left[\frac{s^2+9-6}{s(s^2+9)}\right]=L^{-1}\left[\frac{1}{s}-\frac{6}{s(s^2+9)}\right]$$

$$=1-2\int_0^t \sin 3t\,dt=1+2\times\frac{1}{3}[\cos 3t]_0^t=1+\frac{2}{3}\cos 3t-\frac{2}{3}$$

$$=\frac{2}{3}\cos 3t+\frac{1}{3}=\frac{1}{3}[2\cos 3t+1]$$ **Ans.**

EXERCISE 6.3

Find the Inverse Laplace Transform of the following:

1. $\frac{1}{2s(s-3)}$ **Ans.** $\frac{1}{2}\left[\frac{e^{3t}}{3}-1\right]$
2. $\frac{1}{s(s+2)}$ **Ans.** $\frac{1-e^{-2t}}{2}$
3. $\frac{1}{s(s^2-16)}$ **Ans.** $\frac{1}{16}[\cosh 4t-1]$
4. $\frac{1}{s(s^2+a^2)}$ **Ans.** $\frac{1-\cos at}{a^2}$
5. $\frac{s^2+2}{s(s^2+4)}$ **Ans.** $\cos^2 t$
6. $\frac{1}{s^2(s+1)}$ **Ans.** $t-1+e^{-t}$
7. $\frac{1}{s^3(s^2+1)}$ **Ans.** $\frac{t^2}{2}+\cos t-1$

6.5 FIRST SHIFTING PROPERTY

If $L^{-1}F(s)=f(t)$, then $\boxed{L^{-1}F(s+a)=e^{-at}L^{-1}[F(s)]}$

Example 7. *Find the Inverse Laplace Transform of*

(*i*) $\frac{1}{(s+2)^5}$ (*ii*) $\frac{s}{s^2+4s+13}$ (*iii*) $\frac{1}{9s^2+6s+1}$

Solution.

(*i*) $L^{-1}\frac{1}{s^5}=\frac{t^4}{4!}$

$$\Rightarrow L^{-1}\frac{1}{(s+2)^5}=e^{-2t}\cdot\frac{t^4}{4!}$$ **Ans.**

(*ii*) $L^{-1}\left(\dfrac{s}{s^2+4s+13}\right) = L^{-1}\dfrac{s+2-2}{(s+2)^2+(3)^2} = L^{-1}\dfrac{s+2}{(s+2)^2+(3)^2} - L^{-1}\dfrac{2}{(s+2)^2+(3)^2}$

$$= e^{-2t}\,L^{-1}\frac{s}{s^2+3^2} - e^{-2t}\,L^{-1}\frac{2}{3}\left(\frac{3}{s^2+3^2}\right)$$

$$= e^{-2t}\cos 3t - \frac{2}{3}e^{-2t}\sin 3t$$ **Ans.**

(*iii*) $L^{-1}\dfrac{1}{9s^2+6s+1} = L^{-1}\dfrac{1}{(3s+1)^2} = \dfrac{1}{9}L^{-1}\dfrac{1}{\left(s+\frac{1}{3}\right)^2} = \dfrac{1}{9}e^{-t/3}\,L^{-1}\dfrac{1}{s^2}$

$$= \frac{1}{9}e^{-t/3}\,t = \frac{t\,e^{-t/3}}{9}$$ **Ans.**

Example 8. *Find the Inverse Laplace Transform of* $\dfrac{s+1}{s^2-6s+25}$ (*U.P., II Sem. 2010*)

Solution. $L^{-1}\left(\dfrac{s+1}{s^2-6s+25}\right) = L^{-1}\left[\dfrac{s+1}{(s-3)^2+(4)^2}\right] = L^{-1}\left[\dfrac{s-3+4}{(s-3)^2+(4)^2}\right]$

$$= L^{-1}\left[\frac{s-3}{(s-3)^2+(4)^2}\right] + L^{-1}\left[\frac{4}{(s-3)^2+(4)^2}\right]$$

$$= e^{3t}\cos 4t + e^{3t}\sin 4t.$$ **Ans.**

EXERCISE 6.4

Obtain the Inverse Laplace Transform of the following:

1. $\dfrac{s+8}{s^2+4s+5}$ **Ans.** $e^{-2t}(\cos t + 6\sin t)$
2. $\dfrac{s}{(s+3)^2+4}$ **Ans.** $e^{-3t}(\cos 2t - 1.5\sin 2t)$
3. $\dfrac{s}{(s+7)^4}$ **Ans.** $e^{-7t}\dfrac{t^2}{6}(3-7t)$
4. $\dfrac{s+2}{s^2-2s-8}$ **Ans.** $e^{-t}(\cosh 3t + \sinh 3t)$
5. $\dfrac{s}{s^2+6s+25}$ **Ans.** $e^{-3t}\left[\cos 4t - \dfrac{3}{4}\sin 4t\right]$
6. $\dfrac{1}{2(s-1)^2+32}$ **Ans.** $\dfrac{e^t}{8}\sin 4t$
7. $\dfrac{s-4}{4(s-3)^2+16}$ **Ans.** $\dfrac{1}{4}e^{3t}\cos 2t - \dfrac{1}{8}e^{3t}\sin 2t$

6.6 SECOND SHIFTING PROPERTY

$$L^{-1}\left[e^{-as}F(s)\right] = f(t-a)\,U(t-a)$$

Example 9. *Obtain Inverse Laplace Transform of*

(*i*) $\dfrac{e^{-\pi s}}{(s+3)}$ (*ii*) $\dfrac{e^{-s}}{(s+1)^3}$

Solution.

(*i*) $L^{-1}\dfrac{1}{s+3}=e^{-3t}$

$$L^{-1}\frac{e^{-\pi s}}{s+3}=e^{-3(t-\pi)}\,u(t-\pi)$$ **Ans.**

(*ii*) $L^{-1}\dfrac{1}{s^3}=\dfrac{t^2}{2!} \Rightarrow L^{-1}\dfrac{1}{(s+1)^3}=e^{-t}\dfrac{t^2}{2!}$

$$L^{-1}\frac{e^{-s}}{(s+1)^3}=e^{-(t-1)}\cdot\frac{(t-1)^2}{2!}\cdot u(t-1)$$ **Ans.**

Example 10. *Evaluate*

$$L^{-1}\left[\frac{e^{-s}-3e^{-3s}}{s^2}\right]$$ (*U.P. II Semester, Summer 2002*)

Solution. $L^{-1}\left[\dfrac{e^{-s}-3e^{-3s}}{s^2}\right]=L^{-1}\left[\dfrac{e^{-s}}{s^2}-\dfrac{3e^{-3s}}{s^2}\right]$ ...(1)

We know that $L\,u(t-a)=\dfrac{e^{-as}}{s}$

and $L\,[(t-a)\,u(t-a)]=\dfrac{e^{-as}}{s^2}$

Using these results in (1), we get

$$\therefore \quad L^{-1}\left[\frac{e^{-s}-3e^{-3s}}{s^2}\right]=(t-1)\,u(t-1)-3(t-3)\,u(t-3)$$ **Ans.**

Example 11. *Find the Inverse Laplace Transform of*

$$\frac{se^{-s/2}+\pi e^{-s}}{s^2+\pi^2}$$

in terms of unit step functions.

Solution. $L^{-1}\dfrac{\pi}{s^2+\pi^2}=\sin\pi t$

$$L^{-1}\left[e^{-s}\frac{\pi}{s^2+\pi^2}\right]=\sin\pi(t-1)\cdot u(t-1)=-\sin(\pi t)\cdot u(t-1) \quad ...(1)$$

and $L^{-1}\dfrac{s}{s^2+\pi^2}=\cos\pi t$

$$L^{-1}\left[e^{-s/2}\frac{s}{s^2+\pi^2}\right]=\cos\pi\left(t-\frac{1}{2}\right)\cdot u\left(t-\frac{1}{2}\right)=\sin\pi t\cdot u\left(t-\frac{1}{2}\right) \quad ...(2)$$

On adding (1) and (2), we get

$$L^{-1}\left[\frac{e^{-s/2}\,s+e^{-s}\cdot\pi}{s^2+\pi^2}\right]=\sin(\pi t)\cdot u\left(t-\frac{1}{2}\right)-\sin(\pi t)\cdot u(t-1)$$

$$=\sin\pi t\left[u\left(t-\frac{1}{2}\right)-u(t-1)\right]$$ **Ans.**

Example 12. *Find the value of*

$$L^{-1}\left\{\frac{1}{(s^2+a^2)^2}\right\}$$

Solution. $\dfrac{1}{(s^2+a^2)^2}=\dfrac{1}{s}\dfrac{s}{(s^2+a^2)^2}=-\dfrac{1}{2s}\dfrac{d}{ds}\left(\dfrac{1}{s^2+a^2}\right)$

$$\Rightarrow \quad L^{-1}\left\{\frac{1}{(s^2+a^2)^2}\right\}=L^{-1}\left\{-L^{-1}\frac{1}{2s}\frac{d}{ds}\left(\frac{1}{s^2+a^2}\right)\right\}=L^{-1}\frac{1}{2s}\left\{-t\frac{1}{a}\sin at\right\}$$

$$=\frac{1}{2a}L^{-1}\frac{1}{s}\{t\sin at\}=\frac{1}{2a}\int_0^t t\sin at\,dt$$

$$=\frac{1}{2a}\left[t\left(-\frac{\cos at}{a}\right)_0^t-\int_0^t-\frac{\cos at}{a}dt\right]$$

$$=\frac{1}{2a}\left[-\frac{t}{a}\cos at+\frac{\sin at}{a^2}\right]_0^t=\frac{1}{2a^3}\left[-at\cos at+\sin at\right] \qquad \textbf{Ans.}$$

EXERCISE 6.5

Obtain Inverse Laplace Transform of the following:

1. $\dfrac{e^{-s}}{(s+2)^3}$ **Ans.** $e^{-(t-2)}\dfrac{(t-2)^2}{2}U(t-2)$

2. $\dfrac{e^{-2s}}{(s+1)(s^2+2s+2)}$ **Ans.** $e^{-(t-2)}\left|1-\cos(t-2)\right|U(t-2)$

3. $\dfrac{e^{-s}}{\sqrt{s+1}}$ **Ans.** $\dfrac{e^{-(t-1)}}{\sqrt{\pi(t-1)}}.U(t-1)$

4. $\dfrac{e^{-\frac{\pi}{2}s}+e^{-\frac{3\pi}{2}s}}{s^2+1}$ **Ans.** $\cot t\left[U\left(t-\dfrac{3\pi}{2}\right)-U\left(t-\dfrac{\pi}{2}\right)\right]$

5. $\dfrac{e^{-4s}(s+2)}{s^2+4s+5}$ **Ans.** $e^{-2(t-u)}\cos(t-u)\,U(t-4)$

6. $\dfrac{e^{-as}}{s^2}$ **Ans.** $f(t)=t-a$, when $t>a$
   $=0$, when $t<a$

7. $\dfrac{e^{-\pi s}}{s^2+1}$ **Ans.** $-\sin t.\ u(t-\pi)$

6.7 INVERSE LAPLACE TRANSFORMS OF DERIVATIVES

$$L^{-1}\left[\frac{d}{ds}F(s)\right]=-t\,L^{-1}\left[F(s)\right]=-t\,f(t) \quad \Rightarrow \quad \boxed{L^{-1}\left[F(s)\right]=-\frac{1}{t}L^{-1}\left[\frac{d}{ds}F(s)\right]}$$

Example 13. *Find* $L^{-1}\left\{\log\dfrac{s+1}{s-1}\right\}$. *(Uttarakhand, II Semester, June 2012, 2010, 2009,* 2007)

Solution. $L^{-1}\left\{\log\left(\dfrac{s+1}{s-1}\right)\right\}=-\dfrac{1}{t}L^{-1}\left[\dfrac{d}{ds}\log\left(\dfrac{s+1}{s-1}\right)\right]$

$$=-\frac{1}{t}L^{-1}\left[\frac{d}{ds}\log(s+1)-\frac{d}{ds}\log(s-1)\right]$$

$$=-\frac{1}{t}L^{-1}\left[\frac{1}{s+1}-\frac{1}{s-1}\right]$$

$$=-\frac{1}{t}\left[e^{-t}-e^{t}\right]=\frac{1}{t}\left[e^{t}-e^{-t}\right]$$ **Ans.**

Example 14. *Find the Inverse Laplace Transform of F(s)* $=\log\dfrac{s+a}{s+b}$

(U.P., II Semester, Summer 2003)

Solution. $L^{-1}\log\left(\dfrac{s+a}{s+b}\right)=-\dfrac{1}{t}L^{-1}\left[\dfrac{d}{ds}\log\dfrac{s+a}{s+b}\right]$

$$=-\frac{1}{t}L^{-1}\left[\frac{d}{ds}\log(s+a)-\frac{d}{ds}\log(s+b)\right]$$

$$=-\frac{1}{t}L^{-1}\left[\frac{1}{s+a}-\frac{1}{s+b}\right]$$

$$=-\frac{1}{t}\left[e^{-at}-e^{-bt}\right]=\frac{1}{t}\left(e^{-bt}-e^{-at}\right)$$ **Ans.**

Example 15. *Obtain the Inverse Laplace Transform of log* $\dfrac{s^2-1}{s^2}$.

Solution. $L^{-1}\left[\log\dfrac{s^2-1}{s^2}\right]=-\dfrac{1}{t}L^{-1}\left[\dfrac{d}{ds}\log\dfrac{s^2-1}{s^2}\right]$

$$=-\frac{1}{t}L^{-1}\left[\frac{d}{ds}\left\{\log\left(s^2-1\right)-2\log s\right\}\right]=-\frac{1}{t}L^{-1}\left[\frac{2s}{s^2-1}-\frac{2}{s}\right]=-\frac{1}{t}\left[2\cosh t-2\right]$$

$$=\frac{2}{t}\left[1-\cosh t\right]$$ **Ans.**

Example 16. *Find the function whose Laplace transform is*

$ln\left(1+\dfrac{1}{s}\right)$. *(U.P., II Semester, June 2007)*

Solution. $L^{-1}\left[\log\left(1+\dfrac{1}{s}\right)\right]=\dfrac{1}{t}L^{-1}\left[\dfrac{d}{ds}\log\left(\dfrac{s+1}{s}\right)\right]$

$$=-\frac{1}{t}L^{-1}\left[\left(\frac{s}{s+1}\right)\left(-\frac{1}{s^2}\right)\right]=-\frac{1}{t}L^{-1}\left[-\frac{1}{s(s+1)}\right]$$

$$= -\frac{1}{t} L^{-1}\left[\frac{1}{s+1} - \frac{1}{s}\right] \qquad \text{(Partial fraction)}$$

$$= -\frac{1}{t}[e^{-t} - 1] = \frac{1}{t}[1 - e^{-t}] \qquad \textbf{Ans.}$$

Example 17. *Find the inverse Laplace transform of* $\tan^{-1}\left(\frac{2}{s^2}\right)$ (*Q.Bank U.P.T.U. 2001*)

Solution. Here, we have

$$L^{-1}\left[\tan^{-1}\left(\frac{2}{s^2}\right)\right] = -\frac{1}{t} L^{-1}\left[\frac{d}{ds}\tan^{-1}\frac{2}{s^2}\right]$$

$$= -\frac{1}{t} L^{-1}\left[\frac{1}{1+\frac{4}{s^4}}\left(-\frac{4}{s^3}\right)\right] = -\frac{1}{t} L^{-1}\left[\frac{s^4}{s^4+4}\left(-\frac{4}{s^3}\right)\right]$$

$$= \frac{1}{t} L^{-1}\left[\frac{4s}{s^4+4}\right] = \frac{4}{t} L^{-1}\left[\frac{s}{s^4+4}\right]$$

$$= \frac{4}{t} L^{-1}\left[\frac{s}{(s^2+2s+2)(s^2-2s+2)}\right]$$

$$= \frac{4}{t} L^{-1}\left[-\frac{1}{4}\frac{1}{(s^2+2s+2)} + \frac{1}{4}\frac{1}{(s^2-2s+2)}\right] \qquad \left(\begin{array}{c}\text{By}\\ \text{partial}\\ \text{fraction}\end{array}\right)$$

$$= \frac{1}{t} L^{-1}\left[-\frac{1}{(s^2+2s+2)} + \frac{1}{(s^2-2s+2)}\right]$$

$$= \frac{1}{t} L^{-1}\left[-\frac{1}{(s+1)^2+1} + \frac{1}{(s-1)^2+1}\right]$$

$$= \frac{1}{t}\left[-e^{-t}\sin t + e^{t}\sin t\right] = \frac{\sin t}{t}\left[e^{t} - e^{-t}\right] \qquad \textbf{Ans.}$$

Example 18. *Find Inverse Laplace Transform of* $\tan^{-1}\frac{1}{s}$.

Solution. $$L^{-1}\left(\tan^{-1}\frac{1}{s}\right) = -\frac{1}{t} L^{-1}\left[\frac{d}{ds}\tan^{-1}\frac{1}{s}\right]$$

$$= -\frac{1}{t} L^{-1}\left[\frac{1}{1+\frac{1}{s^2}}\left(-\frac{1}{s^2}\right)\right] = \frac{1}{t} L^{-1}\left[\frac{1}{1+s^2}\right] = \frac{\sin t}{t} \qquad \textbf{Ans.}$$

Example 19. *Find* $L^{-1}\left[\tan^{-1}(1+s)\right]$ (*M.D.U., Dec. 2010*)

Solution. $$L^{-1}\left[\tan^{-1}(1+s)\right] = -\frac{1}{t} L^{-1}\left[\frac{d}{ds}\tan^{-1}(1+s)\right]$$

$$= -\frac{1}{t} L^{-1}\left[\frac{1}{1+(s+1)^2}\right] = -\frac{1}{t} L^{-1}\left[\frac{1}{(s+1)^2+1}\right]$$

$$= -\frac{1}{t} e^{-t}\sin t \qquad \textbf{Ans.}$$

Example 20. *Find the inverse Laplace transform of*

$$\cot^{-1}\left(\frac{s}{2}\right)$$ (*Q.Bank U.P. 2001*)

Solution.

Let $$L^{-1}\left[\cot^{-1}\left(\frac{s}{2}\right)\right] = f(t) \Rightarrow L^{-1}\left[\frac{d}{ds}\cot^{-1}\left(\frac{s}{2}\right)\right] = -t\,f(t)$$

$$\Rightarrow \quad L^{-1}\left[\frac{-1}{1+\frac{s^2}{4}}\;\frac{1}{2}\right] = -t\,f(t) \Rightarrow L^{-1}\left[\frac{2}{s^2+2^2}\right] = t\,f(t)$$

$$\Rightarrow \quad \sin 2t = t\,f(t)$$

$$\Rightarrow \quad f(t) = \frac{1}{t}\sin 2t$$ **Ans.**

Example 21. *Obtain the Inverse Laplace Transform of* $\cot^{-1}\left(\frac{s+3}{2}\right)$

(*U. P., II Semester, Summer* 2002)

Solution. We know that $L^{-1}\left[F(s)\right] = -\frac{1}{t}L^{-1}\left[\frac{d}{ds}F(s)\right]$

$$\therefore \quad L^{-1}\left[\cot^{-1}\left(\frac{s+3}{2}\right)\right] = -\frac{1}{t}L^{-1}\left[\frac{d}{ds}\cot^{-1}\left(\frac{s+3}{2}\right)\right]$$

$$= -\frac{1}{t}L^{-1}\left\{\frac{-\frac{1}{2}}{1+\left(\frac{s+3}{2}\right)^2}\right\} = \frac{1}{2t}L^{-1}\left\{\frac{4}{4+(s+3)^2}\right\}$$

$$= \frac{1}{t}L^{-1}\left\{\frac{2}{2^2+(s+3)^2}\right\} = \frac{1}{t}e^{-3t}L^{-1}\left(\frac{2}{2^2+s^2}\right)$$

$$= \frac{e^{-3t}}{t}\sin 2t$$ **Ans.**

Example 22. *Find the inverse Laplace transform of*

$$\frac{2as}{(s^2+a^2)^2}$$

Solution. $$L^{-1}\left(\frac{a}{s^2+a^2}\right) = \sin at$$

$$L^{-1}\left[\frac{d}{ds}\left\{\frac{a}{s^2+a^2}\right\}\right] = -t\sin at \Rightarrow L^{-1}\left\{\frac{-2as}{(s^2+a^2)^2}\right\} = -t\sin at$$

$$\Rightarrow \quad L^{-1}\left\{\frac{2as}{(s^2+a^2)^2}\right\} = t\sin at$$ **Ans.**

Example 23. *Find the inverse Laplace transform of* $\dfrac{s^2-a^2}{(s^2+a^2)^2}$

Solution. We know that

$$L^{-1}\left\{\frac{s}{s^2+a^2}\right\} = \cos at$$

$$\therefore \quad L^{-1}\left[\frac{d}{ds}\left\{\frac{s}{s^2+a^2}\right\}\right] = -t\cos at$$

$$\Rightarrow \quad L^{-1}\left\{\frac{(s^2+a^2).1-s(2s)}{(s^2+a^2)^2}\right\} = -t\cos at \quad \Rightarrow \quad L^{-1}\left[\frac{a^2-s^2}{(s^2+a^2)^2}\right] = -t\cos at$$

$$\therefore \quad L^{-1}\left[\frac{s^2-a^2}{(s^2+a^2)^2}\right] = t\cos at$$

Ans.

Example 24. *Find the inverse Laplace Transform of* $\dfrac{s+1}{(s^2+6s+13)^2}$

Solution. $L^{-1}\dfrac{s+1}{(s^2+6s+13)^2} = L^{-1}\dfrac{s+1}{\left[(s+3)^2+2^2\right]^2}$

$$= L^{-1}\frac{s+3-2}{\left[(s+3)^2+2^2\right]^2} = L^{-1}\frac{s+3}{[(s+3)^2+2^2]^2} - L^{-1}\frac{2}{[(s+3)^2+2^2]^2}$$

$$= e^{-3t}L^{-1}\frac{s}{(s^2+2^2)^2} - e^{-3t}L^{-1}\frac{2}{(s^2+2^2)^2} \quad \ldots(1)$$

First Part.

We know that

$$L^{-1}\frac{-2s}{(s^2+2^2)^2} = L^{-1}\frac{d}{ds}\left(\frac{1}{s^2+2^2}\right)$$

$$= -t\,L^{-1}\frac{1}{s^2+2^2}$$

$$= -\frac{t}{2}\sin 2t$$

$$\Rightarrow \quad L^{-1}\frac{s}{(s^2+2^2)^2} = \frac{t}{4}\sin 2t$$

$$\Rightarrow \quad e^{-3t}L^{-1}\frac{s}{(s^2+2^2)^2} = \frac{t}{4}e^{-3t}\sin 2t$$

$$\Rightarrow \quad L^{-1}\frac{s+3}{(s+3)^2+2^2} = \frac{t}{4}e^{-3t}\sin 2t$$

Second Part.

We know that from the formula 13 on page 198 or solved example 12 on page 206.

$$L^{-1}\frac{1}{(s^2+a^2)^2} = \frac{1}{2a^3}(\sin at - at\cos at)$$

$$L^{-1}\frac{2}{(s^2+2^2)^2} = \frac{2}{2(2)^3}[\sin 2t - 2t\cos 2t]$$

$$= \frac{1}{8}[\sin 2t - 2t\cos 2t]$$

$$e^{-3t}L^{-1}\frac{-2}{(s^2+2^2)^2} = \frac{e^{-3t}}{8}[\sin 2t - 2t\cos 2t]$$

$$L^{-1}\frac{-2}{[(s+3)^2+2^2]^2} = -\frac{e^{-3t}}{8}[\sin 2t - 2t\cos 2t]$$

Putting the values of both parts in (1), we get

$$L^{-1}\frac{s+1}{(s^2+6s+13)^2} = \frac{t}{4}e^{-3t}\sin 2t - \frac{e^{-3t}}{8}(\sin 2t - 2t\cos 2t)$$

$$= \frac{e^{-3t}}{8}(2t.\sin 2t - \sin 2t + 2t\cos 2t)$$

Ans.

EXERCISE 6.6

Obtain Inverse Laplace Transform of the following:

1. $\log\left(1+\frac{\omega^2}{s^2}\right)$ **Ans.** $-\frac{2}{t}\cos\omega t+2$ **2.** $\log\left(1+\frac{1}{s^2}\right)$ **Ans.** $\frac{2}{t}[1-\cos\omega t]$

3. $\frac{s}{1+s^2+s^4}$ **Ans.** $\frac{2}{\sqrt{3}}\sin\frac{\sqrt{3}}{2}t\sinh\frac{t}{2}$

4. $\frac{s}{\left(s^2+a^2\right)^2}$ **Ans.** $\frac{t\sin at}{2a}$ **5.** $s\log\frac{s}{\sqrt{s^2+1}}+\cot^{-1}s$ **Ans.** $\frac{1-\cos t}{t^2}$

6. $\frac{1}{2}\log\left\{\frac{s^2+b^2}{(s-a)^2}\right\}$ **Ans.** $\frac{e^{-at}-\cos bt}{t}$ **7.** $\tan^{-1}(s+1)$ (*MDU, 2010*) **Ans.** $-\frac{1}{t}e^{-t}\sin t$

6.8 INVERSE LAPLACE TRANSFORM OF INTEGRALS

$$L^{-1}\left[\int_s^\infty F(s)\,ds\right]=\frac{f(t)}{t}=\frac{1}{t}L^{-1}[F(s)] \quad \text{or} \quad \boxed{L^{-1}[F(s)]=t\,L^{-1}\left[\int_s^\infty F(s)\,ds\right]}$$

Example 25. *Obtain* $L^{-1}\frac{2s}{\left(s^2+1\right)^2}$

Solution. $L^{-1}\frac{2s}{\left(s^2+1\right)^2}=t\,L^{-1}\int_s^\infty\frac{2s\,ds}{\left(s^2+1\right)^2}=t\,L^{-1}\left[-\frac{1}{s^2+1}\right]_s^\infty=t\,L^{-1}\left[-0+\frac{1}{s^2+1}\right]$

$= t\sin t$ **Ans.**

6.9 PARTIAL FRACTIONS METHOD

Example 26. *Find the Inverse Laplace Transform of* $\frac{1}{s^2-5s+6}$.

Solution. Let us convert the given function into partial fractions.

$$L^{-1}\left[\frac{1}{s^2-5s+6}\right]=L^{-1}\left[\frac{1}{s-3}-\frac{1}{s-2}\right]$$

$$=L^{-1}\left(\frac{1}{s-3}\right)-L^{-1}\left(\frac{1}{s-2}\right)=e^{3t}-e^{2t}$$ **Ans.**

Example 27. *Find the inverse Laplace transform of*

$$\frac{s^3}{s^4-a^4}$$ (*Q. Bank U.P. 2001*)

Solution. Here, we have

$$L^{-1}\left(\frac{s^3}{s^4-a^4}\right)=L^{-1}\left[s\left\{\frac{s^2}{(s^2-a^2)(s^2+a^2)}\right\}\right]$$

$$=L^{-1}\left[\frac{s}{2}\left(\frac{1}{s^2-a^2}+\frac{1}{s^2+a^2}\right)\right]$$

$$=\frac{1}{2}L^{-1}\left(\frac{s}{s^2-a^2}+\frac{s}{s^2+a^2}\right)$$ (By partial fraction)

$$= \frac{1}{2}(\cosh at + \cos at)$$ **Ans.**

Example 28. *Find the Inverse Laplace Transform of* $\dfrac{s+4}{s(s-1)(s^2+4)}$.

Solution. Let us first resolve $\dfrac{s+4}{s(s-1)(s^2+4)}$ into partial fractions.

$$\frac{s+4}{s(s-1)(s^2+4)} \equiv \frac{A}{s} + \frac{B}{s-1} + \frac{Cs+D}{s^2+4} \quad \text{... (1)}$$

$$s+4 \equiv A(s-1)(s^2+4) + Bs(s^2+4) + (Cs+D)s(s-1)$$

Putting $s = 0$, we get $4 = -4A \quad \Rightarrow \quad A = -1$

Putting $s = 1$, we get $5 = B \cdot 1 \cdot (1+4) \quad \Rightarrow B = 1$

Equating the coefficients of s^3 on both sides of (1), we have

$$0 = A + B + C \quad \Rightarrow \quad 0 = -1 + 1 + C \quad \Rightarrow \quad C = 0.$$

Equating the coefficients of s on both sides of (1), we get

$$1 = 4A + 4B - D \quad \Rightarrow \quad 1 = -4 + 4 - D \quad \Rightarrow \quad D = -1.$$

On putting the values of A, B, C, D in (1), we get

$$\frac{s+4}{s(s-1)(s^2+4)} = -\frac{1}{s} + \frac{1}{s-1} - \frac{1}{s^2+4}$$

$$\therefore \quad L^{-1}\left[\frac{s+4}{s(s-1)(s^2+4)}\right] = L^{-1}\left[-\frac{1}{s} + \frac{1}{s-1} - \frac{1}{s^2+4}\right]$$

$$= -L^{-1}\left(\frac{1}{s}\right) + L^{-1}\left(\frac{1}{s-1}\right) - \frac{1}{2}L^{-1}\left(\frac{2}{s^2+2^2}\right)$$

$$= -1 + e^t - \frac{1}{2}\sin 2t.$$ **Ans.**

Example 29. *Find the inverse Laplace transform of*

$$\frac{1}{s^4+4}$$ [*U.P., II Semester, (SUM) 2007*]

Solution. Here, we have

$$s^4 + 4 = (s^2+2)^2 - (2s)^2 = (s^2 - 2s + 2)(s^2 + 2s + 2)$$

$$\frac{1}{s^4+4} = \frac{1}{(s^2-2s+2)(s^2+2s+2)}$$

$$= \frac{1}{4s}\left[\frac{1}{s^2-2s+2} - \frac{1}{s^2+2s+2}\right] \quad \left(\begin{array}{c}\text{By}\\ \text{partial}\\ \text{fraction}\end{array}\right) \text{...(1)}$$

Now, $$L^{-1}\left(\frac{1}{s^2-2s+2}\right) = L^{-1}\left[\frac{1}{(s-1)^2+1}\right] = e^t \sin t$$

and $$L^{-1}\left(\frac{1}{s^2+2s+2}\right) = L^{-1}\left[\frac{1}{(s+1)^2+1}\right] = e^{-t}\sin t$$

$\therefore$ $$\frac{1}{4}L^{-1}\left(\frac{1}{s^2-2s+2}-\frac{1}{s^2+2s+2}\right) = \frac{1}{4}(e^t-e^{-t})\sin t$$

Hence, $$L^{-1}\left[\frac{1}{4s}\left(\frac{1}{s^2-2s+2}-\frac{1}{s^2+2s+2}\right)\right] = \frac{1}{4}\int_0^t (e^t-e^{-t})\sin t\,dt$$

$$\Rightarrow \quad L^{-1}\left(\frac{1}{s^2+4}\right) = \frac{1}{4}\left[\frac{e^t}{2}(\sin t-\cos t)-\frac{e^{-t}}{2}(-\sin t-\cos t)\right]$$

$$= \frac{1}{4}\left[\sin t\left(\frac{e^t+e^{-t}}{2}\right)-\cos t\left(\frac{e^t-e^{-t}}{2}\right)\right]$$

$$\Rightarrow \quad L^{-1}\left(\frac{1}{s^2+4}\right) = \frac{1}{4}[\sin t\cosh t-\cos t\sinh t]$$ **Ans.**

Example 30. *Find the inverse Laplace transform of*

$$\frac{s}{s^4+4a^4}$$

Solution. $$s^4+4a^4 = (s^2+2a^2)^2-(2as)^2 = (s^2-2as+2a^2)(s^2+2as+2a^2)$$
$$= \{(s-a)^2+a^2\}\{(s+a)^2+a^2\}$$

$$\frac{s}{s^4+4a^4} = \frac{s}{\{(s-a)^2+a^2\}\{(s+a)^2+a^2\}}$$

$$= \frac{1}{4a}\left[\frac{1}{(s-a)^2+a^2}-\frac{1}{(s+a)^2+a^2}\right]$$ (By partial fraction)

$\therefore$ $$L^{-1}\left(\frac{s}{s^4+4a^4}\right) = \frac{1}{4a}\left[L^{-1}\left\{\frac{1}{(s-a)^2+a^2}\right\}-L^{-1}\left\{\frac{1}{(s+a)^2+a^2}\right\}\right]$$

$$= \frac{1}{4a}\left[\frac{1}{a}e^{at}\sin at-e^{-at}.\frac{1}{a}\sin at\right]$$

$$= \frac{1}{2a^2}\sin at\left(\frac{e^{at}-e^{-at}}{2}\right) = \frac{1}{2a^2}\sin at\sinh at.$$ **Ans.**

Example 31. *Find the Inverse Laplace Transform of* $\frac{e^{-cs}}{s^2(s+a)}$, *c > 0.*

(U.P. II Semester, Summer 2002)

Solution. We have,

$$L^{-1}\left[\frac{e^{-cs}}{s^2(s+a)}\right]=L^{-1}\left[-\frac{e^{-cs}}{a^2 s}+\frac{e^{-cs}}{as^2}+\frac{e^{-cs}}{a^2(s+a)}\right] \quad \text{(By Partial fractions)}$$

$$= L^{-1}\left[\left(\frac{-1}{a^2}\frac{e^{-cs}}{s}\right)+\left(\frac{1}{a}\right)\frac{e^{-cs}}{s^2}+\left(\frac{1}{a^2}\right)\frac{e^{-c(s+a)}}{e^{-ca}(s+a)}\right]$$

$$= -\frac{1}{a^2}u(t-c)+\frac{1}{a}(t-c)\,u(t-c)+\frac{1}{a^2e^{-ca}}e^{at}\,u(t-c)$$

$$= u(t-c)\left[\frac{-1}{a^2}+\frac{1}{a}(t-c)+\frac{1}{a^2}e^{a(c+t)}\right]$$, where $u(t-c)$ is unit step function.

Ans.

Example 32. *Find the Inverse Laplace Transform of*

$$\frac{5s+3}{(s-1)(s^2+2s+5)}$$

(U.P. II Semester Summer 2005)

Solution. $L^{-1}\left\{\frac{5s+3}{(s-1)(s^2+2s+5)}\right\}$

Let $$\frac{5s+3}{(s-1)(s^2+2s+5)}=\frac{A}{s-1}+\frac{Bs+C}{s^2+2s+5}$$

$$5s+3 = A(s^2+2s+5)+(Bs+C)(s-1)$$

$$5s+3 = s^2(A+B)+s(2A-B+C)+(5A-C)$$

Comparing the coefficients of s^2, s and constant, we get

$$A+B=0 \quad \ldots (1)$$

$$2A-B+C=5 \quad \ldots (2)$$

$$5A-C=3 \quad \ldots (3)$$

On adding equations (1) and (2), we have

$$3A+C=5 \quad \ldots (4)$$

Adding equations (3) and (4), we get

$$8A=8 \Rightarrow A=1$$

Putting A = 1 in (3), we get

$$C=2$$

Putting $A=1$, $C=2$ in (2), we get

$$B=-1$$

Thus $$\frac{5s+3}{(s-1)(s^2+2s+5)}=\frac{1}{s-1}+\frac{-s+2}{s^2+2s+5}=\frac{1}{s-1}-\frac{s-2}{(s+1)^2+2^2}$$

$$=\frac{1}{s-1}-\frac{s+1}{(s+1)^2+2^2}+\frac{3}{(s+1)^2+2^2}$$

$$L^{-1}\left\{\frac{5s+3}{(s-1)(s^2+2s+5)}\right\}=L^{-1}\left\{\frac{1}{s-1}\right\}+L^{-1}\left\{\frac{3}{(s+1)^2+2^2}\right\}-L^{-1}\left\{\frac{s+1}{(s+1)^2+2^2}\right\}$$

$$= e^t + 3e^{-t} L^{-1}\left\{\frac{1}{s^2+2^2}\right\} - e^{-t} L^{-1}\left\{\frac{s}{s^2+2^2}\right\}$$

$$= e^t + 3e^{-t} \cdot \frac{1}{2} \sin 2t - e^{-t} \cos 2t \qquad \textbf{Ans.}$$

Example 32(a). *Find the Inverse Laplace Transform of* $\frac{s^2}{(s^2+a^2)(s^2+b^2)}$.

Solution. Let us convert the given function into partial fractions.

$$L^{-1}\left[\frac{s^2}{(s^2+a^2)(s^2+b^2)}\right] = L^{-1}\left[\frac{a^2}{a^2-b^2}\cdot\frac{1}{s^2+a^2} - \frac{b^2}{a^2-b^2}\cdot\frac{1}{s^2+b^2}\right]$$

$$= \frac{1}{a^2-b^2} L^{-1}\left[\frac{a^2}{s^2+a^2} - \frac{b^2}{s^2+b^2}\right] = \frac{1}{a^2-b^2}\left[a^2\left(\frac{1}{a}\sin at\right) - b^2\left(\frac{1}{b}\sin bt\right)\right]$$

$$= \frac{1}{a^2-b^2}[\, a \sin at - b \sin bt \,] \qquad \textbf{Ans.}$$

Note: This question is also solved by using the Convolution Theorem as an example 36.

EXERCISE 6.7

Find the Inverse Laplace Transforms of the following by partial fractions method:

1. $\frac{1}{s^2-7s+12}$ **Ans.** $e^{4t} - e^{3t}$
2. $\frac{s+2}{s^2-4s+13}$ **Ans.** $e^{2t}\cos 3t + \frac{4}{3}e^{2t}\sin 3t$
3. $\frac{3s+1}{(s-1)(s^2+1)}$ **Ans.** $e^t - 2\cos t + \sin t$
4. $\frac{11s^2-2s+5}{2s^3-3s^2-3s+2}$ **Ans.** $2e^{-t} + 5e^{2t} - \frac{3}{2}e^{t/2}$
5. $\frac{2s^2-6s+5}{(s-1)(s-2)(s-3)}$ **Ans.** $\frac{1}{2}e^t - e^{2t} + \frac{5}{2}e^{3t}$
6. $\frac{s-4}{(s-4)^2+9}$ **Ans.** $e^{4t}\cos 3t$
7. $\frac{16}{(s^2+2s+5)^2}$ **Ans.** $e^{-t}(\sin 2t - 2t\cos 2t)$
8. $\frac{1}{(s+1)(s^2+2s+2)}$ **Ans.** $e^{-t}(1-\cos t)$
9. $\frac{1}{(s-2)(s^2+1)}$ **Ans.** $\frac{1}{5}e^{2t} - \frac{1}{5}\cos t - \frac{2}{5}\sin t$
10. $\frac{s^2-6s+7}{(s^2-4s+5)^2}$ **Ans.** $te^{2t}\{\cos t - \sin t\}$

6.10 INVERSE LAPLACE TRANSFORM BY CONVOLUTION

$$L\left\{\int_0^t f_1(x).f_2(t-x)\,dx\right\} = F_1(s).F_2(s) \text{ or } \boxed{\int_0^t f_1(x).f_2(t-x)\,dx = L^{-1}[F_1(s).F_2(s)]}$$

Example 33. *Use convolution theorem to evaluate:*

$$L^{-1}\left\{\frac{s}{(s^2+4)^2}\right\}$$

(U.P., II Semester, 2010)

Solution. $$\frac{s}{(s^2+4)^2} = \frac{1}{s^2+4}\cdot\frac{s}{s^2+4}$$

Let $$F_1(s) = \frac{1}{s^2+4} \text{ and } F_2(s) = \frac{s}{s^2+4}$$

and $$L^{-1}[F_1(s)] = L^{-1}\left(\frac{1}{s^2+4}\right) = \frac{1}{2}\sin 2t$$

and $$L^{-1}[F_2(s)] = L^{-1}\left(\frac{s}{s^2+4}\right) = \cos 2t$$

According to Convolution Theorem

$$\begin{aligned}
{}^{-1}[F_1(s).F_2(s)] &= \int_0^t f_1(x).f_2(t-x)dx \\
&= \int_0^t \frac{1}{2}\sin 2x\cos 2(t-x)dx \\
&= \frac{1}{4}\int_0^t [\sin(2x+2t-2x)+\sin(2x-2t+2x)]\,dx \\
&= \frac{1}{4}\int_0^t [\sin 2t+\sin(4x-2t)]dx \\
&= \frac{1}{4}[x\sin 2t - \frac{1}{4}\cos(4x-2t)]_0^t \\
&= \frac{1}{4}\left[t\sin 2t - \frac{1}{4}\cos(4t-2t) + \frac{1}{4}\cos(-2t)\right] \\
&= \frac{1}{4}\left[t\sin 2t - \frac{1}{4}\cos 2t + \frac{1}{4}\cos 2t\right] \\
&= \frac{1}{4}\sin 2t
\end{aligned}$$

Ans.

Example 34. *Use convolution theorem to find the inverse of the function* $\frac{1}{(s^2+a^2)^2}$.

Solution. We know that *(U.P., II Semester, 2009)*

$$L^{-1}\frac{1}{s^2+a^2} = \frac{1}{a}\sin at$$

Hence by convolution theorem

$$L^{-1}\frac{1}{(s^2+a^2)}\frac{1}{(s^2+a^2)} = \int_0^t \frac{1}{a}\sin ax \frac{1}{a}\sin a(t-x)\,dx$$

$$= \frac{1}{a^2}\int_0^t \frac{1}{2}[\cos(ax-at+ax)-\cos(ax+at-ax)]\,dx$$

$$\left\{\sin A\sin B = \frac{1}{2}[\cos(A-B)-\cos(A+B)]\right\}$$

$$= \frac{1}{2a^2}\int_0^t [\cos(2ax-at)-\cos at]\,dx$$

$$= \frac{1}{2a^2}\left[\frac{1}{2a}\sin(2ax-at)-x\cos at\right]_0^t$$

$$= \frac{1}{2a^2}\left[\frac{1}{2a}\sin(2at-at)-t\cos at-\frac{1}{2a}\sin(-at)\right]$$

$$= \frac{1}{2a^2}\left[\frac{1}{2a}\sin at-t\cos at+\frac{1}{2a}\sin at\right]$$

$$= \frac{1}{2a^2}\left[\frac{2}{2a}\sin at-t\cos at\right]$$

$$= \frac{1}{2a^3}[\sin at - at\cos at]$$ **Ans.**

Example 35. *State convolution theorem and hence find*

$$L^{-1}\left\{\frac{1}{(s+2)^2(s-2)}\right\}$$ *(Uttarakhand, II Semester, June 2007)*

Solution. Convolution Theorem (See Art 5.23 on page 187).

Let $L\{f_1(t)\} = F_1(s)$ and Let $L\{f_2(t)\} = F_2(s)$

$$F_1(s) = \frac{1}{(s+2)^2} \quad \text{and} \quad F_2(s) = \frac{1}{s-2}$$

$$f_1(t) = L^{-1}\left[\frac{1}{(s+2)^2}\right] = t\,e^{-2t}$$

$$f_2(t) = L^{-1}\left[\frac{1}{(s-2)}\right] = e^{2t}$$

$$\left[\begin{array}{l} L(e^{-2t}) = \dfrac{1}{s+2} \\ L(t\,e^{-2t}) = -\dfrac{d}{ds}\dfrac{1}{(s+2)} = \dfrac{1}{(s+2)^2} \end{array}\right.$$

According to Convolution Theorem

$$L^{-1}[F_1(s).F_2(s)] = \int_0^t f_1(x)\,f_2(t-x)\,dx$$

$$L^{-1}\left[\frac{1}{(s+2)^2(s-2)}\right] = \int_0^t x\,e^{-2x}\cdot e^{2(t-x)}dx = \int_0^t x\,e^{2t-4x}dx$$

$$= \left[x\frac{e^{2t-4x}}{-4} - \int 1.\frac{e^{(2t-4x)}}{-4}\,dx\right]_0^t = \left[-\frac{x}{4}e^{2t-4x} + \frac{1}{4}\left\{\frac{e^{(2t-4x)}}{-4}\right\}\right]_0^t$$

$$= \frac{-t}{4}e^{2t-4t} - \frac{1}{16}e^{2t-4t} + \frac{1}{16}e^{2t} = \frac{-t}{4}e^{-2t} - \frac{1}{16}e^{-2t} + \frac{1}{16}e^{2t}$$

$$= \frac{e^{2t}}{16} - \frac{1}{16}e^{-2t}[4t+1]$$ **Ans.**

Example 36. *Using the Convolution Theorem find*

$$L^{-1}\left\{\frac{s^2}{(s^2+a^2)(s^2+b^2)}\right\},\ a \neq b.$$

Uttarakhand, IInd Semester, Summer 2010, U.P. II Semester Summer, 2010, 2006, 2004

Solution. We have, $L(\cos at) = \dfrac{s}{s^2+a^2}$ and $L(\cos bt) = \dfrac{s}{s^2+b^2}$

Hence, by the convolution theorem

$$L\left\{\int_0^t \cos ax \cos b(t-x)\,dx\right\} = \frac{s^2}{(s^2+a^2)(s^2+b^2)}$$

Therefore,

$$L^{-1}\left\{\frac{s^2}{(s^2+a^2)(s^2+b^2)}\right\} = \int_0^t \cos ax \cos b(t-x)\,dx$$

$$= \frac{1}{2}\int_0^t \{\cos(ax+bt-bx) + \cos(ax-bt+bx)\}\,dx$$

$$= \frac{1}{2}\int_0^t \cos[(a-b)x+bt]\,dx + \frac{1}{2}\int_0^t \cos[(a+b)x-bt]\,dx$$

$$= \left[\frac{\sin[(a-b)x+bt]}{2(a-b)}\right]_0^t + \left[\frac{\sin[(a+b)x-bt]}{2(a+b)}\right]_0^t$$

$$= \frac{\sin at - \sin bt}{2(a-b)} + \frac{\sin at + \sin bt}{2(a+b)}$$

$$= \frac{a\sin at - b\sin bt}{a^2-b^2}$$ **Ans.**

Example 37. *Evaluate* $L^{-1}\left\{\dfrac{s}{(s^2+1)(s^2+4)}\right\}$ (*U.P., II Semester, Summer* 2002)

Solution. We know that $L^{-1}\dfrac{s}{s^2+1} = \cos x$ and $L^{-1}\dfrac{2}{s^2+2^2} = \sin 2x$

$$\therefore \quad L^{-1}\left(\frac{s}{(s^2+1)(s^2+4)}\right)=\frac{1}{2}L^{-1}\left[\left(\frac{s}{s^2+1}\right)\left(\frac{2}{s^2+4}\right)\right]$$

$$=\frac{1}{2}\int_0^t \sin 2x\cos(t-x)\,dx \qquad \text{[By Convolution Th.]}$$

$$=\int_0^t \sin x\cos x\,\{\cos t\cos x+\sin t\sin x\}\,dx$$

$$=\int_0^t \left[\sin x\cos^2 x\cos t+\sin^2 x\cos x\sin t\right]dx$$

$$=\left[-\frac{\cos^3 x}{3}\cos t+\frac{\sin^3 x}{3}\sin t\right]_0^t$$

$$=-\frac{\cos^4 t}{3}+\frac{\sin^4 t}{3}+\frac{\cos t}{3}=\frac{1}{3}\left[\sin^4 t-\cos^4 t\right]+\frac{\cos t}{3}$$

$$=\frac{1}{3}\left(\sin^2 t+\cos^2 t\right)\left(\sin^2 t-\cos^2 t\right)+\frac{\cos t}{3}$$

$$=\frac{1}{3}(\sin^2 t-\cos^2 t)+\frac{\cos t}{3}$$

$$=-\frac{1}{3}\cos 2t+\frac{\cos t}{3}$$

$$=\frac{1}{3}(\cos t-\cos 2t)$$

Ans.

Example 38. *Obtain* $L^{-1}\dfrac{1}{s\left(s^2+a^2\right)}$.

Solution. $L^{-1}\dfrac{1}{s}=1$ and $L^{-1}\dfrac{1}{s^2+a^2}=\dfrac{\sin at}{a}$.

$$L^{-1}\{F_1(s)\,.\,F_2(s)\}=\int_0^t f_1(t)\,f_2(t-x)\,dx \qquad \text{(Convolution Theorem)}$$

Hence by the Convolution Theorem

$$L^{-1}\left[\frac{1}{s}\cdot\frac{1}{s^2+a^2}\right]$$

$$=\int_0^t \frac{\sin a(t-x)}{a}\,dx=\left[\frac{-\cos(at-ax)}{-a^2}\right]_0^t$$

$$=\frac{1}{a^2}\left[1-\cos at\right]$$

Ans.

Example 39. *Using Convolution Theorem, prove that*

$$L^{-1}\left[\frac{1}{s^3\left(s^2+1\right)}\right]=\frac{t^2}{2}+\cos t-1$$

(U.P., II Semester, Summer 2005)

Solution. We know that,

$$L^{-1}\left\{\frac{1}{s^3}\right\}=\frac{t^2}{2!}\qquad\left[L^{-1}\frac{1}{s}=1\right.$$

$$L^{-1}\left\{\frac{1}{s^2+1}\right\}=\sin t\qquad\left.L^{-1}\frac{1}{s^2+a^2}=\frac{1}{a}\sin at\right]$$

Using Convolution Theorem,

$$L^{-1}\left\{\frac{1}{s^3\left(s^2+1\right)}\right\}=\int_0^t\frac{(t-x)^2}{2!}\sin x\,dx$$

$$=\frac{1}{2}\int_0^t\left(t^2+x^2-2tx\right)\sin x\,dx$$

$$=\frac{1}{2}\left[\left(t^2+x^2-2tx\right)(-\cos x)-\int(2x-2t)(-\cos x)\,dx\right]_0^t$$

$$=\frac{1}{2}\left[(t^2+x^2-2tx)(-\cos x)+2\int(x-t)\cos x\,dx\right]_0^t$$

$$=\frac{1}{2}\left[(t^2+x^2-2tx)(-\cos x)+2(x-t)\sin x+2\cos x\right]_0^t$$

$$=\frac{1}{2}\left[(t^2+t^2-2t^2)(-\cos t)+0+2\cos t+t^2\cos 0-2\cos 0\right]$$

$$=\frac{1}{2}[2\cos t+t^2-2]=\cos t+\frac{t^2}{2}-1=\frac{t^2}{2}+\cos t-1\qquad\textbf{Ans.}$$

EXERCISE 6.8

Obtain the Inverse Laplace Transform by convolution.

1. $\dfrac{s^2}{\left(s^2+a^2\right)^2}$ **Ans.** $\dfrac{1}{2}t\cos at+\dfrac{1}{2a}\sin at$

2. $\dfrac{1}{\left(s^2+1\right)^3}$ **Ans.** $\dfrac{1}{8}\left[\left(3-t^2\right)\sin t-3t\cos t\right]$

3. $\dfrac{s}{\left(s^2+a^2\right)^2}$ **Ans.** $\dfrac{t\sin at}{2a}$

4. $\dfrac{1}{s^2\left(s^2-a^2\right)}$ **Ans.** $\dfrac{1}{a^3}[-at+\sinh at]$

5. $\dfrac{1}{(s+1)\left(s^2+1\right)}$ **Ans.** $\dfrac{1}{2}\left(\cos t-\sin t-e^{-t}\right)$

6.11 HEAVISIDE INVERSE FORMULA OF $\dfrac{F(s)}{G(s)}$

If $F(s)$ and G (s) be two polynomials in S. The degree of F(s) is less than that of G (s). Let $\alpha_1, \alpha_2, \alpha_3, \ldots \alpha_n$ be n roots of the equation $G(s)=0$

Inverse Laplace formula of $\frac{F(s)}{G(s)}$ is given by

$$L^{-1}\left\{\frac{F(s)}{G(s)}\right\}=\sum_{i=1}^{n}\frac{F(\alpha_i)}{G'(\alpha_i)}e^{\alpha_i t}$$

Example 40. *Find* $L^{-1}\left\{\frac{2s^2+5s-4}{s^3+s^2-2s}\right\}$.

Solution. Let $F(s) = 2s^2 + 5s - 4$

and $G(s) = s^3 + s^2 - 2s = s(s^2 + s - 2) = s(s+2)(s-1)$

$G'(s) = 3s^2 + 2s - 2$

$G(s) = 0$ has three roots, $0, 1, -2$

$\Rightarrow$ $\alpha_1 = 0, \quad \alpha_2 = 1, \quad \alpha_3 = -2$

By Heaviside Inverse formula

$$L^{-1}\left\{\frac{F(s)}{G(s)}\right\}=\sum_{i=1}^{n}\frac{F(\alpha_i)}{G'(\alpha_i)}e^{\alpha_i t}$$

$$=\left\{\frac{F(\alpha_1)}{G'(\alpha_1)}\right\}e^{t\alpha_1}+\frac{F(\alpha_2)}{G'(\alpha_2)}e^{t\alpha_2}+\frac{F(\alpha_3)}{G'(\alpha_3)}e^{t\alpha_3}=\frac{F(0)}{G'(0)}e^0+\frac{F(1)}{G'(1)}e^t+\frac{F(-2)}{G'(-2)}e^{-2t}$$

$$=\frac{-4}{-2}e^0+\frac{3}{3}e^t+\frac{(-6)}{(6)}e^{-2t}=2+e^t-e^{-2t}$$ **Ans.**

Example 41. *Find* $L^{-1}\left[\frac{2s^2-6s+5}{s^3-6s^2+11s-6}\right]$ (*U.P. II Semester,* 2004)

Solution. Let $F(s) = 2s^2 - 6s + 5$

$G(s) = s^3 - 6s^2 + 11s - 6 = (s-1)(s-2)(s-3)$

$G(s) = 0$ has three roots, 1, 2, 3.

$\Rightarrow$ $\alpha_1 = 1, \quad \alpha_2 = 2, \quad \alpha_3 = 3$

$G'(s) = 3s^2 - 12s + 11$

By Heaviside Inverse formula, we have $L^{-1}\left\{\frac{F(s)}{G(s)}\right\}=\sum_{i=1}^{n}\frac{F(\alpha_i)}{G'(\alpha_i)}e^{t\alpha_i}$

$$L^{-1}\left\{\frac{2s^2-6s+5}{s^3-6s^2+11s-6}\right\}=\frac{F(\alpha_1)}{G'(\alpha_1)}e^{t\alpha_1}+\frac{F(\alpha_2)}{G'(\alpha_2)}e^{t\alpha_2}+\frac{F(\alpha_3)}{G'(\alpha_3)}e^{t\alpha_3}$$

$$=\frac{F(1)}{G'(1)}e^t+\frac{F(2)}{G'(2)}e^{2t}+\frac{F(3)}{G'(3)}e^{3t}=\frac{(1)}{(2)}e^t+\frac{(1)}{(-1)}e^{2t}+\frac{(5)}{(2)}e^{3t}=\frac{1}{2}e^t-e^{2t}+\frac{5}{2}e^{3t}$$ **Ans.**

EXERCISE 6.9

Using Heaviside expansion formula, find the Inverse Laplace Transform of the following :

1. $\frac{s-1}{s^2+3s+2}$ **Ans.** $-2e^{-t}+3e^{-2t}$

2. $\frac{s}{(s-1)(s-2)(s-3)}$ **Ans.** $\frac{1}{2}e^t-2e^{2t}+\frac{3}{2}e^{3t}$

3. $\dfrac{2s+3}{(s-2)(s-3)(s-4)}$ **Ans.** $\dfrac{7}{2}e^{2t}-9e^{3t}+\dfrac{11}{2}e^{4t}$

4. $\dfrac{11s^2-2s+5}{2s^3-3s^2-3s+2}$ **Ans.** $2e^{-2t}+5\cdot e^{2t}-\dfrac{3}{2}e^{\frac{t}{2}}$

6.12 SOLUTION OF DIFFERENTIAL EQUATIONS BY LAPLACE TRANSFORMS

Ordinary linear differential equations with constant coefficients can be easily solved by the Laplace Transform method, without finding the general solution and the arbitrary constants. The method will be clear from the following examples:

Example 42. *Solve the equations :*

$$\frac{d^3y}{dt^3}+2\frac{d^2y}{dt^2}-\frac{dy}{dt}-2y=0, \text{ where } y=1, \frac{dy}{dt}=2, \frac{d^2y}{dt^2}=2 \text{ at } t=0.$$

Solution. Here we have

$$y'''+2y''-y'-2y=0 \quad ...(1)$$

Taking Laplace transform of both sides of (1), we get

$$[s^3\bar{y}-s^2y(0)-sy'(0)-y''(0)]+2[s^2\bar{y}-sy(0)-y'(0)]-[s\bar{y}-y(0)]-2\bar{y}=0 \quad ...(2)$$

Putting $y(0)=1$, $y'(0)=2$, $y''(0)=2$ in equation (2), we get

$$(s^3+2s^2-s-2)\bar{y}=s^2+4s+5$$

$$\bar{y}=\frac{s^2+4s+5}{s^3+2s^2-s-2}=\frac{s^2+4s+5}{(s-1)(s+1)(s+2)}$$

$$=\frac{5}{3(s-1)}-\frac{1}{s+1}+\frac{1}{3(s+2)} \quad ...(3) \text{ (Partial Fractions)}$$

Taking the Inverse Laplace transforms of both sides, of (3), we get

$$y=\frac{5}{3}L^{-1}\left\{\frac{1}{s-1}\right\}-L^{-1}\left\{\frac{1}{s+1}\right\}+\frac{1}{3}L^{-1}\left\{\frac{1}{s+2}\right\}=\frac{5}{3}e^{t}-e^{-t}+\frac{1}{3}e^{-2t}$$

$$\Rightarrow \quad y=\frac{1}{3}(5e^t+e^{-2t})-e^{-t}$$ **Ans.**

Example 43. *Solve the following equation by Laplace transform*

$$y'''-2y''+5y'=0\ ;\ y=0, y'=1 \text{ at } t=0 \text{ and } y=1 \text{ at } t=\frac{\pi}{8}.$$

(*Q. Bank U.P., II Semester 2001*)

Solution. Here, we have

$$y'''-2y''+5y'=0 \quad ...(1)$$

Taking Laplace transform on both sides of (1), we get

$$L(y''')-2L(y'')+5L(y')=L(0)$$

$$\Rightarrow s^3\bar{y}-s^2y(0)-sy'(0)-y''(0)-2[s^2\bar{y}-sy(0)-y'(0)]+5[s\bar{y}-y(0)]=0$$

$$\Rightarrow (s^3-2s^2+5s)\bar{y}-s-k+2=0 \qquad [\text{Let } y''(0)=k \text{ (say)}]$$

$$\Rightarrow \quad \bar{y} = \frac{(k-2)+s}{s(s^2-2s+5)}$$

$$= \left(\frac{k-2}{5}\right)\left(\frac{1}{s} - \frac{s-2}{s^2-2s+5}\right) + \frac{1}{s^2-2s+5}$$

$$= \left(\frac{k-2}{5}\right)\frac{1}{s} - \left(\frac{k-2}{5}\right)\left\{\frac{(s-1)-1}{(s-1)^2+4}\right\} + \frac{1}{(s-1)^2+4}$$

$$\Rightarrow \quad \bar{y} = \left(\frac{k-2}{5}\right)\frac{1}{s} - \left(\frac{k-2}{5}\right)\left\{\frac{(s-1)}{(s-1)^2+4}\right\} + \left(\frac{k+3}{10}\right).\left\{\frac{2}{(s-1)^2+4}\right\} \quad ...(3)$$

Taking Inverse Laplace transform on both sides of (3), we get

$$y = \left(\frac{k-2}{5}\right) - \left(\frac{k-2}{5}\right)e^t \cos 2t + \left(\frac{k+3}{10}\right)e^t \sin 2t \quad ... (4)$$

Puttting $y\left(\frac{\pi}{8}\right) = 1$, we get

$$1 = \left(\frac{k-2}{5}\right) - \left(\frac{k-2}{5}\right)e^{\pi/8}.\frac{1}{\sqrt{2}} + \left(\frac{k+3}{10}\right)e^{\pi/8}.\frac{1}{\sqrt{2}} \quad ...(5)$$

$$\Rightarrow \quad k = 7$$

(on simplification)

On putting the value of k in (4), we get

Hence required solution is

$$y = 1 + e^t(\sin 2t - \cos 2t)$$ **Ans.**

Example 44. *Using Laplace transforms, find the solution of the initial value problem*

$$y'' - 4y' + 4y = 64 \sin 2t$$
$$y(0) = 0, \quad y'(0) = 1.$$

Solution. Here, we have $y'' - 4y' + 4y = 64 \sin 2t$ ... (1)

$$y(0) = 0, y'(0) = 1.$$

Taking Laplace transform of both sides of (1), we have

$$[s^2\bar{y} - sy(0) - y'(0)] - 4[s\bar{y} - y(0)] + 4\bar{y} = \frac{64\times 2}{s^2+4} \quad ... (2)$$

On putting the values of $y(0)$ and $y'(0)$ in (2), we get

$$s^2\bar{y} - 1 - 4s\bar{y} + 4\bar{y} = \frac{128}{s^2+4}$$

$$\Rightarrow \quad (s^2 - 4s + 4)\bar{y} = 1 + \frac{128}{s^2+4}, \quad \Rightarrow (s-2)^2\bar{y} = 1 + \frac{128}{s^2+4}$$

$$\Rightarrow \quad \bar{y} = \frac{1}{(s-2)^2} + \frac{128}{(s-2)^2(s^2+4)} = \frac{1}{(s-2)^2} - \frac{8}{s-2} + \frac{16}{(s-2)^2} + \frac{8s}{s^2+4}$$

$$y = L^{-1}\left[-\frac{8}{s-2} + \frac{17}{(s-2)^2} + \frac{8s}{s^2+4}\right]$$

$$y = -8e^{2t} + 17te^{2t} + 8\cos 2t$$ **Ans.**

Example 45. *Using the Laplace transforms, find the solution of the initial value problem*

$$y'' + 25\,y = 10\cos 5t$$
$$y(0) = 2,\ \ y'(0) = 0.$$

Solution. Taking Laplace transforms of the given differential equation, we get

$$[s^2\,\bar{y} - sy(0) - y'(0)] + 25\bar{y} = 10\,\frac{s}{s^2+25} \qquad \ldots (1)$$

Putting the value of y (0) and y' (0) in (1), we get

$$s^2\,\bar{y} - 2s + 25\bar{y} = \frac{10s}{s^2+25}$$

$$\Rightarrow \quad (s^2+25)\,\bar{y} = 2s + \frac{10s}{s^2+25}$$

$$\Rightarrow \quad \bar{y} = \frac{2s}{s^2+25} + \frac{10s}{(s^2+25)^2}$$

$$y = L^{-1}\left[\frac{2s}{s^2+25} + \frac{10s}{(s^2+25)^2}\right] = 2\cos 5t + L^{-1}\left[\frac{10s}{(s^2+25)^2}\right]$$

$$= 2\cos 5t + L^{-1}\frac{d}{ds}\left[\frac{-5}{(s^2+25)}\right] \qquad \left[\text{On differentiating } \frac{-5}{s^2+25}, \text{ we get } \frac{10\,s}{(s^2+25)^2}.\right]$$

$$= 2\cos 5t - t\sin 5t$$ **Ans.**

Example 46. *Applying Convolution Theorem, solve the following initial value problem*

$$y'' + y = \sin 3t,$$
$$y(0) = 0,\ y'(0) = 0.$$

Solution. $y'' + y = \sin 3t$,

Taking Laplace transform of both sides, we have

$$\left[s^2\,\bar{y} - sy(0) - y'(0)\right] + \bar{y} = \frac{3}{s^2+9} \qquad \ldots (1)$$

On putting the values of y (0) and $y'(0)$ in (1), we get

$$s^2\bar{y} + \bar{y} = \frac{3}{s^2+9} \quad \Rightarrow \quad (s^2+1)\,\bar{y} = \frac{3}{s^2+9}$$

$$\bar{y} = \frac{3}{(s^2+1)(s^2+9)} = \frac{3}{8}\left[\frac{1}{s^2+1} - \frac{1}{s^2+9}\right]$$

Taking the inverse transform, we get

$$y = \frac{3}{8}L^{-1}\frac{1}{s^2+1} - \frac{3}{8}L^{-1}\frac{1}{s^2+9}$$

$$y = \frac{3}{8}\sin t - \frac{3}{8}\times\frac{1}{3}\sin 3t = \frac{3}{8}\sin t - \frac{1}{8}\sin 3t$$ **Ans.**

Example 47. *Using Laplace transforms, find the solution of the initial value problem*

$$y'' + 9y = 6\cos 3t$$

$$y(0) = 2,\ y'(0) = 0 \qquad \textit{(U. P. II Semester Summer 2006)}$$

Solution. We have, $\quad y'' + 9y = 6\cos 3t \qquad \ldots (1)$

$$y(0) = 2,\ y'(0) = 0$$

Taking Laplace transform of (1), we get

$$[s^2\bar{y} - sy(0) - y'(0)] + 9\bar{y} = 6\frac{s}{s^2+9} \qquad \ldots (2)$$

Putting the values of $y(0)$ and $y'(0)$ in (2), we have

$$s^2\bar{y} - 2s + 9\bar{y} = \frac{6s}{s^2+9}$$

$$\Rightarrow \quad (s^2+9)\bar{y} = 2s + \frac{6s}{s^2+9} \quad \Rightarrow \quad \bar{y} = \frac{2s}{s^2+9} + \frac{6s}{(s^2+9)^2}$$

$$y = L^{-1}\frac{2s}{s^2+9} + L^{-1}\frac{6s}{(s^2+9)^2} = 2\cos 3t + L^{-1}\frac{d}{ds}\left[\frac{-3}{(s^2+9)}\right]$$

$$= 2\cos 3t - t\sin 3t \qquad \textbf{Ans.}$$

Example 48. *Using Laplace transformation solve the following differential equation:*

$$\frac{d^2x}{dt^2} + 9x = \cos 2t, \quad \text{if } x(0) = 1, \quad x\left(\frac{\pi}{2}\right) = -1$$

(Uttarakhand, IInd Semester, Summer 2010, U. P. II Semester, Summer 2002)

Solution. $\dfrac{d^2x}{dt^2} + 9x = \cos 2t \qquad \ldots (1)$

Taking Laplace transform of both the sides of (1), we get

$$L\frac{d^2x}{dt^2} + 9Lx = L\cos 2t$$

$$\Rightarrow \quad s^2\bar{x} - sx(0) - x'(0) + 9\bar{x} = \frac{s}{s^2+4} \qquad \ldots (2)$$

On putting $x(0) = 1$ in (2), we get

$$s^2\bar{x} - s + 9\bar{x} - x'(0) = \frac{s}{s^2+4}$$

$$(s^2+9)\bar{x} = s + \frac{s}{s^2+4} + x'(0) = \frac{s(s^2+4)+s}{s^2+4} + x'(0) = \frac{s^3+5s}{s^2+4} + x'(0)$$

$$\Rightarrow \quad \bar{x} = \frac{(s^3+5s)}{(s^2+4)(s^2+9)} + \frac{x'(0)}{s^2+9} = \frac{1}{5}\frac{s}{s^2+4} + \frac{4}{5}\frac{s}{s^2+9} + \frac{x'(0)}{s^2+9}$$

Taking the Inverse Laplace Transform, we get

$$x(t) = \frac{1}{5}L^{-1}\frac{s}{s^2+4} + \frac{4}{5}L^{-1}\frac{s}{s^2+9} + L^{-1}\frac{x'(0)}{s^2+9}$$

$$x(t) = \frac{1}{5}\cos 2t + \frac{4}{5}\cos 3t + \frac{x'(0)\sin 3t}{3} \qquad \ldots (3)$$

On putting $x\left(\frac{\pi}{2}\right) = -1$ in (3), we get

$$-1 = -\frac{1}{5} + 0 - \frac{x'(0)}{3} \qquad \Rightarrow \qquad x'(0) = \frac{12}{5}$$

On putting the value of $x'(0)$ in (3), we get

$$x = \frac{1}{5}\cos 2t + \frac{4}{5}\cos 3t + \frac{12}{5}\,\frac{\sin 3t}{3} = \frac{1}{5}[\cos 2t + 4\cos 3t + 4\sin 3t] \qquad \textbf{Ans.}$$

Example 49. *Using Laplace transform technique solve the following initial value problem*

$$\frac{d^2y}{dt^2} + 2\frac{dy}{dt} + 2y = 5\sin t, \quad y(0) = y'(0) = 0.$$

Solution. $\quad y'' + 2y' + 2y = 5\sin t$

$$y(0) = y'(0) = 0$$

Taking the Laplace Transform of both sides, we have

$$[s^2\bar{y} - sy(0) - y'(0)] + 2[s\bar{y} - y(0)] + 2\bar{y} = 5 \times \frac{1}{s^2+1} \qquad \ldots (1)$$

On substituting the values of $y(0)$, and $y'(0)$ in (1), we get

$$s^2\bar{y} + 2s\bar{y} + 2\bar{y} = \frac{5}{s^2+1} \qquad \Rightarrow \qquad [s^2 + 2s + 2]\bar{y} = \frac{5}{s^2+1}$$

$$\Rightarrow \qquad \bar{y} = \frac{5}{(s^2+2s+2)(s^2+1)}$$

Resolving into partial fractions, $y = \dfrac{2s+3}{s^2+2s+2} + \left(\dfrac{-2s+1}{s^2+1}\right)$

Taking the inverse transform, we get

$$y = L^{-1}\left(\frac{2s+3}{s^2+2s+2}\right) + L^{-1}\left(\frac{-2s+1}{s^2+1}\right)$$

$$= L^{-1}\left[\frac{2(s+1)+1}{(s+1)^2+1}\right] + L^{-1}\left(\frac{-2s}{s^2+1}\right) + L^{-1}\left(\frac{1}{s^2+1}\right)$$

$$= L^{-1}\left[\frac{2(s+1)}{(s+1)^2+1}\right] + L^{-1}\left[\frac{1}{(s+1)^2+1}\right] - 2\cos t + \sin t$$

$$= 2e^{-t}\cos t + e^{-t}\sin t - 2\cos t + \sin t \qquad \textbf{Ans.}$$

Example 50. *Solve the initial value problem*

$$2y'' + 5y' + 2y = e^{2t}, \qquad y(0) = 1,\ y'(0) = 1,$$

using the Laplace transforms.

Solution. $\quad 2y'' + 5y' + 2y = e^{-2t} \qquad y(0) = 1,\ y'(0) = 1$

Taking the Laplace Transform of both sides, we get

$$2[s^2\,\bar{y} - sy(0) - y'(0)] + 5[s\bar{y} - y(0)] + 2\bar{y} = \frac{1}{s+2} \qquad \text{... (1)}$$

On substituting the values of y (0) and y' (0) in (1), we get

$$2[s^2\,\bar{y} - s - 1] + 5[s\bar{y} - 1] + 2\bar{y} = \frac{1}{s+2}$$

$$[2s^2 + 5s + 2]\,\bar{y} - 2s - 2 - 5 = \frac{1}{s+2} \Rightarrow (2s^2 + 5s + 2)\,\bar{y} = 2s + 7 + \frac{1}{s+2}$$

$$\bar{y} = \frac{1}{(s+2)(2s^2+5s+2)} + \frac{2s+7}{2s^2+5s+2} = \frac{1 + 2s^2 + 7s + 4s + 14}{(2s^2+5s+2)(s+2)} = \frac{2s^2 + 11s + 15}{(2s+1)(s+2)^2}$$

$$= \frac{\frac{40}{9}}{2s+1} - \frac{\frac{11}{9}}{s+2} - \frac{\frac{1}{3}}{(s+2)^2} = \frac{40}{9}\,\frac{1}{2}\,\frac{1}{s+\frac{1}{2}} - \frac{11}{9}\,\frac{1}{s+2} - \frac{1}{3}\,\frac{1}{(s+2)^2}$$

$$y = \frac{20}{9}\,\frac{1}{s+\frac{1}{2}} - \frac{11}{9}\,\frac{1}{(s+2)} + \frac{1}{3}\,\frac{d}{ds}\,\frac{1}{(s+2)}$$

$$\Rightarrow \quad y = \frac{20}{9}\,e^{-\frac{1}{2}t} - \frac{11}{9}\,e^{-2t} - \frac{1}{3}\,te^{-2t}. \qquad \textbf{Ans.}$$

Example 51. *Solve, using Laplace transform method*

$$y(0) = -2,\ y'(0) = 8;\ \ y'' + 4y' + 4y = 6e^{-t} \qquad \textit{(U.P., II Semester, 2007)}$$

Solution. Here, we have

$$y'' + 4y' + 4y = 6e^{-t} \qquad \text{...(1)}$$

Taking Laplace transform on both sides of (1), we get

$$L(y'') + 4L(y') + 4L(y) = 6L(e^{-t})$$

$$\Rightarrow [s^2\bar{y} - sy(0) - y'(0)] + 4[s\,\bar{y} - y(0)] + 4\bar{y} = \frac{6}{s+1} \qquad \text{... (2)}\ \ [\text{Here } \bar{y} = L(y)]$$

Putting the values of $y(0) = -2$ and $y'(0) = 8$ in (2), we get

$$(s^2 + 4s + 4)\,\bar{y} + 2s - 8 + 8 = \frac{6}{s+1}$$

$$\Rightarrow \qquad (s^2 + 4s + 4)\,\bar{y} + 2s = \frac{6}{s+1}$$

$$(s+2)^2\,\bar{y} + 2s = \frac{6}{s+1}$$

$$\bar{y} = \frac{6}{(s+1)(s+2)^2} - \frac{2s}{(s+2)^2}$$

$$= 6\left[\frac{1}{s+1} - \frac{1}{s+2} - \frac{1}{(s+2)^2}\right] - \frac{2\{(s+2)-2\}}{(s+2)^2}$$

$$= \frac{6}{s+1} - \frac{6}{s+2} - \frac{6}{(s+2)^2} - \frac{2}{s+2} + \frac{4}{(s+2)^2}$$

$$= \frac{6}{s+1} - \frac{8}{s+2} - \frac{2}{(s+2)^2} \qquad ...(3)$$

Taking inverse Laplace transform on both sides of (3), we get

$$y = 6e^{-t} - 8e^{-2t} - 2t\, e^{-2t} \qquad \textbf{Ans.}$$

Example 52. *Solve the following differential equation using Laplace transform*

$$\frac{d^3 y}{dt^3} - 3\frac{d^2 y}{dt^2} + 3\frac{dy}{dt} - y = t^2 e^t$$

where $y(0) = 1,\ \left(\frac{dy}{dt}\right)_{t=0} = 0,\ \left(\frac{d^2 y}{dt^2}\right)_{t=0} = -2$ *(U. P., II Semester, (SUM) 2008)*

Solution. Here we have equation

$$y''' - 3y'' + 3y' - y = t^2 e^t \qquad ...(1)$$

Taking Laplace transform on both side of equation (1), we get

$$L(y''') - 3L(y'') + 3L(y') - L(y) = L(t^2 e^t)$$

$$\Rightarrow [s^3 \bar{y} - s^2 y(0) - sy'(0) - y''(0)] - 3[s^2 \bar{y} - sy(0) - y'(0)]$$

$$+ 3[s\bar{y} - y(0)] - \bar{y} = \frac{2}{(s-1)^3} \quad ...(2)\ [\text{where } \bar{y} = L(y)]$$

Putting the values of $y(0)$, $y'(0)$ and y'' at $x = 0$ in (2), we get

$$\Rightarrow (s^3 \bar{y} - s^2 + 2) - 3(s^2 \bar{y} - s) + 3(s\bar{y} - 1) - \bar{y} = \frac{2}{(s-1)^3}$$

$$\Rightarrow (s^3 - 3s^2 + 3s - 1)\bar{y} - s^2 + 3s - 1 = \frac{2}{(s-1)^3}$$

$$\Rightarrow (s-1)^3 \bar{y} = s^2 - 3s + 1 + \frac{2}{(s-1)^3}$$

$$\Rightarrow \bar{y} = \frac{(s-1)^2}{(s-1)^3} - \frac{s}{(s-1)^3} + \frac{2}{(s-1)^6}$$

$$= \frac{1}{s-1} - \frac{(s-1)+1}{(s-1)^3} + \frac{2}{(s-1)^6}$$

$$= \frac{1}{s-1} - \frac{1}{(s-1)^2} - \frac{1}{(s-1)^3} + \frac{2}{(s-1)^6} \qquad ...(3)$$

Taking inverse Laplace transform on both sides of equation (3), we get

$$y = e^t - te^t - \frac{t^2}{2}e^t + \frac{t^5 e^t}{60} = \left(1 - t - \frac{t^2}{2} + \frac{t^5}{60}\right)e^t \qquad \textbf{Ans.}$$

Example 53. *Solve* $\frac{d^2 y}{dx^2} + 2\frac{dy}{dx} + 5y = e^{-x}\sin x$ *where* $y(0) = 0$, $y'(0) = 1$.

(U.P., II Semester, 2004)

Solution. Here, we have $\frac{d^2y}{dx^2} + 2\frac{dy}{dx} + 5y = e^{-x} \sin x$

Taking the Laplace Transform of both the sides, we get

$$[s^2 \bar{y} - sy(0) - \bar{y}(0)] + 2[s\bar{y} - y(0)] + 5\bar{y} = L(e^{-x} \sin x)$$

$$[s^2 \bar{y} - sy(0) - y'(0)] + 2[s\bar{y} - y(0)] + 5\bar{y} = \frac{1}{(s+1)^2+1} \qquad \text{... (1)}$$

On substituting the values of $y(0)$ and $y'(0)$ in (1), we get

$$(s^2\bar{y} - 1) + 2(s\bar{y}) + 5\bar{y} = \frac{1}{s^2+2s+2}$$

$$(s^2+2s+5)\bar{y} = 1 + \frac{1}{s^2+2s+2} = \frac{s^2+2s+3}{s^2+2s+2}$$

$$\bar{y} = \frac{s^2+2s+3}{(s^2+2s+5)(s^2+2s+2)}$$

On resolving the R.H.S. into partial fractions, we get

$$\bar{y} = \frac{2}{3}\frac{1}{s^2+2s+5} + \frac{1}{3}\frac{1}{s^2+2s+2}$$

On inversion, we obtain

$$y = \frac{2}{3}L^{-1}\frac{1}{s^2+2s+5} + \frac{1}{3}L^{-1}\frac{1}{s^2+2s+2}$$

$$\Rightarrow \qquad y = \frac{1}{3}L^{-1}\frac{2}{(s+1)^2+(2)^2} + \frac{1}{3}L^{-1}\frac{1}{(s+1)^2+(1)^2}$$

$$\Rightarrow \qquad y = \frac{1}{3}e^{-x}\sin 2x + \frac{1}{3}e^{-x}\sin x$$

$$\Rightarrow \qquad y = \frac{1}{3}e^{-x}(\sin x + \sin 2x)$$ **Ans.**

Example 54. *Solve the equation by the transform method:*

$$\frac{dy}{dt} + 2y + \int_0^t y\,dt = \sin t, \quad y(0) = 1$$

(*R. G. P. V. Bhopal, June* 2003)

Solution. Taking Laplace transform of the given equation, we get

$$[s\bar{y} - y(0)] + 2\bar{y} + \frac{\bar{y}}{s} = \frac{1}{s^2+1} \qquad \text{... (1)} \qquad \left[\because L\left\{\int_0^t y\,dt = \frac{\bar{y}}{s}\right\}\right]$$

Putting the values of $y(0) = 1$ in (1), we get

$$[s\bar{y} - 1] + 2\bar{y} + \frac{\bar{y}}{s} = \frac{1}{s^2+1}$$

$$\bar{y}\left(s + 2 + \frac{1}{s}\right) = 1 + \frac{1}{s^2+1} \qquad [\because y(0) = 1]$$

$$\frac{1}{s}\,\bar{y}(s^2+2s+1)=\frac{s^2+1+1}{s^2+1}$$

$$\frac{1}{s}\,\bar{y}\,(s+1)^2 = \frac{s^2+2}{s^2+1}$$

$$\Rightarrow \qquad \bar{y}\,(s+1)^2 = \frac{s^3+2s}{s^2+1}$$

$$\Rightarrow \qquad \bar{y}=\frac{s^3+2s}{(s+1)^2\,(s^2+1)}$$

$$=\frac{1}{s+1}-\frac{3}{2\,(s+1)^2}+\frac{1}{2\,(s^2+1)} \qquad \text{[By partial fractions]}$$

Taking Inverse Laplace Transform, we have

$$y=e^{-t}-\frac{3}{2}\,t\,e^{-t}+\frac{1}{2}\sin t \qquad \textbf{Ans.}$$

Example 55. *Apply Laplace transform to solve the equation*

$$\frac{d^2y}{dt^2}+y=t\cos 2t,\ t>0,\ \textit{given that}\ y=\frac{dy}{dt}=0\ \textit{for}\ t=0.$$

(U. P. II. Semester Summer 2012, 2005)

Solution. Given equation is $\frac{d^2y}{dt^2}+y=t\cos 2t,\ t>0,\ y=\frac{dy}{dt}=0$ *for* $t=0.$

Taking Laplace transform of both sides, we get

$$L\left\{\frac{d^2y}{dt^2}\right\}+L\{\,y\,\}=L\{\,t\cos 2t\,\} \qquad \text{... (1)}$$

We know that $\quad L\{\,\cos 2t\,\}=\frac{s}{s^2+4}$

$$\Rightarrow \qquad L\{\,t\cos 2t\}=(-1)\frac{d}{ds}\left\{\frac{s}{s^2+4}\right\}=(-1)\frac{\left(s^2+4\right)(1)-(2s)(s)}{\left(s^2+4\right)^2}$$

$$=(-1)\left\{\frac{s^2+4-2s^2}{\left(s^2+4\right)^2}\right\}=\frac{s^2-4}{\left(s^2+4\right)^2} \qquad \text{...(2)}$$

Putting the Laplace transform of $y\,(0)$, y and $t\cos 2t$ in (1), we get

$$[\,s^2\,\bar{y}-sy\,(0)-y'\,(0)\,]+\bar{y}=\frac{s^2-4}{(s^2+4)^2}$$

$$\Rightarrow \qquad [\,s^2\,\bar{y}-0-0\,]+\bar{y}=\frac{s^2-4}{(s^2+4)^2} \qquad [\,\because\ y(0)=y'(0)=0\,]$$

$$\Rightarrow \qquad (\,s^2+1\,)\,\bar{y}=\frac{s^2-4}{(s^2+4)^2}$$

$$\Rightarrow \qquad \bar{y} = \frac{s^2 - 4}{(s^2+1)(s^2+4)^2}$$

Taking Inverse Laplace Transform

$$y = L^{-1}\left\{\frac{s^2-4}{(s^2+1)(s^2+4)^2}\right\} = L^{-1}\left\{\frac{s^2+4-8}{(s^2+1)(s^2+4)^2}\right\}$$

$$= L^{-1}\left[\frac{s^2+4}{(s^2+1)(s^2+4)^2} - \frac{8}{(s^2+1)(s^2+4)^2}\right]$$

$$= L^{-1}\left\{\frac{1}{(s^2+1)(s^2+4)} - \frac{8}{(s^2+1)(s^2+4)^2}\right\}$$

$$= L^{-1}\left\{\frac{1}{(s^2+1)(s^2+4)}\right\} - 8\,L^{-1}\left\{\frac{1}{(s^2+1)(s^2+4)^2}\right\}$$

$$= L^{-1}\left[\frac{1}{3(s^2+1)} - \frac{1}{3(s^2+4)}\right] - 8\,L^{-1}\left\{\frac{1}{9(s^2+1)} - \frac{1}{9(s^2+4)} - \frac{1}{3(s^2+4)^2}\right\}$$

$$= \frac{1}{3}\sin t - \frac{1}{6}\sin 2t - 8\left[\frac{1}{9}\sin t - \frac{1}{18}\sin 2t - \frac{1}{3}L^{-1}\left\{\frac{1}{(s^2+4)}\cdot\frac{1}{(s^2+4)}\right\}\right]$$

$$= \frac{1}{3}\sin t - \frac{1}{6}\sin 2t - \frac{8}{9}\sin t + \frac{8}{18}\sin 2t + \frac{8}{3}\int_0^t \frac{1}{4}\sin 2(t-x)\sin 2x\,dx$$

[Convolution Theorem]

$$= \frac{1}{3}\sin t - \frac{1}{6}\sin 2t - \frac{8}{9}\sin t + \frac{8}{18}\sin 2t + \frac{8}{12}\times\frac{1}{2}\int_0^t 2\sin 2(t-x)\sin 2x\,dx$$

$$= \frac{1}{3}\sin t - \frac{1}{6}\sin 2t - \frac{8}{9}\sin t + \frac{8}{18}\sin 2t + \frac{8}{24}\int_0^t \{\cos 2(t-2x) - \cos 2t\}\,dx$$

$$= \frac{1}{3}\sin t - \frac{1}{6}\sin 2t - \frac{8}{9}\sin t + \frac{4}{9}\sin 2t + \frac{1}{3}\left[\frac{\sin 2(t-2x)}{-4} - x\cos 2t\right]_0^t$$

$$= \frac{1}{3}\sin t - \frac{1}{6}\sin 2t - \frac{8}{9}\sin t + \frac{4}{9}\sin 2t + \frac{1}{3}\left[\frac{\sin 2t}{4} - t\cos 2t + \frac{\sin 2t}{4}\right]$$

$$= \frac{1}{3}\sin t - \frac{1}{6}\sin 2t - \frac{8}{9}\sin t + \frac{4}{9}\sin 2t + \frac{\sin 2t}{6} - \frac{t}{3}\cos 2t$$

$$= -\frac{5}{9}\sin t + \frac{4}{9}\sin 2t - \frac{t}{3}\cos 2t.$$

Ans.

Example 56. *Solve the following differential equation using Laplace transform:*

$$x\frac{d^2y}{dx^2} + \frac{dy}{dx} + xy = 0, \qquad \text{given } y(0) = 2,\ y'(0) = 0.$$

(*Uttarakhand, II Semester, June* 2007)

Solution. Taking Laplace transform on both sides of the given equation, we have

$$L[xy''] + L[y'] + L[xy] = L(0)$$

$$\Rightarrow \qquad -\frac{d}{ds}L(y'') + L(y') - \frac{d}{ds}L(y) = 0$$

$$\Rightarrow \qquad -\frac{d}{ds}\{s^2\bar{y} - sy(0) - y'(0)\} + \{s\bar{y} - y(0)\} - \frac{d\bar{y}}{ds} = 0 \qquad \text{... (1)}$$

Putting $y(0) = 2$ and $y'(0) = 0$ in (1), we get

$$-\frac{d}{ds}\{s^2\bar{y} - 2s - 0\} + \{s\bar{y} - 2\} - \frac{d\bar{y}}{ds} = 0$$

$$\Rightarrow \qquad -s^2\frac{d\bar{y}}{ds} - (2s)\bar{y} + 2 + s\bar{y} - 2 - \frac{d\bar{y}}{ds} = 0$$

$$\Rightarrow \qquad (s^2+1)\frac{d\bar{y}}{ds} + s\bar{y} = 0 \qquad \text{... (2)}$$

Separating the variables, we have

$$\frac{d\bar{y}}{\bar{y}} + \frac{s\,ds}{s^2+1} = 0 \qquad \text{... (3)}$$

On integrating, we have

$$\log\bar{y} + \frac{1}{2}\log(s^2+1) = \log C$$

$$\Rightarrow \qquad \log\bar{y} = \log C - \log\sqrt{s^2+1} \qquad \Rightarrow \qquad \log\bar{y} = \log\frac{C}{\sqrt{s^2+1}}$$

$$\Rightarrow \qquad \bar{y} = \frac{C}{\sqrt{s^2+1}} \qquad \text{... (4)}$$

Taking Inverse Laplace Transform, we get

$$y = L^{-1}\left[\frac{C}{\sqrt{s^2+1}}\right]$$

$$\Rightarrow \qquad y = C\,J_0(x) \qquad \text{... (5) [See Art. 5.24]}$$

At $x = 0$ $\qquad y(0) = C\,J_0(0) \qquad$ ... (6)

Putting $y(0) = 2$ and $J_0(0) = 1$ in (6), we get

$$2 = C(1) \quad \Rightarrow \quad C = 2$$

On putting the value of C in (5), we get

$$y = 2\,J_0(x)$$

Ans.

Example 57. *Using Laplace transforms, find the solution of the initial value problem*

$$y'' + 9y = 9\,u(t-3), \qquad y(0) = y'(0) = 0$$

where $u(t-3)$ is the unit step functions.

Solution. We have, $\qquad y'' + 9y = 9\,u(t-3) \qquad$ ... (1)

Taking Laplace transform of (1), we have

$$s^2\bar{y} - sy(0) - y'(0) + 9\bar{y} = 9\frac{e^{-3s}}{s} \qquad \text{... (2)}$$

Putting the values of $y(0) = 0$ and $y'(0) = 0$ in (2), we get

$$s^2\,\bar{y}+9\bar{y}=9\,\frac{e^{-3s}}{s}$$

$$(s^2+9)\,\bar{y}=9\,\frac{e^{-3s}}{s}$$

$$\bar{y}=\frac{9\,e^{-3s}}{s\,(s^2+9)} \Rightarrow \qquad y=L^{-1}\,\frac{9\,e^{-3s}}{s\,(s^2+9)} \qquad \dots (3)$$

We know that $\quad L^{-1}\,\dfrac{3}{s^2+9}=\sin 3t$

$$\Rightarrow \quad 3\,L^{-1}\,\frac{3}{s\,(s^2+9)}=3\int_0^t \sin 3t\,dt=-[\,\cos 3t\,]_0^t=1-\cos 3t \qquad \dots (4)$$

[Using second shifting theorem]

Using (4), we get the inverse of (3)

$$y=[\,1-\cos 3\,(t-3)\,]\,u\,(t-3) \qquad \textbf{Ans.}$$

Example 58. *Solve, by the method of Laplace transform, the differential equation*

$$(D^2+n^2)\,x=a\sin(nt+\alpha),$$

$$x=Dx=0 \text{ at } t=0. \qquad \textit{(U.P., II Semester, Summer, 2010, 2002)}$$

Solution. Taking Laplace transform of the given differential equation, we get

$$[\,s^2\,\bar{x}-sx\,(0)-x'(0)\,]+n^2\,\bar{x}=a\,L\sin(\,nt+\alpha\,)$$

$$=a\,L\,[\,\sin nt\cos\alpha+\cos nt\sin\alpha\,]$$

$$=a\left[\frac{n}{s^2+n^2}\cos\alpha+\frac{s}{s^2+n^2}\sin\alpha\right]$$

$$=\frac{a\,[n\cos\alpha+s\sin\alpha\,]}{s^2+n^2} \qquad \dots (1)$$

Putting $x\,(0)=x'\,(0)=0$ in (1), we get

$$s^2\,\bar{x}+n^2\,\bar{x}=a\left[\frac{n\cos\alpha+s\sin\alpha}{s^2+n^2}\right] \Rightarrow \quad (s^2+n^2)\,\bar{x}=\frac{a\,[\,n\cos\alpha+s\sin\alpha\,]}{s^2+n^2}$$

$$\Rightarrow \quad \bar{x}=\frac{a\,n\cos\alpha+as\sin\alpha}{(s^2+n^2)^2}$$

$$\Rightarrow \quad \bar{x}=a\cos\alpha\,.\,\frac{n}{(s^2+n^2)^2}+a\sin\alpha\,.\,\frac{s}{(s^2+n^2)^2} \qquad \dots (2)$$

Taking the Inverse Laplace Transform of (2), we get

$$x=a\cos\alpha\,L^{-1}\left[\frac{n}{(s^2+n^2)^2}\right]+(\,a\sin\alpha\,)\,L^{-1}\left[\frac{s}{(s^2+n^2)^2}\right] \qquad \dots (3)$$

Let us find out the inverse of the term on R.H.S.

But $\quad L^{-1}\left\{\dfrac{2s}{(s^2+n^2)^2}\right\}=L^{-1}\,\dfrac{d}{ds}\,\dfrac{1}{s^2+n^2}=\dfrac{t}{n}\sin nt \qquad \left[-\dfrac{d}{ds}[\,F\,(s)\,]=t\,f(t)\right]$

$$\Rightarrow \quad L^{-1}\left\{\frac{s}{(s^2+n^2)^2}\right\}=\frac{t}{2n}\sin nt \qquad \dots (4)$$

Again $L^{-1}\frac{1}{s}\left\{\frac{s}{(s^2+n^2)^2}\right\}=\frac{1}{2n}\int_0^t t\sin nt\,dt$ $\left[\frac{F(s)}{s}=\int_0^t f(t)\,dt\right]$

$$L^{-1}\left[\frac{1}{(s^2+n^2)^2}\right]=\frac{1}{2n}\left[t\left(\frac{-\cos nt}{n}\right)+\frac{1}{n^2}\sin nt\right]_0^t$$

$$L^{-1}\left[\frac{n}{(s^2+n^2)^2}\right]=\frac{1}{2n^3}\left[-nt\cos nt+\sin nt\right] \qquad \text{... (5)}$$

Putting the values of $L^{-1}\left[\frac{s}{(s^2+n^2)^2}\right]$ from (4) and $L^{-1}\left[\frac{n}{(s^2+n^2)^2}\right]$ from (5) in (3), we get

$$x=(a\cos\alpha)\frac{1}{2n^3}[\sin nt-nt\cos nt]+(a\sin\alpha)\frac{t}{2n}\sin nt$$

$$=\frac{a}{2n^2}[\cos\alpha\sin nt-nt\cos\alpha\cos nt+nt\sin\alpha\sin nt]$$

$$=\frac{a}{2n^2}[\cos\alpha\sin nt-nt(\cos nt\cos\alpha-\sin nt\sin\alpha)]$$

$$=\frac{a}{2n^2}[\sin nt\cos\alpha-nt\cos(nt+\alpha)]$$

Which is the required solution. **Ans.**

Example 59. *Solve* $t\,y''+2\,y'+t\,y=\cos t$, *if* $y(0)=1,\ y'(0)=0$

Solution. We have, $t\,y''+2\,y'+t\,y=\cos t$

$$\Rightarrow\quad L\{t\,y''\}+L\{2y'\}+L\{t\,y\}=L\{\cos t\}$$

$$\Rightarrow\quad -\frac{d}{ds}L\{y''\}+2L\{y'\}-\frac{d}{ds}L\{y\}=L\{\cos t\}$$

$[i=0 \text{ at } t=0]$

$$\Rightarrow\quad -\frac{d}{ds}[s^2\bar{y}-sy(0)-y'(0)]+2[s\bar{y}-y(0)]-\frac{d}{ds}\bar{y}=\frac{s}{s^2+1}$$

$$\Rightarrow\quad -\frac{d}{ds}[s^2\bar{y}-s]+2s\bar{y}-2-\frac{d}{ds}\bar{y}=\frac{s}{s^2+1}$$

$$\Rightarrow\quad -2s\bar{y}-s^2\frac{d\bar{y}}{ds}+1+2s\bar{y}-2-\frac{d\bar{y}}{ds}=\frac{s}{s^2+1}$$

$$\Rightarrow\quad (s^2+1)\frac{d\bar{y}}{ds}+1=\frac{-s}{s^2+1}\quad\Rightarrow\quad \frac{d\bar{y}}{ds}=\frac{-s}{(s^2+1)^2}-\frac{1}{s^2+1}$$

Taking Inverse Laplace Transform, we get

$$L^{-1}\frac{d}{ds}(\bar{y})=L^{-1}\left[\frac{-s}{(s^2+1)^2}-\frac{1}{s^2+1}\right],\qquad \left[L^{-1}\frac{d}{ds}F(s)=(-1)^1t^1f(t)\right]$$

$$\Rightarrow\quad (-1)^1t^1\,y=-\frac{1}{2}L^{-1}\left\{\frac{2s}{(s^2+1)^2}\right\}-L^{-1}\left\{\frac{1}{s^2+1}\right\}=-\frac{1}{2}t\sin t-\sin t$$

$$\Rightarrow\quad y=\frac{1}{2}\left(1+\frac{2}{t}\right)\sin t$$

Ans.

Example 60. *Solve* $[\, t\, D^2 + (1 - 2t)\, D - 2]\, y = 0$, *where* $y(0) = 1$, $y'(0) = 2$.

(*M.D.U., 2010, U. P. II Semester. June* 2002)

Solution. Here, $t\, D^2\, y + (1 - 2t)\, Dy - 2y = 0 \quad \Rightarrow \quad ty'' + y' - 2\, t\, y' - 2y = 0$

Taking Laplace transform of given differential equation, we get

$$L(t\, y'') + L\,(y') - 2\, L\,(t\, y') - 2\, L\,(y) = 0$$

$$\Rightarrow \quad -\frac{d}{ds} L\{\, y''\,\} + L\{\, y'\,\} + 2\frac{d}{ds} L\,(y') - 2\, L\,(y) = 0$$

$$\Rightarrow \quad -\frac{d}{ds}[\, s^2\, \bar{y} - sy\,(0) - y'\,(0)\,] + [\, s\, \bar{y} - y\,(0)\,] + 2\, \frac{d}{ds}[\, s\, \bar{y} - y\,(0)\,] - 2\bar{y} = 0$$

Putting the values of $y(0)$ and $y'\,(0)$, we get

$$-\frac{d}{ds}(\, s^2\, \bar{y} - s - 2\,) + (\, s\, \bar{y} - 1\,) + 2\frac{d}{ds}(s\, \bar{y} - 1) - 2\, \bar{y} = 0 \qquad [\because\ y(0) = 1,\ y'\,(0) = 2]$$

$$\Rightarrow \quad -s^2\frac{d\,\bar{y}}{ds} - 2s\,\bar{y} + 1 + s\,\bar{y} - 1 + 2\left(s\,\frac{d\,\bar{y}}{ds} + \bar{y}\right) - 2\,\bar{y} = 0$$

$$\Rightarrow \quad -(\, s^2 - 2s\,)\frac{d\,\bar{y}}{ds} - s\,\bar{y} = 0$$

$$\Rightarrow \quad -\frac{d\,\bar{y}}{\bar{y}} - \frac{1}{s-2}\, ds = 0 \quad (\textit{Separating the variables})$$

$$\Rightarrow \quad \int\frac{d\,\bar{y}}{\bar{y}} + \int\frac{ds}{s-2} = 0 \Rightarrow \log\bar{y} + \log(\, s - 2\,) = \log C \Rightarrow \log\bar{y}\,(\, s - 2\,) = \log C$$

$$\Rightarrow \quad \bar{y}\,(\, s - 2\,) = C \Rightarrow \bar{y} = \frac{C}{s-2} \quad \Rightarrow \quad y = C\, L^{-1}\left\{\frac{1}{s-2}\right\} \Rightarrow y = C\, e^{2t} \qquad \ldots (1)$$

$$y\,(0) = C\, e^0 \qquad \ldots (2)$$

Putting $y\,(0) = 1$ in (2), we get

$$1 = C\, e^0 \quad \Rightarrow \quad C = 1.$$

Putting $C = 1$ in (1), we get

$$y = e^{2t}$$

This is the required solution. **Ans.**

Example 61. *Solve the problem*

$$t\, y'' + y' + 4ty = 0 \textit{ given that } y = 3 \textit{ and } y' = 0 \textit{ when } t = 0.$$

Solution. Taking the Laplace transform of both sides of the given equation, we have

$$L(ty'') + L(y') + 4L(ty) = L(0)$$

$$\Rightarrow \quad -\frac{d}{ds} L\,(y'') + L\,(y') - 4\frac{d}{ds} L\,(y) = 0$$

$$\Rightarrow \quad -\frac{d}{ds}\left\{s^2\,\bar{y} - sy(0) - y'\,(0)\right\} + \{\, s\bar{y} - y(0)\,\} - 4\frac{d\,\bar{y}}{ds} = 0 \qquad \ldots (1)$$

Putting the value of $y(0)$, $y'(0)$ in (1), we get

$$-\frac{d}{ds}(\, s^2\,\bar{y} - 3s - 0\,) + (\, s\,\bar{y} - 3\,) - 4\frac{d\,\bar{y}}{ds} = 0$$

$$-\left(2s\,\bar{y}+\frac{s^2\,d\,\bar{y}}{ds}-3\right)+(s\,\bar{y}-3)-4\frac{d\,\bar{y}}{ds}=0$$

$$\Rightarrow \qquad (s^2+4)\frac{d\,\bar{y}}{ds}+s\,\bar{y}=0$$

Separating the variables, we have

$$\frac{d\,\bar{y}}{\bar{y}}+\frac{s\,ds}{s^2+4}=0$$

$$\Rightarrow \qquad \int\frac{d\,\bar{y}}{\bar{y}}+\int\frac{s\,ds}{s^2+4}=0 \qquad \ldots (2)$$

$$\Rightarrow \qquad \log\bar{y}+\frac{1}{2}\log(s^2+4)=\log C \qquad \Rightarrow \qquad \log\bar{y}\sqrt{s^2+4}=\log C$$

$$\Rightarrow \qquad \bar{y}\sqrt{s^2+4}=C$$

$$\Rightarrow \qquad \bar{y}=\frac{C}{\sqrt{s^2+4}} \qquad \ldots (3)$$

Taking Inverse Laplace Transform, we get

$$y=C\,J_0(2t) \qquad \ldots (4)$$

$$\Rightarrow \qquad y(0)=C\,J_0(0) \qquad \ldots (5)$$

Putting $y(0) = 3$ and $J_0(0) = 0$ in (5), we get

$$\Rightarrow \qquad 3=C\times 1 \qquad \Rightarrow \qquad C=3$$

On putting the value of C in (4), the required solution is

$$y=3\,J_0(2t)$$ **Ans.**

Example 62. *Using Laplace transform, solve the following differential equation: -*

$$y''+2t\,y'-y=t$$

when $y(0) = 0$ *and* $y'(0) = 1$ *(U.P., II Semester, Summer* 2003)

Solution. We have, $y''+2t\,y'-y=t$ ... (1)

Taking Laplace transform of (1), we get

$$[s^2\,\bar{y}-sy(0)-y'(0)]-2\frac{d}{ds}[s\,\bar{y}-y(0)]-\bar{y}=\frac{1}{s^2} \qquad \ldots (2)$$

On putting $y(0) = 0$ and $y'(0) = 1$ in (2), we get

$$(s^2\,\bar{y}-1)-2\frac{d}{ds}(s\,\bar{y}-0)-\bar{y}=\frac{1}{s^2}$$

$$\Rightarrow \qquad (s^2\,\bar{y}-1)-2\,\bar{y}-2s\frac{d\,\bar{y}}{ds}-\bar{y}=\frac{1}{s^2}$$

$$\Rightarrow \qquad -2s\frac{d\,\bar{y}}{ds}+(s^2-3)\,\bar{y}=\frac{1}{s^2}+1=\frac{1+s^2}{s^2}$$

$$\Rightarrow \qquad \frac{d\,\bar{y}}{ds}-\frac{s^2-3}{2s}\,\bar{y}=\frac{1+s^2}{-2s^3} \qquad \Rightarrow \qquad \frac{d\,\bar{y}}{ds}-\left(\frac{s}{2}-\frac{3}{2s}\right)\bar{y}=-\frac{1}{2s^3}-\frac{1}{2s} \qquad \ldots (3)$$

Thus (3) is a linear differential equation.

$$\text{I.F.} = e^{\frac{1}{2}\int\left(\frac{3}{s}-s\right)ds} = e^{\frac{1}{2}\left(3\log s - \frac{s^2}{2}\right)} = e^{-\frac{s^2}{4}} \cdot s^{\frac{3}{2}}$$

Solution of differential equation (3) is

$$\bar{y}\, e^{-\frac{s^2}{4}} \cdot s^{\frac{3}{2}} = -\frac{1}{2}\int\left(\frac{1}{s^3}+\frac{1}{s}\right) s^{\frac{3}{2}} \cdot e^{-\frac{s^2}{4}}\, ds = -\frac{1}{2}\int\left(\sqrt{s}+\frac{1}{s^{\frac{3}{2}}}\right) e^{-\frac{s^2}{4}}\, ds$$

Put $s^2 = 4z \quad \Rightarrow \quad s = 2\sqrt{z}$ so that $ds = \dfrac{dz}{\sqrt{z}}$

$$\bar{y}\, s^{\frac{3}{2}} \cdot e^{-\frac{s^2}{4}} = -\frac{1}{2}\int\left(\sqrt{2}\, z^{\frac{1}{4}} + \frac{1}{2\sqrt{2}}\, z^{-\frac{3}{4}}\right) e^{-z}\, \frac{dz}{\sqrt{z}}$$

$$= -\frac{1}{\sqrt{2}}\int\left(z^{-\frac{1}{4}} + \frac{1}{4} z^{-\frac{5}{4}}\right) e^{-z}\, dz = -\frac{1}{\sqrt{2}}\int z^{-\frac{1}{4}} e^{-z}\, dz - \frac{1}{4\sqrt{2}}\int z^{-\frac{5}{4}} e^{-z}\, dz$$

$$= -\frac{1}{\sqrt{2}}\left[z^{-\frac{1}{4}}\frac{e^{-z}}{-1} + \int\left(-\frac{1}{4}\right) z^{-\frac{5}{4}} e^{-z}\, dz\right] - \frac{1}{4\sqrt{2}}\int z^{-\frac{5}{4}} e^{-z}\, dz + C$$

$$= \frac{1}{\sqrt{2}} e^{-z} z^{-\frac{1}{4}} + C = \frac{1}{\sqrt{2}} e^{-\frac{s^2}{4}}\left(\frac{s^2}{4}\right)^{-\frac{1}{4}} + C = \frac{1}{\sqrt{s}} e^{-\frac{s^2}{4}} + C$$

(Particular case)

$$\Rightarrow \quad \bar{y} = \frac{1}{s^2} + C \qquad \Rightarrow \qquad y = L^{-1}\left(\frac{1}{s^2} + C\right) = t + C$$

$y = t$ (c must vanish if $\bar{y}$ is a transform since $\bar{y} \to 0$ as $s \to \infty$) **Ans.**

Example 63. *Solve* $y'' + t\, y' - y = 0$, *if* $y(0) = 0,\ y'(0) = 1$.

Solution. Taking the Laplace transform of both sides of the given equation, we get

$$L(y'') + L(t\, y') - L(y) = L(0)$$

$$\left[s^2\, \bar{y} - sy(0) - y'(0)\right] - \frac{d}{ds} L(y') - \bar{y} = 0 \qquad \dots (1)$$

Putting the values of $y(0)$ and $y'(0)$ in (1), we get

$$\Rightarrow \qquad \left[s^2\, \bar{y} - 0 - 1\right] - \frac{d}{ds}[s\bar{y} - y(0)] - \bar{y} = 0$$

$$(s^2\bar{y} - 1) - \left[\bar{y} + s\frac{d\bar{y}}{ds}\right] - \bar{y} = 0$$

$$s^2\bar{y} - 2\bar{y} - s\frac{d\bar{y}}{ds} - 1 = 0$$

$$-s\frac{d\bar{y}}{ds} + (s^2 - 2)\,\bar{y} - 1 = 0$$

$$\frac{d\bar{y}}{ds} + \left(\frac{2}{s} - s\right)\bar{y} = -\frac{1}{s}$$

which is a linear differential equation in $\bar{y}$.

I. F. $= e^{\int\left(\frac{2}{s}-s\right)ds} = e^{2\log s - \frac{s^2}{2}} = s^2 e^{-\frac{s^2}{2}}$

Solution of the equation is

$$\bar{y}\left(s^2 e^{-\frac{s^2}{2}}\right) = \int\left(-\frac{1}{s}\right)\left(s^2 e^{-\frac{s^2}{2}}\right) ds$$

$$= -\int s e^{-\frac{s^2}{2}} ds$$

$$= e^{-\frac{s^2}{2}} + C$$

$\Rightarrow$ $$\bar{y} = \frac{1}{s^2} + \frac{C}{s^2 e^{-\frac{s^2}{2}}}$$

$\Rightarrow$ $$\bar{y} = \frac{1}{s^2} + \frac{C}{s^2} e^{\frac{s^2}{2}}$$

Taking Inverse Laplace Transform, we get

$\Rightarrow$ $$y = L^{-1}\frac{1}{s^2} + C\, L^{-1}\left[\frac{1}{s^2} e^{\frac{s^2}{2}}\right]$$

$\Rightarrow$ $$y = t + C\, L^{-1}\left[\frac{1}{s^2}\left(1 + \frac{s^2}{2} + \frac{1}{2!}\frac{s^4}{4} + \ldots\right)\right]$$

$\Rightarrow$ $$y = t + C\, L^{-1}\left[\frac{1}{s^2} + \frac{1}{2} + \frac{s^2}{8} + \ldots\right]$$

$\Rightarrow$ $$y = t + Ct + C\, L^{-1}\left[\frac{1}{2} + \frac{s^2}{8} + \ldots\right]$$

$[L^{-1} s^n = 0, \text{ for } n = 0, 1, 2, \ldots]$

$\Rightarrow$ $y = t + C\,t + 0,$ $\Rightarrow$ $y = t + C\,t$ ... (2)

Differentiating (2) *w.r.t.* '*t*', we get

$$y' = 1 + C \qquad \ldots (3)$$

Putting $y'(0) = 1$ in (3), we get

$$1 = 1 + C \quad \Rightarrow \quad C = 0$$

Putting the value of $C = 0$ in (2), we get

$$y = t$$ **Ans.**

Example 64. *Solve:* $\frac{d^2y}{dt^2} - t\frac{dy}{dt} + y = 1$, *when* $y(0) = 1$, $y'(0) = 2$.

Solution. The given equation is

$$y'' - t\,y' + y = 1$$

Taking Laplace transform, we have

$$L\{y''\} - L\{ty'\} + L\{y\} = L\{1\}$$

$$\Rightarrow \quad [s^2\bar{y} - sy(0) - y'(0)] + \frac{d}{ds}[L\{y'\}] + \bar{y} = \frac{1}{s}$$

$$\Rightarrow \quad s^2\bar{y} - s - 2 + \frac{d}{ds}[s\bar{y} - y(0)] + \bar{y} = \frac{1}{s}$$

$$\Rightarrow \quad s^2\bar{y} + \frac{d}{ds}(s\bar{y} - 1) + \bar{y} = s + 2 + \frac{1}{s}$$

$$\Rightarrow \quad s^2\bar{y} + \bar{y} + s\frac{d\bar{y}}{ds} + \bar{y} = s + 2 + \frac{1}{s}$$

$$\Rightarrow \quad s\frac{d\bar{y}}{ds} + (s^2 + 2)\bar{y} = s + 2 + \frac{1}{s}$$

$$\Rightarrow \quad \frac{d\bar{y}}{ds} + \left(s + \frac{2}{s}\right)\bar{y} = 1 + \frac{2}{s} + \frac{1}{s^2} \quad \ldots (1)$$

Equation (1) is linear differential equation in $\bar{y}$.

Its I. F. $= e^{\int\left(s + \frac{2}{s}\right)ds} = e^{\frac{s^2}{2} + 2\log s} = e^{\frac{s^2}{2}} \cdot e^{2\log s} = s^2 e^{\frac{s^2}{2}}$

Therefore, solution of (1) is

$$s^2 e^{\frac{s^2}{2}} \cdot \bar{y} = \int\left(1 + \frac{2}{s} + \frac{1}{s^2}\right)s^2 e^{\frac{s^2}{2}}\, ds$$

$$= \int(s^2 + 2s + 1)\, e^{\frac{s^2}{2}}\, ds$$

$$= \int(s^2 + 1)\, e^{\frac{s^2}{2}}\, ds + 2\int s\, e^{\frac{s^2}{2}}\, ds$$

Putting $\frac{s^2}{2} = v$, so that $s\,ds = dv$, $ds = \frac{dv}{s} = \frac{dv}{\sqrt{2v}}$, we get

$$s^2 e^{\frac{s^2}{2}}\, \bar{y} = \int(2v + 1)\, e^v \cdot \frac{dv}{\sqrt{2v}} + 2\int e^v\, dv$$

$$= \int(\sqrt{2v})\, e^v\, dv + \int\frac{e^v}{\sqrt{2v}}\, dv + 2\int e^v\, dv$$

$$= (\sqrt{2v})\, e^v - \sqrt{2}\int\frac{1}{2}\frac{1}{\sqrt{v}}\, e^v\, dv + \int\frac{e^v}{\sqrt{2v}}\, dv + 2e^v + C_1$$

$$= \sqrt{2v}\, e^v - \int\frac{e^v}{\sqrt{2v}}\, dv + \int\frac{e^v}{\sqrt{2v}}\, dv + 2e^v + C_1$$

$$= (\sqrt{2v})\, e^v + 2e^v + C_1 = s\, e^{\frac{s^2}{2}} + 2\, e^{\frac{s^2}{2}} + C_1$$

$$= C_1 + (s + 2)\, e^{\frac{s^2}{2}}$$

$$\Rightarrow \qquad \bar{y} = \frac{C_1}{s^2} e^{-\frac{s^2}{2}} + \left(\frac{1}{s} + \frac{2}{s^2}\right) = \frac{C_1}{s^2}\left(1 - \frac{s^2}{2} + \frac{s^4}{2!4} - \ldots\right) + \frac{1}{s} + \frac{2}{s^2}$$

$$\Rightarrow \qquad \bar{y} = \frac{(2 + C_1)}{s^2} + \frac{1}{s} - \left(\frac{C_1}{2}\right) + \left(\frac{C_1}{8}\right) s^2 + \ldots$$

Taking Inverse Laplace Transform and noting that $\left[L^{-1}\frac{C_1}{2} = 0 \text{ and } L^{-1}\frac{C_1}{8}s^2 = 0\right]$, **we get**

$$y = (2 + C_1)\, t + 1 \qquad \ldots (2)$$

On differentiating (2), we get $y' = 2 + C_1$ **... (3)**

Now given that $y'(0) = 2$, **so putting** $t = 0$, $y' = 2$ **in (3), we get**

$$2 = 2 + C_1 \;\; i.e., \; C_1 = 0$$

Putting $C_1 = 0$ **in (2), we get**

$$y = 2t + 1$$

This is the required solution. **Ans.**

Example 65. *A particle moves in a line so that its displacement x from a fixed point O at any time t, is given by*

$$\frac{d^2x}{dt^2} + 4\frac{dx}{dt} + 5x = 80 \sin 5t$$

Using Laplace transform, find its displacement at any time t if initially particle is at rest at x = 0 *(U.P., II Semester 2009)*

Solution. Here, we have

$$\frac{d^2x}{dt^2} + 4\frac{dx}{dt} + 5x = 80 \sin 5t, \;\; x(0) = 0, \; x'(0) = 0 \qquad \ldots(1)$$

Taking Laplace transform of both sides of (1), we get

$$[s^2\bar{x} - sx(0) - x'(0)] + 4[s\bar{x} - x(0)] + 5\bar{x} = L[80 \sin 5t]$$

$$[s^2\bar{x} - 0 - 0] + 4s\bar{x} + 5\bar{x} = 80\left(\frac{5}{s^2 + 25}\right)$$

$$(s^2 + 4s + 5)\,\bar{x} = \frac{400}{s^2 + 25}$$

$$\bar{x} = \left(\frac{1}{s^2 + 4s + 5}\right)\left(\frac{400}{s^2 + 25}\right)$$

$$\bar{x} = \frac{2s + 18}{s^2 + 4s + 5} - \frac{2s + 10}{s^2 + 25} \qquad \text{[By Partial fraction]}$$

$$= \frac{2(s + 2) + 14}{(s + 2)^2 + 1} - \frac{2s}{s^2 + (5)^2} - \frac{10}{s^2 + (5)^2}$$

$$= \frac{2(s + 2)}{(s + 2)^2 + 1} + \frac{14}{(s + 2)^2 + 1} - \frac{2s}{s^2 + (5)^2} - \frac{10}{s^2 + (5)^2}$$

$$\Rightarrow x = 2\,L^{-1}\frac{(s+2)}{(s+2)^2+1} + 14\,L^{-1}\frac{1}{(s+2)^2+1}$$

$$-\,2L^{-1}\frac{s}{s^2+(5)^2} - 2L^{-1}\frac{5}{s^2+(5)^2}$$

$$= 2e^{-2t}\cos t + 14e^{-2t}\sin t - 2\cos 5t - 2\sin 5t$$

$$= 2e^{-2t}(\cos t + 7\sin t) - 2(\cos 5t + \sin 5t) \qquad \textbf{Ans.}$$

6.13 ELECTRIC CIRCUIT

Consider an electric circuit consisting of a resistance R, inductance L, a condensor of **capacity** C and electromotive power of voltage E in a series. A switch is also connected in the circuit. Here,

$$i = \frac{dq}{dt}$$

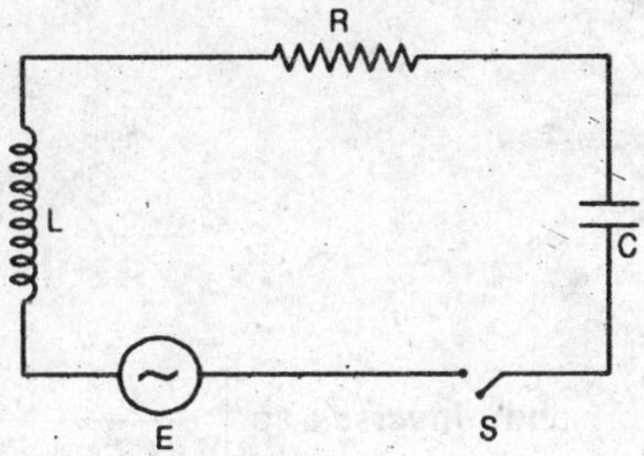

Voltage developed by Ri, $L\dfrac{di}{dt}$ and $\dfrac{q}{C}$

By Kirchhoff low $\boxed{L\dfrac{di}{dt} + Ri + \dfrac{q}{C} = E}$

Example 66. *A resistance R in series with inductance L is connected with e.m.f. E (t). The current i is given by*

$$L\frac{di}{dt} + Ri = E$$

If the switch is connected at t = 0 and disconnected at t = a, find the current i in terms of t.

(*U.P. II Semester, Summer* 2001)

Solution. Conditions under which current i flows are

$$E(t) = \begin{cases} E, & 0 < t < a \\ 0, & t > a \end{cases} \qquad [\,i = 0 \text{ at } t = 0]$$

Given equation is $L\dfrac{di}{dt} + Ri = E$ ... (1)

Taking Laplace transform of (1), we get

$$L[s\,\bar{i} - i(0)] + R\,\bar{i} = \int_0^\infty e^{-st}\,E\,dt$$

$$L\,s\,\bar{i} + R\,\bar{i} = \int_0^\infty e^{-st}\,E\,dt \qquad [\,i(0) = 0]$$

$$(Ls + R)\,\bar{i} = \int_0^\infty e^{-st}\,.\,E\,dt = \int_0^a e^{-st}E\,dt + \int_a^\infty e^{-st}\,E\,dt$$

$$= E\left[\frac{e^{-st}}{-s}\right]_0^a + 0 = \frac{E}{s}[1 - e^{-as}] = \frac{E}{s} - \frac{E}{s}e^{-as}$$

$$\Rightarrow \qquad \bar{i} = \frac{E}{s(Ls+R)} - \frac{Ee^{-as}}{s(Ls+R)}$$

Taking Inverse Laplace Transform, we obtain

$$i = \text{Inverse Lap.}\left[\frac{E}{s(Ls+R)}\right] - \text{Inverse Lap.}\left[\frac{E\,e^{-as}}{s(Ls+R)}\right] \quad \ldots (2)$$

Now we have to find the value of Inverse Lap. $\left[\dfrac{E}{s(Ls+R)}\right]$

$$\text{Inverse Lap.}\left[\frac{E}{s(Ls+R)}\right] = \frac{E}{L}\ \text{Inverse Lap.}\left[\frac{1}{s\left(s+\frac{R}{L}\right)}\right]$$

$$= \frac{E}{L}\frac{L}{R}\ \text{Inverse Lap.}\left[\frac{1}{s} - \frac{1}{s+\frac{R}{L}}\right] = \frac{E}{R}\left[1 - e^{-\frac{R}{L}t}\right]$$

(*Resolving into partial fractions*)

and $\quad \text{Inverse Lap.}\left[\dfrac{E\,e^{-as}}{s(Ls+R)}\right] = \dfrac{E}{R}\left[1 - e^{-\frac{R}{L}(t-a)}\right] u(t-a)$

(*By the Second Shifting Theorem*)

On substituting the values of the inverse transforms in (2), we get

$$i = \frac{E}{R}\left[1 - e^{-\frac{R}{L}t}\right] - \frac{E}{R}\left[1 - e^{-\frac{R}{L}(t-a)}\right] u(t-a)$$

Hence $\quad i = \dfrac{E}{R}\left[1 - e^{-\frac{R}{L}t}\right]$ for $0 < t < a$, $\quad [u(t-a) = 0]$

$$i = \frac{E}{R}\left[1 - e^{-\frac{R}{L}t}\right] - \frac{E}{R}\left[1 - e^{-\frac{R}{L}(t-a)}\right]$$

for $t > a$
$[u(t-a) = 1]$

$$= \frac{E}{R}\left[e^{-\frac{R}{L}(t-a)} - e^{-\frac{R}{L}t}\right] = \frac{E}{R}e^{-\frac{R}{L}t}\left[e^{\frac{Ra}{L}} - 1\right] \qquad \textbf{Ans.}$$

Example 67. *Voltage Ee^{-at} is applied at $t = 0$ to a circuit of inductance L and resistance R. Show that the current at time t is* $\dfrac{E}{R-aL}(e^{-at} - e^{-Rt/L})$. [*U.P., II Semester, (SUM) 2007*]

Solution. We know that

$$L\frac{dI}{dt} + RI = Ee^{-at}, \quad \ldots (1)$$

where $\quad I(0) = 0$

Taking Laplace transform of both sides, of (1) we get

$$L[s\bar{I} - I(0)] + R\bar{I} = \frac{E}{s+a} \qquad ...(2)$$

Putting I (0) = 0 in (2), we get

$$(Ls + R)\bar{I} = \frac{E}{s+a}$$

$$\Rightarrow \qquad \bar{I} = \frac{E}{(s+a)(Ls+R)}$$

$$= \frac{E}{R-aL}\left(\frac{1}{s+a} - \frac{1}{Ls+R}\right)$$

$$= \frac{E}{R-aL}\left(\frac{1}{s+a} - \frac{1}{s+R/L}\right) \qquad ...(3)$$

Taking the Inverse Laplace transform of both sides of (3), we get

$$I = \frac{E}{R-aL} L^{-1}\left\{\frac{1}{s+a} - \frac{1}{s+R/L}\right\}$$

$$= \frac{E}{R-aL}[e^{-at} - e^{-Rt/L}] \qquad \textbf{Ans.}$$

Example 68. *Using the Laplace transform, find the current i (t) in the LC - circuit. Assuming L = 1 henry, C = 1 farad, zero initial current and charge on the capacitor, and*

v (t) = t when 0 < t < 1

= 0 otherwise.

i

C

L

V(t)

Solution. The differential equation for L and C circuit is given by

$$L\frac{d^2q}{dt^2} + \frac{q}{C} = E \qquad ... (1)$$

Putting $L = 1$, $C = 1$, $E = v(t)$ in (1), we get

$$\frac{d^2q}{dt^2} + q = v(t) \qquad ... (2)$$

Taking Laplace Transform of (2), we have

$$s^2\bar{q} - sq(0) - q'(0) + \bar{q} = \int_0^\infty v(t)e^{-st}\,dt$$

Substituting q (0) = 0, and q' (0) = 0, we get

$$s^2\bar{q} + \bar{q} = \int_0^1 te^{-st}\,dt + \int_1^\infty 0\,e^{-st}\,dt$$

$$(s^2+1)\bar{q} = \left[t\frac{e^{-st}}{-s}\right]_0^1 - \int_0^1 \frac{e^{-st}}{-s}\,dt = \frac{e^{-s}}{-s} - \left[\frac{e^{-st}}{s^2}\right]_0^1 = -\frac{e^{-s}}{s} - \frac{e^{-s}}{s^2} + \frac{1}{s^2}$$

$$\bar{q} = \frac{1}{s^2+1}\left[-\frac{e^{-s}}{s} - \frac{e^{-s}}{s^2} + \frac{1}{s^2}\right]$$

$$\bar{q} = \frac{-e^{-s}}{s\left(s^2+1\right)} - \frac{e^{-s}}{s^2\left(s^2+1\right)} + \frac{1}{s^2\left(s^2+1\right)}$$

Taking Inverse Laplace Transform, we get

$$q = \text{Inverse Lap.} \frac{-e^{-s}}{s\,(s^2+1)} - \text{Inverse Lap.} \frac{e^{-s}}{s^2\,(s^2+1)} + \text{Inverse Lap.} \frac{1}{s^2\,(s^2+1)} \qquad \dots (3)$$

We know that

$$\text{Inverse Lap.}\ [\,e^{-as}\ F(s)\,] = f(t-a)\,u\,(t-a)$$

$$\text{Inverse Lap.}\ \frac{1}{s(s^2+1)} = \int_0^t \sin t\ dt = [\,-\cos t\,]_0^t = 1 - \cos t \qquad \dots (4)$$

$$\text{Inverse Lap.}\ \frac{1}{s^2\,(s^2+1)} = \int_0^t (1-\cos t)\ dt = t - \sin t \qquad \dots (5)$$

In view of this, we have

$$\text{Inverse Lap.} \left[\frac{-e^{-s}}{s\,(s^2+1)}\right] = -[\,1-\cos\,(t-1)\,]\,u\,(t-1) \qquad \text{[From (4)]}$$

$$\text{Inverse Lap.}\ \frac{e^{-s}}{s^2\,(s^2+1)} = [\,(t-1) - \sin\,(t-1)\,]\,u\,(t-1) \qquad \text{[From (5)]}$$

Putting the above values in (3), we get

$$q = -[1-\cos\,(t-1)]\ u(t-1) - [(t-1) - \sin(t-1)]\ u(t-1) + t - \sin t \qquad \textbf{Ans.}$$

EXERCISE 6.10

Solve the following differential equations:

1. $\frac{d^2y}{dx^2} + y = 0$ where $y = 1$ and $\frac{dy}{dx} = -1$ at $x = 0$. **Ans.** $y = \cos x - \sin x$

2. $\frac{d^2y}{dx^2} - 4y = 0$, where $y = 0$ and $\frac{dy}{dx} = -6$ at $x = 0$ **Ans.** $y = -\frac{3}{2}e^{2x} + \frac{3}{2}e^{-2x}$

3. $\frac{d^2y}{dx^2} + y = 0$, where $y = 1$, $\frac{dy}{dx} = 1$ at $x = 0$. **Ans.** $y = \sin x + \cos x$

4. $\frac{d^2y}{dx^2} + 2\frac{dy}{dx} + 5y = 0$, where $y = 2$, $\frac{dy}{dx} = -4$ at $x = 0$. **Ans.** $y = e^{-x}\,(2\cos 2x - \sin 2x)$

5. $\frac{d^3y}{dx^3} + 2\frac{d^2y}{dx^2} - \frac{dy}{dx} - 2y = 0$, given $y = \frac{dy}{dx} = 0$, $\frac{d^2y}{dx^2} = 6$ at $x = 0$.
Ans. $y = e^x - 3e^{-x} + 2e^{-2x}$

6. $\frac{d^2y}{dx^2} + y = 3\cos 2x$, where $y = \frac{dy}{dx} = 0$ at $x = 0$. **Ans.** $y = \cos x - \cos 2x$.

7. $\frac{d^2y}{dx^2} + \frac{dy}{dx} - 2y = 1 - 2x$, given $y = 0$, $\frac{dy}{dx} = 4$ at $x = 0$. **Ans.** $y = e^x - e^{-2x} + x$

8. $\dfrac{d^2y}{dx^2} - 3\dfrac{dy}{dx} + 2y = 4e^{2x}$, given $y = -3$, and $\dfrac{dy}{dx} = 5$ at $x = 0$

Ans. $y = -7e^x + 4e^{2x} + 4x\,e^{2x}$

9. $\dfrac{d^2y}{dx^2} - 3\dfrac{dy}{dx} + 2y = 4x + e^{2x}$, where $y = 1$, $\dfrac{dy}{dx} = -1$ at $x = 0$.

Ans. $y = 3 + 2x + \dfrac{1}{2}e^{3x} - 2e^{2x} - \dfrac{1}{2}e^x$.

10. $(D^2 - D - 2)x = 20\sin 2t, x_0 = -1, x_1 = 2$ **Ans.** $x = 2e^{2t} - 4e^{-t} + \cos 2t - 3\sin 2t$

11. $(D^3 + D^2)\,x = 6t^2 + 4$, $x(0) = 0$, $x'(0) = 2$, $x''(0) = 0$

Ans. $x = \dfrac{1}{2}t^4 - 2t^3 + 8t^2 - 16t + 16 - 16\,e^{-t}$

12. $\dfrac{d^2x}{dt^2} - 2\dfrac{dx}{dt} + x = e^t$, where $x(0) = 2$, $\dfrac{dx}{dt} = -1$ at $t = 0$

Ans. $x = 2e^t - 3\,t\,e^t + \dfrac{1}{2}t^2e^t$

13. $y'' + 2y' + y = t\,e^{-t}$ if $y(0) = 1$, $y'(0) = -2$. **Ans.** $y = \left(1 - t + \dfrac{t^3}{6}\right)e^{-t}$

14. $\dfrac{d^2y}{dx^2} + y = x\cos 2x$, where $y = \dfrac{dy}{dx} = 0$ at $x = 0$. **Ans.** $y = \dfrac{4}{9}\sin 2x - \dfrac{5}{9}\sin x - \dfrac{x}{3}\cos 2x$

15. $\dfrac{d^3y}{dx^3} - 3\dfrac{d^2y}{dx^2} + 3\dfrac{dy}{dx} - y = x^2e^{2x}$, where $y = 1$, $\dfrac{dy}{dx} = 0$, $\dfrac{d^2y}{dx^2} = -2$ at $x = 0$.

Ans. $y = e^{2x}(x^2 - 6x + 12) - e^x(15x^2 + 7x + 11)$.

16. $y'' + 4y' + 3y = t$, $t > 0$; given that $y(0) = 0$ and $y'(0) = 1$.

Ans. $y = -\dfrac{4}{9} + \dfrac{t}{6} + e^{-t} - \dfrac{5}{9}e^{-3t}$

17. $[\,t D^2 + (1 - 2t)D - 2]\,y = 0$, if $y(0) = 1$, $y'(0) = 2$. **Ans.** $y = e^{2t}$

18. $t\,y'' + at\,y' - 2ay = 1$; $y(0) = y'(0) = 0$, $a > 0$. **Ans.** $y = \dfrac{1}{2}t^2$

19. $y'' + 2y = r(t), y(0) = 0, y'(0) = 0$, where $r(t) = \begin{cases} 0, & t \geq 1 \\ 1, & 0 \leq t < t \end{cases}$ **Ans.** $y = \dfrac{1}{2} - \dfrac{1}{2}\cos\sqrt{2}\,t$.

20. $\dfrac{d^2y}{dt^2} + 4y = u(t - 2)$, where u is unit step function.

$y(0) = 0$ and $y'(0) = 1$. **Ans.** $y = \dfrac{1}{2}\sin 2t$ *for* $t < 2$

21. $\dfrac{d^2y}{dx^2} + y\,u(t - \pi) - u(t - 2\pi)$, $y(0) = y'(0) = 0$

Ans. $y = (1 + \cos t)\,u(t - \pi) - (1 - \cos t)\,u(t - 2\pi)$

22. A condenser of capacity C is charged to potential E and discharged at $t = 0$ through an inductance L and resistance R. The charge q at time t is governed by the differential equation

$$L\frac{d^2q}{dt^2} + R\frac{dq}{dt} + \frac{q}{C} = E$$

Using Laplace transforms, show that the charge q is given by

$$q = \frac{CE}{n}e^{-\mu t}[\mu \sin nt + n\cos nt], \text{ where } \mu = \frac{R}{2L} \text{ and } n^2 = \frac{1}{LC} - \frac{R^2}{4L^2}$$

6.14 SOLUTION OF SIMULTANEOUS DIFFERENTIAL EQUATIONS BY LAPLACE TRANSFORMS

Simultaneous differential equations can also be solved by Laplace Transform method.

Example 69. *Solve* $\frac{dx}{dt} + y = 0$ *and* $\frac{dy}{dt} - x = 0$ *under the condition* $x(0) = 1,\ y(0) = 0.$

Solution. We have, $x' + y = 0$ ... (1)

and $y' - x = 0$ ... (2)

Taking the Laplace transform of (1) and (2), we get

$$[s\bar{x} - x(0)] + \bar{y} = 0 \quad \text{... (3)}$$

$$[s\bar{y} - y(0)] - \bar{x} = 0 \quad \text{... (4)}$$

On substituting the values of $x(0)$ and $y(0)$ in (3) and (4), we get

$$s\bar{x} - 1 + \bar{y} = 0 \quad \text{... (5)}$$

$$s\bar{y} - \bar{x} = 0 \quad \text{... (6)}$$

Solving (5) and (6) for $\bar{x}$ and $\bar{y}$, we get

$$\bar{x} = \frac{s}{s^2+1}, \quad \bar{y} = \frac{1}{s^2+1},$$

On inversion, we obtain $x = L^{-1}\left(\frac{s}{s^2+1}\right)$, $y = L^{-1}\left(\frac{1}{s^2+1}\right)$,

$x = \cos t,$ $y = \sin t$ **Ans.**

Example 70. *Solve the following simultaneous differential equations by Laplace transform*

$$3\frac{dx}{dt} - y = 2t, \quad \frac{dx}{dt} + \frac{dy}{dt} - y = 0$$

with the conditions $x(0) = y(0) = 0.$ *[U.P., II Semester, (SUM) 2008]*

Solution. Here, we have

$$3\frac{dx}{dt} - y = 2t, \quad \text{...(1)}$$

and $$\frac{dx}{dt} + \frac{dy}{dt} - y = 0 \quad \text{...(2)}$$

Taking Laplace transform on both sides of equation (1), we get

$$3L(x') - L(y) = L\,(2t)$$

$$3[s\bar{x} - x(0)] - \bar{y} = \frac{2}{s^2} \qquad \text{[where } L(x) = \bar{x} \text{ and } L\,(y) = \bar{y}]$$

$$3s\bar{x} - \bar{y} = \frac{2}{s^2} \qquad ...(3)$$

Again taking Laplace transform on both sides of equation (2), we get

$$L(x') + L(y') - L(y) = L\,(0)$$

$$\Rightarrow \quad [s\bar{x} - x(0)] + [s\bar{y} - y\,(0)] - \bar{y} = 0$$

$$\Rightarrow \quad s\bar{x} + (s-1)\,\bar{y} = 0 \qquad ...(4)$$

Multiplying equation (4) by 3, we get

$$3s\bar{x} + 3(s-1)\,\bar{y} = 0 \qquad ...(5)$$

Substracting equation (3) from (5), we get

$$(3s-2)\,\bar{y} = -\frac{2}{s^2}$$

$$\Rightarrow \quad \bar{y} = -\frac{2}{s^2\,(3s-2)} = \frac{1}{s^2} + \frac{3}{2s} - \frac{3}{2\left(s - \frac{2}{3}\right)}$$

Taking inverse Laplace transform on both sides, we get

$$y = t + \frac{3}{2} - \frac{3}{2}\,e^{\frac{2t}{3}} \qquad ...(6)$$

Substituting $\bar{y}$ in (3), we get

$$3s\bar{x} - \frac{1}{s^2} - \frac{3}{2s} + \frac{3}{2\left(s - \frac{2}{3}\right)} = \frac{2}{s^2}$$

$$\Rightarrow \quad 3s\bar{x} = \frac{3}{s^2} + \frac{3}{2s} - \frac{3}{2\left(s - \frac{2}{3}\right)}$$

$$\Rightarrow \quad \bar{x} = \frac{1}{s^3} + \frac{1}{2s^2} - \frac{1}{2s\left(s - \frac{2}{3}\right)}$$

$$\Rightarrow \quad \bar{x} = \frac{1}{s^3} + \frac{1}{2s^2} - \frac{3}{4}\left(\frac{1}{s - \frac{2}{3}} - \frac{1}{s}\right)$$

Taking inverse Laplace transform on both sides, we get

$$x = \frac{t^2}{2} + \frac{t}{2} - \frac{3}{4}\,e^{\frac{2t}{3}} + \frac{3}{4} \qquad ...(7)$$

Equation (6) and (7) when taken together, give the complete solution. **Ans.**

Example 71. *Solve the simultaneous equations:*

$$\frac{dx}{dt} - y = e^t$$

$$\frac{dy}{dt} + x = \sin t, \text{ given } x(0) = 1,\ y(0) = 0,$$ (*U.P.,II Semester, Summer* 2006)

Solution. $$\frac{dx}{dt} - y = e^t \qquad \dots (1)$$

$$\frac{dy}{dt} + x = \sin t \qquad \dots (2)$$

Taking Laplace transform of (1), we get

$$[s\bar{x} - x(0)] - \bar{y} = \frac{1}{s-1}$$

i.e., $$s\bar{x} - 1 - \bar{y} = \frac{1}{s-1} \qquad [\because x(0) = 1]$$

$$s\bar{x} - \bar{y} = 1 + \frac{1}{s-1}$$

$$s\bar{x} - \bar{y} = \frac{s}{s-1} \qquad \dots (3)$$

Taking Laplace Transform of (2), we get

$$[s\bar{y} - y(0)] + \bar{x} = \frac{1}{s^2+1} \qquad [y(0) = 0)$$

$$\bar{x} + s\bar{y} = \frac{1}{s^2+1} \qquad \dots (4)$$

Solving (3) and (4) for $\bar{x}$ and $\bar{y}$, we have

$$\bar{x} = \frac{s^2}{(s-1)(s^2+1)} + \frac{1}{(s^2+1)^2} = \frac{1}{2}\left[\frac{1}{s-1} + \frac{s}{s^2+1} + \frac{1}{s^2+1}\right] + \frac{1}{(s^2+1)^2} \qquad \dots (5)$$

and $$\bar{y} = \frac{s}{(s^2+1)^2} - \frac{s}{(s-1)(s^2+1)} = \frac{s}{(s^2+1)^2} - \frac{1}{2}\left[\frac{1}{s-1} - \frac{s}{s^2+1} + \frac{1}{s^2+1}\right] \qquad \dots (6)$$

Taking Inverse Laplace Transform of (5), we get

$$x = \frac{1}{2}L^{-1}\left[\frac{1}{s-1} + \frac{s}{s^2+1} + \frac{1}{s^2+1}\right] + L^{-1}\left[\frac{1}{(s^2+1)^2}\right]$$

$$= \frac{1}{2}[e^t + \cos t + \sin t] + \frac{1}{2}(\sin t - t\cos t) \qquad \left[\because L^{-1}\left[\frac{1}{(s^2+a^2)^2}\right] = \frac{1}{2a^2}(\sin at - at\cos at)\right]$$

$$= \frac{1}{2}[e^t + \cos t + 2\sin t - t\cos t]$$

Taking Inverse Laplace Transform of (6), we get

$$y = L^{-1}\left[\frac{s}{(s^2+1)^2}\right] - \frac{1}{2}L^{-1}\left[\frac{1}{s-1} - \frac{s}{s^2+1} + \frac{1}{s^2+1}\right]$$

$$= \frac{1}{2} t \sin t - \frac{1}{2}[e^t - \cos t + \sin t] \qquad \left[\because L^{-1}\left[\frac{s}{(s^2+a^2)^2}\right] = \frac{1}{2a^2} t \sin at\right]$$

$$= \frac{1}{2}[t \sin t - e^t + \cos t - \sin t]$$

Hence, $x = \frac{1}{2}(e^t + \cos t + 2\sin t - t\cos t)$

$$y = \frac{1}{2}(t \sin t - e^t + \cos t - \sin t)$$ **Ans.**

Example 72. *Use Laplace transform to solve:*

$$\frac{dx}{dt} + y = \sin t, \quad \frac{dy}{dt} + x = \cos t$$

given that $x = 2$, $y = 0$ at $t = 0$. [*U.P., II Semester, 2004*]

Solution. Here, we have

$$\frac{dx}{dt} + y = \sin t \qquad ...(1)$$

$$\frac{dy}{dt} + x = \cos t \qquad ...(2)$$

Taking Laplace transform of (1) and (2), we get

$$s\bar{x} - x(0) + \bar{y} = \frac{1}{s^2+1} \Rightarrow s\bar{x} + \bar{y} = \frac{1}{s^2+1} + 2 \qquad ...(3)$$

and $$s\bar{y} - y(0) + \bar{x} = \frac{s}{s^2+1} \Rightarrow \bar{x} + s\bar{y} = \frac{s}{s^2+1} \qquad ...(4)$$

Solving (3) and (4) for $\bar{x}$ and $\bar{y}$, we get

$$\bar{x} = \frac{2s}{s^2-1} \text{ and } \bar{y} = \frac{1}{1+s^2} + \frac{2}{1-s^2}$$

$$\bar{x} = \frac{1}{s+1} + \frac{1}{s-1} \qquad ...(5)$$

and $$\bar{y} = \frac{1}{1+s^2} + \frac{1}{s+1} - \frac{1}{s-1} \qquad ...(6) \text{ (By partial fractions)}$$

Taking Inverse Laplace transform on both sides of (5) and (6), we get

$$x = e^{-t} + e^t$$

and $$y = \sin t + e^{-t} - e^t$$ **Ans.**

Example 73. *The co-ordinates (x, y) of a particle moving along a plane curve at any time t are given by*

$$\frac{dy}{dt} + 2x = \sin 2t, \quad \frac{dx}{dt} - 2y = \cos 2t; \ (t > 0)$$

It is given that at $t = 0$, $x = 1$ and $y = 0$. Show using transforms that the particle moves along the curve $4x^2 + 4xy + 5y^2 = 4$. [*U.P. II Semester 2003*]

Solution. Here, we have

$$\begin{bmatrix} \dfrac{dy}{dt} + 2x = \sin 2t \\ \dfrac{dx}{dt} - 2y = \cos 2t \end{bmatrix}$$

$$\Rightarrow \quad \begin{bmatrix} 2x + Dy = \sin 2t & \quad ...(1) \\ Dx - 2y = \cos 2t, & \quad ...(2) \end{bmatrix}$$

Taking Laplace transform of (1) on both sides, we get

$$2\bar{x} + s\bar{y} - y(0) = \frac{2}{s^2+4} \qquad \text{where } \bar{x} = L(x) \text{ and } \bar{y} = L(y)$$

$$2\bar{x} + s\bar{y} = \frac{2}{s^2+4} \qquad ...(3)\ [\therefore y(0) = 0]$$

Again, taking Laplace transform of equation, (2) on both sides, we get

$$s\bar{x} - x(0) - 2\bar{y} = \frac{s}{s^2+4}, \text{ where } \bar{x} = L(x) \text{ and } \bar{y} = L(y)$$

$$\Rightarrow \quad s\bar{x} - 2\bar{y} = \frac{s}{s^2+4} + 1 \qquad ...(4)\ [\because x(0) = 0]$$

Multiplying equation (3) bv 2 and equation (4) by s and then adding, we get

$$s\bar{x} + s^2\bar{x} = \frac{4}{s^2+4} + \frac{s^2}{s^2+4} + s$$

$$(4+s^2)\,\bar{x} = 1 + s$$

$$\bar{x} = \frac{1+s}{4+s^2} = \frac{1}{s^2+4} + \frac{s}{s^2+4} \qquad ...(5)$$

Taking Inverse Laplace transform of (5), we get

$$x = \frac{1}{2}\sin 2t + \cos 2t \qquad ...(6)$$

Again, multiplying (3) by s and equation (4) by 2 then subtracting equation (6) from (3), we get

$$s^2\bar{y} + 4\bar{y} = \frac{2s}{s^2+4} - \frac{2s}{s^2+4} - 2$$

$$\Rightarrow \quad \bar{y} = \frac{-2}{s^2+4} \qquad ...(7)$$

Taking Inverse Laplace transform of (7), we get $y = -\sin 2\text{t}$

Now,
$$4x^2 = 4\left[\frac{1}{4}\sin^2 2t + \cos^2 2t + \sin 2t \cos 2t\right]$$

$$5y^2 = 5\sin^2 2\text{t}$$

$$4xy = 4\left[\left(\frac{1}{2}\sin 2t + \cos 2t\right).(-\sin 2t)\right]$$

$$= -(2\sin^2 2t + 4\sin 2t \cos 2t)$$

$$\therefore \quad 4x^2 + 5y^2 + 4xy = 4\sin^2 2t + 4\cos^2 2t = 4$$ **Ans.**

Example 74. *Solve the following simultaneous differential equations by Laplace transform*

$$\frac{dx}{dt}+4\frac{dy}{dt}-y=0;\quad \frac{dx}{dt}+2y=e^{-t}$$

with conditions $x(0)=y(0)=0.$ [*U.P., II Semester, 2008*]

Solution. Here, we have

$$\frac{dx}{dt}+4\frac{dy}{dt}-y=0 \qquad ...(1)$$

and

$$\frac{dx}{dt}+2y=e^{-t} \qquad ...(2)$$

Taking Laplace transform on both sides of equation (1), we get

$$L(x')+4L(y')-L(y)=L(0)$$

$$\Rightarrow \quad s\bar{x}-x(0)+4[s\bar{y}-y(0)]-\bar{y}=0$$

$$\Rightarrow \quad s\bar{x}+(4s-1)\,\bar{y}=0 \qquad ...(3)$$

Again, taking Laplace transform on both sides of equation (2), we get

$$L(x')+2L(y)=L(e^{-t})$$

$$\Rightarrow \quad s\bar{x}-x(0)+2\bar{y}=\frac{1}{s+1}$$

$$\Rightarrow \quad s\bar{x}+2\bar{y}=\frac{1}{s+1} \qquad ...(4)$$

Subtracting (4) from (3), we get

$$(4s-3)\bar{y}=-\frac{1}{s+1}$$

$$\bar{y}=-\frac{1}{(s+1)(4s-3)}$$

$$=-\frac{1}{7}\left(\frac{-1}{s+1}+\frac{1}{s-3/4}\right)$$

$$=\frac{1}{7}\left(\frac{1}{s+1}-\frac{1}{s-3/4}\right) \qquad ...(5)$$

Taking inverse Laplace transform on both sides of (5), we get

$$y=\frac{1}{7}\left(e^{-t}-e^{\frac{3t}{4}}\right) \qquad ...(6)$$

Substituting $\bar{y}$ in (4), we get

$$s\bar{x}+\frac{2}{7}\left(\frac{1}{s+1}-\frac{1}{s-3/4}\right)=\frac{1}{s+1}$$

$$\Rightarrow \quad s\bar{x}=\frac{5}{7(s+1)}+\frac{2}{7(s-3/4)}$$

$$\Rightarrow \quad \bar{x}=\frac{5}{7s(s+1)}+\frac{2}{7s(s-3/4)}$$

$$= \frac{5}{7}\left(\frac{1}{s}-\frac{1}{s+1}\right)+\frac{8}{21}\left(\frac{1}{s-3/4}-\frac{1}{s}\right)$$

$$= \frac{1}{3s}-\frac{5}{7(s+1)}+\frac{8}{21(s-3/4)} \qquad \ldots(7)$$

Taking Inverse Laplace transform on both sides of (7), we get

$$x = \frac{1}{3}-\frac{5}{7}e^{-t}+\frac{8}{21}e^{\frac{3t}{4}} \qquad \textbf{Ans.}$$

Example 75. *Using Laplace transforms, solve the differential equations*

$$(D+1)\,y_1+(D-1)\,y_2=e^{-t}$$

$$(D+2)\,y_1+(D+1)\,y_2=e^{t}, \qquad \text{where } D=\frac{d}{dt} \text{ and } y_1(0)=1, y_2(0)=0.$$

Solution. $(D+1)\,y_1+(D-1)\,y_2=e^{-t}$ ...(1)

$(D+2)\,y_1+(D+1)\,y_2=e^{t}$ ... (2)

Multiplying (1) by $(D+1)$ and (2) by $(D-1)$, we get

$$(D+1)^2\,y_1+(D^2-1)\,y_2=(D+1)\,e^{-t} \qquad \ldots(3)$$

$$(D-1)(D+2)y_1+(D^2-1)\,y_2=(D-1)\,e^{t} \qquad \ldots(4)$$

Subtracting (4) from (3), we get

$$(D^2+2D+1-D^2-D+2)\,y_1=(-e^{-t}+e^{-t})-(e^{t}-e^{t})$$

$$\Rightarrow \qquad (D+3)y_1=0 \qquad \Rightarrow \qquad Dy_1+3y_1=0 \qquad \ldots(5)$$

Taking Laplace transform of (5), we have

$$s\,\bar{y}_1-y_1(0)+3\,\bar{y}_1=0$$

$$s\,\bar{y}_1-1+3\,\bar{y}_1=0$$

$$(s+3)\,\bar{y}_1=1 \quad \Rightarrow \quad \bar{y}_1=\frac{1}{s+3} \quad \Rightarrow \quad y_1=e^{-3t}$$

Putting the value of y_1 in (1), we get

$$(D+1)\,e^{-3t}+(D-1)\,y_2=e^{-t}$$

$$-3\,e^{-3t}+e^{-3t}+(D-1)\,y_2=e^{-t}$$

$$(D-1)\,y_2=e^{-t}+2e^{-3t} \quad \Rightarrow \quad Dy_2-y_2=e^{-t}+2e^{-3t}$$

Taking Laplace transform, we get

$$s\,\bar{y}_2-y_2(0)-\bar{y}_2=\frac{1}{s+1}+\frac{2}{s+3}$$

$$(s-1)\,\bar{y}_2=\frac{1}{s+1}+\frac{2}{s+3} \qquad \text{or} \qquad \bar{y}_2=\frac{1}{s^2-1}+\frac{2}{s^2+2s-3}$$

$$y_2=L^{-1}\left[\frac{1}{s^2-1}+\frac{2}{(s+1)^2-(2)^2}\right] \qquad \text{or} \qquad y_2=\sinh t+e^{-t}\sinh 2t$$

$$y_1=e^{-3t} \text{ and } y_2=\sinh t+e^{-t}\sinh 2t \qquad \textbf{Ans.}$$

Example 76. *Using Laplace Transformation, solve*

$$(D-2)\,x-(D+1)\,y=6\,e^{3t} \qquad \text{... (1)}$$
$$(2D-3)\,x+(D-3)\,y=6e^{3t}$$

Given $x=3,\ y=0$ *when* $t=0$. (*U.P., II Semester Summer* 2001)

Solution. Taking Laplace transformation of the given equations, we get

$$\begin{bmatrix} L\,Dx-2\,Lx-LDy-Ly=6\,Le^{3t} \\ 2L\,Dx-3Lx+LDy-3Ly=6\,Le^{3t} \end{bmatrix}$$

$$\Rightarrow \begin{bmatrix} s\,\bar{x}-x(0)-2\,\bar{x}-s\bar{y}+y(0)-\bar{y}=6\dfrac{1}{s-3} \\ 2s\,\bar{x}-2x(0)-3\,\bar{x}+s\,\bar{y}-y(0)-3\,\bar{y}=\dfrac{6}{s-3} \end{bmatrix}$$

$$\Rightarrow \begin{bmatrix} (s-2)\,\bar{x}-(s+1)\,\bar{y}-3=\dfrac{6}{s-3} \\ (2s-3)\,\bar{x}+(s-3)\,\bar{y}-6=\dfrac{6}{s-3} \end{bmatrix} \Rightarrow \begin{bmatrix} (s-2)\,\bar{x}-(s+1)\,\bar{y}=\dfrac{3s-3}{s-3} \\ (2s-3)\,x+(s-3)\,\bar{y}=\dfrac{6s-12}{s-3} \end{bmatrix}$$

$$\Rightarrow \begin{bmatrix} (s-3)\,(s-2)\,\bar{x}-(s-3)\,(s+1)\,\bar{y}=3s-3 \\ (s+1)\,(2s-3)\,\bar{x}+(s+1)\,(s-3)\,\bar{y}=\dfrac{(s+1)\,(6s-12)}{s-3} \end{bmatrix} \text{ on adding, we get:}$$

$$(3s^2-6s+3)\,x=3\,(s-1)+\frac{6\,(s^2-s-2)}{s-3} \Rightarrow \bar{x}=\frac{3\,(s-1)}{3\,(s-1)^2}+\frac{6\,(s^2-s-2)}{3\,(s-1)^2\,(s-3)}$$

$$x=L^{-1}\left[\frac{1}{s-1}+\frac{2}{(s-1)^2}+\frac{2}{s-3}\right]=e^t+2t\,e^t+2e^{3t}$$

Putting the value of x in (1), we get

$$(D-2)\,(e^t+2te^t+2e^{3t})-(D+1)\,y=6e^{3t}$$

$$\Rightarrow \quad e^t+2te^t+2e^t+6e^{3t}-2e^t-4te^t-4e^{3t}-(D+1)\,y=6e^{3t}$$

$$\Rightarrow \quad (D+1)\,y=e^t-2te^t-4e^{3t} \qquad \text{... (2)}$$

Taking Laplace transform of (2), we get

$$s\,\bar{y}-y(0)+\bar{y}=\frac{1}{s-1}-\frac{2}{(s-1)^2}-\frac{4}{s-3}$$

$$\Rightarrow \quad (s+1)\,\bar{y}=\frac{1}{s-1}-\frac{2}{(s-1)^2}-\frac{4}{s-3} \qquad [\because y(0)=0]$$

$$\Rightarrow \quad \bar{y}=\frac{1}{s^2-1}-\frac{2}{(s+1)\,(s-1)^2}-\frac{4}{(s+1)\,(s-3)}$$

$$=\frac{1}{s^2-1}-\frac{\frac{1}{2}}{s+1}+\frac{\frac{1}{2}}{s-1}-\frac{1}{(s-1)^2}+\frac{1}{s+1}-\frac{1}{s-3}$$

$$\Rightarrow \quad \bar{y} = \frac{-1}{s^2-1} + \frac{\frac{1}{2}}{s+1} + \frac{\frac{1}{2}}{s-1} - \frac{1}{(s-1)^2} - \frac{1}{s-3}$$

$$\Rightarrow \quad y = L^{-1}\left[\frac{1}{s^2-1} + \frac{1}{2}\frac{1}{s+1} + \frac{1}{2}\frac{1}{s-1} - \frac{1}{s-3} - \frac{1}{(s-1)^2}\right]$$

$$\Rightarrow \quad y = \sinh t + \frac{1}{2}e^{-t} + \frac{1}{2}e^{t} - e^{3t} - te^{t}$$

$$y = \sinh t + \cosh t - e^{3t} - te^{t}$$ **Ans.**

Example 77. *Solve the simultaneous equations*

$$(D^2 - 3)x - 4y = 0$$
$$x + (D^2 + 1)y = 0$$

for $t > 0$, *given that* $x = y = \frac{dy}{dt} = 0$ *and* $\frac{dx}{dt} = 2$ *at* $t = 0$. *[U.P., II Semester, 2004]*

Solution. Here, we have

$$(D^2 - 3)x - 4y = 0 \qquad \text{...(1)}$$
$$x + (D^2 + 1)y = 0 \qquad \text{...(2)}$$

Taking Laplace transform of (1) and (2), we get

$$s^2\bar{x} - sx(0) - x'(0) - 3\bar{x} - 4\bar{y} = 0$$

i.e., $$(s^2 - 3)\bar{x} - 4\bar{y} = 2 \qquad \text{...(3)}[\because x(0) = 0, x'(0) = 2]$$

and $$\bar{x} + s^2\bar{y} - sy(0) - y'(0) + \bar{y} = 0$$

i.e., $$\bar{x} + (s^2 + 1)\bar{y} = 0 \qquad \text{...(4)}[\because y(0) = 0, y'(0) = 0]$$

Solving (3) and (4) for $\bar{x}$ and $\bar{y}$, we get

$$\bar{x} = \frac{2(s^2+1)}{(s^2-1)^2} = \frac{1}{(s-1)^2} + \frac{1}{(s+1)^2} \qquad \text{...(5)}$$

and $$\bar{y} = -\frac{2}{(s^2-1)^2} = -\frac{1}{2}\left[\frac{1}{s+1} - \frac{1}{s-1} - \frac{1}{(s+1)^2} + \frac{1}{(s-1)^2}\right] \qquad \text{...(6)}$$

Taking Inverse Laplace transform of both sides of (5) and (6), we get

$$x = L^{-1}\left[\frac{1}{(s-1)^2} + \frac{1}{(s+1)^2}\right] = te^{t} + te^{-t} = 2t\left(\frac{e^{t} + e^{-t}}{2}\right) = 2t\cosh t$$

and $$y = -\frac{1}{2}L^{-1}\left(\frac{1}{s+1} - \frac{1}{s-1} - \frac{1}{(s+1)^2} + \frac{1}{(s-1)^2}\right)$$

$$= -\frac{1}{2}(e^{-t} - e^{t} - te^{-t} + te^{t}) = \frac{e^{t} - e^{-t}}{2} - t\left(\frac{e^{t} - e^{-t}}{2}\right) = (1 - t)\sinh t$$

Hence, $$x = 2t\cosh t, \; y = (1 - t)\sinh t.$$ **Ans.**

Example 78. *Using the Laplace transforms, solve the initial value problem*

$$y_1'' = y_1 + 3y_2$$

$$y_2'' = 4y_1 - 4e^t$$

$$y_1(0) = 2, \quad y'_1(0) = 3, \quad y_2(0) = 1, \quad y'_2(0) = 2$$

Solution. We have,

$$y_1'' = y_1 + 3y_2 \qquad \text{... (1)}$$

$$y_2'' = 4y_1 - 4e^t \qquad \text{... (2)}$$

Taking the Laplace transform of (1) and (2), we get

$$s^2\,\bar{y}_1 - sy_1(0) - y'_1(0) = \bar{y}_1 + 3\,\bar{y}_2 \qquad \text{... (3)}$$

$$s^2\,\bar{y}_2 - sy_2(0) - y'_2(0) = 4\bar{y}_1 - \frac{4}{s-1} \qquad \text{... (4)}$$

Putting the values of $y_1(0)$, $y'_1(0)$, $y_2(0)$, $y'_2(0)$ in (3) and (4), we get

$$s^2\,\bar{y}_1 - 2s - 3 = \bar{y}_1 + 3\,\bar{y}_2 \quad \text{or} \quad (s^2-1)\,\bar{y}_1 - 3\,\bar{y}_2 = 2s + 3 \qquad \text{... (5)}$$

$$s^2\,\bar{y}_2 - s - 2 = 4\bar{y}_1 - \frac{4}{s-1} \quad \text{or} \quad 4\,\bar{y}_1 - s^2\,\bar{y}_2 = \frac{4}{s-1} - s - 2 \qquad \text{... (6)}$$

On solving (5) and (6), we get

$$\bar{y}_1 = \frac{(2s-3)(s^2+3)(s+2)}{(s-1)(s^2+3)(s^2-4)} = \frac{2s-3}{(s-1)(s-2)} = \frac{1}{s-1} + \frac{1}{s-2}$$

$$y_1 = e^t + e^{2t}$$

$$\bar{y}_2 = \frac{(s+2)(s^2+3)}{(s^2+3)(s^2-4)} = \frac{1}{s-2} \quad \Rightarrow \quad y_2 = e^{2t} \qquad \textbf{Ans.}$$

EXERCISE 6.11

Solve the following:

1. $\dfrac{dx}{dt} + 4y = 0$, $\dfrac{dy}{dt} - 9x = 0$. Given $x = 2$ and $y = 1$ at $t = 0$.

   Ans. $x = -\dfrac{2}{3}\sin 6t + 2\cos 6t,\ y = \cos 6t + 3\sin 6t$

2. $4\dfrac{dy}{dt} + \dfrac{dx}{dt} + 3y = 0,\ 3\dfrac{dx}{dt} + 2x + \dfrac{dy}{dt} = 1$ under the condition $x = y = 0$ at $t = 0$.

   Ans. $x = \dfrac{1}{2} - \dfrac{1}{5}e^{-t} - \dfrac{3}{10}e^{-\frac{6}{11}t},\ y = \dfrac{1}{5}e^{-t} - \dfrac{1}{5}e^{-\frac{6}{11}t}$

3. $\dfrac{dx}{dt} + 5x - 2y = t,\ \dfrac{dy}{dt} + 2x + y = 0$ being given $x = y = 0$ when $t = 0$.

   Ans. $x = -\dfrac{1}{27}(1+6t)\,e^{-3t} + \dfrac{1}{27}(1+3t),\ y = -\dfrac{2}{27}(2+3t)\,e^{-3t} - \dfrac{2t}{9} + \dfrac{4}{27}$

4. $(D-1)x - 2y = t,$

$-2x + (D-1)y = t, \quad t > 0 \quad$ where $D = \dfrac{d}{dt}$ and $x(0) = 2, \quad y(0) = 4$

5. The small oscillations of a certain system with two degrees of freedom are given by the equations.
$D^2x + 3x - 2y = 0,\ D^2x + D^2y - 3x + 5y = 0$
If $x = 0,\ y = 0,\ Dx = 3,\ Dy = 2$ when $t = 0$.

Ans. $x = -\dfrac{11}{4}\sin t + \dfrac{1}{12}\sin 3t,\ y = \dfrac{11}{4}\sin t - \dfrac{1}{4}\sin 3t.$

6. $3\dfrac{dx}{dt} + 3\dfrac{dy}{dt} + 5x = 25\cos t,\ 2\dfrac{dx}{dt} - 3\dfrac{dy}{dt} = 5\sin t,$ with $x(0) = 2,\ y(0) = 3.$

Ans. $x = 2\cos t + 3\sin t,\ y = 3\cos t + 2\sin 2t$

6.17 SOLUTION OF PARTIAL DIFFERENTIAL EQUATION BY LAPLACE TRANSFORM

Example 79. *Solve the differential equation using Laplace transform method:*

$$\frac{\partial y}{\partial t} = 3\frac{\partial^2 y}{\partial t^2} \quad \text{where} \quad y\left(\frac{\pi}{2}, t\right) = 0, \left(\frac{\partial y}{\partial x}\right)_{x=0} = 0 \text{ and } y(x, 0) = 30\cos 5x.$$

(*U.P., II Semester Summer* 2005)

Solution. Given equation is

$$\frac{\partial y}{\partial t} = 3\frac{\partial^2 y}{\partial t^2}$$

Taking Laplace transform of both sides, we get

$$sL\{y\} - y(x, 0) = 3\frac{d^2}{dx^2}L\{y\}, \qquad [\text{Let } L\{y\} = \bar{y}]$$

$$s\bar{y} - y(x, 0) = 3\frac{d^2\bar{y}}{dx^2}$$

$$3\frac{d^2\bar{y}}{dx^2} - s\bar{y} = -30\cos 5x$$

$$\left(D^2 - \frac{s}{3}\right)\bar{y} = -10\cos 5x$$

A.E. is $\quad m^2 - \dfrac{s}{3} = 0 \quad \Rightarrow \quad m = \pm\sqrt{\dfrac{s}{3}}$

$$C.F. = C_1 e^{x\sqrt{s/3}} + C_2 e^{-x\sqrt{s/3}}$$

$$P.I. = \frac{1}{\left(D^2 - \frac{s}{3}\right)}(-10\cos 5x)$$

$$P.I. = \frac{30\cos 5x}{75 + s}$$

Thus
$$\bar{y} = C_1 e^{-x\sqrt{s/3}} + C_2 e^{x\sqrt{s/3}} + \frac{30\cos 5x}{75 + s} \qquad \ldots (1)$$

$$\frac{\partial y}{\partial x} = 0 \qquad \text{when} \qquad x = 0$$

$$\Rightarrow \qquad L\left\{\frac{\partial y}{\partial x}\right\} = 0 \qquad \text{at} \qquad x = 0$$

$$\Rightarrow \qquad \frac{d\bar{y}}{dx} = 0 \qquad \text{at} \qquad x = 0$$

Again, $\quad y\left(\frac{\pi}{2}, t\right) = 0 \quad \Rightarrow \quad L\left\{y\left(\frac{\pi}{2}, t\right)\right\} = 0 \quad \Rightarrow \quad \bar{y}\left(\frac{\pi}{2}, s\right) = 0$

$$\frac{d\,\bar{y}}{dx} = \sqrt{\frac{s}{3}}\left[Ae^{x\sqrt{s/3}} - Be^{-x\sqrt{s/3}} - \frac{150 \sin 5x}{75+s}\right]$$

Putting $\quad \frac{d\,\bar{y}}{dx} = 0 \quad \text{at} \quad x = 0$

$$0 = \sqrt{\frac{s}{3}}\,[A - B] \qquad \Rightarrow \qquad A = B$$

Equation (1) becomes,

$$\bar{y} = A\left[e^{x\sqrt{s/3}} + e^{-x\sqrt{s/3}}\right] + \frac{30 \cos 5x}{75 + s} \qquad \dots (2)$$

Subjecting this to the condition

$$\bar{y}\left(\frac{\pi}{2}, s\right) = 0$$

$$0 = A\left[e^{\frac{\pi}{2}\sqrt{s/3}} + e^{-\frac{\pi}{2}\sqrt{s/3}}\right] + \frac{30 \cos (5\pi/2)}{75+s}$$

$$24 \cosh\left[\frac{\pi}{2}\sqrt{\frac{s}{3}}\right] = 0 \qquad \Rightarrow \quad A = 0$$

From equation (2), $\bar{y} = \frac{30 \cos 5x}{75 + s}$, taking inverse Laplace, we get

$$y = 30e^{-75t} \cos 5x \qquad \textbf{Ans.}$$

EXERCISE 6.12

Solve the following differential equation using Laplace transform method:

1. $\frac{\partial u}{\partial t} = \frac{\partial^2 u}{\partial x^2}$ if $u(x, 0) = \sin \pi x$ **Ans.** $u = \sin \pi x.\, e^{-p^2 t}$

2. $\frac{\partial u}{\partial t} = 4\frac{\partial^2 u}{\partial x^2}$ if $u(x, 0) = 4x - \frac{1}{2}x^2$ **Ans.** $u = \left(4x - \frac{x^2}{2}\right)e^{-p^2 t}$

3. $\frac{\partial^2 u}{\partial t^2} = \frac{\partial^2 u}{\partial x^2}$ if $u(x, 0) = \frac{1}{2}x(1-x)$

Ans. $u = \frac{x}{2}(1-x)\cos pt + C_2 \sin pt\,(C_3 \cos px + C_4 \sin px)$

4. $16\frac{\partial^2 u}{\partial t^2} = \frac{\partial^2 u}{\partial x^2}$ if $u(x, 0) = x^2(5 - x)$

Ans. $u = x^2(5 - x)\cos pt + C_4 \sin pt\left(C_1 \cos\frac{px}{4} + C_2 \sin\frac{px}{4}\right)$

5. $\frac{\partial u}{\partial t} = \frac{\partial^2 u}{\partial x^2}$ if $u(x, 0) = \begin{cases} 2x, & \text{when } 0 \le x \le \frac{1}{2} \\ 2(1 - x), & \text{when } \frac{1}{2} \le x \le 1 \end{cases}$

OBJECTIVE TYPE QUESTIONS

Choose the correct alternative

1. $L^{-1}\left(\frac{1}{s}\right) =$

(*i*) -1 (*ii*) $\frac{1}{\sqrt{2}}$ (*iii*) 1 (*iv*) 2! **Ans.** (*iii*)

2. $L^{-1}\left(\frac{1}{2s + 3}\right) =$

(*i*) $-\frac{1}{2}e^{-\frac{3t}{2}}$ (*ii*) $\frac{1}{2}e^{\frac{3t}{2}}$ (*iii*) $\frac{1}{2}e^{-\frac{3t}{2}}$ (*iv*) $\frac{1}{\sqrt{\pi t}}$ **Ans.** (*iii*)

3. $L^{-1}\left(\frac{1}{\sqrt{s}}\right) =$

(*i*) $\frac{1}{\pi t}$ (*ii*) $\frac{1}{\sqrt{\pi t}}$ (*iii*) $-\frac{1}{\sqrt{\pi t}}$ (*iv*) $-\frac{1}{\pi t}$ **Ans.** (*ii*)

4. $L^{-1}\left(\frac{s}{s^2 + 1}\right) =$

(*i*) $-\cos t$ (*ii*) $\sinh t$ (*iii*) $\cos t$ (*iv*) $\cosh t$ **Ans.** (*iii*)

5. $L^{-1}\left(\frac{e^{-s}}{s}\right) =$

(*i*) $u(t - 1)$ (*ii*) $u(t + 1)$ (*iii*) $-u(t - 1)$ (*iv*) $-u(t + 1)$ **Ans.** (*i*)

6. $L^{-1}(e^{-s}) =$

(*i*) $-\delta(t - 1)$ (*ii*) $\delta(t - 1)$ (*iii*) $\delta(t + 1)$ (*iv*) $-\delta(t + 1)$ **Ans.** (*ii*)

7. $L^{-1}\left[\frac{1}{s(s^2 + 1)}\right]$ is equal to

(*i*) $1 - \cos t$ (*ii*) $1 + \sin t$ (*iii*) $1 - \sin t$ (*iv*) $1 + \cos t$ **Ans.** (*i*)

8. $L^{-1}\left[\frac{1}{(s + 1)^2}\right]$ is

(*i*) te^t (*ii*) $\frac{1}{t}e^{-t}$ (*iii*) te^{-t} (*iv*) $\frac{1}{t}e^t$ **Ans.** (*iii*)

9. $L^{-1}\{F(s)\} = f(t)$ and $L^{-1}\{\phi(s)\} = g(t)$, then $L^{-1}[F(s).\ \phi(s)]$ is given by

(*i*) $\int_0^t f(x)\, g(t-x)\, dx$ (*ii*) $\int_0^t f(x)\, g(t+x)\, dx$

(*iii*) $\int_0^t f(t)\, g(x-t)\, dt$ (*iv*) $\int_0^t f(x)\, g(x)\, dx$ **Ans.** (*i*)

10. The Inverse Laplace Transform of $\frac{(e^{-3s})}{s^3}$, is

(*i*) $(t-3)\ u_3(t)$ (*ii*) $(t-3)^2\ u_3(t)$ (*iii*) $(t-3)^2\ u_3(t)$ (*iv*) $(t+3)\ u_3(t)$ **Ans.** (*iv*)

11. If Laplace transform of a function $f(t)$ equals $\frac{(e^{-2s} - e^{-s})}{s}$, then

(*i*) $f(t) = 1,\ t > 1$;
(*ii*) $f(t) = 1$, when $1 < t < 2$, and 0 otherwise ;
(*iii*) $f(t) = -1$, when $1 < t < 3$, and 0 otherwise ;
(*iv*) $f(t) = -1$, when $1 < t < 2$, and 0 otherwise. **Ans.** (*iv*)

12. The Laplace inverse transform of $\left[\frac{2}{s}\left(e^{-2s} - e^{-4s}\right)\right]$ equals

(*i*) 2, if $0 < t < 4$; 0 otherwise, (*ii*) 2, if $t > 0$
(*iii*) 2, if $0 < t < 2$; 0 otherwise, (*iv*) 2, if $2 < t < 4$; 0 otherwise **Ans.** (*iv*)

13. The Laplace transform of $t\ u_2(t)$ is

(*i*) $\left(\frac{1}{s^2} + \frac{2}{s}\right) e^{-2s}$ (*ii*) $\frac{1}{s^2} e^{-2s}$ (*iii*) $\left(\frac{1}{s^2} - \frac{2}{s}\right) e^{-2s}$ (*iv*) $\frac{1}{s^2} e^{2s}$ **Ans.** (*i*)

14. The Inverse Laplace Transform of $\frac{Ke^{-as}}{s^2 + k^2}$ is

(*i*) $\sin kt$ (*ii*) $\cos kt$ (*iii*) $u(t-a) \sin kt$ (*iv*) none of these. **Ans.** (*iv*)

15. Inverse Laplace Transform of 1 is:

(*i*) 1 (*ii*) $\delta(t)$ (*iii*) $\delta(t-1)$ (*iv*) $u(t)$ **Ans.** (*ii*)

16. $L^{-1}\left(\frac{e^{-2s}}{s^5}\right) =$

(*i*) $\frac{(t+2)^2}{2} u(t+2)$ (*ii*) $\frac{(t+2)^4}{4} u(t-2)$

(*iii*) $\frac{(t-2)^3}{3} u(t-4)$ (*iv*) $\frac{(t-2)^4}{4} u(t-2)$ **Ans.** (*iv*)

17. $L^{-1}\left[\frac{s^2}{(s^2+4)^2}\right] =$

(*i*) $\frac{1}{4}\sin 2t + \frac{t}{2}\cos 2t$ (*ii*) $\sin 2t + \cos 2t$ (*iii*) $\frac{1}{2}\sin 2t + t\cos t$ (*iv*) $\frac{1}{2}\sin t + \frac{1}{4}\cos t$ **Ans.** (*i*)

18. $L^{-1} \log\left(\frac{s+1}{s-1}\right)$ equal to (*U.P., II Semester, 2009*)

(i) $\frac{2}{t}\cos t$ (*ii*) $\frac{2}{t}\sin t$ (*iii*) $2t \cos t$ (*iv*) $2t \sin t$ **Ans.** (*ii*)

19. $L^{-1}\left(\frac{e^{-s}}{s^2}\right)$ is equal to

(i) $(t-1)\,u_1\,(t)$ (ii) $(t-1)\,u_0\,(t-1)$

(iii) $t\,u_1\,(t)$ (iv) $t\,u_0\,(t)$ **Ans.** (i)

Fill in the blanks:

20. $L^{-1}\left[\frac{s^4-2s^2+8}{s^3(s^2-4)}\right]=$ ...... **Ans.** $\cos 2t - t^2$

21. $L^{-1}\left[\frac{6}{(s^2-1)(s^2+4)}\right]=$ ...... **Ans.** $\sinh t - \sin 2t$

22. $L^{-1}\left[\frac{6}{(s+3)^2+16}\right]=$ ...... **Ans.** $e^{-3t}\sin 4t$

23. $L^{-1}\left(\frac{2}{(s-3)^3}\right)=$ .... **Ans.** $t^2\,e^{3t}$

24. $L^{-1}\left[\frac{2}{s^2-6s+8}\right]=$ ...... **Ans.** $e^{4t}-e^{2t}$

25. $L^{-1}\left[\frac{-5}{s^2+s-6}\right]=$ .... **Ans.** $e^{-3t}-e^{2t}$

26. $L^{-1}\left[\frac{s+1}{s^2+1}\right]=$ .... **Ans.** $\cos t+\sin t$

27. $L^{-1}\left[\frac{1}{2}F\left(\frac{s}{2}\right)\right]=$ ..... **Ans.** $f(2t)$

28. $L^{-1}\left[\sqrt{\frac{\pi}{s}}\right]=$ ... **Ans.** $t^{-\frac{1}{2}}$

29. $L^{-1}\left[\frac{1}{s}F(s)\right]=$ ...... **Ans.** $\int_0^t f(t)$

30. $L^{-1}\left[(-1)^n\frac{d^n}{ds^n}[F(s)]\right]=$ ...... **Ans.** $t^n f(t)$

31. $L^{-1}\left[\int_o^\infty F(s)\,ds\right]=$ ...... **Ans.** $\frac{1}{t}f(t)$

32. $L^{-1}\left[\cos^{-1}\left(\frac{s}{2}\right)\right]=$ ...... **Ans.** $\frac{\sin 2t}{t}$

33. $L^{-1}\left[\frac{e^{-2s}}{s}\right]=$ ...... **Ans.** $u(t-2)$

34. $L^{-1}\left[\frac{\int_o^T e^{-at}f(t)\,dt}{1-e^{-st}}\right]=$ ...... **Ans.** Periodic function $f(t)$

35. $L^{-1}[F_1(s).\,F_2(s)]=$ ...... **Ans.** $\int_o^t f_1(x)\,f_2(t-x)\,dx$

36. $L^{-1}[F(s+a)] = \ldots\ldots$ **Ans.** $e^{-at} f(t)$

37. $L^{-1}[s^2\bar{y} - sy(0) - y'(0)] = \ldots\ldots$ **Ans.** $\dfrac{d^2y}{dx^2}$

38. $L^{-1}[s\bar{y} - y(0)] = \ldots\ldots$ **Ans.** $\dfrac{dy}{dx}$

39. $L^{-1}[\{s\bar{x} - x(0)\} + \bar{y}] = \ldots\ldots..$ **Ans.** $\dfrac{dx}{dt} + y$

Indicate True or False for the following statement

40. $L^{-1}\left(\dfrac{1}{\sqrt{s+7}}\right) = e^{-5t/2}$ **Ans.** False

41. $L^{-1}\dfrac{1}{(s+2)^2+1} = e^{-2t}\sin t$ **Ans.** True

42. $L^{-1}\left(\dfrac{e^{-2s}}{s+3}\right) = e^{-3(t-2)}\,u(t-2)$ **Ans.** True

43. If $L^{-1}\{f(s)\} = F(t)$ then $L^{-1}\left\{\dfrac{F(s)}{s}\right\} = \int_0^t f(u)\,dt$ **Ans.** True

UNIT–III

7 CHAPTER

INFINITE SERIES

7.1 SEQUENCE

A *sequence* is a succession of numbers or terms formed according to some definite rule. The *n*th term in a sequence is denoted by u_n.

For example, if $u_n = 2n + 1$.

By giving different values of n in u_n, we get different terms of the sequence.

Thus, $u_1 = 3$, $u_2 = 5$, $u_3 = 7$, ...

A sequence having unlimited number of terms is known as an *infinite sequence.*

7.2 LIMIT

If a sequence tends to a limit l, then we write $\lim_{n \to \infty} (u_n) = l$

7.3 CONVERGENT SEQUENCE

If the limit of a sequence is finite, the sequence is *convergent*. If the limit of a sequence does not tend to a finite number, the sequence is said to be *divergent.*

e.g., $1, \frac{1}{4}, \frac{1}{9}, \frac{1}{16}, \ldots, \frac{1}{n^2} + \ldots$. is a convergent sequence.

$3, 5, 7, \ldots, (2n + 1), \ldots$ is a divergent sequence.

7.4 BOUNDED SEQUENCE

$u_1, u_2, u_3 \ldots, u_n \ldots$ is a bounded sequence if $u_n < k$ for every n.

7.5 MONOTONIC SEQUENCE

The sequence is either increasing or decreasing, such sequences are called *monotonic.*

e.g., $1, 4, 7, 10, \ldots$ is a monotonic sequence.

$1, \frac{1}{2}, \frac{1}{3}, \frac{1}{4}, \ldots$ is also a monotonic sequenece.

$1, -1, 1, -1, 1, \ldots$ is not a monotonic sequence.

A sequence which is monotonic and bounded is a convergent sequence.

EXERCISE 7.1

Determine the general term of each of the following sequence. Prove that the following sequences are convergent.

1. $\frac{1}{2}, \frac{1}{4}, \frac{1}{8}, \frac{1}{16}, \ldots$ **Ans.** $\frac{1}{2^n}$

2. $\frac{1}{2}, \frac{2}{3}, \frac{3}{4}, \frac{4}{5}, \ldots$ **Ans.** $\frac{n}{n+1}$

3. $1, -1, 1, -1, \ldots$ **Ans.** $(-1)^{n-1}$

4. $\frac{1^2}{1!}, \frac{2^2}{2!}, \frac{3^2}{3!}, \frac{4^2}{4!}, \frac{5^2}{5!}, \ldots$ **Ans.** $\frac{n^2}{n!}$

Which of the following sequences are convergent ?

5. $u_n = \frac{n+1}{n}$ **Ans.** Convergent

6. $u_n = 3n$ **Ans.** Divergent

7. $u_n = n^2$ **Ans.** Divergent

8. $u_n = \frac{1}{n}$ **Ans.** Convergent

7.6 REMEMBER THE FOLLOWING LIMITS

(*i*) $\lim_{n \to \infty} x^n = 0$ if $x < 1$ and $\lim_{n \to \infty} x^n = \infty$ if $x > 1$

(*ii*) $\lim_{n \to \infty} \frac{x^n}{n!} = 0$ for all values of x

(*iii*) $\lim_{n \to \infty} \frac{\log n}{n} = 0$

(*iv*) $\lim_{n \to \infty} \left(1 + \frac{1}{n}\right)^n = e$

(*v*) $\lim_{n \to \infty} (n)^{1/n} = 1$

(*vi*) $\lim_{n \to \infty} [n!]^{1/n} = \infty$

(*vii*) $\lim_{n \to \infty} \left[\frac{(n!)}{n}\right]^{1/n} = \frac{1}{e}$

(*viii*) $\lim_{n \to \infty} n x^n = 0$ if $x < 1$

(*ix*) $\lim_{n \to \infty} n^h = \infty$

(*x*) $\lim_{n \to \infty} \frac{1}{n^h} = 0$

(*xi*) $\lim_{x \to \infty} \left[\frac{a^x - 1}{x}\right] = \log a$ or $\lim_{n \to \infty} \frac{a^{1/n} - 1}{1/n} = \log a$

(*xii*) $\lim_{x \to 0} \frac{\sin x}{x} = 1$

(*xiii*) $\lim_{x \to 0} \frac{\tan x}{x} = 1$

7.7 SERIES

A *series* is the sum of a sequence.

Let $u_1, u_2, u_3, \ldots\ldots\ldots, u_n, \ldots\ldots\ldots$ be a given sequence. Then, the expression $u_1 + u_2 + u_3 + \ldots\ldots\ldots + u_n + \ldots\ldots\ldots$ is called the series associated with the given sequence.

For example, $1 + 3 + 5 + 7 + \ldots$ is a series.

If the number of terms of a series is limited, the series is called *finite.* When the number of terms of a series are unlimited, it is called an *infinite series.*

$$u_1 + u_2 + u_3 + u_4 + \ldots + u_n + \ldots \infty$$

is called an infinite series and it is denoted by $\sum_{n=1}^{\infty} u_n$ or $\Sigma\, u_n$. The sum of the first n terms of a series is denoted by S_n.

7.8 CONVERGENT, DIVERGENT AND OSCILLATORY SERIES

Consider the infinite series $\Sigma u_n = u_1 + u_2 + u_3 + \ldots + u_n + \ldots \infty$

$$S_n = u_1 + u_2 + u_3 + \ldots + u_n$$

Three cases arise:

(*i*) If S_n tends to a finite number as $n \to \infty$, the series Σu_n is said to be *convergent*.

(*ii*) If S_n tends to infinity as $n \to \infty$, the series Σu_n is said to be *divergent*.

(*iii*) If S_n does not tend to a unique limit, finite or infinite, the series Σu_n is called *oscillatory*.

7.9 PROPERTIES OF INFINITE SERIES

1. The nature of an infinite series does not change:
   (*i*) by multiplication of all terms by a constant k.
   (*ii*) by addition or deletion of a finite number of terms.
2. If two series Σu_n and Σv_n are convergent, then $\Sigma (u_n + v_n)$ is also convergent.

Example 1. *Examine the nature of the series $1 + 2 + 3 + 4 + \ldots + n + \ldots \infty$.*

Solution. Let $\quad S_n = 1 + 2 + 3 + 4 + \ldots + n = \dfrac{n(n+1)}{2}$ [Series in A.P.]

Since $\quad \lim\limits_{n \to \infty} S_n = \lim\limits_{n \to \infty} \dfrac{n(n+1)}{2} \Rightarrow \infty$

Hence, this series is divergent. **Ans.**

Example 2. *Test the convergence of the series $1 + \dfrac{1}{2} + \dfrac{1}{4} + \dfrac{1}{8} + \ldots \infty$*

Solution. Let $\quad S_n = 1 + \dfrac{1}{2} + \dfrac{1}{4} + \dfrac{1}{8} + \ldots \infty$ [Series in G.P.]

$$= \frac{1}{1 - \frac{1}{2}} = 2 \qquad \left(S_n = \frac{a}{1-r}\right)$$

$$\lim_{n \to \infty} S_n = 2$$

Hence, the series is convergent. **Ans.**

Example 3. *Prove that the following series:*

$\dfrac{2}{3!} + \dfrac{3}{4!} + \dfrac{4}{5!} + \ldots$ *is convergent and find its sum.* (*M.U. 2008*)

Solution. Here, $\quad u_n = \dfrac{n+1}{(n+2)!} = \dfrac{n+2-1}{(n+2)!} = \dfrac{n+2}{(n+2)!} - \dfrac{1}{(n+2)!}$

$$= \frac{1}{(n+1)!} - \frac{1}{(n+2)!}$$

$$S_n = \left(\frac{1}{2!} - \frac{1}{3!}\right) + \left(\frac{1}{3!} - \frac{1}{4!}\right) + \left(\frac{1}{4!} - \frac{1}{5!}\right) + \ldots$$

$$+ \left(\frac{1}{(n+1)!} - \frac{1}{(n+2)!}\right) = \frac{1}{2!} - \frac{1}{(n+2)!}$$

$$\lim_{n \to \infty} S_n = \lim_{n \to \infty} \left[\frac{1}{2!} - \frac{1}{(n+2)!}\right] = \frac{1}{2}$$

$\therefore \quad \Sigma u_n$ converges and its limit is $\dfrac{1}{2}$. **Ans.**

Example 4. *Discuss the nature of the series* $2 - 2 + 2 - 2 + 2 - \ldots \infty$.

Solution. Let

$$S_n = 2 - 2 + 2 - 2 + 2 - \ldots \infty$$
$$= 0 \text{ if } n \text{ is even}$$
$$= 2 \text{ if } n \text{ is odd.}$$

Hence, S_n does not tend to a unique limit, and, therefore, the given series is oscillatory. **Ans.**

EXERCISE 7.2

Discuss the nature of the following series:

1. $1 + 4 + 7 + 10 + \ldots \infty$ **Ans.** Divergent

2. $1 + \frac{5}{4} + \frac{6}{4} + \frac{7}{4} + \ldots \infty$ **Ans.** Divergent

3. $6 - 5 - 1 + 6 - 5 - 1 + 6 - 5 - 1 + \ldots \infty$ **Ans.** Oscillatory

4. $3 + \frac{3}{2} + \frac{3}{2^2} + \ldots \infty$ **Ans.** Convergent

5. $1^2 + 2^2 + 3^2 + 4^2 + \ldots \infty$ **Ans.** Divergent

6. $1 - \frac{1}{2} + \frac{1}{4} - \frac{1}{8} + \ldots \infty$ **Ans.** Convergent

7. $\frac{1}{1.3} + \frac{1}{3.5} + \frac{1}{5.7} + \ldots \infty$ **Ans.** Convergent

8. $\frac{1}{2!} + \frac{2}{3!} + \frac{3}{4!} + \ldots \infty$ **Ans.** Convergent

9. $\log 3 + \log \frac{4}{3} + \log \frac{5}{4} + \ldots \infty$ **Ans.** Divergent

10. $\sum \log \frac{n}{n+1}$ **Ans.** Divergent

11. $\Sigma\left(\sqrt{n+1} - \sqrt{n}\right)$ **Ans.** Divergent

12. $\sum \frac{1}{n(n+2)}$ **Ans.** Convergent

13. $\sum \frac{1}{n(n+1)(n+2)(n+3)}$ **Ans.** Convergent

14. $\sum \frac{n}{(n+1)(n+2)(n+3)}$ **Ans.** Convergent

15. $\sum \frac{2n+1}{n^2(n+1)^2}$ **Ans.** Convergent

7.10 PROPERTIES OF GEOMETRIC SERIES

The series $1 + r + r^2 + r^3 + \ldots \infty$ is

(*i*) convergent if $|r| < 1$ *(Uttarakhand, IInd Semester, Summer 2012, 2010)*

(*ii*) divergent if $r \geq 1$ (*iii*) oscillatory if $r \leq -1$.

Proof.

$$S_n = 1 + r + r^2 + \ldots + r^{n-1} = \frac{1 - r^n}{1 - r}$$

(*i*) When $|r| < 1$, $\lim\limits_{n \to \infty} r^n = 0$

$$\lim_{n \to \infty} S_n = \lim_{n \to \infty} \frac{1 - r^n}{1 - r} = \frac{1 - 0}{1 - r} = \frac{1}{1 - r}$$

Hence, the series is convergent.

(*ii*) (*a*) When $r > 1$, $\lim\limits_{n \to \infty} r^n = \infty$ $\therefore$ $\lim\limits_{n \to \infty} S_n = \lim\limits_{n \to \infty} \frac{r^n - 1}{r - 1} \Rightarrow \infty$

Hence, the series is divergent.

(*b*) When $r = 1$, the series becomes $1 + 1 + 1 + 1 + \ldots \infty$

$$S_n = 1 + 1 + 1 + 1 + \ldots = n$$

$$\lim_{n\to\infty} S_n = \lim_{n\to\infty} n = \infty$$

Hence, the series is divergent.

(*iii*) (*a*) When $r = -1$, the series becomes $1 - 1 + 1 - 1 + 1 - \ldots \infty$

$$S_n = 0 \text{ if } n \text{ is even} = 1 \text{ if } n \text{ is odd}$$

Hence, the series is oscillatory.

(*b*) When $r < -1$, let $r = -k$ where $k > 1$.

$$r^n = (-k)^n = (-1)^n k^n$$

$$\lim_{n\to\infty} S_n = \lim_{n\to\infty} \frac{1-r^n}{1-r} = \lim_{n\to\infty} \frac{1-(-1)^n k^n}{1-(-k)}$$

$$= +\infty \text{ if } n \text{ is odd} = -\infty \text{ if } n \text{ is even}$$

Hence, the series is oscillatory. **Proved.**

EXERCISE 7.3

Test the nature of the following series :

1. $1+\frac{1}{2}+\frac{1}{2^2}+\frac{1}{2^3}+\ldots\infty$ **Ans.** Convergent
2. $1+\frac{3}{4}+\frac{9}{16}+\frac{27}{64}+\ldots\infty$ **Ans.** Convergent
3. $1-\frac{1}{3}+\frac{1}{9}-\frac{1}{27}+\ldots\infty$ **Ans.** Convergent
4. $1-2+4-8+\ldots\infty$ **Ans.** Oscillatory
5. $2+3+\frac{9}{2}+\frac{27}{4}+\ldots\infty$ **Ans.** Divergent
6. $1+\frac{4}{3}+\left(\frac{4}{3}\right)^2+\left(\frac{4}{3}\right)^3+\ldots\infty$ **Ans.** Divergent
7. State, which one of the alternatives in the following is correct:

   The series $1 - 1 + 1 - 1 + \ldots$ is

   (*i*) Convergent with its sum equal to 0. (*ii*) Convergent with its sum equal to 1.

   (*iii*) Divergent. (*iv*) Oscillatory. **Ans.** Oscillatory series

7.11 POSITIVE TERM SERIES

If all terms after few negative terms in an infinite series are positive, such a series is a positive term series.

e.g., $-10 - 6 - 1 + 5 + 12 + 20 + \ldots$ is a positive term series.

By omitting the negative terms, the nature of a positive term series remains unchanged.

7.12 NECESSARY CONDITIONS FOR CONVERGENT SERIES

For every convergent series $\Sigma\, u_n$.

$$\lim_{n\to\infty} u_n = 0$$

Solution. Let $S_n = u_1 + u_2 + u_3 + \ldots + u_n$

$$\lim_{n\to\infty} S_n = k \quad \text{(a finite quantity)}$$

Also $$\lim_{n\to\infty} S_{n-1} = k \quad \text{(a finite quantity)}$$

$$S_n = S_{n-1} + u_n$$

$$u_n = S_n - S_{n-1}$$

$$\lim_{n\to\infty} u_n = \lim_{n\to\infty} [S_n - S_{n-1}] = 0$$

$$\lim_{n\to\infty} u_n = 0$$

Corollary. Converse of the above theorem is not true.

e.g., $$1+\frac{1}{\sqrt{2}}+\frac{1}{\sqrt{3}}+\frac{1}{\sqrt{4}}+\ldots+\frac{1}{\sqrt{n}}+\ldots\infty \text{ is divergent.}$$

$$S_n = 1+\frac{1}{\sqrt{2}}+\frac{1}{\sqrt{3}}+\frac{1}{\sqrt{4}}+\ldots+\frac{1}{\sqrt{n}}$$

$$> \frac{1}{\sqrt{n}}+\frac{1}{\sqrt{n}}+\frac{1}{\sqrt{n}}+\frac{1}{\sqrt{n}}+\ldots+\frac{1}{\sqrt{n}}$$

$$> \frac{n}{\sqrt{n}} > \sqrt{n}$$

$$\lim_{n\to\infty} S_n = \lim_{n\to\infty} \sqrt{n} = \infty$$

Thus, the series is divergent, although $\lim_{n\to\infty} u_n = \lim_{n\to\infty} \frac{1}{\sqrt{n}} = 0$

So $\lim_{n\to\infty} u_n = 0$ is a necessary condition but not a sufficient condition for convergence.

Note: 1. Test for divergence

If $\lim_{n\to\infty} u_n \neq 0$, the series $\Sigma\, u_n$ must be divergent.

2. ***To determine the nature of a series we have to find S_n. Since it is not possible to find S_n for every series, we have to devise tests for convergence without involving S_n.***

7.13 CAUCHY'S FUNDAMENTAL TEST FOR DIVERGENCE

If $\lim_{n\to\infty} u_n \neq 0$, the series is divergent.

Example 5. *Test for convergence of the series* $1+\frac{2}{3}+\frac{3}{4}+\frac{4}{5}+\ldots+\frac{n}{n+1}+\ldots\infty$

Solution. Here, $$\lim_{n\to\infty} u_n = \lim_{n\to\infty}\frac{n}{n+1} = \lim_{n\to\infty}\frac{1}{1+\frac{1}{n}} = 1 \neq 0$$

Hence, by **Cauchy's Fundamental Test** for divergence, the series is divergent. **Ans.**

Example 6. *Test for convergence the series* $1+\frac{3}{5}+\frac{8}{10}+\frac{15}{17}+\ldots+\frac{2^n-1}{2^n+1}+\ldots\infty$

Solution. Here, $$\lim_{n\to\infty} u_n = \lim_{n\to\infty}\frac{2^n-1}{2^n+1} = \lim_{n\to\infty}\frac{1-\frac{1}{2^n}}{1+\frac{1}{2^n}} = 1 \neq 0$$

Hence, by **Cauchy's Fundamental Test** for divergence the series is divergent. **Ans.**

Example 7. Test the convergence of the following series:

$$\sqrt{\frac{1}{4}}+\sqrt{\frac{2}{6}}+\sqrt{\frac{3}{8}}+\cdots+\sqrt{\frac{n}{2(n+1)}}+\cdots$$ *(M.D.U., 2000)*

Solution. Here, we have

$$\sqrt{\frac{1}{4}}+\sqrt{\frac{2}{6}}+\sqrt{\frac{3}{8}}+\cdots+\sqrt{\frac{n}{2(n+1)}}+\cdots$$

$$u_n = \sqrt{\frac{n}{2(n+1)}} = \sqrt{\frac{1}{2\left(1+\frac{1}{n}\right)}}$$

$$\lim_{n\to\infty} u_n = \lim_{n\to\infty} \frac{1}{\sqrt{2\left(1+\frac{1}{n}\right)}} = \frac{1}{\sqrt{2}} \neq 0$$

$\Rightarrow \Sigma u_n$ does not converge.

The given series is a series of + ve terms,

Hence by Cauchy fundamental test for divergence, the series is divergent. **Ans.**

Example 8. *Test the convergence of* $\sum_{n=1}^{\infty} \cos(1/n)$ *(Uttarakhand, IInd Semester Summer 2010)*

Solution. Here we have

$$u_n = \cos\frac{1}{n}$$

$$= \left[1 - \frac{1}{2!n^2} + \frac{1}{4!n^4} - \frac{1}{6!n^6} + \ldots\ldots\ldots\right]$$

$$\lim_{n\to\infty} u_n = \lim_{n\to\infty}\left[1 - \frac{1}{2n^2} + \frac{1}{24n^4} - \frac{1}{720n^6} + \ldots\ldots\right]$$

$$= 1 \neq 0$$

Hence by Cauchys' Fundamental test for divergence, the series is divergent. **Ans.**

Example 9. *Test whether the following series is convergent*

or divergent $\sum_{n=1}^{\infty} \frac{1}{n}\sin\left(\frac{1}{n}\right)$ *(Uttarakhand, IInd Semester Summer 2010)*

Solution. Here we have

$$u_n = \frac{1}{n}\sin\frac{1}{n}$$

$$= \frac{1}{n}\left(\frac{1}{n} - \frac{1}{3!n^3} + \frac{1}{5!n^5} - \frac{1}{7!n^7} \ldots\ldots\ldots\right)$$

$$= \frac{1}{n}\left(\frac{1}{n} - \frac{1}{6n^3} + \frac{1}{120n^5} - \frac{1}{5040n^7} \ldots\ldots\ldots\right)$$

$$= \frac{1}{n^2} - \frac{1}{6n^4} + \frac{1}{120n^6} - \frac{1}{5040n^8} \ldots\ldots$$

$$\lim_{n\to\infty} u_n = \lim_{n\to\infty}\left[\frac{1}{n^2} - \frac{1}{6n^4} + \frac{1}{120n^6} - \frac{1}{5040n^8} \ldots\ldots\right]$$

$$= 0$$

Hence by Canchy's Fundamental test for divergence, the series is convergent. **Ans.**

EXERCISE 7.4

Examine for convergence:

1. $\frac{1}{\sqrt{2}}+\frac{2}{\sqrt{5}}+\frac{4}{\sqrt{17}}+\dots+\frac{2^n}{\sqrt{4^n+1}}+\dots\infty$ **Ans. Divergent**
2. $\sum_{n=1}^{\infty}\sqrt{\frac{n}{n+1}}$ **Ans. Divergent**
3. $1+\frac{1}{2}+2+\frac{1}{3}+3+\frac{1}{4}+4+\dots$ **Ans. Divergent**
4. $\Sigma\,(6-n^2)$ **Ans. Divergent**
5. $\Sigma\,(-2^n)$ **Ans. Divergent**
6. $\Sigma\,3^{n+1}$ **Ans. Divergent**

7.14 *p*-SERIES

The series $\frac{1}{1^p}+\frac{1}{2^p}+\frac{1}{3^p}+\dots\infty$ *is*

(*i*) *convergent if* $p>1$ (*ii*) *Divergent if* $p\le 1$. (*MDU, Dec. 2010*)

Solution. Case 1: ($p>1$)

The given series can be grouped as

$$\frac{1}{1^p}+\left(\frac{1}{2^p}+\frac{1}{3^p}\right)+\left(\frac{1}{4^p}+\frac{1}{5^p}+\frac{1}{6^p}+\frac{1}{7^p}\right)+\left(\frac{1}{8^p}+\frac{1}{9^p}+\frac{1}{10^p}+\frac{1}{11^p}+\frac{1}{12^p}+\frac{1}{13^p}+\frac{1}{14^p}+\frac{1}{15^p}\right)+\dots$$

Now

$$\frac{1}{1^p}=1 \qquad \dots(1)$$

$$\frac{1}{2^p}+\frac{1}{3^p}<\frac{1}{2^p}+\frac{1}{2^p}=\frac{2}{2^p} \qquad \dots(2)$$

$$\frac{1}{4^p}+\frac{1}{5^p}+\frac{1}{6^p}+\frac{1}{7^p}<\frac{1}{4^p}+\frac{1}{4^p}+\frac{1}{4^p}+\frac{1}{4^p}=\frac{4}{4^p} \qquad \dots(3)$$

$$\frac{1}{8^p}+\frac{1}{9^p}+\dots+\frac{1}{15^p}<\frac{1}{8^p}+\frac{1}{8^p}+\dots+\frac{1}{8^p}=\frac{8}{8^p} \qquad \dots(4)$$

On adding (1), (2), (3) and (4), we get:

$$\frac{1}{1^p}+\left(\frac{1}{2^p}+\frac{1}{3^p}\right)+\left(\frac{1}{4^p}+\frac{1}{5^p}+\frac{1}{6^p}+\frac{1}{7^p}\right)+\left(\frac{1}{8^p}+\frac{1}{9^p}+\dots+\frac{1}{15^p}\right)+\dots$$

$$<\frac{1}{1^p}+\frac{2}{2^p}+\frac{4}{4^p}+\frac{8}{8^p}+\dots$$

$$<1+\left(\frac{1}{2}\right)^{p-1}+\left(\frac{1}{2}\right)^{2p-2}+\left(\frac{1}{2}\right)^{3p-3}+\dots$$

$$<\frac{1}{1-\left(\frac{1}{2}\right)^{p-1}} \qquad \left[\text{G.P., } r=\left(\frac{1}{2}\right)^{p-1},\ S=\frac{1}{1-r}\right]$$

$$<\text{Finite number if } p>1$$

Hence, the given series is convergent when $p>1$.

Case 2: $p = 1$

When $p = 1$, the given series becomes

$$1+\frac{1}{2}+\left(\frac{1}{3}+\frac{1}{4}\right)+\left(\frac{1}{5}+\frac{1}{6}+\frac{1}{7}+\frac{1}{8}\right)+\left(\frac{1}{9}+\frac{1}{10}+\ldots+\frac{1}{16}\right)+\ldots$$

$$1+\frac{1}{2} = 1+\frac{1}{2} \qquad \ldots(1)$$

$$\frac{1}{3}+\frac{1}{4} > \frac{1}{4}+\frac{1}{4}=\frac{1}{2} \qquad \ldots(2)$$

$$\frac{1}{5}+\frac{1}{6}+\frac{1}{7}+\frac{1}{8} > \frac{1}{8}+\frac{1}{8}+\frac{1}{8}+\frac{1}{8}=\frac{4}{8}=\frac{1}{2} \qquad \ldots(3)$$

$$\frac{1}{9}+\frac{1}{10}+\ldots+\frac{1}{16} > \frac{1}{16}+\frac{1}{16}+\ldots+\frac{1}{16}=\frac{8}{16}=\frac{1}{2} \qquad \ldots(4)$$

On adding (1), (2), (3) and (4), we get

$$1+\frac{1}{2}+\left(\frac{1}{3}+\frac{1}{4}\right)+\left(\frac{1}{5}+\frac{1}{6}+\frac{1}{7}+\frac{1}{8}\right)+\left(\frac{1}{9}+\frac{1}{10}+\ldots+\frac{1}{16}\right)+\ldots$$

$$> 1+\frac{1}{2}+\frac{1}{2}+\frac{1}{2}+\frac{1}{2}+\ldots$$

$$> 1+\frac{n}{2} \quad (n \to \infty)$$

$$> \infty$$

Hence, the given series is divergent when $p = 1$.

Case 3: $p < 1$

$$\frac{1}{2^p} > \frac{1}{2}, \qquad \frac{1}{3^p} > \frac{1}{3}, \qquad \frac{1}{4^p} > \frac{1}{4} \text{ and so on}$$

Therefore, $\frac{1}{1^p}+\frac{1}{2^p}+\frac{1}{3^p}+\frac{1}{4^p}+\ldots > 1+\frac{1}{2}+\frac{1}{3}+\frac{1}{4}+\ldots$

$>$ divergent series ($p = 1$) [From Case 2]

$$\left[\text{As the series on R.H.S.} \left(1+\frac{1}{2}+\frac{1}{3}+\frac{1}{4}+\ldots\right) \text{is divergent}\right]$$

Hence, the given series is divergent when $p < 1$.

7.15 COMPARISON TEST

If two positive terms $\Sigma\, u_n$ and $\Sigma\, v_n$ be such that

$\lim\limits_{n\to\infty} \dfrac{u_n}{v_n} = k$ (finite number), then both series converge or diverge together.

Proof. By definition of limit there exists a positive number ε, however small, such that

$$\left|\frac{u_n}{v_n}-k\right| < \varepsilon \text{ for } n > m \qquad i.e., -\varepsilon < \frac{u_n}{v_n}-k < +\varepsilon$$

$$k-\varepsilon < \frac{u_n}{v_n} < k+\varepsilon \text{ for } n > m$$

Ignoring the first m terms of both series, we have

$$k-\varepsilon < \frac{u_n}{v_n} < k+\varepsilon \text{ for all } n. \qquad \ldots(1)$$

Case 1. Σv_n is convergent, then

$$\lim_{n \to \infty} (v_1 + v_2 + \dots v_n) = h \text{ (say)} \qquad \text{where } h \text{ is a finite number.}$$

From (1), $u_n < (k + \varepsilon) v_n$ for all n.

$$\lim_{n \to \infty} (u_1 + u_2 + \dots + u_n) < (k + \varepsilon) \lim_{n \to \infty} (v_1 + v_2 + \dots + v_n) = (k + \varepsilon)h$$

Hence, Σu_n is also convergent.

Case 2. Σv_n is divergent, then

$$\lim_{n \to \infty} (v_1 + v_2 + \dots + v_n) \to \infty \qquad \dots(2)$$

Now from (1) $\qquad k - \varepsilon < \dfrac{u_n}{v_n}$

$$u_n > (k - \varepsilon) v_n \text{ for all } n$$

$$\lim_{n \to \infty} (u_1 + u_2 + \dots + u_n) > (k - \varepsilon) \lim_{n \to \infty} (v_1 + v_2 + \dots + v_n)$$

From (2), $\lim_{n \to \infty} (u_1 + u_2 + \dots + u_n) \to \infty$

Hence, Σu_n is also divergent.

Note. For testing the convergence of a series, this Comparison Test is very useful. We choose Σv_n (p-series) in such a way that

$$\boxed{\lim_{n \to \infty} \frac{u_n}{v_n} = \text{finite number.}}$$

Then the nature of both the series is the same. The nature of Σv_n (p-series) is already known, so the nature of Σu_n is also known.

Example 10. *Test the series* $\displaystyle\sum_{n=1}^{\infty} \frac{1}{n + 10}$ *for convergence or divergence.*

Solution. Here, $\qquad u_n = \dfrac{1}{n + 10}$

Let $\qquad v_n = \dfrac{1}{n}$

$$\lim_{n \to \infty} \frac{u_n}{v_n} = \lim_{n \to \infty} \frac{n}{n + 10} = \lim_{n \to \infty} \frac{1}{1 + \frac{10}{n}} = 1 = \text{finite number.}$$

According to Comparison Test both series converge or diverge together, but Σv_n is divergent as $p = 1$.

$\therefore \quad \Sigma u_n$ is also divergent. **Ans.**

Example 11. *Test the convergence of the following series:*

$$\frac{1}{\sqrt{1} + \sqrt{2}} + \frac{1}{\sqrt{2} + \sqrt{3}} + \frac{1}{\sqrt{3} + \sqrt{4}} + \cdots \qquad \textit{(M.D.U., 2000)}$$

Solution. Here, we have

$$\frac{1}{\sqrt{1} + \sqrt{2}} + \frac{1}{\sqrt{2} + \sqrt{3}} + \frac{1}{\sqrt{3} + \sqrt{4}} + \cdots$$

$$u_n = \frac{1}{\sqrt{n} + \sqrt{n+1}} = \frac{1}{\sqrt{n}\left[1 + \sqrt{1 + \frac{1}{n}}\right]}$$

Let us compare Σu_n with Σv_n, where

$$v_n = \frac{1}{\sqrt{n}}$$

$$\frac{u_n}{v_n} = \frac{1}{\sqrt{n}\left[1+\sqrt{1+\frac{1}{n}}\right]} \cdot \frac{\sqrt{n}}{1} = \frac{1}{1+\sqrt{1+\frac{1}{n}}}$$

$$\lim_{n\to\infty} \frac{u_n}{v_n} = \lim_{n\to\infty} \frac{1}{1+\sqrt{1+\frac{1}{n}}} = \frac{1}{1+1} = \frac{1}{2}$$

Which is finite and non-zero.

$\therefore \Sigma u_n$ and Σv_n, converge or diverge together since $\Sigma v_n = \Sigma \frac{1}{n^{\frac{1}{2}}}$ is of the form $\Sigma \frac{1}{n^P}$.

$$P = \frac{1}{2} < 1$$

$\therefore \Sigma v_n$ is divergent $\Rightarrow \Sigma u_n$ is also divergent. **Ans.**

Example 12. *Examine the convergence of the series:*

$$\sum\left(\sqrt[3]{n^3+1}-n\right)$$ (*M.D.U. 2003*)

Solution. Here, we have $\sum\left(\sqrt[3]{n^3+1}-n\right)$

$$u_n = (n^3+1)^{\frac{1}{3}} - n = \left[n^3\left(1+\frac{1}{n^3}\right)\right]^{\frac{1}{3}} - n$$

$$= n\left[\left(1+\frac{1}{n^3}\right)^{\frac{1}{3}} - 1\right] = n\left[1+\frac{1}{3}.\frac{1}{n^3}+\frac{\frac{1}{3}\left(\frac{1}{3}-1\right)}{2!}\left(\frac{1}{n^3}\right)^2+\ldots-1\right]$$

$$= \frac{n}{n^3}\left[\frac{1}{3}-\frac{1}{9}.\frac{1}{n^3}\ldots\right] = \frac{1}{n^2}\left[\frac{1}{3}-\frac{1}{9}.\frac{1}{n^3}\ldots\right]$$

Let $$v_n = \frac{1}{n^2}$$

$$\frac{u_n}{v_n} = \frac{1}{n^2}\left[\frac{1}{3}-\frac{1}{9}.\frac{1}{n^3}\ldots\right].\frac{n^2}{1} = \left[\frac{1}{3}-\frac{1}{9}.\frac{1}{n^3}\ldots\right]$$

$$\Rightarrow \lim_{n\to\infty}\frac{u_n}{v_n} = \lim_{n\to\infty}\left(\frac{1}{3}-\frac{1}{9}.\frac{1}{n^3}\ldots\right) = \frac{1}{3}$$

which is finite and non-zero.

$\therefore \sum u_n$ and $\sum v_n$ converge or diverge together.

Since $\sum v_n = \sum \frac{1}{n^2}$ is of the form $\sum \frac{1}{n^p}$ with $p = 2 > 1$

$\therefore \sum v_n$ is convergent $\Rightarrow \sum u_n$ is convergent. **Ans.**

Example 13. *Test the convergence of the following series*

$$\frac{1}{1+2^{-1}}+\frac{2}{1+2^{-2}}+\frac{3}{1+2^{-3}}+\cdots$$ (*M.D.U., 2000*)

Solution. Here, we have

$$\frac{1}{1+2^{-1}}+\frac{2}{1+2^{-2}}+\frac{3}{1+2^{-3}}+\cdots$$

Here $$u_n = \frac{n}{1+2^{-n}} = \frac{n}{1+\frac{1}{2^n}}$$

Let $$u_n = n$$

Let us compare $\sum u_n$ with $\sum v_n$,

$$\frac{u_n}{v_n} = \frac{n}{1+\frac{1}{2^n}} \cdot \frac{1}{n} = \frac{1}{1+\frac{1}{2^n}}$$

$$\lim_{n\to\infty} \frac{u_n}{v_n} = \lim_{n\to\infty} \frac{1}{1+\frac{1}{2^n}} = \frac{1}{1+0} = 1$$

Which is finite and non-zero.

$\therefore$ $\sum u_n$ and $\sum v_n$, converge or diverge together since $\sum v_n = \sum \frac{1}{n}$ is of the form $\sum \frac{1}{n^p}$ with $p = 1$.

$\therefore$ $\sum v_n$ is divergent $\Rightarrow$ $\sum u_n$ is also divergent. **Ans.**

Example 14. *Examine the convergence of the series* $\frac{\sqrt{2}-1}{3^3-1}+\frac{\sqrt{3}-1}{4^3-1}+\frac{\sqrt{4}-1}{5^3-1}+\cdots$

(M.D.U., 2000)

Solution. Here, we have

$$\frac{\sqrt{2}-1}{3^3-1}+\frac{\sqrt{3}-1}{4^3-1}+\frac{\sqrt{4}-1}{5^3-1}+\cdots$$

Here $$u_n = \frac{\sqrt{n+1}-1}{(n+2)^3-1} = \frac{\sqrt{n}\left(\sqrt{1+\frac{1}{n}}-\frac{1}{\sqrt{n}}\right)}{n^3\left[\left(1+\frac{2}{n}\right)^3-\frac{1}{n^3}\right]} = \frac{\sqrt{1+\frac{1}{n}}-\frac{1}{\sqrt{n}}}{n^{\frac{5}{2}}\left[\left(1+\frac{2}{n}\right)^3-\frac{1}{n^3}\right]}$$

Let $$v_n = \frac{1}{n^{\frac{5}{2}}}$$

Let us compare $\sum u_n$ with $\sum v_n$,

$$\frac{u_n}{v_n} = \frac{\sqrt{1+\frac{1}{n}}-\frac{1}{\sqrt{n}}}{n^{\frac{5}{2}}\left[\left(1+\frac{2}{n}\right)^3-\frac{1}{n^3}\right]} \times \frac{n^{\frac{5}{2}}}{1} = \frac{\sqrt{1+\frac{1}{n}}-\frac{1}{\sqrt{n}}}{\left[\left(1+\frac{2}{n}\right)^3-\frac{1}{n^3}\right]}$$

$$\lim_{n\to\infty} \frac{u_n}{v_n} = \lim_{n\to\infty} \frac{\sqrt{1+\frac{1}{n}}-\frac{1}{\sqrt{n}}}{\left[\left(1+\frac{2}{n}\right)^3-\frac{1}{n^3}\right]} = \frac{\sqrt{1+0}-0}{(1-0)^3-0} = 1$$

Which is finite and non-zero.

$\therefore \sum u_n$ and $\sum v_n$, converge or diverge together since $\sum v_n = \sum \frac{1}{n^{\frac{5}{2}}}$ is of the form

$\sum \frac{1}{n^p}$.
$$P = \frac{5}{2} > 1.$$

$\therefore \sum v_n$ is convergent $\Rightarrow \sum u_n$ is convergent. **Ans.**

Example 15. *Test the convergence and divergence of the following series.*

$$\sum_{n=1}^{\infty} \frac{2n^2 + 3n}{5 + n^5}$$ *(Gujarat, I Semester, Jan. 2009)*

Solution. Here,
$$u_n = \frac{2n^2 + 3n}{5 + n^5} = \frac{n^2\left(2 + \frac{3}{n}\right)}{n^5\left(\frac{5}{n^5} + 1\right)} = \frac{1}{n^3} \frac{2 + \frac{3}{n}}{\frac{5}{n^5} + 1}$$

Let
$$v_n = \frac{1}{n^3}$$

By Comparison Test

$$\lim_{n \to \infty} \frac{u_n}{v_n} = \lim_{n \to \infty} \frac{n^3\left(2 + \frac{3}{n}\right)}{n^3\left(\frac{5}{n^5} + 1\right)} = \lim_{n \to \infty} \frac{2 + \frac{3}{n}}{\frac{5}{n^5} + 1} = 2 = \text{Finite number.}$$

According to comparison test both series converge or diverge together but $\Sigma\, v_n$ is convergent as $p = 2$.

Hence, the given series is convergent. **Ans.**

Example 16. *Test for convergence the series* $\frac{1}{1.2.3} + \frac{3}{2.3.4} + \frac{5}{3.4.5} + \dots \infty$

(Uttarakhand, II Semester, 2012)

Solution. Here,
$$u_n = \frac{2n - 1}{n(n + 1)(n + 2)} = \frac{1}{n^2} \frac{2 - \frac{1}{n}}{\left(1 + \frac{1}{n}\right)\left(1 + \frac{2}{n}\right)}$$

Let $v_n = \frac{1}{n^2}$,
$$\lim_{n \to \infty} \frac{u_n}{v_n} = \lim_{n \to \infty} \frac{2 - \frac{1}{n}}{\left(1 + \frac{1}{n}\right)\left(1 + \frac{2}{n}\right)} = 2$$
$$= \text{finite number.}$$

According to **Comparison Test** both series converge or diverge together, but $\Sigma\, v_n$ is convergent as $p = 2$.

$\therefore \quad \Sigma\, u_n$ is also convergent. **Ans.**

Example 17. *Test the following for convergence :* $\sum_{1}^{\infty} \frac{\sqrt[3]{3n^2 + 1}}{\sqrt[4]{4n^3 + 2n + 7}}$

Solution. Here,
$$u_n = \frac{(3n^2 + 1)^{1/3}}{(4n^3 + 2n + 7)^{1/4}} = \frac{3^{1/3} n^{2/3}\left[1 + \frac{1}{3n^2}\right]^{1/3}}{n^{3/4}\left[4 + \frac{2}{n^2} + \frac{7}{n^3}\right]^{1/4}}$$

$$= \frac{3^{1/3}\left[1+\frac{1}{3n^2}\right]^{1/3}}{n^{1/12}\left[4+\frac{2}{n^2}+\frac{7}{n^3}\right]^{1/4}}$$

Let $$v_n = \frac{1}{n^{1/12}}$$

By Comparison Test

$$\lim_{n\to\infty}\frac{u_n}{v_n} = \lim_{n\to\infty}\frac{3^{1/3}\left[1+\frac{1}{3n^2}\right]^{1/3}}{\left[4+\frac{2}{n^2}+\frac{7}{n^3}\right]^{1/4}} = \frac{3^{1/3}}{\sqrt{2}} = \text{finite number.}$$

According to Comparison Test both series converge or diverge together; but $\Sigma\, v_n$ is divergent as $p = \frac{1}{12} < 1$.

$\therefore\quad \Sigma u_n$ is also divergent. **Ans.**

Example 18. *Test for convergence the series whose nth term is* $n^{\log x}$.

Solution. The series is $1^{\log x} + 2^{\log x} + 3^{\log x} + 4^{\log x} + \ldots\ \infty$

$$= \frac{1}{1^{-\log x}} + \frac{1}{2^{-\log x}} + \frac{1}{3^{-\log x}} + \frac{1}{4^{-\log x}} + \ldots\ \infty$$

Put $-\log x = p$

$$= \frac{1}{1^p} + \frac{1}{2^p} + \frac{1}{3^p} + \frac{1}{4^p} + \ldots\ \infty$$

(*a*) which is convergent $p > 1$, *i.e.*, $-\log x > 1$

$$\Rightarrow\quad \log_e\frac{1}{x} > \log_e e \quad\Rightarrow\quad \frac{1}{x} > e \ \text{ or }\ x < \frac{1}{e}$$

(*b*) which is divergent if $p \le 1 \ \Rightarrow\ -\log x \le 1$

$$\Rightarrow\quad \log_e\frac{1}{x} \le \log_e e \quad\Rightarrow\quad \frac{1}{x} \le e \ \Rightarrow\ \frac{1}{e} \le x.$$

The given series is convergent if $x < \frac{1}{e}$

The given series is divergent if $x \ge \frac{1}{e}$. **Ans.**

Example 19. *Test the following series for convergence* $\frac{2}{1^p} + \frac{3}{2^p} + \frac{4}{3^p} + \frac{5}{4^p} + \ldots$

Solution. Given series is $\frac{2}{1^p} + \frac{3}{2^p} + \frac{4}{3^p} + \frac{5}{4^p} + \ldots$

Here $$u_n = \frac{n+1}{n^p} = \frac{1+\frac{1}{n}}{n^{p-1}}$$

Let $$v_n = \frac{1}{n^{p-1}} \quad\therefore\quad \frac{u_n}{v_n} = \frac{1+\frac{1}{n}}{n^{p-1}} \times \frac{n^{p-1}}{1} = 1 + \frac{1}{n}$$

$$\lim_{n\to\infty}\frac{u_n}{v_n} = 1$$

Therefore, both the series are either convergent or divergent.

But Σv_n is convergent if $p - 1 > 1$, *i.e.,* if $p > 2$ (P series)

and is divergent if $p - 1 \le 1$, *i.e.,* if $p \le 2$

$\therefore$ The given series is convergent if $p > 2$ and divergent if $p \le 2$. **Ans.**

Example 20. *Test the series for convergence* $1 + \frac{1}{2^2} + \frac{2^2}{3^3} + \frac{3^3}{4^4} + \ldots$

Solution. Let us consider $\frac{1}{2^2} + \frac{2^2}{3^3} + \frac{3^3}{4^4} + \ldots$

Here
$$u_n = \frac{n^n}{(n+1)^{n+1}} = \frac{n^n}{(n+1)(n+1)^n}$$

$$= \frac{n^n}{n\left(1+\frac{1}{n}\right) \times n^n\left(1+\frac{1}{n}\right)^n} = \frac{1}{n\left(1+\frac{1}{n}\right)\left(1+\frac{1}{n}\right)^n}$$

$$\frac{u_n}{v_n} = \frac{1}{\left(1+\frac{1}{n}\right)\left(1+\frac{1}{n}\right)^n}. \quad \text{Let } v_n = \frac{1}{n}$$

By Comparison Test

$$\lim_{n \to \infty} \frac{u_n}{v_n} = \lim_{n \to \infty} \frac{1}{\left(1+\frac{1}{n}\right)\left(1+\frac{1}{n}\right)^n} = \frac{1}{e} = \text{finite quantity.}$$

Hence, either both the series are convergent or both the series are divergent. But Σv_n is divergent as $p = 1$.

$\therefore$ The given series is divergent. **Ans.**

EXERCISE 7.5

Examine the convergence or divergence of the following series:

1. $2 + \frac{3}{2}.\frac{1}{4} + \frac{4}{3}.\frac{1}{4^2} + \frac{5}{4}.\frac{1}{4^3} + \ldots \infty$ **Ans.** Convergent
2. $1 + \frac{1.2}{1.3} + \frac{1.2.3}{1.3.5} + \frac{1.2.3.4}{1.3.5.7} + \ldots \infty$ **Ans.** Convergent
3. $1 + \frac{3}{2!} + \frac{5}{3!} + \frac{7}{4!} + \ldots \infty$ **Ans.** Convergent
4. $\frac{1}{1.2} + \frac{2}{3.4} + \frac{3}{5.6} + \ldots \infty$ **Ans.** Divergent
5. $\frac{1}{1.3} + \frac{2}{3.5} + \frac{3}{5.7} + \ldots \infty$ **Ans.** Divergent
6. $\frac{1}{1.2.3} + \frac{1}{2.3.4} + \frac{1}{3.4.5} + \ldots\ldots\ldots \infty$ **Ans.** Convergent (*M.D. University, Dec. 2004*)
7. $1 + \frac{2^2}{2!} + \frac{3^2}{3!} + \frac{4^2}{4!} + \ldots \infty$ **Ans.** Convergent
8. $1 + \frac{1}{2^2} + \frac{2^2}{3^3} + \frac{3^3}{4^4} + \ldots \infty$ **Ans.** Divergent
9. $\frac{1}{3} + \frac{2!}{3^2} + \frac{3!}{3^3} + \ldots \infty$ **Ans.** Convergent

10. $\sum_{n=1}^{\infty} \frac{1}{\sqrt{n}+\sqrt{n+1}}$ **Ans.** Divergent

11. $\sum_{n=1}^{\infty} \frac{2n^3+5}{4n^5+1}$ **Ans.** Convergent

12. $\sum_{n=1}^{\infty} \frac{a^n}{x^n+n^a}$ **Ans.** If $x > a$, convergent; if $x \le a$, Divergent

13. $\sum_{n=1}^{\infty} \frac{\sqrt{n}}{n^2+1}$ **Ans.** Convergent

14. $\sum_{n=1}^{\infty} \sqrt{(n^2+1)} - n$ **Ans.** Divergent

15. $\sum_{n=1}^{\infty} \left[\sqrt{(n^4+1)} - \sqrt{(n^4-1)}\right]$ **Ans.** Convergent

16. $\sum_{n=1}^{\infty} \frac{n!}{5^n}$ **Ans.** Divergent

17. $\sum_{n=1}^{\infty} \frac{2^n+1}{3^n+n}$ **Ans.** Convergent

18. $\sum_{n=1}^{\infty} \frac{n^n}{n!}$ **Ans.** Convergent

19. $\sum_{n=1}^{\infty} \frac{n^2}{e^n}$ **Ans.** Convergent

20. $\sum_{n=1}^{\infty} \frac{x^n}{1+x^{2n}}$ $(x > 0)$ **Ans.** If $x > 1$, convergent; if $x \le 1$, divergent

21. $\sum_{n=1}^{\infty} \frac{1}{x^n+x^{-n}}$ **Ans.** If $x \le 1$, convergent; divergent if $x = 1$

22. $\sum_{n=1}^{\infty} \sin\frac{1}{n}$ **Ans.** Divergent

23. $\frac{(n^2+n-1)^{1/2}}{(n^5-2)^{1/2}}$ **Ans.** Convergent

7.16 D'ALEMBERT'S RATIO TEST

Statement. ***If Σu_n is a positive term series such that $\lim_{n \to \infty} \frac{u_{n+1}}{u_n} = k$ then***

(i) the series is convergent if $k < 1$ (ii) the series is divergent if $k > 1$

Solution.

Case I. When $\lim_{n \to \infty} \frac{u_{n+1}}{u_n} = k < 1$

By definition of a limit, we can find a number r (< 1) such that

$$\frac{u_{n+1}}{u_n} < r \text{ for all } n \ge m \qquad \left[\frac{u_2}{u_1} < r, \frac{u_3}{u_2} < r, \frac{u_4}{u_3} < r \ldots\right]$$

Omitting the first m terms, let the series be

$$u_1 + u_2 + u_3 + u_4 + \ldots \infty$$

$$= u_1\left(1 + \frac{u_2}{u_1} + \frac{u_3}{u_1} + \frac{u_4}{u_1} + \ldots\right) = u_1\left(1 + \frac{u_2}{u_1} + \frac{u_3}{u_2}.\frac{u_2}{u_1} + \frac{u_4}{u_3}.\frac{u_3}{u_2}.\frac{u_2}{u_1} + \ldots \infty\right)$$

$$< u_1 (1 + r + r^2 + r^3 + \ldots \infty) \qquad (r < 1)$$

$$= \frac{u_1}{1-r}, \text{ which is a finite quantity.}$$

Hence, Σu_n is convergent.

Case 2. When $\lim_{n \to \infty} \frac{u_{n+1}}{u_n} = k > 1$

By definition of limit, we can find a number m such that $\dfrac{u_{n+1}}{u_n} \geq 1$ for all $n \geq m$

$$\frac{u_2}{u_1} > 1, \qquad \frac{u_3}{u_2} > 1, \qquad \frac{u_4}{u_3} > 1$$

Ignoring the first m terms, let the series be

$$u_1 + u_2 + u_3 + u_4 + \ldots \infty$$

$$= u_1\left(1 + \frac{u_2}{u_1} + \frac{u_3}{u_1} + \frac{u_4}{u_1} + \ldots\right) = u_1\left(1 + \frac{u_2}{u_1} + \frac{u_3}{u_2}.\frac{u_2}{u_1} + \frac{u_4}{u_3}.\frac{u_3}{u_2}.\frac{u_2}{u_1} + \ldots \infty\right)$$

$$\geq u_1 (1 + 1 + 1 + 1 \ldots \text{ to } n \text{ terms}) = nu_1$$

$$\left[\because \lim_{n \to \infty} (u_1 + u_2 + \ldots + u_n) = nu_1\right]$$

$$\lim_{n \to \infty} S_n \geq \lim_{n \to \infty} nu_1 = \infty$$

Hence, $\Sigma\, u_n$ is divergent.

Note. When $\dfrac{u_{n+1}}{u_n} = 1 \quad (k = 1)$

The ratio test fails.

For Example. Consider the series whose nth term $= \dfrac{1}{n}$

$$\lim_{n \to \infty} \frac{u_{n+1}}{u_n} = \lim_{n \to \infty} \frac{\frac{1}{n+1}}{\frac{1}{n}} = \lim_{n \to \infty} \frac{n}{n+1} = \lim_{n \to \infty} \frac{1}{1 + \frac{1}{n}} = 1 \qquad \ldots(1)$$

Consider the second series whose nth term is $\dfrac{1}{n^2}$.

$$\lim_{n \to \infty} \frac{u_{n+1}}{u_n} = \lim_{n \to \infty} \frac{\frac{1}{(n+1)^2}}{\frac{1}{n^2}} = \lim_{n \to \infty} \left(\frac{n}{n+1}\right)^2 = 1 \qquad \ldots(2)$$

Thus, from (1) and (2) in both cases $\displaystyle\lim_{n \to \infty} \frac{u_{n+1}}{u_n} = 1$

ut we know that the first series is divergent as $p = 1$.

The second series is convergent as $p = 2$.

Hence, when $\displaystyle\lim_{n \to \infty} \frac{u_{n+1}}{u_n} = 1$, the series may be convergent or divergent.

Thus, ratio test fails when $k = 1$.

Example 21. *Test for convergence of the series whose nth term is* $\dfrac{n^2}{2^n}$.

Solution. Here, we have $u_n = \dfrac{n^2}{2^n}, \qquad u_{n+1} = \dfrac{(n+1)^2}{2^{n+1}}$

By D'Alembert's Test

$$\lim_{n \to \infty} \frac{u_{n+1}}{u_n} = \lim_{n \to \infty} \frac{(n+1)^2}{2^{n+1}}.\frac{2^n}{n^2} = \lim_{n \to \infty} \frac{1}{2}\left(1 + \frac{1}{n}\right)^2 = \frac{1}{2} < 1$$

Hence, the series is convergent by D'Alembert's Ratio Test. **Ans.**

Example 22. *Test for convergence the series whose nth term is* $\dfrac{2^n}{n^3}$

Solution. Here, we have $u_n = \dfrac{2^n}{n^3}$, $\quad u_{n+1} = \dfrac{2^{n+1}}{(n+1)^3}$

By D'Alembert's Ratio Test

$$\frac{u_{n+1}}{u_n} = \frac{2^{n+1}}{(n+1)^3}\cdot\frac{n^3}{2^n} = \frac{2}{\left(1+\dfrac{1}{n}\right)^3} \Rightarrow \lim_{n\to\infty}\frac{u_{n+1}}{u_n} = \lim_{n\to\infty}\frac{2}{\left(1+\dfrac{1}{n}\right)^3} = 2 > 1$$

Hence, the series is divergent. **Ans.**

Example 23. *Discuss the convergence of the series:*

$$\sum \frac{\sqrt{n}}{\sqrt{n^2+1}}x^n \qquad (x > 0)$$ *(M.D.University, Dec., 2001)*

Solution. Here, we have

$$u_n = \sqrt{\frac{n}{n^2+1}}\,x^n$$

$$\therefore \qquad u_{n+1} = \sqrt{\frac{n+1}{(n+1)^2+1}}.x^{n+1}$$

$$\frac{u_n}{u_{n+1}} = \sqrt{\frac{n}{n+1}}.\sqrt{\frac{n^2+2n+2}{n^2+1}}.\frac{1}{x} = \sqrt{\frac{1}{1+\dfrac{1}{n}}.\frac{\left(1+\dfrac{2}{n}+\dfrac{2}{n^2}\right)}{\left(1+\dfrac{1}{n^2}\right)}}.\frac{1}{x}$$

$$\lim_{n\to\infty}\frac{u_n}{u_{n+1}} = \lim_{n\to\infty}\sqrt{\frac{1}{1+\dfrac{1}{n}}.\frac{\left(1+\dfrac{2}{n}+\dfrac{2}{n^2}\right)}{\left(1+\dfrac{1}{n^2}\right)}}.\frac{1}{x} = \frac{1}{x}$$

$\therefore$ By D' Alembert's Ratio Test, $\sum u_n$ converges if $\dfrac{1}{x} > 1$, *i.e.* $x < 1$ and diverges if $\dfrac{1}{x} < 1$ *i.e.*, $x > 1$.

When $x = 1$, the Ratio Test fails.

When $x = 1$, $u_n = \sqrt{\dfrac{n}{n^2+1}} = \sqrt{\dfrac{n}{n^2\left(1+\dfrac{1}{n^2}\right)}} = \dfrac{1}{\sqrt{n}}.\dfrac{1}{\sqrt{1+\dfrac{1}{n^2}}}$

$$v_n = \frac{1}{\sqrt{n}},$$

$$\frac{u_n}{v_n} = \frac{1}{\sqrt{n}}.\frac{1}{\sqrt{1+\dfrac{1}{n^2}}}.\frac{\sqrt{n}}{1} = \frac{1}{\sqrt{1+\dfrac{1}{n^2}}}$$

$$\lim_{n\to\infty}\frac{u_n}{v_n} = \lim_{n\to\infty}\frac{1}{\sqrt{1+\dfrac{1}{n^2}}} = 1$$

Which is finite and non-zero.

$\therefore$ By comparison test, $\sum u_n$ and $\sum v_n$ converge or diverge together.

Since $\sum v_n = \sum \frac{1}{\sqrt{n}}$ is of the form $\sum \frac{1}{n^p}$ with $p = \frac{1}{2} < 1$.

$\sum v_n$ diverges $\Rightarrow \sum u_n$ diverges.

Hence, the given series $\sum u_n$ converges as $x < 1$ and diverges if $x \geq 1$. **Ans**

Example 24. *By D'Alembert's ratio test, discuss the convergence of the series,*

$$\frac{1^2 . 2^2}{1!} + \frac{2^2 . 3^2}{2!} + \frac{3^2 . 4^2}{3!} + ...$$

(Uttarakhand, IInd Semester, June 2010)

Solution. Here we have

$$\Sigma u_n = \frac{1^2 . 2^2}{1!} + \frac{2^2 . 3^2}{2!} + \frac{3^2 . 4^2}{3!} + ... + \frac{n^2 (n+1)^2}{n!} = ...$$

$$u_n = \frac{n^2 (n+1)^2}{n!}$$

$$u_{n+1} = \frac{(n+1)^2 (n+2)^2}{(n+1)!}$$

By D' Alembert's Ratio test

$$\frac{u_{n+1}}{u_n} = \frac{(n+1)^2 (n+2)^2}{(n+1)!} \cdot \frac{n!}{n^2 .(n+1)^2} = \frac{(n+2)^2}{(n+1)!} \frac{n!}{n^2} = \frac{(n+2)^2}{(n+1)n^2} = \frac{\left(1+\frac{2}{n}\right)^2}{n+1}$$

$$\lim_{n \to \infty} \frac{u_{n+1}}{u_n} = \lim_{n \to \infty} \frac{\left(1+\frac{1}{n}\right)^2}{n+1}$$

$= 0 < 1$

Hence, the series is convergent by D' Alemberts Ratio test. **Ans.**

EXERCISE 7.6

Test the convergence for series:

1. $\sum_{n=1}^{n} \frac{n^2}{3^n}$ **Ans.** Convergent

2. $\sum_{n=1}^{n} \frac{n!}{n^n}$ **Ans.** Convergent

3. $\left(\frac{1}{3}\right)^2 + \left(\frac{1.2}{3.5}\right)^2 + \left(\frac{1.2.3}{3.5.7}\right)^2 + ... \infty$ **Ans.** Convergent

4. $\frac{2}{1} + \frac{2.5.8}{1.5.9} + \frac{2.5.8.11}{1.5.9.13} + ... \infty$ **Ans.** Convergent

5. $\sum_{n=1}^{n} \frac{n!.2^n}{n^n}$ **Ans.** Convergent

6. $\sum_{n=1}^{\infty} \frac{x^{n-1}}{n.3^n}$ **Ans.** Convergent if $x > 3$, Divergent if $x < 3$

7. $\frac{1}{1+2} + \frac{2}{1+2^2} + \frac{3}{1+2^3} + \ldots$ **Ans.** Convergent *(M.D. University, 2001)*

8. Prove that, if $u_{n+1} = \frac{k}{1+u_n}$, where $k > 0$, $u_1 > 0$, then the series $\Sigma\, u_n$ converges to the positive root of the equation $x^2 + x = k$.

9. Prove that, if $u_{n+1} = \sqrt{k + u_n}$, where $k > 0$, $u_1 > 0$, then the series $\Sigma\, u_n$ converges to the positive root of the equation $x^2 - x = k$

10. Prove that, if $u_{n+1} = u_n^2 + k - k^2$, where $u_1 > 0$ and u_n tends to a finite limit then l must be either k or $1 - k$.

7.17 RAABE'S TEST (HIGHER RATIO TEST)

If $\Sigma\, u_n$ is a positive term series such that $\lim_{n \to \infty} n\left(\frac{u_n}{u_{n+1}} - 1\right) = k$, then

(i) the series is convergent if $k > 1$ (ii) the series is divergent if $k < 1$.

Proof. Case I. $k > 1$

Let p be such that $k > p > 1$ and compare the given series $\Sigma\, u_n$ with $\sum \frac{1}{n^p}$ which is convergent as $p > 1$.

$$\frac{u_n}{u_{n+1}} > \frac{(n+1)^p}{n^p} \quad \text{or} \quad \left(\frac{u_n}{u_{n+1}}\right) > \left(1 + \frac{1}{n}\right)^p > 1 + \frac{p}{n} + \frac{p(p-1)}{2!}\frac{1}{n^2} + \ldots$$

(Binomial Theorem)

$$n\left(\frac{u_n}{u_{n+1}} - 1\right) > p + \frac{p\,(p-1)}{2!}\frac{1}{n} + \ldots.$$

If $$\lim_{n \to \infty} n\left(\frac{u_n}{u_{n+1}} - 1\right) > p$$

and $k > p$ which is true as $k > p > 1$; Σu_n is convergent when $k > 1$.

Case II. $k < 1$ Same steps as in Case I.

Notes:

1. Raabe's Test fails if $k = 1$
2. Raabe's Test is applied only when D'Alembert's Ratio Test fails.

Example 25. *Test the convergence for the series* $\frac{x}{1.2} + \frac{x^2}{3.4} + \frac{x^3}{5.6} + \frac{x^4}{7.8} + \ldots$ *(M.U. 2009)*

Solution. Here, $$u_n = \frac{x^n}{(2n-1)\,2n} \quad \text{and} \quad u_{n+1} = \frac{x^{n+1}}{(2n+1)\,(2n+2)}$$

By D'Alembert's Test

$$\frac{u_{n+1}}{u_n} = \frac{x^{n+1}}{(2n+1)\,(2n+2)} \times \frac{(2n-1)2n}{x^n} = \frac{x\left(1 - \frac{1}{2n}\right)}{\left(1 + \frac{1}{2n}\right)\left(1 + \frac{2}{2n}\right)}$$

$$\lim_{n \to \infty} \frac{u_{n+1}}{u_n} = x$$

(*i*) If $x < 1$, $\Sigma\, u_n$ is convergent (*ii*) If $x > 1$, $\Sigma\, u_n$ is divergent (*iii*) If $x = 1$, Test fails.

Let us apply **Raabe's Test** when $x = 1$

$$\lim_{n\to\infty} n\left(\frac{u_n}{u_{n+1}}-1\right) = \lim_{n\to\infty} n\left[\frac{(2n+1)(2n+2)}{2n(2n-1)}-1\right]$$

$$= \lim_{n\to\infty} n\left[\frac{(2n+1)(2n+2)-2n(2n-1)}{2n\,(2n-1)}\right]$$

$$= \lim_{n\to\infty} n\left[\frac{(8n+2)}{(2n)\,(2n-1)}\right] = \lim_{n\to\infty} \frac{\left(2+\frac{2}{4n}\right)}{1\left(1-\frac{1}{2n}\right)} = 2$$

So the series is convergent.

Hence we can say that the given series is convergent if $x \le 1$ and divergent, if $x > 1$. **Ans.**

Example 26. *Test the following series for convergence* $\sum \frac{1}{\sqrt{n+1}-1}$.

Solution. Here, $u_n = \frac{1}{\sqrt{n+1}-1}$, $u_{n+1} = \frac{1}{\sqrt{n+2}-1}$

$$\frac{u_{n+1}}{u_n} = \frac{\sqrt{n+1}-1}{\sqrt{n+2}-1}$$

$$\lim_{n\to\infty} \frac{u_{n+1}}{u_n} = \lim_{n\to\infty} \frac{\sqrt{1+\frac{1}{n}-\frac{1}{n^2}}}{\sqrt{1+\frac{2}{n}-\frac{1}{n^2}}} = 1$$

D'Alembert's test fails.

By Raabe's test.

$$\lim_{n\to\infty} n\left(\frac{u_n}{u_{n+1}}-1\right) = \lim_{n\to\infty} n\left(\frac{\sqrt{n+2}-1}{\sqrt{n+1}-1}-1\right)$$

By D'Alembert's Test

$$= \lim_{n\to\infty} n\left[\frac{\sqrt{n+2}-1-\sqrt{n+1}+1}{\sqrt{n+1}-1}\right] = \lim_{n\to\infty} n\left[\frac{\sqrt{n+2}-\sqrt{n+1}}{\sqrt{n+1}-1}\right]$$

$$= \lim_{n\to\infty} n\left[\frac{\sqrt{1+\frac{2}{n}}-\sqrt{1+\frac{1}{n}}}{\sqrt{1+\frac{1}{n}-\frac{1}{n^2}}}\right] = 0 < 1$$

Hence, $\Sigma\, u_n$ is divergent. **Ans.**

Example 27. *Discuss the convergence of the series,*

$$\frac{1}{2}+\frac{1\cdot 3}{2\cdot 4}+\frac{1\cdot 3\cdot 5}{2\cdot 4\cdot 6}+\ldots\ldots$$ *Uttarakhand, IInd Semester, Summer 2010*

Solution. Here we have

$$\Sigma u_n = \frac{1}{2}+\frac{1\cdot 3}{2\cdot 4}+\frac{1\cdot 3\cdot 5}{2\cdot 4\cdot 6}+\ldots\ldots$$

$$u_n = \frac{1\cdot 3\cdot 5\ldots\ldots(2n-1)}{2\cdot 4\cdot 6\ldots\ldots.2n}$$

$$u_{n+1} = \frac{1\cdot 3\cdot 5......(2n-1)(2n+1)}{2\cdot 4\cdot 6.......(2n).(2n+2)}$$

$$\frac{u_{n+1}}{u_n} = \frac{1\cdot 3\cdot 5......(2n-1)(2n+1)}{2\cdot 4\cdot 6.......(2n).(2n+2)} \times \frac{2\cdot 4\cdot 6......2n}{1\cdot 3\cdot 5.......(2n-1)} = \frac{(2n+1)}{(2n+2)}$$

$$\lim_{n\to\infty} \frac{u_{n+1}}{u_n} = \lim_{n\to\infty} \frac{2n+1}{2n+2} = \lim_{n\to\infty} \frac{1+\frac{1}{2n}}{1+\frac{2}{2n}} = 1$$

D; Alembert's test fails. Let us apply Raabe's test.

$$\lim_{n\to\infty}\left(\frac{u_n}{u_n+1}-1\right) = \lim_{n\to\infty} n\left(\frac{2n+2}{2n+1}-1\right) = \lim_{n\to\infty} \frac{n(2n+2-2n-1)}{2n+1}$$

$$= \lim_{n\to\infty} \frac{n}{2n+1} = \lim_{n\to\infty} \frac{1}{2+\frac{1}{n}} = \frac{1}{2} < 1$$

Hence the series is convergent by Raabe's test. **Ans.**

Example 28. *Discuss the convergence of the series:*

$$\frac{x}{1} + \frac{1}{2}\cdot\frac{x^3}{3} + \frac{1\cdot 3}{2\cdot 4}\frac{x^5}{5} + \frac{1\cdot 3\cdot 5}{2\cdot 4\cdot 6}.\frac{x^7}{7} + ...\ (x > 0)$$ *(M.D.U. Dec., 2001)*

Solution. Here, we have

$$\frac{x}{1} + \frac{1}{2}\cdot\frac{x^3}{3} + \frac{1\cdot 3}{2\cdot 4}\frac{x^5}{5} + \frac{1\cdot 3\cdot 5}{2\cdot 4\cdot 6}.\frac{x^7}{7} + ...$$

Neglecting the first term, we have

$$u_n = \frac{1\cdot 3\cdot 5...(2n-1)}{2\cdot 4\cdot 5...(2n)} \frac{x^{2n+1}}{2n+1}$$

and $$u_{n+1} = \frac{1\cdot 3\cdot 5...(2n-1)(2n+1)}{2\cdot 4\cdot 5...(2n)(2n+2)}.\frac{x^{2n+3}}{2n+3}$$

$$\frac{u_n}{u_{n+1}} = \frac{(2n-1)\,x^{2n+1}}{2n\,(2n+1)} \times \frac{2n\,(2n+2)\,(2n+3)}{(2n-1)\,(2n+1)\,x^{2n+3}}$$

$$= \frac{(2n+2)\,(2n+3)}{(2n+1)\,(2n+1)}\,\frac{1}{x^2}$$

$$\therefore \quad = \frac{2n+2}{2n+1}.\frac{2n+3}{2n+1}.\frac{1}{x^2} = \frac{2n\left(1+\frac{1}{n}\right).2n\left(1+\frac{3}{2n}\right)}{2n\left(1+\frac{1}{2n}\right).2n\left(1+\frac{1}{2n}\right)}.\frac{1}{x^2} = \frac{\left(1+\frac{1}{n}\right)\left(1+\frac{3}{2n}\right)}{\left(1+\frac{1}{2n}\right)^2}.\frac{1}{x^2}$$

$$\lim_{n\to\infty} \frac{u_n}{u_{n+1}} = \lim_{n\to\infty} \frac{\left(1+\frac{1}{n}\right)\left(1+\frac{3}{2n}\right)}{\left(1+\frac{1}{2n}\right)^2}.\frac{1}{x^2} = \frac{1}{x^2}$$

$\therefore$ Ratio Test, $\sum u_n$ is convergent if $\dfrac{1}{x^2} > 1$.

i.e.; $x^2 < 1$ and divergent if $\dfrac{1}{x^2} < 1$. *i.e.,* $x^2 > 1$.

If $x^2 = 1$, then Ratio Test fails.

Now Raabe's test

When $x^2 = 1$, we have $\quad \dfrac{u_n}{u_{n+1}} = \dfrac{(2n+2)(2n+3)}{(2n+1)^2} = \dfrac{4n^2+10n+6}{4n^2+4n+1}$

$$\lim_{n\to\infty} n\left(\frac{u_n}{u_{n+1}} - 1\right) = \lim_{n\to\infty} n\left(\frac{4n^2+10n+6}{4n^2+4n+1} - 1\right)$$

$$= \lim_{n\to\infty} \frac{6n^2+5n}{4n^2+4n+1} = \lim_{n\to\infty} \frac{6+\frac{5}{n}}{4+\frac{4}{n}+\frac{1}{n^2}} = \frac{6}{4} = \frac{3}{2} > 1$$

$\therefore$ By Raabe's Test, the series converges.

Hence, $\sum u_n$ is convergent if $x^2 \le 1$ and divergent if $x^2 > 1$. **Ans.**

Example 29. *Test for convergence the series for positive values of x:*

$$1 + \frac{2}{5}x + \frac{6}{9}x^2 + \frac{14}{17}x^3 + \dots + \frac{2^n-2}{2^n+1}x^{n-1} + \dots$$

Solution. Here, $\quad u_n = \dfrac{2^n-2}{2^n+1}x^{n-1}$ and $u_{n+1} = \dfrac{2^{n+1}-2}{2^{n+1}+1}x^n$

By D'Alembert's Ratio Test

$$\frac{u_{n+1}}{u_n} = \left(\frac{2^{n+1}-2}{2^{n+1}+1}x^n\right)\left(\frac{2^n+1}{2^n-2}\frac{1}{x^{n-1}}\right)$$

$$\lim_{n\to\infty} \frac{u_{n+1}}{u_n} = \lim_{n\to\infty} \frac{2^{n+1}\left(1-\frac{1}{2^n}\right)x}{2^{n+1}\left(1+\frac{1}{2^{n+1}}\right)}\left[\frac{2^n\left(1+\frac{1}{2^n}\right)}{2^n\left(1-\frac{1}{2^{n-1}}\right)}\right]$$

$$\lim_{n\to\infty} \frac{u_{n+1}}{u_n} = x$$

(*i*) If $x < 1$, then Σu_n is convergent. (*ii*) If $x > 1$, then Σu_n is divergent.

(*iii*) If $x = 1$, then test fails.

Let us apply **Raabe's Test** when $x = 1$

$$\lim_{n\to\infty} n\left(\frac{u_n}{u_{n+1}} - 1\right) = \lim_{n\to\infty} n\left[\left(\frac{2^n-2}{2^n+1}\right)\left(\frac{2^{n+1}+1}{2^{n+1}-2}\right) - 1\right]$$

$$= \lim_{n\to\infty} n[1-1] = 0 < 1$$

so the series is divergent.

Hence, the given series is divergent if $x \ge 1$ and convergent if $x < 1$. **Ans.**

Example 30. *Test the following series for convergence*

$$1+\frac{3}{7}x+\frac{3.6}{7.10}x^2+\frac{3.6.9}{7.10.13}x^3+\frac{3.6.9.12}{7.10.13.16}x^4+\ldots$$

Solution. Here,
$$u_n = \frac{3.6.9.12\ldots 3(n-1)}{7.10.13.16\ldots(3n+1)}x^{n-1}$$
$$u_{n+1} = \frac{3.6.9.12\ldots 3(n-1).3n}{7.10.13.16\ldots(3n+1)(3n+4)}x^n$$

By D'Alembert's Test

$$\frac{u_{n+1}}{u_n} = \frac{3n}{3n+4}x, \qquad \lim_{n\to\infty}\frac{u_{n+1}}{u_n} = \lim_{n\to\infty}\frac{3}{3+\frac{4}{n}}x = x$$

If $x < 1$, then Σu_n is convergent. If $x > 1$, then Σu_n is divergent.
If $x = 1$, then the test fails.

By Raabe's Test

$$\lim_{n\to\infty} n\left(\frac{u_n}{u_{n+1}}-1\right) = \lim_{n\to\infty} n\left(\frac{3n+4}{3n}-1\right) = \lim_{n\to\infty} n\frac{4}{3n} = \frac{4}{3} > 1$$

Hence, the series is convergent if $x \le 1$ and divergent if $x > 1$. **Ans.**

Example 31. *Test the following series for convergence*

$$\frac{1}{2}x+x^2+\frac{9}{8}x^3+x^4+\frac{25}{32}x^5+\ldots\infty$$

Solution. Here,
$$u = \frac{n^2.x^n}{2^n}, \qquad u_{n+1} = \frac{(n+1)^2.x^{n+1}}{2^{n+1}}$$

By D'Alembert's Test

$$\frac{u_{n+1}}{u_n} = \frac{(n+1)^2x^{n+1}}{2^{n+1}}.\frac{2^n}{n^2x^n} = \left(\frac{n+1}{n}\right)^2\frac{x}{2}$$

$$\lim_{n\to\infty}\frac{u_{n+1}}{u_n} = \lim_{n\to\infty}\left(1+\frac{1}{n}\right)^2\frac{x}{2} = \frac{x}{2}$$

(*i*) If $\frac{x}{2} < 1$ or $x < 2$, then Σu_n is convergent. (*ii*) If $\frac{x}{2} > 1$ or $x > 2$, then Σu_n is divergent.

(*iii*) If $\frac{x}{2} = 1$ or $x = 2$, then the test fails

Let us apply **Raabe's test**

$$n\left(\frac{u_n}{u_{n+1}}-1\right) = n\left[\frac{n^2}{(n+1)^2}\frac{2}{2}-1\right] = n\left[\frac{n^2-n^2-2n-1}{(n+1)^2}\right] = \frac{-2n^2-n}{(n+1)^2}$$

$$\lim_{n\to\infty} n\left(\frac{u_n}{u_{n+1}}-1\right) = \lim_{n\to\infty}\frac{-2-\frac{1}{n}}{\left(1+\frac{1}{n}\right)^2} = -2 < 1$$

Hence, Σu_n is divergent if $x \ge 2$, and convergent if $x < 2$. **Ans.**

Example 32. *Show that the series* $\frac{1}{x} + \frac{2!}{x(x+1)} + \frac{3!}{x(x+1)(x+2)} + \ldots$ *converges if* $x > 2$ *and diverges if* $x < 2$.

Solution. Here,
$$u_n = \frac{n!}{x(x+1)(x+2)\ldots(x+n-1)}$$
$$u_{n+1} = \frac{(n+1)!}{x(x+1)(x+2)\ldots(x+n-1)(x+n)}$$

By D'Alembert's test

$$\frac{u_{n+1}}{u_n} = \frac{n+1}{(x+n)}, \qquad \lim_{n\to\infty}\frac{u_{n+1}}{u_n} = \frac{1+\frac{1}{n}}{1+\frac{x}{n}} = 1$$

Test fails. Let us apply **Raabe's Test.**

$$\lim_{n\to\infty} n\left(\frac{u_n}{u_{n+1}} - 1\right) = \lim_{n\to\infty} n\left(\frac{x+n}{n+1} - 1\right) = \lim_{n\to\infty} n\left(\frac{x-1}{n+1}\right) = \lim_{n\to\infty}\frac{x-1}{1+\frac{1}{n}} = x-1$$

If $x - 1 > 1$ or $x > 2$, then $\Sigma\, u_n$ is convergent.
If $x - 1 < 1$ or $x < 2$, then $\Sigma\, u_n$ is divergent. **Ans.**

Example 33. *Discuss the convergence of the series* $\frac{x^2}{2\log 2} + \frac{x^3}{3\log 3} + \frac{x^4}{4\log 4} + \ldots$

Solution. Here, we have $\frac{x^2}{2\log 2} + \frac{x^3}{3\log 3} + \frac{x^4}{4\log 4} + \ldots$

$$u_n = \frac{x^{n+1}}{(n+1)\log(n+1)}, \qquad u_{n+1} = \frac{x^{n+2}}{(n+2)\log(n+2)}$$

By D'Alembert's Test

$$\lim_{n\to\infty}\frac{u_{n+1}}{u_n} = \lim_{n\to\infty}\frac{x^{n+2}}{(n+2)\log(n+2)} \times \frac{(n+1)\log(n+1)}{x^{n+1}}$$
$$= \lim_{n\to\infty} x\left(\frac{n+1}{n+2}\right)\frac{\log(n+1)}{\log(n+2)}$$
$$= \lim_{n\to\infty} x\left(\frac{1+\frac{1}{n}}{1+\frac{2}{n}}\right)\frac{\log n + \log\left(1+\frac{1}{n}\right)}{\log n + \log\left(1+\frac{2}{n}\right)}$$
$$= \lim_{n\to\infty} x\left(\frac{1+\frac{1}{n}}{1+\frac{2}{n}}\right)\left[\frac{\log n + \frac{1}{n} - \frac{1}{2}\cdot\frac{1}{n^2} + \ldots}{\log n + \frac{2}{n} - \frac{1}{2}\cdot\frac{4}{n^2} + \ldots}\right]$$
$$= \lim_{n\to\infty} x\left(\frac{1+\frac{1}{n}}{1+\frac{2}{n}}\right)\left[\frac{1+\frac{1}{n\log n}+\ldots}{1+\frac{2}{n\log n}+\ldots}\right]$$
$$= x$$

(*i*) When $x < 1$, the series is convergent

(*ii*) When $x > 1$, the series is divergent.

(*iii*) When $x = 1$, the test fails.

Let us apply Raabe's Test

$$\frac{u_n}{u_{n+1}} = \left(\frac{n+2}{n+1}\right)\frac{\log(n+2)}{\log(n+1)} = \left(\frac{n+2}{n+1}\right)\frac{\log n + \log\left(1+\frac{2}{n}\right)}{\log n + \log\left(1+\frac{1}{n}\right)}$$

$$= \left(\frac{n+2}{n+1}\right)\frac{\log n + \frac{2}{n} - \frac{1}{2}\cdot\frac{4}{n^2} + \ldots}{\log n + \frac{1}{n} - \frac{1}{2}\cdot\frac{1}{n^2} + \ldots} = \left(\frac{n+2}{n+1}\right)\frac{1 + \frac{2}{n\log n} + \ldots}{1 + \frac{1}{n\log n} + \ldots}$$

$$= \frac{n+2}{n+1}\left(1+\frac{2}{n\log n}\right)\left(1+\frac{1}{n\log n}\right)^{-1} = \frac{n+2}{n+1}\left(1+\frac{2}{n\log n}\right)\left(1-\frac{1}{n\log n}\right)$$

$$= \frac{n+2}{n+1}\left(1+\frac{2}{n\log n}-\frac{1}{n\log n}+\ldots\right) = \left(\frac{n+2}{n+1}\right)\left[1+\frac{1}{n\log n}\right]$$

$$\lim_{n\to\infty}\frac{u_n}{u_{n+1}} = \lim_{n\to\infty}\left[\frac{1+\frac{2}{n}}{1+\frac{1}{n}}\right]\left[1 + \frac{1}{n\log n}\right] = 1 + \frac{1}{n\log n}$$

$$n\left[\frac{u_n}{u_{n+1}} - 1\right] = n\left[1 + \frac{1}{n\log n} - 1\right] = \frac{1}{\log n} = 0 < 1$$

Thus the series is divergent when $x = 1$.

Hence, the series converges if $x < 1$ and diverges if $x \geq 1$. **Ans.**

Example 34. *Test the series for convergence*

$$1 + \frac{\alpha.\beta}{1.\gamma}x + \frac{\alpha(\alpha+1).\beta(\beta+1)}{1.2.\gamma(\gamma+1)}x^2 + \frac{\alpha(\alpha+1)(\alpha+2).\beta(\beta+1)(\beta+2)}{1.2.3.\gamma(\gamma+1)(\gamma+2)}x^3 + \ldots$$

Solution. $$u_n = \frac{\alpha(\alpha+1)(\alpha+2)\ldots[\alpha+(n-1)].\beta(\beta+1)\ldots[\beta+(n-1)]}{n!\,\gamma(\gamma+1)\ldots[\gamma+(n-1)]}x^n$$

$$u_{n+1} = \frac{\alpha(\alpha+1)(\alpha+2)\ldots[\alpha+(n-1)](\alpha+n).\beta(\beta+1)\ldots[\beta+(n-1)](\beta+n)}{(n+1)!\,\gamma(\gamma+1)\ldots[\gamma+(n-1)](\gamma+n)}x^{n+1}$$

By D'Alembert's Test

$$\lim_{n\to\infty}\frac{u_{n+1}}{u_n} = \lim_{n\to\infty}\frac{(\alpha+n)(\beta+n)}{(n+1)(\gamma+n)}x = \lim_{n\to\infty}\frac{\left(1+\frac{\alpha}{n}\right)\left(1+\frac{\beta}{n}\right)}{\left(1+\frac{1}{n}\right)\left(1+\frac{\gamma}{n}\right)}.x$$

$$\lim_{n\to\infty}\frac{u_{n+1}}{u_n} = x$$

(*i*) If $x < 1$, the series is convergent. (*ii*) If $x > 1$, the series is divergent.

(*iii*) If $x = 1$, the test fails.

Let us apply Raabe's Test

$$n\left(\frac{u_n}{u_{n+1}} - 1\right) = n\left[\frac{(n+1)(\gamma+n)}{(\alpha+n)(\beta+n)} - 1\right] = n\left[\frac{n\gamma+n^2+\gamma+n-\alpha\beta-n\alpha-n\beta-n^2}{(\alpha+n)(\beta+n)}\right]$$

$$\lim_{n\to\infty} n\left(\frac{u_n}{u_{n+1}} - 1\right) = \lim_{n\to\infty} \frac{\gamma + \frac{\gamma}{n} + 1 - \frac{\alpha\beta}{n} - \alpha - \beta}{\left(\frac{\alpha}{n}+1\right)\left(\frac{\beta}{n}+1\right)} = \gamma + 1 - \alpha - \beta$$

(*i*) If $\gamma + 1 - \alpha - \beta > 1$ or $\gamma > \alpha + \beta$, then $\Sigma\, u_n$ is convergent.

(*ii*) If $\gamma + 1 - \alpha - \beta < 1$ or $\gamma < \alpha + \beta$, then $\Sigma\, u_n$ is divergent. **Ans.**

Example 35. *Test for convergence the series* $\sum \frac{4.7. \ldots .(3n+1)}{1.2. \ldots .n} x^n$

Solution. $u_{n+1} = \sum \frac{4.7. \ldots (3n+1)(3n+4)}{1.2. \ldots n\,(n+1)} x^{n+1}$

By D'Alembert's Test

$$\frac{u_{n+1}}{u_n} = \frac{(3n+4)}{n+1}x, \quad \lim_{n\to\infty}\frac{u_{n+1}}{u_n} = \lim_{n\to\infty}\frac{\left(3+\frac{4}{n}\right)x}{1+\frac{1}{n}} = 3x$$

If $3x < 1 \Rightarrow x < \frac{1}{3}$, then $\Sigma\, u_n$ is convergent. If $3x > 1 \Rightarrow x > \frac{1}{3}$, then $\Sigma\, u_n$ is divergent.

If $3x = 1 \Rightarrow x = \frac{1}{3}$, the test fails.

By Raabe's Test

$$\lim_{n\to\infty} n\left(\frac{u_n}{u_{n+1}} - 1\right) = \lim_{n\to\infty} n\left(\frac{n+1}{n+\frac{4}{3}} - 1\right) = \lim_{n\to\infty} n\left(\frac{n+1-n-4/3}{n+4/3}\right)$$

$$= \lim_{n\to\infty}\frac{-1/3}{1+\frac{4}{3n}} = -\frac{1}{3} < 1$$

Thus, $\Sigma\, u_n$ is divergent.

Hence, the series is convergent if $x < \frac{1}{3}$ and divergent if $x \geq \frac{1}{3}$. **Ans.**

Example 36. *Test the series for convergence* $2x + \frac{3}{8}x^2 + \frac{4}{27}x^3 + \ldots + \frac{n+1}{n^3}x^n + \ldots \infty$

Solution. Here, $u_n = \frac{n+1}{n^3}x^n, \quad u_{n+1} = \frac{n+2}{(n+1)^3}x^{n+1}$

By D'Alembert's Test

$$\lim_{n\to\infty}\frac{u_{n+1}}{u_n} = \lim_{n\to\infty}\frac{(n+2)\,x^{n+1}}{(n+1)^3}\cdot\frac{n^3}{(n+1)x^n}$$

$$= \lim_{n\to\infty}\frac{n^3(n+2)x}{(n+1)^4} = \lim_{n\to\infty}\frac{1.\left(1+\frac{2}{n}\right)}{\left(1+\frac{1}{n}\right)^4}x = x$$

(*i*) If $x < 1$, $\Sigma\, u_n$ is convergent.

(*ii*) If $x > 1$, $\Sigma\, u_n$ is divergent.

(*iii*) If $x = 1$, the test fails.

Let us apply Raabe's Test when $x = 1$.

$$\lim_{n \to \infty} n\left(\frac{u_n}{u_{n+1}} - 1\right) = \lim_{n \to \infty} n\left[\frac{(n+1)^4}{n^3\,(n+2)} - 1\right]$$

$$= \lim_{n \to \infty} n\left[\frac{n^4 + 4n^3 + 6n^2 + 4n + 1 - n^4 - 2n^3}{n^4 + 2n^3}\right]$$

$$= \lim_{n \to \infty} \frac{2n^3 + 6n^2 + 4n + 1}{n^3 + 2n^2} = \lim_{n \to \infty} \frac{2 + \frac{6}{n} + \frac{4}{n^2} + \frac{1}{n^3}}{1 + \frac{2}{n}} = 2 > 1$$

Thus, the series is convergent for $x = 1$.
Hence, the series is convergent for $x \le 1$ and divergent if $x > 1$. **Ans.**

EXERCISE 7.7

Determine the nature of the following series:

1. $1 + \frac{2!}{2^2} + \frac{3!}{3^2} + \frac{4!}{4^2} + \ldots \infty$ **Ans.** Divergent
2. $\frac{1}{1} + \frac{1.3}{1.4} + \frac{1.3.5}{1.4.7} + \frac{1.3.5.7}{1.4.7.10} + \ldots \infty$ **Ans.** Convergent
3. $1 + \frac{1+\alpha}{1+\beta} + \frac{(1+\alpha)(2+\alpha)}{(1+\beta)(2+\beta)} + \ldots \infty$ **Ans.** If $\beta - \alpha > 1$, convergent. If $\beta - \alpha \le 1$, Divergent.
4. $\sum_{n=1}^{\infty} \frac{n^3}{e^n}$ **Ans.** Convergent
5. $x + \frac{2x^2}{2!} + \frac{3x^3}{3!} + \frac{4x^4}{4!} + \ldots \infty$ **Ans.** Convergent
6. $1 + \frac{x}{2} + \frac{x^2}{3} + \frac{x^3}{4} + \ldots$ **Ans.** Divergent
7. $1 + \frac{1}{2}x + \frac{1}{5}x^2 + \frac{1}{10}x^3 + \ldots$ **Ans.** Convergent if $-1 \le x < 1$ and divergent if $|x| > 1$
8. $1 + \frac{x}{2} + \frac{1.3}{2.4}x^2 + \frac{1.3.5}{2.4.6}x^3 + \ldots \infty$ **Ans.** If $x < 1$, convergent; and divergent if $x \ge 1$
9. $x + \frac{(2x)^2}{2!} + \frac{(3x)^3}{3!} + \frac{(4x)^4}{4!} + \frac{(5x)^5}{5!} + \ldots \infty$ **Ans.** If $x < \frac{1}{e}$, convergent; and divergent if $x \ge \frac{1}{e}$.
10. $1 + \frac{(1!)^2}{2!}x^2 + \frac{(2!)^2}{4!}x^4 + \frac{(3!)^2}{6!}x^6 + \ldots \infty \ (x > 0)$
    Ans. If $x^2 < 4$, convergent; and divergent if $x^2 \ge 4$

Find the values of x for which the following series converges:

11. $x^2 (\log 2)^q + x^3 (\log 3)^q + x^4 (\log 4)^q + \ldots$
    Ans. If $x < 1$, convergent; and divergent if $x \ge 1$
12. $\sum_{n=0}^{\infty} \frac{(-1)^n\, n!\, x^n}{10^n}$
13. $\sum \frac{x^n}{2n\,(2n+1)}$ **Ans.** If $x \le 1$, convergent; and if $x > 1$, divergent
14. $\sum \frac{1.2 \ldots n}{4.7 \ldots (3n+1)} x^n$ **Ans.** If $0 < x < 3$, convergent and divergent if $x \ge 3$.
15. $\sum \frac{1.2 \ldots n}{7.10 \ldots (3n+4)} x^n$ **Ans.** If $0 < x < 3$, convergent and divergent if $x \ge 3$.

7.18. GAUSS'S TEST

Statement. *If Σu_n is a positive term series such that*

$$\frac{u_n}{u_{n+1}} = \alpha + \frac{\beta}{n} + \frac{\gamma}{n^2} \quad \textit{where } \alpha > 0$$

(i) if $\alpha > 1$, convergent if $\alpha < 1$, divergent, whatever β may be

(ii) if $\alpha = 1$ and $\begin{cases} \beta > 1, \textit{ convergent} \\ \beta \le 1, \textit{ divergent} \end{cases}$

Example 37. *Test for convergence the series* $\frac{2^2.4^2}{3^2.5^2} + \frac{2^2.4^2.6^2}{3^2.5^2.7^2} + \frac{2^2.4^2.6^2.8^2}{3^2.5^2.7^2.9^2} + \ldots$

Solution. The given series is $\frac{2^2.4^2}{3^2.5^2} + \frac{2^2.4^2.6^2}{3^2.5^2.7^2} + \frac{2^2.4^2.6^2.8^2}{3^2.5^2.7^2.9^2} + \ldots$

$$+ \frac{2^2.4^2.6^2.8^2 \ldots (2n+2)^2}{3^2.5^2.7^2.9^2 \ldots (2n+3)^2} + \ldots \infty$$

$$u_{n+1} = \frac{2^2.4^2.6^2.8^2 \ldots (2n+2)^2\,(2n+4)^2}{3^2.5^2.7^2.9^2 \ldots (2n+3)^2\,(2n+5)^2}$$

By D'Alembert's Test

$$\frac{u_{n+1}}{u_n} = \frac{(2n+4)^2}{(2n+5)^2} = \frac{4n^2+16n+16}{4n^2+20n+25} = \frac{4 + \frac{16}{n} + \frac{16}{n^2}}{4 + \frac{20}{n} + \frac{25}{n^2}}$$

$$\lim_{n \to \infty} \frac{u_{n+1}}{u_n} = \lim_{n \to \infty} \frac{4 + \frac{16}{n} + \frac{16}{n^2}}{4 + \frac{20}{n} + \frac{25}{n^2}} = 1$$

D'Alembert's Test fails. Let us apply **Raabe's Test.**

$$\lim_{n \to \infty} n\left(\frac{u_n}{u_{n+1}} - 1\right) = \lim_{n \to \infty} n\left(\frac{4n^2+20n+25}{4n^2+16n+16} - 1\right)$$

$$= \lim_{n \to \infty}\left(\frac{4n^2+9n}{4n^2+16n+16}\right) = \lim_{n \to \infty}\left[\frac{4+\frac{9}{n}}{4+\frac{16}{n}+\frac{16}{n^2}}\right] = 1, \text{ Raabe's Test fails}$$

Let us apply **Gauss's Test**

$$\frac{u_n}{u_{n+1}} = \frac{(2n+5)^2}{(2n+4)^2} = \frac{\left(1+\frac{5}{2n}\right)^2}{\left(1+\frac{2}{n}\right)^2} = \left(1+\frac{5}{n}+\frac{25}{4n^2}\right)\left(1+\frac{2}{n}\right)^{-2}$$

$$= \left(1+\frac{5}{n}+\frac{25}{4n^2}\right)\left(1-\frac{4}{n}+\frac{(-2)\times(-3)}{2!}\frac{4}{n^2}+\ldots\right) = \left(1+\frac{5}{n}+\frac{25}{4n^2}\right)\left(1-\frac{4}{n}+\frac{12}{n^2}+\ldots\right)$$

$$= 1 - \frac{4}{n} + \frac{12}{n^2} + \frac{5}{n} - \frac{20}{n^2} + \frac{25}{4n^2} + \ldots = 1 + \frac{1}{n} - \frac{7}{n^2} \qquad \left(\frac{u_n}{u_{n+1}} = \alpha + \frac{\beta}{n} + \frac{\gamma}{n^2}\right)$$

Hence, $\alpha = 1$, $\beta = 1$. Thus, the series is divergent. **Ans.**

7.19 CAUCHY'S INTEGRAL TEST

Statement. *A positive term series* $f(1) + f(2) + f(3) + \dots + f(n) + \dots$ *where* $f(n)$ *decreases as n increases, converges or diverges according to the integral*

$$\int_1^\infty f(x)\, dx$$

is finite or infinite.

Proof. In the figure, the area under the curve from $x = 1$ to $x = n + 1$ lies between the sum of the areas of small rectangles (small height) and sum of the areas of large rectangles (large height).

[$f(1), f(2)$ ... represent the height of the rectangles]

$$\Rightarrow \quad f(1) + f(2) + \dots + f(n) \geq \int_1^{n+1} f(x)\, dx \geq f(2) + f(3) + \dots + f(n+1)$$

$$S_n \geq \int_1^{n+1} f(x)\, dx \geq S_{n+1} - f(1)$$

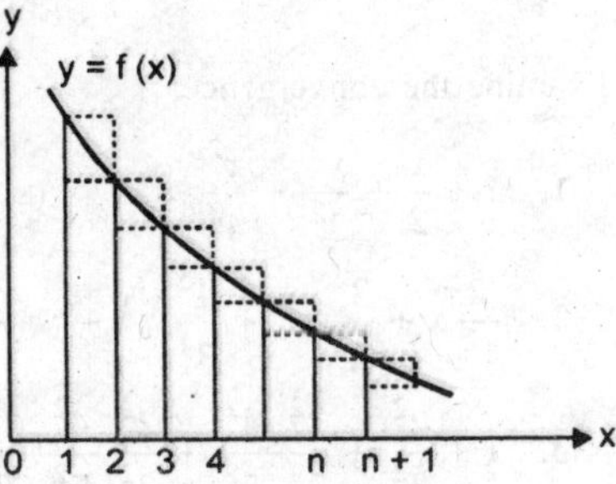

As $n \to \infty$, from the second inequality that if the integral has a finite value then $\lim_{n \to \infty} S_{n+1}$ is also finite, so $\Sigma f(n)$ is convergent.

Similarly, if the integral is infinite, then from the first inequality that $\lim_{n \to \infty} S_n \to \infty$, so the series is divergent.

Example 38. *Apply the integral test to determine the convergence of the p-series*

$$\frac{1}{1^p} + \frac{1}{2^p} + \frac{1}{3^p} + \dots + \frac{1}{n^p} + \dots \infty$$

Solution. (*i*) When $p > 1$, $\quad f(x) = \dfrac{1}{x^p}$

$$\int_1^\infty f(x)\, dx = \lim_{m \to \infty} \int_1^m \frac{1}{x^p}\, dx = \lim_{m \to \infty} \left[\frac{x^{1-p}}{1-p}\right]_1^m = \lim_{m \to \infty} \frac{1}{1-p}(m^{1-p} - 1)$$

$$= \lim_{m \to \infty} \frac{1}{1-p}\left[\frac{1}{m^{p-1}} - 1\right] = \frac{1}{p-1}, \text{ which is finite.}$$

By Cauchy's Integral Test, the series is convergent for $p > 1$.

(*ii*) When $p < 1$,

$$\int_1^\infty f(x)\, dx = \frac{1}{1-p}\left[\lim_{m \to \infty} (m^{1-p} - 1)\right] \to \infty$$

Thus, the series is divergent, if $p < 1$.

(*iii*) When $p = 1$,

$$\int_1^\infty \frac{1}{x}\, dx = [\log x]_1^\infty \to \infty$$

Thus, the series is divergent.

Hence, $\Sigma \dfrac{1}{n^p}$ is convergent if $p > 1$ and divergent if $p \leq 1$. **Ans.**

Example 39. *Examine the convergence of* $\displaystyle\sum_{n=2}^{\infty} \frac{1}{n \log n}$.

Solution. Here $f(x) = \dfrac{1}{x \log x}$

$$\int_2^{\infty} \frac{1}{x \log x}\, dx = \lim_{m \to \infty} \left[\log \log x \right]_2^m = \lim_{m \to \infty} [\log \log m - \log \log 2] \to \infty$$

By Cauchy's Integral Test the series is divergent. **Ans.**

Example 40. *Examine the convergence of* $\sum_{x=1}^{\infty} x\, e^{-x^2}$

Solution. Here $f(x) = x\, e^{-x^2}$

Now, $$\int_1^{\infty} x\, e^{-x^2}\, dx = \lim_{m \to \infty} \left[\frac{e^{-x^2}}{-2} \right]_1^m = \lim_{m \to \infty} \left[\frac{e^{-m^2}}{-2} + \frac{e^{-1}}{2} \right] = \frac{e^{-1}}{2} = \frac{1}{2e},$$ which is finite.

Hence, the given series is convergent. **Ans.**

EXERCISE 7.8

Examine the convergence:

1. $1 + \frac{x}{2} + \frac{x^2}{3^2} + \frac{x^2}{4^3} + \ldots \infty \quad (x > 0)$ **Ans.** Convergent

2. $\frac{2}{1^2} x + \frac{3^2}{2^3} x^2 + \frac{4^3}{3^4} x^3 + \ldots + \frac{(n+1)^n}{n^{n+1}} x^n + \ldots$ **Ans.** $x < 1$, convergent; $x \geq 1$, divergent

3. $1 + \frac{2^2}{3^2} + \frac{2^2 . 4^2}{3^2 . 5^2} + \frac{2^2 . 4^2 . 6^2}{3^2 . 5^2 . 7^2} + \ldots \infty$ **Ans.** Divergent

4. $\sum_{n=1}^{\infty} \frac{1}{\sqrt{n}}$ **Ans.** Divergent

5. $\sum_{n=1}^{\infty} \frac{1}{n^2 + 1}$ **Ans.** Convergent

6. $\sum_{n=1}^{\infty} \frac{1}{n^n}$ **Ans.** Convergent

7. $\sum_{n=1}^{\infty} n^2 e^{-n^3}$ **Ans.** Convergent

8. $\sum_{n=1}^{\infty} \frac{1}{n (\log n)^2}$ **Ans.** Convergent

7.20 CAUCHY'S ROOT TEST

Statement. *If* $\Sigma\, u_n$ *is positive term series such that* $\lim_{n \to \infty} (u_n)^{1/n} = k$, *then*

(i) if $k < 1$, *the series converges.* *(ii) if* $k > 1$, *the series diverges.*

Proof. By definition of limit

$$|(u_n)^{1/n} - k| < \varepsilon \text{ for } n > m$$

$$k - \varepsilon < (u_n)^{1/n} < k + \varepsilon \text{ for } n > m$$

(*i*) $$k < 1$$

$$k + \varepsilon < r < 1$$

$$(u_n)^{1/n} < k$$

$\Rightarrow$ $$u_n < k^n$$

$$u_1 + u_2 + \ldots \infty < k + k^2 + \ldots + k^n + \ldots \infty$$

$$< \frac{1}{1-k} \text{ (a finite quantity)}$$

$\therefore$ The series is convergent.

(*ii*) $$k > 1$$

$$k - \varepsilon > 1$$

$$(u_n)^{1/n} > k - \varepsilon > 1$$
$$u_n > 1$$
$$S_n = u_1 + u_2 + \dots u_n > n$$
$$\lim_{n \to \infty} S_n \to \infty$$

$\therefore$ The series is divergent.

(*iii*) $k = 1$

If $\lim_{n \to \infty} (u_n)^{1/n} = 1$, the test fails.

For example, $\Sigma u_n = \sum \frac{1}{n^p}$

$$\lim_{n \to \infty} (u_n)^{1/n} = \lim_{n \to \infty} \left(\frac{1}{n^p}\right)^{1/n} = \lim_{n \to \infty} \left(\frac{1}{n^{1/n}}\right)^{-p} = 1 \text{ for all } p,\ k = 1$$

But $\sum \frac{1}{n^p}$ is convergent for $p > 1$ and divergent for $p \le 1$.

Thus, we cannot say whether Σu_n is convergent or divergent for $k = 1$.

Example 41. *Examine the convergence of the series* $\sum \left(\frac{n}{n+1}\right)^{n^2}$

(*Uttarakhand, IInd Semester, Summer 2010*)

Solution. Here, $u_n = \left(\frac{n}{n+1}\right)^{n^2} = \frac{1}{\left(1+\frac{1}{n}\right)^{n^2}} \Rightarrow (u_n)^{1/n} = \left[\frac{1}{\left(1+\frac{1}{n}\right)^{n^2}}\right]^{\frac{1}{n}} = \frac{1}{\left(1+\frac{1}{n}\right)^n}$

$$\lim_{n \to \infty} (u_n)^{1/n} = \lim_{n \to \infty} \frac{1}{\left(1+\frac{1}{n}\right)^n} = \frac{1}{e} < 1$$

Hence, the given series is convergent. **Ans.**

Example 42. *Test the following series for convergence*

$$\sum_{n=1}^{\infty} \frac{(n+1)^n x^n}{n^{n+1}}$$

(*M.D.U. Dec., 2001*)

Solution. Here, we have

$$\sum_{n=1}^{\infty} \frac{(n+1)^n x^n}{n^{n+1}}$$

Here, $u_n = \frac{(n+1)^n x^n}{n^{n+1}} = \left[\frac{(n+1)x}{n}\right]^n \cdot \frac{1}{n}$

$$\Rightarrow \quad (u_n)^{\frac{1}{n}} = \frac{(n+1)x}{n} \cdot \frac{1}{\frac{1}{n^n}} = \left(1+\frac{1}{n}\right)x \cdot \frac{1}{\frac{1}{n^n}}$$

$$\Rightarrow \quad \lim_{n\to\infty} (u_n)^{\frac{1}{n}} = \left[\lim_{n\to\infty}\left(1+\frac{1}{n}\right)x\right]\left[\lim_{n\to\infty} \frac{1}{\frac{1}{n^n}}\right]$$

$$= (1+0)\,x \cdot \frac{1}{1} = x \qquad \left[\because \lim_{n\to\infty} n^{\frac{1}{n}} = 1\right]$$

$\therefore$ By Cauchy's root test, $\sum u_n$ is convergent if $x < 1$ and divergent if $x > 1$. The test fails when $x = 1$.

When $x = 1$, $$u_n = \frac{(n+1)^n}{n^{n+1}} = \frac{1}{n}\cdot\frac{(n+1)^n}{n^n} = \frac{1}{n}\left(1+\frac{1}{n}\right)^n$$

Let $$v_n = \frac{1}{n},$$

$$\frac{u_n}{v_n} = \left(1+\frac{1}{n}\right)^n$$

$$\lim_{n\to\infty}\frac{u_n}{v_n} = \lim_{n\to\infty}\left(1+\frac{1}{n}\right)^n = e,$$ which is finite and non-zero.

$\therefore$ By camparison test, $\sum u_n$ and $\sum v_n$ converge or diverge together.

Since $\sum v_n = \sum \frac{1}{n}$ is of the form $\sum \frac{1}{n^p}$ with $p = 1$,

$\sum v_n$ divergent $\Rightarrow \sum u_n$ is also divergent.

Hence, $\sum u_n$ is convergent as $x < 1$ and divergent if $x \geq$ **Ans.**

Example 43. *Discuss the convergence of the following series:*

$$\left(\frac{2^2}{1^2}-\frac{2}{1}\right)^{-1} + \left(\frac{3^3}{2^3}-\frac{3}{2}\right)^{-2} + \left(\frac{4^4}{3^4}-\frac{4}{3}\right)^{-3} + \dots \infty$$

Solution. Here, $$u_n = \left[\frac{(n+1)^{n+1}}{n^{n+1}} - \frac{(n+1)}{n}\right]^{-n}$$

$$[u_n]^{1/n} = \left[\left[\frac{(n+1)^{n+1}}{n^{n+1}} - \frac{n+1}{n}\right]^{-n}\right]^{\frac{1}{n}} = \left[\frac{(n+1)^{n+1}}{n^{n+1}} - \frac{n+1}{n}\right]^{-1}$$

$$\lim_{n\to\infty}(u_n)^{1/n} = \lim_{n\to\infty}\left[\left(1+\frac{1}{n}\right)^{n+1} - \left(1+\frac{1}{n}\right)\right]^{-1} = (e-1)^{-1} = \frac{1}{e-1} < 1$$

Hence, the given series is convergent. **Ans.**

Example 44. *Discuss the convergence of the series* $\frac{1}{2} + \frac{2}{3}x + \left(\frac{3}{4}\right)^2 x^2 + \left(\frac{4}{5}\right)^3 x^3 + \dots \infty$

Solution. Ignoring the first term, we have:

$$u_n = \left(\frac{n+1}{n+2}\right)^n x^n$$

$$(u_n)^{1/n} = \left(\frac{n+1}{n+2}\right)x \qquad \text{[Cauchy Root Test]}$$

$$\lim_{n\to\infty}(u_n)^{1/n} = \lim_{n\to\infty}\left(\frac{1+\frac{1}{n}}{1+\frac{2}{n}}\right)x = x$$

Hence, the series converges if $x < 1$ and diverges if $x > 1$.
If $x = 1$

$$u_n = \left(\frac{n+1}{n+2}\right)^n = \left(\frac{1+\frac{1}{n}}{1+\frac{2}{n}}\right)^n$$

$$\lim_{n\to\infty} u_n = \lim_{n\to\infty} \frac{\left(1+\frac{1}{n}\right)^n}{\left[\left(1+\frac{2}{n}\right)^{\frac{n}{2}}\right]^2} = \frac{e}{e^2} = \frac{1}{e} \neq 0$$

$\therefore$ The series is divergent by Cauchy Fundamental test. **Ans.**

EXERCISE 7.9

Discuss the convergence of the following series:

1. $\sum_{n=1}^{\infty} \frac{1}{n^n}$ **Ans.** Convergent
2. $\sum_{n=1}^{\infty} \left(1+\frac{1}{n}\right)^{n^2}$ **Ans.** Divergent
3. $\sum_{n=1}^{\infty} \left(1+\frac{1}{\sqrt{n}}\right)^{-n^{3/2}}$ **Ans.** Convergent
4. $\sum_{n=1}^{\infty} \frac{1}{(\log n)^n}$ **Ans.** Convergent
5. $\Sigma\, n^{-k}$ **Ans.** If $k > 1$, convergent
6. $\sum_{n=2}^{\infty} \frac{1}{n(\log n)^k}$ **Ans.** If $k > 1$, convergent; and divergent if $k \leq 1$.
7. $\Sigma\, (n \log n)^{-1} (\log \log n)^{-k}$ **Ans.** If $k > 1$, convergent; and divergent if $k \leq 1$.
8. $\sum \left(1-\frac{1}{n}\right)^{n^2}$ **Ans.** Convergent
9. $\sum \frac{x^n}{n^n}$ **Ans.** Convergent
10. $(a + b) + (a^2 + b^2) + (a^3 + b^3) + \ldots$ **Ans.** Convergent if $a < 1, b < 1$; divergent if $a \geq 1, b \geq 1$

7.21 LOGARITHMIC TEST

If $\Sigma\, u_n$ is a positive term series such that $\lim_{n\to\infty}\left(n \log \frac{u_n}{u_{n+1}}\right) = k$

(*i*) If $k > 1$, then the series is convergent. (*ii*) If $k < 1$, then the series is divergent.

Proof. (*i*) If $k > 1$

Compare $\Sigma\, u_n$ with $\sum \frac{1}{n^p}$, if $k > p > 1$, then $\Sigma\, u_n$ converges.

$$\frac{u_n}{u_{n+1}} = \frac{(n+1)^p}{n^p} = \left(1+\frac{1}{n}\right)^p \qquad \ldots(1)$$

Taking logarithm of both sides of (1), we have:

$$\log \frac{u_n}{u_{n+1}} > p \log\left(1+\frac{1}{n}\right) \quad \left[\log(1+x) = x - \frac{x^2}{2} + \frac{x^3}{3} - \frac{x^4}{4} + \ldots\right]$$

if $$\log \frac{u_n}{u_{n+1}} > p\left(\frac{1}{n} - \frac{1}{2n^2} + \frac{1}{3n^3} - \frac{1}{4n^4} + \ldots\right)$$

if $$n \log \frac{u_n}{u_{n+1}} > p\left(1 - \frac{1}{2n} + \frac{1}{3n^2} - \frac{1}{4n^3} + \ldots\right)$$

$$\lim_{n\to\infty} n \log \frac{u_n}{u_{n+1}} > p$$

i.e., $k > p$ which is true as $k > p > 1$. $\left[\lim_{n \to \infty} n \log \frac{u_n}{u_{n+1}} = k\right]$

Hence, $\Sigma\, u_n$ is convergent.
When $p < 1$
Similarly, when $p < 1$, $\Sigma\, u_n$ is divergent.
When $p = 1$, the test fails.

Example 45. *Test the convergence of the series* $x + \frac{2^2.x^2}{2!} + \frac{3^3.x^3}{3!} + \frac{4^4.x^4}{4!} + \ldots \infty$

(MDU, Dec. 2010)

Solution. Here the series is $x + \frac{2^2.x^2}{2!} + \frac{3^3.x^3}{3!} + \frac{4^4.x^4}{4!} + \ldots + \frac{n^n.x^n}{n!} + \ldots \infty$

$$u_n = \frac{n^n.x^n}{n!} \text{ and } u_{n+1} = \frac{(n+1)^{n+1}.x^{n+1}}{(n+1)!}$$

$$\frac{u_n}{u_{n+1}} = \frac{n^n.x^n}{n!} \frac{(n+1)!}{(n+1)^{n+1}.x^{n+1}} = \frac{n^n}{(n+1)^n} \frac{1}{x} = \frac{1}{\left(1+\frac{1}{n}\right)^n} \frac{1}{x}$$

$$\lim_{n \to \infty} \frac{u_n}{u_{n+1}} = \lim_{n \to \infty} \frac{1}{\left(1+\frac{1}{n}\right)^n} \frac{1}{x} = \frac{1}{e}.\frac{1}{x}$$

If $\frac{1}{e\,x} > 1$ or $x < \frac{1}{e}$, the series is convergent.

If $\frac{1}{e\,x} < 1$ or $\frac{1}{e} < x$, the series is divergent. If $\frac{1}{e\,x} = 1$ or $x = \frac{1}{e}$, the test fails.

$$\log \frac{u_n}{u_{n+1}} = \log \frac{1}{\left(1+\frac{1}{n}\right)^n}.e = \log e - \log\left(1+\frac{1}{n}\right)^n$$

$$= 1 - n\log\left(1+\frac{1}{n}\right) = 1 - n\left[\frac{1}{n} - \frac{1}{2n^2} + \frac{1}{3n^3} + \ldots\right]$$

$$= 1 - 1 + \frac{1}{2n} - \frac{1}{3n^2} + \ldots = \frac{1}{2n} - \frac{1}{3n^2}$$

$$\lim_{n \to \infty} n \log \frac{u_n}{u_{n+1}} = \lim_{n \to \infty} \left[\frac{1}{2} - \frac{1}{3n}\right] = \frac{1}{2} < 1.$$

Thus, the series is divergent. **Ans.**

Example 46. *Test the convergence of the series* $1 + \frac{2x}{2!} + \frac{3^2x^2}{3!} + \frac{4^3x^3}{4!} + \ldots$

Solution. The given series is

$$1 + \frac{2x}{2!} + \frac{3^2x^2}{3!} + \frac{4^3x^3}{4!} + \ldots$$

Here $u_n = \frac{n^{n-1}\,x^{n-1}}{n!}$ $\quad \therefore \quad u_{n+1} = \frac{(n+1)^n\,x^n}{(n+1)!}$

$$\therefore \quad \frac{u_{n+1}}{u_n} = \frac{(n+1)^n\,x^n}{(n+1)!} \times \frac{n!}{n^{n-1}\,x^{n-1}} = \frac{(n+1)^n}{(n+1)\,n^{n-1}}\,x$$

$$= \frac{(n+1)^{n-1}}{n^{n-1}} x = \left[\frac{n+1}{n}\right]^{n-1} x = \left[1+\frac{1}{n}\right]^{n-1} x = \left(1+\frac{1}{n}\right)^n \times \left(1+\frac{1}{n}\right)^{-1} x$$

$$\therefore \quad \lim_{n\to\infty} \frac{u_{n+1}}{u_n} = e\,x \qquad \left[\because \lim_{n\to\infty}\left(1+\frac{1}{n}\right)^n = e\right]$$

By **D'Alembert's Ratio Test**,

(*i*) $\Sigma\, u_n$ is convergent if $|ex| < 1$, *i.e.*, $|x| < \frac{1}{e}$ and

(*ii*) is divergent if $|ex| > 1$, *i.e.*, $|x| > \frac{1}{e}$.

(*iii*) If $ex = 1$ *i.e.*, $x = \frac{1}{e}$, this test fails.

If $x = \frac{1}{e}$ then $\frac{u_{n+1}}{u_n} = \left(1+\frac{1}{n}\right)^{n-1} \times \frac{1}{e}$

$$\therefore \quad \frac{u_n}{u_{n+1}} = \frac{e}{\left(1+\frac{1}{n}\right)^{n-1}}$$

$$\therefore \quad \log\left[\frac{u_n}{u_{n+1}}\right] = \log\left[\frac{e}{\left(1+\frac{1}{n}\right)^{n-1}}\right]$$

$$= \log e - (n-1)\log\left(1+\frac{1}{n}\right) = \log e - (n-1)\left[\frac{1}{n} - \frac{1}{2n^2} + \frac{1}{3n^3} - \ldots\right]$$

$$= 1 - \left[\left(1 - \frac{1}{2n} + \frac{1}{3n^2} - \ldots\right) - \left(\frac{1}{n} - \frac{1}{2n^2} + \ldots\right)\right] = 1 - 1 + \frac{1}{2n} - \frac{1}{3n^2} + \frac{1}{n} - \frac{1}{2n^2} - \ldots$$

$$= \frac{3}{2n} - \frac{5}{6n^2}$$

By Logarithmic Test

$$\therefore \quad n \log \frac{u_n}{u_{n+1}} = \frac{3}{2} - \frac{5}{6n}$$

$$\therefore \quad \lim_{n\to\infty}\left[n \log \frac{u_n}{u_{n+1}}\right] = \frac{3}{2} > 1$$

Thus, if $x = \frac{1}{e}$, then $\Sigma\, u_n$ is convergent.

Hence, if $x \le \frac{1}{e}$, then $\Sigma\, u_n$ is convergent and if $x > \frac{1}{e}$, then $\Sigma\, u_n$ is divergent.

Ans.

Example 47. *Discuss the convergence of the series:*

$$1 + \frac{x}{2} + \frac{2!}{3^2}x^2 + \frac{3!}{4^3}x^3 + \frac{4!}{5^4}x^4 + \ldots\ldots\ldots \infty \; (x > 0)$$ *(M.D.University, I Semester, 2009)*

Solution. Here, we have

$$1+\frac{x}{2}+\frac{2!}{3^2}x^2+\frac{3!}{4^3}x^3+\frac{4!}{5^4}x^4+.........\infty$$

Neglecting the first term, we get

$$u_n = \frac{n!}{(n+1)^n}x^n$$

and
$$u_{n+1} = \frac{(n+1)!}{(n+2)^{n+1}}x^{n+1}$$

$$\frac{u_n}{u_{n+1}} = \frac{n!}{(n+1)^n}.x^n.\frac{(n+2)^{n+1}}{(n+1)!.x^{n+1}} = \frac{(n+2)^{n+1}}{(n+1)^n.(n+1)}.\frac{1}{x}$$

$$\lim_{n\to\infty}\frac{u_n}{u_{n+1}} = \lim_{n\to\infty}\frac{(n+2)^{n+1}}{(n+1)^n.(n+1)}\frac{1}{x}$$

$$= \lim_{n\to\infty}\frac{n^{n+1}\left(1+\frac{2}{n}\right)^{n+1}}{n^n\left(1+\frac{1}{n}\right)^n.n\left(1+\frac{1}{n}\right)}.\frac{1}{x}$$

$$= \lim_{n\to\infty}\frac{\left(1+\frac{2}{n}\right)^n\left(1+\frac{2}{n}\right)}{\left(1+\frac{1}{n}\right)^n\left(1+\frac{1}{n}\right)}.\frac{1}{x} \qquad ...(1)$$

$$= \frac{e^2}{e}.\frac{1}{x}=\frac{e}{x}. \qquad \left[\because \lim_{n\to\infty}\left(1+\frac{a}{n}\right)^n = \lim_{n\to\infty}\left\{\left(1+\frac{a}{n}\right)^{\frac{n}{a}}\right\}^a = e \right.$$

$$\left. \Rightarrow \lim_{n\to\infty}\left(1+\frac{2}{n}\right)^n = \lim_{n\to\infty}\left[\left(1+\frac{2}{n}\right)^{\frac{n}{2}}\right]^2 = e^2\right]$$

$\therefore$ By D' Alembert's ratio test, the series converges if $1<\frac{e}{x}$ or if $x<e$ and diverges if $\frac{e}{x}<1$ or if $e<x$.

If $x = e$, the ratio test fails, $\because \lim_{n\to\infty}\frac{u_n}{u_{n+1}}=1$

Now when $x = e$

Putting the value of x in (1), we get

$$\frac{u_n}{u_{n+1}} = \frac{\left(1+\frac{2}{n}\right)^{n+1}}{\left(1+\frac{1}{n}\right)^{n+1}}.\frac{1}{e}$$

Since the expression $\frac{u_n}{u_{n+1}}$ involves the number e, so we do not apply Raabe's test but apply logarithmic test.

$$\therefore \quad \log \frac{u_n}{u_{n+1}} = (n+1)\log\left(1+\frac{2}{n}\right)-(n+1)\log\left(1+\frac{1}{n}\right)-\log e$$

$$= (n+1)\left[\log\left(1+\frac{2}{n}\right)-\log\left(1+\frac{1}{n}\right)\right]-1$$

$$= (n+1)\left[\left(\frac{2}{n}-\frac{1}{2}\cdot\frac{4}{n^2}+\frac{1}{3}\cdot\frac{8}{n^3}\ldots\right)-\left(\frac{1}{n}-\frac{1}{2n^2}+\frac{1}{3n^3}+\ldots\right)\right]-1$$

$$= (n+1)\left[\frac{1}{n}-\frac{3}{2n^2}+\frac{7}{3n^3}-\ldots\ldots\right]-1$$

$$= \left(1-\frac{3}{2n}\right)+\left(\frac{1}{n}-\frac{3}{n^2}+\ldots\ldots\right)-1=\frac{1}{n}-\frac{3}{n^2}-\frac{3}{2n^2}$$

$$= 1-\frac{1}{2n}-\frac{3}{2n^2}+\ldots\ldots-1=-\frac{1}{2n}-\frac{3}{2n^2}$$

$$\therefore \quad \lim_{n\to\infty} n\log\frac{u_n}{u_{n+1}} = \lim_{n\to\infty} n\left[-\frac{1}{2n}-\frac{3}{2n^2}+\ldots\ldots\right] = \lim_{n\to\infty}\left(-\frac{1}{2}-\frac{3}{2n}+\ldots\ldots\right)=-\frac{1}{2}<1$$

$\therefore$ By log test, the series diverges.

Hence, the given series $\sum u_n$ converges if $x < e$ and diverges if $x \geq e$. **Ans.**

EXERCISE 7.10

Examine the convergence for the following series :

1. $\frac{1^2}{4^2}+\frac{5^2}{8^2}+\frac{9^2}{12^2}+\frac{13^2}{16^2}+\ldots\infty$ **Ans.** Convergent

2. $1+\frac{1!}{2}x+\frac{2!}{3^2}x^2+\frac{3!}{4^3}x^3+\ldots\infty$ **Ans.** If $x < e$, convergent and divergent if $x \geq e$

3. $\frac{1^2}{2^2}+\frac{1^2.3^2}{2^2.4^2}x+\frac{1^2.3^2.5^2}{2^2.4^2.6^2}x^2+\ldots\infty$ **Ans.** Convergent if $x < 1$, and divergent if $x \geq 1$

4. $\frac{a+x}{1!}+\frac{(a+2x)^2}{2!}+\frac{(a+3x)^3}{3!}+\ldots$ **Ans.** Convergent if $x < \frac{1}{e}$, divergent if $x \geq \frac{1}{e}$

7.22 DE MORGAN'S AND BERTRAND'S TEST

If Σu_n is a positive term series such that

$$\lim_{n\to\infty}\left[\left\{n\left(\frac{u_n}{u_{n+1}}-1\right)-1\right\}\log n\right]=k$$

then the series is convergent if $k > 1$ and divergent if $k < 1$.

Example 48. *Test for convergence the series* $1^p + \left(\frac{1}{2}\right)^p + \left(\frac{1 \times 3}{2 \times 4}\right)^p + \left(\frac{1 \times 3 \times 5}{2 \times 4 \times 6}\right)^p + \ldots$

Solution. The given series is :

$$1^p + \left(\frac{1}{2}\right)^p + \left(\frac{1\times 3}{2\times 4}\right)^p + \left(\frac{1\times 3\times 5}{2\times 4\times 6}\right)^p + \ldots$$

Here $$u_n = \left[\frac{1\times 3\times 5\times \ldots \times (2n-3)}{2\times 4\times 6\times \ldots \times (2n-2)}\right]^p$$

$\therefore$ $$u_{n+1} = \left[\frac{1\times 3\times 5\times \ldots (2n-3)(2n-1)}{2\times 4\times 6\times \ldots \times (2n-2)(2n)}\right]^p$$

$\therefore$ $$\frac{u_{n+1}}{u_n} = \left(\frac{2n-1}{2n}\right)^p = \left(1 - \frac{1}{2n}\right)^p$$

$$\lim_{n\to\infty} \frac{u_{n+1}}{u_n} = 1$$

$\therefore$ **D'Alembert's Test fails.**

Now let us apply Raabe's Test.

Here $n\left(\frac{u_n}{u_{n+1}} - 1\right) = n\left[\left(1 - \frac{1}{2n}\right)^{-p} - 1\right] = n\left[1 + \frac{p}{2n} + \frac{p(p+1)}{8n^2} + \ldots - 1\right] = \frac{p}{2} + \frac{p(p+1)}{8n} + \ldots$

$\therefore$ $$\lim_{n\to\infty} n\left[\frac{u_n}{u_{n+1}} - 1\right] = \frac{p}{2}$$

If $\frac{p}{2} > 1, i.e., p > 2$, the series is convergent and divergent if $\frac{p}{2} < 1, i.e., p < 2$.

This test fails if $\frac{p}{2} = 1, i.e., p = 2$.

Now let us apply **De Morgan's Test**. When $p = 2$

$$n\left[\frac{u_n}{u_{n+1}} - 1\right] = 1 + \frac{3}{4n} + \ldots$$

Now, $$\lim_{n\to\infty}\left[n\left(\frac{u_n}{u_{n+1}} - 1\right) - 1\right]\log n = \lim_{n\to\infty}\left[1 + \frac{3}{4n} + \ldots - 1\right]\log n$$

$$= \lim_{n\to\infty} \frac{3}{4}\left[\frac{\log n}{n} - \ldots\right] = 0 < 1 \qquad \left[\lim_{x\to\infty} \frac{\log n}{n} = 0\right]$$

$\therefore$ $\Sigma\, u_n$ is divergent when $p = 2$. **Ans.**

7.23 CAUCHY'S CONDENSATION TEST

If $\phi(n)$ is positive for all positive integral values of n and continually diminishes as n increases and if a be a positive integer greater than 1, then the two series $\Sigma\, \phi(n)$ and $\Sigma\, a^n \phi(a^n)$ are either both convergent or both divergent.

Example 49. *Show that the series*

$$1 + \frac{1}{2(\log 2)^p} + \frac{1}{3(\log 3)^p} + \ldots + \frac{1}{n(\log n)^p} + \ldots$$

is convergent if $p > 1$ and divergent if $p = 1$ or $p < 1$.

Solution. We apply Cauchy's Condensation Test.

Here $$\phi(n) = \frac{1}{n(\log n)^p}$$

$\therefore$ nth term of the second series $\Sigma\, a^n \phi(a^n)$ is :

$$a^n\left[\frac{1}{a^n\left(\log a^n\right)^p}\right] \text{ i.e., } \frac{1}{\left(\log a^n\right)^p} \text{ i.e., } \frac{1}{(n\log a)^p} \text{ i.e., } \frac{1}{(\log a)^p}\times\frac{1}{n^p}$$

$\therefore$ The given series will be convergent or divergent if $\sum\left[\frac{1}{(\log a)^p}\times\frac{1}{n^p}\right]$ is convergent or divergent, i.e., if $\sum\frac{1}{n^p}$ is convergent or divergent.

But we know the $\sum\frac{1}{n^p}$ is convergent when $p > 1$ and divergent if $p = 1$ or < 1

Hence, the given series is convergent if $p > 1$ and divergent if $p = 1$ or < 1. **Proved.**

7.24 ALTERNATING SERIES

A series in which the terms are alternately negative is called the alternating series.

e.g., $$u_1 - u_2 + u_3 - u_4 + \ldots\infty$$

7.25 LEIBNITZ'S RULE FOR CONVERGENCE OF AN ALTERNATING SERIES

(i) Each term is numerically less than its preceeding term.

(ii) $\lim\limits_{n\to\infty} u_n = 0$

Exmaple 50. *Discuss the convergence of the series* $\sum\limits_{n=1}^{\infty}(-1)^n\frac{n}{n^2+1}$.

Solution. The terms of the given series are alternately positive and negative;

(i) $$|u_n| = \frac{n}{n^2+1} \quad\text{and}\quad |u_{n+1}| = \frac{(n+1)}{(n+1)^2+1}$$

$$|u_n| - |u_{n+1}| = \frac{n}{n^2+1} - \frac{(n+1)}{(n+1)^2+1} = \frac{n(n+1)^2 + n - (n+1)(n^2+1)}{(n^2+1)[(n+1)^2+1]}$$

$$= \frac{n^2+n-1}{(n^2+1)[(n+1)^2+1]} = +\text{ve}$$

and each term is numerically less than its preceeding term.

(ii) $$\lim_{n\to\infty}|u_n| = \lim_{n\to\infty}\frac{n}{n^2+1} = \lim_{n\to\infty}\frac{1}{n+\frac{1}{n}} = 0$$

Both conditions are satisfied.

Hence, by Leibnitz's rule, the given series is convergent. **Ans.**

Example 51. *Test the convergence of the series* $\frac{1}{6}-\frac{2}{11}+\frac{3}{16}-\frac{4}{21}+\frac{5}{26}-\ldots\infty$

Solution. The terms of the given series are alternately positive and negative.

$$u_n = (-1)^{n-1}\frac{n}{5n+1}$$

$$|u_n| = \frac{n}{5n+1} \text{ and } |u_{n+1}| = \frac{n+1}{5(n+1)+1}$$

$$(i)\ |u_n| - |u_{n+1}| = \frac{n}{5n+1} - \frac{n+1}{5(n+1)+1} = \frac{5n^2+6n-5n^2-5n-n-1}{(5n+1)(5n+6)}$$

$$= \frac{-1}{(5n+1)(5n+6)}$$

$$\therefore \quad |u_n| > |u_{n+1}|$$

Thus each term is not numerically less than its preceeding terms.

$$(ii) \quad \lim_{n\to\infty} u_n = \lim_{n\to\infty} \frac{n}{5n+1} = \lim_{n\to\infty} \frac{1}{5+\frac{1}{n}} = \frac{1}{5} \neq 0$$

$$\Rightarrow \quad \lim_{n\to\infty} u_n \neq 0$$

Both conditions for convergence are not satisfied.

Hence, the series is not convergent. It is oscillatory. **Ans.**

Example 52. *Test the following series for convergence and absolute convergence:*

$$1 - \frac{1}{2\sqrt{2}} + \frac{1}{3\sqrt{3}} - \frac{1}{4\sqrt{4}} + \ldots$$ *(M.D.U. Dec., 2002)*

Solution. The given series is

$$\sum_{n=1}^{\infty} u_n = \sum_{n=1}^{\infty} (-1)^{n-1} \cdot \frac{1}{n\sqrt{n}} = \sum_{n=1}^{\infty} (-1)^{n-1} \frac{1}{n^{\frac{3}{2}}} = \sum_{n=1}^{\infty} (-1)^{n-1} a_n$$

It is an alternating series.

Here,
$$a_n = \frac{1}{n^{\frac{3}{2}}}$$

$$a_{n+1} = \frac{1}{(n+1)^{\frac{3}{2}}}$$

Since,
$$\frac{1}{n^{\frac{3}{2}}} > \frac{1}{(n+1)^{\frac{3}{2}}} \ \forall\, n \qquad (\therefore a_n > a_{n+1}\ \forall\, n)$$

Also,
$$\lim_{n\to\infty} a_n = \lim_{n\to\infty} \frac{1}{n^{\frac{3}{2}}} = 0$$

$\therefore$ By Leibnitz's test, the series $\sum u_n$ is convergent.

$$|u_n| = \frac{1}{n^{\frac{3}{2}}}$$

Now $\sum |u_n| = \sum \frac{1}{n^{\frac{3}{2}}}$ is convergent $\left(\because p = \frac{3}{2} > 1\right)$

Hence, the series $\sum u_n$ is absolutely convergent. **Ans.**

Example 53. *Test the convergence of the series:*

$$\sum_{n=1}^{\infty} \frac{(-1)^n (x+1)^n}{2^n n^2}$$ *(M.D. University, I Semester, 2009)*

Solution. Here, we have

$$\sum_{n=1}^{\infty} \frac{(-1)^n (x+1)^n}{2^n n^2} \quad ...(1)$$

Here, $u_n = \dfrac{(-1)^n (x+1)^n}{2^n . n^2}$ and $u_{n+1} = \dfrac{(-1)^{n+1} (x+1)^{n+1}}{2^{n+1} . (n+1)^2}$

$$\frac{|u_n|}{|u_{n+1}|} = \frac{|x+1|^n}{2^n . n^2} \cdot \frac{2^{n+1} . (n+1)^2}{|x+1|^{n+1}}$$

$$= 2\left(\frac{n+1}{n}\right)^2 \cdot \frac{1}{|x+1|} = 2\left(1+\frac{1}{n}\right)^2 \cdot \frac{1}{|x+1|}$$

$$\therefore \quad \lim_{n\to\infty} \frac{|u_n|}{|u_{n+1}|} = \lim_{n\to\infty} 2\left(1+\frac{1}{n}\right)^2 \cdot \frac{1}{|x+1|} = \frac{2}{|x+1|}$$

$\therefore$ By ratio test, the series $\sum |u_n|$ is convergent if

$1 < \dfrac{2}{|x+1|}$ *i.e.*, if $|x+1| < 2$

i.e., if $-2 < x+1 < 2$ *i.e.*, if $-3 < x < 1$

Also $\sum |u_n|$ is divergent if $\dfrac{2}{|x+1|} < 1$

i.e., if $|x+1| > 2$ *i.e.*, if $x+1 > 2$ or $x+1 < -2$ *i.e.*, if $x > 1$ or $x < -3$.

Ratio test fails when $x = 1$ or -3.

When $x = 1$, $\sum u_n = \sum \dfrac{(-1)^n . 2^n}{2^n . n^2} = \sum \dfrac{(-1)^n}{n^2} = \sum (-1)^n . v_n$ From (1)

It is an alternating series.

Here, $v_n = \dfrac{1}{n^2}$, $v_{n+1} = \dfrac{1}{(n+1)^2}$

Clearly $v_n > v_{n+1}$ $\forall n$

Also $\lim_{n\to\infty} v_n = \lim_{n\to\infty} \dfrac{1}{n^2} = 0$

$\therefore$ By Leibnitz's test, $\sum u_n$ is convergent.

When $x = -3$, $\sum u_n = \sum \dfrac{(-1)^n . (-2)^n}{2^n . n^2} = \sum \dfrac{(-1)^{2n} 2^n}{2^n . n^2} = \sum \dfrac{1}{n^2}$

Which is convergent.

Hence, the given series is convergent if $-3 \le x \le 1$ and divergent if $x > 1$ or $x < -3$

EXERCISE 7.11

Discuss the convergence of the following series ·

1. $1 - \dfrac{1}{\sqrt{2}} + \dfrac{1}{\sqrt{3}} - \dfrac{1}{\sqrt{4}} + ...$ **Ans.** Convergent
2. $1 - 2x + 3x^2 - 4x^3 + ... \infty \ (x < 1)$ **Ans.** Convergent
3. $\dfrac{x}{1+x} - \dfrac{x^2}{1+x^2} + \dfrac{x^3}{1+x^3} - \dfrac{x^4}{1+x^4} + ...\infty \ (0 < x < 1)$ **Ans.** Convergent

4. $\frac{1}{1^p}-\frac{1}{2^p}+\frac{1}{3^p}-\frac{1}{4^p}+\ldots$ **Ans.** If $p>0$, convergent; oscillatory if $p<0$.

5. $\sum_{n=1}^{\infty}(-1)^{n-1}\frac{n}{2n-1}$ **Ans.** Oscillatory

6. Show that the series $\frac{1}{x+1}-\frac{1}{x+2}+\frac{1}{x+3}-\frac{1}{x+4}+\ldots$ is convergent for all real values of x other than negative integers.

7. Prove that the series $x-\frac{x^2}{2}+\frac{x^3}{3}-\ldots$ converges if $-1\ x\leq 1$.

7.26 ALTERNATING CONVERGENT SERIES

There are two types of alternating convergent series :

(1) Absolutely convergent series. (2) Conditionally convergent series.
Absolutely convergent series. If $u_1+u_2+u_3+\ldots$ be such that $|u_1|+|u_2|+|u_3|+\ldots$ be convergent then $u_1+u_2+u_3+\ldots\infty$ is called absolutely convergent.

Example 54. *Show that the series* $1-\frac{1}{2}+\frac{1}{3}-\frac{1}{4}+\frac{1}{5}-\frac{1}{6}+\ldots$ *is convergent but not absolutely convergent.*

Solution. $1-\frac{1}{2}+\frac{1}{3}-\frac{1}{4}+\ldots$

The terms of the series are alternately positive and negative.

(*i*) $|u_{n+1}|<|u_n|$ as $\frac{1}{n+1}<\frac{1}{n}$ (*ii*) $\lim_{n\to\infty}u_n=\lim_{n\to\infty}\frac{1}{n}=0$

Both conditions are satisfied. Hence, the given series is convergent.

But $1+\frac{1}{2}+\frac{1}{3}+\frac{1}{4}+\ldots\infty$ is divergent since in *p-series*, $p=1$.

Hence, the given series is conditionally convergent. **Ans.**

Example 55. *What can you say about the series* $1-\frac{1}{2^2}+\frac{1}{3^2}-\frac{1}{4^2}+\ldots\infty$?

Solution. $1-\frac{1}{2^2}+\frac{1}{3^2}-\frac{1}{4^2}+\ldots$

$$|u_n|=\frac{1}{n^2},\quad\text{and}\quad|u_{n+1}|=\frac{1}{(n+1)^2}$$

(*i*) $|u_{n+1}|<|u_n|$ (*ii*) $\lim_{n\to\infty}u_n=\lim_{n\to\infty}\frac{1}{n^2}=0$

Thus, the given series is convergent by Leibnitz's rule.

And $1+\frac{1}{2^2}+\frac{1}{3^2}+\frac{1}{4^2}+\ldots.$ is also convergent since in *p*-series, $p=2>1$.

Both the conditions are satisfied.

Hence, the given series is absolutely convergent. **Ans.**

Example 56. *Discuss the series for convergence* $1-\frac{1}{3}+\frac{1}{2}-\frac{1}{3^3}+\frac{1}{2^2}-\frac{1}{3^5}+\frac{1}{2^3}-\frac{1}{3^7}+\ldots$

Solution. The given series is rewritten as

$$1+\frac{1}{2}+\frac{1}{2^2}+\frac{1}{2^3}+...-\left(\frac{1}{3}+\frac{1}{3^3}+\frac{1}{3^5}+\frac{1}{3^7}+...\right)$$

$$\lim_{n\to\infty} S_n = \frac{1}{1-\frac{1}{2}}-\frac{\frac{1}{3}}{1-\frac{1}{3^2}}=2-\frac{3}{8}=1\frac{5}{8}$$

The given series is convergent.

Again $1+\frac{1}{2}+\frac{1}{2^2}+\frac{1}{2^3}+...+\frac{1}{3}+\frac{1}{3^3}+\frac{1}{3^5}+...$

$$\lim_{n\to\infty} S_n = \frac{1}{1-\frac{1}{2}}+\frac{\frac{1}{3}}{1-\frac{1}{3}} \qquad \left(\lim_{n\to\infty} S_n = \frac{a}{1-r}\right)$$

$$\lim_{n\to\infty} S_n = 2+\frac{3}{8}=\frac{19}{8}$$

Both the conditions are satisfied.

This series is also convergent.

Hence, the given series is absolutely convergent. **Ans.**

Example 57. *Test the convergence and divergence of the series*

$$5-\frac{10}{3}+\frac{20}{9}-\frac{40}{27}+.........$$ *(Gujarat, I Semester, Jan. 2009)*

Solution. The terms of the given series are alternately positive and negative and the given series is geometric infinite series.

(*i*) $$S = 5-\frac{10}{3}+\frac{20}{9}-\frac{40}{27}+\frac{80}{81}+.........$$

Here $a = 5$ and $r = -\frac{2}{3}$

$$S = \frac{a}{1-r}$$

$$S = \frac{5}{1-\left(-\frac{2}{3}\right)}=\frac{5}{1+\frac{2}{3}}=\frac{5}{\frac{5}{3}}=3$$

Sum of the series is finite.

Hence, the given series is convergent.

(*ii*) Again $5+\frac{10}{3}+\frac{20}{9}+\frac{40}{27}+\frac{80}{81}+...$

This is also G.P.

Here, $a = 5$ and $r = \frac{2}{3}$

$$S = \frac{a}{1-r}$$

$$S = \frac{5}{1-\frac{2}{3}}=\frac{5}{\frac{1}{3}}=15$$

Again sum of the positive terms is finite.
Thus the series is also convergent.
Both of the conditions are satisfied.
Hence, the given series is absolutely convergent. **Ans.**

Example 58. *Test the following series for convergence and divergence.*

$$\sum_{n=1}^{\infty} \tan^{-1}\left(\frac{1}{n^2+n+1}\right)$$ (*Gujarat, I Semester, Jan. 2009*)

Solution. Let $u_n = \tan^{-1}\dfrac{1}{1+n(n+1)}$

$$u_n = \tan^{-1}\frac{(n+1)-n}{1+n(n+1)}$$

$$u_n = \tan^{-1}(n+1) - \tan^{-1}(n)$$

$$u_{n+1} = \tan^{-1}(n+2) - \tan^{-1}(n+1)$$

By D' Alembert's test

$$\lim_{n\to\infty}\frac{u_{n+1}}{u_n} = \lim_{n\to\infty}\frac{\tan^{-1}(n+2)-\tan^{-1}(n+1)}{\tan^{-1}(n+1)-\tan^{-1} n}$$

$$= \lim_{n\to\infty}\frac{\dfrac{1}{1+(n+2)^2}-\dfrac{1}{1+(n+1)^2}}{\dfrac{1}{1+(n+1)^2}-\dfrac{1}{1+n^2}} \quad \text{[L'Hopital Rule]}$$

$$= \lim_{n\to\infty}\frac{\dfrac{1+(n+1)^2-1-(n+2)^2}{[1+(n+2)^2]\,[1+(n+1)^2]}}{\dfrac{1+n^2-1-(n+1)^2}{[1+(n+1)^2]\,[1+n^2]}}$$

$$= \lim_{n\to\infty}\frac{(2n+3)(-1)}{(2n+1)(-1)}\times\frac{1+n^2}{1+(n+2)^2} = \lim_{n\to\infty}\frac{(2n+3)(1+n^2)}{(2n+1)[n^2+4n+5]}$$

$$= \lim_{n\to\infty}\frac{\left(2+\frac{3}{n}\right)\left(\frac{1}{n^2}+1\right)}{\left(2+\frac{1}{n}\right)\left(1+\frac{4}{n}+\frac{5}{n^2}\right)} = \frac{2}{2} = 1$$

Test fails.

Let us apply **Raabe's Test.**

$$\lim_{n\to\infty} n\left(\frac{u_n}{u_{n+1}}-1\right) = \lim_{n\to\infty} n\left[\frac{(2n+1)(n^2+4n+5)}{(2n+3)(n^2+1)}-1\right]$$

$$= \lim_{n\to\infty} n\left[\frac{(2n+1)(n^2+4n+5)-(2n+3)(n^2+1)}{(2n+3)(n^2+1)}\right]$$

$$= \lim_{n\to\infty} n\left[\frac{2n^3+8n^2+10n+n^2+4n+5-2n^3-2n-3n^2-3}{(2n+3)\,(n^2+1)}\right]$$

$$= \lim_{n\to\infty} n\left[\frac{6n^2+12n+2}{(2n+3)(n^2+1)}\right] = \lim_{n\to\infty}\frac{6+\frac{12}{n}+\frac{2}{n^2}}{\left(2+\frac{3}{n}\right)\left(1+\frac{1}{n^2}\right)} = \frac{6}{2} = 3 > 1$$

Hence, the given series is convergent. **Ans.**

EXERCISE 7.12

Discuss the convergence of the following series :

1. $1-\frac{1}{2}+\frac{1}{4}-\frac{1}{8}+\ldots\infty$ **Ans.** Absolutely convergent
2. $1-\frac{2}{5}+\frac{3}{10}-\frac{4}{17}+\ldots+\frac{(-1)^{n-1}n}{n^2+1}+\ldots\infty$ **Ans.** Conditionally convergent
3. $1-\frac{2}{3}+\frac{3}{3^2}-\frac{4}{3^3}+\ldots$ **Ans.** Absolutely convergent
4. $1+\frac{1}{2^2}-\frac{1}{3^2}-\frac{1}{4^2}+\frac{1}{5^2}+\frac{1}{6^2}-\frac{1}{7^2}-\frac{1}{8^2}+\ldots$ **Ans.** Absolutely convergent
5. $1-\frac{1}{\sqrt{2}}+\frac{1}{\sqrt{3}}-\frac{1}{\sqrt{4}}+\ldots\infty$ **Ans.** Conditionally convergent
6. $\frac{\sin x}{1^3}-\frac{\sin 2x}{2^3}+\frac{\sin 3x}{3^3}+\ldots$ **Ans.** Absolutely convergent

7.27 POWER SERIES IN *x*

$$a_0 + a_1x + a_2x^2 + \ldots + a_nx^n + \ldots$$

is power series in x, here a's are independent of x.

Proof. $u_n = a_nx^n$ and $u_{n+1} = a_{n+1}x^{n+1}$

D'Alembert's Ratio Test

$$\lim_{n\to\infty}\frac{u_{n+1}}{u_n} = \lim_{n\to\infty}\frac{a_{n+1}x^{n+1}}{a_nx^n} = \lim_{n\to\infty}\frac{a_{n+1}}{a_n}x$$

If $\frac{a_{n+1}}{a_n} = k$,

If $|kx| < 1 \Rightarrow |x| < \frac{1}{k}$, then the series is convergent.

Thus, the power series is convergent if $-\frac{1}{k} < x < \frac{1}{k}$.

Thus, the interval of the power series is $-\frac{1}{k}$ to $\frac{1}{k}$ for convergence. Outside this interval the series is divergent. **Ans.**

Example 59. *Find the values of x for which the series* $x-\frac{x^2}{2^2}+\frac{x^3}{3^2}-\frac{x^4}{4^2}+\ldots\infty$ *converges.*

Solution. Here $u_n = (-1)^{n-1}\frac{x^n}{n^2}$, and $u_{n+1} = (-1)^n\frac{x^{n+1}}{(n+1)^2}$

$$\frac{u_{n+1}}{u_n} = -\frac{n^2}{(n+1)^2}x \quad \Rightarrow \quad \lim_{n\to\infty}\frac{u_{n+1}}{u_n} = -\lim_{n\to\infty}\frac{1}{\left(1+\frac{1}{n}\right)^2}x = -x$$

By **D'Alembert's Test** the given series is convergent for $|x| < 1$ and divergent if $|x| > 1$.

At $x = 1$. The series becomes

$$1 - \frac{1}{2^2} + \frac{1}{3^2} - \frac{1}{4^2} + \ldots$$

This is an alternately convergent series.

At $x = -1$. The series becomes

$$-1 - \frac{1}{2^2} - \frac{1}{3^2} - \frac{1}{4^2} + \ldots$$

This is also convergent series, $p = 2$

Hence, the interval of convergence is $-1 \le x \le 1$. **Ans.**

7.28 EXPONENTIAL SERIES

$$e^x = 1 + x + \frac{x^2}{2!} + \frac{x^3}{3!} + \ldots + \frac{x^n}{n!} + \ldots$$

is convergent for all values of x.

Proof. Here, we have $\quad u_n = \dfrac{x^{n-1}}{(n-1)!}, \quad$ and $\quad u_{n+1} = \dfrac{x^n}{n!}$

$$\lim_{n\to\infty}\frac{u_{n+1}}{u_n} = \lim_{n\to\infty}\frac{x^n}{n!}\,\frac{(n-1)!}{x^{n-1}} = \lim_{n\to\infty}\frac{x}{n} = 0 < 1$$

Hence, by D'Alembert's Test the exponential series is convergent for all values of x.

7.29 LOGARITHMIC SERIES

$$\log(1+x) = x - \frac{x^2}{2} + \frac{x^3}{3} - \frac{x^4}{4} + \ldots + (-1)^{n-1}\frac{x^n}{n} + \ldots$$

is convergent for $-1 < x \le 1$.

Proof. Here, $u_n = (-1)^{n-1}\dfrac{x^n}{n}, \quad$ and $\quad u_{n+1} = (-1)^n\dfrac{x^{n+1}}{n+1}$

$$\lim_{n\to\infty}\frac{u_{n+1}}{u_n} = \lim_{n\to\infty}\frac{(-1)^n x^{n+1}}{n+1}\cdot\frac{n}{(-1)^{n-1}x^n} = -\lim_{n\to\infty}\left(\frac{n}{n+1}\right)x = -\lim_{n\to\infty}\left(\frac{1}{1+\frac{1}{n}}\right)x = -x$$

Thus, the series is convergent for $|x| < 1$ and divergent for $|x| > 1$

At $x = 1$. The series becomes

$$1 - \frac{1}{2} + \frac{1}{3} - \frac{1}{4} + \ldots \infty \text{ which is convergent.}$$

At $x = -1$. The series becomes

$$-1 - \frac{1}{2} - \frac{1}{3} - \frac{1}{4} + \ldots \infty \text{ which is divergent.}$$

7.30 BINOMIAL SERIES

$$(1+x)^n = 1 + nx + \frac{n(n-1)}{2!}x^2 + \frac{n(n-1)(n-2)}{3!}x^3 + \ldots$$

is convergent for $|x| < 1$.

Proof. $$u_r = \frac{n(n-1)\ldots(n-r+2)}{(r-1)!}x^{r-1}$$

$$u_{r+1} = \frac{n(n-1)\ldots(n-r+1)}{r!}x^r$$

$$\lim_{r\to\infty}\frac{u_{r+1}}{u_r} = \lim_{r\to\infty}\frac{n-r+1}{r}x = \lim_{r\to\infty}\left(\frac{n+1}{r}-1\right)x = -x \text{ for } r > n+1$$

If $|x| < 1$, the series is convergent by D'Alembert's Test.

EXERCISE 7.13

Test the convergence of the following series

1. $1 + x + 2x^2 + 3x^3 + \ldots + nx^n + \ldots\infty$ **Ans.** Convergent if $-1 < x > 1$
2. $\frac{1}{2}+\frac{x}{3}+\frac{x^2}{4}+\frac{x^3}{5}+\ldots+\frac{x^n}{n+2}+\ldots\infty$ **Ans.** Convergent for $-1 < x < 1$
3. $1+\frac{1}{1!}+\frac{1}{2!}+\frac{1}{3!}+\ldots\infty$ **Ans.** Convergent
4. $1+\frac{4}{1!}+\frac{4^2}{2!}+\frac{4^3}{3!}+\ldots\infty$ **Ans.** Convergent
5. $3-\frac{3^2}{2}+\frac{3^3}{3}-\frac{3^4}{4}+\ldots\infty$ **Ans.** Convergent
6. $\sqrt{2}-\frac{2}{2}+\frac{2^{\frac{3}{2}}}{3}-\frac{2^2}{4}+\ldots\infty$ **Ans.** Convergent
7. $\frac{1}{1-x}+\frac{1}{2(1-x)^2}+\frac{1}{3(1-x)^3}+\ldots\infty$ **Ans.** Convergent, if $x \neq 1$

7.31 UNIFORM CONVERGENCE

If for a given ε > 0, a number N can be found *independent of x*, such that for every x in the interval (a, b), the series is said to be uniformly convergent in the interval (a, b).

Example 60. *Discuss the uniform convergence of the series*

$$1 + x + x^2 + \ldots\infty$$

Solution. $$S_n = 1 + x + x^2 + \ldots + x^n = \frac{1-x^n}{1-x}$$

$$S_\infty(x) = \lim_{n\to\infty}\frac{1-x^n}{1-x} = \frac{1}{1-x} \qquad \text{for } |x| < 1$$

$$|S_\infty(x) - S_n(x)| = \left|\frac{1}{1-x}-\frac{1-x^n}{1-x}\right| = \left|\frac{x^n}{1-x}\right| = \frac{|x|^n}{1-x} < \varepsilon$$

if $$|x|^n < \varepsilon(1-x)$$

Let $\qquad |x|^N = \varepsilon\,(1 - x) \Rightarrow N = \dfrac{\log \varepsilon (1-x)}{\log|x|}$

In the interval $\left(-\dfrac{1}{2}, \dfrac{1}{2}\right)$, N has maximum value. $N = \dfrac{\log \dfrac{\varepsilon}{2}}{\log \dfrac{1}{2}}$

Hence, the given series is uniformly convergent in the interval $\left(-\dfrac{1}{2}, \dfrac{1}{2}\right)$.

Note. The series is convergent in (– 1, 1) but not uniformly convergent.

7.32 ABEL'S TEST

If $v_n(x)$ be either monotonic decreasing in n for each fixed x in (a, b) or monotonic increasing in n for each fixed x in (a, b), $\Sigma\, a_n(x)\, v_n(x)$ is uniformly convergent in (a, b) if

(*i*) $\Sigma\, a_n(x)$ is uniformly convergent in (a, b).

(*ii*) There exists k such that $|v_n(x)| < k$ for all n when $a \le x < b$.

Example 61. *Prove that* $\dfrac{x^n}{n^3}$ *is uniformly convergent in* (– *1, 1*).

Solution. (*i*) $\sum \dfrac{1}{n^3}$ is uniformly convergent.

(*ii*) $|x^n| < k$ for all n when $-1 < k < 1$

Hence, $\sum \dfrac{x^n}{n^3}$ is uniformly convergent by Abel's Test. **Ans.**

EXERCISE 7.14

Prove that the following series are uniformly convergent in (– 1, 1).

1. $\sum \dfrac{x^n}{n^2}$ 2. $\sum \dfrac{x^n}{n(n+1)}$ 3. $\sum \dfrac{x^{2n}}{x^{2n}+n^2}$

4. Prove that the series $1 + \dfrac{e^{-2x}}{2^2-1} - \dfrac{e^{-4x}}{4^2-1} + \dfrac{e^{-6x}}{6^2-1} + \ldots\infty$ is uniformly convergent with regard to x in $x \ge 0$.

7.33 BRIEF PROCEDURE FOR TESTING A SERIES FOR CONVERGENCE

There are four types of series :

1. Alternating series
2. Positive term series
3. Power series
4. Geometric series

1. Alternating Series. Apply Leibnitz's Test.

2. Positive Term Series.

(*a*) *Cauchy's Fundamental Test.* Find u_n if $\lim\limits_{n\to\infty} u_n \ne 0$, the series is divergent.

(*b*) *Comparison Test.* Apply Comparison Test $\lim\limits_{n\to\infty} \dfrac{u_n}{v_n}$ where $\Sigma\, v_n$ is $\dfrac{1}{n^p}$

(*c*) *D'Alembert's Ratio Test.* If Comparison Test is not applicable, apply D'Alembert's Ratio Test, $\lim\limits_{n\to\infty} \dfrac{u_{n+1}}{u_n}$.

(*d*) *Raabe's Test.* If Ratio Test fails, apply Raabe's Test $\lim_{n\to\infty} n\left(\frac{u_n}{u_{n+1}}-1\right)$.

(*e*) *Cauchy's Root Test.* If Raabe's Test also fails apply Cauchy's Root Test $\lim_{n\to\infty} (u_n)^{1/n}$

(*f*) *Logarithmic Test.* If Cauchy's Root Test fails, apply $\lim_{n\to\infty}\left[n\log\frac{u_n}{u_{n+1}}\right]$

(*g*) *De Morgan's And Bertrand's Test.* Apply De Morgan's and Bertrand's Test.

3. **Power Series.** Apply D'Alembert's Ratio Test. If this test fails, apply all the tests of (2).
4. **Geometric Series**

Note. Gauss's Test and Integral Test are also useful, if they can be applied easily.

7.34 LIST OF THE TESTS FOR CONVERGENCE

1. Cauchy's Fundamental Test

$$\lim_{n\to\infty} u_n \neq 0, \Sigma\, u_n \text{ is divergent.}$$

2. Comparison Test

$$\lim_{n\to\infty}\frac{u_n}{v_n} = k, \Sigma\, u_n \text{ and } \Sigma\, v_n \text{ are of the same nature.}$$

$$\sum\frac{1}{n^p},\ p > 1 \text{ convergent; } p \le 1, \text{ divergent}$$

3. D'Alembert's Test

$$\lim_{n\to\infty}\frac{u_{n+1}}{u_n} = k, \text{ if } k < 1, \Sigma\, u_n \text{ is convergent; if } k > 1, \Sigma\, u_n \text{ is divergent.}$$

4. Raabe's Test

$$\lim_{n\to\infty} n\left(\frac{u_n}{v_{n+1}}-1\right) = k, \text{ if } k > 1, \Sigma\, u_n \text{ is convergent; if } k < 1, \Sigma\, u_n \text{ is divergent.}$$

5. Cauchy's Root Test

$$\lim_{n\to\infty}(u_n)^{1/n} = k, \text{ if } k < 1, \Sigma\, u_n \text{ is convergent; if } k > 1, \Sigma\, u_n \text{ is divergent.}$$

6. Logarithmic Test

$$\lim_{n\to\infty} n\log\frac{u_n}{u_{n+1}} = k, \text{ if } k > 1, \Sigma\, u_n \text{ is convergent; if } k < 1, \Sigma\, u_n \text{ is divergent.}$$

7. De Morgan's and Bertrand's Test

$$\lim_{n\to\infty}\left[\left\{n\left(\frac{u_n}{u_{n+1}}-1\right)-1\right\}\log n\right] = k.$$

If $k > 1$, convergent; $k < 1$, divergent.

8. Gauss's Test

$$\frac{u_n}{u_{n+1}} = \alpha + \frac{\beta}{n} + \frac{\gamma}{n^2}$$

(*i*) (*a*) If $\alpha > 1$, $\Sigma\, u_n$ is convergent

(*b*) If $\alpha < 1$, $\Sigma\, u_n$ is divergent.

(*ii*) If $\alpha = 1$ and $\left[\begin{array}{l}\beta > 1, \Sigma u_n \text{ is convergent} \\ \beta < 1, \Sigma u_n \text{ is divergent}\end{array}\right.$

9. Cauchy's Condensation Test

$\Sigma\,\phi\,(n)$ and $\Sigma\, a^n\,\phi\,(a^n)$ are either both convergent or both divergent.

10. Leibnitz's Test

Alternately positive and negative is convergent if

(*i*) $|u_{n+1}| > |u_n|$ (*ii*) $\lim_{n \to \infty} u_n = 0$

EXERCISE 7.15

Test the following series for convergence:

1. $\dfrac{1}{1\times 2}+\dfrac{1}{2\times 3}+\dfrac{1}{3\times 4}+\ldots+\dfrac{1}{n(n+1)}$ **Ans.** Convergent

2. $\dfrac{1}{1\times 2\times 3}+\dfrac{1}{2\times 3\times 4}+\dfrac{1}{3\times 4\times 5}+\ldots+\dfrac{1}{n(n+1)(n+2)}+\ldots$ **Ans.** Convergent

3. $\sqrt{\dfrac{1}{2}}+\sqrt{\dfrac{2}{3}}+\sqrt{\dfrac{3}{4}}+\sqrt{\dfrac{4}{5}}+\ldots$ **Ans.** Divergent

4. $\sum \tan\dfrac{1}{n}$ **Ans.** Divergent

5. $\sum n\tan\dfrac{1}{n}$ **Ans.** Divergent

6. $\dfrac{1}{1}+\dfrac{1}{2\sqrt{2}}+\dfrac{1}{3\sqrt{3}}+\dfrac{1}{4\sqrt{4}}+\ldots$ **Ans.** Convergent

7. $\Sigma\,(10^{1/n}-1)$ **Ans.** Divergent

8. $\Sigma\,[\sqrt{n^3+1}-\sqrt{n^3}\,]$ **Ans.** Convergent

9. $\sum \dfrac{\sqrt[3]{2n^2-1}}{\sqrt[4]{3n^3+2n+5}}$ **Ans.** Divergent

10. $\sum \dfrac{1}{n^{1+\frac{1}{n}}}$ **Ans.** Divergent

11. $\sum \sin^{-1}\dfrac{1}{n}$ **Ans.** Divergent

12. $\sum \sin^2\dfrac{1}{n}$ $\left[\textbf{Hint. } \sin^2\dfrac{1}{n}=\left[\dfrac{1}{n}-\dfrac{1}{3\,!\,n^3}+\ldots\right]^2\right]$ **Ans.** Convergent

13. $\sum \dfrac{1}{(1)^2\,a+b}+\dfrac{2}{(2)^2\,a+b}+\dfrac{3}{(3)^2\,a+b}+\ldots\dfrac{n}{(n)^2\,a+b}+\ldots$ **Ans.** Divergent

14. $\sum \dfrac{1}{(x-d)^2}+\dfrac{1}{(x-2d)^2}+\dfrac{1}{(x-3d)^2}+\ldots$ **Ans.** Convergent

15. $\sum \dfrac{1}{2^{n-1}}\left(\dfrac{x+1}{x-2}\right)^n$ **Ans.** Convergent if $x < 1$ and $x > 5$; Divergent if $1 \le x \ge 5$

16. $\dfrac{2x}{x+4}+\dfrac{1}{2}\left(\dfrac{2x}{x+4}\right)^2+\dfrac{1}{3}\left(\dfrac{2x}{x+4}\right)^3+\ldots$ **Ans.** Convergent if $-\dfrac{4}{3} \le x < 4$; divergent if $x = 4$.

17. $\dfrac{1}{1\times 3\times 5}+\dfrac{(x+2)^2}{2\times 4\times 6}+\dfrac{(x+2)^4}{3\times 5\times 7}+\dfrac{(x+2)^6}{4\times 6\times 8}+\ldots$ **Ans.** Convergent if $-3 \le x < 1$ and $x = -1$.

18. $\frac{1}{10}+\frac{2!}{10^2}+\frac{3!}{10^3}+\frac{4!}{10^4}+\dots$ **Ans.** Divergent

19. $1+\frac{2!}{2^2}+\frac{3!}{3^2}+\dots$ **Ans.** Divergent

20. $\frac{ax}{2}+\frac{a^2x^2}{5}+\frac{a^3x^3}{10}+\dots+\frac{a^nx^n}{n^2+1}+\dots$ **Ans.** Convergent if $-1 \le ax \le 1$; divergent if $|ax|>1$

21. $\sum \frac{1}{n\,x^n}$ **Ans.** Convergent if $1 < x$; divergent if $x \ge 1$

22. $\sum (-1)^{n-1}\,\frac{x^{2n-1}}{(2n-1)!}$ **Ans.** Convergent

23. $\sum \left(\frac{n^2-1}{n^2+1}\right)x^n$ **Ans.** Convergent if $x < 1$; divergent if $x \ge 1$

24. $\sum \frac{n^3+a}{2^n+a}$ **Ans.** Convergent

25. $\frac{1}{2}+\frac{x^2}{3\sqrt{2}}+\frac{x^4}{4\sqrt{3}}+\dots$ *(Uttarakhand, II Semester, 2012)*

Ans. Convergent if $x^2 \le 1$; divergent if $x^2 > 1$

26. $\sum \frac{n^p}{n!}$ **Ans.** Convergent

27. $1+\frac{3}{2!}+\frac{5}{3!}+\frac{7}{4!}+\dots$ **Ans.** Convergent

28. $1+\frac{2^2}{2!}+\frac{3^2}{3!}+\frac{4^2}{4!}+\dots$ **Ans.** Convergent

29. $\frac{1\times 2\times 3\times\dots\times n}{3\times 5\times\dots\times(2n+1)}\,x^n$ **Ans.** Convergent if $x < 2$; divergent if $x \ge 2$

30. $x^2+\frac{2^2}{3\times 4}x^4+\frac{2^2\times 4^2}{3\times 4\times 5\times 6}x^6+\frac{2^2\times 4^2\times 6^2}{3\times 4\times 5\times 6\times 7\times 8}x^8+\dots$

Ans. Convergent if $x^2 \le 1$; divergent if $x^2 > 1$

31. $\left(\frac{1}{3}\right)^2+\left(\frac{1\times 4}{3\times 6}\right)^2+\left(\frac{1\times 4\times 7}{3\times 6\times 9}\right)^2+\dots+\left(\frac{1\times 4\times 7\times\dots\times \overline{3n-2}}{3\times 6\times 9\times\dots\times 3n}\right)^2+\dots$ **Ans.** Convergent

32. $\frac{1^2}{4^2}+\frac{1^2\times 5^2}{4^2\times 8^2}+\frac{1^2\times 5^2\times 9^2}{4^2\times 8^2\times 12^2}+\dots$ **Ans.** Convergent

33. $1+\frac{(1!)^2}{2!}x+\frac{(2!)^2}{4!}x^2+\frac{(3!)^2}{6!}x^3+\dots$ *(MDU, Dec. 2010)*

Ans. Convergent if $x < 4$; divergent if $x \ge 4$

OBJECTIVE TYPE QUESTIONS

Choose the correct answers :

1. The series $\Sigma\left(n+\frac{1}{n}\right)^{-n^2}$ is
   (*a*) Oscillatory (*b*) Convergent (*c*) Divergent to + ∞ (*d*) None **Ans.** (*b*)
2. The series $\Sigma(\sqrt{n^4+1}-\sqrt{n^4-1})$ is
   (*a*) Divergent (*b*) Convergent (*c*) Oscillatory (*d*) None **Ans.** (*b*)
3. The series Σx^n converges absolutely if
   (*a*) $|x| < 1$ (*b*) $|x| \le 1$ (*c*) $|x| > 1$ (*d*) None **Ans.** (*a*)
4. The series $\sum \frac{1}{n^k}$ is convergent if
   (*a*) $k > 1$ (*b*) $k = 1$ (*c*) $k < 1$ (*d*) $k = -1$ **Ans.** (*a*)
5. The series $\frac{2}{1^2}-\frac{3}{2^2}+\frac{4}{3^2}-\frac{5}{4^2}+ \ldots\ldots\ldots$ is
   (*a*) Divergent (*b*) Absolutely convergent
   (*c*) Conditionally convergent (*d*) Convergent **Ans.** (*c*)
6. An absolutely convergent series is
   (*a*) Convergent but not conditionally (*b*) has its nth term non-zero
   (c) Convergent conditionally as well (*d*) None **Ans.** (*c*)
7. The series which is convergent but not absolutely convergent is called
   (*a*) Oscillatory (*b*) Absolutely convergent
   (*c*) Divergent (*d*) Conditionally convergent **Ans.** (*d*)
8. Every absolutely convergent series is
   (*a*) Convergent (*b*) Oscillatory
   (c) Divergent (*d*) Conditionally convergent **Ans.** (*a*)
9. Σu_n is said to be absolutely convergent if
   (*a*) $|u_n|$ is divergent (*b*) $\Sigma|u_n|$ is convergent (*c*) $\Sigma|u_n|$ is oscillating (*d*) None of these **Ans.** (*b*)
10. The series whose n^{th} term is $\frac{(-1)^{n+1}}{3n-2}$
    (*a*) Conditionally convergent (*b*) Divergent
    (*c*) Absolutely convergent (*d*) Oscillates **Ans.** (*c*)
11. The series $\Sigma\,(-1)^{n-1}\,n^2$ is
    (*a*) Oscillatory (*b*) Absolutely convergent
    (*c*) Divergent (*d*) Conditionally convergent **Ans.** (*d*)
12. $\int_0^\infty f(x)\,dx$ is infinite then the series $\sum_{x=1}^{\infty} f(x)$ is
    (*a*) Divergent (*b*) Convergent (*c*) Oscillates infinity (*d*) None **Ans.** (*a*)
13. Let Σu_n be a series of positive terms and $\lim_{n\to\infty}\frac{u_{n+1}}{u_n} = L$
    (*a*) Divergent (*b*) May be or may not be convergent
    (*c*) Convergent (*d*) Oscillates **Ans.** (*b*)
14. The series $\Sigma\,(-1)^n$
    (*a*) Oscillates (*b*) diverges to ∞ (*c*) Oscillates finitely (*d*) Convergent **Ans.** (*c*)

15. The series $1 + 2e + 2^2 e^2 + 2^3 e^3 + \dots \infty$ is
(*a*) Oscillates (*b*) Divergent to ∞ (*c*) Divergent to $-\infty$ (*d*) is convergent **Ans.** (*d*)

16. The sum of the series is $\frac{1}{1.2} + \frac{1}{2.3} + \frac{1}{3.4} + \dots$
(*a*) Finite (*b*) Infinite (*c*) Not unique for all values of n (*d*) None of these **Ans.** (*a*)

17. If $\lim_{n \to \infty} n\left[\frac{u_n}{u_{n+1}} - 1\right] = 0$ then +ve termed series Σu_n is
(*a*) Oscillating infinitely (*b*) Divergent
(*c*) Oscillates finitely (*d*) Convergent **Ans.** (*b*)

18. $\Sigma \frac{\left(1 + \frac{1}{n}\right)^{2n}}{e^n}$ is
(*a*) Divergent (*b*) Convergent (*c*) Oscillatory (*d*) Cannot say. **Ans.** (*b*)

19. $\sum_{n=0}^{\infty} \frac{x^n}{n!}$ is ... for all values of R.
(*a*) Divergent (*b*) Convergent
(*c*) Absolutely convergent (*d*) None **Ans.** (*c*)

20. $\sum_{n=1}^{\infty} (\sqrt{n^2 + 1} - \sqrt{n^2})$
(*a*) Converges conditionally (*b*) Converges
(*c*) Diverges (*d*) None of them **Ans.** (*b*)

21. The series $1 - \frac{1}{2} + \frac{1}{3} - \frac{1}{4} + \dots$
(*a*) Conditionally convergent (*b*) Absolutely convergent
(*c*) Divergent (*d*) None of them **Ans.** (*b*)

22. Let Σu_n be a series of positive terms with $\lim_{n \to \infty} \frac{u_{n+1}}{u_n} = \frac{2}{3}$ then series is
(*a*) Oscillatory (*b*) May be or may not be convergent
(*c*) Convergent (*d*) Divergent **Ans.** (*c*)

23. The series $\Sigma \frac{\log n}{n^k}$ is convergent when
(*a*) $k > 1$ (*b*) $k < 1$ (*c*) $k = 1$ (*d*) $k = 0$ **Ans.** (*a*)

24. $\sum_{n=1}^{\infty} (\sqrt{n+1} - \sqrt{n})$
(*a*) Converges to 1 (*b*) Converges to 0 (*c*) Diverges to ∞ (*d*) Converges to -1 **Ans.** (*c*)

25. The series $\Sigma \frac{1}{n \log n}$ and the series $\Sigma \frac{1}{n^2}$
(*a*) Both converges (*b*) The first converges and the second diverges
(*c*) The first diverges and the second converges **Ans.** (*c*)

26. $\{e^n\}$ sequence is
(*a*) Oscillates infinity (*c*) Convergent (*c*) Oscillates flintily (*d*) Divergent **Ans.** (*d*)

27. $\Sigma(-1)^n \left[\frac{n^2}{n^3 - 1}\right]$ is
(*a*) Convergent (*b*) Divergent (*c*) Oscillatory (*d*) None **Ans.** (*a*)

28. The series $\sum_{n=0}^{\infty} (2x)^n$ is convergent if

(*a*) –1 £ *x* £ 1 (*b*) $-\frac{1}{2} \le x \le \frac{1}{2}$ (*c*) $-2 < x < 2$ (*d*) None **Ans.** (*b*)

29. The infinite series $\Sigma \frac{1}{2^n + 3^n}$ is

(*a*) Oscillates flintily (*b*) Oscillates infinity (*c*) Convergent (*d*) Divergent **Ans.** (*c*)

30. The series $\sum_{n=2}^{\infty} \frac{(-1)^n}{n \log n}$ is

(*a*) Convergent (*b*) Absolutely convergent (*c*) Oscillatory(*d*) None **Ans.** (*d*)

31. $\sum_{n=1}^{\infty} [(-1)^{2n+1}]$

(*a*) Oscillating between –1, 1 (*b*) Converges to 1
(*c*) Diverges to ∞ (*d*) Converges to 0 **Ans.** (*c*)

32. Σu_n is a series of positive terms such that $\lim_{n \to \infty} \frac{u_n}{u_{n+1}} = l$ then Σu_n diverges if

(*a*) $l < 1$ (*b*) $l > 1$ (*c*) $l = 1$ (*d*) None of them **Ans.** (*a*)

33. $\sum_{n \to 0}^{\infty} \frac{1}{n} \tan \frac{1}{n}$ is

(*a*) Convergent (*b*) Divergent (*c*) Can not say (*d*) None of them **Ans.** (*a*)

34. Σu_n is a series of positive terms such that $\lim u_n^{1/n} = l$, then Σu_n converges if

(*a*) $l < 1$ (*b*) $l > 1$ (*c*) $l = 1$ (*d*) None of them **Ans.** (*b*)

35. $\sum_{n=0}^{\infty} \frac{x^n}{n}$ is ... in every bounded interval of *R*

(*a*) Convergent (*b*) Divergent
(*c*) Absolutely convergent (*d*) None of them **Ans.** (*c*)

36. The series $\sum_{n=1}^{\infty} \frac{(-1)^{n-1}}{n}$

(*a*) Converges (*b*) Absolutely Converges
(*c*) Conditionally converges (*d*) None of them **Ans.** (*c*)

37. $\sum_{n=1}^{\infty} (-1)^{n+1} \frac{\sin^2 n}{2^n}$

(*a*) Divergent (*c*) Conditionally convergent
(*c*) Absolutely convergent (*d*) Convergent **Ans.** (*c*)

38. Which of the following series is not convergent.

(*a*) $\frac{1}{2\sqrt{2}} + \frac{1}{3\sqrt{3}} + \frac{1}{4\sqrt{4}} + ...$ (*b*) $1\frac{1}{2} - 1\frac{1}{3} + 1\frac{1}{4} - 1\frac{1}{5} + ...$

(*c*) $\frac{1}{2} - \frac{1}{3} + \frac{1}{4} - \frac{1}{5} + ...$ (*d*) $x + x^2 + x^3 + x^4$ .... $|x| < 1$ **Ans.** (*a*)

39. $\Sigma \frac{1}{(\log n)^n}$ is

(*a*) Divergent (*b*) Oscillates
(*c*) Convergent (*d*) Conditionally convergent **Ans.** (*c*)

40. Sum of the series $\frac{1}{1.2} - \frac{1}{3.4} + \frac{1}{4.5} \ldots$ is

(*a*) $\log e$ (*b*) $\log \frac{1}{e}$ (*c*) $\log_e 4$ (*d*) None of them **Ans.** (*c*)

41. $\sum_{n=1}^{\infty} \sin\left(\frac{1}{n}\right)$ is

(*a*) Convergent (*b*) Divergent (*c*) Oscillatory (*d*) None of them **Ans.** (*b*)

42. The series with *n*th term equal to $(-1)^{n+1} \frac{\log n}{n}$ is

(*a*) Not convergent (*b*) Absolutely convergent
(*c*) Conditionally convergent (*d*) Convergent **Ans.** (*c*)

43. $\sum_{n=2}^{\infty} \frac{1}{n(\log n)^p}$

(*a*) Converges for all values of p (*b*) Diverges for all values of p
(*c*) Convergent if $p > 1$ (*c*) None **Ans.** (*c*)

44. The series $1 - \frac{1}{3} + \frac{1}{5} - \frac{1}{7} + \ldots$

(*a*) Divergent (*b*) Absolutely convergent
(*c*) Conditionally convergent (*d*) None **Ans.** (*c*)

45. The series $\Sigma \frac{1}{4n+2}$ is

(*a*) Convergent (*b*) Oscillates flintily (*c*) Oscillates infinitely (*d*) Divergent **Ans.** (*d*)

46. The series $\Sigma(-1)^n \left[\frac{n}{2n-1}\right]$ is

(*a*) Convergent (*b*) Divergent (*c*) Oscillatory (*d*) None **Ans.** (*c*)

47. The series $\sum_{n=1}^{\infty} \frac{(-1)^n}{n^2}$

(*a*) Absolutely convergent (*b*) Conditionally convergent (*c*)Divergent (*d*) None of these **Ans.** (*a*)

48. The series $1 - \frac{1}{2!} + \frac{1}{3!} - \frac{1}{4!} + \ldots$ is

(*a*) Divergent (*b*) Conditionally convergent
(*c*) Absolutely convergent (*d*) Convergent **Ans.** (*c*)

49. $\sum \frac{\log n}{n^p}$ is convergent when

(*a*) $p > 1$ (*b*) $p < 1$ (*c*) $p = 1$ (*d*) None **Ans.** (*a*)

50. Sum of the series $\frac{1}{2.3} + \frac{1}{4.5} + \frac{1}{6.7} + \ldots \infty$

(*a*) $\log_e 2$ (*b*) $\log_2 (2/e)$ (*c*) $1 - \log_e 2$ (*d*) None **Ans.** (*c*)

51. $\sum_{n=1}^{\infty} \frac{1}{\sqrt{n+2}}$ is

(*a*) Convergent (*b*) Divergent (*c*) Oscillates (*d*) None **Ans.** (*b*)

52. Σu is a series of positive terms such that $\underset{n \to \infty}{Lt} u_n^{1/n} = p$, then Σu diverges if

(*a*) $p < 1$ (*b*) $p > 1$ (*c*) $p = 1$ (*d*) $p = 0$ **Ans.** (*b*)

53. The series $\sum \frac{1}{2^n + 1}$ is

(*a*) Convergent (*b*) Divergent
(*c*) Conditionally convergent (*d*) None of them **Ans.** (*a*)

54. $\sum_{n=1}^{\infty} \frac{1}{\sqrt{n+1}}$ is

(*a*) Convergent (*b*) Divergent (*c*) Oscillatory (*d*) None of them **Ans.** (*b*)

55. The series $\sum_{n=0}^{\infty} \frac{(-1)^{n-1}}{n}$ is

(*a*) Convergent (*b*) Absolutely divergent
(*c*) Conditionally convergent (*d*) Divergent **Ans.** (*c*)

56. Σu_n is a positive terms series with $\underset{n \to \infty}{Lt} \frac{u_n}{u_{n+1}} = k$ then Σu_n converses if

(*a*) $k < 1$ (*b*) $k > 1$ (*c*) $k = 1$ (*d*) $k = 0$ **Ans.** (*b*)

57. If $\lim_{n \to \infty} u_n = 0$ then Σu_n is

(*a*) Convergent (*b*) Divergent (*c*) May or may not be convergent (*d*) None of these **Ans.** (*b*)

58. $\sum \frac{1}{\log n}$ is

(*a*) Convergent (*b*) Divergent (*c*) Absolutely convergent (*d*) None of these **Ans.** (*b*)

59. The series $\Sigma \left(\frac{n+1}{n+3}\right)$ is

(*a*) Convergent (*b*) Divergent (*c*) Oscillatory (*d*) Can not say **Ans.** (*b*)

60. The sum of the alternating harmonic series $1 - \frac{1}{2} + \frac{1}{3} - \frac{1}{4} + \ldots$ is

(*a*) 0 (*b*) ∞ (*c*) log 2 (*d*) None **Ans.** (*a*)

61. Let Σu_n be a series of positive integers and let $\lim_{n \to \infty} \frac{u_{n+1}}{u_n} = l$, then if $l = 1$ then the series is

(*a*) Oscillatory (*b*) Divergent
(*c*) May be or may not be convergent (*d*) Convergent **Ans.** (*a*)

62. $\sum \left(\frac{n+1}{2n+3}\right)^n$ is

(*a*) Convergent (*b*) Divergent (*c*) Oscillatory (*d*) None **Ans.** (*a*)

63. The series $\frac{2}{1} + \frac{3}{2} + \frac{4}{3} + \frac{5}{4} + \ldots \infty$ is

(*a*) Divergent (*b*) Oscillates finitely (*c*) Oscillates infinitely (*d*) Convergent **Ans.** (*a*)

64. If Σu_n is a series of positive terms and Σu_n is convergent then Σu_n^2 is

(*a*) Convergent (*b*) Oscillatory (*c*) Divergent (*d*) Need not be convergent **Ans.** (*a*)

65. If Lt $S_n \to \infty$ where S_n is the sum of first *n* terms of Σu_n, then Σu_n is

(*a*) Divergent (*b*) Convergent
(*c*) Oscillatory (*d*) Conditionally convergent **Ans.** (*a*)

66. If $\Sigma\, |u_n|$ is convergent then Σu_n is

(*a*) Convergent (*b*) Divergent (*c*) May be or may not be convergent **Ans.** (*a*)

67. The series whose *n*th term is $2^n c^{-n}$ is

(*a*) Oscillatory (*b*) Convergent (*c*) Divergent (*d*) None **Ans.** ()

UNIT–IV

CHAPTER

FOURIER SERIES

8.1 PERIODIC FUNCTIONS

If the value of each ordinate $f(t)$ repeats itself at equal intervals in the abscissa, then $f(t)$ is said to be a periodic function.

If $f(t) = f(t + T) = f(t + 2T) = \ldots$ then T is called the period of the function $f(t)$.

For example:

The period of $\sin x$, $\cos x$, $\sec x$, and $\operatorname{cosec} x$ is 2π.

The period of $\tan x$ and $\cot x$ is π.

$\sin x = \sin(x + 2\pi) = \sin(x + 4\pi) = \ldots$ so $\sin x$ is a periodic function with the period 2π.

$$\sin 5x = \sin(5x + 2\pi) = \sin 5\left(x + \frac{2\pi}{5}\right), \text{ Period} = \frac{2\pi}{5}$$

$$\cos 3x = \cos(3x + 2\pi) = \cos 3\left(x + \frac{2\pi}{3}\right), \text{Period} = \frac{2\pi}{3}$$

$$\cos\frac{2n\pi x}{k} = \cos\left(\frac{2n\pi x}{k} + 2\pi\right) = \cos\frac{2n\pi}{k}\left(x + \frac{2\pi k}{2n\pi}\right)$$

$$= \cos\frac{2n\pi}{k}\left(x + \frac{k}{n}\right), \text{ Period} = \frac{k}{n}$$

$$\tan 2x = \tan(2x + \pi) = \tan 2\left(x + \frac{\pi}{2}\right), \text{ period} = \frac{\pi}{2}$$

This is also called sinusoidal periodic function.

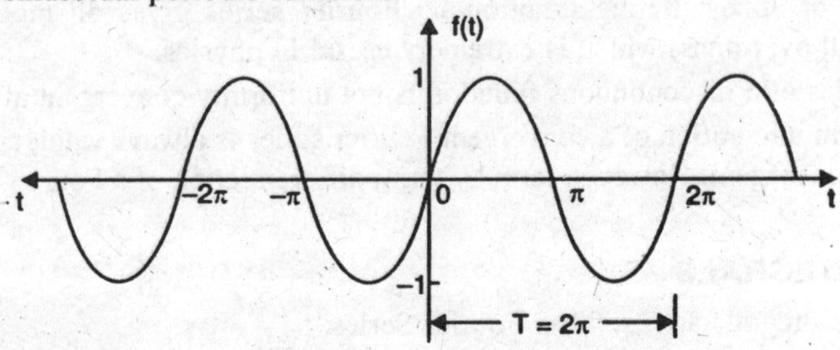

8.2 FOURIER SERIES

Here we will express a non-sinusoidal periodic function into a fundamental and its harmonics. A series of sines and cosines of an angle and its multiples of the form.

$$\frac{a_0}{2} + a_1 \cos x + a_2 \cos 2x + a_3 \cos 3x + \ldots + a_n \cos nx + \ldots$$
$$+ b_1 \sin x + b_2 \sin 2x + b_3 \sin 3x + \ldots + b_n \sin nx + \ldots$$
$$= \frac{a_0}{2} + \sum_{n=1}^{\infty} a_n \cos nx + \sum_{n=1}^{\infty} b_n \sin nx.$$

is called the Fourier series, where $a_0, a_1, a_2, \ldots a_n, \ldots b_1, b_2, b_3 \ldots b_n \ldots$ are constants.

A periodic function $f(x)$ can be expanded in a Fourier Series. The series consists of the following:

(*i*) A constant term a_0 (called d.c. component in electrical work).

(*ii*) A component at the fundamental frequency determined by the values of a_1, b_1.

(*iii*) Components of the harmonics (multiples of the fundamental frequency) determined by $a_2, a_3 \ldots b_2, b_3 \ldots$. And $a_0, a_1, a_2 \ldots, b_1, b_2 \ldots$ are known as *Fourier coefficients* or Fourier constants.

Note. (1) When the function and its derivatives are continuous then the function can be expanded in powers of x by Maclaurin's theorem.

(2) But by Fourier series we can expand continuous and discontinuous both types of functions under certain conditions.

8.3 DIRICHLET'S CONDITIONS FOR A FOURIER SERIES

If the function $f(x)$ for the interval $(-\pi, \pi)$

(1) is single-valued **(2)** is bounded

(3) has at most a finite number of maxima and minima.

(4) has only a finite number of discontinuous

(5) is $f(x + 2\pi) = f(x)$ for values of x outside $[-\pi, \pi]$, then

$$S_P(x) = \frac{a_0}{2} + \sum_{n=1}^{P} a_n \cos nx + \sum_{n=1}^{P} b_n \sin nx$$

converges to $f(x)$ as $P \to \infty$ at values of x for which $f(x)$ is continuous and the sum of the series is equal to $\frac{1}{2}[f(x+0) + f(x-0)]$ at points of discontinuity.

8.4 ADVANTAGES OF FOURIER SERIES

1. Discontinuous function can be represented by Fourier series. Although derivatives of the discontinuous functions do not exist. (This is not true for Taylor's series).

2. The Fourier series is useful in expanding the periodic functions since outside the closed interval, there exists a periodic extension of the function.

3. Expansion of an oscillating function by Fourier series gives all modes of oscillation (fundamental and all overtones) which is extremely useful in physics.

4. Fourier series of a discontinuous function is not uniformly convergent at all points.

5. Term by term integration of a convergent Fourier series is always valid, and it may be valid if the series is not convergent. However, term by term, differentiation of a Fourier series is not valid in most cases.

8.5 USEFUL INTEGRALS

The following integrals are useful in Fourier Series.

(*i*) $\int_0^{2\pi} \sin nx \, dx = 0$ (*ii*) $\int_0^{2\pi} \cos nx \, dx = 0$

(*iii*) $\int_0^{2\pi} \sin^2 nx \, dx = \pi$ (*iv*) $\int_0^{2\pi} \cos^2 nx \, dx = \pi$

(v) $\int_0^{2\pi} \sin nx \cdot \sin mx \, dx = 0$ (vi) $\int_0^{2\pi} \cos nx \cos mx \, dx = 0$

(vii) $\int_0^{2\pi} \sin nx \cdot \cos mx \, dx = 0$ (viii) $\int_0^{2\pi} \sin nx \cdot \cos nx \, dx = 0$

(ix) $[uv]_1 = uv_1 - u'v_2 + u''v_3 - u'''v_4 + \ldots$

where $v_1 = \int v \, dx, v_2 = \int v_1 \, dx$ and so on $u' = \frac{du}{dx}$, $u'' = \frac{d^2u}{dx^2}$ and so on and

(x) $\sin n\pi = 0$, $\cos n\pi = (-1)^n$ where $n \in I$

8.6 DETERMINATION OF FOURIER COEFFICIENTS (EULER'S FORMULAE)

$$f(x) = \frac{a_0}{2} + a_1 \cos x + a_2 \cos 2x + \ldots + a_n \cos nx + \ldots$$
$$+ b_1 \sin x + b_2 \sin 2x + \ldots + b_n \sin nx + \ldots \quad \ldots(1)$$

(i) **To find a_0 :** Integrate both sides of (1) from $x = 0$ to $x = 2\pi$.

$$\int_0^{2\pi} f(x)\,dx = \frac{a_0}{2}\int_0^{2\pi} dx + a_1 \int_0^{2\pi} \cos x\,dx + a_2 \int_0^{2\pi} \cos 2x\,dx + \ldots + a_n \int_0^{2\pi} \cos nx\,dx + \ldots$$
$$+ b_1 \int_0^{2\pi} \sin x\,dx + b_2 \int_0^{2\pi} \sin 2x\,dx + \ldots + b_n \int_0^{2\pi} \sin nx\,dx + \ldots$$

$$= \frac{a_0}{2}\int_0^{2\pi} dx$$ (other integrals = 0 by formulae (i) and (ii) of Art 8.5)

$$\int_0^{2\pi} f(x)\,dx = \frac{a_0}{2} 2\pi \quad \Rightarrow \quad \boxed{a_0 = \frac{1}{\pi}\int_0^{2\pi} f(x)\,dx} \quad \ldots(2)$$

(ii) **To find a_n :** Multiply each side of (1) by cos nx and integrate from $x = 0$ to $x = 2\pi$.

$$\int_0^{2\pi} f(x) \cos nx\,dx = \frac{a_0}{2}\int_0^{2\pi} \cos nx\,dx + a_1 \int_0^{2\pi} \cos x \cos nx\,dx + \ldots + a_n \int_0^{2\pi} \cos^2 nx\,dx \ldots$$
$$+ b_1 \int_0^{2\pi} \sin x \cos nx\,dx + b_2 \int_0^{2\pi} \sin 2x \cos nx\,dx + \ldots$$

$$= a_n \int_0^{2\pi} \cos^2 nx\,dx = a_n \pi$$ (Other integrals = 0, by formulae Art. 8.5)

$$\therefore \quad \boxed{a_n = \frac{1}{\pi}\int_0^{2\pi} f(x) \cos nx\,dx} \quad \ldots(3)$$

By taking $n = 1, 2 \ldots$ we can find the values of $a_1, a_2 \ldots$.

(iii) **To find b_n :** Multiply each side of (1) by sin nx and integrate from $x = 0$ to $x = 2\pi$.

$$\int_0^{2\pi} f(x) \sin nx\,dx = \frac{a_0}{2}\int_0^{2\pi} \sin nx\,dx + a_1 \int_0^{2\pi} \cos x \sin nx\,dx + \ldots + a_n \int_0^{2\pi} \cos nx \sin nx\,dx + \ldots$$
$$\ldots + b_1 \int_0^{2\pi} \sin x \sin nx\,dx + \ldots + b_n \int_0^{2\pi} \sin^2 nx\,dx + \ldots$$

$$= b_n \int_0^{2\pi} \sin^2 nx\,dx$$

$$= b_n \pi$$ (All other integrals = 0, Article No. 8.5)

$$\therefore \quad \boxed{b_n = \frac{1}{\pi}\int_0^{2\pi} f(x) \sin nx\,dx} \quad \ldots(4)$$

Note : To get similar formula of a_0, $\frac{1}{2}$ has been written with a_0 in Fourier series.

Example 1. *Find the Fourier series representing*

$$f(x) = x, \; 0 < x < 2\pi$$

and sketch its graph from $x = -4\pi$ *to* $x = 4\pi$.

Solution. Let $f(x) = \frac{a_0}{2} + a_1 \cos x + a_2 \cos 2x + ... + b_1 \sin x + b_2 \sin 2x + ...$ ...(1)

Hence
$$a_0 = \frac{1}{\pi}\int_0^{2\pi} f(x)\,dx = \frac{1}{\pi}\int_0^{2\pi} x\,dx = \frac{1}{\pi}\left[\frac{x^2}{2}\right]_0^{2\pi} = 2\pi$$

$$a_n = \frac{1}{\pi}\int_0^{2\pi} f(x)\cos nx\,dx = \frac{1}{\pi}\int_0^{2\pi} x\cos nx\,dx$$

$$= \frac{1}{\pi}\left[x\frac{\sin nx}{n} - 1.\left(-\frac{\cos nx}{n^2}\right)\right]_0^{2\pi} = \frac{1}{\pi}\left[\frac{\cos 2n\pi}{n^2} - \frac{1}{n^2}\right] = \frac{1}{n^2\pi}(1-1) = 0$$

$$b_n = \frac{1}{\pi}\int_0^{2\pi} f(x)\sin nx\,dx = \frac{1}{\pi}\int_0^{2\pi} x\sin nx\,dx$$

$$= \frac{1}{\pi}\left[x\left(-\frac{\cos nx}{n}\right) - 1.\left(\frac{-\sin nx}{n^2}\right)\right]_0^{2\pi} = \frac{1}{\pi}\left[\frac{-2\pi\cos 2n\pi}{n}\right] = -\frac{2}{n}$$

Substituting the values of $a_0, a_1, a_2 ..., b_1, b_2 ...$ in (1), we get

$$x = \pi - 2\left[\sin x + \frac{1}{2}\sin 2x + \frac{1}{3}\sin 3x + ...\right]$$ **Ans.**

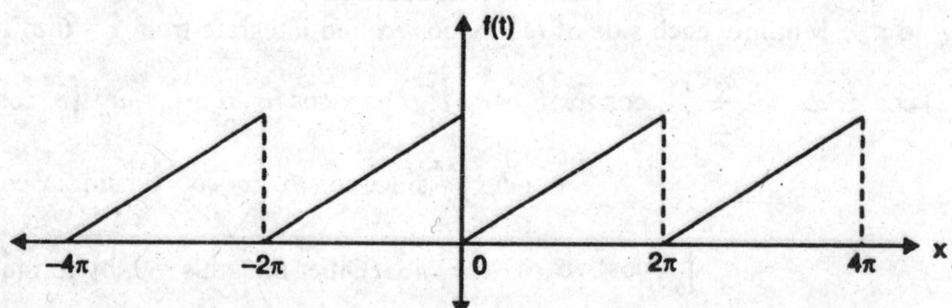

Example 2. *Given that* $f(x) = x + x^2$ *for* $-\pi < x < \pi$, *find the Fourier expression of* $f(x)$.

Deduce that $\frac{\pi^2}{6} = 1 + \frac{1}{2^2} + \frac{1}{3^2} + \frac{1}{4^2} + ...$

(*U.P., II Semester, Summer 2003, Uttarakhand, June 2009*)

Solution. Let $x + x^2 = \frac{a_0}{2} + a_1\cos x + a_2\cos 2x + ... + b_1\sin x + b_2\sin 2x + ...$ ...(1)

$$a_0 = \frac{1}{\pi}\int_{-\pi}^{\pi} f(x)\,dx = \frac{1}{\pi}\int_{-\pi}^{\pi}(x + x^2)\,dx$$

$$= \frac{1}{\pi}\left[\frac{x^2}{2} + \frac{x^3}{3}\right]_{-\pi}^{\pi} = \frac{1}{\pi}\left[\frac{\pi^2}{2} + \frac{\pi^3}{3} - \frac{\pi^2}{2} + \frac{\pi^3}{3}\right] = \frac{2\pi^2}{3}$$

$$a_n = \frac{1}{\pi}\int_{-\pi}^{\pi} f(x)\cos nx\,dx = \int_{-\pi}^{\pi}(x + x^2)\cos nx\,dx$$

$$= \frac{1}{\pi}\left[(x+x^2)\frac{\sin nx}{n} - (2x+1)\frac{(-\cos nx)}{n^2} + (2)\left(-\frac{\sin nx}{n^3}\right)\right]_{-\pi}^{\pi}$$

$$= \frac{1}{\pi}\left[(2\pi+1)\frac{\cos n\pi}{n^2} - (-2\pi+1)\frac{\cos(-n\pi)}{n^2}\right] = \frac{1}{\pi}\left[4\pi\frac{\cos n\pi}{n^2}\right] = \frac{4(-1)^n}{n^2}$$

$$b_n = \frac{1}{\pi}\int_{-\pi}^{\pi} f(x)\sin nx\,dx = \frac{1}{\pi}\int_{-\pi}^{\pi}(x+x^2)\sin nx\,dx$$

$$= \frac{1}{\pi}\left[(x+x^2)\left(-\frac{\cos nx}{n}\right) - (2x+1)\left(\frac{-\sin nx}{n^2}\right) + 2\frac{\cos nx}{n^3}\right]_{-\pi}^{\pi}$$

$$= \frac{1}{\pi}\left[-(\pi+\pi^2)\frac{\cos n\pi}{n} + 2\frac{\cos n\pi}{n^3} + (-\pi+\pi^2)\frac{\cos n\pi}{n} - 2\frac{\cos n\pi}{n^3}\right]$$

$$= \frac{1}{\pi}\left[-\frac{2\pi}{n}\cos n\pi\right] = -\frac{2}{n}(-1)^n$$

Substituting the values of a_0, a_n, b_n, in (1), we get

$$x + x^2 = \frac{\pi^2}{3} + 4\left[-\cos x + \frac{1}{2^2}\cos 2x - \frac{1}{3^2}\cos 3x + \ldots\right]$$

$$- 2\left[-\sin x + \frac{1}{2}\sin 2x - \frac{1}{3}\sin 3x + \ldots\right] \quad \ldots(2)$$

Here, $f(x) = x + x^2$ is valid for all values of x between $-\pi$ and π but not at the end points $-\pi$ and π due to open interval.

$$f(-\pi) = \frac{1}{2}\,[f(-\pi-0) + f(-\pi+0)]$$

$$= \frac{1}{2}\,[f(\pi-0) + f(-\pi+0)] \qquad [f(x) \text{ is periodic with period } 2\pi]$$

$$= \frac{1}{2}\,[(\pi+\pi^2) + \{(-\pi) + (-\pi)^2\}] = \pi^2 \quad \ldots(3)$$

Putting the value of $f(-\pi)$ from (3) and $x = -\pi$ in (2), we get

$$\pi^2 = \frac{\pi^2}{3} + 4\left[\frac{1}{1^2} + \frac{1}{2^2} + \frac{1}{3^2} + \frac{1}{4^2} + \ldots\ldots\right]$$

$$\frac{\pi^2}{6} = \frac{1}{1^2} + \frac{1}{2^2} + \frac{1}{3^2} + \frac{1}{4^2} + \ldots.. \qquad \textbf{Proved.}$$

Example 3. *Find the Fourier series expansion for f (x)* $= x + \frac{x^2}{4}$, $-\pi \le x \le \pi$

(U.P. II Semester, 2009)

Solution. Let $x + \frac{x^2}{4} = \frac{a_0}{2} + a_1\cos x + a_2\cos 2x + \ldots\ldots + b_1\sin x + b_2\sin 2x + \ldots..$

where $$a_0 = \frac{1}{\pi}\int_{-\pi}^{\pi} f(x)\,dx = \frac{1}{\pi}\int_{-\pi}^{\pi}\left(x + \frac{x^2}{4}\right)dx$$

$$= \frac{1}{\pi}\left[\frac{x^2}{2} + \frac{x^3}{12}\right]_{-\pi}^{\pi} = \frac{1}{\pi}\left[\frac{\pi^2}{2} + \frac{\pi^3}{12} - \frac{\pi^2}{2} + \frac{\pi^3}{12}\right] = \frac{1}{\pi}\left[\frac{2\pi^3}{12}\right] = \frac{\pi^2}{6}$$

$$a_n = \frac{1}{\pi}\int_{-\pi}^{\pi} f(x)\cos n\,x\,dx = \frac{1}{\pi}\int_{-\pi}^{\pi}\left(x+\frac{x^2}{4}\right)\cos n\,x\,dx$$

$$= \frac{1}{\pi}\left[\left(x+\frac{x^2}{4}\right)\left(\frac{\sin nx}{n}\right)-\left(1+\frac{2x}{4}\right)\left(\frac{-\cos nx}{n^2}\right)+\frac{1}{2}\left(-\frac{\sin nx}{n^3}\right)\right]_{-\pi}^{\pi}$$

$$= \frac{1}{\pi}\left[\left(\pi+\frac{\pi^2}{4}\right)\left(\frac{\sin n\pi}{n}\right)+\left(1+\frac{2\pi}{4}\right)\left(\frac{\cos n\pi}{n^2}\right)-\frac{1}{2}\left(\frac{\sin n\pi}{n^3}\right)-\left(-\pi+\frac{\pi^2}{4}\right)\left(\frac{\sin(-n\pi)}{n}\right)\right.$$

$$\left.-\left(1-\frac{2\pi}{4}\right)\frac{\cos(-n\pi)}{n^2}+\frac{1}{2}\frac{\sin(-n\pi)}{n^3}\right]$$

$$= \frac{1}{\pi}\left[\left(1+\frac{\pi}{2}\right)\frac{(-1)^n}{n^2}-\left(1-\frac{\pi}{2}\right)\frac{(-1)^n}{n^2}\right]$$

$$= \frac{(-1)^n}{n^2\pi}\left[1+\frac{\pi}{2}-1+\frac{\pi}{2}\right] = \frac{(-1)^n}{n^2\pi}(\pi) = \frac{(-1)^n}{n^2}$$

$$a_1 = -1$$

$$a_2 = \frac{1}{4}$$

$$a_3 = -\frac{1}{9}$$

$$a_4 = \frac{1}{16}$$

....

....

$$b_n = \frac{1}{\pi}\int_{-\pi}^{\pi} f(x)\sin nx\,dx$$

$$= \frac{1}{\pi}\int_{-\pi}^{\pi}\left(x+\frac{x^2}{4}\right)\sin nx\,dx$$

$$= \frac{1}{\pi}\int_{-\pi}^{\pi} x\sin nx\,dx + \frac{1}{\pi}\int_{-\pi}^{\pi}\frac{x^2}{4}\sin nx\,dx$$

Even function Odd function

$$= \frac{2}{\pi}\int_0^{\pi} x\sin n\,x\,dx + 0$$

$$= \frac{2}{\pi}\left[x\left(-\frac{\cos n\,x}{n}\right)-(1)\left(-\frac{\sin n\,x}{n^2}\right)\right]_0^{\pi}$$

$$= \frac{2}{\pi}\left[-\pi\frac{\cos n\pi}{n}+\frac{\sin n\,\pi}{n^2}\right]$$

$$= \frac{2}{\pi}\left[-\pi\frac{(-1)^n}{n}\right] = -\frac{2(-1)^n}{n}$$

$$b_1 = \frac{2}{1}, \quad b_2 = -1, \quad b_3 = \frac{2}{3}, \quad b_4 = -\frac{1}{2} \ldots$$

Hence, Fourier series of the given function is

$$f(x) = \frac{a_0}{2} + \sum_{n=1}^{\infty} a_n \cos nx + \sum_{n=1}^{\infty} b_n \sin nx$$

$$= \frac{\pi^2}{12} + \sum \frac{(-1)^n}{n^2} \cos nx - \sum_{n=1}^{\infty} \frac{2}{n}(-1)^n \sin nx$$

$$f(x) = \frac{\pi^2}{12} - \cos x + \frac{1}{4}\cos 2x - \frac{1}{9}\cos 3x + \frac{1}{16}\cos 4x + \ldots + 2\sin x - \sin 3x + \frac{2}{3}\sin 4x - \frac{1}{2}\sin 4x + \ldots$$

Ans.

EXERCISE 8.1

1. Find a Fourier series to represent, $f(x) = \pi - x$ for $0 < x < 2\pi$.

Ans. $2\left[\sin x + \frac{1}{2}\sin 2x + \frac{1}{3}\sin 3x + \ldots + \frac{1}{n}\sin nx + \ldots\right]$

2. Find a Fourier series to represent $x - x^2$ from $x = -\pi$ to π and show that

$$\frac{\pi^2}{12} = \frac{1}{1^2} - \frac{1}{2^2} + \frac{1}{3^2} - \frac{1}{4^2} + \ldots$$

Ans. $-\frac{\pi^2}{3} + 4\left[\frac{\cos x}{1^2} - \frac{\cos 2x}{2^2} + \frac{\cos 3x}{3^2} - \frac{\cos 4x}{4^2} + \ldots\right] + 2\left[\frac{\sin x}{1} - \frac{\sin 2x}{2} + \frac{\sin 3x}{3} - \frac{\sin 4x}{4} + \ldots\right]$

3. Find a Fourier series to represent : $f(x) = x \sin x$, for $0 < x < 2\pi$. *(U.P., II Semester, Summer 2001)*

Ans. $-1 + \pi \sin x - \frac{1}{2}\cos x + 2\left[\frac{\cos 2x}{2^2 - 1} + \frac{\cos 3x}{3^2 - 1} + \frac{\cos 4x}{4^2 - 1} + \ldots\right]$

4. Find a Fourier series to represent the function $f(x) = e^x$, for $-\pi < x < \pi$ and hence derive a series for $\frac{\pi}{\sinh \pi}$.

Ans. $\frac{2\sinh \pi}{\pi}\left[\frac{1}{2} - \frac{1}{1^2+1}\cos x + \frac{1}{2^2+1}\cos 2x - \frac{1}{3^2+1}\cos 3x + \ldots\right.$

$\left. + \frac{1}{1^2+1}\sin x - \frac{2}{2^2+1}\sin 2x - \frac{3}{3^2+1}\sin 3x \ldots,\right]$

$$\frac{\pi}{\sinh \pi} = 1 + 2\left[-\frac{1}{2} + \frac{1}{5} - \frac{1}{10} + \ldots\right]$$

5. Obtain the Fourier series for $f(x) = e^{-x}$ in the interval $0 \le x < 2\pi$.

Ans. $\frac{1 - e^{-2\pi}}{\pi}\left[\frac{1}{2} + \frac{1}{2}\cos x + \frac{1}{5}\cos 2x + \frac{1}{10}\cos 3x + \frac{1}{2}\sin x + \frac{2}{5}\sin 2x + \frac{3}{10}\sin 3x + \ldots\right]$

6. If $f(x) = \left(\frac{\pi - x}{2}\right)^2, 0 < x < 2\pi$, show that $f(x) = \frac{\pi^2}{12} + \sum_{n=1}^{\infty} \frac{\cos nx}{n^2}$

7. Prove that $x^2 = \frac{\pi^2}{3} + 4\sum_{n=1}^{\infty} (-1)^n \frac{\cos nx}{n^2}, -\pi < x < \pi$.

Hence show that

(*i*) $\Sigma \frac{1}{(2n-1)^2} = \frac{\pi^2}{8}$ (*ii*) $\Sigma \frac{1}{n^4} = \frac{\pi^4}{90}$

8. If $f(x)$ is a periodic function defined over a period $(0, 2\pi)$ by $f(x) = \dfrac{(3x^2 - 6x\pi + 2\pi^2)}{12}$.

Prove that $f(x) = \sum_{n=1}^{\infty} \dfrac{\cos nx}{n^2}$ and hence show that $\dfrac{\pi^2}{6} = 1 + \dfrac{1}{2^2} + \dfrac{1}{3^2} + \ldots$

8.7 FOURIER SERIES FOR DISCONTINUOUS FUNCTIONS

Let the function $f(x)$ be defined by

$$f(x) = f_1(x), \quad c < x < x_0$$
$$= f_2(x), \quad x_0 < x < c + 2\pi,$$

where x_0 is the point of discontinuity in the interval $(c, c + 2\pi)$.

In such cases also, we obtain the Fourier series for $f(x)$ in the usual way. The values of a_0, a_n, b_n are evaluated by

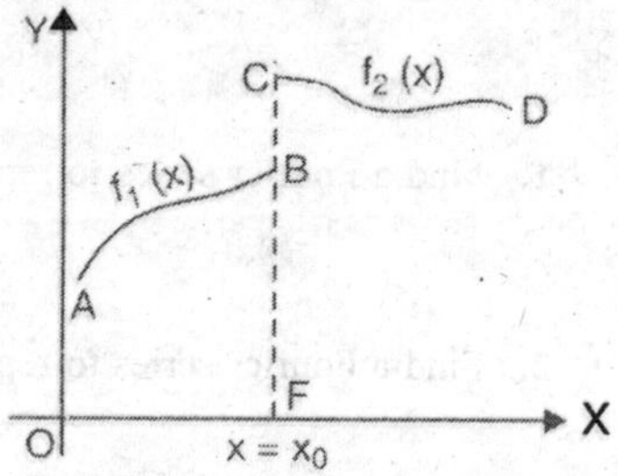

$$a_0 = \frac{1}{\pi}\left[\int_c^{x_0} f_1(x)\, dx + \int_{x_0}^{2+2\pi} f_2(x)\, dx\right];$$

$$a_n = \frac{1}{\pi}\left[\int_c^{x_0} f_1(x) \cos n x\, dx + \int_{x_0}^{2+2\pi} f_2(x) \cos n x\, dx\right]$$

$$b_n = \frac{1}{\pi}\left[\int_c^{x_0} f_1(x) \sin n x\, dx + \int_{x_0}^{c+2\pi} f_2(x) \cos n x\, dx\right]$$

If $x = x_0$ is the point of finite discontinuity, then the sum of the Fourier series

$$= \frac{1}{2}\left[\lim_{h \to 0} f(x_0 - h) + \lim_{h \to 0} f(x_0 + h)\right]$$

$$= \frac{1}{2}[f(x_0 - 0) + f(x_0 + 0)] = \frac{1}{2}(FB + FC)$$

Remarks.

1. It may be seen from the graph, that at a point of finite discontinuity $x = x_0$, there is a finite jump equal to BC in the value of the function $f(x)$ at $x = x_0$.

2. A given function $f(x)$ may be defined by different formulae in different regions. Such types of functions are quite common in Fourier Series.

3. At a point of discontinuity the sum of the series is equal to the mean of the limits on the right and left.

8.8 FUNCTION DEFINED IN TWO OR MORE SUB-RANGES

Example 4. *Find the fourier series to represent the function f (x) given by :*

$$f(x) = \begin{cases} -k & \text{for } -\pi < x < 0 \\ k & \text{for } 0 < x < \pi \end{cases}$$

Hence show that :

$$1 - \frac{1}{3} + \frac{1}{5} - \frac{1}{7} + \ldots = \frac{\pi}{4}.$$ (*U.P. II Semester 2010*)

Solution. $f(x) = \begin{cases} -k, & -\pi < x < 0 \\ k, & 0 < x < \pi \end{cases}$

$$a_0 = \frac{1}{\pi}\int_{-\pi}^{\pi} f(x)\, dx$$

$$= \frac{1}{\pi}\left[\int_{-\pi}^{0} -k\,dx + \int_{0}^{\pi} k\,dx\right] = \frac{1}{\pi}\left[\,[-kx]_{-\pi}^{0} + [kx]_{0}^{\pi}\right]$$

$$= \frac{1}{\pi}\,k\,[0 - \pi + \pi - 0] = 0$$

$$a_n = \frac{1}{\pi}\int_{-\pi}^{\pi} f(x)\;\cos nx\,dx$$

$$= \frac{1}{\pi}\left[\int_{-\pi}^{0} -k\cos nx\,dx + \int_{0}^{\pi} k\cos nx\,dx\right]$$

$$= \frac{1}{\pi}\,k\left[-\left\{\frac{\sin nx}{n}\right\}_{-\pi}^{0} + \left\{\frac{\sin nx}{n}\right\}_{0}^{\pi}\right] = \frac{1}{\pi}\,k\,[-0+0] = 0$$

$$b_n = \frac{1}{\pi}\int_{-\pi}^{\pi} f(x)\sin nx\,dx$$

$$= \frac{1}{\pi}\left[\int_{-\pi}^{0} -k\sin nx\,dx + \int_{0}^{\pi} k\sin nx\,dx\right]$$

$$= \frac{1}{\pi}\,k\left[\left\{\frac{\cos nx}{n}\right\}_{-\pi}^{0} - \left\{\frac{\cos nx}{n}\right\}_{0}^{\pi}\right]$$

$$= \frac{1}{\pi}\,k\left[\frac{1}{n} - \frac{(-1)^n}{n} - \frac{(-1)^n}{n} + \frac{1}{n}\right] = \frac{1}{\pi}\,k\left[\frac{2}{n} - \frac{2\,(-1)^n}{n}\right]$$

If n is even $\quad b_n = 0$

If n is odd $\quad b_n = \dfrac{4k}{n\pi}$

$$f(x) = a_0 + \sum_{n=1}^{\infty} a_n\cos nx + \sum_{n=1}^{\infty} b_n\sin nx \qquad \text{... (1)}$$

$$f(x) = \sum_{n=1}^{\infty} b_n\sin nx$$

$$= b_1\sin x + b_2\sin 2x + b_3\sin 3x + b_4\sin 4x + \ldots$$

Thus required Fourier sine series is

$$f(x) = \frac{4k}{\pi}\sin x + \frac{4k}{3\pi}\sin 3x + \frac{4k}{5\pi}\sin 5x + \ldots$$

$$\Rightarrow \qquad f(x) = \frac{4k}{\pi}\left[\sin x + \frac{1}{3}\sin 3x + \frac{1}{5}\sin 5x + \ldots\right] \qquad \text{... (2)}$$

Putting $x = \dfrac{\pi}{2}$ in (2), we get

$$k = \frac{4k}{\pi}\left[\sin\frac{\pi}{2} + \frac{1}{3}\sin\frac{3\pi}{2} + \frac{1}{5}\sin\frac{5\pi}{2} + \ldots.\right]$$

$$\Rightarrow \qquad 1 = \frac{4}{\pi}\left[1 + \frac{1}{3}(-1) + \frac{1}{5}(1) + \frac{1}{7}(-1) + \ldots\right]$$

$$= \frac{4}{\pi}\left[1 - \frac{1}{3} + \frac{1}{5} - \frac{1}{7} + \ldots\right]$$

$$\Rightarrow \qquad \frac{\pi}{4} = 1 - \frac{1}{3} + \frac{1}{5} - \frac{1}{7} + \ldots$$

Proved.

Example 5. *Find the Fourier series of the function*

$$f(x) = \begin{cases} -1 & \text{for} \quad -\pi < x < -\dfrac{\pi}{2} \\ 0 & \text{for} \quad -\dfrac{\pi}{2} < x < \dfrac{\pi}{2} \\ +1 & \text{for} \quad \dfrac{\pi}{2} < x < \pi. \end{cases}$$

Solution. Let $f(x) = \dfrac{a_0}{2} + a_1 \cos x + a_2 \cos 2x + \ldots + b_1 \sin x + b_2 \sin 2x + \ldots$ ...(1)

$$a_0 = \frac{1}{\pi}\int_{-\pi}^{\pi} f(x)\,dx = \frac{1}{\pi}\int_{-\pi}^{-\pi/2}(-1)\,dx + \frac{1}{\pi}\int_{-\pi/2}^{\pi/2} 0\,dx + \frac{1}{\pi}\int_{\pi/2}^{\pi} 1\,dx$$

$$= \frac{1}{\pi}[-x]_{-\pi}^{-\pi/2} + \frac{1}{\pi}[x]_{\pi/2}^{\pi} = \frac{1}{\pi}\left[\frac{\pi}{2} - \pi + \pi - \frac{\pi}{2}\right] = 0$$

$$a_n = \frac{1}{\pi}\int_{-\pi}^{\pi} f(x)\cos nx\,dx$$

$$= \frac{1}{\pi}\int_{-\pi}^{-\pi/2}(-1)\cos nx\,dx + \frac{1}{\pi}\int_{-\pi/2}^{\pi/2}(0)\cos nx\,dx + \frac{1}{\pi}\int_{\pi/2}^{\pi}(1)\cos nx\,dx$$

$$= -\frac{1}{\pi}\left[\frac{\sin nx}{n}\right]_{-\pi}^{-\pi/2} + \frac{1}{\pi}\left[\frac{\sin nx}{n}\right]_{\pi/2}^{\pi}$$

$$= -\frac{1}{\pi}\left[-\frac{\sin\dfrac{n\pi}{2}}{n}\right] + \frac{1}{\pi}\left[-\frac{\sin\dfrac{n\pi}{2}}{n}\right] = 0$$

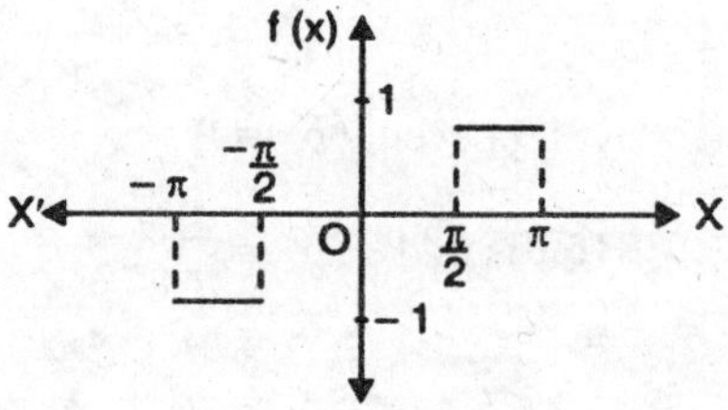

$$b_n = \frac{1}{\pi}\int_{-\pi}^{\pi} f(x)\sin nx\,dx$$

$$= \frac{1}{\pi}\int_{-\pi}^{-\pi/2}(-1)\sin nx\,dx + \frac{1}{\pi}\int_{-\pi/2}^{\pi/2}(0)\sin nx\,dx$$

$$+\frac{1}{\pi}\int_{\pi/2}^{\pi}(1)\sin nx\,dx = \frac{1}{\pi}\left[\frac{\cos nx}{n}\right]_{-\pi}^{-\pi/2} - \frac{1}{\pi}\left[\frac{\cos nx}{n}\right]_{\pi/2}^{\pi}$$

$$= \frac{1}{n\pi}\left[\cos\frac{n\pi}{2} - \cos n\pi\right] - \frac{1}{n\pi}\left(\cos n\pi - \cos\frac{n\pi}{2}\right) = \frac{2}{n\pi}\left[\cos\frac{n\pi}{2} - \cos n\pi\right]$$

$$b_1 = \frac{2}{\pi}, \qquad b_2 = -\frac{2}{\pi}, \qquad b_3 = \frac{2}{3\pi}$$

Putting the values of a_0, a_n, b_n in (1), we get

$$f(x) = \frac{1}{\pi}\left[2\sin x - 2\sin 2x + \frac{2}{3}\sin 3x + \ldots\right]$$ **Ans.**

Example 6. *Find the Fourier series for the periodic function*

$$f(x) = \begin{cases} 0, & -\pi < x < 0 \\ x, & 0 < x < \pi \end{cases}$$

$$f(x + 2\pi) = f(x)$$

Solution. Let $f(x) = \dfrac{a_0}{2} + a_1 \cos x + a_2 \cos 2x + \ldots + b_1 \sin x + b_2 \sin 2x + \ldots$ ...(1)

$$a_0 = \frac{1}{\pi}\int_{-\pi}^{0} 0.\,dx + \frac{1}{\pi}\int_0^{\pi} x\,dx = \frac{1}{\pi}\left[\frac{x^2}{2}\right]_0^{\pi} = \frac{1}{\pi}\left(\frac{\pi^2}{2}\right) = \frac{\pi}{2}$$

$$a_n = \frac{1}{\pi}\int_0^{\pi} x\cos nx\,dx = \frac{1}{\pi}\left[x\cdot\frac{\sin nx}{n} - (1)\left(-\frac{\cos nx}{n^2}\right)\right]_0^{\pi}$$

$$= \frac{1}{\pi}\left[\frac{(-1)^n}{n^2} - \frac{1}{n^2}\right] = -\frac{2}{n^2\pi}, \text{ when } n \text{ is odd}$$

$= 0,$ when n is even.

$$b_n = \frac{1}{\pi}\int_0^{\pi} x\sin nx\,dx = \frac{1}{\pi}\left[x\left(-\frac{\cos nx}{n}\right) - (1)\left(-\frac{\sin nx}{n^2}\right)\right]_0^{\pi} = \frac{1}{\pi}\left[-\pi\frac{(-1)^n}{n}\right] = -\frac{(-1)^n}{n}$$

Substituting the values of $a_0, a_1, a_2 \ldots b_1, b_2, \ldots$ in (1), we get

$$f(x) = \frac{\pi}{4} - \frac{2}{\pi}\left[\frac{\cos x}{1^2} + \frac{\cos 3x}{3^2} + \frac{\cos 5x}{5^2} + \ldots\right] + \left[\frac{\sin x}{1} - \frac{\sin 2x}{2} + \frac{\sin 3x}{3} + \ldots\right]$$ **Ans.**

8.9 DISCONTINUOUS FUNCTIONS

At a point of discontinuity, Fourier series gives the value of $f(x)$ as the arithmetic mean of left and right limits.

At the point of discontinuity, $x = c$

$$\text{At } x = c,\ f(x) = \frac{1}{2}\,[f(c-0) + f(c+0)]$$

Example 7. *Find the Fourier series for $f(x)$, if* $f(x) = \begin{cases} -\pi, & -\pi < x < 0 \\ x, & 0 < x < \pi \end{cases}$

Deduce that $\dfrac{1}{1^2} + \dfrac{1}{3^2} + \dfrac{1}{5^2} + \ldots = \dfrac{\pi^2}{8}$

Solution. Let $f(x) = \dfrac{a_0}{2} + a_1 \cos x + a_2 \cos 2x + \ldots + a_n \cos nx + \ldots$

$+ b_1 \sin x + b_2 \sin 2x + \ldots + b_n \sin nx + \ldots$ ...(1)

$$a_0 = \frac{1}{\pi}\int_{-\pi}^{\pi} f(x)\,dx$$

Then $a_0 = \dfrac{1}{\pi}\left[\int_{-\pi}^{0}(-\pi)\,dx + \int_0^{\pi} x\,dx\right] = \dfrac{1}{\pi}\left[-\pi\,(x)_{-\pi}^{0} + (x^2/2)_0^{\pi}\right] = \dfrac{1}{\pi}(-\pi^2 + \pi^2/2) = -\dfrac{\pi}{2};$

$$a_n = \frac{1}{\pi}\int_{-\pi}^{\pi} f(x)\cos nx\,dx$$

$$a_n = \frac{1}{\pi}\left[\int_{-\pi}^{0}(-\pi)\cos nx\,dx + \int_0^{\pi} x\cos nx\,dx\right] = \frac{1}{\pi}\left[-\pi\left(\frac{\sin nx}{n}\right)_{-\pi}^{0} + \left(\frac{x\sin nx}{n} + \frac{\cos nx}{n^2}\right)_0^{\pi}\right]$$

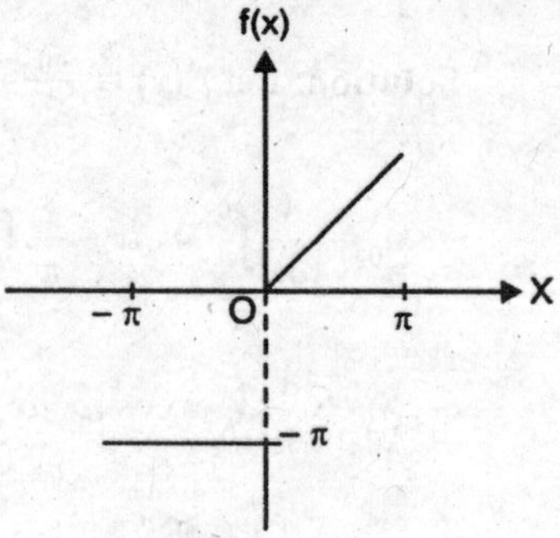

$$= \frac{1}{\pi}\left[0 + \frac{1}{n^2}\cos n\pi - \frac{1}{n^2}\right] = \frac{1}{\pi n^2}(\cos n\pi - 1)$$

$$b_n = \frac{1}{\pi}\int_{-\pi}^{\pi} f(x)\sin nx\, dx$$

$$= \frac{1}{\pi}\left[\int_{-\pi}^{0}(-\pi)\sin nx\, dx + \int_0^{\pi} x\sin nx\, dx\right]$$

$$= \frac{1}{\pi}\left[\left(\frac{\pi\cos nx}{n}\right)_{-\pi}^{0} + \left(-x\frac{\cos nx}{n} + \frac{\sin nx}{n^2}\right)_0^{\pi}\right]$$

$$= \frac{1}{\pi}\left[\frac{\pi}{n}(1 - \cos n\pi) - \frac{\pi}{n}\cos n\pi\right] = \frac{1}{n}(1 - 2\cos n\pi)$$

$$f(x) = -\frac{\pi}{4} - \frac{2}{\pi}\left(\cos x + \frac{\cos 3x}{3^2} + \frac{\cos 5x}{5^2} + \ldots\right) + 3\sin x - \frac{\sin 2x}{2} + \frac{3\sin 3x}{3} - \frac{\sin 4x}{4} \quad \ldots(2)$$

Putting $x = 0$ in (2), we get $f(0) = -\frac{\pi}{4} - \frac{2}{\pi}\left(1 + \frac{1}{3^2} + \frac{1}{5^2} + \ldots\infty\right)$ ...(3)

Now, $f(x)$ is discontinuous at $x = 0$.
But $f(0 - 0) = -\pi$ and $f(0 + 0) = 0$

$$\therefore \quad f(0) = \frac{1}{2}[f(0 - 0) + f(0 + 0)] = -\pi/2 \quad \ldots(4)$$

Form (3) and (4), we get $-\frac{\pi}{2} = -\frac{\pi}{4} - \frac{2}{\pi}\left[\frac{1}{1^2} + \frac{1}{3^2} + \frac{1}{5^2} + \ldots\right] \Rightarrow \frac{\pi^2}{8} = \frac{1}{1^2} + \frac{1}{3^2} + \frac{1}{5^2} + \ldots$ **Proved.**

Example 8. *Obtain Fourier Series of the function*

$$f(x) = \begin{cases} x, & -\pi < x < 0 \\ -x, & 0 < x < \pi \end{cases}$$

and hence show that

$$\frac{1}{1^2} + \frac{1}{3^2} + \frac{1}{5^2} + \ldots\ldots\ldots\ldots = \frac{\pi^2}{8}$$

(U.P., II Semester, June 2008, 2002)

Solution. We have, $f(x) = \begin{cases} x, & -\pi < x < 0 \\ -x, & 0 < x < \pi \end{cases}$

Here $f(x)$ is an even function so $b_n = 0$

$$a_0 = \frac{2}{\pi}\int_0^{\pi} f(x)\, dx = \frac{2}{\pi}\int_0^{\pi} -x\, dx = -\frac{2}{\pi}\left[\frac{x^2}{2}\right]_0^{\pi} = -\frac{2}{\pi}\left[\frac{\pi^2}{2}\right] = -\pi$$

$$a_n = \frac{2}{\pi}\int_0^{\pi} f(x)\cos nx\, dx = \frac{2}{\pi}\int_0^{\pi} -x\cos nx\, dx = -\frac{2}{\pi}\left[x\frac{\sin nx}{n} + \frac{\cos nx}{n^2}\right]_0^{\pi}$$

$$= -\frac{2}{\pi}\left[\frac{(-1)^n}{n^2} - \frac{1}{n^2}\right] = \frac{2}{\pi n^2}[1 - (-1)^n] = \begin{cases} 0, & n \text{ is even} \\ \frac{4}{\pi n^2}, & n \text{ is odd} \end{cases}$$

Fourier series

$$f(x) = -\frac{\pi}{2}+\frac{4}{\pi}\left[\frac{\cos x}{1^2}+\frac{\cos 3x}{3^2}+\frac{\cos 5x}{5^2}+....\right]$$ **Ans.**

At the point of discontinuity

$$f(0) = \frac{1}{2}\ [f(0-0)+f(0+0)] = \frac{1}{2}\ (0-0)=0$$

Putting $x = 0$ in the above, we get

$$0 = -\frac{\pi}{2}+\frac{4}{\pi}\left(\frac{1}{1^2}+\frac{1}{3^2}+\frac{1}{5^2}+.........\right)$$

$$\Rightarrow \quad \frac{1}{1^2}+\frac{1}{3^2}+\frac{1}{5^2}+.........=\frac{\pi^2}{8}$$ **Ans.**

Example 9. *Find the Fourier series expansion of the periodic function of period 2π, defined by*

$$f(x) = \begin{cases} x, & \text{if } -\dfrac{\pi}{2}<x<\dfrac{\pi}{2} \\ \pi - x, & \text{if } \dfrac{\pi}{2}<x<\dfrac{3\pi}{2} \end{cases}$$

Solution. Let $f(x) = \dfrac{a_0}{2} + a_1 \cos x + a_2 \cos 2x + ... + b_1 \sin x + b_2 \sin 2x + ...$ ... (1)

$$a_0 = \frac{1}{\pi}\int_{-\pi/2}^{\pi/2} x\,dx + \frac{1}{\pi}\int_{\pi/2}^{\frac{3\pi}{2}} (\pi - x)\,dx = \frac{1}{\pi}\left(\frac{x^2}{2}\right)_{-\pi/2}^{\pi/2} + \frac{1}{\pi}\left(\pi x - \frac{x^2}{2}\right)_{\pi/2}^{\frac{3\pi}{2}}$$

$$= \frac{1}{\pi}\left(\frac{\pi^2}{8}-\frac{\pi^2}{8}\right)+\frac{1}{\pi}\left(\frac{3\pi^2}{2}-\frac{9\pi^2}{8}-\frac{\pi^2}{2}+\frac{\pi^2}{8}\right) = \pi\left(\frac{3}{2}-\frac{9}{8}-\frac{1}{2}+\frac{1}{8}\right)=0$$

$$a_n = \frac{1}{\pi}\int_{-\pi/2}^{\pi/2} x \cos nx\,dx + \frac{1}{\pi}\int_{\pi/2}^{\frac{3\pi}{2}} (\pi - x)\cos nx\,dx$$

$$= \frac{1}{\pi}\left[x\frac{\sin nx}{n} - (1)\left(-\frac{\cos nx}{n^2}\right)\right]_{-\pi/2}^{\pi/2} + \frac{1}{\pi}\left[(\pi - x)\frac{\sin nx}{n} - (-1)\left(-\frac{\cos nx}{n^2}\right)\right]_{\pi/2}^{\frac{3\pi}{2}}$$

$$= \frac{1}{\pi}\left[\frac{\pi}{2}\frac{\sin\frac{n\pi}{2}}{n}+\frac{\cos\frac{n\pi}{2}}{n^2}-\frac{\pi}{2}\frac{\sin\frac{n\pi}{2}}{n}-\frac{\cos\frac{n\pi}{2}}{n^2}\right]$$

$$+\frac{1}{\pi}\left[-\frac{\pi}{2}\frac{\sin\frac{3n\pi}{2}}{n}+\frac{\cos\frac{3n\pi}{2}}{n^2}-\frac{\pi}{2}\frac{\sin\frac{n\pi}{2}}{n}-\frac{\cos\frac{n\pi}{2}}{n^2}\right]$$

$$= \frac{1}{\pi}\left[-\frac{\pi}{2n}\left(\sin\frac{3n\pi}{2}+\sin\frac{n\pi}{2}\right)-\frac{1}{n^2}\left(\cos\frac{3n\pi}{2}-\cos\frac{n\pi}{2}\right)\right]$$

$$= \frac{1}{\pi}\left[-\frac{\pi}{n}\sin n\pi\cos\frac{n\pi}{2}+\frac{2}{n^2}\sin\frac{n\pi}{2}\sin n\pi\right]=0$$

$$b_n = \frac{1}{\pi}\int_{-\pi/2}^{\pi/2} x \sin nx\, dx + \frac{1}{\pi}\int_{\pi/2}^{3\pi/2} (\pi - x)\sin nx\, dx$$

$$= \frac{2}{\pi}\int_{0}^{\pi/2} x \sin nx\, dx + \frac{1}{\pi}\int_{\pi/2}^{\frac{3\pi}{2}} (\pi - x)\sin nx\, dx$$

$$= \frac{2}{\pi}\left[x\left(-\frac{\cos nx}{n}\right) - (1)\left(-\frac{\sin nx}{n^2}\right)\right]_0^{\pi/2} + \frac{1}{\pi}\left[(\pi - x)\left(-\frac{\cos nx}{n}\right) - (-1)\left(-\frac{\sin nx}{n^2}\right)\right]_{\pi/2}^{\frac{3\pi}{2}}$$

$$= \frac{2}{\pi}\left[-\frac{\pi}{2}\frac{\cos\frac{n\pi}{2}}{n} + \frac{\sin\frac{n\pi}{2}}{n^2}\right] + \frac{1}{\pi}\left[\frac{\pi}{2}\frac{\cos\frac{3n\pi}{2}}{n} - \frac{\sin\frac{3nx}{2}}{n^2} + \frac{\pi}{2}\frac{\cos\frac{n\pi}{2}}{n} + \frac{\sin\frac{n\pi}{2}}{n^2}\right]$$

$$= \frac{1}{\pi}\left[-\frac{\pi}{2}\frac{\cos\frac{n\pi}{2}}{n} + \frac{3\sin\frac{n\pi}{2}}{n^2} + \frac{\pi}{2}\frac{\cos 3\frac{n\pi}{2}}{n} - \frac{\sin\frac{3n\pi}{2}}{n^2}\right]$$

$$= \frac{1}{\pi}\left[\frac{\pi}{2n}\left(\cos\frac{3n\pi}{2} - \cos\frac{n\pi}{2}\right) + \frac{3}{n^2}\sin\frac{n\pi}{2} - \frac{1}{n^2}\sin\frac{3n\pi}{2}\right]$$

$$= \frac{1}{\pi}\left[-\frac{\pi}{n}\sin\frac{n\pi}{2}\sin n\pi + \frac{3}{n^2}\sin\frac{n\pi}{2} - \frac{1}{n^2}\sin\frac{3n\pi}{2}\right] = \frac{1}{n^2\pi}\left[3\sin\frac{n\pi}{2} - \sin\frac{3n\pi}{2}\right]$$

Substituting the values of $a_0, a_1, a_2 \ldots b_1, b_2, \ldots$ in (1), we get

$$f(x) = \frac{4}{\pi}\left[\frac{\sin x}{1^2} - \frac{\sin 3x}{3^2} + \frac{\sin 5x}{5^2} - \ldots\right] \qquad \textbf{Ans.}$$

Example 10. *Find the Fourier series of the function defined as*

$$f(x) = \begin{cases} x + \pi, & \text{for } 0 \le x \le \pi, \\ -x - \pi, & \text{for } -\pi \le x < 0 \end{cases} \qquad \textit{and} \qquad f(x + 2\pi) = f(x).$$

(U.P., II Semester Summer 2006)

Solution. $a_0 = \dfrac{1}{\pi}\displaystyle\int_{-\pi}^{\pi} f(x)\,dx = \dfrac{1}{\pi}\int_{-\pi}^{0} f(x)\,dx + \dfrac{1}{\pi}\int_{0}^{\pi} f(x)\,dx$

$$= \frac{1}{\pi}\int_{-\pi}^{0}(-x-\pi)\,dx + \frac{1}{\pi}\int_0^{\pi}(x+\pi)\,dx = \frac{1}{\pi}\left(-\frac{x^2}{2} - \pi x\right)_{-\pi}^{0} + \frac{1}{\pi}\left(\frac{x^2}{2} + \pi x\right)_0^{\pi}$$

$$= \frac{1}{\pi}\left(\frac{\pi^2}{2} - \pi^2\right) + \frac{1}{\pi}\left(\frac{\pi^2}{2} + \pi^2\right) = \pi\left(\frac{1}{2} - 1\right) + \pi\left(\frac{1}{2} + 1\right) = \pi$$

$$a_n = \frac{1}{\pi}\int_{-\pi}^{0} f(x)\cos nx\,dx = \frac{1}{\pi}\int_{-\pi}^{0} f(x)\cos nx\,dx + \frac{1}{\pi}\int_0^{\pi} f(x)\cos nx\,dx$$

$$= \frac{1}{\pi}\int_{-\pi}^{0}(-x-\pi)\cos nx\,dx + \frac{1}{\pi}\int_0^{\pi}(x+\pi)\cos nx\,dx$$

$$= \frac{1}{\pi}\left[(-x-\pi)\frac{\sin nx}{n} - (-1)\left\{-\frac{\cos nx}{n^2}\right\}\right]_{-\pi}^{0} + \frac{1}{\pi}\left[(x+\pi)\frac{\sin nx}{n} - (-1)\left\{-\frac{\cos nx}{n^2}\right\}\right]_0^{\pi}$$

$$= \frac{1}{\pi}\left[-\frac{1}{n^2}+\frac{(-1)^n}{n^2}\right]+\frac{1}{\pi}\left[\frac{(-1)^n}{n^2}-\frac{1}{n^2}\right]=\frac{2}{n^2\pi}[(-1)^n-1]=\frac{-4}{n^2\pi},\ \text{if } n \text{ is odd.}$$

$= 0$, if n is even.

$$b_n = \frac{1}{\pi}\int_{-\pi}^{\pi} f(x)\sin nx\,dx = \frac{1}{\pi}\int_{-\pi}^{0} f(x)\sin nx\,dx + \frac{1}{\pi}\int_{0}^{\pi} f(x)\sin nx\,dx$$

$$= \frac{1}{\pi}\int_{-\pi}^{0}(-x-\pi)\sin nx\,dx + \frac{1}{\pi}\int_{0}^{\pi}(x+\pi)\sin nx\,dx$$

$$= \frac{1}{\pi}\left[(-x-\pi)\left(-\frac{\cos nx}{n}\right)-(-1)\left(-\frac{\sin nx}{n^2}\right)\right]_{-\pi}^{0}$$

$$+\frac{1}{\pi}\left[(x+\pi)\left(-\frac{\cos nx}{n}\right)-(-1)\left(-\frac{\sin nx}{n^2}\right)\right]_{0}^{\pi}$$

$$= \frac{1}{\pi}\left[\frac{\pi}{n}\right]+\frac{1}{\pi}\left[-\frac{2\pi}{n}(-1)^n+\frac{\pi}{n}\right]=\frac{1}{n}[(1)-2(-1)^n+(1)]=\frac{2}{n}[1-(-1)^n]$$

$= \frac{4}{n}$, if n is odd.

$= 0$, if n is even.

Fourier series is

$$f(x) = \frac{a_0}{2}+a_1\cos x+a_2\cos 2x+\ldots+b_1\sin x+b_2\sin 2x+\ldots$$

$$f(x) = \frac{\pi}{2}-\frac{4}{\pi}\left(\frac{\cos x}{1^2}+\frac{\cos 3x}{3^2}+\ldots\right)+4\left(\frac{\sin x}{1}+\frac{\sin 3x}{3}+\ldots\right)$$ **Ans.**

EXERCISE 8.2

1. Find the Fourier series of the function

$$f(x) = \begin{cases} -1 & \text{for } -\pi < x < 0 \\ 1 & \text{for } 0 < x < \pi \end{cases}$$

(*U.P. II Semester 2005*)

where $f(x+2\pi) = f(x)$. **Ans.** $\frac{4}{\pi}\left[\frac{1}{1}\sin x+\frac{1}{3}\sin 3x+\frac{1}{5}\sin 5x+\frac{1}{7}\sin 7x+\ldots\right]$

2. Find the Fourier series for the function

$$f(x) = \begin{cases} -\frac{\pi}{4} & \text{for } -\pi < x < 0 \\ \frac{\pi}{4} & \text{for } 0 < x < \pi \end{cases}$$ and $f(-\pi) = f(0) = f(\pi) = 0$, $f(x) = f(x+2\pi)$ for all x.

Deduce that $\frac{\pi}{4} = 1-\frac{1}{3}+\frac{1}{5}-\frac{1}{7}+\ldots$ **Ans.** $\frac{\sin x}{1}+\frac{\sin 3x}{3}+\frac{\sin 5x}{5}+\frac{\sin 7x}{7}+\ldots$

3. Find the Fourier series of the function

$$f(x) = \begin{cases} 0 & \text{for } -\pi \le x \le 0 \\ 1 & \text{for } 0 < x < \frac{\pi}{2} \\ 0 & \text{for } \frac{\pi}{2} \le x \le \pi \end{cases}$$

Ans. $\frac{1}{4}+\frac{1}{\pi}\left[\frac{\cos x}{1}-\frac{\cos 3x}{3}+\frac{\cos 5x}{5}+\ldots+\frac{\sin x}{1}+\frac{\sin 2x}{2}+\frac{\sin 3x}{3}+\ldots\right]$

4. Obtain a Fourier series to represent the following periodic function
$f(x) = 0$ when $0 < x < \pi$

$f(x) = 1$ when $\pi < x < 2\pi$ **Ans.** $\frac{1}{2} - \frac{2}{\pi}\left(\sin x + \frac{1}{3}\sin 3x + \frac{1}{5}\sin 5x + ...\right)$

5. Find the Fourier expansion of the function defined in a single period by the relations.

$$f(x) = \begin{cases} 1 & \text{for} \quad 0 < x < \pi \\ 2 & \text{for} \quad \pi < x < 2\pi \end{cases}$$

and from it deduce that $\frac{\pi}{4} = 1 - \frac{1}{3} + \frac{1}{5} - \frac{1}{7} + ...$ **Ans.** $\frac{3}{2} - \frac{2}{\pi}\left(\sin x + \frac{1}{3}\sin 3x + \frac{1}{5}\sin 5x + ...\right)$

6. Find a Fourier series to represent the function

$$f(x) = \begin{cases} 0 & \text{for} \quad -\pi < x \le 0 \\ \frac{1}{4}\pi x & \text{for} \quad 0 < x < \pi \end{cases}$$

and hence deduce that $\frac{\pi^2}{8} = \frac{1}{1^2} + \frac{1}{3^2} + \frac{1}{5^2} + ...$

Ans. $\frac{\pi^2}{16} + \sum_{n=1}^{\infty}\left(\frac{[(-1)^n - 1]}{4n^2}\cos nx - \frac{(-1)^n \pi}{4n}\sin nx + ...\right)$

7. Find the Fourier series for $f(x)$, if

$$f(x) = -\pi \quad \text{for} \quad -\pi < x \le 0$$
$$= x \quad \text{for} \quad 0 < x < \pi$$
$$= \frac{-\pi}{2} \quad \text{for} \quad x = 0$$

Deduce that $\frac{1}{1^2} + \frac{1}{3^2} + \frac{1}{5^2} + ... = \frac{\pi^2}{8}$

Ans. $-\frac{\pi}{4} - \frac{2}{\pi}\left(\cos x + \frac{\cos 3x}{3^2} + \frac{\cos 5x}{5^2} + ...\right) + 3\sin x - \frac{1}{2}\sin 2x + \frac{3}{3}\sin 3x - \frac{1}{4}\sin 4x + ...$

8. Obtain a Fourier series to represent the function
$f(x) = |x|$ for $-\pi < x < \pi$

and hence deduce $\frac{\pi^2}{8} = \frac{1}{1^2} + \frac{1}{3^2} + \frac{1}{5^2} + ...$ **Ans.** $\frac{\pi}{2} - \frac{4}{\pi}\left[\cos x + \frac{1}{3^2}\cos 3x + \frac{1}{5^2}\cos 5x + ...\right]$

9. Expand as a Fourier series, the function $f(x)$ defined as

$$f(x) = \pi + x \quad \text{for} \quad -\pi < x < -\frac{\pi}{2}$$
$$= \frac{\pi}{2} \quad \text{for} \quad -\frac{\pi}{2} < x < \frac{\pi}{2}$$
$$= \pi - x \quad \text{for} \quad \frac{\pi}{2} < x < \pi$$

Ans. $\frac{3\pi}{8} + \frac{2}{\pi}\left[\frac{1}{1^2}\cos x - \frac{2}{2^2}\cos 2x + \frac{1}{3^2}\cos 3x + ...\right]$

10. Obtain a Fourier series to represent the function

$f(x) = |\sin x|$ for $-\pi < x < \pi$ $\left\{\textbf{Hint}\ \begin{matrix} f(x) = -\sin x & \text{for} & -\pi < x < 0 \\ = \sin x & \text{for} & 0 < x < \pi \end{matrix}\right\}$

Ans. $\frac{2}{\pi} - \frac{4}{\pi}\left[\frac{1}{3}\cos 2x + \frac{1}{15}\cos 4x + \frac{1}{35}\cos 6x + ...\right]$

11. An alternating current after passing through a rectifier has the form

$$i = I \sin\theta \quad \text{for} \quad 0 < \theta < \pi$$
$$= 0 \quad \text{for} \quad \pi < \theta < 2\pi$$

Find the Fourier series of the function.

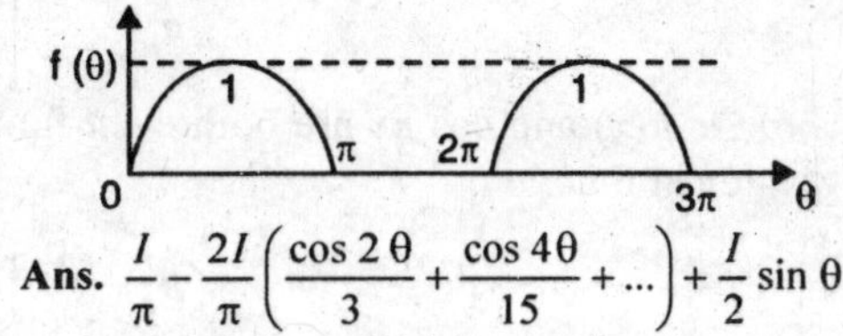

Ans. $\dfrac{I}{\pi} - \dfrac{2I}{\pi}\left(\dfrac{\cos 2\theta}{3} + \dfrac{\cos 4\theta}{15} + \ldots\right) + \dfrac{I}{2}\sin\theta$

12. If $f(x) = 0$ for $-\pi < x < 0$
$= \sin x$ for $0 < x < \pi$.

Prove that $f(x) = \dfrac{1}{\pi} + \dfrac{\sin x}{2} - \dfrac{2}{\pi}\sum_{m=1}^{\infty}\dfrac{\cos 2mx}{4m^2 - 1}$. Hence show that $\dfrac{1}{1.3} - \dfrac{1}{3.5} + \dfrac{1}{5.7} \ldots \infty = \dfrac{1}{4}(\pi - 2)$

8.10 EVEN FUNCTION AND ODD FUNCTION

(a) Even Function

A function $f(x)$ is said to be even (or symmetric) function if, $f(-x) = f(x)$

The graph of such a function is symmetrical with respect to y-axis [$f(x)$ axis]. Here y-axis is a mirror for the reflection of the curve.

The area under such a curve from $-\pi$ to π is double the area from 0 to π.

$$\therefore \quad \int_{-\pi}^{\pi} f(x)\,dx = 2\int_0^{\pi} f(x)\,dx$$

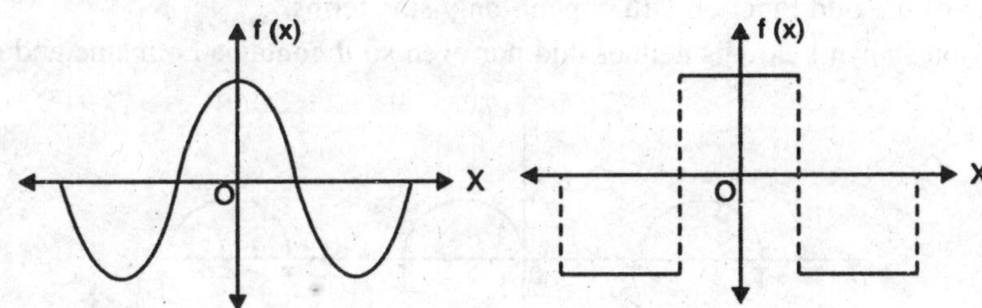

(b) Odd Function

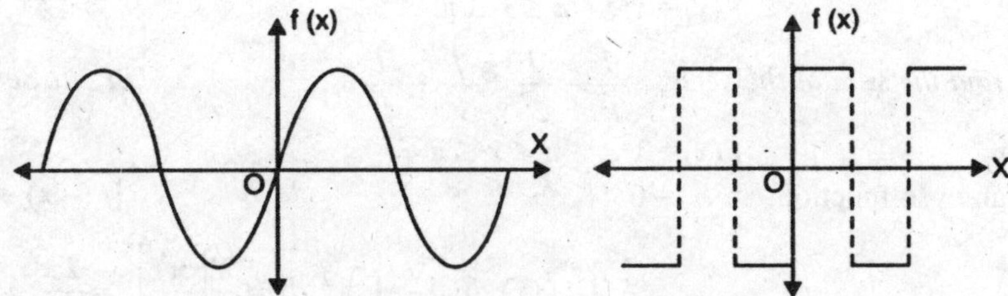

A function $f(x)$ is called odd (or skew symmetric) function if

$$f(-x) = -f(x)$$

Here the area under the curve from $-\pi$ to π is zero.

$$\int_{-\pi}^{\pi} f(x)\,dx = 0$$

Expansion of an even function:

$$a_0 = \frac{1}{\pi}\int_{-\pi}^{\pi} f(x)\,dx = \frac{2}{\pi}\int_0^{\pi} f(x)\,dx$$

$$a_n = \frac{1}{\pi}\int_{-\pi}^{\pi} f(x)\cos nx\,dx = \frac{2}{\pi}\int_0^{\pi} f(x)\cos nx\,dx$$

As $f(x)$ and $\cos nx$ are both even functions, therefore, the product of $f(x).\cos nx$ is also an even function.

$$b_n = \frac{1}{\pi}\int_{-\pi}^{\pi} f(x)\sin nx\,dx = 0$$

As $\sin nx$ is an odd function so $f(x).\sin nx$ is also an odd function. We need not to calculate b_n. It saves our labour a lot.

The series of the even function will contain only cosine terms. *(U.P. II Semester 2010)*

Expansion of an odd function :

$$a_0 = \frac{1}{\pi}\int_{-\pi}^{\pi} f(x)\,dx = 0$$

$$a_n = \frac{1}{\pi}\int_{-\pi}^{\pi} f(x)\cos nx\,dx = 0 \qquad [f(x).\cos nx \text{ is odd function.}]$$

$$b_n = \frac{1}{\pi}\int_{-\pi}^{\pi} f(x)\sin nx\,dx = \frac{2}{\pi}\int_0^{\pi} f(x)\sin nx\,dx$$

[$f(x)$. sin nx is even function.]

The series of the odd function will contain only sine terms.

The function shown below is neither odd nor even so it contains both sine and cosine terms

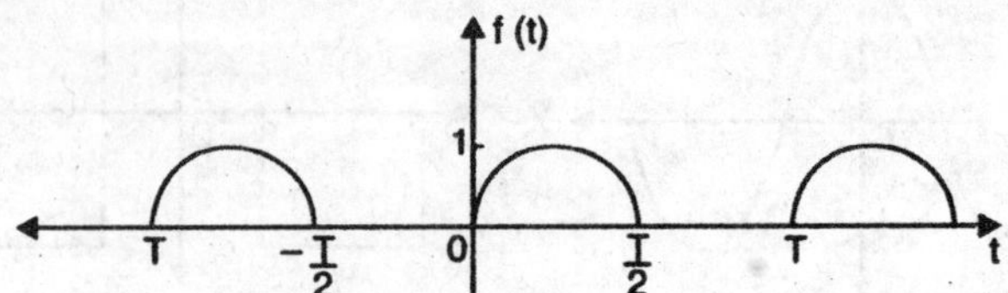

Example 11. *Find the Fourier series expansion of the periodic function of period 2π*

$$f(x) = x^2, -\pi \le x \le \pi.$$

Hence, find the sum of the series $\frac{1}{1^2} - \frac{1}{2^2} + \frac{1}{3^2} - \frac{1}{4^2} + \ldots$ *(U.P., II Semester 2004)*

Solution. $f(x) = x^2, \; -\pi \le x \le \pi$

This is an even function. $\therefore\; b_n = 0$ $\qquad [f(-x) = f(x)]$

$$a_0 = \frac{2}{\pi}\int_0^{\pi} f(x)\,dx = \frac{2}{\pi}\int_0^{\pi} x^2\,dx = \frac{2}{\pi}\left[\frac{x^3}{3}\right]_0^{\pi} = \frac{2\pi^2}{3}$$

$$a_n = \frac{2}{\pi}\int_0^{\pi} f(x)\cos nx\,dx = \frac{2}{\pi}\int_0^{\pi} x^2\cos nx\,dx$$

$$= \frac{2}{\pi}\left[x^2\left(\frac{\sin nx}{n}\right) - (2x)\left(-\frac{\cos nx}{n^2}\right) + (2)\left(-\frac{\sin nx}{n^3}\right)\right]_0^{\pi}$$

$$= \frac{2}{\pi}\left[\frac{\pi^2 \sin n\pi}{n} + \frac{2\pi\cos n\pi}{n^2} - \frac{2\sin n\pi}{n^3}\right] = \frac{4(-1)^n}{n^2}$$

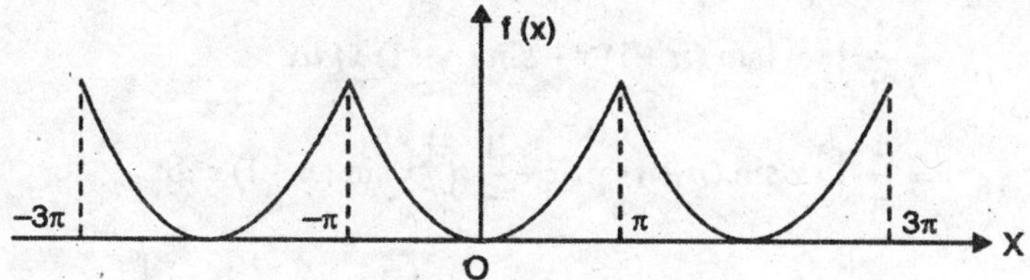

Fourier series is $$f(x)=\frac{a_0}{2}+a_1\cos x+a_2\cos 2x+a_3\cos 3x+\ldots+a_n\cos nx+\ldots$$

$$x^2=\frac{\pi^2}{3}-4\left[\frac{\cos x}{1^2}-\frac{\cos 2x}{2^2}+\frac{\cos 3x}{3^2}-\frac{\cos 4x}{4^2}+\ldots\right]$$

On putting $x = 0$, we have

$$0=\frac{\pi^2}{3}-4\left[\frac{1}{1^2}-\frac{1}{2^2}+\frac{1}{3^2}-\frac{1}{4^2}\ldots\right]$$

$$\Rightarrow \quad \frac{1}{1^2}-\frac{1}{2^2}+\frac{1}{3^2}-\frac{1}{4^2}\ldots=\frac{\pi^2}{12}$$ **Ans.**

Example 12. *Obtain a Fourier expression for*

$$f(x)=x^3 \quad for \quad -\pi<x<\pi.$$

Solution. $f(x) = x^3$ is an odd function.

$$\therefore \quad a_0=0 \text{ and } a_n=0 \qquad [f(-x)=-f(x)]$$

$$b_n=\frac{2}{\pi}\int_0^\pi f(x)\sin nx\,dx=\frac{2}{\pi}\int_0^\pi x^3\sin nx\,dx$$

$$\left[\int uv=uv_1-u'v_2+u''v_3-u'''v_4+\ldots\right]$$

$$=\frac{2}{\pi}\left[x^3\left(-\frac{\cos nx}{n}\right)-3x^2\left(-\frac{\sin nx}{n^2}\right)+6x\left(\frac{\cos nx}{n^3}\right)-6\left(\frac{\sin nx}{n^4}\right)\right]_0^\pi$$

$$=\frac{2}{\pi}\left[-\frac{\pi^3\cos n\pi}{n}+\frac{6\pi\cos n\pi}{n^3}\right]=2.(-1)^n\left[-\frac{\pi^2}{n}+\frac{6}{n^3}\right]$$

$$f(x)=b_1\sin x+b_2\sin 2x+b_3\sin 3x+\ldots.$$

$$\therefore \quad x^3=2\left[-\left(-\frac{\pi^2}{1}+\frac{6}{1^3}\right)\sin x+\left(-\frac{\pi^2}{2}+\frac{6}{2^3}\right)\sin 2x-\left(-\frac{\pi^2}{3}+\frac{6}{3^3}\right)\sin 3x+\ldots\right]$$

Ans.

Example 13. *Expand the function* $f(x) = x \sin x$, *as a Fourier series in the interval* $-\pi \le x \le \pi$.

Hence deduce that $\frac{1}{1.3}-\frac{1}{3.5}+\frac{1}{5.7}-\frac{1}{7.9}+\ldots=\frac{\pi-2}{4}$

(U.P., II Semester, Summer 2008, 2001, Uttarakhand, II Semester, June 2007)

Solution. $f(x) = x \sin x$

$$a_0=\frac{2}{\pi}\int_0^\pi f(x)\,dx=\frac{2}{\pi}\int_0^\pi x\sin x\,dx \qquad \text{(Here } x\sin x \text{ is an even function)}$$

$$=\frac{2}{\pi}[x(-\cos x)-(1)(-\sin x)]_0^\pi=\frac{2}{\pi}(\pi)=2$$

$$a_n=\frac{2}{\pi}\int_0^\pi f(x)\cos nx\,dx=\frac{2}{\pi}\int_0^\pi x\sin x\cos nx\,dx$$

$$= \frac{1}{\pi}\int_0^\pi x\{\sin(n+1)x - \sin(n-1)x\}\,dx$$

$$= \frac{1}{\pi}\int_0^\pi x\sin(n+1)x\,dx - \frac{1}{\pi}\int_0^\pi x\sin(n-1)x\,dx$$

$$= \frac{1}{\pi}\left[x\left(-\frac{\cos(n+1)x}{n+1}\right)_0^\pi - (1)\left\{-\frac{\sin(n+1)x}{(n+1)^2}\right\}\right]_0^\pi$$

$$-\frac{1}{\pi}\left[x\left(-\frac{\cos(n-1)x}{(n-1)}\right) - (1)\left\{-\frac{\sin(n-1)x}{(n-1)^2}\right\}\right]_0^\pi$$

$$= \frac{1}{\pi}\left[-\pi\frac{(-1)^{(n+1)}}{n+1} + 0\right] - \frac{1}{\pi}\left[-\pi\frac{(-1)^{(n-1)}}{n-1} - 0\right]$$

$$= -\frac{(-1)^{n+1}}{n+1} + \frac{(-1)^{n-1}}{n-1} = (-1)^{n+1}\left[-\frac{1}{n+1} + \frac{1}{n-1}\right] = \frac{2(-1)^{n+1}}{n^2-1}$$

$$a_1 = \frac{2}{\pi}\int_0^\pi x\sin x\cos x\,dx = \frac{1}{\pi}\int_0^\pi x\sin 2x\,dx$$

$$= \frac{1}{\pi}\left[x\left(-\frac{\cos 2x}{2}\right) - (1)\left(\frac{-\sin 2x}{4}\right)\right]_0^\pi = \frac{1}{\pi}\left[-\frac{\pi}{2}\right] = -\frac{1}{2}$$

$b_n = 0$ [As $x \sin x \sin nx$ is an odd function]

Hence $f(x) = 1 - \frac{1}{2}\cos x + 2\sum_{n=2}^{\infty}\frac{(-1)^{n+1}}{n^2-1}\cos nx = 1 - \frac{1}{2}\cos x + 2\sum_{n=2}^{\infty}\frac{(-1)^{n+1}}{(n-1)(n+1)}\cos nx$

$$x\sin x = 1 + 2\left[-\frac{1}{4}\cos x - \frac{1}{1.3}\cos 2x + \frac{1}{2.4}\cos 3x - \frac{1}{3.5}\cos 4x + \dots\right]$$

Putting $x = \frac{\pi}{2}$, we get $\quad \frac{\pi}{2} = 1 + 2\left\{\frac{1}{1.3} - \frac{1}{3.5} + \frac{1}{5.7} - \dots\right\}$

or $\quad \frac{\pi}{4} = \frac{1}{2} + \frac{1}{1.3} - \frac{1}{3.5} + \frac{1}{5.7} - \dots$

$$\Rightarrow \quad \frac{\pi}{4} - \frac{1}{2} = \frac{1}{1.3} - \frac{1}{3.5} + \frac{1}{5.7} - \dots$$

$$\Rightarrow \quad \frac{\pi - 2}{4} = \frac{1}{1.3} - \frac{1}{3.5} + \frac{1}{5.7} - \dots$$

Proved.

Example 14. *Find the Fourier Series expansion for the function*

$f(x) = x\cos x,\ -\pi < x < \pi.$ *(U.P., II Semester, Summer 2002)*

Solution. Since $x \cos x$ is an odd function therefore, $a_0 = a_n = 0$.

Let $x\cos x = \Sigma\, b_n \sin bx$, where

$$b_n = \frac{2}{\pi}\int_0^\pi x\cos x.\sin nx\,dx = \frac{1}{\pi}\int_0^\pi x\{\sin(n+1)x + \sin(n-1)x\}\,dx$$

$$= \frac{1}{\pi}\int_0^\pi x\sin(n+1)x\,dx + \frac{1}{\pi}\int_0^\pi x\sin(n-1)x\,dx$$

$$= \frac{1}{\pi}\left[x\left(\frac{-\cos(n+1)x}{n+1}\right)+\frac{\sin(n+1)x}{(n+1)^2}\right]_0^\pi + \frac{1}{\pi}\left[-x\frac{\cos(n-1)x}{n-1}+\frac{\sin(n-1)x}{(n-1)^2}\right]_0^\pi$$

$$= \frac{1}{\pi}\left[x.\left\{-\frac{\cos(n+1)x}{(n+1)}-\frac{\cos(n-1)x}{(n-1)}\right\}+1.\left\{\frac{\sin(n+1)x}{(n+1)^2}+\frac{\sin(n-1)x}{(n-1)^2}\right\}\right]_0^\pi$$

$$= \frac{1}{\pi}\left[\pi.\left\{-\frac{\cos(n+1)\pi}{(n+1)}-\frac{\cos(n-1)\pi}{(n-1)}\right\}\right]$$

$$\Rightarrow \quad b_n = \left\{-\frac{(-1)^{n+1}}{n+1}-\frac{(-1)^{n-1}}{n-1}\right\}, \; n \neq 1$$

$$b_n = -(-1)^{n+1}\left[\frac{1}{n+1}+\frac{1}{n-1}\right]$$

$$= -\left\{\frac{1}{(n+1)}+\frac{1}{(n-1)}\right\} = \frac{-2n}{n^2-1}, \qquad \text{If } n \text{ is odd; } n \neq 1.$$

$$\text{But } b_n = \left\{\frac{1}{n+1}+\frac{1}{n-1}\right\} = \frac{2n}{n^2-1}, \qquad \text{If } n \text{ is even; } n \neq 1$$

If $\quad n = 1$, then $b_1 = \frac{2}{\pi}\int_0^\pi x\cos x.\sin x\,dx = \frac{1}{\pi}\int_0^\pi x.\sin 2x\,dx$

$$= \frac{1}{\pi}\left[x.\left(-\frac{\cos 2x}{2}\right)-1\left(-\frac{\sin 2x}{4}\right)\right]_0^\pi = \frac{1}{\pi}\left[\pi\left(-\frac{1}{2}\right)\right] = -\frac{1}{2}$$

$\therefore\; x\cos x = b_1\sin x + b_2\sin 2x + b_3\sin 3x + \ldots$

$$= -\frac{1}{2}\sin x + \frac{4\sin 2x}{2^2-1} - \frac{6\sin 3x}{3^2-1} + \ldots$$ **Ans.**

8.11 HALF-RANGE SERIES, PERIOD 0 TO π

The given function is defined in the interval $(0, \pi)$ and it is immaterial whatever the function may be outside the interval $(0, \pi)$. To get the series of cosines only we assume that $f(x)$ is an even function in the interval $(-\pi, \pi)$.

$$a_n = \frac{2}{\pi}\int_0^\pi f(x)\cos nx\,dx \text{ and } b_n = 0$$

To expand $f(x)$ as a sine series we extend the function in the interval $(-\pi, \pi)$ as an odd function.

$$b_n = \frac{2}{\pi}\int_0^\pi f(x)\sin nx\,dx \text{ and } a_n = 0$$

Example 15. *Find the Haff-range cosine series of the following function*

$$f(x) = |\cos x|, \; 0 < x < \pi$$

Draw its diagram also. (*Uttarakhand, II Semester, Summer 2010*)

Solution. Here $f(-x) = |\cos(-x)| = |\cos x| = \cos x = f(x)$

$f(x)$ is an even function of x.

$$a_0 = \frac{2}{\pi}\int_0^{\pi} |\cos x|\, dx = \frac{2}{\pi}\left[\int_0^{\frac{\pi}{2}} \cos x\, dx + \int_{\frac{\pi}{2}}^{\pi} (-\cos x)\, dx\right]$$

$$= \frac{2}{\pi}\left([\sin x]_0^{\frac{\pi}{2}} - (\sin x)_{\frac{\pi}{2}}^{\pi}\right)$$

$$= \frac{2}{\pi}\left[\left(\sin\frac{\pi}{2}\right) - \left(\sin \pi - \sin\frac{\pi}{2}\right)\right] = \frac{2}{\pi}[1+1] = \frac{4}{\pi}$$

$$a_1 = \frac{2}{\pi}\int_0^{\pi} |\cos x| \cos nx\, dx = \frac{2}{\pi}\left[\int_0^{\frac{\pi}{2}} \cos x \cos nx\, dx + \int_{\frac{\pi}{2}}^{\pi} -\cos x \cos nx\, dx\right]$$

$$= \frac{2}{2\pi}\left[\int_0^{\frac{\pi}{2}} [\cos (n+1)x + \cos (n-1)x]\, dx - \int_{\frac{\pi}{2}}^{\pi} (\cos (n+1)x + \cos (n-1)x)\, dx\right]$$

$$= \frac{2}{2\pi}\left[\frac{\sin (n+1)x}{n+1} + \frac{\sin (n-1)x}{n-1}\right]_0^{\frac{\pi}{2}} - \frac{2}{2\pi}\left[\frac{\sin (n+1)x}{n+1} + \frac{\sin (n-1)x}{n-1}\right]_{\frac{\pi}{2}}^{\pi}$$

$$= \frac{1}{\pi}\left[\frac{\sin (n+1)\frac{\pi}{2}}{n+1} + \frac{\sin (n-1)\frac{\pi}{2}}{n-1}\right] - \frac{1}{\pi}\left[\frac{\sin (n+1)\pi}{n+1} + \frac{\sin (n-1)\pi}{n-1} - \frac{\sin (n+1)\frac{\pi}{2}}{n+1} - \frac{\sin (n-1)\frac{\pi}{2}}{n-1}\right]$$

$$= \frac{2}{\pi}\left[\frac{\sin (n+1)\frac{\pi}{2}}{n+1} + \frac{\sin (n-1)\frac{\pi}{2}}{n-1}\right] - \frac{1}{\pi}\left[\frac{\sin (n+1)\pi}{n+1} + \frac{\sin (n-1)\pi}{n-1}\right]$$

$$= \frac{2}{\pi}\left[\frac{\cos n\frac{\pi}{2}}{n+1} - \frac{\cos\frac{n\pi}{2}}{n-1}\right]$$

$$= \frac{2\cos\frac{n\pi}{2}}{\pi}\left[\frac{1}{n+1} - \frac{1}{n-1}\right] = \frac{2\cos\frac{n\pi}{2}}{\pi}\left[\frac{n-1-n-1}{n^2-1}\right]$$

$$= \frac{\cos\frac{n\pi}{2}}{\pi}\frac{(-4)}{n^2-1} = \frac{-4\cos N\pi}{\pi(4N^2-1)} = \frac{-4(-1)^N}{(4N^2-1)\pi} \qquad [n \to 2N = \text{ever number}]$$

$$|\cos x| = \frac{2}{\pi} + \frac{4}{\pi}\left[\frac{\cos 2x}{3} - \frac{\cos 4x}{15} + \ldots\ldots\right]$$ **Ans.**

Example 16. *Represent the following function by a Fourier sine series :*

$$f(t) = \begin{cases} t, & 0 < t \le \dfrac{\pi}{2} \\ \dfrac{\pi}{2}, & \dfrac{\pi}{2} < t \le \pi \end{cases}$$

f (t)
t
– π
$-\frac{\pi}{2}$
O
$\frac{\pi}{2}$
π

Solution. $b_n = \frac{2}{\pi}\int_0^{\pi} f(t)\sin nt\,dt$

$$= \frac{2}{\pi}\int_0^{\pi/2} t\sin nt\,dt + \frac{2}{\pi}\int_{\pi/2}^{\pi}\frac{\pi}{2}\sin nt\,dt$$

$$= \frac{2}{\pi}\left[t\left(-\frac{\cos nt}{n}\right)-(1)\left(-\frac{\sin nt}{n^2}\right)\right]_0^{\pi/2} + \frac{2}{\pi}\frac{\pi}{2}\left[-\frac{\cos nt}{n}\right]_{\pi/2}^{\pi}$$

$$= \frac{2}{\pi}\left[-\frac{\pi}{2}\frac{\cos\frac{n\pi}{2}}{n}+\frac{\sin\frac{n\pi}{2}}{n^2}\right]+\left[-\frac{\cos n\pi}{n}+\frac{\cos\frac{n\pi}{2}}{n}\right]$$

$$b_1 = \frac{2}{\pi}\left[-\frac{\pi}{2}\cos\frac{\pi}{2}+\sin\frac{\pi}{2}\right]+\left[-\cos\pi+\cos\frac{\pi}{2}\right] = \frac{2}{\pi}[0+1]+[1] = \frac{2}{\pi}+1$$

$$b_2 = \frac{2}{\pi}\left[-\frac{\pi}{2}\frac{\cos\pi}{2}+\frac{\sin\pi}{2^2}\right]+\left[-\frac{\cos 2\pi}{2}+\frac{\cos\pi}{2}\right] = \frac{2}{\pi}\left[-\frac{\pi}{2}\frac{(-1)}{2}+0\right]+\left[-\frac{1}{2}-\frac{1}{2}\right]$$

$$= \frac{2}{\pi}\left[\frac{\pi}{4}\right]-1 = \frac{1}{2}-1 = -\frac{1}{2}$$

$$b_3 = \frac{2}{\pi}\left[-\frac{\pi}{2}\frac{\cos\frac{3\pi}{2}}{3}+\frac{\sin\frac{3\pi}{2}}{3^2}\right]+\left[-\frac{\cos 3\pi}{3}+\frac{\cos\frac{3\pi}{2}}{3}\right]$$

$$= \frac{2}{\pi}\left[-\frac{\pi}{2}(0)-\frac{1}{9}\right]+\left[\frac{1}{3}+0\right] = -\frac{2}{9\pi}+\frac{1}{3}$$

$$f(t) = \left(\frac{2}{\pi}+1\right)\sin t-\frac{1}{2}\sin 2t+\left(-\frac{2}{9\pi}+\frac{1}{3}\right)\sin 3t+\ldots$$ **Ans.**

Example 17. *Find the Fourier sine series for the function*

$$f(x) = e^{ax} \quad \textit{for} \quad 0 < x < \pi$$

where a is constant.

Solution. $b_n = \frac{2}{\pi}\int_0^{\pi} e^{ax}\sin nx\,dx \qquad \left[\int e^{ax}\sin bx\,dx = \frac{e^{ax}}{a^2+b^2}[a\sin bx - b\cos bx]\right]$

$$= \frac{2}{\pi}\left[\frac{e^{ax}}{a^2+n^2}(a\sin nx - n\cos nx)\right]_0^{\pi}$$

$$= \frac{2}{\pi}\left[\frac{e^{a\pi}}{a^2+n^2}(a\sin n\pi - n\cos n\pi)+\frac{n}{a^2+n^2}\right]$$

$$= \frac{2}{\pi}\frac{n}{a^2+n^2}\left[-(-1)^n e^{a\pi}+1\right] = \frac{2n}{(a^2+n^2)\pi}[1-(-1)^n e^{a\pi}]$$

$$b_1 = \frac{2(1+e^{a\pi})}{(a^2+1^2)\pi}, \qquad b_2 = \frac{2.2.(1-e^{a\pi})}{.(a^2+2^2)\pi}$$

$$e^{ax} = \frac{2}{\pi}\left[\frac{1+e^{a\pi}}{a^2+1^2}\sin x+\frac{2(1-e^{a\pi})}{a^2+2^2}\sin 2x+\ldots\right]$$ **Ans.**

EXERCISE 8.3

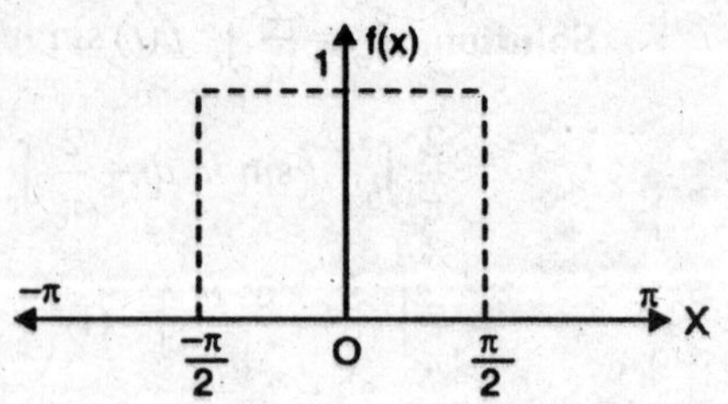

1. Find the Fourier cosine series for the function

$$f(x)=\begin{cases}1 & \text{for } 0<x<\dfrac{\pi}{2}\\[2mm] 0 & \text{for } \dfrac{\pi}{2}<x<\pi.\end{cases}$$

Ans. $\dfrac{1}{2}+\dfrac{2}{\pi}\left[\cos x-\dfrac{1}{3}\cos 3x+\dfrac{1}{5}\cos 5x-\ldots\right]$

2. Find a series of cosine of multiples of x which will represent $f(x)$ in $(0, \pi)$ where

$f(x)=0$ for $0<x<\dfrac{\pi}{2}$

$f(x)=\dfrac{\pi}{2}$ for $\dfrac{\pi}{2}<x<\pi$

Deduce that $1-\dfrac{1}{3}+\dfrac{1}{5}-\dfrac{1}{7}+\ldots\infty=\dfrac{\pi}{4}$ **Ans.** $\dfrac{\pi}{4}-\cos x+\dfrac{1}{3}\cos 3x-\dfrac{1}{5}\cos 5x+\ldots$

3. Express $f(x)=x$ as a sine series in $0<x<\pi$. **Ans.** $2\left[\sin x-\dfrac{1}{2}\sin 2x+\dfrac{1}{3}\sin 3x-\ldots\right]$

4. Find the cosine series for $f(x)=\pi-x$ in the interval $0<x<\pi$.

Ans. $\dfrac{\pi}{2}+\dfrac{4}{\pi}\left(\cos x+\dfrac{\cos 3x}{3^2}+\dfrac{\cos 5x}{5^2}+\ldots\right)$

5. If $f(x)=x$, for $0<x<\dfrac{\pi}{2}$

$=\pi-x$, for $\dfrac{\pi}{2}<x<\pi$

Show that:

(i) $f(x)=\dfrac{4}{\pi}\left(\sin x-\dfrac{1}{3^2}\sin 3x+\dfrac{1}{5^2}\sin 5x-\ldots\right)$

(ii) $f(x)=\dfrac{\pi}{4}-\dfrac{2}{\pi}\left(\dfrac{1}{1^2}\cos 2x+\dfrac{1}{3^2}\cos 6x+\dfrac{1}{5^2}\cos 10x+\ldots\right)$

6. Obtain the half-range cosine series for $f(x)=x^2$ in $0<x<\pi$.

Ans. $\dfrac{\pi^2}{3}-\dfrac{4}{\pi}\left(\cos x-\dfrac{1}{2^2}\cos 2x+\dfrac{1}{3^2}\cos 3x-\ldots\right)$

7. Find (i) sine series and (ii) cosine series for the function
$f(x)=e^x$ for $0<x<\pi$.

Ans. (i) $\dfrac{2}{\pi}\sum_1^{\infty} n\left[\dfrac{1-(-1)^n e^{\pi}}{n^2+1}\right]\sin nx$ (ii) $\dfrac{e^{\pi}-1}{\pi}-\dfrac{2}{\pi}\sum_1^{\infty}\dfrac{1-(-1)^n e^{\pi}}{n^2+1}\cos nx$

8. If $f(x)=x+1$, for $0<x<\pi$, find its Fourier (i) sine series (ii) cosine series. Hence deduce that

(i) $1-\dfrac{1}{3}+\dfrac{1}{5}-\dfrac{1}{7}+\ldots=\dfrac{\pi}{4}$ (ii) $1+\dfrac{1}{3^2}+\dfrac{1}{5^2}+\dfrac{1}{7^2}+\ldots=\dfrac{\pi^2}{8}$

Ans. (i) $\dfrac{2}{\pi}\left[(\pi+2)\sin x-\dfrac{\pi}{2}\sin 2x+\dfrac{1}{3}(\pi+2)\sin 3x-\dfrac{\pi}{4}\sin 4x+\ldots\right]$

(ii) $\dfrac{\pi}{2}+1-4\left(\cos x+\dfrac{\cos 3x}{3^2}+\dfrac{\cos 5x}{5^2}+\ldots\right)$

9. Find the Fourier series expansion of the function
$f(x) = \cos(sx),\ -\pi \le x \le \pi$

where s is a fraction. Hence, show that $\cot\theta = \dfrac{1}{\theta} + \dfrac{2\theta}{\theta^2 - \pi^2} + \dfrac{2\theta}{\theta^2 - 4\pi^2} + \ldots$

Ans. $\dfrac{\sin \pi x}{\pi s} + \dfrac{1}{\pi}\Sigma\left(\dfrac{\sin(s\pi + n\pi)}{s+n} + \dfrac{\sin(s\pi - n\pi)}{s-n}\right)\cos nx$

8.12 CHANGE OF INTERVAL AND FUNCTIONS HAVING ARBITRARY PERIOD

In electrical engineering problems, the period of the function is not always 2π but T or $2c$. This period must be converted to the length 2π. The independent variable x is also to be changed proportionally.

Let the function $f(x)$ be defined in the interval $(-c, c)$. Now we want to change the function to the period of 2π so that we can use the formulae of a_n, b_n as discussed in Article 8.6.

$\therefore$ $2c$ is the interval for the variable x.

$\therefore$ 1 is the interval for the variable $= \dfrac{x}{2c}$

$\therefore$ 2π is the interval for the variable $= \dfrac{x\,2\pi}{2c} = \dfrac{\pi x}{c}$

so put $z = \dfrac{\pi x}{c}$ or $x = \dfrac{zc}{\pi}$

Thus the function $f(x)$ of period $2c$ is transformed to the function

$f\left(\dfrac{cz}{\pi}\right)$ or $F(z)$ of period 2π.

$F(z)$ can be expanded in the Fourier series.

$$F(z) = f\left(\frac{cz}{\pi}\right) = \frac{a_0}{2} + a_1\cos z + a_2\cos 2z + \ldots + b_1\sin z + b_2\sin 2z + \ldots$$

where
$$a_0 = \frac{1}{\pi}\int_0^{2\pi} F(z)\,dz = \frac{1}{\pi}\int_0^{2\pi} f\left(\frac{cz}{\pi}\right)dz$$

$$= \frac{1}{\pi}\int_0^{2c} f(x)\,d\left(\frac{\pi x}{c}\right) = \frac{1}{c}\int_0^{2c} f(x)\,dx \qquad \left(\text{put } z = \frac{\pi x}{c}\right)$$

$$\boxed{a_0 = \frac{1}{c}\int_0^{2c} f(x)\,dx}$$

$$a_n = \frac{1}{\pi}\int_0^{2\pi} F(z)\cos nz\,dz = \frac{1}{\pi}\int_0^{2\pi} f\left(\frac{cz}{\pi}\right)\cos nz\,dz$$

$$= \frac{1}{\pi}\int_0^{2c} f(x)\cos\frac{n\pi x}{c}\,d\left(\frac{\pi x}{c}\right) = \frac{1}{c}\int_0^{2c} f(x)\cos\frac{n\pi x}{c}\,dx. \qquad \left[\text{Put } z = \frac{\pi x}{c}\right]$$

$$\boxed{a_n = \frac{1}{c}\int_0^{2c} f(x)\cos\frac{n\pi x}{c}\,dx}$$

Similarly,
$$\boxed{b_n = \frac{1}{c}\int_0^{2c} f(x)\sin\frac{n\pi x}{c}\,dx}$$

Example 18. *Find the Fourier series corresponding to the function f (x) defined in* (– 2, 2) *as follows*

$$f(x) = \begin{cases} 2 & in \quad -2 \le x \le 0 \\ x & in \quad 0 < x < 2 \end{cases}$$

Solution. Here the interval is (– 2, 2) and $c = 2$

$$a_0 = \frac{1}{c}\int_{-c}^{c} f(x)\,dx = \frac{1}{2}\left[\int_{-2}^{0} 2.dx + \int_0^2 x.dx\right]$$

$$= \frac{1}{2}\left[[2x]_{-2}^{0} + \left(\frac{x^2}{2}\right)_0^2\right] = \frac{1}{2}[4+2] = 3$$

$$a_n = \frac{1}{c}\int_{-c}^{c} f(x)\cos\left(\frac{n\pi x}{c}\right)dx = \frac{1}{2}\left[\int_{-2}^{0} 2\cos\frac{n\pi x}{2}\,dx + \int_0^2 x\cos\frac{n\pi x}{2}\,dx\right]$$

$$= \frac{1}{2}\left[\frac{4}{n\pi}\left(\sin\frac{n\pi x}{2}\right)_{-2}^{0} + \left(x\frac{2}{n\pi}\sin\frac{n\pi x}{2} + \frac{4}{n^2\pi^2}\cos\frac{n\pi x}{2}\right)_0^2\right]$$

$$= \frac{1}{2}\left[\frac{4}{n^2\pi^2}\cos n\pi - \frac{4}{n^2\pi^2}\right] = \frac{2}{n^2\pi^2}[(-1)^n - 1]$$

$$= -\frac{4}{n^2\pi^2}, \qquad \text{when } n \text{ is odd}$$

$$= 0, \qquad \text{when } n \text{ is even.}$$

$$b_n = \frac{1}{c}\int_{-c}^{c} f(x)\sin\frac{n\pi x}{c}\,dx = \frac{1}{2}\int_{-2}^{0} 2\sin\frac{n\pi x}{2}\,dx + \frac{1}{2}\int_0^2 x\sin\frac{n\pi x}{2}\,dx$$

$$= \frac{1}{2}\left[2\left(-\frac{2}{n\pi}\cos\frac{n\pi x}{2}\right)\right]_{-2}^{0} + \frac{1}{2}\left[x\left(-\frac{2}{n\pi}\cos\frac{n\pi x}{2}\right) + (1)\frac{4}{n^2\pi^2}\sin\frac{n\pi x}{2}\right]_0^2$$

$$= \frac{1}{2}\left[-\frac{4}{n\pi} + \frac{4}{n\pi}\cos n\pi\right] + \frac{1}{2}\left[-\frac{4}{n\pi}\cos n\pi + \frac{4}{n^2\pi^2}\sin n\pi\right] = \frac{1}{2}\left[-\frac{4}{n\pi}\right] = -\frac{2}{n\pi}$$

$$f(x) = \frac{a_0}{2} + a_1\cos\frac{\pi x}{c} + a_2\cos\frac{2\pi x}{c} + a_3\cos\frac{3\pi x}{c} + \ldots$$

$$+ b_1\sin\frac{\pi x}{c} + b_2\sin\frac{2\pi x}{c} + b_3\sin\frac{3\pi x}{c} + \ldots$$

$$= \frac{3}{2} - \frac{4}{\pi^2}\left\{\frac{1}{1^2}\cos\frac{\pi x}{2} + \frac{1}{3^2}\cos\frac{3\pi x}{2} + \ldots\right\}$$

$$-\frac{2}{\pi}\left\{\frac{1}{1}\sin\frac{\pi x}{2} + \frac{1}{2}\sin\frac{2\pi x}{2} + \frac{1}{3}\sin\frac{3\pi x}{2} + \ldots\right\}$$ **Ans.**

Example 19. *A periodic function of period 4 is defined as*

$$f(x) = |\,x\,|, -2 < x < 2.$$

Find its Fourier series expansion.

Solution. $f(x) = |x| \quad -2 < x < 2$

$$\Rightarrow \quad f(x) = \begin{cases} x, & 0 < x < 2 \\ -x, & -2 < x < 0 \end{cases}$$

f(x), 2, X', X, −2, O, 2

$$a_0 = \frac{1}{c}\int_{-c}^{c} f(x)\,dx = \frac{1}{2}\int_0^2 x\,dx + \frac{1}{2}\int_{-2}^{0}(-x)\,dx$$

$$= \frac{1}{2}\left[\frac{x^2}{2}\right]_0^2 + \frac{1}{2}\left[\frac{-x^2}{2}\right]_{-2}^{0} = \frac{1}{4}(4-0) + \frac{1}{4}(0+4) = 2$$

$$a_n = \frac{1}{c}\int_{-c}^{c} f(x)\cos\frac{n\pi x}{c}\,dx = \frac{1}{2}\int_0^2 x\cos\frac{n\pi x}{2}\,dx + \frac{1}{2}\int_{-2}^{0}(-x)\cos\frac{n\pi x}{2}\,dx$$

$$= \frac{1}{2}\left[x\left(\frac{2}{n\pi}\sin\frac{n\pi x}{2}\right) - (1)\left(-\frac{4}{n^2\pi^2}\cos\frac{n\pi x}{2}\right)\right]_0^2$$

$$+ \frac{1}{2}\left[(-x)\left(\frac{2}{n\pi}\sin\frac{n\pi x}{2}\right) - (-1)\left(-\frac{4}{n^2\pi^2}\right)\cos\frac{n\pi x}{2}\right]_{-2}^{0}$$

$$= \frac{1}{2}\left[0 + \frac{4}{n^2\pi^2}(-1)^n - \frac{4}{n^2\pi^2}\right] + \frac{1}{2}\left[0 - \frac{4}{n^2\pi^2} + \frac{4}{n^2\pi^2}(-1)^n\right]$$

$$= \frac{1}{2}\frac{4}{n^2\pi^2}[(-1)^n - 1 - 1 + (-1)^n] = \frac{4}{n^2\pi^2}[(-1)^n - 1]$$

$$= -\frac{8}{n^2\pi^2} \qquad \text{(If } n \text{ is odd.)}$$

$$= 0 \qquad \text{(If } n \text{ is even)}$$

$b_n = 0$ as $f(x)$ is even function.

Fourier series is

$$f(x) = \frac{a_0}{2} + a_1\cos\frac{\pi x}{c} + a_2\cos\frac{2\pi x}{c} + \ldots + b_1\sin\frac{\pi x}{c} + b_2\sin\frac{2\pi x}{c} + \ldots$$

$$f(x) = 1 - \frac{8}{\pi^2}\left[\frac{\cos\frac{\pi x}{2}}{1^2} + \frac{\cos\frac{3\pi x}{2}}{3^2} + \frac{\cos\frac{5\pi x}{2}}{5^2} + \ldots\right] \qquad \textbf{Ans.}$$

8.13 HALF PERIOD SERIES

Cosine series: $$f(x) = \frac{a_0}{2} + a_1\cos\frac{\pi x}{c} + a_2\cos\frac{2\pi x}{c} + \ldots + a_n\cos\frac{n\pi x}{c} + \ldots$$

where $$a_0 = \frac{2}{c}\int_0^c f(x)\,dx,\quad a_n = \frac{2}{c}\int_0^c f(x)\cos\frac{n\pi x}{c}\,dx$$

Sine series: $$f(x) = b_1\sin\frac{\pi x}{c} + b_2\sin\frac{2\pi x}{c} + \ldots + b_n\sin\frac{n\pi x}{c} + \ldots$$

where $$b_n = \frac{2}{c}\int_0^c f(x)\sin\frac{n\pi x}{c}\,dx.$$

Example 20. *Expand for $f(x) = k$ for $0 < x < 2$ in a half range sine series*

(*U.P., II Semester, June 2007*)

Solution. $f(x) = k$

$$b_n = \frac{2}{c}\int_0^c f(x)\cdot \sin\frac{n\pi x}{c}\,dx \text{ in half range } (0, c) = \frac{2}{2}\int_0^2 k\sin\frac{n\pi x}{2}\,dx$$

$$= k\frac{2}{n\pi}\left(-\cos\frac{n\pi x}{2}\right)_0^2 = \frac{2k}{n\pi}[-\cos n\pi + 1]$$

Half range sine series is

$$f(x) = \sum_{n=1}^{\infty} b_n \sin\frac{n\pi x}{2}$$

$$k = \sum_{n=1}^{\infty}\frac{2k}{n\pi}[1-\cos n\pi]\sin\frac{n\pi x}{2}$$ **Ans.**

Example 21. *Obtain the half-range sine series for the function $f(x) = x^2$ in the interval $0 < x < 3$.*

(*U.P., II Semester, Summer 2002*)

Solution. We know that half range sine series is given by $f(x) = \Sigma b_n \sin nx$

Where $$b_n = \frac{2}{c}\int_0^c f(x)\sin\frac{n\pi x}{c}\,dx \text{ in the half-range } (0, c).$$

Here, we have half range $0 < x < 3$ and $f(x) = x^2$

$$b_n = \frac{2}{3}\int_0^3 x^2\sin\frac{n\pi x}{3}\,dx$$

$$= \frac{2}{3}\left[x^2\left(\frac{3}{n\pi}\right)\left(-\cos\frac{n\pi x}{3}\right) + 2x\times\left(\frac{3}{n\pi}\right)\left(\frac{3}{n\pi}\right)\sin\frac{n\pi x}{3} - 2\left(\frac{3}{n\pi}\right)\left(\frac{3}{n\pi}\right)\left(\frac{3}{n\pi}\right)\left(-\cos\frac{n\pi x}{3}\right)\right]_0^3$$

$$\Rightarrow \quad b_n = \frac{2}{3}\left[\left\{-\frac{27}{n\pi}(-1)^n - \frac{54}{n^3\pi^3}(-1)^n\right\} + \frac{54}{n^3\pi^3}\right]$$

$$\Rightarrow \quad b_n = \frac{2}{3}\left[\frac{54}{n^3\pi^3}\{1-(-1)^n\} - \frac{27}{n\pi}(-1)^n\right] \Rightarrow b_n = \frac{2}{3}\left[\frac{108}{n^3\pi^3} + \frac{27}{n\pi}\right] \text{ when } n \text{ is odd}$$

And $b_n = -\dfrac{18}{n\pi}$ when n is even

$\therefore$ Half range sine series

$$f(x) = b_1\sin x + b_2\sin 2x + b_3\sin 3x + \ldots$$

$$= \frac{2}{3}\left[\frac{108}{\pi^3}\left(\frac{\sin x}{1^3} + \frac{\sin 3x}{3^3} + \frac{\sin 5x}{5^3} + \ldots\right) + \frac{27}{\pi}\left(\frac{\sin x}{1} + \frac{\sin 3x}{3} + \frac{\sin 5x}{5} + \ldots\right)\right]$$

$$- \frac{18}{\pi}\left(\frac{\sin 2x}{2} + \frac{\sin 4x}{4} + \ldots\right)$$ **Ans.**

Example 22. *Expand $f(x) = e^x$ in a cosine series over (0, 1).*

Solution. Here, we have $f(x) = e^x$ and $c = 1$

$$a_0 = \frac{2}{c}\int_0^c f(x)\,dx = \frac{2}{1}\int_0^1 e^x\,dx = 2\,(e-1)$$

$$a_n = \frac{2}{c}\int_0^c f(x)\cos\frac{n\pi x}{c}\,dx = \frac{2}{1}\int_0^1 e^x\cos\frac{n\pi x}{1}\,dx$$

$$= 2\left[\frac{e^x}{n^2\pi^2+1}(n\,\pi\sin n\,\pi\,x + \cos n\,\pi\,x)\right]_0^1$$

$$= 2\left[\frac{e^1}{n^2\pi^2+1}(n\,\pi\sin n\,\pi + \cos n\,\pi) - \frac{1}{n^2\pi^2+1}\right]$$

$$= \frac{2}{n^2\pi^2+1}[(-1)^n\,e-1]$$

$$f(x) = \frac{a_0}{2} + a_1\cos\pi x + a_2\cos 2\pi x + a_3\cos 3\pi x + \ldots$$

$$e^x = e-1+2\left[\frac{-e-1}{\pi^2+1}\cos\pi x + \frac{e-1}{4\pi^2+1}\cos 2\pi x + \frac{-e-1}{9\pi^2+1}\cos 3\pi x + \ldots\right]$$ **Ans.**

Example 23. *Find the Fourier half-range cosine series of the function*

$$f(t) = \begin{cases} 2t, & 0 < t < 1 \\ 2\,(2-t), & 1 < t < 2 \end{cases}$$ *(U.P., II Semester, Summer 2007, 2006, 2001)*

Solution. $f(t) = \begin{cases} 2t, & 0 < t < 1 \\ 2\,(2-t), & 1 < t < 2 \end{cases}$

Let $$f(t) = \frac{a_0}{2} + a_1\cos\frac{\pi t}{c} + a_2\cos\frac{2\pi t}{c} + a_3\cos\frac{3\pi t}{c} + \ldots$$
$$+ b_1\sin\frac{\pi t}{c} + b_2\sin\frac{2\pi t}{c} + b_3\sin\frac{3\pi t}{c} + \ldots \quad \ldots(1)$$

Here, $c = 2$, because it is half range series.

Hence, $$a_0 = \frac{2}{c}\int_0^c f(t)\,dt = \frac{2}{2}\int_0^1 2t\,dt + \frac{2}{2}\int_1^2 2\,(2-t)\,dt$$

$$= [t^2]_0^1 + \left[2\left(2t - \frac{t^2}{2}\right)\right]_1^2 = 1 + [4t - t^2]_1^2 = 1 + (8-4-4+1) = 2$$

$$a_n = \frac{2}{c}\int_0^c f(t)\cos\frac{n\pi t}{c}\,dt = \frac{2}{2}\int_0^1 2t\cos\frac{n\pi t}{2}\,dt + \frac{2}{2}\int_1^2 2(2-t)\cos\frac{n\pi t}{2}\,dt$$

$$= \left[2t\left(\frac{2}{n\pi}\sin\frac{n\pi t}{2}\right) - (2)\left(-\frac{4}{n^2\pi^2}\cos\frac{n\pi t}{2}\right)\right]_0^1$$

$$+ \left[(4-2t)\left(\frac{2}{n\pi}\sin\frac{n\pi t}{2}\right) - (-2)\left(-\frac{4}{n^2\pi^2}\cos\frac{n\pi t}{2}\right)\right]_1^2$$

$$= \left[\frac{4}{n\pi}\sin\frac{n\pi}{2} + \frac{8}{n^2\pi^2}\cos\frac{n\pi}{2} - \frac{8}{n^2\pi^2}\right] + \left[0 - \frac{8}{n^2\pi^2}\cos n\pi - \frac{4}{n\pi}\sin\frac{n\pi}{2} + \frac{8}{n^2\pi^2}\cos\frac{n\pi}{2}\right]$$

$$= \frac{16}{n^2\pi^2}\cos\frac{n\pi}{2} - \frac{8}{n^2\pi^2} - \frac{8}{n^2\pi^2}\cos n\pi = \frac{8}{n^2\pi^2}\left[2\cos\frac{n\pi}{2} - 1 - \cos n\pi\right]$$

$$f(t) = 1 + \frac{8}{\pi^2}\sum_{n=1}^{\infty}\frac{1}{n^2}\left[2\cos\frac{n\pi}{2} - 1 - \cos n\pi\right]\cos\frac{n\pi t}{2}$$ **Ans.**

Example 24. *Obtain the Fourier cosine series expansion of the periodic function defined by*

$f(t) = \sin\left(\frac{\pi t}{l}\right)$, $0 < t < l$ *(U.P., II Semester, Summer 2001)*

Solution. We have, $f(t) = \sin\left(\frac{\pi t}{l}\right)$, $0 < t < l$

$$a_0 = \frac{2}{l}\int_0^l \sin\left(\frac{\pi t}{l}\right)dt = \frac{2}{l}\left(-\frac{l}{\pi}\cos\frac{\pi t}{l}\right)_0^l = -\frac{2}{\pi}(\cos\pi - \cos 0) = -\frac{2}{\pi}(-1-1) = \frac{4}{\pi}$$

$$a_n = \frac{2}{l}\int_0^l \sin\left(\frac{\pi t}{l}\right)\cos\frac{n\pi t}{l}\,dt = \frac{1}{l}\int_0^l\left[\sin\left(\frac{\pi t}{l} + \frac{n\pi t}{l}\right) - \sin\left(\frac{n\pi t}{l} - \frac{\pi t}{l}\right)\right]dt$$

$$= \frac{1}{l}\int_0^l \sin(n+1)\frac{\pi t}{l}\,dt - \frac{1}{l}\int_0^l \sin(n-1)\frac{\pi t}{l}\,dt$$

$$= \frac{1}{l}\left[-\frac{l}{(n+1)\pi}\cos\frac{(n+1)\pi t}{l}\right]_0^l - \frac{1}{l}\left[\frac{l}{(n-1)\pi}\cos\frac{(n-1)\pi t}{l}\right]_0^l$$

$$= \frac{-1}{(n+1)\pi}[\cos(n+1)\pi - \cos 0] + \frac{1}{(n-1)\pi}[\cos(n-1)\pi - \cos 0]$$

$$= \frac{1}{(n+1)\pi}[(-1)^{n+1} - 1] + \frac{1}{(n-1)\pi}[(-1)^{n+1} - 1]$$

$$= (-1)^{n+1}\left[\frac{1}{(n+1)\pi} + \frac{1}{(n-1)\pi}\right] + \frac{1}{(n+1)\pi} - \frac{1}{(n-1)\pi}$$

$$= (-1)^{n+1}\frac{2}{(n^2-1)\pi} - \frac{2}{(n^2-1)\pi} = \frac{2}{(n^2-1)\pi}[(-1)^{n+1} - 1]$$

$$= \frac{-4}{(n^2-1)\pi}, \qquad \text{when } n \text{ is even}$$

$$= 0, \qquad \text{when } n \text{ is odd.}$$

The above formula for finding the value of a_1 is not applicable.

$$a_1 = \frac{2}{l}\int_0^l \sin\frac{\pi t}{l}\cos\frac{\pi t}{l}\,dt = \frac{1}{l}\int_0^l \sin\frac{2\pi t}{l}\,dt$$

$$= \frac{1}{l}\left(-\frac{l}{2\pi}\cos\frac{2\pi t}{l}\right)_0^l = -\frac{l}{2\pi l}(\cos 2\pi - \cos 0) = -\frac{1}{2\pi}(1-1) = 0$$

$$f(t) = \frac{a_0}{2} + a_1\cos\frac{\pi t}{l} + a_2\cos\frac{2\pi t}{l} + a_3\cos\frac{3\pi t}{l} + a_4\cos\frac{4\pi t}{l} + \ldots$$

$$= \frac{2}{\pi} - \frac{4}{\pi}\left[\frac{1}{3}\cos\frac{2\pi t}{l} + \frac{1}{15}\cos\frac{4\pi t}{l} + \frac{1}{35}\cos\frac{6\pi t}{l} + \ldots\right]$$ **Ans.**

Example 25. *Find the Fourier cosine series expansion of the periodic function of period 1*

$$f(x) = \begin{cases} \frac{1}{2} + x, & -\frac{1}{2} < x \le 0 \\ \frac{1}{2} - x, & 0 < x < \frac{1}{2} \end{cases}$$

Solution. Let $f(x) = \frac{a_0}{2} + a_1\cos\frac{\pi x}{c} + a_2\cos\frac{2\pi x}{c} + \ldots$ ...(1)

as $f(x)$ is a cosine series.

Here $2c = 1 \Rightarrow c = \frac{1}{2}$

$$a_0 = \frac{1}{c}\int_{-c}^{c} f(x)\,dx = \frac{1}{1/2}\int_{-1/2}^{0}\left(\frac{1}{2}+x\right)dx + \frac{1}{1/2}\int_0^{1/2}\left(\frac{1}{2}-x\right)dx$$

$$= 2\left[\frac{x}{2}+\frac{x^2}{2}\right]_{-1/2}^{0} + 2\left[\frac{x}{2}-\frac{x^2}{2}\right]_0^{1/2} = 2\left[\frac{1}{4}-\frac{1}{8}\right] + 2\left[\frac{1}{4}-\frac{1}{8}\right] = \frac{1}{2}$$

$$a_n = \frac{1}{c}\int_{-c}^{c} f(x)\cos\frac{n\pi x}{c}\,dx$$

$$= \frac{1}{1/2}\int_{-1/2}^{0}\left(\frac{1}{2}+x\right)\cos\frac{n\pi x}{1/2}\,dx + \frac{1}{1/2}\int_0^{1/2}\left(\frac{1}{2}-x\right)\cos\frac{n\pi x}{1/2}\,dx$$

$$= 2\int_{-1/2}^{0}\left(\frac{1}{2}+x\right)\cos 2n\pi x\,dx + 2\int_0^{1/2}\left(\frac{1}{2}-x\right)\cos 2n\pi x\,dx$$

$$= 2\left[\left(\frac{1}{2}+x\right)\frac{\sin 2n\pi x}{2n\pi} - (1)\left(-\frac{\cos 2n\pi x}{4n^2\pi^2}\right)\right]_{-1/2}^{0}$$

$$+ 2\left[\left(\frac{1}{2}-x\right)\frac{\sin 2n\pi x}{2n\pi} - (-1)\left(\frac{-\cos 2n\pi x}{4n^2\pi^2}\right)\right]_0^{1/2}$$

$$= 2\left[0 + \frac{1}{4n^2\pi^2} - \frac{(-1)^n}{4n^2\pi^2}\right] + 2\left[0 - \frac{(-1)^n}{4n^2\pi^2} + \frac{1}{4n^2\pi^2}\right] = \frac{1}{\pi^2}\left[\frac{1}{n^2} - \frac{(-1)^n}{n^2}\right]$$

$$= \frac{2}{n^2\pi^2} \qquad \text{(if } n \text{ is odd)}$$
$$= 0 \qquad \text{(if } n \text{ is even)}$$

Substituting the values of a_0, a_1, a_2, a_3, ... in (1), we have

$$f(x) = \frac{1}{4} + \frac{2}{\pi^2}\left[\frac{\cos 2\pi x}{1^2} + \frac{\cos 6\pi x}{3^2} + \frac{\cos 10\pi x}{5^2} + \dots\right]$$ **Ans.**

Example 26. *Let* $f(x) = \begin{cases} wx, & \text{where } 0 \le x \le \frac{l}{2} \\ w(l-x), & \text{where } \frac{l}{2} \le x \le l \end{cases}$

Show that $$f(x) = \frac{4wl}{\pi^2} \sum_{n=0}^{\infty} \frac{(-1)^n}{(2n+1)^2} \sin \frac{(2n+1)\pi x}{l}$$

Hence, obtain the sum of the series

$$1 + \frac{1}{3^2} + \frac{1}{5^2} + \dots$$ *(U.P., Second Semester 2003)*

Solution. Half range sine series is

$$f(x) = \sum_{n=1}^{\infty} b_n \sin \frac{n\pi x}{l} \qquad \dots(1)$$

where, $b_n = \frac{2}{l}\int_0^l f(x) \sin \frac{n\pi x}{l}\, dx$

$$b_n = \frac{2}{l}\left[\int_0^{\frac{l}{2}} f(x) \sin \frac{n\pi x}{l}\, dx + \int_{\frac{l}{2}}^{l} f(x) \sin \frac{n\pi x}{l}\, dx\right]$$

$$= \frac{2}{l}\left[\int_0^{\frac{l}{2}} wx \sin \frac{n\pi x}{l}\, dx + \int_{\frac{l}{2}}^{l} w(l-x) \sin \frac{n\pi x}{l}\, dx\right]$$

$$= \frac{2}{l}\left[\left\{wx\frac{\left(-\cos\frac{n\pi x}{l}\right)}{\left(\frac{n\pi}{l}\right)} - w\frac{\left(-\sin\frac{n\pi x}{l}\right)}{\left(\frac{n\pi}{l}\right)^2}\right\}_0^{\frac{l}{2}} + \left\{w(l-x)\frac{\left(-\cos\frac{n\pi x}{l}\right)}{\left(\frac{n\pi}{l}\right)} - w(-1)\frac{\left(-\sin\frac{n\pi x}{l}\right)}{\left(\frac{n\pi}{l}\right)^2}\right\}_{\frac{l}{2}}^{l}\right]$$

$$= \frac{2}{l}\left[\left\{\frac{wl}{2}\frac{\left(-\cos\frac{n\pi}{2}\right)}{\left(\frac{n\pi}{l}\right)} + \frac{w\sin\frac{n\pi}{2}}{\left(\frac{n\pi}{l}\right)^2} - 0 - 0\right\} - \left\{\frac{wl}{2}\frac{\left(-\cos\frac{n\pi}{2}\right)}{\left(\frac{n\pi}{l}\right)} - \frac{w\sin\frac{n\pi}{2}}{\left(\frac{n\pi}{l}\right)^2}\right\}\right]$$

$$= \frac{2w}{l}\left[-\frac{l^2}{2n\pi}\cos\frac{n\pi}{2} + \frac{l^2}{n^2\pi^2}\sin\frac{n\pi}{2} + \frac{l^2}{2n\pi}\cos\frac{n\pi}{2} + \frac{l^2}{n^2\pi^2}\sin\frac{n\pi}{2}\right]$$

$$= \frac{2w}{l}\left[\frac{l^2}{n^2\pi^2}\sin\frac{n\pi}{2} + \frac{l^2}{n^2\pi^2}\sin\frac{n\pi}{2}\right] = \frac{4wl^2}{ln^2\pi^2}\sin\frac{n\pi}{2} = \frac{4wl}{(2n+1)^2\pi^2}(-1)^n,$$

$$b_n = \frac{4wl}{n^2\pi^2} \sin \frac{n\pi}{2}, \text{ when } n \text{ is odd.}$$

$$b_n = 0, \quad \text{when } n \text{ is even.}$$

Now, putting the value of b_n in (1), we get

$$f(x) = \sum_{n=1}^{\infty} \frac{4\,wl}{n^2\pi^2} \sin \frac{n\pi}{2} \sin \frac{n\pi x}{l}, \text{ when } n \text{ is odd}$$

$$f(x) = \sum_{n=1}^{\infty} \frac{4\,wl}{(2n+1)^2\pi^2} \sin \frac{(2n+1)\pi}{2} \sin \frac{(2n+1)\,\pi x}{l}$$

$$\boxed{f(x) = \frac{4wl}{\pi^2} \sum_{n=0}^{\infty} \frac{(-1)^n \sin \frac{(2n+1)\pi x}{l}}{(2n+1)^2}}$$

$$f(x) = \frac{4wl}{\pi^2} \left[\frac{\sin \frac{\pi x}{l}}{1^2} - \frac{1}{3^2} \sin \frac{3\pi x}{l} + \frac{1}{5^2} \sin \frac{5\pi x}{l} + \ldots \right] \quad \ldots(2)$$

Putting $x = \frac{l}{2}$, $f(x) = wx$ and $f\left(\frac{l}{2}\right) = \frac{wl}{2}$ in (2), we get

$$\frac{wl}{2} = \frac{4wl}{\pi^2} \left[\frac{\sin \frac{\pi}{2}}{1^2} - \frac{1}{3^2} \sin \frac{3\pi}{2} + \frac{1}{5^2} \sin \frac{5\pi}{2} + \ldots \right]$$

$$\Rightarrow \frac{\pi^2}{8} = \frac{1}{1^2} + \frac{1}{3^2} + \frac{1}{5^2} + \ldots \qquad \textbf{Ans.}$$

Example 27. *Prove that*

$$\frac{l}{2} - x = \frac{l}{\pi} \sum_{l}^{\infty} \frac{1}{n} \sin \frac{2n\pi x}{l}, \; 0 < x < l.$$

Solution. $$f(x) = \frac{1}{2} - x$$

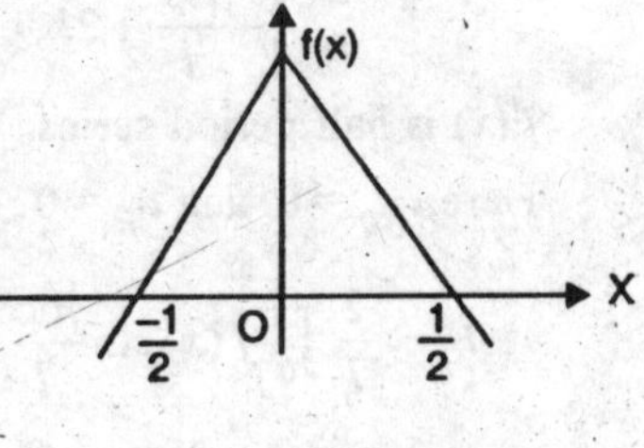

$$a_0 = \frac{1}{l/2} \int_0^l f(x)\, dx = \frac{2}{l} \int_0^l \left(\frac{l}{2} - x\right) dx = \frac{2}{l} \left[\frac{lx}{2} - \frac{x^2}{2} \right]_0^l = 0$$

$$a_n = \frac{1}{l/2} \int_0^l f(x) \cos \frac{n\pi x}{l/2}\, dx = \frac{2}{l} \int_0^l \left(\frac{l}{2} - x\right) \cos \frac{2n\pi x}{l}\, dx$$

$$= \frac{2}{l} \left[\left(\frac{l}{2} - x\right) \frac{l}{2n\pi} \sin \frac{2n\pi x}{l} + (-1) \frac{l^2}{4n^2\pi^2} \cos \frac{2n\pi x}{l} \right]_0^l$$

$$= \frac{2}{l} \left[0 - \frac{l^2}{4n^2\pi^2} \cos 2n\pi + \frac{l^2}{4n^2\pi^2} \right]$$

$$= \frac{2}{l} \frac{l^2}{4n^2\pi^2} (-\cos 2n\pi + 1) = \frac{l}{2n^2\pi^2} (-1 + 1) = 0$$

$$b_n = \frac{1}{l/2}\int_0^l f(x)\sin\frac{n\pi x}{l/2}\,dx = \frac{2}{l}\int_0^l\left(\frac{1}{2}-x\right)\sin\frac{2n\pi x}{l}\,dx$$

$$= \frac{2}{l}\left[\left(\frac{1}{2}-x\right)\left(-\frac{1}{2n\pi}\cos\frac{2n\pi x}{l}\right)-(-1)\left(-\frac{l^2}{4n^2\pi^2}\sin\frac{2n\pi x}{l}\right)\right]_0^l$$

$$= \frac{2}{l}\left[\frac{l}{2}\frac{l}{2n\pi}\cos 2n\pi - 0 + \frac{l}{2}\cdot\frac{l}{2n\pi}(1)\right] = \frac{2}{l}\left[\frac{l^2}{2n\pi}\right] = \frac{l}{n\pi}$$

Fourier series is

$$f(x) = \frac{a_0}{2} + a_1\cos\frac{n\pi x}{l/2} + a_2\cos\frac{2n\pi x}{l/2} + a_3\cos\frac{3n\pi x}{l/2} + \ldots$$

$$+ b_1\sin\frac{n\pi x}{l/2} + b_2\sin\frac{2n\pi x}{l/2} + b_3\sin\frac{3n\pi x}{l/2} + \ldots$$

$$\frac{l}{2} - x = \frac{l}{\pi}\sin\frac{2\pi x}{l} + \frac{l}{2\pi}\sin\frac{4\pi x}{l} + \frac{l}{3\pi}\sin\frac{6\pi x}{l} + \ldots$$

$$= \frac{l}{\pi}\sum_1^\infty \frac{1}{n}\sin\frac{2n\pi x}{l}$$ **Proved.**

Example 28. *Find the half period sine series for f (x) given in the range (0, l) by the graph OPQ as shown in figure.* *(U.P. II semester, 2009)*

Solution. The equation of line *OP* is $y = \dfrac{kx}{\frac{l}{2}} \Rightarrow y = \dfrac{2kx}{l}$

and the equation of the line *PQ* is $y - 0 = \dfrac{0-k}{l-\frac{l}{2}}(x-l) \quad\Rightarrow\quad y = -\dfrac{2kx}{l} + 2k$

$f(x)$ is the half period

$$f(x) = \begin{cases} \dfrac{2kx}{l}, & 0 < x < \dfrac{l}{2} \\ -\dfrac{2kx}{l} + 2k, & \dfrac{l}{2} < x < l \end{cases}$$

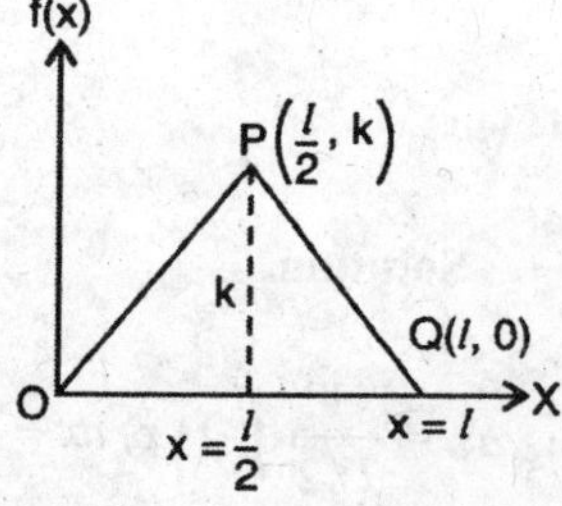

$f(x)$ is half period series. It is to be expanded as sine series.

Here, $a_0 = 0$ and $a_n = 0$

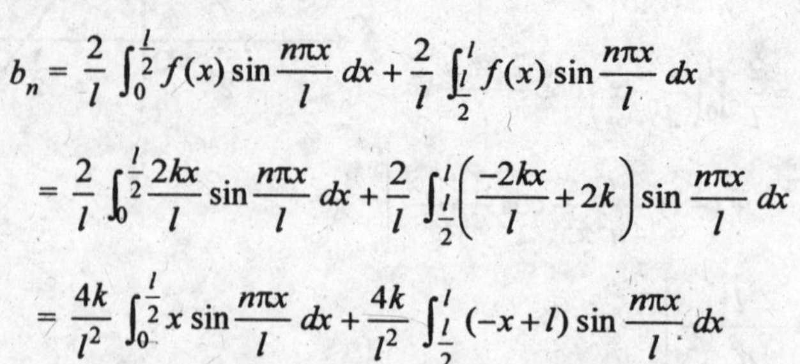

$$b_n = \frac{2}{l}\int_0^{\frac{l}{2}} f(x)\sin\frac{n\pi x}{l}\,dx + \frac{2}{l}\int_{\frac{l}{2}}^l f(x)\sin\frac{n\pi x}{l}\,dx$$

$$= \frac{2}{l}\int_0^{\frac{l}{2}}\frac{2kx}{l}\sin\frac{n\pi x}{l}\,dx + \frac{2}{l}\int_{\frac{l}{2}}^l\left(\frac{-2kx}{l}+2k\right)\sin\frac{n\pi x}{l}\,dx$$

$$= \frac{4k}{l^2}\int_0^{\frac{l}{2}} x\sin\frac{n\pi x}{l}\,dx + \frac{4k}{l^2}\int_{\frac{l}{2}}^l(-x+l)\sin\frac{n\pi x}{l}\,dx$$

$$= \frac{4k}{l^2}\left[x\left(\frac{-l}{n\pi}\cos\frac{n\pi x}{l}\right)-\left(\frac{-l^2}{n^2\pi^2}\sin\frac{n\pi x}{l}\right)\right]_0^{\frac{l}{2}} + \frac{4k}{l^2}\left[(-x+l)\left(\frac{-l}{n\pi}\cos\frac{n\pi x}{l}\right)-(-1)\left(\frac{-l^2}{n^2\pi^2}\sin\frac{n\pi x}{l}\right)\right]_{\frac{l}{2}}^l$$

$$= \frac{4k}{l^2}\left[-\frac{l}{2}\left(\frac{l}{n\pi}\right)\cos\frac{n\pi\frac{l}{2}}{l}+\frac{l^2}{n^2\pi^2}\sin\frac{n\pi\frac{l}{2}}{l}\right]$$

$$+\frac{4k}{l^2}\left[(-l+l)\left(-\frac{l}{n\pi}\cos n\pi\right)-\frac{l^2}{n^2\pi^2}\sin n\pi-\left(-\frac{l}{2}+l\right)\left(-\frac{l}{n\pi}\cos\frac{n\pi\frac{l}{2}}{l}\right)+\frac{l^2}{n^2\pi^2}\sin\frac{n\pi\frac{l}{2}}{l}\right]$$

$$= \frac{4k}{l^2}\left[-\frac{l^2}{2n\pi}\cos\frac{n\pi}{2}+\frac{l^2}{n^2\pi^2}\sin\frac{n\pi}{2}\right]+\frac{4k}{l^2}\left[0-0+\frac{l^2}{2n\pi}\cos\frac{n\pi}{2}+\frac{l^2}{n^2\pi^2}\sin\frac{n\pi}{2}\right]$$

$$= \frac{4k}{l^2}\left(\frac{l^2}{2n\pi}\right)\left[-\cos\frac{n\pi}{2}+\frac{2}{n\pi}\sin\frac{n\pi}{2}\right]+\frac{4k}{l^2}\left[\frac{l^2}{2n\pi}\cos\frac{n\pi}{2}+\frac{l^2}{n^2\pi^2}\sin\frac{n\pi}{2}\right]$$

$$= \frac{4k}{l^2}\left(\frac{l^2}{2n\pi}\right)\left[-\cos\frac{n\pi}{2}+\frac{2}{n\pi}\sin\frac{n\pi}{2}+\cos\frac{n\pi}{2}+\frac{2}{n\pi}\sin\frac{n\pi}{2}\right]$$

$$= \frac{2k}{n\pi}\left[\frac{4}{n\pi}\sin\frac{n\pi}{2}\right]=\frac{8k}{n^2\pi^2}\sin\frac{n\pi}{2}$$

Hence, Fourier series of $f(x)$ is

$$f(x)=\frac{8k}{\pi^2}\sum_{n=1}^{\infty}\frac{1}{n^2}\sin\frac{n\pi}{2}\sin\frac{n\pi x}{l}$$ **Ans.**

EXERCISE 8.4

1. Find the Fourier series to represent $f(x)$, where

$$f(x)=\begin{cases}-a, & -c<x<0\\ a, & 0<x<c\end{cases}$$ **Ans.** $\frac{4a}{\pi}\left[\sin\frac{\pi x}{c}+\frac{1}{3}\sin\frac{3\pi x}{c}+\frac{1}{5}\sin\frac{5\pi x}{c}+...\right]$

2. Find the half-range sine series for the function

$f(x)=2x-1 \quad 0<x<1$ **Ans.** $-\frac{2}{\pi}\left[\sin 2\pi x+\frac{1}{2}\sin 4\pi x+\frac{1}{3}\sin 6\pi x+...\right]$

3. Express $f(x)=x$ as a cosine, half range series in $0<x<2$.

Ans. $1-\frac{8}{\pi^2}\left[\frac{1}{1^2}\cos\frac{\pi x}{2}+\frac{1}{3^2}\cos\frac{3\pi x}{2}+\frac{1}{5^2}\cos\frac{5\pi x}{2}+...\right]$

4. Find the Fourier series of the function

$$f(x)=\begin{cases}-2 & \text{for } -4<x<-2\\ x & \text{for } -2<x<2\\ 2 & \text{for } 2<x<4\end{cases}$$

Ans. $\frac{4}{\pi}+\frac{8}{\pi^2}\sin\frac{\pi x}{4}-\frac{2}{\pi}\sin\frac{2\pi x}{4}+\left(\frac{4}{3\pi}-\frac{8}{3^2\pi}\right)\sin\frac{3\pi x}{4}-\frac{2}{2\pi}\sin\frac{4\pi x}{4}+...$

5. Find the Fourier series to represent
$f(x)=x^2-2$ from $-2<x<2$.

Ans. $-\frac{2}{3}-\frac{16}{\pi^2}\left[\cos\frac{\pi x}{2^2}-\frac{1}{4}\cos\pi x+\frac{1}{9}\cos\frac{3\pi x}{2}+...\right]$

6. If $f(x) = e^{-x}, -c < x < c$, show that

$$f(x) = (e^c - e^{-c})\left\{\frac{1}{2c} - c\left(\frac{1}{c^2+\pi^2}\cos\frac{\pi x}{c} - \frac{1}{c^2+4\pi^2}\cos\frac{2\pi x}{c} + \ldots\right)\right.$$
$$\left. - \pi\left(\frac{1}{c^2+\pi^2}\sin\frac{\pi x}{c} - \frac{2}{c^2+4\pi^2}\sin\frac{2\pi x}{c}\ldots\right)\right\}$$

7. A sinusoidal voltage E sin ωt is passed through a half wave rectifier which clips the negative portion of the wave. Develop the resulting portion of the function

$$u(t) = \begin{cases} 0, & \text{when } -\frac{T}{2} < t < 0 \\ E \sin \omega t, & \text{when } 0 < t < \frac{T}{2} \end{cases} \qquad \left(T = \frac{2\pi}{\omega}\right)$$

Ans. $\frac{E}{\pi} + \frac{E}{2}\sin \omega t - \frac{2E}{\pi}\left[\frac{1}{1.3}\cos 2\omega t + \frac{1}{3.5}\cos 4\omega t + \frac{1}{5.7}\cos 6\omega t + \ldots\right]$

8. A periodic square wave has a period 4. The function generating the square is

$$f(t) = \begin{cases} 0 & \text{for } -2 < t < -1 \\ k & \text{for } -1 < t < 1 \\ 0 & \text{for } 1 < t < 2 \end{cases}$$

Find the Fourier series of the function. **Ans.** $f(t) = \frac{k}{2} + \frac{2k}{\pi}\left[\cos\frac{\pi t}{2} - \frac{1}{3}\cos\frac{3\pi t}{2} + \ldots\right]$

9. Find a Fourier series to represent x^2 in the interval $(-l, l)$.

Ans. $\frac{l^2}{3} - \frac{4l^2}{\pi^2}\left[\cos \pi x - \frac{\cos \pi x}{2^2} + \frac{\cos 3\pi x}{3^2}\right]$

8.14. PARSEVAL'S FORMULA

$$\int_{-c}^{c} [f(x)]^2\, dx = c\left\{\frac{1}{2}a_0^2 + \sum_{n=1}^{\infty}(a_n^2 + b_n^2)\right\}$$

We know that $f(x) = \frac{a_0}{2} + \sum_{n=1}^{\infty}\left(a_n \cos\frac{n\pi x}{c} + b_n \sin\frac{n\pi x}{c}\right)$ ...(1)

Multiplying (1) by $f(x)$, we get

$$[f(x)]^2 = \frac{a_0}{2} f(x) + \sum_{n=1}^{\infty} a_n f(x)\cos\frac{n\pi x}{c} + \sum_{n=1}^{\infty} b_n f(x)\sin\frac{n\pi x}{c} \quad \text{..(2)}$$

Intergrating term by term from $-c$ to c, we have

$$\int_{-c}^{c}[f(x)]^2\, dx = \frac{a_0}{2}\int_{-c}^{c} f(x)\, dx + \sum_{n=1}^{\infty} a_n \int_{-c}^{c} f(x)\cos\frac{n\pi x}{c}\, dx$$
$$+ \sum_{n=1}^{\infty} b_n \int_{-c}^{c} f(x)\sin\frac{n\pi x}{c}\, dx \quad \text{...(3)}$$

In article 8.12, we have the following results

$$\int_{-c}^{c} f(x)\, dx = c\, a_0$$

$$\int_{-c}^{c} f(x)\cos\frac{n\pi x}{c}\, dx = c\, a_n$$

$$\int_{-c}^{c} f(x)\sin\frac{n\pi x}{c}\, dx = c\, b_n$$

On putting these integrals in (3), we get

$$\int_{-c}^{c}[f(x)]^2\,dx = c\frac{a_0^2}{2}+\sum_{n=1}^{\infty}c\,a_n^2+\sum_{n=1}^{\infty}c\,b_n^2 = c\left[\frac{a_0^2}{2}+\sum_{n=1}^{\infty}(a_n^2+b_n^2)\right]$$

This is the Parseval's formula

Note. 1. If $0 < x < 2c$, then $\int_0^{2c}[f(x)]^2\,dx = c\left[\frac{a_0^2}{2}+\sum_{n=1}^{\infty}(a_n^2+b_n^2)\right]$

2. If $0 < x < c$ (Half range consine series), $\int_0^{c}[f(x)]^2\,dx = \frac{c}{2}\left[\frac{a_0^2}{2}+\sum_{n=1}^{\infty}a_n^2\right]$

3. If $0 < x < c$ (Half range sine series), $\int_0^{c}[f(x)]^2\,dx = \frac{c}{2}\left[\sum_{n=1}^{\infty}b_n^2\right]$

4. R.M.S. $= \left\{\frac{\int_a^b[f(x)]^2\,dx}{b-a}\right\}^{\frac{1}{2}}$

Example 29. *By using the series for $f(x) = 1$ in $0 < x < \pi$ show that*

$$\frac{\pi^2}{8} = 1+\frac{1}{3^2}+\frac{1}{5^2}+\frac{1}{7^2}+\ldots$$

Solution. Sine series is $f(x) = \Sigma\, b_n \sin nx$

$$b_n = \frac{2}{\pi}\int_0^{\pi} f(x)\sin nx\,dx$$

$$= \frac{2}{\pi}\int_0^{\pi}(1)\sin nx\,dx = \frac{2}{\pi}\left(\frac{-\cos nx}{n}\right)_0^{\pi} = \frac{-2}{n\pi}[\cos n\pi - 1] = \frac{-2}{n\pi}[(-1)^n-1]$$

$$= \frac{4}{n\pi} \quad \text{if } n \text{ is odd}$$

$$= 0 \quad \text{if } n \text{ is even}$$

Then the sine series is

$$1 = \frac{4}{\pi}\sin x+\frac{4}{3\pi}\sin 3x+\frac{4}{5\pi}\sin 5x+\frac{4}{7\pi}\sin 7x+\ldots$$

$$\int_0^{c}[f(x)]^2\,dx = \frac{c}{2}\left[b_1^2+b_2^2+b_3^2+b_4^2+b_5^2+\ldots\right]$$

$$\int_0^{\pi}(1)^2\,dx = \frac{\pi}{2}\left[\left(\frac{4}{\pi}\right)^2+\left(\frac{4}{3\pi}\right)^2+\left(\frac{4}{5\pi}\right)^2+\left(\frac{4}{7\pi}\right)^2+\ldots\right]$$

$$[x]_0^{\pi} = \left(\frac{\pi}{2}\right)\left(\frac{16}{\pi^2}\right)\left[1+\frac{1}{3^2}+\frac{1}{5^2}+\frac{1}{7^2}+\ldots\right]$$

$$\pi = \frac{\pi}{2}\left(\frac{16}{\pi^2}\right)\left[1+\frac{1}{3^2}+\frac{1}{5^2}+\frac{1}{7^2}+\ldots\right]$$

$$\frac{\pi^2}{8} = 1+\frac{1}{3^2}+\frac{1}{5^2}+\frac{1}{7^2}+\ldots$$ **Proved.**

Example 30. *If* $f(x) = \begin{cases} \pi x & 0 < x < 1 \\ \pi(2-x), & 1 < x < 2 \end{cases}$

using half range cosine series, show that

$$\frac{1}{1^4} + \frac{1}{3^4} + \frac{1}{5^4} + \ldots = \frac{\pi^4}{96}$$

Solution. Half range consine series is

$$f(x) = \frac{a_0}{2} + \Sigma\, a_n \cos \frac{n\pi x}{c}$$

$$\text{where } a_0 = \frac{2}{c}\int_0^c f(x)\, dx = \frac{2}{2}\left[\int_0^1 \pi\, x\, dx + \int_1^2 \pi\,(2-x)\, dx\right]$$

$$= \pi\left(\frac{x^2}{2}\right)_0^1 + \pi\left(2x - \frac{x^2}{2}\right)_1^2 = \frac{\pi}{2} + \pi\left[(4-2) - \left(2 - \frac{1}{2}\right)\right] = \pi$$

$$a_n = \frac{2}{c}\int_0^c f(x) \cos \frac{n\pi x}{c}\, dx$$

$$= \frac{2}{2}\left[\int_0^1 \pi x \cos \frac{n\pi x}{2}\, dx + \int_1^2 \pi\,(2-x) \cos \frac{n\pi x}{2}\, dx\right]$$

$$= \pi\left[x\,\frac{\sin\frac{n\pi x}{2}}{\frac{n\pi}{2}} - \left(\frac{-\cos\frac{n\pi x}{2}}{\frac{n^2\pi^2}{4}}\right)\right]_0^1 + \pi\left[(2-x)\frac{\sin\frac{n\pi x}{2}}{\frac{n\pi}{2}} - (-1)\left(\frac{-\cos\frac{n\pi x}{2}}{\frac{n^2\pi^2}{4}}\right)\right]_1^2$$

$$= \pi\left[\frac{2}{n\pi}\sin\frac{n\pi}{2} + \frac{4}{n^2\pi^2}\cos\frac{n\pi}{2} - \frac{4}{n^2\pi^2}\right] + \pi\left[0 - \frac{4}{n^2\pi^2}\cos n\pi - \frac{2}{n\pi}\sin\frac{n\pi}{2} + \frac{4}{n^2\pi^2}\cos\frac{n\pi}{2}\right]$$

$$= \pi\left[\frac{8}{n^2\pi^2}\cos\frac{n\pi}{2} - \frac{4}{n^2\pi^2} - \frac{4}{n^2\pi^2}\cos n\pi\right] = \frac{4}{n^2\pi}\left[2\cos\frac{n\pi}{2} - 1 - \cos n\pi\right]$$

$$a_1 = 0,\ a_2 = \frac{-4}{\pi},\ a_3 = 0,\ a_4 = 0,\ a_5 = 0,\ a_6 = \frac{-4}{9\pi}\ \ldots$$

$$\int_0^c [f(x)]^2\, dx = \frac{c}{2}\left[\frac{a_0^2}{2} + a_1^2 + a_2^2 + a_3^2 + \ldots\right]$$

$$\int_0^1 (\pi x)^2\, dx + \int_1^2 \pi^2\,(2-x)^2\, dx = \frac{2}{2}\left[\frac{\pi^2}{2} + \frac{16}{\pi^2} + \frac{16}{81\pi^2} + \ldots\right]$$

$$\pi^2\left[\frac{x^3}{3}\right]_0^1 - \pi^2\left[\frac{(2-x)^3}{3}\right]_1^2 = \frac{\pi^2}{2} + \frac{16}{\pi^2} + \frac{16}{81\pi^2} + \ldots$$

$$\frac{\pi^2}{3} - \pi^2\left(0 - \frac{1}{3}\right) = \frac{\pi^2}{2} + \frac{16}{\pi^2}\left[1 + \frac{1}{81} + \ldots\right]$$

$$\frac{2\pi^2}{3} - \frac{\pi^2}{2} = \frac{16}{\pi^2}\left[1 + \frac{1}{3^4} + \frac{1}{5^4} + \ldots\right]$$

$$\frac{\pi^2}{6} = \frac{16}{\pi^2}\left[1+\frac{1}{3^4}+\frac{1}{5^4}+...\right]$$

$$\frac{\pi^4}{96} = 1+\frac{1}{3^4}+\frac{1}{5^4}+...$$ **Ans.**

Example 31. *Prove that for* $0 < x < \pi$

$$(a)\ x(\pi - x) = \frac{\pi^2}{6} - \left[\frac{\cos 2x}{1^2}+\frac{\cos 4x}{2^2}+\frac{\cos 6x}{3^2}+...\right]$$

$$(b)\ x(\pi - x) = \frac{8}{\pi}\left[\frac{\sin x}{1^2}+\frac{\sin 3x}{3^2}+\frac{\sin 5x}{5^2}+...\right]$$

Deduce from (a) and (b) respectively that

$$(c)\ \sum_{n=1}^{\infty}\frac{1}{n^4} = \frac{\pi^4}{90} \qquad (d)\ \sum_{n=1}^{\infty}\frac{1}{n^6} = \frac{\pi^4}{945}$$

Solution. (*a*) Half range cosine series

$$a_0 = \frac{2}{\pi}\int_0^{\pi} x(\pi - x) = \frac{2}{\pi}\left[\frac{\pi x^2}{2}-\frac{x^3}{3}\right]_0^{\pi} = \frac{2}{\pi}\left[\frac{\pi^3}{2}-\frac{\pi^3}{3}\right] = \frac{\pi^2}{3}$$

$$a_n = \frac{2}{\pi}\int_0^{\pi} x(\pi - x)\cos nx\, dx$$

$$= \frac{2}{\pi}\left[(\pi x - x^2)\frac{\sin nx}{n} - (\pi - 2x)\left(\frac{-\cos nx}{n^2}\right) + (-2)\left(\frac{-\sin nx}{n^3}\right)\right]_0^{\pi}$$

$$= \frac{2}{\pi}\left[0 - \frac{\pi(-1)^n}{n^2} + 0 - \frac{\pi}{n^2}\right] = \frac{2}{\pi}\left(\frac{\pi}{n^2}\right)[-(-1)^n - 1]$$

$$= -\frac{4}{n^2} \qquad \text{(when } n \text{ is even)}$$

$$= 0 \qquad \text{(when } n \text{ is odd)}$$

Hence,
$$x(\pi - x) = \frac{\pi^2}{6} - 4\left[\frac{\cos 2x}{2^2}+\frac{\cos 4x}{4^2}+...\right]$$

By Parseval's formula
$$\frac{2}{\pi}\int_0^{\pi} x^2(\pi - x)^2\, dx = \frac{a_0^2}{2} + \Sigma a_n^2$$

$$\frac{2}{\pi}\int_0^{\pi}(\pi^2 x^2 - 2\pi x^3 + x^4)\, dx = \frac{1}{2}\left(\frac{\pi^4}{9}\right) + 16\left[\frac{1}{2^4}+\frac{1}{4^4}+\frac{1}{6^4}+...\right]$$

$$\frac{2}{\pi}\left[\frac{\pi^2 x^3}{3}-\frac{2\pi x^4}{4}+\frac{x^5}{5}\right]_0^{\pi} = \frac{\pi^4}{18}+\left[\frac{1}{1^4}+\frac{1}{2^4}+\frac{1}{3^4}+...\right]$$

$$\frac{2}{\pi}\left[\frac{\pi^5}{3}-\frac{2\pi^5}{4}+\frac{\pi^5}{5}\right] = \frac{\pi^4}{18}+\left[\frac{1}{1^4}+\frac{1}{2^4}+\frac{1}{3^4}+...\right]$$

$$\frac{\pi^4}{15} = \frac{\pi^4}{18}+\sum_{n=1}^{\infty}\frac{1}{n^4} \quad \text{or} \quad \sum_{n=1}^{\infty}\frac{1}{n^4} = \frac{\pi^4}{90}$$

(*b*) Half range sine series

$$b_n = \frac{2}{\pi}\int_0^{\pi} x(\pi - x)\sin nx\, dx$$

$$= \frac{2}{\pi}\left[(\pi x - x^2)\left(\frac{-\cos nx}{n}\right) - (\pi - 2x)\left(\frac{-\sin nx}{n^2}\right) + (-2)\frac{\cos nx}{n^3}\right]_0^{\pi}$$

$$= \frac{2}{\pi}\left[-2\frac{(-1)^n}{n^3} + \frac{2}{n^3}\right] = \frac{4}{\pi n^3}[-(-1)^n + 1]$$

$$= \frac{8}{n^3\pi} \qquad \textbf{(when } n \textbf{ is odd)}$$

$$= 0 \qquad \textbf{(when } n \textbf{ is even)}$$

$$\therefore \quad x(\pi - x) = \frac{8}{\pi}\left[\frac{\sin x}{1^3} + \frac{\sin 3x}{3^3} + \frac{\sin 5x}{5^3} + \ldots\right]$$

By Parseval's formula

$$\frac{2}{\pi}\int_0^{\pi} x^2(\pi - x^2)\, dx = \Sigma b_n^2$$

$$\frac{\pi^2}{15} = \frac{64}{\pi^2}\left[\frac{1}{1^6} + \frac{1}{3^6} + \frac{1}{5^6} + \ldots\right]$$

$$\frac{\pi^4}{960} = \frac{1}{1^6} + \frac{1}{3^6} + \frac{1}{5^6} + \ldots.$$

$$\text{Let } S = \frac{1}{1^6} + \frac{1}{2^6} + \frac{1}{3^6} + \frac{1}{4^6} + \ldots = \left(\frac{1}{1^6} + \frac{1}{3^6} + \frac{1}{5^6}\ldots\right) + \left(\frac{1}{2^6} + \frac{1}{4^6} + \frac{1}{6^6} + \ldots\right)$$

$$S = \frac{\pi^4}{960} + \left(\frac{1}{2^6} + \frac{1}{4^6} + \frac{1}{6^6} + \ldots\right) = \frac{\pi^4}{960} + \frac{1}{2^6}\left[\frac{1}{1^6} + \frac{1}{2^6} + \frac{1}{3^6} + \ldots\right]$$

$$S = \frac{\pi^4}{960} + \frac{S}{64}$$

$$S - \frac{S}{64} = \frac{\pi^4}{960} \quad \Rightarrow \quad \frac{63}{64}S = \frac{\pi^4}{960}$$

$$S = \frac{\pi^4}{960} \times \frac{64}{63} = \frac{\pi^4}{945}$$

$$\sum_{n=1}^{\infty}\frac{1}{n^6} = \frac{\pi^4}{945}$$

Proved.

EXERCISE 8.5

1. Prove that in $0 < x < c$,

$$x = \frac{c}{2} - \frac{4c}{\pi^2}\left(\cos\frac{\pi x}{c} + \frac{1}{3^2}\cos\frac{3\pi x}{c} + \frac{1}{5^2}\cos\frac{5\pi x}{c} + \ldots\right) \text{ and deduce that}$$

(*i*) $\frac{1}{1^4} + \frac{1}{3^4} + \frac{1}{5^4} + \ldots = \frac{\pi^4}{96}$ (*ii*) $\frac{1}{1^4} + \frac{1}{2^4} + \frac{1}{3^4} + \frac{1}{4^4} + \ldots = \frac{\pi^4}{90}$

8.15 FOURIER SERIES IN COMPLEX FORM

Fourier series of a function $f(x)$ of period $2l$ is

$$f(x) = \frac{a_0}{2} + a_1 \cos\frac{\pi x}{l} + a_2 \cos\frac{2\pi x}{l} + \dots + a_n \cos\frac{n\pi x}{l} + \dots$$

$$+ b_1 \sin\frac{\pi x}{l} + b_2 \sin\frac{2\pi x}{l} + \dots + b_n \sin\frac{n\pi x}{l} + \dots \quad \dots(1)$$

We know that $\cos x = \dfrac{e^{ix} + e^{-ix}}{2}$ and $\sin x = \dfrac{e^{ix} - e^{-ix}}{2i}$

On putting the values of $\cos x$ and $\sin x$ in (1), we get

$$f(x) = \frac{a_0}{2} + a_1 \frac{e^{\frac{i\pi x}{l}} + e^{\frac{-i\pi x}{l}}}{2} + a_2 \frac{e^{\frac{2i\pi x}{l}} + e^{\frac{-2i\pi x}{l}}}{2} + \dots + b_1 \frac{e^{\frac{i\pi x}{l}} - e^{\frac{i\pi x}{l}}}{2i} + b_2 \frac{e^{\frac{2i\pi x}{l}} - e^{\frac{-2i\pi x}{l}}}{2i} + \dots$$

$$= \frac{a_0}{2} + (a_1 - ib_1)\, e^{\frac{i\pi x}{l}} + (a_2 - ib_2)\, e^{\frac{2i\pi x}{l}} + \dots + (a_1 + ib_1)\, e^{\frac{-i\pi x}{l}} + (a_2 + ib_2)\, e^{\frac{-2i\pi x}{l}} + \dots$$

$$= c_0 + c_1 e^{\frac{i\pi x}{l}} + c_2 e^{\frac{2i\pi x}{l}} + \dots + c_{-1} e^{\frac{-i\pi x}{l}} + c_{-2} e^{\frac{-2i\pi x}{l}} + \dots$$

$$= c_0 + \sum_{n=1}^{\infty} c_n e^{\frac{in\pi x}{l}} + \sum_{n=1}^{\infty} c_{-n} e^{\frac{-in\pi x}{l}}$$

$$c_n = \frac{1}{2}(a_n - ib_n),\ c_{-n} = \frac{1}{2}(a_n + ib_n)$$

where $c_0 = \dfrac{a_0}{2} = \dfrac{1}{2}\dfrac{1}{l}\displaystyle\int_0^{2l} f(x)\, dx$

$$c_n = \frac{1}{2}\left[\frac{1}{l}\int_0^{2l} f(x)\cos\frac{n\pi x}{l}\, dx - \frac{i}{l}\int_0^{2l} f(x)\sin\frac{n\pi x}{l}\, dx\right]$$

$$\Rightarrow \quad c_n = \frac{1}{2}\frac{1}{l}\int_0^{2l} f(x)\left\{\cos\frac{n\pi x}{l} - i\sin\frac{n\pi x}{l}\right\} dx$$

$$c_n = \frac{1}{2l}\int_0^{2l} f(x)\, e^{\frac{-in\pi x}{l}}\, dx,$$

$$c_{-n} = \frac{1}{2l}\int_0^{2l} f(x)\, e^{\frac{in\pi x}{l}}\, dx$$

Example 32. *Obtain the complex form of the Fourier series of the function*

$$f(x) = \begin{cases} 0, & -\pi \le x \le 0 \\ 1, & 0 \le x \le \pi \end{cases}$$

Solution. $\quad c_0 = \dfrac{1}{2\pi}\displaystyle\int_0^{\pi} dx = \dfrac{1}{2}$

$$c_n = \frac{1}{2\pi}\int_{-\pi}^{\pi} f(x)\, e^{-inx}\, dx$$

$$= \frac{1}{2\pi}\left[\int_{-\pi}^{0} 0\cdot e^{-inx}\, dx + \int_0^{\pi} 1\cdot e^{-inx}\, dx\right] = \frac{1}{2\pi}\int_0^{\pi} e^{-inx}\, dx = \frac{1}{2\pi}\left[\frac{e^{-inx}}{-in}\right]_0^{\pi}$$

$$= -\frac{1}{2n\pi i}[e^{-inx} - 1] = -\frac{1}{2n\pi i}[\cos n\pi - i \sin n\pi - 1] = -\frac{1}{2n\pi i}[(-1)^n - 0 - 1]$$

$$= \begin{cases} \frac{1}{in\pi}, & n \text{ is odd} \\ 0, & n \text{ is even} \end{cases}$$

$$f(x) = \frac{1}{2} + \frac{1}{i\pi}\left[\frac{e^{ix}}{1} + \frac{e^{3ix}}{3} + \frac{e^{5ix}}{5} + \ldots\right] + \frac{1}{i\pi}\left[\frac{e^{-ix}}{-1} + \frac{e^{-3ix}}{-3} + \frac{e^{-5ix}}{-5} + \ldots\right]$$

$$= \frac{1}{2} - \frac{1}{i\pi}\left[\left(e^{ix} - e^{-ix}\right) + \frac{1}{3}\left(e^{3ix} - e^{-3ix}\right) + \frac{1}{5}\left(e^{5ix} - e^{-5ix}\right) + \ldots\right]$$

$$= \frac{1}{2} - \frac{2}{\pi}\left[\sin x + \frac{1}{3}\sin 3x + \frac{1}{5}\sin 5x + \ldots\ldots\right]$$ **Ans.**

EXERCISE 8.6

Find the complex form of the Fourier series of

1. $f(x) = e^{-x}, \ -1 \le x \le 1.$ **Ans.** $\sum_{n=-\infty}^{\infty} \frac{(-1)^n (1 - in\pi)}{1 + n^2\pi^2} \sinh 1 . e^{in\pi x}$

2. $f(x) = e^{ax}, \ -l < x < l$ **Ans.** $\frac{2}{\pi} - \frac{2}{\pi}\left[\frac{e^{2it} + e^{-2it}}{1 \cdot 3} + \frac{e^{4it} + e^{-4it}}{3 \cdot 5} + \frac{e^{6it} + e^{-6it}}{5 - 7} + \ldots\right]$

3. $f(x) = \cos ax, \ -\pi < x < \pi$ **Ans.** $\frac{a}{\pi} \sin a\pi \sum_{-\infty}^{\infty} \frac{(-1)^n e^{inx}}{a^2 - n^2}$

8.16 PRACTICAL HARMONIC ANALYSIS

Some times the function is not given by a formula, but by a graph or by a table of corresponding values. The process of finding the Fourier series for a function given by such values of the function and independent variable is known as Harmonic Analysis. The Fourier constants are evaluated by the following formulae :

(1) $$a_0 = \frac{1}{\pi}\int_0^{2\pi} f(x)\,dx$$

$$= 2\frac{1}{2\pi - 0}\int_0^{2\pi} f(x)\,dx \qquad \left[\text{Mean} = \frac{1}{b-a}\int_a^b f(x)\,dx\right]$$

$\Rightarrow$ $$\boldsymbol{a_0 = 2\ [\text{mean value of } f(x) \text{ in } (0, 2\pi)]}$$

(2) $$a_n = \frac{1}{\pi}\int_0^{2\pi} f(x)\cos nx\,dx = 2\frac{1}{2\pi - 0}\int_0^{2\pi} f(x)\cos nx\,dx$$

$$\boldsymbol{= 2\ [\text{mean value of } f(x) \cos nx \text{ in } (0, 2\pi)]}$$

(3) $$b_n = \frac{1}{\pi}\int_0^{2\pi} f(x)\sin nx\,dx = 2\frac{1}{2\pi - 0}\int_0^{2\pi} f(x)\sin nx\,dx$$

$$\boldsymbol{= 2\ [\text{mean value of } f(x) \sin nx \text{ in } (0, 2\pi)]}$$

Fundamental or first harmonic. The term $(a_1 \cos x + b_1 \sin x)$ in Fourier series is called the fundamental or first harmonic.

Second harmonic. The term $(a_2 \cos 2x + b_2 \sin 2x)$ in Fourier series is called the second harmonic and so on.

Example 33. *Find the Fourier series as far as the second harmonic to represent the function given by table below :*

x	0°	30°	60°	90°	120°	150°	180°	210°	240°	270°	300°	330°
$f(x)$	2.34	3.01	3.69	4.15	3.69	2.20	0.83	0.51	0.88	1.09	1.19	1.64

Solution.

$x°$	$\sin x$	$\sin 2x$	$\cos x$	$\cos 2x$	$f(x)$	$f(x) \sin x$	$f(x) \sin 2x$	$f(x) \cos x$	$f(x) \cos 2x$
0°	0	0	1	1	2.34	0	0	2.340	2.340
30°	0.50	0.87	0.87	0.50	3.01	1.505	2.619	2.619	1.505
60°	0.87	0.87	0.50	– 0.50	3.69	3.210	3.210	1.845	– 1.845
90°	1.00	0	0	– 1.00	4.15	4.150	0	0	– 4.150
120°	0.87	– 0.87	– 0.50	– 0.50	3.69	3.210	– 3.210	– 1.845	– 1.845
150°	0.50	– 0.87	– 0.87	0.50	2.20	1.100	– 1.914	– 1.914	1.100
180°	0	0	– 1	1.00	0.83	0	0	– 0.830	0.830
210°	– 0.50	0.87	– 0.87	0.50	0.51	0.255	0.444	– 0.444	0.255
240°	– 0.87	0.87	– 0.50	– 0.50	0.88	– 0.766	0.766	–0.440	– 0.440
270°	– 1.00	0	0	– 1.00	1.09	– 1.090	0	0	– 1.090
300°	– 0.87	– 0.87	0.50	– 0.50	1.19	– 1.035	– 1.035	0.595	– 0.595
330°	– 0.50	– 0.87	0.87	0.50	1.64	– 0.820	– 1.427	1.427	0.820
					25.22	9.209	– 0.547	3.353	– 3.115

$$a_0 = 2 \times \text{Mean of } f(x) = 2 \times \frac{25.22}{12} = 4.203$$

$$a_1 = 2 \times \text{Mean of } f(x) \cos x = 2 \times \frac{3.353}{12} = 0.559$$

$$a_2 = 2 \times \text{Mean of } f(x) \cos 2x = 2 \times \frac{-3.115}{12} = -0.519$$

$$b_1 = 2 \times \text{Mean of } f(x) \sin x = 2 \times \frac{9.209}{12} = 1.535$$

$$b_2 = 2 \times \text{Mean of } f(x) \sin 2x = 2 \times \frac{-0.547}{12} = -0.091$$

Fourier series is

$$f(x) = \frac{a_0}{2} + a_1 \cos x + a_2 \cos 2x + \ldots + b_1 \sin x + b_2 \sin 2x + \ldots$$

$$= 2.1015 + 0.559 \cos x - 0.519 \cos 2x + \ldots + 1.535 \sin x - 0.091 \sin 2x + \ldots$$ **Ans.**

Example 34. *A machine completes its cycle of operations every time as certain pulley completes a revolution. The displacement $f(x)$ of a point on a certain portion of the machine is given in the table given below for twelve positions of the pulley, x being the angle in degree turned through by the pulley. Find a Fourier series to represent $f(x)$ for all values of x.*

x	30°	60°	90°	120°	150°	180°	210°	240°	270°	300°	330°	360°
$f(x)$	7.976	8.026	7.204	5.676	3.674	1.764	0.552	0.262	0.904	2.492	4.736	6.824

Solution.

x	$\sin x$	$\sin 2x$	$\sin 3x$	$\cos x$	$\cos 2x$	$\cos 3x$	$f(x)$	$f(x) \times \sin x$	$f(x) \times \sin 2x$	$f(x) \times \sin 3x$	$f(x) \times \cos x$	$f(x) \times \cos 2x$	$f(x) \times \cos 3x$
30°	0.50	0.87	1	0.87	0.50	0	7.976	3.988	6.939	7.976	6.939	3.988	0
60°	0.87	0.87	0	0.50	– 0.50	– 1	8.026	6.983	6.983	0	4.013	4.013	– 8.026
90°	1.00	0	– 1	0	– 1	0	7.204	7.204	0	– 7.204	0	– 7.204	0
120°	0.87	– 0.87	0	– 0.50	– 0.50	1	5.676	4.938	– 4.939	0	– 2.838	– 2.838	5.676
150°	0.50	– 0.87	1	– 0.87	0.50	0	3.674	1.837	– 3.196	– 3.196	– 3.196	1.837	0
180°	0	0	0	– 1	1	– 1	1.764	0	0	– 1.764	– 1.764	1.764	– 1.764
210°	– 0.50	0.87	– 1	– 0.87	0.50	0	0.552	– 0.276	0.480	0.480	–0.480	0.276	0
240°	– 0.87	0.87	0	– 0.50	– 0.50	1	0.262	– 0.228	0.228	– 0.131	– 0.131	0.131	0.262
270°	– 1.00	0	1	0	– 1.00	0	0.904	– 0.904	0	0	0	– 0.904	0
300°	– 0.87	– 0.87	0	0.50	– 0.50	– 1	2.492	– 2.168	– 2.168	1.246	1.246	–1.296	– 2.492
330°	– 0.50	– 0.87	– 1	0.87	0.50	0	4.736	–2.368	– 4.120	4.120	4.120	2.368	0
360°	0	0	0	1	1	1	6.824	0	0	0	6.824	6.824	6.824
						Σ	50.09	19.206	0.207	0.062	14.733	0.721	0.460

$$a_0 = 2 \times \text{Mean value of } f(x) = 2 \times \frac{50.09}{12} = 8.34$$

$$a_1 = 2 \times \text{Mean value of } f(x) \cos x = 2 \times \frac{14.733}{12} = 2.45$$

$$a_2 = 2 \times \text{Mean value of } f(x) \cos 2x = 2 \times \frac{0.721}{12} = 0.12$$

$$a_3 = 2 \times \text{Mean value of } f(x) \cos 3x = 2 \times \frac{0.460}{12} = 0.08$$

$$b_1 = 2 \times \text{Mean value of } f(x) \sin x = 2 \times \frac{19.206}{12} = 3.16$$

$$b_2 = 2 \times \text{Mean value of } f(x) \sin 2x = 2 \times \frac{0.207}{12} = 0.03$$

$$b_3 = 2 \times \text{Mean value of } f(x) \sin 3x = 2 \times \frac{0.062}{12} = 0.01$$

Fourier series is

$$f(x) = \frac{a_0}{2} + a_1 \cos x + a_2 \cos 2x + a_3 \cos 3x + \ldots + b_1 \sin x + b_2 \sin 2x + b_3 \sin 3x + \ldots$$

$$= 4.17 + 2.45 \cos x + 0.12 \cos 2x + 0.08 \cos 3x + \ldots$$

$$+ 3.16 \sin x + 0.03 \sin 2x + 0.01 \sin 3x + \ldots$$ **Ans.**

Example 35. *Obtain the constant term and the coefficient of the first sine and cosine terms in the Fourier series of f (x) as given in the following table.*

x	0	1	2	3	4	5
$f(x)$	9	18	24	28	26	20

Solution.

x	$\frac{x\pi}{3}$	$\sin\frac{\pi x}{3}$	$\cos\frac{\pi x}{3}$	$f(x)$	$f(x)\sin\frac{\pi x}{3}$	$f(x)\cos\frac{\pi x}{3}$
0	0	0	1.0	9	0	9
1	$\frac{\pi}{3}$	0.867	0.5	18	15.606	9
2	$\frac{2\pi}{3}$	0.867	– 0.5	24	20.808	– 12
3	$\frac{3\pi}{3}$	0	– 1.0	28	0	– 28
4	$\frac{4\pi}{3}$	– 0.867	– 0.5	26	– 22.542	– 13
5	$\frac{5\pi}{3}$	– 0.867	0.5	20	– 17.340	10
			$\Sigma =$	125	– 3.468	– 25

$$a_0 = 2 \text{ Mean value of } f(x) = 2 \times \frac{125}{6} = 41.66$$

$$a_1 = 2 \text{ Mean value of } f(x) \cos\frac{\pi x}{3} = 2 \times \frac{-25}{6} = -8.33$$

$$b_1 = 2 \text{ Mean value of } f(x) \sin\frac{\pi x}{3} = 2 \times \frac{-3.468}{6} = -1.156$$

Fourier series is
$$f(x) = \frac{a_0}{2} + a_1 \cos\frac{\pi x}{3} + \ldots + b_1 \sin\frac{\pi x}{3} + \ldots$$

$$= 20.83 - 8.33 \cos\frac{\pi x}{3} + \ldots - 1.156 \sin\frac{\pi x}{3} + \ldots$$ **Ans.**

EXERCISE 8.7

1. In a machine the displacement $f(x)$ of a given point is given for a certain angle $x°$ as follows:

$x°$	0°	30°	60°	90°	120°	150°	180°	210°	240°	270°	300°	330°
$f(x)$	7.9	8.0	7.2	5.6	3.6	1.7	0.5	0.2	0.9	2.5	4.7	6.8

Find the coefficient of sin 2 x in the Fourier series representing the above variations. **Ans.** – 0.072

2. The displacement $f(x)$ of a part of a machine is tabulated with corresponding angular moment 'x' of the crank. Express $f(x)$ as a Fourier series upto third harmonic.

$x°$	0°	30°	60°	90°	120°	150°	180°	210°	240°	270°	300°	330°
$f(x)$	1.80	1.10	0.30	0.16	0.50	1.30	2.16	1.25	1.30	1.52	1.76	2.00

Ans. $f(x) = 1.26 + 0.04 \cos x + 0.53 \cos 2x - 0.1 \cos 3x + \ldots$
$- 0.63 \sin x - 0.23 \sin 2x + 0.085 \sin 3x + \ldots$

3. The following values of y give the displacement in cms of a certain machine part of the rotation x of the flywheel. Expand $f(x)$ in the form of a Fourier series.

x	0	$\frac{\pi}{6}$	$\frac{2\pi}{6}$	$\frac{3\pi}{6}$	$\frac{4\pi}{6}$	$\frac{5\pi}{6}$
$f(x)$	0	9.2	14.4	17.8	17.3	11.7

Ans. $f(x) = 11.733 - 7.733 \cos 2x - 2.833 \cos 4x + \ldots$
$- 1.566 \sin 2x - 0.116 \sin 4x + \ldots$

4. Analyse harmonically the data given below and express y in Fourier series upto the second harmonic.

x	0	$\frac{\pi}{3}$	$\frac{2\pi}{3}$	π	$\frac{4\pi}{3}$	$\frac{5\pi}{3}$	2π
y	1.0	1.4	1.9	1.7	1.5	1.2	1.0

Ans. $a_0 = 2.9,\ a_1 = -0.37,\ a_2 = -0.1,\ a_3 = 0.03,\ b_1 = 0.17,\ b_2 = -0.06,\ b_3 = 0$

OBJECTIVE TYPE QUESTIONS

Choose the correct alternative:

1. For a function given by a graph or by a table of corresponding value, a_n =
   (*i*) 2 [Mean value of $f(x)$ sin nx in $(0, 2\pi)$]
   (*ii*) 2 [Mean value of $f(x)$ cos nx in $(0, 2\pi)$]
   (*iii*) [Mean value of $f(x)$ cos nx in $(0, 2\pi)$]
   (*iv*) n [Mean value of $f(x)$ cos mx in $(0, 2\pi)$] **Ans.** (*ii*)
2. For a function not given by formula but by a graph or by a table of corresponding values b_n =
   (*i*) 2 [Mean value of $f(x)$ cos nx in $(0, 2\pi)$]
   (*ii*) 2 [Mean value of $f(x)$ cosec nx in $(0, 2\pi)$]
   (*iii*) 2 [Mean value of $f(x)$ sin nx in $(0, 2\pi)$]
   (*iv*) n [Mean value of $f(x)$ sin nx in $(0, 2\pi)$] **Ans.** (*iii*)
3. Fourier coefficient 'a_0' in Fourier series expansion of a function represents the:
   (*i*) maximum value of the function (*ii*) 2 mean value of the function
   (*iii*) minimum value of the function (*iv*) None of these **Ans.** (*ii*)
   (*U.P. II Semester 2010*)
4. If the Fourier series of $f(x)$ has only cosine terms then $f(x)$ must be :
   (*i*) odd function (*ii*) even function **Ans.** (*ii*) (*U.P. II Semester 2010*)

UNIT–IV

CHAPTER 9

INTRODUCTION TO PARTIAL DIFFERENTIAL EQUATIONS

9.1. PARTIAL DIFFERENTIAL EQUATIONS

Partial differential equations are those equations which contain partial derivatives, independent variables and dependent variables.

The independent variables will be denoted by x and y and the dependent variable by z. The partial differential coefficients are denoted as follows .

$$\frac{\partial z}{\partial x} = p, \qquad \frac{\partial z}{\partial y} = q. \qquad \frac{\partial^2 z}{\partial x^2} = r, \qquad \frac{\partial^2 z}{\partial x \partial y} = s, \qquad \frac{\partial^2 z}{\partial y^2} = t$$

9.2. ORDER

Order of a partial differential equation is the same as that of the order of the highest differential coefficient in it.

9.3. METHOD OF FORMING PARTIAL DIFFERENTIAL EQUATIONS

A partial differential equation is formed by two methods.

(*i*) By eliminating arbitrary constants.

(*ii*) By eliminating arbitrary functions.

(*i*) Method of elimination of arbitrary constants

Example 1. *Find the PDE of all sphere whose centre lie on z-axis and given by equations* $x^2 + y^2 + (z - a)^2 = b^2$, *a and b being constants.* (*U.P., II Semester, 2009*)

Solution. We have, $x^2 + y^2 + (z - a)^2 = b^2$ ... (1)

(1) contains two arbitrary constants a and b.

Differentiating (1) partially w.r.t. x, we get

$$2x + 2(z - a)\frac{\partial z}{\partial x} = 0$$

$$\Rightarrow \qquad x + (z - a)\,p = 0 \qquad \text{... (2)}$$

Differentiating (1) partially w.r.t. y, we get

$$2y + 2(z - a)\frac{\partial z}{\partial y} = 0$$

$$y + (z - a)\,q = 0 \qquad \text{... (3)}$$

Let us eliminate a from (2) and (3).

From (2) $$(z - a) = -\frac{x}{p}$$

Putting this value of $z - a$ in (3), we get

$$y - \frac{x}{p} q = 0$$

$$\Rightarrow \quad y\,p - x\,q = 0 \qquad \textbf{Ans.}$$

(*ii*) **Method of elimination of arbitrary functions**

Example 2. *Form the partial differential equation from*

$$z = f(x^2 - y^2)$$

Solution. We have, $\quad z = f(x^2 - y^2) \quad \ldots (1)$

Differentiating (1) w.r.t. x and y, we get

$$p = \frac{\partial z}{\partial x} = f'(x^2 - y^2)\, 2x \quad \ldots (2)$$

$$q = \frac{\partial z}{\partial y} = f'(x^2 - y^2)\,(-2y) \quad \ldots (3)$$

Dividing (2) by (3), we get $\quad \frac{p}{q} = \frac{-x}{y} \quad \Rightarrow \quad py = -qx$

$$\Rightarrow \quad yp + xq = 0 \qquad \textbf{Ans.}$$

EXERCISE 9.1

Form the partial differential equations from:

1. $z = (x + a)(y + b)$ **Ans.** $pq = z$
2. $(x - h)^2 + (y - k)^2 + z^2 = a^2$ **Ans.** $z^2 (p^2 + q^2 + 1) = a^2$
3. $2z = (a\,x + y)^2 + b$ **Ans.** $p\,x + q\,y = q^2$
4. $a\,x^2 + b\,y^2 + z^2 = 1$ **Ans.** $z\,(p\,x + q\,y) = z^2 - 1$
5. $x^2 + y^2 = (z - c)^2 \tan^2 \alpha$ **Ans.** $y\,p - x\,q = 0$
6. $z = f(x^2 + y^2)$ **Ans.** $yp - xq = 0$
7. $2z = \frac{x^2}{a^2} + \frac{y^2}{b^2}$ *(A.M.I.E., Winter 2001)* **Ans.** $2z = xp + yq$
8. $f(x + y + z, x^2 + y^2 + z^2) = 0$ **Ans.** $(y - z)\,p + (z - x)\,q = x - y$

9.4 SOLUTION OF EQUATION BY DIRECT INTEGRATION

Example 3. *Solve* $\quad \frac{\partial^3 z}{\partial x^2\, \partial y} = \cos(2x + 3y)$

Solution. We have, $\quad \frac{\partial^3 z}{\partial x^2 \partial y} = \cos(2x + 3y)$

Integrating w.r.t. 'x', we get $\quad \frac{\partial^2 z}{\partial x\, \partial y} = \frac{1}{2} \sin(2x + 3y) + f(y)$

Again, integrating w.r.t. x, we get $\quad \frac{\partial z}{\partial y} = -\frac{1}{4} \cos(2x + 3y) + x \int f(y)\, dx + g(y)$

$$= -\frac{1}{4} \cos(2x + 3y) + x\phi(y) + g(y)$$

Integrating w.r.t. 'y', we get $z = -\frac{1}{12}\sin(2x+3y) + x\int \phi(y)\,dy + \int g(y)\,dy$

$$z = -\frac{1}{12}\sin(2x+3y) + x\phi_1(y) + \phi_2(y) \qquad \textbf{Ans.}$$

Example 4. *Solve* $\frac{\partial^2 z}{\partial x\, \partial y} = x^2 y$

subject to the conditions $z(x, 0) = x^2$ *and* $z(1, y) = \cos y$.

Solution. We have, $\frac{\partial}{\partial x}\left(\frac{\partial z}{\partial y}\right) = x^2 y$

On integrating w.r.t. x, we obtain

$$\frac{\partial z}{\partial y} = \frac{x^3}{3} y + f(y)$$

Integrating w.r.t. y, we obtain

$$z = \frac{x^3}{3}\cdot\frac{y^2}{2} + \int f(y)\,dy + g(x) \qquad \left[F(y) = \int f(y)\,dy\right]$$

$$\Rightarrow \qquad z = \frac{x^3 y^2}{6} + F(y) + g(x) \qquad \ldots (1)$$

Condition 1 : Putting $z = x^2$ and $y = 0$ in (1), we get

$$x^2 = 0 + F(0) + g(x)$$

Putting the value of $g(x)$ in (1), we get

$$z = \frac{x^3 y^2}{6} + F(y) + x^2 - F(0) \qquad \ldots (2)$$

Condition 2 : $z(1, y) = \cos y$

Putting $x = 1$ and $z = \cos y$ in (2), we get

$$\cos y = \frac{y^2}{6} + F(y) + 1 - F(0) \Rightarrow F(y) = \cos y - \frac{1}{6} y^2 - 1 + F(0)$$

Putting the value of $F(y)$ in (2), we get

$$z = \frac{1}{6} x^3 y^2 + \cos y - \frac{1}{6} y^2 - 1 + F(0) + x^2 - F(0)$$

$$\Rightarrow \qquad z = \frac{1}{6} x^3 y^2 + \cos y - \frac{1}{6} y^2 - 1 + x^2 \qquad \textbf{Ans.}$$

Example 5. *Solve* $\frac{\partial^2 z}{\partial y^2} = z$, *if* $y = 0$, $z = e^x$ *and* $\frac{\partial z}{\partial y} = e^{-x}$

Solution. If z is a function of y alone, then

$$z = A \sinh y + B \cosh y$$

where A and B are constants, since z is a function of x and y, A and B are the functions of x

Then $\qquad z = \sinh y \,.\, f(x) + \cosh y \,.\, \phi(x) \qquad \ldots (1)$

On putting $y = 0$ and $z = e^x$ in (1), we obtain

$$e^x = \phi(x)$$

(1) becomes $\qquad z = \sinh y \,.\, f(x) + \cosh y \,.\, e^x \qquad \ldots (2)$

On differentiating (2) w.r.t. y, we get

$$\frac{\partial z}{\partial y} = \cosh y \,.\, f(x) + \sinh y \,.\, e^x \quad \text{... (3)}$$

On putting $y = 0$ and $\frac{\partial z}{\partial y} = e^{-x}$ in (3), we obtain

$$e^{-x} = f(x)$$

Thus, (2) becomes, $z = e^{-x} \sinh y + e^x \cosh y$ **Ans.**

EXERCISE 9.2

Solve the following:

1. $\frac{\partial^2 z}{\partial x \partial y} = xy^2$ **Ans.** $z = \frac{x^2 y^3}{6} + f(y) + \phi(x)$

2. $\frac{\partial^2 z}{\partial x \partial y} = e^y \cos x$ **Ans.** $z = e^y \sin x + f(y) + \phi(x)$

3. $\frac{\partial^2 z}{\partial x \partial y} = \frac{y}{x} + 2$ **Ans.** $z = \frac{y^2}{2} \log x + 2xy + f(y) + \phi(x)$

4. $\frac{\partial^2 z}{\partial x^2} = a^2 z$, when $x = 0$, $\frac{\partial z}{\partial x} = a \sin y$ and $\frac{\partial z}{\partial y} = 0$ **Ans.** $z = \sin x + e^y \cos x$

5. $\frac{\partial^2 z}{\partial x \partial y} = \sin x \sin y$ if $\frac{\partial z}{\partial y} = -2 \sin y$ and $z = 0$ when $x = 0$

9.5 LAGRANGE'S LINEAR EQUATION IS AN EQUATION OF THE TYPE

$$Pp + Qq = R$$

where P, Q, R are the function of x, y, z and $p = \frac{\partial z}{\partial x}$, $q = \frac{\partial z}{\partial y}$

Solution. $Pp + Qq = R$ ... (1)

This form of the equation is obtained by eliminating an arbitrary function f from

$$f(u, v) = 0 \quad \text{... (2)}$$

where u, v are functions of x, y, z.

Differentiating (2) partially w.r.t. to x and y.

$$\frac{\partial f}{\partial u}\left(\frac{\partial u}{\partial x} + \frac{\partial u}{\partial z}\frac{\partial z}{\partial x}\right) + \frac{\partial f}{\partial v}\left(\frac{\partial v}{\partial x} + \frac{\partial v}{\partial z}\frac{\partial z}{\partial x}\right) = 0 \quad \text{... (3)}$$

and

$$\frac{\partial f}{\partial u}\left(\frac{\partial u}{\partial y} + \frac{\partial u}{\partial z}\frac{\partial z}{\partial y}\right) + \frac{\partial f}{\partial v}\left(\frac{\partial v}{\partial y} + \frac{\partial v}{\partial z}\frac{\partial z}{\partial y}\right) = 0 \quad \text{... (4)}$$

Let us eliminate $\frac{\partial f}{\partial u}$ and $\frac{\partial f}{\partial v}$ from (3) and (4).

From (3),
$$\frac{\partial f}{\partial u}\left[\frac{\partial u}{\partial x} + \frac{\partial u}{\partial z} p\right] = -\frac{\partial f}{\partial v}\left[\frac{\partial v}{\partial x} + \frac{\partial v}{\partial z} p\right] \quad \text{... (5)}$$

From (4),
$$\frac{\partial f}{\partial u}\left[\frac{\partial u}{\partial y} + \frac{\partial u}{\partial z} q\right] = -\frac{\partial f}{\partial v}\left[\frac{\partial v}{\partial y} + \frac{\partial v}{\partial z} q\right] \quad \text{... (6)}$$

Dividing (5) by (6), we get

$$\frac{\dfrac{\partial u}{\partial x}+\dfrac{\partial u}{\partial z}\cdot p}{\dfrac{\partial u}{\partial y}+\dfrac{\partial u}{\partial z}\cdot q}=\frac{\dfrac{\partial v}{\partial x}+\dfrac{\partial v}{\partial z}\cdot p}{\dfrac{\partial v}{\partial y}+\dfrac{\partial v}{\partial z}\cdot q}$$

$$\Rightarrow \quad \left[\frac{\partial u}{\partial x}+\frac{\partial u}{\partial z}\cdot p\right]\left[\frac{\partial v}{\partial y}+\frac{\partial v}{\partial z}\cdot q\right]=\left[\frac{\partial u}{\partial y}+\frac{\partial u}{\partial z}\cdot q\right]\left[\frac{\partial v}{\partial x}+\frac{\partial v}{\partial z}\cdot p\right]$$

$$\frac{\partial u}{\partial x}\times\frac{\partial v}{\partial y}+\frac{\partial u}{\partial x}\times\frac{\partial v}{\partial z}\cdot q+\frac{\partial u}{\partial z}\times p\times\frac{\partial v}{\partial y}+\frac{\partial u}{\partial z}\times\frac{\partial v}{\partial z}\cdot pq$$

$$=\frac{\partial u}{\partial y}\times\frac{\partial v}{\partial x}+\frac{\partial u}{\partial y}\times\frac{\partial v}{\partial z}p+\frac{\partial u}{\partial z}\cdot q\times\frac{\partial v}{\partial x}+\frac{\partial u}{\partial z}\times\frac{\partial v}{\partial z}\cdot pq$$

$$\Rightarrow \quad \left[\frac{\partial u}{\partial y}\times\frac{\partial v}{\partial z}-\frac{\partial u}{\partial z}\times\frac{\partial v}{\partial y}\right]p+\left[\frac{\partial u}{\partial z}\times\frac{\partial v}{\partial x}-\frac{\partial u}{\partial x}\times\frac{\partial v}{\partial z}\right]q$$

$$=\frac{\partial u}{\partial x}\times\frac{\partial v}{\partial y}-\frac{\partial u}{\partial y}\times\frac{\partial v}{\partial x} \quad \text{... (7)}$$

If (1) and (7) are the same, then the coefficients of p, q are equal

$$P=\frac{\partial u}{\partial y}\times\frac{\partial v}{\partial z}-\frac{\partial u}{\partial z}\times\frac{\partial v}{\partial y}$$

$$Q=\frac{\partial u}{\partial z}\times\frac{\partial v}{\partial x}-\frac{\partial u}{\partial x}\times\frac{\partial v}{\partial z} \quad \text{... (8)}$$

$$R=\frac{\partial u}{\partial x}\times\frac{\partial v}{\partial y}-\frac{\partial u}{\partial y}\times\frac{\partial v}{\partial x}$$

Now suppose $u = c_1$ and $v = c_2$ are two solutions, where c_1, c_2 are constants. Differentiating $u = c_1$ and $v = c_2$

$$\frac{\partial u}{\partial x}dx+\frac{\partial u}{\partial y}dy+\frac{\partial u}{\partial z}dz=0 \quad \text{... (9)}$$

and
$$\frac{\partial v}{\partial x}dx+\frac{\partial v}{\partial y}dy+\frac{\partial v}{\partial z}dz=0 \quad \text{... (10)}$$

Solving (9) and (10), we get

$$\frac{dx}{\dfrac{\partial u}{\partial y}\times\dfrac{\partial v}{\partial z}-\dfrac{\partial u}{\partial z}\times\dfrac{\partial v}{\partial y}}=\frac{dy}{\dfrac{\partial u}{\partial z}\times\dfrac{\partial v}{\partial x}-\dfrac{\partial u}{\partial x}\times\dfrac{\partial v}{\partial z}}=\frac{dz}{\dfrac{\partial u}{\partial x}\times\dfrac{\partial v}{\partial y}-\dfrac{\partial u}{\partial y}\times\dfrac{\partial v}{\partial x}} \quad \text{... (11)}$$

From (8) and (11), we have

$$\frac{dx}{P}=\frac{dy}{Q}=\frac{dz}{R}$$

Solutions of these equations are

$$u = c_1 \quad \text{and} \quad v = c_2$$

$\therefore$ $f(u, v) = 0$ is the required solution of (1).

9.6. WORKING RULE TO SOLVE Pp + Qq = R

Step 1. Write down the auxiliary equations

$$\frac{dx}{P}=\frac{dy}{Q}=\frac{dz}{R}$$

Step 2. Solve the above auxiliary equations.

Let the two solutions be $u = c_1$ and $v = c_2$.

Step 3. Then $f(u, v) = 0$ or $u = \phi(v)$ is the required solution of

$$Pp + Qq = R.$$

Example 6. *Solve the following partial differential equation*

$$yq - xp = z, \qquad \textit{where } p=\frac{\partial z}{\partial x},\ q=\frac{\partial z}{\partial y}.$$

Solution. We have, $y\,q - x\,p = z$

Here the auxiliary equations are

$$\frac{dx}{-x}=\frac{dy}{y}=\frac{dz}{z}$$

$$-\log x = \log y - \log a \quad \text{(From first two equations)}$$

$$x\,y = a \qquad \text{... (1)}$$

$$\log y = \log z + \log b \quad \text{(From last two equations)}$$

$$\frac{y}{z} = b \qquad \text{... (2)}$$

From (1) and (2) we get the solution

$$f\left(xy, \frac{y}{z}\right) = 0. \qquad \textbf{Ans.}$$

Example 7. *Solve* $y^2p - xyq = x(z - 2y)$. (*A.M.I.E., Summer 2001*)

Solution. We have, $y^2p - xyq = x(z - 2y)$

The auxiliary equations are

$$\frac{dx}{y^2}=\frac{dy}{-xy}=\frac{dz}{x(z-2y)} \qquad \text{... (1)}$$

Considering first two members of the equations

$$\frac{dx}{y}=\frac{dy}{-x} \quad \Rightarrow \quad x\,dx = -y\,dy$$

Integrating $\dfrac{x^2}{2}=-\dfrac{y^2}{2}+\dfrac{C_1}{2}$

$$\Rightarrow \quad x^2 + y^2 = C_1 \qquad \text{... (2)}$$

From last two equations of (1), we have

$$-\frac{dy}{y}=\frac{dz}{z-2y}$$

$$\Rightarrow \quad -z\,dy + 2y\,dy = y\,dz \quad \Rightarrow \quad 2y\,dy = y\,dz + z\,dy$$

On integration, we get

$$y^2 = yz + C_2$$

$$y^2 - yz = C_2 \qquad \text{... (3)}$$

From (2) and (3), we have

$$x^2 + y^2 = f(y^2 - yz) \qquad \textbf{Ans.}$$

Example 8. *Solve* $(x^2 - yz)\,p + (y^2 - zx)\,q = z^2 - xy$

(A.M.I.E., Summer 2001)

Solution. $(x^2 - y\,z)\,p + (y^2 - z\,x)\,q = z^2 - x\,y$ ... (1)

The auxiliary equations are

$$\frac{dx}{x^2 - yz} = \frac{dy}{y^2 - zx} = \frac{dz}{z^2 - xy}$$

$$\Rightarrow \quad \frac{dx - dy}{x^2 - yz - y^2 + zx} = \frac{dy - dz}{y^2 - zx - z^2 + xy} = \frac{dz - dx}{z^2 - xy - x^2 + yz}$$

$$\Rightarrow \quad \frac{dx - dy}{(x-y)(x+y+z)} = \frac{dy - dz}{(y-z)(x+y+z)} = \frac{dz - dx}{(z-x)(x+y+z)}$$

$$\Rightarrow \quad \frac{dx - dy}{x - y} = \frac{dy - dz}{y - z} = \frac{dz - dx}{z - x} \quad \text{... (2)}$$

Integrating first two members of (2), we have

$$\log(x - y) = \log(y - z) + \log c_1$$

$$\log\frac{x - y}{y - z} = \log c_1 \quad \Rightarrow \quad \frac{x - y}{y - z} = c_1 \quad \text{... (3)}$$

Similarly from last two members of (2), we have

$$\frac{y - z}{z - x} = c_2 \quad \text{.... (4)}$$

From (3) and (4), the required solution is

$$f\left[\frac{x - y}{y - z}, \frac{y - z}{z - x}\right] = 0$$ **Ans.**

9.7 METHOD OF MULTIPLIERS

Let the auxiliary equations be

$$\frac{dx}{P} = \frac{dy}{Q} = \frac{dz}{R}$$

l, m, n may be constants or functions of x, y, z then, we have

$$\frac{dx}{P} = \frac{dy}{Q} = \frac{dz}{R} = \frac{l\,dx + m\,dy + n\,dz}{lP + mQ + nR}$$

l, m, n are chosen in such a way that

$$lP + mQ + nR = 0$$

Thus $\quad l\,dx + m\,dy + n\,dz = 0$

Solve this differential equation, if the solution is $u = c_1$.

Similarly, choose another set of multipliers (l_1, m_1, n_1) and if the second solution is $v = c_2$.

$\therefore$ Required solution is $f(u, v) = 0$.

Example 9. *Solve*

$$(mz - ny)\frac{\partial z}{\partial x} + (nx - lz)\frac{\partial z}{\partial y} = l\,y - mx.$$ *(A.M.I.E. Winter 2001)*

Solution. We have, $(mz - ny)\dfrac{\partial z}{\partial x} + (nx - lz)\dfrac{\partial z}{\partial y} = l\,y - mx$

Here, the auxiliary equations, are

$$\frac{dx}{mz-ny}=\frac{dy}{nx-lz}=\frac{dz}{ly-mx}$$

Using multipliers $x, y, z,$ we get

$$\text{each fraction}=\frac{x\,dx+y\,dy+z\,dz}{x(mz-ny)+y(nx-lz)+z(ly-mx)}=\frac{x\,dx+y\,dy+z\,dz}{0}$$

$\therefore \quad x\,dx + y\,dy + z\,dz = 0$

which on integration gives

$$x^2+y^2+z^2=c_1 \qquad \text{... (1)}$$

Again using multipliers, $l, m, n,$ we get

$$\text{each fraction}=\frac{l\,dx+m\,dy+n\,dz}{l(mz-ny)+m(nx-lz)+n(ly-mx)}=\frac{l\,dx+m\,dy+n\,dz}{0}$$

$\therefore \quad l\,dx + m\,dy + n\,dz = 0$

which on integration gives

$$l\,x+m\,y+n\,z=c_2 \qquad \text{... (2)}$$

Hence from (1) and (2), the required solution is

$$x^2+y^2+z^2=f(lx+my+nz) \qquad \textbf{Ans.}$$

Example 10. *Solve the partial differential equation* $x(y^2+z)\,p-y(x^2+z)\,q=z(x^2-y^2)$

where, $p=\dfrac{\partial z}{\partial x}$ and $q=\dfrac{\partial z}{\partial y}$. *(U.P., II Semester, 2008)*

Solution. Lagrange's subsidiary equations are

$$\frac{dx}{x(y^2+z)}=\frac{dy}{-y(x^2+z)}=\frac{dz}{z(x^2-y^2)} \qquad \text{...(1)}$$

Using $x, y, -1$ as multipliers, we get

$$\text{each fraction}=\frac{x\,dx+y\,dy-dz}{x^2\left(y^2+z\right)-y^2\left(x^2+z\right)-z\left(x^2-y^2\right)}$$

$$=\frac{x\,dx+y\,dy-dz}{0}$$

$\therefore \quad x\,dx + y\,dy - dz = 0$

Integrating, we get

$$\frac{x^2}{2}+\frac{y^2}{2}-z=\frac{C_1}{2}$$

$$\Rightarrow \quad x^2+y^2-2z=C_1 \qquad \text{...(2)}$$

Again, using $\dfrac{1}{x}$, $\dfrac{1}{y}$ and $\dfrac{1}{z}$ as multipliers, we get

$$\text{each fraction}=\frac{\frac{1}{x}dx+\frac{1}{y}dy+\frac{1}{z}dz}{y^2+z-x^2-z+x^2-y^2}$$

$$=\frac{\frac{1}{x}dx+\frac{1}{y}dy+\frac{1}{z}dz}{0}$$

$$\therefore \quad \frac{1}{x}dx + \frac{1}{y}dy + \frac{1}{z}dz = 0$$

Integrating, we get

$\log x + \log y + \log z = \log C_2$

$\Rightarrow \quad xyz = C_2$ ...(3)

Hence the general solution is

$\phi(x^2 + y^2 - 2z, xyz) = 0$ **Ans.**

Example 11. *Find the general solution of*

$$x(z^2 - y^2)\frac{\partial z}{\partial x} + y(x^2 - z^2)\frac{\partial z}{\partial y} = z(y^2 - x^2)$$

Solution. $x(z^2 - y^2)\dfrac{\partial z}{\partial x} + y(x^2 - z^2)\dfrac{\partial z}{\partial y} = z(y^2 - x^2)$ ... (1)

$\therefore$ The auxiliary simultaneous equations are

$$\frac{dx}{x(z^2 - y^2)} = \frac{dy}{y(x^2 - z^2)} = \frac{dz}{z(y^2 - x^2)} \quad \text{... (2)}$$

Using multipliers x, y, z, we get

$$\text{Each term of (2)} = \frac{x\,dx + y\,dy + z\,dz}{x^2(z^2 - y^2) + y^2(x^2 - z^2) + z^2(y^2 - x^2)} = \frac{x\,dx + y\,dy + z\,dz}{0}$$

$\therefore \quad x\,dx + y\,dy + z\,dz = 0$ ... (3)

On integration $x^2 + y^2 + z^2 = C_1$

Again (2) can be written as

$$\frac{\frac{dx}{x}}{z^2 - y^2} = \frac{\frac{dy}{y}}{x^2 - z^2} = \frac{\frac{dz}{z}}{y^2 - x^2} = \frac{\frac{dx}{x} + \frac{dy}{y} + \frac{dz}{z}}{(z^2 - y^2) + (x^2 - z^2) + (y^2 - x^2)} \quad \text{... (4)}$$

$$= \frac{\frac{dx}{x} + \frac{dy}{y} + \frac{dz}{z}}{0} \quad \Rightarrow \quad \frac{dx}{x} + \frac{dy}{y} + \frac{dz}{z} = 0$$

On integration, we get

$\Rightarrow \quad \log x + \log y + \log z = \log C_2$

$\log x\,y\,z = \log C_2 \;\Rightarrow\; x\,y\,z = C_2$ .... (5)

From (3) and (5), the general solution is

$xyz = f(x^2 + y^2 + z^2)$ **Ans.**

Example 12. *Solve the partial differential equation*

$$\frac{y - z}{yz}p + \frac{z - x}{zx}q = \frac{x - y}{xy}$$

Solution. We have, $\dfrac{y - z}{yz}p + \dfrac{z - x}{zx}q = \dfrac{x - y}{xy}$

Multiplying by xyz, we get

$x(y - z)p + y(z - x)q = z(x - y)$

$$\frac{dx}{x(y-z)}=\frac{dy}{y(z-x)}=\frac{dz}{z(x-y)}=\frac{dx+dy+dz}{x(y-z)+y(z-x)+z(x-y)} \quad \text{.. (1)}$$

$$=\frac{dx+dy+dz}{0}$$

$\therefore$ $dx + dy + dz = 0$

Which on integration gives

$$x+y+z=a \quad \text{...(2)}$$

Again (1) can be written as

$$\frac{\frac{dx}{x}}{y-z}=\frac{\frac{dy}{y}}{z-x}=\frac{\frac{dz}{z}}{x-y}=\frac{\frac{dx}{x}+\frac{dy}{y}+\frac{dz}{z}}{(y-z)+(z-x)+(x-y)}=\frac{\frac{dx}{x}+\frac{dy}{y}+\frac{dz}{z}}{0}$$

$\Rightarrow$ $$\frac{dx}{x}+\frac{dy}{y}+\frac{dz}{z}=0$$

On integration, we get

$$\log x+\log y+\log z=\log b \;\Rightarrow\; \log xyz=\log b \;\Rightarrow\; xyz=b \quad \text{... (3)}$$

From (2) and (3) the general solution is

$$xyz=f(x+y+z)$$ **Ans.**

Example 13. *Solve* $(x^2-y^2-z^2)\,p+2xy\,q=2xz.$ *(A.M.I.E., Summer 2004, 2000)*

Solution. We have, $(x^2-y^2-z^2)\,p+2xy\,q=2xz$ ... (1)

Here the auxiliary equations are

$$\frac{dx}{x^2-y^2-z^2}=\frac{dy}{2xy}=\frac{dz}{2xz} \quad \text{... (2)}$$

From the last two members of (2), we have

$$\frac{dy}{y}=\frac{dz}{z}$$

which on integration gives

$$\log y=\log z+\log a \;\Rightarrow\; \log\frac{y}{z}=\log a$$

$\Rightarrow$ $$\frac{y}{z}=a \quad \text{... (3)}$$

Using multipliers $x, y, z,$ we have

$$\frac{dx}{x^2-y^2-z^2}=\frac{dy}{2xy}=\frac{dz}{2xz}=\frac{x\,dx+y\,dy+z\,dz}{x(x^2+y^2+z^2)}$$

$$\frac{2x\,dx+2y\,dy+2z\,dz}{(x^2+y^2+z^2)}=\frac{dz}{z}$$

which on integration gives

$$\log(x^2+y^2+z^2)=\log z+\log b$$

$$\frac{x^2+y^2+z^2}{z}=b \quad \text{... (4)}$$

Hence from (3) and (4), the required solution is

$$x^2+y^2+z^2=z\,f\left(\frac{y}{z}\right)$$ **Ans.**

Example 14. *Solve the differential equation*

$$x^2\frac{\partial z}{\partial x}+y^2\frac{\partial z}{\partial y}=(x+y)z$$

Solution. We have, $$x^2\frac{\partial z}{\partial x}+y^2\frac{\partial z}{\partial y}=(x+y)z \quad \text{... (1)}$$

The auxiliary equations of (1) are

$$\frac{dx}{x^2}=\frac{dy}{y^2}=\frac{dz}{(x+y)z} \quad \text{... (2)}$$

Take first two members of (2) and integrate them

$$-\frac{1}{x}=-\frac{1}{y}+c$$

$$\frac{1}{x}-\frac{1}{y}=c_1 \quad \text{... (3)}$$

(2) can be written as

$$\frac{\frac{dx}{x}}{x}=\frac{\frac{dy}{y}}{y}=\frac{\frac{dz}{z}}{x+y}=\frac{\frac{dx}{x}+\frac{dy}{y}-\frac{dz}{z}}{(x+y)-(x+y)}$$

$$\Rightarrow \quad \frac{dx}{x}+\frac{dy}{y}-\frac{dz}{z}=0$$

On integration, we get

$$\Rightarrow \quad \log x+\log y-\log z=\log c_2$$

$$\Rightarrow \quad \log\frac{xy}{z}=\log c_2 \quad \Rightarrow \quad \frac{xy}{z}=c_2 \quad \text{... (4)}$$

From (3) and (4), we have

$$f\left[\frac{1}{x}-\frac{1}{y},\frac{xy}{z}\right]=0.$$ **Ans.**

Example 15. *Find the general solution of*

$$x\frac{\partial z}{\partial x}+y\frac{\partial z}{\partial y}+t\frac{\partial z}{\partial t}=xyt$$

Solution. The auxiliary equations are

$$\frac{dx}{x}=\frac{dy}{y}=\frac{dt}{t}=\frac{dz}{xyt} \quad \text{... (1)}$$

Taking the first two members and integrating, we get

$$\log x=\log y+\log a=\log ay$$

$$\Rightarrow \quad x=ay, \quad i.e. \quad x/y=a \quad \text{... (2)}$$

Similarly, from the 2nd and 3rd members

$$\frac{t}{y}=b \quad \text{... (3)}$$

Multiplying the equation (1) by xyt, we get

$$dz=\frac{ty\,dx}{1}=\frac{tx\,dy}{1}=\frac{xy\,dt}{1}=\frac{ty\,dx+tx\,dy+xy\,dt}{3}$$

Integrating, we get

$$z = \frac{1}{3} x\,y\,t + c \;\Rightarrow\; z - \frac{1}{3} xy\,t = c \qquad \dots (4)$$

From (2), (3) and (4) the solution is

$$z - \frac{1}{3} xyt = f\left(\frac{y}{x}\right) + \phi\left(\frac{t}{y}\right) \qquad \textbf{Ans.}$$

Example 16. *Solve* $(y + z)\,p - (x + z)\,q = x - y$

Solution. $(y + z)\,p - (x + z)\,q = x - y \qquad \dots (1)$

$\therefore$ The auxiliary equations are

$$\frac{dx}{y+z} = \frac{dy}{-(x+z)} = \frac{dz}{x-y} \qquad \dots (2)$$

$$\Rightarrow \quad \frac{dx}{y+z} = \frac{dy}{-(x+z)} = \frac{dz}{x-y} = \frac{dx+dy+dz}{y+z-(x+z)+x-y}$$

$$\Rightarrow \quad \frac{dx}{y+z} = \frac{dy}{-(x+z)} = \frac{dz}{x-y} = \frac{dx+dy+dz}{0}$$

Thus, we have $\quad dx + dy + dz = 0$

Which on integration gives $\quad x + y + z = c_1 \qquad \dots (3)$

Using multipliers $x, y, -z$ for (2), we get

$$\frac{dx}{y+z} = \frac{dy}{-(x+z)} = \frac{dz}{x-y} = \frac{x\,dx + y\,dy - z\,dz}{x(y+z) - y(x+z) - z(x-y)}$$

$$\Rightarrow \quad \frac{dx}{y+z} = \frac{dy}{-(x+z)} = \frac{dz}{x-y} = \frac{x\,dx + y\,dy - z\,dz}{0}$$

Integrating $x\,dx + y\,dy - z\,dz = 0$, we get

$$\frac{x^2}{2} + \frac{y^2}{2} - \frac{z^2}{2} = c_2$$

$$\Rightarrow \quad x^2 + y^2 - z^2 = 2c_2 \qquad \dots (4)$$

From (3) and (4), we get the required solution

$$f(x + y + z,\; x^2 + y^2 - z^2) = 0 \qquad \textbf{Ans.}$$

Example 17. *Solve* $z\,p + yq = x.$

Solution. The auxiliary equations are

$$\underset{(i)}{\frac{dx}{z}} = \underset{(ii)}{\frac{dy}{y}} = \underset{(iii)}{\frac{dz}{x}} \qquad \dots (1)$$

From (i) and (iii)

$$\frac{dx}{z} = \frac{dz}{x} \qquad \Rightarrow \qquad x\,.\,dx = z\,.\,dz$$

$$\Rightarrow \quad \frac{x^2}{2} = \frac{z^2}{2} - \frac{c_1}{2} \qquad \Rightarrow \qquad x^2 = z^2 - c_1 \qquad \dots (2)$$

$$\Rightarrow \qquad z = \sqrt{x^2 + c_1}$$

Putting the value of z in (1), we get

$$\frac{dx}{\sqrt{x^2 + c_1}} = \frac{dy}{y}$$

$$\sinh^{-1} \frac{x}{\sqrt{c_1}} = \log y + c_2 \qquad \text{... (3)}$$

From (2) and (3), the required solution is

$$f(z^2 - x^2) = \sinh^{-1} \frac{x}{\sqrt{c_1}} - \log y \qquad \textbf{Ans.}$$

Example 18. *Solve* $px(z - 2y^2) = (z - qy)(z - y^2 - 2x^3)$.

(A.M.I.E., Summer 2000)

Solution. $px(z - 2y^2) = (z - qy)(z - y^2 - 2x^3)$

$$\Rightarrow \qquad px(z - 2y^2) + qy(z - y^2 - 2x^3) = z(z - y^2 - 2x^3) \qquad \text{... (1)}$$

Here the auxiliary equations are

$$\frac{dx}{x(z - 2y^2)} = \frac{dy}{y(z - y^2 - 2x^3)} = \frac{dz}{z(z - y^2 - 2x^3)} \qquad \text{... (2)}$$

From the last two members of (2), we have

$$\frac{dy}{y} = \frac{dz}{z}$$

which gives on integration

$$\log y = \log z + \log a \quad \Rightarrow \quad y = az \qquad \text{... (3)}$$

From the first and third members of (2), we have

$$\frac{dx}{x(z - 2y^2)} = \frac{dz}{z(z - y^2 - 2x^3)} \qquad \text{[Using (3), } y = az]$$

$$\Rightarrow \qquad \frac{dx}{x(z - 2a^2z^2)} = \frac{dz}{z(z - a^2z^2 - 2x^3)}$$

$$\Rightarrow \qquad \frac{dx}{x(1 - 2a^2z)} = \frac{dz}{z - a^2z^2 - 2x^3}$$

$$\Rightarrow \qquad z\,dx - a^2z^2dx - 2x^3dx = x\,dz - 2a^2\,x\,z\,dz$$

$$\Rightarrow \qquad (x\,dz - z\,dx) - a^2(2\,x\,z\,dz - z^2dx) + 2x^3\,dx = 0$$

$$\Rightarrow \qquad \frac{x\,dz - z\,dx}{x^2} - a^2\frac{(2\,x\,z\,dz - z^2\,dx)}{x^2} + 2\,x\,dx = 0$$

On integrating,

$$\frac{z}{x} - a^2\frac{z^2}{x} + x^2 = b \qquad \text{... (4)}$$

From (3) and (4), we have the required solution:

$$\frac{y}{z} = f\left(\frac{z}{x} - \frac{a^2z^2}{x} + x^2\right) \qquad \textbf{Ans.}$$

EXERCISE 9.3

Solve the following partial differential equations

1. $p \tan x + q \tan y = \tan z$ **Ans.** $f\left(\frac{\sin x}{\sin y}, \frac{\sin y}{\sin z}\right) = 0$
2. $(y - z)\, p + (x - y)\, q = z - x$ **Ans.** $f(x + y + z, x^2 + 2yz)$
3. $(y + zx)\, p - (x + yz)\, q = x^2 - y^2$ **Ans.** $f(x^2 + y^2 - z^2) = (x - y)^2 - (z + 1)^2$
4. $zx\frac{\partial z}{\partial x} - zy\frac{\partial z}{\partial y} = y^2 - x^2$ **Ans.** $f(x^2 + y^2 + z^2, xy) = 0$
5. $pz - qz = z^2 + (x + y)^2$ **Ans.** $[z^2 + (x + y)^2]\, e^{-2x} = f(x + y)$
6. $p + q + 2xz = 0$ **Ans.** $f(x - y) = x^2 + \log z$
7. $x^2p + y^2q + z^2 = 0$ **Ans.** $f\left(\frac{1}{y} - \frac{1}{x}, \frac{1}{y} + \frac{1}{z}\right) = 0$
8. $(x^2 + y^2)\, p + 2\, x\, y\, q = (x + y)\, z$ (*A.M.I.E., Summer 2000*) **Ans.** $f\left(\frac{x+y}{z}, \frac{2y}{x^2 - y^2}\right) = 0$
9. $\frac{\partial z}{\partial x} - 2\frac{\partial z}{\partial y} = 2x - e^y + 1$ **Ans.** $f(2x + y) = z - \frac{(2x+1)^2}{4} - \frac{e^y}{2}$
10. $p + 3q = 5\, z + \tan(y - 3x)$ **Ans.** $f(y - 3x) = \frac{e^{5x}}{5z + \tan(y - 3x)}$
11. $xp - yq + x^2 - y^2 = 0$ **Ans.** $f(xy) = \frac{x^2}{2} + \frac{y^2}{2} + z$
12. $(x + y)\left(\frac{\partial z}{\partial x} + \frac{\partial z}{\partial y}\right) = z - 1$ **Ans.** $f(x - y) = \frac{x + y}{(z - 1)^2}$.
13. $(x^3 + 3xy^2)\frac{\partial z}{\partial x} + (y^3 + 3x^2y)\frac{\partial z}{\partial y} = 2(x^2 + y^2)z$ (*A.M.I.E.T.E., Summer 2000*)
14. $(z^2 - 2\, y\, z - y^2)\, P + (x\, y + z\, x)\, q = x\, y - z\, x$
15. Find the solution of the equation $\frac{x\partial z}{\partial y} - \frac{y\partial z}{\partial x} = 0$, which passes through the curve $z = 1, x^2 + y^2 = 4$

    Ans. $f(x^2 + y^2 - 4, z - 1) = 0$

OBJECTIVE TYPE QUESTIONS

Choose the correct alternative :

1. The partial differential equation from $z = (a + x)^2 + y$ is

   (*i*) $z = \frac{1}{4}\left(\frac{\partial z}{\partial x}\right)^2 + y$ (*ii*) $z = \left(\frac{\partial z}{\partial x}\right)^2 + y$

   (*iii*) $z = \frac{1}{4}\left(\frac{\partial z}{\partial y}\right)^2 + y$ (*iv*) $z = \left(\frac{\partial z}{\partial y}\right)^2 + y$ **Ans.** (*i*)

2. The solution of $xp + yq = z$ is

   (*i*) $f(x^2, y^2) = 0$ (*ii*) $f(x, y) = 0$

   (*iii*) $f\left(\frac{x}{y}, \frac{y}{z}\right) = 0$ (*iv*) $f(xy, yz) = 0$ **Ans.** (*iii*)

3. The solution of $\frac{y-z}{yz}p + \frac{z-x}{zx}q = \frac{x-y}{xy}$ is

(i) $x - y - z = f(x + y + z)$ (ii) $x + y + z = f(x + y + z)$

(iii) $x - y - z = f(x - y - z)$ (iv) $xyz = f(x + y + z)$ **Ans.** (iv)

4. The solution of $(y + z)p - (x + z)q = x - y$ is

(i) $f(x + y + z, x^2 y^2 z^2)$ (ii) $f(x + y + z, x^2 + y^2 - z^2) = 0$

(iii) $f(x - y - z, x^2 + y^2 - z^2)$ (iv) $f(xyz, x^2 + y^2 - z^2) = 0$ **Ans.** (ii)

Indicate True or False for the following statements

5. The auxiliary equation of $Pp + Qq = R$ can be written as

$$\frac{dp}{P} = \frac{dq}{Q} = \frac{dz}{R}$$ **Ans.** True

6. From the equation $z = (x + 4)(y + 6)$ a partial differential equation $pq = 2$ is formed. **Ans.** False

7. The auxiliary equation of

$$(mz - ny)\frac{\partial z}{\partial x} + (nx - lz)\frac{\partial z}{\partial y} = ly - mx \text{ is } \frac{dx}{mz - ny} = \frac{dy}{nx - lz} = \frac{dz}{ly + mx}.$$

Ans. False

8. The auxiliary equation of $x\frac{\partial z}{\partial x} + y\frac{\partial z}{\partial y} + t\frac{\partial z}{\partial t} = xyt$ is $\frac{dx}{x} = \frac{dy}{y} = \frac{dt}{t}$ **Ans.** False

9. The *A.E.* of $(y + z)p - (x + z)q = x - y$ is $\frac{dx}{y+z} = \frac{dy}{x+z} = \frac{dz}{x-y}$ **Ans.** False

10. With usual symbols, the *P.D.E.* $u_{xx} + u^2 u_{yy} = f(xy)$ is non-linear in 'u' and is of second order. **Ans.** True

(*U.P., II Semester, 2009*)

11. Solution of the *P.D.E.* $\frac{\partial^2 z}{\partial x \partial y} = xy^2$ is $z = \frac{x^2 y^3}{6} + f(y) + \phi(x)$ **Ans.** True

UNIT–IV

CHAPTER 10

Linear Partial Differential Equations with Constant Coefficients of 2nd Order

10.1 LINEAR HOMOGENEOUS PARTIAL DIFFERENTIAL EQUATIONS OF *n*TH ORDER WITH CONSTANT COEFFICIENTS

An equation of the type

$$a_0 \frac{\partial^n z}{\partial x^n} + a_1 \frac{\partial^n z}{\partial x^{n-1} \partial y} + ... + a_n \frac{\partial^n z}{\partial y^n} = F(x, y) \quad ... (1)$$

is called a homogeneous linear partial differential equation of *n*th order with constant coefficients. It is called homogeneous because all the terms contain derivatives of the same order.

Putting $\frac{\partial}{\partial x} = D$ and $\frac{\partial}{\partial y} = D'$ in (1), we get

$$(a_0 D^n + a_1 D^{n-1} D' + ... + a_n D'^n) z = F(x, y)$$

$$\Rightarrow \qquad f(D, D')z = F(x, y)$$

10.2 RULES FOR FINDING THE COMPLEMENTARY FUNCTION

Consider the equation

$$a_0 \frac{\partial^2 z}{\partial x^2} + a_1 \frac{\partial^2 z}{\partial x \partial y} + a_2 \frac{\partial^2 z}{\partial y^2} = 0 \Rightarrow \qquad (a_0 D^2 + a_1 DD' + a_2 D'^2) z = 0$$

Step 1. Put $D = m$ and $D' = 1$

$$a_0 m^2 + a_1 m + a_2 = 0$$

This is the auxiliary equation.

Step 2. Solve the auxiliary equation.

Case 1. If the roots of the auxiliary equation are real and different; say m_1, m_2.

Then $\quad$ C.F. $= f_1(y + m_1 x) + f_2(y + m_2 x)$

Theory: $(D - m_1 D')(D - m_2 D')z = 0 \quad ...(1)$

(1) will be satisfied by the solution of

$$(D - m_2 D') z = 0 \quad \Rightarrow \quad p - m_2 q = 0 \quad ... (2)$$

This is a Lagrange's linear equation. Its subsidiary equations are

$$\frac{dx}{1} = \frac{dy}{-m_2} = \frac{dz}{0} \Rightarrow y + m_2 x = C_1 \text{ and } z = C_2$$

$\therefore$ Solution of (2) is $\qquad z = f_2\,(y + m_2\,x)$

Similarly the solution of $(D - m_1 D')\,z = 0$ is

$$z = f_1(y + m_1 x)$$

Hence the complete solution of (1) is

$$z = f_1\,(y + m_1 x) + f_2\,(y + m_2 x)$$

Case 2. If the roots are equal; say m

Then $\qquad C.F. = f_1(y + mx) + x\,f_2(y + mx)$

Theory:

$$(D - m\,D')(D - m\,D')z = 0 \quad \text{...(1)}$$

Let $\qquad (D - mD')\,z = u \quad \text{... (2)}$

(1) becomes $\qquad (D - m\,D')u = 0 \quad \text{... (3)}$

Solution of (3) is $\qquad u = f\,(y + mx)$

(2) becomes $\qquad (D - mD')\,z = f\,(y + mx) \quad \Rightarrow p - mq = f\,(y + mx)$

This is Lagrange's equation and its subsidiary equations are

$$\underset{(i)}{\frac{dx}{1}} = \underset{(ii)}{\frac{dy}{-m}} = \underset{(iii)}{\frac{dz}{f(y + mx)}}$$

From (*i*) and (*ii*), $y + mx = C_1$ and $dz = f\,(y + mx)\,dx$

$$dz = f\,(C_1)\,dx \Rightarrow z = f(y + mx).x + f_1(y + mx)$$

S. No.	***Roots of A.E.***	*C.F.*
1.	m_1, m_2, m_3 (different)	$f_1\,(y + m_1 x) + f_2(y + m_2 x) + f_3(y + m_3 x)$
2.	$m_1, m_2, m_3 \begin{bmatrix} m_2 = m_1 \\ m_3 \neq m_1 \end{bmatrix}$	$f_1(y + m_1 x) + x\; f_2(y + m_1 x) + f_3(y + m_3 x)$
3.	m_1, m_2, m_3 $m_1 = m_2 = m_3$	$f_1(y + m_1 x) + x\; f_2(y + m_1 x) + x^2\,f_3(y + m_1 x)$

Example 1. *Solve* $\quad (D^3 - 4D^2 D' + 3DD'^2\,)z = 0.$

Solution. $\qquad (D^3 - 4D^2 D' + 3DD'^2)z = 0. \qquad [\,D = m,\;\; D' = 1\,]$

Its auxiliary equation is $m^3 - 4m^2 + 3m = 0$

$$m(m^2 - 4m + 3) = 0$$

$$m\,(m - 1)(m - 3) = 0 \quad \Rightarrow \quad m = 0,\,1,\,3$$

The required solution is

$$z = f_1(y) + f_2(y + x) + f_3(y + 3x) \qquad \textbf{Ans.}$$

Example 2. *Solve* $\quad \dfrac{\partial^2 z}{\partial x^2} - 4\dfrac{\partial^2 z}{\partial x\,\partial y} + 4\dfrac{\partial^2 z}{\partial y^2} = 0.$

Solution. $(D^2 - 4DD' + 4D'^2)z = 0$

Its auxiliary equation is $m^2 - 4m + 4 = 0$ $[D = m, D' = 1]$

$\Rightarrow$ $(m-2)^2 = 0 \Rightarrow m = 2, 2$

The required solution is

$$z = f_1(y+2x) + x f_2(y+2x)$$ **Ans.**

Example 3. *Solve the linear partial differential equation*

$$\frac{\partial^4 z}{\partial x^4} - 2\frac{\partial^4 z}{\partial x^3 \partial y} + 2\frac{\partial^4 z}{\partial x \partial y^3} - \frac{\partial^4 z}{\partial y^4} = 0.$$ *(Q.Bank U.P. II semester 2002)*

Solution. Here, we have

$$(D^4 - 2D^3D' + 2DD'^3 - D'^4)\, z = 0, \text{ where } D \equiv \frac{\partial}{\partial x} \text{ and } D' \equiv \frac{\partial}{\partial y}$$

Auxiliary equation is

$$m^4 - 2m^3 + 2m - 1 = 0$$

$$m^3 (m-1) - m^2 (m-1) - m(m-1) + 1(m-1) = 0$$

$\Rightarrow$ $(m^3 - m^2 - m + 1)(m-1) = 0$

$\Rightarrow$ $(m+1)(m-1)^3 = 0$

$\Rightarrow$ $m = -1, 1, 1, 1$

$\therefore$ $z = f_1(y-x) + f_2(y+x) + x f_3(y+x) + x^2 f_4(y+x)$ **Ans.**

Example 4. *Solve the linear partial differential equation* $\dfrac{\partial^4 z}{\partial x^4} + \dfrac{\partial^4 z}{\partial y^4} = 0.$

Solution. $(D^4 + D'^4)\, z = 0$

Auxiliary equation is $m^4 + 1 = 0$

$\Rightarrow$ $m^4 + 1 + 2m^2 = 2m^2$

$\Rightarrow$ $(m^2+1)^2 - (m\sqrt{2})^2 = 0$

$\Rightarrow$ $(m^2 + \sqrt{2}m + 1)(m^2 - \sqrt{2}m + 1) = 0$

so that $m^2 + \sqrt{2}m + 1 = 0$ or $m^2 - \sqrt{2}m + 1 = 0$

$\Rightarrow$ $m = \dfrac{-1+i}{\sqrt{2}}, \dfrac{1+i}{\sqrt{2}}, \dfrac{-1-i}{\sqrt{2}}, \dfrac{1-i}{\sqrt{2}},$

Hence

$$z = f_1\left\{y + \left(\frac{-1+i}{\sqrt{2}}\right)x\right\} + f_2\left\{y + \left(\frac{1+i}{\sqrt{2}}\right)x\right\}$$

$$+ f_3\left\{y + \left(\frac{-1-i}{\sqrt{2}}\right)x\right\} + f_4\left\{y + \left(\frac{1-i}{\sqrt{2}}\right)x\right\}$$ **Ans.**

EXERCISE 10.1

Solve the following equations:

1. $\dfrac{\partial^2 z}{\partial x^2} + \dfrac{4\partial^2 z}{\partial x \partial y} - 5\dfrac{\partial^2 z}{\partial y^2} = 0$ **Ans.** $z = f_1(y+x) + f_2(y-5x)$

2. $2\dfrac{\partial^2 z}{\partial x^2} + 5\dfrac{\partial^2 z}{\partial x \partial y} + 2\dfrac{\partial^2 z}{\partial y^2} = 0$ **Ans.** $z = f_1(2y-x) + f_2(y-2x)$

3. $(D^3-6D^2D'+11DD'^2-6D'^3)z=0$ **Ans.** $z=f_1(y+x)+f_2(y+2x)+f_3(y+3x)$

4. $\frac{\partial^2 z}{\partial x^2}-2\frac{\partial^2 z}{\partial x\partial y}+\frac{\partial^2 z}{\partial y^2}=0$ **Ans.** $z=f_1(y+x)+xf_2(y+x)$

5. $(D^3-6D^2D'+12DD'^2-8D'^3)z=0$ **Ans.** $z=f_1(y+2x)+xf_2(y+2x)+x^2f_3(y+2x)$

6. $\frac{\partial^4 z}{\partial x^4}-\frac{\partial^4 z}{\partial y^4}=0$ **Ans.** $z=f_1(y+x)+f_2(y-x)+f_3(y+ix)+f_4(y-ix)$

7. $\frac{\partial^3 z}{\partial x^3}-4\frac{\partial^3 z}{\partial x^2\partial y}+4\frac{\partial^3 z}{\partial x\partial y^2}=0$ **Ans.** $z=f_1(y)+f_2(y+2x)+xf_3(y+2x)$

8. $\frac{\partial^3 z}{\partial x^3}-7\frac{\partial^3 z}{\partial x\partial y^2}+6\frac{\partial^3 z}{dy^3}=0$ **Ans.** $z=f_1(y+x)+f_2(y+2x)+f_3(y-3x)$

9. $\frac{\partial^4 z}{\partial x^4}+\frac{\partial^4 z}{\partial y^4}=2\frac{\partial^4 z}{\partial x^2\partial y^2}$ **Ans.** $z=f_1(y-x)+xf_2(y-2x)+f_3(y+x)+xf_4(y+x)$

10. $\frac{\partial^2 u}{\partial x^2}+\frac{\partial^2 u}{\partial y^2}=0$, when $u=\sin y$, $x=0$ for all y and $u\to 0$ when $x\to\infty$. *(A.M.I.E., Summer 2000)*

10.3 GENERAL RULES FOR FINDING THE PARTICULAR INTEGRAL

Given partial differential equation is

$$f(D, D')\,z = F(x, y)$$

$$\text{P.I.} = \frac{1}{f(D, D')}F(x, y)$$

If $f(D, D')$ is a homogeneous function of D and D′ of degree n and R.H.S. function $\phi(ax+by)$, $ax+by$, $e^{(ax+by)}$, $\sin(ax+by)$. Then the particular integral

$$\text{P.I.} = \frac{1}{F(D, D')}\phi(ax+by)$$

$$\text{P.I. of } \frac{1}{F(D, D')}F(x, y) = \frac{1}{F(a, b)}\iiint\cdots\int\phi(u)\,du\,du\ldots du \ (n \text{ times}), \text{ where } u = ax+by$$

GENERAL RULE

Integrate $f(u)$ w.r.t. u, n times and after integration replace u by $ax+by$.

Case of failure:

To find P.I.

The given equation is

$$F(D, D')\,z = \phi(ax+by) \text{ and } F(a, b) = 0$$

Procedure: Let $F(D, D')$ is a homogeneous function of degree n.

Differentiating $F(D, D')$ partially w.r.t D and multiply L.H.S by x we get

$$x\frac{1}{\frac{\partial}{\partial D}F(D, D')}\phi(ax+by)$$

If F (a, b) is again zero.

Differentiate it second time and multiply by x to get $x^2\frac{1}{\frac{\partial^2}{\partial D^2}F(D, D')}\phi(ax+by)$

If $F(a, b) \neq 0$ then stop.

If $F(a, b) = 0$, then repeat the above procedure. After m times differentiating and multiplying x by m times, we get

$$x^m \frac{1}{\frac{\partial^m}{\partial D^m} F(D, D')} \phi(ax + by)$$

Let $F(a, b) \neq 0$, then $\left[\text{Put } \frac{\partial^m}{\partial D^m} F(D, D') = \psi(D, D')\right]$

$$\text{P.I.} = x^m \frac{1}{(a, b)} \phi(ax + by)$$

SHORT FORMULAE:

We can find P.I. of function on the R.H.S. of the form $e^{ax + by}$, sin $(ax + by)$, cos $(ax + by)$ either by general formula or by the short formula given below:

(*i*) When $F(x, y) = e^{ax+by}$

$$\text{P.I.} = \frac{1}{f(D, D')} e^{ax+by} = \frac{e^{ax+by}}{f(a, b)} \quad [\text{Put } D = a,\ D' = b]$$

Theory. $De^{ax+by} = a e^{ax+by}$, $D' e^{ax+by} = b e^{ax+by}$

$$D^2 e^{ax+by} = a^2 e^{ax+by},\ DD' e^{ax+by} = ab e^{ax+by},\ D'^2 e^{ax+by} = b^2 e^{ax+by}$$

$$\therefore \quad (D^2 + k_1 DD' + k_2 D'^2) e^{ax+by} = (a^2 + k_1 ab + k_2 b^2) e^{ax+by}$$

i.e., $\boxed{f(D, D') e^{ax+by} = f(a, b) e^{ax+by}}$

$$\Rightarrow \quad \frac{1}{f(D, D')} f(D, D') e^{ax+by} = \frac{1}{f(D, D')} f(a, b) e^{ax+by}$$

$$\Rightarrow \quad \frac{1}{f(D, D')} e^{ax+by} = \frac{1}{f(a, b)} e^{ax+by}$$

(*ii*) When $F(x, y) = \sin(ax + by)$ or $F(x, y) = \cos(ax + by)$

$$\text{P.I.} = \frac{1}{f(D^2, DD', D'^2)} \sin(ax + by) = \frac{1}{f(-a^2, -ab, -b^2)} \sin(ax + by)$$

Put $D^2 = -a^2, \quad DD' = -ab, \quad D'^2 = -b^2$

$$\text{P.I.} = \frac{1}{f(D^2, DD', D'^2)} \cos(ax + by) = \frac{1}{f(-a^2, -ab, -b^2)} \cos(ax + by)$$

Theory. $D^2 \sin(ax + by) = -a^2 \sin(ax + by)$

$$DD' \sin(ax + by) = -ab \sin(ax + by)$$

$$D'^2 \sin(ax + by) = -b^2 \sin(ax + by)$$

$$f(D^2, DD', D'^2) \sin(ax + by) = f(-a^2, -ab, -b^2) \sin(ax + by)$$

$$\Rightarrow \quad \frac{1}{f(D^2, DD', D'^2)} f(D^2, DD', D'^2) \sin(ax + by)$$

$$= \frac{1}{f(D^2, DD', D'^2)} f(-a^2, -ab, -b^2) \sin(ax+by)$$

$$\boxed{\frac{1}{f(-a^2, -ab, -b^2)} \sin(ax+by) = \frac{1}{f(D^2, DD', D'^2)} \sin(ax+by)}$$

Case 1. When R.H.S. $= e^{ax+by}$

Example 5. *Solve:* $\frac{\partial^3 z}{\partial x^3} - 3\frac{\partial^3 z}{\partial x^2 \partial y} + 4\frac{\partial^3 z}{\partial y^3} = e^{x+2y}$ *(U.P. II Semester, June 2007)*

Solution. $\frac{\partial^3 z}{\partial x^3} - 3\frac{\partial^3 z}{\partial x^2 \partial y} + 4\frac{\partial^3 z}{\partial y^3} = e^{x+2y}$

Given equation in symbolic form is

$$(D^3 - 3D^2 D' + 4D'^3) z = e^{x+2y}$$

Its A.E. is $m^3 - 3m^2 + 4 = 0 \Rightarrow m = -1, 2, 2.$

$\therefore$ C.F. $= f_1(y-x) + f_2(y+2x) + x f_3(y+2x)$

$$\text{P.I.} = \frac{1}{D^3 - 3D^2 D' + 4D'^3} e^{x+2y}$$

Put $D = 1,\ D' = 2$ $= \frac{1}{1-6+32} e^{x+2y} = \frac{e^{x+2y}}{27}$

Hence, complete solution is

$$z = f_1(y-x) + f_2(y+2x) + x f_3(y+2x) + \frac{e^{x+2y}}{27}$$ **Ans.**

Example 6. *Solve*:

$$4\frac{\partial^2 z}{\partial x^2} - 4\frac{\partial^2 z}{\partial x \partial y} + \frac{\partial^2 z}{\partial y^2} = e^{x+2y}$$

Solution. $(4D^2 - 4DD' + D'^2) z = e^{x+2y}$

$$4D^2 - 4DD' + D'^2 = 0 \qquad \text{... (1)}$$

Put $D = m,\ D' = 1$ in (1), we get

$$4m^2 - 4m + 1 = 0 \Rightarrow (2m-1)^2 = 0 \Rightarrow m = \frac{1}{2}, \frac{1}{2}$$

$$\text{C.F.} = f_1\left(y + \frac{x}{2}\right) + x f_2\left(y + \frac{x}{2}\right)$$

$\Rightarrow$ C.F. $= f_1(2y+x) + x f_2(2y+x)$

$$\text{P.I.} = \frac{1}{4D^2 - 4DD' + D'^2} e^{x+2y}$$

On putting $D = 1,\ D' = 2$, in $4D^2 - 4DD' + D'^2$ it becomes zero.

Since the denominator becomes zero, we differentiate the denominator w.r.t. D and multiplying by x, we get

$$\text{P.I.} = x \frac{1}{8D - 4D'} e^{x+2y}$$

On putting $D = 1$, $D' = 2$ in $8D - 4D'$ it becomes zero.

Since the denominator becomes zero, we differentiate the denominator w.r.t. D and multiplying by x, we get

$$\text{P.I.} = \frac{x^2}{8} e^{x+2y}$$

Hence, the complete solution is

$$z = f_1(2y+x) + x f_2(2y+x) + \frac{x^2 e^{x+2y}}{8} \qquad \textbf{Ans.}$$

EXERCISE 10.2

Solve the following equations:

1. $\frac{\partial^2 z}{\partial x^2} - \frac{\partial^2 z}{\partial y^2} = e^{x+2y}$ **Ans.** $z = f_1(y+x) + f_2(y-x) - \frac{e^{x+2y}}{3}$

2. $\frac{\partial^2 z}{\partial x^2} - 5\frac{\partial^2 z}{\partial x \partial y} + 6\frac{\partial^2 z}{\partial y^2} = e^{x+y}$ **Ans.** $z = f_1(y+2x) + f_2(y+3x) + \frac{1}{2}e^{x+y}$

3. $\frac{\partial^2 z}{\partial x^2} - 4\frac{\partial^2 z}{\partial x \partial y} + 4\frac{\partial^2 z}{\partial y^2} = e^{2x+y}$ **Ans.** $z = f_1(y+2x) + x f_2(y+2x) + \frac{x^2}{2}e^{2x+y}$

4. $\frac{\partial^2 z}{\partial x^2} - 7\frac{\partial^2 z}{\partial x \partial y} + 12\frac{\partial^2 z}{\partial y^2} = e^{x-y}$ **Ans.** $z = f_1(y+3x) + f_2(y+4x) + \frac{1}{20}e^{x-y}$

5. $\frac{\partial^3 z}{\partial x^3} - 2\frac{\partial^3 z}{\partial x^2 \partial y} = 2e^{2x-y}$ **Ans.** $z = f_1(y) + x f_2(y) + f_3(y+2x) + \frac{1}{8}e^{2x-y}$

6. $(D^2 - 2DD' + D'^2)z = e^{x+2y}$ **Ans.** $z = f_1(y+x) + x f_2(y+x) + e^{x+2y}$

7. $\frac{\partial^2 z}{\partial x^2} - \frac{\partial^2 z}{\partial y^2} - 2\frac{\partial z}{\partial x} + 2\frac{\partial z}{\partial y} = e^{2x+3y}$ **Ans.** $z = f_1(y+x) + e^{2x} f_2(y-x) - \frac{1}{3}e^{2x+3y}$

8. $\frac{\partial^2 z}{\partial x^2} - 5\frac{\partial^2 z}{\partial x \partial y} + 6\frac{\partial^2 z}{\partial y^2} = \exp(3x - 2y)$ **Ans.** $z = f_1(y+2x) + f_2(y+3x) + \frac{1}{63}e^{3x-2y}$

Case II. When R.H.S. = sin $(ax + by)$ or cos $(ax + by)$

Example 7. *Solve* $(D+1)(D+D'-1)z = \sin(x+2y)$ *(U.P., II Semester 2010)*

Solution. $\text{C.F.} = e^{-x}\phi(y) + e^{x}\phi_2(y-x)$

$$\text{P.I.} = \frac{1}{(D+1)(D+D'-1)}\sin(x+2y) = \frac{1}{D^2 + DD' + D' - 1}\sin(x+2y)$$

$$= \frac{1}{-1+(-2)+D'-1}\sin(x+2y) = \frac{1}{D'-4}\sin(x+2y)$$

$$= \frac{D'+4}{(D'^2-16)}\sin(x+2y) = \frac{D'+4}{(-4-16)}\sin(x+2y)$$

$$= -\frac{1}{20}(D'+4)\sin(x+2y) = -\frac{1}{20}[D'\sin(x+2y) + 4\sin(x+2y)]$$

$$= -\frac{1}{20}[2\cos(x+2y)+4\sin(x+2y)]$$

Hence, the solution is

$$z = e^{-x}\phi(y)+e^{x}\phi_2(y-x)-\frac{1}{10}[\cos(x+2y)+2\sin(x+2y)] \qquad \textbf{Ans.}$$

Example 8. *Solve the linear partial differential equation*

$$\frac{\partial^2 z}{\partial x^2}+2\frac{\partial^2 z}{\partial x\,\partial y}+\frac{\partial^2 z}{\partial y^2}=\sin(2x+3y)$$ *(U.P. II Semester Summer 2006)*

Solution. We have,

$$\frac{\partial^2 z}{\partial x^2}+2\frac{\partial^2 z}{\partial x\,\partial y}+\frac{\partial^2 z}{\partial y^2}=\sin(2x+3y)$$

$$(D^2+2DD'+D'^2)z=\sin(2x+3y) \quad \text{where } D=\frac{\partial}{\partial x} \text{ and } D'=\frac{\partial}{\partial y}$$

Put $D = m, \quad D' = 1$

The auxiliary equation is

$$m^2+2m+1=0$$

$$\Rightarrow \quad (m+1)^2=0 \Rightarrow m=-1,-1$$

$$\Rightarrow \quad \text{C.F.} = f_1(y-x)+x\,f_2(y-x)$$

$$\text{P.I.} = \frac{1}{D^2+2DD'+D'^2}\sin(2x+3y)$$

$$= \frac{1}{-4+2(-6)-9}\sin(2x+3y)$$

$$= \frac{1}{-25}\sin(2x+3y) \qquad \left[\begin{array}{l} D^2=-2^2=-4 \\ D'^2=-3^2=-9 \\ DD'=-2\times 3=-6 \end{array}\right]$$

Hence, the complete solution is

$$z = C.F. + P.I.$$

$$\Rightarrow \quad z = f_1(y-x)+x\,f_2(y-x)+\frac{1}{-25}\sin(2x+3y)$$

$$\Rightarrow \quad z = f_1(y-x)+x\,f_2(y-x)-\frac{1}{25}\sin(2x+3y) \qquad \textbf{Ans.}$$

Example 9. *Solve*

$$\frac{\partial^3 z}{\partial x^3}-4\frac{\partial^3 z}{\partial x^2\partial y}+4\frac{\partial^3 z}{\partial x\partial y^2}=2\sin(3x+2y)$$ *(Uttarakhand, II Semester, 2012)*

Solution. We have, $\frac{\partial^3 z}{\partial x^3}-4\frac{\partial^3 z}{\partial x^2\partial y}+4\frac{\partial^3 z}{\partial x\partial y^2}=2\sin(3x+2y)$

Putting $\dfrac{\partial}{\partial x} = D, \dfrac{\partial}{\partial y} = D'$

$$D^3 z - 4D^2 D' z + 4DD'^2 z = 2\sin(3x+2y)$$

A.E. is $D^3 - 4D^2 D' + 4DD'^2 = 0 \Rightarrow D(D^2 - 4DD' + 4D'^2) = 0$

Put $D = m,\ D' = 1$

$$m(m^2 - 4m + 4) = 0 \quad \Rightarrow \quad m(m-2)^2 = 0 \Rightarrow m = 0, 2, 2$$

$$\text{C.F.} = f_1(y) + f_2(y+2x) + x f_3(y+2x)$$

$$\text{P.I.} = \frac{1}{D^3 - 4D^2 D' + 4DD'^2} 2\sin(3x+2y) = 2\frac{1}{D(D^2 - 4DD' + 4D'^2)}\sin(3x+2y)$$

$$= 2.\frac{1}{D[-9 - 4(-6) + 4(-4)]}\sin(3x+2y) = -\frac{2}{D}\sin(3x+2y)$$

$$= -\frac{2}{3}[-\cos(3x+2y)] = \frac{2}{3}\cos(3x+2y) \qquad \left[\frac{1}{D} = \int\right]$$

General solution is

$$z = f_1(y) + f_2(y+2x) + x f_3(y+2x) + \frac{2}{3}\cos(3x+2y)$$ **Ans.**

Example 10. *Solve* $\dfrac{\partial^2 z}{\partial x^2} - \dfrac{\partial^2 z}{\partial x \partial y} = \sin x \cos 2y$ *(U.P., II Semester, June, 2010, 2008)*

Solution. Here, we have $\dfrac{\partial^2 z}{\partial x^2} - \dfrac{\partial^2 z}{\partial x \partial y} = \sin x \cos 2y$

The given equation can be written in the form

$$(D^2 - DD')z = \sin x \cos 2y \text{ where } D = \frac{\partial}{\partial x},\ D' = \frac{\partial}{\partial y}$$

Writing $D = m$ and $D' = 1$, the auxiliary equation is

$$m^2 - m = 0 \quad \Rightarrow \quad m(m-1) = 0, \Rightarrow m = 0, 1$$

$\therefore$ $\text{C.F.} = f_1(y) + f_2(y+x)$

$$\text{P.I.} = \frac{1}{D^2 - DD'}\sin x \cos 2y$$

$$= \frac{1}{D^2 - DD'}\frac{1}{2}[\sin(x+2y) + \sin(x-2y)]$$

$$= \frac{1}{2}\frac{1}{D^2 - DD'}\sin(x+2y) + \frac{1}{2}\frac{1}{D^2 - DD'}\sin(x-2y)$$

Put $D^2 = -1,\ DD' = -2$ in the first integral and $D^2 = -1,\ DD' = 2$ in the second integral.

$$= \frac{1}{2}\left[\frac{\sin(x+2y)}{-1-(-2)}\right] + \frac{1}{2}\left[\frac{\sin(x-2y)}{-1-(2)}\right] = \frac{1}{2}\sin(x+2y) - \frac{1}{6}\sin(x-2y)$$

Hence the complete solution is z = C.F. + P.I.

i.e., $$z = f_1(y) + f_2(y+x) + \frac{1}{2}\sin(x+2y) - \frac{1}{6}\sin(x-2y)$$ **Ans.**

Example 11. *Solve* $(D^2 + DD' - 6D'^2)z = \cos(2x+y)$

Solution. $(D^2 + DD' - 6D'^2)z = \cos(2x+y)$

A.E. is $\quad m^2 + m - 6 = 0 \quad$ or $\quad m = 2, -3$

$$\text{C.F.} = f_1(y+2x) + f_2(y-3x)$$

$$\text{P.I.} = \frac{1}{D^2 + DD' - 6D'^2}\cos(2x+y)$$

$$D^2 + DD' - 6D'^2 = -4 - 2 - 6(-1) = 0$$

$\therefore$ It is a case of failure.

Now
$$\text{P.I.} = \frac{1}{D^2 + DD' - 6D'^2}\cos(2x+y)$$

$$= x\frac{1}{2D + D'}\cos(2x+y) = x\frac{D}{2D^2 + DD'}\cos(2x+y)$$

$$= x\frac{D}{2\times(-4) - 2}\cos(2x+y) = -\frac{x}{10}D\cos(2x+y)$$

$$= 2\frac{x}{10}\sin(2x+y) = \frac{x}{5}\sin(2x+y)$$

Hence, the complete solution is

$$z = f_1(y+2x) + f_2(y-3x) + \frac{x}{5}\sin(2x+y) \qquad \textbf{Ans.}$$

Example 12. *Solve the partial differential equation ;*

$$\frac{\partial^2 z}{\partial x^2} - 3\frac{\partial^2 z}{\partial x\,\partial y} + 2\frac{\partial^2 z}{\partial y^2} = e^{2x-y} + e^{x+y} + \cos(x+2y)$$ *(U.P. II Semester Summer 2006)*

Solution. Given equation is

$$\frac{\partial^2 z}{\partial x^2} - 3\frac{\partial^2 z}{\partial x\,\partial y} + 2\frac{\partial^2 z}{\partial y^2} = e^{2x-y} + e^{x+y} + \cos(x+2y)$$

Given equation can be written as :

$$(D^2 - 3DD' + 2D'^2)z = e^{2x-y} + e^{x+y} + \cos(x+2y)$$

The auxiliary equation is

$$m^2 - 3m + 2 = 0$$
$$m^2 - 2m - m + 2 = 0$$
$$m(m-2) - 1(m-2) = 0$$
$$(m-1)(m-2) = 0$$
$$m = 1, 2$$

Hence, $\quad \text{C.F.} = \phi_1(y+x) + \phi_2(y+2x)$

Now
$$\text{P.I.} = \frac{1}{(D-D')(D-2D')}\{e^{2x-y} + e^{x+y} + \cos(x+2y)\}$$

$$= \frac{1}{(D-D')(D-2D')}e^{2x-y} + \frac{1}{(D-D')(D-2D')}e^{x+y} + \frac{1}{(D-D')(D-2D')}\cos(x+2y)$$

$$= I_1 + I_2 + I_3$$

Let $I_1 = \dfrac{1}{(D-D')(D-2D')} e^{2x-y}$ (Replacing D by 2 and D' by – 1.)

$$I_1 = \frac{1}{(2+1)(2+2)} e^{2x-y} = \frac{1}{12} e^{2x-y}$$

Now, $I_2 = \dfrac{1}{(D-D')(D-2D')} e^{x+y}$, (Replacing D by 1 and D' by 1)

$$= \frac{1}{(D-D')(-1)} e^{x+y} = -\frac{1}{(D-D')} e^{x+y}$$

$$= -x\frac{1}{1} e^{x+y} = -x\, e^{x+y}$$

Now, $I_3 = \dfrac{1}{(D-D')(D-2D')} \cos(x+2y)$

$$= \frac{1}{D^2 - 3DD' + 2D'^2} \cos(x+2y)$$

$I_3 = \dfrac{1}{-1-3(-2)+2(-4)} \cos(x+2y)$ (Replacing D^2 by –1; DD' by –2; D^2 by –4)

$$= \frac{1}{-1+6-8} \cos(x+2y)$$

$$= -\frac{1}{3} \cos(x+2y)$$

P.I. $= I_1 + I_2 + I_3$

Thus required $P.I. = \dfrac{1}{12} e^{2x-y} - xe^{x+y} - \dfrac{1}{3}\cos(x+2y)$

Hence, the complete solution is

$$z = C.F. + P.I.$$

$$= \phi_1(y+x) + \phi_2(y+2x) + \frac{1}{12} e^{2x-y} - xe^{x+y} - \frac{1}{3}\cos(x+2y)$$ **Ans.**

Example 13. *Solve the equation*

$$(D^3 - 7DD'^2 - 6D'^3)z = \sin(x+2y) + e^{2x+y}$$

Solution. $(D^3 - 7DD'^2 - 6D'^3)z = \sin(x+2y) + e^{2x+y}$ ... (1)

Its auxiliary equation is

$$m^3 - 7m - 6 = 0 \quad \text{or} \quad (m+1)(m+2)(m-3) = 0 \Rightarrow m = -1, -2, 3$$

$\therefore$ C.F. $= f_1(y-x) + f_2(y-2x) + f_3(y+3x)$

$$\text{P.I.} = \frac{1}{D^3 - 7DD'^2 - 6D'^3}[\sin(x+2y) + e^{2x+y}]$$

$$= \frac{1}{D^3 - 7DD'^2 - 6D'^3}\sin(x+2y) + \frac{1}{D^3 - 7DD'^2 - 6D'^3} e^{2x+y}$$

$$= \frac{1}{D^2D - 7DD'^2 - 6D'^3}\sin(x+2y) + \frac{e^{2x+y}}{(2)^3 - 7(2)(1)^2 - 6(1)^3}$$

Put $D^2 = -1$, $D'^2 = -2^2$

$$= \frac{1}{-D-7D(-4)-6(-4)D'}\sin(x+2y)+\frac{e^{2x+y}}{8-14-6}$$

$$= \frac{1}{27D+24D'}\sin(x+2y)-\frac{1}{12}e^{2x+y} = \frac{1}{3}\frac{1}{9D+8D'}\sin(x+2y)-\frac{1}{12}e^{2x+y}$$

$$= \frac{1}{3}\frac{D}{9D^2+8DD'}\sin(x+2y)-\frac{1}{12}e^{2x+y} = \frac{1}{3}\frac{D}{9(-1)+8(-2)}\sin(x+2y)-\frac{1}{12}e^{2x+y}$$

$$= -\frac{1}{75}D\sin(x+2y)-\frac{1}{12}e^{2x+y} = -\frac{1}{75}\cos(x+2y)-\frac{1}{12}e^{2x+y}$$

Hence, the complete solution is

$$z = f_1(y-x)+f_2(y-2x)+f_3(y+3x)-\frac{1}{75}\cos(x+2y)-\frac{1}{12}e^{2x+y}$$ **Ans.**

Example 14. Solve the P.D.E $\frac{\partial^2 z}{\partial x^2}-2\frac{\partial^2 z}{\partial x \partial y}+\frac{\partial^2 z}{\partial y^2} = \sin x$.

(*U.P II Semester 2009, 2004*)

Solution. Here, we have

$$\frac{\partial^2 z}{\partial x^2}-2\frac{\partial^2 z}{\partial x \partial y}+\frac{\partial^2 z}{\partial y^2} = \sin x$$

$$\Rightarrow \quad (D^2 - 2DD' + D'^2)z = \sin x$$

Its auxiliary equation is

$$(m^2 - 2m + 1) = 0 \quad \Rightarrow \quad (m-1)^2 = 0 \Rightarrow m = 1, 1$$

$$\text{C.F.} = f_1(x+y) + x f_2(x+y)$$

$$\text{P.I.} = \frac{1}{D^2 - 2DD' + D'^2}\sin x$$

$$= \left[\frac{1}{-1-0+0}\right]\sin x \qquad \begin{bmatrix} D=1 \\ D'=0 \end{bmatrix}$$

$$= -\sin x$$

Hence, the complete solution is

$$z = \text{C.F.} + \text{P.I.}$$

$$= f_1(x+y) + x f_2(x+y) - \sin x$$ **Ans.**

EXERCISE 10.3

Solve the following equations :

1. $[2D^2 - 5DD' + 2D'^2]z = 5\sin(2x+y)$ **Ans.** $z = f_1(y+2x)+f_2(2y+x)-\frac{5}{3}x\cos(2x+y)$

2. $\frac{\partial^2 z}{\partial x^2}-\frac{\partial^2 z}{\partial x \partial y} = \cos(x+2y)$ **Ans.** $z = f_1(y)+f_2(y+x)+\cos(x+2y)$

3. $(D^2 - DD')z = \cos x \cos 2y$ **Ans.** $z = f_1(y)+f_2(y+x)+\frac{1}{2}\cos(x+2y)-\frac{1}{6}\cos(x-2y)$

4. $(D^2 + 2D'D + D'^2)z = \sin(x+2y)$ **Ans.** $z = f_1(y-x)+xf_2(y-x)-\frac{1}{9}\sin(x+2y)$

5. $\frac{\partial^2 z}{\partial x^2}-3\frac{\partial^2 z}{\partial x\partial y}+2\frac{\partial^2 z}{\partial y^2}=e^{2x+3y}+\sin(x+2y)$

Ans. $z=c_1 f(y+x)+f_2(y+2x)+\frac{1}{4}e^{2x+3y}-\frac{1}{15}\sin(x-2y)$

6. $r-2s=\sin x \cos$ **Ans.** $z=f_1(y)+f_2(y+2x)+\frac{1}{15}(\sin x\cos 2y)+4\sin 2y\cos x$

7. $(D^2+D'^2)z=\cos mx.\cos ny$ **Ans.** $z=f_1(y+1x)+f_2(y-1x)-\frac{\cos mx.\cos xy}{(m^2+n^2)}$

8. $(D^3-7DD'^2-6D')z=\sin(x+2y)+e^{3x+y}$

Ans. $z=f_1(y-x)+f_2(y-2x)+f_3(y+3x)-\frac{1}{75}\cos(x+2y)+\frac{x}{20}e^{3x+y}$

9. $(D^2-DD')z=\cos 2y(\sin x+\cos x)$ *(U.P., II Semester, 2003)*

Ans. $z=f_1(y)+f_2(y+x)+\frac{1}{2}[\sin(x+2y)+\cos(x+2y)]-\frac{1}{6}[\sin(x-2y)+\cos(x-2y)]$

Case III. When R.H.S. = ϕ $(ax+by)$ polynomial

Example 15. *Find the general integral of the equation*

$$\frac{\partial^2 z}{\partial x^2}+3\frac{\partial^2 z}{\partial x\partial y}+2\frac{\partial^2 z}{\partial y^2}=x+y$$

Solution. $$\frac{\partial^2 z}{\partial x^2}+3\frac{\partial^2 z}{\partial x\partial y}+2\frac{\partial^2 z}{\partial y^2}=x+y$$

with $D=\frac{\partial}{\partial x}, D'=\frac{\partial}{\partial y}$, the given equation can be written in the form

$$(D^2+3DD'+2D'^2)z=x+y$$

Writing $D=m$ and $D'=1$, the auxiliary equation is

$$m^2+3m+2=0 \quad \text{or} \quad (m+1)(m+2)=0, \quad \text{or} \quad m=-1,-2$$

$$\therefore \quad \text{C.F.} = f_1(y-x)+f_2(y-2x)$$

$$\text{P.I.} = \frac{1}{D^2+3DD'+2D'^2}(x+y)$$

$$=\frac{1}{(1)^2+3(1)(1)+2(1)^2}\iint u\,du\,du \quad \text{[where } x+y=u\text{]}$$

$$=\frac{1}{6}.\frac{u^3}{6}=\frac{u^3}{36}=\frac{(x+y)^3}{36}$$

Hence, the complete solution is

$$z=\text{C.F.}+\text{P.I.}=f_1(y-x)+f_2(y-2x)+\frac{(x+y)^3}{36}$$ **Ans.**

Example 16. *Solve* $\frac{\partial^2 z}{\partial x^2}+\frac{\partial^2 z}{\partial x \partial y}-6\frac{\partial^2 z}{\partial y^2}=x+y.$

Solution. With $D=\frac{\partial}{\partial x}$, $D'=\frac{\partial}{\partial y}$, the given equation can be written in the form

$$(D^2+DD'-6D'^2)z=x+y$$

Writing $D=m$ and $D'=1$, the auxiliary equation is $m^2+m-6=0$

$\Rightarrow$ $$(m+3)(m-2)=0 \Rightarrow m=-3,\ 2$$

$\therefore$ $$\text{C.F.}=f_1(y-3x)+f_2(y+2x)$$

$\therefore$ $$\text{P.I.}=\frac{1}{D^2+DD'-6D'^2}(x+y)$$

$$=\frac{1}{(1)^2+(1)(1)-6(1)^2}\iint u\,du\,du \qquad \text{[where } u=x+y\text{]}$$

$$=\frac{1}{-4}\frac{u^3}{6}=-\frac{u^3}{24}$$

$$=\frac{(x+y)^3}{-24}$$

The complete solution is

$$z=f_1(y-3x)+f_2(y+2x)-\frac{(x+y)^3}{24} \qquad \textbf{Ans.}$$

Example 17. *Solve* $(D^2-D'^2)\,z=x-y$

Solution. Here, we have

$$(D^2-D'^2)\,z=x-y \qquad ...(1)$$

Putting $D=m$ and $D'=1$ in (1), we get

A.E. is $$m^2-1=0$$

$\Rightarrow$ $$m^2=1 \quad\Rightarrow\quad m=1,-1$$

$\therefore$ $$\text{C.F.}=f_1(y+x)+f_2(y-x)$$

$$\text{P.I.}=\frac{1}{D^2-D'^2}(x-y)$$

$$\text{P.I.}=\frac{1}{(D-D')(D+D')}(x-y)$$

$$=x\frac{1}{D-D'}(x-y) \qquad \text{(case of failure)}$$

$$=x\frac{1}{1-(-1)}\int u\,du \qquad \text{where } u=x-y$$

$$=\frac{x}{2}\frac{u^2}{2}=\frac{x}{4}(x-y)^2$$

Complete solution is $z=\text{C.F.}+\text{P.I.}=f_1(y+x)+f_2(y-x)+\frac{x}{4}(x-y)^2$ **Ans.**

Example 17. *Solve :* $(D^2 + 2DD' - 8D'^2)z = \sqrt{2x+3y}$

Solution. Here, we have

$$(D^2 + 2DD' - 8D'^2)\, z = \sqrt{2x+3y}$$

A.E. is $m^2 + 2m - 8 = 0 \Rightarrow (m+4)(m-2) = 0 \Rightarrow m = 2,\ m = -4$

$$C.F. = f_1(y+2x) + f_2(y-4x)$$

$$P.I = \frac{1}{D^2 + 2DD' - 8D'^2}\sqrt{2x+3y}$$

$$= \frac{1}{D^2 + 2DD' - 8D'^2}(2x+3y)^{\frac{1}{2}}$$

$$= \frac{1}{(2)^2 + 2(2)(3) - 8(3)^2}\int\int u^{\frac{1}{2}}\,du\,du, \text{ where } u = 2x+3y$$

$$= \frac{1}{-56}\left[\frac{u^{5/2}}{\frac{3}{2}\cdot\frac{5}{2}}\right] = -\frac{1}{56}\frac{4}{15}(2x+3y)^{\frac{5}{2}}$$

$$= -\frac{1}{210}(2x+3y)^{\frac{5}{2}}$$

Hence, the complete solution = *C.F.* + *P.I.*

$$= f_1(y+2x) + f_2(y-4x) - \frac{1}{210}(2x+3y)^{\frac{5}{2}}$$ **Ans.**

Example 19. *Solve* $(D^3 - 3D^2D' - 4DD'^2 + 12D'^3)z = \sin(y+2x)$.

Solution. Here, we have

$(D^3 - 3D^2D' - 4DD'^2 + 12D'^3)\, z = 0$ ...(1)

Putting $D = m$ and $D' = 1$ in (1) ; we get

A.E. is $m^3 - 3m^2 - 4m + 12 = 0$

$\Rightarrow\quad m^2(m-3) - 4(m-3) = 0$

$\Rightarrow\quad (m^2-4)(m-3) = 0 \Rightarrow m = \pm 2, 3$

$\therefore\quad$ C.F. $= f_1(y+2x) + f_2(y-2x) + f_3(y+3x)$

$$P.I. = \frac{1}{D^3 - 3D^2D' - 4DD'^2 + 12D'^3}\sin(y+2x)$$

$$= \frac{1}{2^3 - 3(2)^2(1) - 4(2)(1)^2 + 12(1)^3}\int\int\int \sin u\,du\,du\,du \quad \left(\begin{array}{l}\text{where } u = y+2x\\ \text{case of failure}\end{array}\right)$$

$$= x\frac{1}{3D^2 - 6DD' - 4D'^2}\int\int \sin u\,du\,du = \frac{x}{3(2)^2 - 6(2)(1) - 4(1)^2}(-\sin u)$$

$$= \frac{x}{4}\sin(y+2x)$$

Complete solution is

$$z = \text{C.F.} + \text{P.I} = f_1(y+2x) + f_2(y-2x) + f_3(y+3x) + \frac{x}{4}\sin(y+2x).$$ **Ans.**

Example 20. *Solve : $(4D^2 - 4DD' + D'^2)z = 16 \log (x + 2y)$*

Solution. Here we have

$$(4D^2 - 4DD' + D'^2)z = 16 \log (x + 2y)$$

Auxiliary equation is

$$4m^2 - 4m + 1 = 0$$

$$\Rightarrow \quad (2m - 1)^2 = 0 \Rightarrow \quad m = \frac{1}{2}, \frac{1}{2}$$

$$C.F. = f_1\left(y + \frac{x}{2}\right) + x\, f_2\left(y + \frac{x}{2}\right)$$

$$P.I. = \frac{1}{4D^2 - 4DD' + D'^2} 16 \log (x + 2y)$$

$$= \frac{1}{4(1)^2 - 4(1)(2) + (2)^2} 16 \int\int \log u \, du \, du, \text{ where } u = x + 2y$$

$$= x \frac{1}{8D - 4D'} 16 \int \log u \, du \qquad \text{(case of failure)}$$

$$= x \frac{1}{8(1) - 4(2)} 16 \int \log u \, du \qquad \text{(case of failure)}$$

$$= 16\, x^2 \left(\frac{1}{8}\right) \log u$$

$$= 16 \frac{x^2}{8} \log (x + 2y)$$

$$= 2x^2 \log (x + 2y)$$

The complete solution $= C.F. + P.I.$

$$= f_1\left(y + \frac{x}{2}\right) + x f_2\left(y + \frac{x}{2}\right) + 2x^2 \log (x + 2y) \qquad \textbf{Ans.}$$

EXERCISE 10.4

Solve the following equations

1. $\dfrac{\partial^2 z}{\partial x^2} + 3\dfrac{\partial^2 z}{\partial x \partial y} - 4\dfrac{\partial^2 z}{\partial y^2} = x + \sin y$ **Ans.** $z = f_1(y + x) + f_2(y - 4x) + \dfrac{x^3}{6} + \dfrac{1}{4}\sin y$

2. $\dfrac{\partial^3 z}{\partial x^2 \partial y} - 2\dfrac{\partial^3 z}{\partial x \partial y^2} + \dfrac{\partial^3 z}{\partial y^3} = \dfrac{1}{x^2}$ **Ans.** $z = f_1(x) + f_2(y + x) + x f_3(y + x) - y \log x$

3. $(D^3 - 4D^2D' + 5DD'^2 - 2D'^3)z = e^{y+2x} + (y + x)^{\frac{1}{2}}$

   Ans. $z = f_1(y + x) + x f_2(y + z) + f_3(y + 2x) + x e^{y+x} - \dfrac{x^2}{3}(y + x)^{\frac{3}{2}}$

4. $\dfrac{\partial^3 z}{\partial x^3} - 4\dfrac{\partial^3 z}{\partial x^2 \partial y} + 4\dfrac{\partial^3 z}{\partial x \partial y^2} = 2 \sin (3x + 2y)$ **Ans.** $z = f_1(y + 2x) + f_2(2y + x) - \dfrac{5}{3} x \cos (2x + y)$

5. $\dfrac{\partial^2 z}{\partial x^2} - 4\dfrac{\partial^2 z}{\partial x \partial y} + 3\dfrac{\partial^2 z}{\partial y^2} = \sqrt{x+3y}$ **Ans.** $z = f_1(y+x) + f_2(y+3x) + \dfrac{1}{60}(x+3y)^{\frac{5}{2}}$

Case IV. *When* $F(x, y) = x^m y^n$

$$P.I. = \frac{1}{f(D, D')} x^m y^n = [f(D, D')]^{-1} x^m y^n$$

(*a*) If $m > n$, then $\dfrac{1}{f(D, D')}$ is expanded in the powers of $\dfrac{D'}{D}$.

(*b*) If $m < n$, then $\dfrac{1}{f(D, D')}$ is expanded in the powers of $\dfrac{D}{D'}$.

Example 21. *Solve* : $(D^2 + D'^2) z = x^2y^2$

Solution. Here, we have

$$(D^2 + D'^2)z = x^2y^2$$

Putting $D = m$ and $D' = 1$, we get the A.E. as

$$m^2 + 1 = 0 \quad \Rightarrow \quad m^2 = -1 \quad \Rightarrow \quad m = \pm i.$$

$\therefore$ C.F. $= f_1(y + ix) + f_2(y - ix)$

$$P.I. = \frac{1}{D^2 + D'^2}(x^2y^2) = \frac{1}{D^2} \cdot \frac{1}{\left(1 + \dfrac{D'^2}{D^2}\right)}(x^2y^2)$$

$$= \frac{1}{D^2}\left(1 + \frac{D'^2}{D^2}\right)^{-1}(x^2y^2)$$

$$= \frac{1}{D^2}\left(1 - \frac{D'^2}{D^2}\right)(x^2y^2)$$

$$= \frac{1}{D^2}(x^2y^2) - \frac{D'^2}{D^4}(x^2y^2)$$

$$= \frac{x^4}{12}y^2 - \frac{1}{D^4}(2x^2)$$

$$= \frac{x^4}{12}y^2 - 2 \cdot \frac{x^6}{3 \cdot 4 \cdot 5 \cdot 6}$$

$$= \frac{1}{180}(15x^4y^2 - x^6)$$

Thus, the complete solution is

$$z = f_1(y + ix) + f_2(y - ix) + \frac{1}{180}(15x^4y^2 - x^6)$$ **Ans.**

Example 22. *Solve* : $\dfrac{\partial^3 z}{\partial x^3} - \dfrac{\partial^3 z}{\partial y^3} = x^3y^3$ (*Q. Bank U.P. 2002*)

Solution. Here, we have

$$\frac{\partial^3 z}{\partial x^3} - \frac{\partial^3 z}{\partial y^3} = x^3y^3$$

$\Rightarrow \qquad (D^3 - D'^3)z = x^3y^3$

Putting $D = m$ and $D' = 1$ in above, we have

$$A.E. \text{ is } m^3 - 1 = 0 \Rightarrow m = 1, w, w^2$$

Where w is one of the cube roots, of unity.

$$\therefore \qquad C.F. = f_1(y + x) + f_2(y + wx) + f_3(y + w^2x)$$

$$P.I. = \frac{1}{D^3 - D'^3}(x^3y^3) = \frac{1}{D^3\left(1 - \frac{D'^3}{D^3}\right)}x^3y^3$$

$$= \frac{1}{D^3}\cdot\left(1 - \frac{D'^3}{D^3}\right)^{-1}(x^3y^3) = \frac{1}{D^3}\left(1 + \frac{D'^3}{D^3}\right)(x^3y^3)$$

$$= \frac{1}{D^3}\left[x^3y^3 + \frac{1}{D^3}D'^3(x^3y^3)\right] = \frac{1}{D^3}\left[x^3y^3 + \frac{1}{D^3}(6x^3)\right]$$

$$= \frac{1}{D^3}(x^3y^3) + \frac{1}{D^6}(6x^3) = \frac{x^6y^3}{6\cdot5\cdot4} + \frac{6x^9}{4\cdot5\cdot6\cdot7\cdot8\cdot9}$$

$$= \frac{x^6y^3}{120} + \frac{x^9}{10080}$$

Hence, the complete solution is

$$z = C.F. + P.I. = f_1(y + x) + f_2(y + wx) + f_3(y + w^2x) + \frac{x^6y^3}{120} + \frac{x^9}{10080} \qquad \textbf{Ans.}$$

Example 23. *Solve* $\frac{\partial^2 z}{\partial x^2} + 2\frac{\partial^2 z}{\partial x\,\partial y} + \frac{\partial^2 z}{\partial y^2} = x^2 + xy + y^2$

Solution. Here, we have

$$\frac{\partial^2 z}{\partial x^2} + 2\frac{\partial^2 z}{\partial x\partial y} + \frac{\partial^2 z}{\partial y^2} = x^2 + xy + y^2$$

A.E. is $\qquad m^2 + 2m + 1 = 0 \Rightarrow (m + 1)^2 = 0 \quad \Rightarrow \quad m = -1, -1$

$$C.F. = f_1(y - x) + xf_2(y - x)$$

$$P.I. = \frac{1}{D^2 + 2DD' + D'^2}(x^2 + xy + y^2)$$

$$= \frac{1}{D^2}\frac{1}{\left(1 + \frac{2D'}{D} + \frac{D'^2}{D^2}\right)}(x^2 + xy + y^2)$$

$$= \frac{1}{D^2}\left(1 + \frac{2D'}{D} + \frac{D'^2}{D^2}\right)^{-1}(x^2 + xy + y^2)$$

$$= \frac{1}{D^2}\left(1 - \frac{2D'}{D} - \frac{D'^2}{D^2} + \frac{4D'^2}{D^2} + ...\right)(x^2 + xy + y^2)$$

$$= \left(\frac{1}{D^2} - \frac{2D'}{D^3} + \frac{3D'^2}{D^4}\right)(x^2 + xy + y^2)$$

$$= \frac{1}{D^2}(x^2+xy+y^2) - \frac{2D'}{D^3}(x^2+xy+y^2) + \frac{3D'^2}{D^4}(x^2+xy+y^2)$$

$$= \left(\frac{x^4}{12}+\frac{x^3y}{6}+\frac{x^2y^2}{2}\right) - \left(\frac{2}{D^3}\right)(x+2y) + \frac{3}{D^4}(2)$$

$$= \frac{x^4}{12}+\frac{x^3y}{6}+\frac{x^2y^2}{2} - \frac{x^4}{12} - \frac{2x^3y}{3} + \frac{6x^4}{2\cdot3\cdot4}$$

$$= \frac{x^4}{12}+\frac{x^3y}{6}+\frac{x^2y^2}{2} - \frac{x^4}{12} - \frac{2x^3y}{3} + \frac{x^4}{4}$$

$$= \frac{x^4}{4} - \frac{1}{2}x^3y + \frac{x^2y^2}{2}$$

Hence, the complete solution is

$$z = f_1(y-x) + xf_2(y-x) + \frac{x^4}{4} - \frac{x^3y}{2} + \frac{x^2y^2}{2}$$ **Ans.**

Example 24. *Solve* $\frac{\partial^3 z}{\partial x^3} - 2\frac{\partial^3 z}{\partial x^2 \partial y} = 2e^{2x} + 3x^2y$

Solution. Here, we have, $\frac{\partial^3 z}{\partial x^3} - 2\frac{\partial^3 z}{\partial x^2 \partial y} = 2e^{2x} + 3x^2y$

$$(D^3 - 2D^2D')z = 2e^{2x} + 3x^2y$$

Its auxiliary equation is

$$m^3 - 2m^2 = 0$$

$$m^2(m-2) = 0$$

$$m = 0, 0, 2.$$

$$\therefore \quad \text{C.F.} = f_1(y) + x f_2(y) + f_3(y+2x)$$

$$\text{P.I.} = \frac{1}{D^3 - 2D^2D'}(2e^{2x} + 3x^2y)$$

$$= \frac{1}{D^3-2D^2D'}2e^{2x} + \frac{1}{D^3-2D^2D'}3x^2y$$

$$= 2\frac{e^{2x}}{(2)^3 - 2(2)^2(0)} + 3\frac{1}{D^3\left(1-\frac{2D'}{D}\right)}x^2y = \frac{2e^{2x}}{8} + \frac{3}{D^3}(1-\frac{2D'}{D})^{-1}x^2y$$

$$= \frac{e^{2x}}{4} + \frac{3}{D^3}\left(1+\frac{2D'}{D}....\right)x^2y = \frac{e^{2x}}{4} + \frac{3}{D^3}\left[x^2y + \frac{2}{D}x^2\right] = \frac{e^{2x}}{4} + \frac{3}{D^3}\left(x^2y + \frac{2x^3}{3}\right)$$

$$= \frac{e^{2x}}{4} + 3y\frac{1}{D^3}x^2 + \frac{2}{D^3}x^3 = \frac{e^{2x}}{4} + 3y\frac{x^5}{3.4.5} + 2\frac{x^6}{4.5.6} = \frac{e^{2x}}{4} + \frac{x^5y}{20} + \frac{x^6}{60}$$

$$= \frac{1}{60}(15e^{2x} + 3x^5y + x^6)$$

Hence, the complete solution is

$$z = f_1(y) + x f_2(y) + f_3(y+2x) + \frac{1}{60}(15e^{2x} + 3x^5y + x^6)$$ **Ans.**

Example 25. *Solve* $(D^2 + 4D'^2)\, z = x^2y^4$

Solution. Here, we have

$$(D^2 + 4D'^2)z = x^2y^4$$

A.E. is $\quad m^2 + 4 = 0 \quad \Rightarrow \quad (m + 2i)\,(m - 2i) = 0 \quad \Rightarrow \quad m = 2i, -2i$

$$C.F. = f_1(y + 2ix) + f_2(y - 2ix)$$

$$P.I. = \frac{1}{D^2 + 4D'^2}\, x^2y^4$$

$$= \frac{1}{4D'^2\left(1 + \frac{D^2}{4D'^2}\right)}\, x^2y^4$$

$\left[\text{Here the power of } y > \text{the power of } x \text{ in } x^2y^u \text{ so we take } 4D'^2 \text{ as common}\right]$

$$= \frac{1}{4D'^2}\left[1 + \frac{D^2}{4D'^2}\right]^{-1} x^2y^4$$

$$= \frac{1}{4D'^2}\left[1 - \frac{D^2}{4D'^2} + \ldots\ldots\right] x^2y^4$$

$$= \frac{1}{4D'^2}\left[x^2y^4 - \frac{D^2}{4D'^2}(x^2y^4) + \ldots\ldots\right]$$

$$= \frac{1}{4D'^2}\left[x^2y^4 - \frac{2}{4D'^2}(y^4) + \ldots\ldots\right]$$

$$= \frac{1}{4D'^2}\left[x^2y^4 - \frac{1}{2}\frac{y^6}{5\times 6}\right]$$

$$= \frac{1}{4}\left[\frac{x^2y^6}{5\times 6} - \frac{1}{2}\frac{y^8}{5\times 6\times 7\times 8}\right]$$

$$= \frac{1}{4\times 5\times 6\times 14\times 8}[14\times 8\, x^2y^6 - y^8]$$

$$= \frac{1}{13440}[112\, x^2y^6 - y^8]$$

The complete solution = C.F. + P.I.

$$= f_1(y + 2ix) + f_2(y - 2ix) + \frac{1}{13440}(112\, x^2y^6 - y^8) \qquad \textbf{Ans.}$$

EXERCISE 10.5

Solve the following equations :

1. $\dfrac{\partial^2 z}{\partial x^2} + 3\dfrac{\partial^2 z}{\partial x\,\partial y} + 2\dfrac{\partial^2 z}{\partial y^2} = 12xy$ *(A.M.I.E., Winter 2001)*

Ans. $z = f_1(y - x) + f_2(y - 2x) + 2x^3y - \dfrac{3x^4}{2}$

2. $\dfrac{\partial^2 z}{\partial x^2} - \dfrac{\partial^2 z}{\partial x\partial y} - 6\dfrac{\partial^2 z}{dy^2} = xy$ **Ans.** $z = f_1(y - 2x) + f_2(y + 3x) + \dfrac{x^3y}{6} + \dfrac{x^4}{24}$

3. $\frac{\partial^2 z}{\partial x^2} - \frac{\partial^2 z}{\partial y^2} - 3\frac{\partial z}{\partial x} + 3\frac{\partial z}{\partial y} = xy + e^{x+2y}$ *(Uttarakhand, June 2009, U.P. III Semester, Summer 2002)*

Ans. $z = f_1(y+x) + e^{3x} f_2(y-x) - \frac{1}{3}\left(\frac{x^2 y}{2} + \frac{x^3}{6} + \frac{x^2}{3} + \frac{xy}{3} + \frac{2x}{9}\right) - xe^{x+2y}$

4. $(D^3 - 3D^2D')z = x^2y$ **Ans.** $z = \phi_1(y+x) + \phi_2(y-x) + \frac{1}{12}e^{2x-y} - xe^{x+y} - \frac{1}{3}\cos(x+2y)$

5. $(D^2 - 6DD' + 9D'^2)z = 12x^2 + 36xy$ **Ans.** $z = f_1(y+3x) + xf_2(y+3x) + 6x^3y + 10x^4$

6. $(D^2 - D'^2 + D + 3D' - 2)z = e^{x-y} - x^2y$

Ans. $z = e^{-2x} f_1(y+x) + e^x f_2(y-x) - \frac{1}{4}e^{x-y} + \frac{1}{2}\left(x^2 y + xy + \frac{3}{2}x^2 + \frac{3y}{2} + 3x + \frac{21}{4}\right)$

10.5 CASE V. P.I. OF ANY FUNCTION

If the function on the R.H.S. of the P.D.E. is not of the form given previous cases

$$\text{P.I.} = \frac{1}{F(D, D')}\phi(x, y)$$

$F(D, D')$ is factorized to get

$$F(D, D') = (D - m_1D')(D - m_2D') \ldots\ldots (D - m_nD')$$

$$\text{P.I.} = \frac{1}{(D - m_1D')(D - m_2D')\ldots(D - m_nD')}\phi(x, y)$$

Let us consider

$$\text{P.I.} = \frac{1}{D - m_1D'}\phi(x, y) \qquad \text{(Taking only one term)}$$

$$\Rightarrow \qquad p - m_1 q = \phi(x, y)$$

Subsidiary equations are (Lagrange's equations)

$$\frac{dx}{1} = \frac{dy}{-m_1} = \frac{dz}{\phi(x, y)}$$

From the first two

$$dy + m_1 dx = 0 \quad \Rightarrow \quad y + m_1 x = c$$

From the first and last equations, we get

$$dz = \phi(x, y)dx = \phi(x, c - m_1x)\,dx$$

$$\Rightarrow \qquad z = \int \phi(x, c - m_1x)dx$$

$$\boxed{\text{P.I.} = \frac{1}{D - mD'}F(x, y) = \int \phi(x, c - mx)dx}$$

where c is replaced by $y + mx$ after integration.

Similarly we repeat the above method to get *P.I.*

Example 26. *Solve* $(D^2 - DD' - 2D'^2)z = (y-1)e^x$

Solution. $(D^2 - DD' - 2D'^2)z = (y-1)e^x$

A.E. is $(D^2 - DD' - 2D'^2) = 0$ or $m^2 - m - 2 = 0$

$$(m-2)(m+1)=0, \quad \text{or} \quad m=2,-1$$

$$\text{C.F.} = f_1(y+2x)+f_2(y-x)$$

$$\text{P.I.} = \frac{1}{D^2-DD'-2D'^2}(y-1)e^x$$

$$= \frac{1}{(D+D')(D-2D')}(y-1)e^x = \frac{1}{D+D'}\int(c-2x-1)e^x dx \qquad [\text{Put } y=c-2x]$$

$$= \frac{1}{D+D'}[(c-2x-1)e^x+2e^x]$$

$$= \frac{1}{D+D'}[ce^x-2xe^x+e^x] \qquad [\text{Put } c=y+2x]$$

$$= \frac{1}{D+D'}[(y+2x)e^x-2xe^x+e^x] = \frac{1}{D+D'}[ye^x+e^x]$$

$$= \int[(c+x)e^x+e^x]\,dx$$

$$= (c+x)e^x-e^x+e^x$$

$$= ce^x+xe^x=(y-x)e^x+xe^x \qquad [\text{Put } y=c+x]$$

$$= ye^x$$

Hence complete solution is

$$\therefore \quad z = f_1(y+2x)+f_2(y-x)+ye^x$$ **Ans.**

Example 27. *Solve* $\frac{\partial^2 z}{\partial x^2}+\frac{\partial^2 z}{\partial x\partial y}-6\frac{\partial^2 z}{\partial y^2} = y\cos x$. *(Uttarakhand, IInd Sem. Summer 2010)*

Solution. We have, $\frac{\partial^2 z}{\partial x^2}+\frac{\partial^2 z}{\partial x\partial y}-6\frac{\partial^2 z}{\partial y^2} = y\cos x$

$$(D^2+DD'-6D'^2) = y\cos x$$

Its auxiliary equation is

$$m^2+m-6=0 \Rightarrow (m+3)(m-2)=0 \Rightarrow m=2,-3$$

$$\therefore \qquad \text{C.F.} = f_1(y+2x)+f_2(y-3x)$$

$$\text{P.I.} = \frac{1}{D^2+DD'-6D'^2}y\cos x = \frac{1}{(D-2D')(D+3D')}y\cos x$$

$$= \frac{1}{D-2D'}\int(c+3x)\cos x\,dx \qquad [\text{Put } y=c+3x]$$

$$= \frac{1}{D-2D'}[(c+3x)\sin x+3\cos x] = \frac{1}{D-2D'}[y\sin x+3\cos x] \qquad [\text{Put } c+3x=y]$$

$$= \int[(c-2x)\sin x+3\cos x]dx \qquad [\text{Put } y=c-2x]$$

$$= (c-2x)(-\cos x)-2\sin x+3\sin x = y\cos x+\sin x \qquad [\text{Put } c-2x=y]$$

Hence, the complete solution is

$$z = f_1(y+2x)+f_2(y-3x)+\sin x-y\cos x \qquad [\text{Put } c-2x=y]$$

Ans.

Example 28. *Solve :* $(D^2 + D D' - 6 D'^2) z = x^2 \sin (x + y)$

Solution. Here, we have

$$(D^2 + D D' - 6 D'^2) z = x^2 \sin (x + y)$$

Putting $\quad D = m \quad$ and $\quad D' = 1$, we have

A.E. is $\quad m^2 + m - 6 = 0 \Rightarrow (m + 3)(m-2) = 0 \Rightarrow m = 2, -3$

$$\text{C.F.} = f_1 (y + 2x) + f_2 (y - 3x)$$

$$\text{P.I.} = \frac{1}{D^2 + D D' - 6 D'^2}\left[x^2 \sin (x+y)\right]$$

$$= \frac{1}{(D-2D')(D+3D')}\left[x^2 \sin (x+y)\right]$$

Let $\frac{1}{D+3D'}[x^2 \sin (x+y)] = u$

$$\Rightarrow \qquad (D + 3D') u = x^2 \sin (x + y)$$

$$u = \int x^2 \sin (x+c+3x)\, dx$$

$$= \int x^2 \sin (4x+c)\, dx \qquad [y = c + 3x]$$

$$= x^2\left(\frac{-\cos (4x+c)}{4}\right) - (2x)\left(\frac{-\sin (4x+c)}{16}\right) + 2\frac{\cos (4x+c)}{64}$$

$$= \left[\frac{-x^2}{4} + \frac{1}{32}\right]\cos (4x+c) + \frac{x}{8}\sin (4x+c) \qquad \ldots (1)$$

On eliminating c, we put $c = y - 3x$ in (1) and get

$$u = \left(\frac{-x^2}{4} + \frac{1}{32}\right)\cos (4x+y-3x) + \frac{x}{8}\sin (4x+y-3x)$$

$$= \left(\frac{-x^2}{4} + \frac{1}{32}\right)\cos (x+y) + \frac{x}{8}\sin (x+y)$$

$$\text{P.I.} = \frac{1}{(D-2D')(D+3D')}x^2 \sin (x+y)$$

$$= \frac{1}{(D-2D')}u$$

$$= \frac{1}{D-2D'}\left[\left(\frac{-x^2}{4} + \frac{1}{32}\right)\cos (x+y) + \frac{x}{8}\sin (x+y)\right]$$

$$= \int \left[\left(\frac{-x^2}{4} + \frac{1}{32}\right)\cos (x+c-2x) + \frac{x}{8}\sin (x+c-2x)\right] dx \qquad (y = c - 2x).$$

$$= \int \left[\left(\frac{-x^2}{4} + \frac{1}{32}\right)\cos (c-x) + \frac{x}{8}\sin (c-x)\right] dx$$

$$= \left(\frac{-x^2}{4} + \frac{1}{32}\right)\{-\sin (c-x)\} - \left(\frac{-x}{2}\right)[-\cos (c-x)]$$

$$+\left(\frac{-1}{2}\right)\sin(c-x)+\frac{x}{8}\cos(c-x)-\frac{1}{8}[-\sin(c-x)]$$

$$=\left(\frac{x^2}{4}-\frac{1}{32}-\frac{1}{2}+\frac{1}{8}\right)\sin(c-x)+\left(\frac{-x}{2}+\frac{x}{8}\right)\cos(c-x)$$

$$=\left(\frac{x^2}{4}-\frac{13}{32}\right)\sin(c-x)-\frac{3x}{8}\cos(c-x) \qquad \ldots(2)(c=2x+y)$$

On eliminating c, we put $c = 2x + y$ in (2) and get

$$\text{P.I.} = \left(\frac{x^2}{4}-\frac{13}{32}\right)\sin(2x+y-x)-\frac{3x}{8}\cos(2x+y-x)$$

$$=\left(\frac{x^2}{4}-\frac{13}{32}\right)\sin(x+y)-\frac{3x}{8}\cos(x+y)$$

The complete solution = *C.F. + P.I.*

$$= f_1(y+2x)+f_2(y-3x)+\left(\frac{x^2}{4}-\frac{13}{32}\right)\sin(x+y)-\frac{3x}{8}\cos(x+y) \textbf{ Ans.}$$

Example 29. *Solve* : $(r-4t)=\dfrac{4x}{y^2}-\dfrac{y}{x^2}$

Solution. Here, we have

$$(D^2-4D'^2)z=\frac{4x}{y^2}-\frac{y}{x^2}$$

A.E. is $m^2-4=0 \Rightarrow (m+2)(m-2)=0 \Rightarrow m=2,-2$

$$\text{C.F.}=f_1(y+2x)+f_2(y-2x)$$

$$\text{P.I.}=\frac{1}{D^2-4D'^2}\left(\frac{4x}{y^2}-\frac{y}{x^2}\right)$$

$$=\frac{1}{(D+2D')(D-2D')}\left(\frac{4x}{y^2}-\frac{y}{x^2}\right)$$

Let $\quad u=\dfrac{1}{D-2D'}\left(\dfrac{4x}{y^2}-\dfrac{y}{x^2}\right)$ $\qquad$ [Put $y=c-2x$]

$$\Rightarrow \quad (D-2D')u=\left(\frac{4x}{y^2}-\frac{y}{x^2}\right)$$

$$\Rightarrow \quad u=\int\left[\frac{4x}{(c-2x)^2}-\frac{c-2x}{x^2}\right]dx$$

$$=\int\left[\frac{-2(c-2x)+2c}{(c-2x)^2}-\frac{c}{x^2}+\frac{2}{x}\right]dx$$

$$=\int\left[\frac{-2}{(c-2x)}+\frac{2c}{(c-2x)^2}-\frac{c}{x^2}+\frac{2}{x}\right]dx$$

$$=\log(c-2x)+\frac{c}{c-2x}+\frac{c}{x}+2\log x$$

On eliminating c, replace c by $2x + y$ and have

$$u = \log(2x+y-2x) + \frac{2x+y}{2x+y-2x} + \frac{2x+y}{x} + 2\log x$$

$$= \log y + \frac{2x+y}{y} + 2 + \frac{y}{x} + 2\log x \qquad \dots (1)$$

Now P.I. $= \dfrac{1}{(D+2D')(D-2D')}\left(\dfrac{4x}{y^2} - \dfrac{y}{x^2}\right) = \dfrac{1}{D+2D'}\left[\dfrac{1}{D-2D'}\left(\dfrac{4x}{y^2} - \dfrac{y}{x^2}\right)\right]$

$$= \frac{1}{(D+2D')} u$$

Putting the value of u from (1), we get

$$= \frac{1}{D+2D'}\left[\log y + \frac{2x+y}{y} + 2 + \frac{y}{x} + 2\log x\right]$$

$$= \int\left[\log(c+2x) + \frac{2x+c+2x}{c+2x} + 2 + \frac{c+2x}{x} + 2\log x\right] dx \qquad [y = c+2x]$$

$$= \int\left[\log(c+2x) + \frac{2x}{c+2x} + 1 + 2 + \frac{c}{x} + 2 + 2\log x\right] dx$$

$$= \int\left[\log(c+2x) + \frac{2x+c-c}{2x+c} + 5 + \frac{c}{x} + 2\log x\right] dx$$

$$= \int\left[\log(c+2x) + 1 - \frac{c}{2x+c} + 5 + \frac{c}{x} + 1.\log x^2\right] dx$$

$$= \left[x\log(c+2x) - \int x\frac{1}{c+2x}2\,dx + 6x - \frac{c}{2}\log(c+2x) + c\log x + x\log x^2 - \int\frac{2x}{x^2}.x\,dx\right]$$

$$= x\log(c+2x) - \int\frac{c+2x-c}{c+2x}dx + 6x - \frac{c}{2}\log(c+2x) + c\log x + x\log x^2 - \int 2\,dx$$

$$= x\log(c+2x) - x + \frac{c}{2}\log(c+2x) + 6x - \frac{c}{2}\log(c+2x) + c\log x + x\log x^2 - 2x$$

$$= x\log(c+2x) + 3x + c\log x + x\log x^2$$

On eliminating c, replacing c by $y - 2x$ and have

$$= x\log(y-2x+2x) + 3x + (y-2x)\log x + x\log x^2$$

$$= x\log y + 3x + y\log x - 2x\log x + x\log x^2$$

$$= x\log y + 3x + y\log x - x\log x^2 + x\log x^2$$

$$= x\log y + 3x + y\log x$$

Hence, the complete solution is

$y = C.F + P.I. = f_1(y+2x) + f_2(y-2x) + x\log y + 3x + y\log x$ **Ans.**

Example 30. *Solve* : $[D^3 + D^2D' - DD'^2 - D'^3]z = e^x\cos 2y$

Solution. We have

$$[D^3 + D^2D' - DD'^2 - D'^3]z = e^x\cos 2y$$

A.E. is $\quad m^3 + m^2 - m - 1 = 0 \Rightarrow (m+1)^2(m-1) = 0, \quad m = 1, -1, -1$

$$\text{C.F.} \quad f_1(y+x) + f_2(y-x) + x f_3(y-x)$$

$$\text{P. I.} = \frac{1}{D^3 + D^2 D' - D D'^2 - D'^3} e^x \cos 2y$$

$$= \frac{1}{(D+D')^2 (D-D')} e^x \cos 2y$$

Let $$u = \frac{1}{D-D'} e^x \cos 2y \qquad (y = c - x)$$

$$= \int e^x \cos 2(c-x)\, dx$$

$$= \frac{e^x}{1+4} [\cos 2(c-x) - 2 \sin 2(c-x)] \qquad \ldots (1)$$

On eliminating c, replace c by $x + y$ in (1), and have

$$u = \frac{e^x}{5}[\cos 2(x+y-x) - 2 \sin 2(x+y-x)]$$

$$= \frac{e^x}{5}[\cos 2y - 2 \sin 2y]$$

Now $$\frac{1}{(D+D')} \cdot \frac{1}{(D-D')} e^x \cos 2y = \frac{1}{(D+D')} u$$

$$= \frac{1}{D+D'} \left[\frac{e^x}{5} (\cos 2y - 2 \sin 2y) \right] \qquad (y = c + x)$$

$$= \int \left[\frac{e^x}{5} \{\cos(2c+2x) - 2 \sin(2c+2x)\} \right] dx$$

$$= \int \frac{e^x}{5} \cos(2c+2x)\, dx - 2 \int \frac{e^x}{5} \sin(2c+2x)\, dx$$

$$= \frac{e^x}{5(1+4)} [\cos(2c+2x) + 2 \sin(2c+2x)] - \frac{2e^x}{5(1+4)} [\sin(2c+2x) - 2\cos(2c+2x)]$$

$$= \frac{e^x}{25} [\cos(2c+2x) + 2 \sin(2c+2x) - 2 \sin(2c+2x) + 4 \cos(2c+2x)]$$

$$= \frac{e^x}{25} [5 \cos(2c+2x)] = \frac{e^x}{5} \cos(2c+2x)$$

On eliminating c, replace c by $y - x$ and have

$$= \frac{e^x}{5} \cos(2y - 2x + 2x) = \frac{e^x}{5} \cos 2y$$

$$\text{P.I.} = \frac{1}{(D+D')} \left(\frac{1}{D+D'} \frac{1}{D-D'} \right) (e^x \cos 2y)$$

$$= \frac{1}{D+D'} \frac{e^x}{5} \cos 2y \qquad (y = c + x)$$

$$= \int \frac{e^x}{5} \cos 2(c+x)\, dx$$

$$= \frac{e^x}{5(1+4)}[\cos 2(c+x) + 2\sin 2(c+x)]$$

On eliminating c, replace c by $(y-x)$ and get

$$= \frac{e^x}{25}[\cos 2(y-x+x) + 2\sin 2(y-x+x)]$$

$$= \frac{e^x}{25}[\cos 2y + 2\sin 2y]$$

The complete solution is $z = C.F. + P.I.$

$$= f_1(y+x) + f_2(y-x) + x\, f_3(y-x) + \frac{e^x}{25}(\cos 2y + 2\sin 2y) \quad \textbf{Ans.}$$

EXERCISE 10.6

Solve the following equations:

1. $(D-D')(D+2D')z = (y+1)e^x$ **Ans.** $z = f_1(y+x) + f_2(y-2x) + ye^x$
2. $\frac{\partial^2 z}{\partial x^2} - \frac{\partial^2 z}{\partial y^2} = \tan^3 x \tan y - \tan x \tan^3 y$ **Ans.** $z = f_1(y+x) + f_2(x-y) + \frac{1}{2}\tan x \tan y$
3. $(D^2 - DD' - 2D'^2)z = (2x^2 + xy - y^2)\sin xy - \cos xy$ **Ans.** $z = f_1(y+2x) + f_2(y-x) + \sin xy$
4. $\frac{\partial^2 z}{\partial x^2} + \frac{\partial^2 z}{\partial x\,\partial y} - 2\frac{\partial^2 z}{\partial y^2} = (y-1)e^x$ *(Q. Bank, U.P. 2002)*
   Ans. $z = f_1(y+x) + f_2(y-2x) + (y-2)e^x$
5. $(D^2 - 6DD' + 9D'^2)z = 12x^2 + 36xy$ **Ans.** $z = f_1(y+3x) + x f_2(y+3x) + 10x^4 + 6x^3 y$

10.6 NON-HOMOGENEOUS LINEAR EQUATIONS

The linear differential equations which are not homogeneous are called Non-homogeneous Linear Equations.

For example, $3\frac{\partial^2 z}{\partial x^2} + 2\frac{\partial^2 z}{\partial x\partial y} + 4\frac{\partial^2 z}{\partial y^2} + 5\frac{\partial z}{\partial x} + \frac{\partial z}{\partial y} + z = 0$

$$f(D, D') = f_1(x, y)$$

Its solution, $z = \text{C.F.} + \text{P.I.}$

Complementary Function: Let the non-homogeneous equation be

$$(D - mD' - a)z = 0 \quad \text{or} \quad \frac{\partial z}{\partial x} - m\frac{\partial z}{\partial y} - az = 0$$

$$p - mq = az$$

The Lagrange's subsidiary equations are

$$\frac{dx}{1} = \frac{dy}{-m} = \frac{dz}{az}$$

From first two relations, we have $-mdx = dy$

$$dy + mdx = 0 \Rightarrow \quad y + mx = c_1 \quad \text{... (1)}$$

and from first and third relation, we have

$$dx = \frac{dz}{az} \Rightarrow x = \frac{1}{a}\log z + c_2 \Rightarrow z = c_3\, e^{ax} \qquad \text{... (2)}$$

From (1) and (2), we have $z = e^{ax}\phi(y+mx)$

Similarly the solution of $(D - mD' - a)^2 z = 0$ is

$$z = e^{ax}\phi_1(y+mx) + x e^{ax}\,\phi_2(y+mx)$$

10.7 IF THE EQUATION IS OF THE FORM

$$(\alpha D + \beta D' + \gamma)\, z = 0$$

$$\Rightarrow \qquad \alpha p + \beta q = -\gamma z$$

It is of Lagrange's form.

Lagrange's subsidiary equations are

$$\frac{dx}{\alpha} = \frac{dy}{\beta} = \frac{dz}{-\gamma z}$$

From first two, we have $\alpha\gamma - \beta x = C_1$

From first and last, we have $\dfrac{dz}{z} = -\dfrac{\gamma}{\alpha}\,dx$

$$\Rightarrow \qquad \log z = -\frac{\gamma}{\alpha}x + \log c_2$$

$$\Rightarrow \qquad z = e^{-\frac{\gamma}{\alpha}x} = \phi(C_1)\, e^{-\frac{\gamma}{\alpha}x}$$

$$\Rightarrow \qquad z = e^{-\frac{\gamma}{\alpha}x} = \varphi(\alpha y - \beta x)$$

where φ is an arbitrary function.

Example 31. *Solve* $(D + D' - 2)(D + 4D' - 3)z = 0$

Solution. The equation can be rewritten as

$$\{D - (-D') - 2\}\,\{D - (-4D') - 3\}\, z = 0$$

Hence the solution is

$$z = e^{2x}\phi_1(y - mx) + e^{3x}\phi_2(y - 4mx)$$ **Ans.**

Example 32. *Solve* $(D + 3D' + 4)^2 z = 0$

Solution. The equation is rewritten as

$$[D - (-3D') - (-4^2)]z = 0$$

Hence the solution is

$$z = e^{-4x}\phi_1(y - 3x) + x e^{-4x}\phi_2(y - 3x)$$ **Ans.**

Example 33. *Solve* $r + 2s + t + 2p + 2q + z = 0$

Solution. The equation is rewritten as

$$(D^2 + 2DD' + D'^2 + 2D + 2D' + 1)\, z = 0$$

$$\Rightarrow \qquad [(D + D')^2 + 2(D + D') + 1]z = 0$$

$$\Rightarrow \qquad (D+D'+1)^2 z = 0$$

$$\Rightarrow \qquad [D-(-D')-(-1)]^2 z = 0$$

Hence the solution is

$$z = e^{-x}\phi_1 y(y-x) + xe^{-x}\phi_2(y-x)$$ **Ans.**

Example 34. *Solve* $r - t + p - q = 0$

Solution. The equation is rewritten as

$$(D^2 - D'^2 + D - D')z = 0$$

$$\Rightarrow \qquad [(D-D')(D+D') + 1(D-D')]z = 0$$

$$\Rightarrow \qquad (D-D')(D+D'+1)z = 0$$

Hence the solution is

$$z = \phi_1(y+x) + e^{-x}\phi_2(y-x)$$ **Ans.**

EXERCISE 10.7

Solve the following equations

1. $(D-D')(D+D'-3)z = 0$ **Ans.** $z = \phi(y+x) + e^{3x}\phi_2(y-x)$
2. $(D-D'-1)(D-D'-2)z = 0$ **Ans.** $z = e^{x}\phi_1(y+x) + e^{2x}\phi_2(y+x)$
3. $(D+D'-1)(D+2D'-2)z = 0$ **Ans.** $z = e^{x}\phi_1(y-x) + e^{2x}\phi_2(y-2x)$
4. $(D^2 + DD' + D' - 1)z = 0$ **Ans.** $z = e^{-x}\phi_1(y) + e^{x}\phi_2(y-x)$
5. $(D^2 - DD' - 2D'^2 + 2D + 2D')z = 0$ **Ans.** $z = \phi_1(y-x) + e^{-2x}\phi_2(y+2x)$
6. $[D^2 - D'^2 + D + 3D' - 2]z = 0$ **Ans.** $z = e^{-2x}\phi_1(y+x) + e^{x}\phi_2(y-x)$
7. $(D^2 - a^2 D'^2 + 2ab\,D + 2ab\,D')z = 0$ **Ans.** $z = \phi_1(y-ax) + e^{-2abx}\phi_2(y+ax)$
8. $t + s + q = 0$ **Ans.** $z = \phi_1(x) + e^{-x}\phi_2(y-x)$
9. $(D+D'-1)(D+2D'-3)z = 0$ **Ans.** $z = e^{x}\phi_1(y-x) + e^{3x}\phi_2(y-2x)$
10. $(D-2D'+5)^2 z = 0$ **Ans.** $z = e^{-5x}\phi_1(y+2x) + xe^{-5x}\phi_2(y+2x)$

Particular Integral

Case 1. $$\frac{1}{F(D, D')}e^{ax+by} = \frac{1}{F(a, b)}e^{ax+by}$$

Example 35. *Solve* $(D-D'-2)(D-D'-3)z = e^{3x-2y}$

Solution. The complementary function is

$$e^{2x}\phi_1(y+x) + e^{3x}\phi_2(y+x).$$

$$\text{P.I.} = \frac{1}{(D-D'-2)(D-D'-3)}e^{3x-2y} = \frac{1}{[3-(-2)-2][3-(-2)-3]}e^{3x-2y} = \frac{1}{6}e^{3x-2y}$$

Hence the complete solution is

$$z = e^{2x}\phi_1(y+x) + e^{3x}\phi_2(y+x) + \frac{1}{6}e^{3x-2y}$$ **Ans.**

Case II. $\dfrac{1}{F(D^2, DD', D'^2)}\sin(ax+by) = \dfrac{1}{F(-a^2, -ab, -b^2)}\sin(ax+by)$

Example 35(*a*). *Find the particular integral of* $(D^2 - DD' + D'^2)\, z = \sin(2x + 3y)$

Solution. $\text{P.I.} = \dfrac{1}{D^2 - DD' + D'^2}\sin(2x+3y) = \dfrac{1}{-4+16-9}\sin(2x+3y) = -\dfrac{1}{7}\sin(2x+3y)$

Case III. $\dfrac{1}{F(D, D')}x^m y^n = [F(D, D')^{-1}]x^m y^n$

Example 36. *Solve* $[D^2 - D'^2 + D + 3D' - 2]z = x^2 y$

Solution. $(D - D' + 2)(D + D' - 1)z = 0$

$\text{C.F.} = e^{-2x}\phi_1(y+x) + e^x\phi_2(y-x)$

$$\text{P.I.} = \frac{1}{(D-D'+2)(D+D'-1)}x^2y = \frac{1}{D^2 - D'^2 + D + 3D' - 2}x^2y = -\frac{1}{2}\left[\frac{1}{1 - \frac{3D'}{2} - \frac{D}{2} + \frac{D'^2}{2} - \frac{D^2}{2}}\right]x^2y$$

$$= -\frac{1}{2}[1 - \frac{1}{2}(3D' + D - D'^2 + D^2)]^{-1}x^2y$$

$$= -\frac{1}{2}[1 + \frac{1}{2}(3D' + D - D'^2 + D^2) + \frac{1}{4}(3D' + D - D'^2 + D^2)^2$$

$$+ \frac{1}{8}(3D' + D - D'^2 + D^2)^3 + \ldots]x^2y$$

$$= -\frac{1}{2}[1 + \frac{1}{2}(3D' + D - D'^2 + D^2) + \frac{1}{4}(9D'^2 + D^2 + 6DD' + 6D^2D') + \frac{1}{8}(9D^2D') + \ldots]x^2y$$

$$= -\frac{1}{2}[x^2y + \frac{1}{2}(3x^2 + 2xy - 0 + 2y) + \frac{1}{4}(0 + 2y + 12x + 12) + \frac{1}{8}(18)]$$

$$= -\frac{1}{2}\left[x^2y + \frac{3x^2}{2} + xy + y + \frac{y}{2} + 3x + 3 + \frac{9}{4}\right] = -\frac{1}{2}\left[x^2y + \frac{3x^2}{2} + xy + \frac{3y}{2} + 3x + \frac{21}{4}\right]$$

Hence, the complete solution is

$$z = e^{-2x}\phi_1(y+x) + e^x\phi_2(y-x) - \frac{1}{2}\left(x^2y + \frac{3x^2}{2} + xy + \frac{3y}{2} + 3x + \frac{21}{4}\right)$$ **Ans.**

Case IV. $\dfrac{1}{F(D, D')}[e^{ax+by}\phi(x, y)] = e^{ax+by}\dfrac{1}{F(D+a, D'+b)}\phi(x, y)$

Example 37. *Solve* $(D - 3D' - 2)^2 z = 2e^{2x}\sin(y + 3x)$

Solution. A.E. is $(D - 3D' - 2)^2 = 0$

$\text{C.F.} = e^{2x}\phi_1(y+3x) + xe^{2x}\phi_2(y+3x)$

$\text{P.I.} = \dfrac{1}{(D - 3D' - 2)^2}2e^{2x}.\sin(y+3x)$

$$= 2e^{2x}\frac{1}{(D+2-3D'^2-2)^2}\sin(y+3x) = 2e^{2x}\frac{1}{(D-3D')^2}\sin(y+3x)$$

$$= 2e^{2x}\frac{1}{D^2+9D'^2-60D'}\sin(y+3x)$$

$$[\, D^2+9D'^2-60D' = -9-9-6(-3)=0\,]$$

$$= 2e^{2x}\,x\frac{1}{2(D-3D')}\sin(y+3x) \qquad \text{(As denominator becomes zero)}$$

$$= 2x^2e^{2x}\frac{1}{2}\sin(y+3x) \qquad \text{(Again differentiate)}$$

$$= x^2e^{2x}\sin(y+3x)$$

Hence, the complete solution is

$$z = e^{2x}\phi_1(y+3x) + xe^{2x}\phi_2(y+3x) + x^2e^{2x}\sin(y+3x) \qquad \textbf{Ans.}$$

Example 38. *Solve* $(D^2+DD'-6D'^2)z = x^2 \sin(x+y)$

Solution. $(D^2+DD'-6D'^2)z = x^2\sin(x+y)$

For complementary function

$$(D^2+DD'-6D'^2)=0 \text{ or } (D-2D')(D+3D')=0$$

$$\text{C.F.} = \phi_1(y+2x)+\phi_2(y-3x)$$

$$\text{P.I.} = \frac{1}{D^2+DD'-6D'^2}x^2\sin(x+y)$$

$$= \text{Imaginary part of } \frac{1}{D^2+DD'-6D'^2}x^2[\cos(x+y)+i\sin(x+y)]$$

$$= \text{Imaginary part of } \frac{1}{D^2+DD'-6D'^2}x^2\, e^{i(x+y)}$$

$$= \text{Imaginary part of } e^{iy}\frac{1}{D^2+Di-6(i)^2}x^2e^{ix}$$

$$= \text{Imaginary part of } e^{i(x+y)}\frac{1}{(D+i)^2+(D+i)i+6}x^2$$

$$= \text{Imaginary part of } e^{i(x+y)}\frac{1}{D^2+3iD+4}x^2$$

$$= \text{Imaginary part of } \frac{e^{i(x+y)}}{4}\frac{1}{1+\frac{3iD}{4}+\frac{D^2}{4}}x^2$$

$$= \text{Imaginary part of } \frac{e^{i(x+y)}}{4}\left[1+\frac{3iD}{4}+\frac{D^2}{4}\right]^{-1}x^2$$

$$= \text{Imaginary part of } \frac{e^{i(x+y)}}{4}\left[1-\frac{3iD}{4}-\frac{D^2}{4}-\frac{9D^2}{16}\dots\right]x^2$$

$$= \text{Imaginary part of } \frac{e^{i(x+y)}}{4}\left[x^2-\frac{3ix}{2}-\frac{2}{4}-\frac{9}{16}(2)\right]$$

$$= \text{Imaginary part of } \frac{1}{4}[\cos(x+y)+i\sin(x+y)]\left[x^2-\frac{3ix}{2}-\frac{13}{8}\right]$$

$$= \frac{1}{4}\left[\sin(x+y)\left(x^2-\frac{13}{8}\right)-\frac{3}{2}x\cos(x+y)\right]$$

$$= \frac{1}{4}\sin(x+y)\left(x^2-\frac{13}{8}\right)-\frac{3x}{8}\cos(x+y)$$

Hence, the complete solution is

$$z = \phi_1(y+2x)+\phi_2(y-3x)+\frac{1}{4}\sin(x+y)\left(x^2-\frac{13}{8}\right)-\frac{3x}{8}\cos(x+y) \quad \textbf{Ans.}$$

EXERCISE 10.8

Solve the following equations:

1. $[D^2+2DD'+D'^2-2D-2D']z=0$ **Ans.** $z=\phi_1(x-y)+e^{2x}\phi_2(x-y)$

2. $(D^2-D'^2-3D+3D')z=e^{x-2y}$ **Ans.** $z=\phi_1(y+x)+e^{3x}\phi_2(y-x)-\frac{1}{12}e^{x-2y}$

3. $(D-D'-1)(D+D'-2)z=e^{2x-y}$ **Ans.** $z=e^{x}\phi_1(x+y)+e^{2x}\phi_2(y-x)-\frac{1}{2}e^{2x-y}$

4. $(D^2-D'^2-3D+3D')z=e^{x+2y}$ **Ans.** $z=\phi_1(y+x)+e^{3x}\phi_1(x-y)-xe^{x+2y}$

5. $(D+D')(D+D'-2)z=\sin(x+2y)$

   Ans. $z=\phi_1(y-x)+e^{2x}(y-x)+\frac{1}{117}[6\cos(x+2y)-9\sin(x+2y)]$

6. $(D^2-DD'-2D)z=\cos(3x+4y)$ **Ans.** $z=\phi_1(y)+e^{2x}\phi_1(y+x)+\frac{1}{15}[\cos(3x+4y)-2\sin(3x+4y)]$

7. $(DD'+D-D'-1)z=xy$ **Ans.** $z=e^{-y}\phi_1(x)+e^{x}\phi_2(y)-(xy+y-x-1)$

8. $(D+D'-1)(D+2D'-3)z=4+3x+6y$ **Ans.** $z=e^{x}\phi_1(x-y)+e^{3x}\phi_2(2x-y)+6+x+2y$

9. $D(D+D'-1)(D+3D'-2)z=x^2-4xy+2y^2$

   Ans. $z=\phi_1(y)+e^{x}\phi_2(x-y)+e^{2x}\phi_3(3x-y)+\frac{1}{2}\left[\frac{x^3}{3}-2x^2y+2xy^2-\frac{7}{2}x^2+4xy+\frac{x}{2}\right]$

10. $(D-D'+2)(D+D'-1)z=e^{x-y}-x^2y$

    Ans. $z=e^{2y}\phi_1(x+y)+e^{x}\phi_2(x-y)-\frac{e^{x-y}}{4}+\frac{1}{2}\left[x^2y+xy+\frac{3x^2}{2}+\frac{3}{2}y+3x+\frac{21}{4}\right]$

11. $(D^2-DD'-2D'^2+2D'+2D)z=e^{2x+3y}+\sin(2x+y)+xy$

    Ans. $z=\phi_1(x-y)+e^{y}\phi_2(2x+y)-\frac{1}{10}e^{2x+3y}-\frac{1}{6}\cos(2x+y)+\frac{x}{24}(6xy-6y+9x-2x^2-12)$

10.8 CLASSIFICATION OF PARTIAL DIFFERENTIAL EQUATIONS

(U.P., II Semester, June 2007)

In practical problems, the following types of equations are generally used :

(*i*) *Wave equation*: $$\frac{\partial^2 u}{\partial t^2} = c^2\frac{\partial^2 u}{\partial x^2}$$

(*ii*) *One-dimensional heat flow*: $\frac{\partial u}{\partial t} = c^2 \frac{\partial^2 u}{\partial y^2}$

(*iii*) *Two-diemensional heat flow*: $\frac{\partial^2 u}{\partial x^2} + \frac{\partial^2 u}{\partial y^2} = 0$

(*iv*) *Radio equations*: $-\frac{\partial V}{\partial x} = L\frac{\partial I}{\partial t}, -\frac{\partial I}{\partial x} = C\frac{\partial V}{\partial t}$

Consider the Equation. $A\frac{\partial^2 u}{\partial x^2} + B\frac{\partial^2 u}{\partial x \partial y} + C\frac{\partial^2 u}{\partial y^2} + F(x, y, u, p, q) = 0$ ...(1)

Where A, B, C may be constants or functions of x and y. Now the equation (1) is

1. Parabolic; if $B^2 - 4AC = 0$
2. Elliptic; if $B^2 - 4AC < 0$
3. Hyperbolic; if $B^2 - 4AC > 0$

1. Parabolic Equation. The one-dimensional heat conduction equation

$\frac{\partial u}{\partial t} = C^2 \frac{\partial^2 u}{\partial x^2}$ is an example of parabolic partial differential equation e.g.

One dimensional heat flow equation $\frac{\partial u}{\partial t} = \frac{\partial^2 u}{\partial x^2}$ is parabolic type.

2. Elliptic Equations

The following are the examples of elliptic equations.

Two dimensional heat flow equation in steady state given by $\frac{\partial^2 z}{\partial x^2} + \frac{\partial^2 z}{\partial y^2} = 0$ is elliptic in nature.

(*i*) Laplace equation: $\frac{\partial^2 u}{\partial x^2} + \frac{\partial^2 u}{\partial y^2} = 0$

(*ii*) Poisson equation : $\frac{\partial^2 u}{\partial x^2} + \frac{\partial^2 u}{\partial y^2} = f(x, y)$

Laplace equation arrises in study-state flow and potential problems.

Poisson equations arrises in fluid mechanics electricity and magnetism and torsion problems.

3. Hyperbolic Equations

The wave equation $\frac{\partial^2 u}{\partial t^2} = c^2 \frac{\partial^2 u}{\partial x^2}$

is the simplest example of hyperbolic partial differential equation.

Remark 1. If A, B, C in (1) are constants, then nature of equation (1) will be the same for all values of x and y.

Remark 2. If A, B, C are functions of x and y in (1), then nature of equation (1) will not for all values of x and y.

Example 39. *Classify the following equations.*

$(a)\ 2\dfrac{\partial^2 u}{\partial x^2}+4\dfrac{\partial^2 u}{\partial x\partial y}+3\dfrac{\partial^2 u}{\partial y^2}=0$ (*Q. Bank U.P. II Semester 2002*)

$(b)\ \dfrac{\partial^2 u}{\partial x^2}+4\dfrac{\partial^2 u}{\partial x\partial y}+4\dfrac{\partial^2 u}{\partial y^2}=0$ $\qquad (c)\ 2\dfrac{\partial^2 u}{\partial x^2}+6\dfrac{\partial^2 u}{\partial x\partial y}+3\dfrac{\partial^2 u}{\partial y^2}=0$

Solution. (*a*) $A = 2,\ B = 4,\ C = 3$

$$B^2 - 4AC = (4)^2 - 4(2)(3) < 0$$ **Ans.** Elliptic

(*b*) $A = 1,\ B = 4,\ C = 4$

$$B^2 - 4AC = (4)^2 - 4(1)(4) = 0$$ **Ans.** Parabolic

(*c*) $A = 2,\ B = 6,\ C = 3$

$$B^2 - 4AC = (6)^2 - 4(2)(3) = +12 > 0$$ **Ans.** Hyperbolic

Example 40. *Determine wether the following equations are hyperbolic, parabolic and elliptic ?*

$(a)\ x^2\dfrac{\partial^2 u}{\partial t^2}-\dfrac{\partial^2 u}{\partial x^2}=u$ $\qquad (b)\ t\dfrac{\partial^2 u}{\partial t^2}+2\dfrac{\partial^2 u}{\partial x\partial t}+x\dfrac{\partial^2 u}{\partial x^2}+\dfrac{\partial u}{\partial x}=0$

$(c)\ x\dfrac{\partial^2 u}{\partial t^2}+t\dfrac{\partial^2 u}{\partial x\partial t}+\dfrac{\partial^2 u}{\partial t^2}=0$

Solution. (*a*) Here $A = x^2,\ B = 0,\ C = -1$

Now, $B^2 - 4AC = (0)^2 - 4x^2(-1) = 4x^2$

$\therefore$ It is hyperbolic if $4x^2 > 0$ *i.e.*, $x > 0$

parabolic if $4x^2 = 0$ *i.e.*, $x = 0$

Since $4x^2$ being a square, cannot be negative hence it cannot be elliptic.

(*b*) Here $A = t,\ B = 2,\ C = x$

Now, $B^2 - 4AC = 4 - 4tx$

It is hyperbolic if $4 - 4tx > 0$ *i.e.*, $tx < 1$

elliptic if $4 - 4tx < 0$ *i.e.*, $tx > 1$

and parabolic if $4 - 4tx = 0$ *i.e.*, $tx = 1$.

(*c*) Here $A = x,\ B = t,\ C = 1$

Now, $B^2 - 4AC = (t)^2 - 4(x)(1) = t^2 - 4x$

$\therefore$ It is hyperbolic if $t^2 - 4x > 0$ *i.e.*, $t^2 > 4x$

elliptic if $t^2 - 4x < 0$ *i.e.*, $t^2 < 4x$

and parabolic if $t^2 - 4x = 0$ *i.e.*, $t^2 = 4x$. **Ans.**

Example 41. *Classify the partial differential equation*

$$\frac{\partial^2 u}{\partial t^2}+t\frac{\partial^2 u}{\partial x\partial t}+x\frac{\partial^2 u}{\partial x^2}+2\frac{\partial u}{\partial t}+\frac{\partial u}{\partial x}+6u=0$$ (*Q. Bank U.P. II Semester 2002*)

Solution. We have,

$$\frac{\partial^2 u}{\partial t^2}+t\frac{\partial^2 u}{\partial x\partial t}+x\frac{\partial^2 u}{\partial x^2}+2\frac{\partial u}{\partial t}+\frac{\partial u}{\partial x}+6u=0$$

Here $A = 1,\ B = t,\ C = x$

Now, $B^2 - 4AC = t^2 - 4(1)(x) = t^2 - 4x$

The equation is elliptic if $t^2 - 4x < 0$.

The equation is parabolic if $t^2 - 4x = 0$.

The equation is hyperbolic if $t^2 - 4x > 0.$ **Ans.**

Example 42. *Classify the partial differential equations*

$$x^2 \frac{\partial^2 u}{\partial t^2} + 3\frac{\partial^2 u}{\partial x \partial t} + x\frac{\partial^2 u}{\partial x^2} + 17\frac{\partial u}{\partial t} = 100u$$ (*Q. Bank U.P. II Semester 2002*)

Solution. We have,

$$x^2 \frac{\partial^2 u}{\partial t^2} + 3\frac{\partial^2 u}{\partial x \partial t} + x\frac{\partial^2 u}{\partial x^2} + 17\frac{\partial u}{\partial t} = 100u$$

Here $A = x^2, B = 3, C = x$

Now, $B^2 - 4AC = (3)^2 - 4x^2 x = 9 - 4x^3$

The equation is elliptic if $9 - 4x^3 < 0.$

The equation is parabolic if $9 - 4x^3 = 0.$

The equation is hyperbolic if $9 - 4x^3 > 0.$ **Ans.**

Example 43. *Show that the equation.*

$$\frac{\partial^2 u}{\partial t^2} = c^2 \frac{\partial^2 u}{\partial x^2} \text{ is hyperbolic.}$$

Solution. We have,

$$\frac{\partial^2 u}{\partial t^2} - c^2 \frac{\partial^2 u}{\partial x^2} = 0.$$

Here $A = 1, B = 0, C = -c^2$

Now, $B^2 - 4AC = (0)^2 - 4(1)(-c^2) = 4c^2$

Hence, $B^2 - 4AC > 0$

Thus the given equation is hyperbolic. **Proved.**

Example 44. *Classify the following differential equation as to type in the second quadrant of xy-plane* $\sqrt{y^2 + x^2}\, u_{xx} + 2(x - y)\, u_{xy} + \sqrt{y^2 + x^2}\, u_{yy} = 0$

Solution. Here $A = \sqrt{y^2 + x^2}, \quad B = 2(x - y), \quad C = \sqrt{y^2 + x^2}$

Now $B^2 - 4AC = 4(x - y)^2 - 4(y^2 + x^2) = -8xy$

$\because$ In second quadrant, y is positive and x is – ve.

$\Rightarrow$ $B^2 - 4AC = +\text{ ve} > 0$

Hence, differential equation is hyperbolic. **Ans.**

Example 45. *Match the column for the items of the Left side to that of right side :*

A second order P.D.E. in the function 'u' of two independent variables x, y given with usual symbols $Au_{xx} + Bu_{xy} + Cu_{yy} + F(u) = 0$ *then*

(*i*) *Hyperbolic* (*a*) $B^2 - 4AC = 0$

(*ii*) *Parabolic* (*b*) $B^2 - 4AC < 0$

(*iii*) *Elliptic* (*c*) $B^2 - 4AC > 0$

(*iv*) *Not classifies* (*d*) $A = B = C = 0$ (*U.P., II Semester, June 2009*)

Solution.

(*i*) Hyperbolic (*c*) $B^2 - 4AC > 0$

(*ii*) Parabolic (*a*) $B^2 - 4AC = 0$

(*iii*) Elliptic (*b*) $B^2 - 4AC < 0$

(*iv*) Not classifies (*d*) $A = B = C = 0$ **Ans.**

EXERCISE 10.8

Classify the following partial differential equations:

1. $9\frac{\partial^2 u}{\partial x^2} - 6\frac{\partial^2 u}{\partial x \partial t} + \frac{\partial^2 u}{\partial t^2} = 0$ **Ans.** Parabolic

2. $3\frac{\partial^2 z}{\partial x^2} + 2\frac{\partial^2 z}{\partial x \partial y} + 4\frac{\partial^2 z}{\partial y^2} = 0$ **Ans.** Elliptic

3. $2\frac{\partial^2 z}{\partial x^2} - 6\frac{\partial^2 z}{\partial x \partial y} + 3\frac{\partial^2 z}{\partial y^2} = 0$ **Ans.** Hyperbolic

4. $t\frac{\partial^2 u}{\partial t^2} + 3\frac{\partial^2 u}{\partial x \partial t} + x\frac{\partial^2 u}{\partial x^2} + 17\frac{\partial u}{\partial x} = 0$

Ans. Hyperbolic if $xt < \frac{9}{4}$, parabolic if $xt = \frac{9}{4}$, elliptic if $xt > \frac{9}{4}$

5. $\frac{\partial^2 z}{\partial x^2} = \frac{\partial x}{\partial y}$ *(U.P. II Semester Summer 2003)* **Ans.** Parabolic

6. $\frac{\partial^2 z}{\partial x^2} = \frac{\partial^2 z}{\partial y^2}$ **Ans.** Hyperbolic

7. $\frac{\partial^2 z}{\partial x^2} + \frac{\partial^2 z}{\partial y^2} = 0$ *(U.P. II Semester Summer 2010)* **Ans.** Elliptic

OBJECTIVE TYPE QUESTIONS

Choose the correct alternative

1. The complementary function of $(D^2 - 6DD' + 9D'^2)\, z = 0$ is
   (*i*) $f_1(y + 3x) + x f_2(y + 3x)$
   (*ii*) $f_1(y - 3x) + x f_2(y - 3x)$
   (*iii*) $f_1(y + 3x) + x f_2(y - 3x)$
   (*iv*) $f_1(y + x) + x f_2(y + x)$ **Ans.** (*i*)

2. The complementary function of $(D^2 - DD' - 6D')^2 z = 0$ is
   (*i*) $f_1(y - 3x) + f_2(y - 2x)$ (*ii*) $f_1(y + 3x) + f_2(y - 2x)$
   (*iii*) $f_1(y + 3x) - f_2(y - 2x)$ (*iv*) $f_1(y - 3x) - f_2(y - 2x)$ **Ans.** (*ii*)

3. The C.F. of $r = C^2 t$ is
   (*i*) $f_1(y - cx) + f_2(y - cx)$
   (*ii*) $f_1(y - cx) + f_2(y + 2x)$
   (*iii*) $f_1(y + cx) + f_2(y - cx)$
   (*iv*) $f_1(y - cx) - f_2(y - cx)$ **Ans.** (*iii*)

4. The C. F. of $\frac{\partial^4 z}{\partial x^4} - \frac{\partial^4 z}{\partial y^4} = 0$ is
   (*i*) $f_1(y - x) + f_2(y - x) + f_3(y - ix) + f_4(y - ix)$
   (*ii*) $f_1(y + x) - f_2(y - x) + f_3(y + ix) + f_4(y - ix)$
   (*iii*) $f_1(y - x) + f_2(y - x) + f_3(y + ix) + f_4(y - ix)$
   (*iv*) $f_1(y - cx) - f_2(y - cx) + f_3(y + ix) + f_4(y - ix)$ **Ans.** (*iv*)

5. The C. F. of $(D'^3 - 6D^2D' + 11DD' - 6D'^3)\, z = 0$ is

(*i*) $f_1(y - x) + f_2(y + 2x) + f_3\,(y - 3x)$

(*ii*) $f_1(y + x) + f_2(y + 2x) + f_3\,(y + 3x)$

(*iii*) $f_1\,(y - x) + f_2(y - 2x) + f_3\,(y - 3x)$

(*iv*) $f_1\,(y + x) + f_2(y - 2x) - f_3\,(y + 3x)$ **Ans.** (*ii*)

6. If the roots of the *A.E.* are $m_1, m_1, -m_2$ then the C. F. is

(*i*) $f_1(y + m_1x) + x\,f_2(y + m_1x) + f_3\,(y - m_2x)$

(*ii*) $f_1(y - m_1x) + x\,f_2(y + m_1x) + f_3\,(y + m_2x)$

(*iii*) $f_1\,(y + m_1x) + x\,f_2(y + m_1x) + f_3\,(y - m_2x)$

(*iv*) $f_1\,(y + m_1x) + x\,f_2(y + m_1x) - f_3\,(y + m_2x)$ **Ans.** (*i*)

7. The solution of $D^4z = 0$ is

(*i*) $z = f(y) + x\,f_2(y) - x^2 f_3\,(y) - x^3 f_4(y)$

(*ii*) $z = f(y) - x\,f_2(y) + x^2 f_3\,(y) - x^3 f_4(y)$

(*iii*) $z = f_1\,(x + y) + f_2(y + 2x) + f_3\,(y) + f_4(y)$

(*iv*) $z = f(y) + x\,f_2(y) + x^2 f_3\,(y) - x^3 f_4(y)$ **Ans.** (*iii*)

8. The complementary function of :

$$\frac{\partial^3 z}{\partial x^3} - 3\frac{\partial^3 z}{\partial^2 x\,\partial y} + 4\frac{\partial^3 z}{\partial y^3} = e^{x + 2y}$$

(*i*) $f_1(y - x) + f_2(y + x) + f_3\,(y + 2x)$

(*ii*) $f_1(y - x) + f_2(y + 2x) + f_3\,(y + x)$

(*iii*) $f_1\,(y + 2x) + x\,f_2(y + 2x) + f_3\,(y + x)$

(*iv*) $f_1\,(y - x) + f_2(y + 2x) + x\,f_3\,(y + 2x)$ **Ans.** (*iv*)

9. $\dfrac{1}{D - 2D'}e^{2x + y}$ is equal to

(*i*) $2^{2x + y}$ (*ii*) $\dfrac{1}{2}xe^{2x + y}$

(*iii*) $\dfrac{1}{2}x^2e^{2x + y}$ (*iv*) $xe^{2x + y}$ **Ans.** (*iv*)

10. The rule for finding the *P.I.* of $f(D, D')\, z = F(x, y)$ is

(*i*) $\dfrac{1}{f(D, D')}F(x, y)$ (*ii*) $\dfrac{1}{f(D, D')}f(x, y)$

(*iii*) $\dfrac{1}{f(D, 1)}F(x, y)$ (*iv*) $\dfrac{1}{f(D, D')}F(x, m)$ **Ans.** (*i*)

11. The rule for finding the *P.I.* of $F(D, D')\, z = e^{ax + by}$ is

(*i*) $\dfrac{e^{ax - by}}{f(a, b)}$ (*ii*) $\dfrac{e^{ax + by}}{f(a, b)}$

(*iii*) $\dfrac{f(a, b)}{e^{ax + by}}$ (*iv*) $\dfrac{e^{ax + by}}{f(a^2, b^2)}$ **Ans.** (*ii*)

12. The rule for finding the *P.I.* of $(D^2 + DD' + D'^2)\, z = \sin(ax + by)$ is

(i) $\dfrac{1}{f(a^2, ab, b^2)} \sin(ax + by)$

(ii) $\dfrac{1}{f(-a^2, -ab, -b^2)} \sin(ax - by)$

(iii) $\dfrac{1}{f(a, ab, b)} \sin(ax + by)$

(iv) $\dfrac{1}{f(-a^2, -ab, -b^2)} \sin(ax + by)$ **Ans.** (iv)

13. P.I. of $(D^2 + DD' - 6D'^2)\, z = e^{2x+3y}$ is

(i) $\dfrac{1}{-3} e^{2x+3y}$ (ii) $\dfrac{1}{-44} e^{2x+3y}$

(iii) $+\dfrac{1}{44} e^{2x+3y}$ (iv) $\dfrac{1}{-8} e^{2x+3y}$ **Ans.** (ii)

14. The P.I. of $(D^2 + 6DD' + D'^2)\, z = e^x$ is

(i) e^x (ii) e^y

(iii) e^{x+y} (iv) e^{x-y} **Ans.** (i)

15. The P.I. of $(2D^2 - DD' + 4D'^2)\, z = \cos(2x + 3y)$ is

(i) $\dfrac{1}{38} \cos(2x - 3y)$ (ii) $\dfrac{1}{-38} \cos(2x - 3y)$

(iii) $\dfrac{1}{-38} \cos(2x + 3y)$ (iv) $\dfrac{1}{38} \cos(2x + 3y)$ **Ans.** (iii)

16. The *P.I.* of $(D^2 - DD' - 6D'^2)\, z = \mathbf{x} + y$ is

(i) $\dfrac{x^2y^2}{2}$ (ii) $\dfrac{x^2y}{2}$

(iii) $\dfrac{xy^2}{2}$ (iv) x^2y **Ans.** (ii)

17. The *P.I.* of $(D^2 - D'^2)\, z = \mathbf{x} - y$ is

(i) $\dfrac{x^3}{6} - \dfrac{x^2y}{2}$ (ii) $\dfrac{x^3}{6} + \dfrac{x^2y}{2}$

(iii) $\dfrac{x^3}{2} - \dfrac{x^2y}{6}$ (iv) $x^3 - x^2y$ **Ans.** (i)

18. The *P.I.* of $(D^2 - DD' - 6D'^2)\, z = \mathbf{xy}$ is

(i) $\dfrac{xy}{6} + \dfrac{x^4}{24}$ (ii) $\dfrac{xy^3}{6} + \dfrac{x^4}{24}$

(iii) $\dfrac{x^3y}{6} - \dfrac{x^4}{24}$ (iv) $\dfrac{x^3y}{6} + \dfrac{x^4}{24}$ **Ans.** (iv)

19. The solution of $\dfrac{\partial^3 z}{\partial x^3} = 0$ is

(i) $z = f_1(y) + x\, f_2(y) + x^2 f_3(y)$

(ii) $z = (1 + x + x^2) f(y)$

(iii) $z = f_1(x) + y f_2(x) + y^2 f_3(x)$

(iv) $z = (1 + y + y^2) f(x)$ **Ans.** (i)

20. The solution of $\frac{\partial^2 z}{\partial x^2} - \frac{\partial^2 z}{\partial y^2} = 0$ is

(i) $z = f_1(y+x) + f_1(y-x)$ (ii) $z = f_1(y+x) + f_2(y-x)$

(iii) $z = f_2(y+x) + f_2(y-x)$ (iv) $z = f(x^2 - y^2)$ **Ans.** (ii)

21. Particular integral of $(2D^2 - 3DD' + D'^2)z = e^{x+2y}$ is

(i) xe^{x+2y} (ii) $\frac{1}{2}e^{x+2y}$ (iii) $-\frac{x}{2}e^{x+2y}$ (iv) $\frac{x^2}{2}e^{x+2y}$ **Ans.** (iii)

22. Particular integral of $(D^2 - D'^2)z = \cos(x+y)$ is

(i) $\frac{x}{2}\cos(x+y)$ (ii) $x\sin(x+y)$ (iii) $x\cos(x+y)$ (iv) $\frac{x}{2}\sin(x+y)$ **Ans.** (iv)

23. The complementary function of $r - 7s + 6t = e^{x+y}$ is :

(i) $f_1(y-x) + f_2(y-6x)$

(ii) $f_1(y+x) + f_2(y+6x)$

(iii) $f_1(y+2x) + f_2(y-2x)$

(iv) $f_1(y+3x) + f_2(y-4x)$ **Ans.** (ii)

(R.G.P.V Bhopal, II Semester Feb. 2006)

24. $\frac{1}{f(D, D')}x^m y^n =$

(i) $[f(D, D')]^{-1}x^m y^n$ (ii) $[f(D, D')]\, x^m y^n$

(iii) $[f(D, D')]\, x^m$ (iv) $[f(D)]\, x^m y^n$ **Ans.** (i)

25. $\frac{1}{(D - mD')}F(x, y) =$

(i) $\int f(x, c - mx)dx$ (ii) $\int F(c, c + mx)dx$

(iii) $\int F(x, mx)\, dx$ (iv) $\frac{d}{dx}F(x, mx - c)$ **Ans.** (i)

Indicate True and false for the following equations

26. The solution of the partial differential equation $(D^2 + DD' + 6D'^2)\, z = 0$ is

$$z = f_1(x) + f_2(x + y)$$ **Ans.** False

27. The C.F. of $(D^2 - 3DD' - 4D'^2)\, z = 0$ is

$$z = f_1(y + x) + f_2(y - x)$$ **Ans.** True

28. The solution of $(D^2 - 6DD' - 8D'^2)\, z = 0$ is

$$z = f_1(y + 2x) + f_2(y + 4x)$$ **Ans.** True

29. The solution of the partial differential equation $(D^2 - 5DD' - 6D'^2)\, z = 0$ is

$$z = f_1(y) + f_2(y + 2x) + f_3(y + 3x)$$ **Ans.** True

30. The solution of the partial differential equation $(D^4 + 2D^2D'^2 + 4D'^4)\, z = 0$ is

$$z = f_1(y + x) + x\, f_2(y + 2x)$$ **Ans.** False

31. The solution of the partial differential equation $(D^3 - 3D^2D' + 4D'^3)\, z = 0$ is

$$z = f_1(y - x) + f_2(y + 2x) + x\, f_3(y + 2x)$$ **Ans.** True

32. The solution of the partial differential equation $\left(\frac{\partial^2 z}{\partial x^2}+\frac{\partial^2 z}{\partial y^2}\right) = 0$ is

$z = f_1(y + x) + f_2(y - x)$ **Ans.** False

33. The solution of the partial differential equation $(D^3 + 4D^2 + 4D)z = 0$ is

$z = f_1(y) + f_2(y + 2x) + x f_3(y + 2x)$ **Ans.** True

34. The C.F. of the partial differential equation

$2r - s - 3t = 5\frac{e^x}{e^y}$ is $f_1\left(y-\frac{1}{2}x\right)+x\,f_2\left(y-\frac{1}{2}x\right)$ **Ans.** False

35. The solution of the linear partial differential equation $\frac{\partial^4 z}{\partial x^4}-2\frac{\partial^4 z}{\partial x^3\partial y}+2\frac{\partial^4 z}{\partial x\partial y^3}-\frac{\partial z^4}{\partial y^4}=0$ is

$z = f_1(y - x) + f_2(y + x) + x f_3(y + x) + x^2 f_4(y + x)$ **Ans.** True

Fill in the blanks

36. The solution of the partial differential equation $\frac{\partial^2 z}{\partial x^2}+5\frac{\partial^2 z}{\partial x\partial y}+2\frac{\partial^2 z}{\partial y^2}=0$ is ....

Ans. $f_1(y-2x)+f_2\left(y-\frac{x}{2}\right)$

37. The solution of $25r - 40s + 16t = 0$ is ..... **Ans.** $f_1\left(y+\frac{4}{5}x\right)+x\,f_2\left(y+\frac{4}{5}x\right)$

38. The solution of $(D^3 - 6D^2D' + 12DD'^2 + 8D'^3)z = 0$ is .....

Ans. $f_1(y + 2x) + x f_2(y + 2x) + x^2 f_3(y + 2x)$

39. The solution of $4r - 4s + t = 0$ is .... **Ans.** $f_1\left(y+\frac{1}{2}x\right)+x\,f_2\left(y+\frac{1}{2}x\right)$

40. The solution of $[D^2 + (\alpha + \beta)DD' + \alpha\beta D'^2]z = 0$ is ..... **Ans.** $f_1(y-\alpha x)+f_2(y-\beta x)+\frac{1}{6}x^3 y$

UNIT–V

11 CHAPTER

APPLICATIONS OF PARTIAL DIFFERENTIAL EQUATIONS

11.1 INTRODUCTION

In applied mathematics, the partial differential equations generally arise from the mathematical formulation of physical problems. Subject to certain given conditions, called boundary conditions solving such an equation is known as solving a boundary value problem.

The method of solution of such equations differs from that used in the case of ordinary differential equations. We first find out the general solution of the ordinary differential equation and determine the particular solution with the help of given conditions. Here, from the start, we try to find particular solutions of the partial differential equations which satisfy all the boundary conditions. Method of separation of variables is employed to solve the applied partial differential equation.

11.2 METHOD OF SEPARATION OF VARIABLES *(U.P., II Semester, June 2007)*

In this method, we assume that the dependent variable is the product of two functions, each of which involves only one of the independent variables. So two ordinary differential equations are formed.

Example 1. *Applying the method of separation of variables techniques, find the solution to the P.D.E.*

$$3u_x + 2u_y = 0 \ldots, \text{ where } u_x = \frac{\partial u}{\partial x}, u_y = \frac{\partial u}{\partial y}.$$

Solution. Here we have

$$\frac{3\,\partial u}{\partial x} + \frac{2\,\partial u}{\partial y} = 0 \qquad \ldots(1)$$

Let $$u = X(x)\,Y(y) \qquad \ldots(2)$$

Where X is a function of x only and Y is a function of y only.

On differentiating (2) partially w.r.t. x, we get

$$\frac{\partial u}{\partial x} = \frac{\partial X}{\partial x}.Y \qquad \ldots(3)$$

On differentiating (2) partially w.r.t. y, we get

$$\frac{\partial u}{\partial y} = X.\frac{\partial Y}{\partial y} \qquad \ldots(4)$$

Putting the values of $\frac{\partial u}{\partial x}$ and $\frac{\partial u}{\partial y}$ from (3) and (4) in (1), we get

$$3\frac{\partial X}{\partial x}.Y + 2X\frac{\partial Y}{\partial y} = 0 \qquad ...(5)$$

Dividing (5) by XY, we get

$$\frac{3}{X}\frac{\partial X}{\partial x} + \frac{2}{Y}\frac{\partial Y}{\partial y} = 0$$

[For L.H.S, R.H.S is constant
So we take
both equation are equal k (constant)]

$$\Rightarrow \quad \frac{3}{X}\frac{\partial X}{\partial x} = -\frac{2}{Y}\frac{\partial Y}{\partial y} = k$$

$$\Rightarrow \quad \frac{3}{X}\frac{\partial X}{\partial x} = k \text{ and } -\frac{2}{Y}\frac{\partial Y}{\partial y} = k$$

$$\Rightarrow \quad \frac{\partial X}{X} = \frac{k}{3}\partial x \text{ and } \frac{\partial Y}{Y} = -\frac{k}{2}\partial y$$

$$\Rightarrow \quad \log X = \frac{k}{3}x + c_1 \text{ and } \log Y = \frac{-k}{2}y + c_2$$

$$\Rightarrow \quad X = e^{\frac{k}{3}x + c_1} \text{ and } Y = e^{-\frac{k}{2}y + c_2}$$

Putting the value of X and Y in (2), we get

$$u = e^{\frac{k}{3}x + c_1}\, e^{\frac{-k}{2}y + c_2}$$

$$= e^{k\left(\frac{x}{3} - \frac{y}{2}\right) + c_1 + c_2}$$

$$= e^{k\left(\frac{x}{3} - \frac{y}{2}\right)} . e^{C_1 + C_2}$$

$$u = A\, e^{k\left(\frac{x}{3} - \frac{y}{2}\right)} \qquad [A = e^{C_1 + C_2}]$$

Hence $$u = A\, e^{k\left(\frac{x}{3} - \frac{y}{2}\right)}$$ **Ans.**

Example 2. *Solve the following equation* $\frac{\partial^2 z}{\partial x^2} - 2\frac{\partial z}{\partial x} + \frac{\partial z}{\partial y} = 0$ *by the method of separation of variables.* *(U.P., II Semester, Summer 2009, 2005)*

Solution. Given equation is

$$\frac{\partial^2 z}{\partial x^2} - 2\frac{\partial z}{\partial x} + \frac{\partial z}{\partial y} = 0 \qquad ...(1)$$

Let $$z = X(x)\, Y(y) \qquad ...(2)$$

where X is a function of x only and Y is a function of y only.

$$\frac{\partial z}{\partial x} = Y\frac{dX}{dx}, \qquad \frac{\partial^2 z}{\partial x^2} = Y\frac{d^2 X}{dx^2}$$

$$\frac{\partial z}{\partial y} = X\frac{dY}{dy}$$

Putting all values in equation (1), we have

$$Y\frac{d^2X}{dx^2}-2Y\frac{dX}{dx}+X\frac{dY}{dy}=0$$

Dividing by XY, we have

$$\frac{1}{X}\frac{d^2X}{dx^2}-\frac{2}{X}\frac{dX}{dx}+\frac{1}{Y}\frac{dY}{dy}=0$$

Separating the variables, we have

$$\frac{1}{X}\frac{d^2X}{dx^2}-\frac{2}{X}\frac{dX}{dx}=-\frac{1}{Y}\frac{dY}{dy}=K\ (let)$$

where K is a constant.

$$\frac{1}{X}\frac{d^2X}{dx^2}-\frac{2}{X}\frac{dX}{dx}=K$$

$$\frac{d^2X}{dx^2}-2\frac{dX}{dx}=KX$$

$\Rightarrow (D^2-2D-K)X=0$

A. E. is

$$m^2-2m-K=0$$

$$m=\frac{2\pm\sqrt{4+4K}}{2}$$

$$m=1\pm\sqrt{1+K}$$

$$-\frac{1}{Y}\frac{dY}{dy}=K$$

$$\frac{dY}{dy}+KY=0$$

$$(D+K)Y=0$$

A.E. is $m+k=0$

$$m=-K$$

$$Y=C_3e^{-Ky} \quad ...(4)$$

Thus $\quad X=C_1e^{(1+\sqrt{1+K})x}+C_2e^{(1-\sqrt{1+K})x} \quad ...(3)$

Putting the values of X and Y from (3) and (4) in (2), we get

$$z=\left\{C_1e^{(1+\sqrt{1+K})x}+C_2e^{(1-\sqrt{1+K})x}\right\}C_3e^{-Ky}$$ **Ans.**

Example 3. *Using the method of separation of variables, solve*

$$\frac{\partial u}{\partial x}=2\frac{\partial u}{\partial t}+u$$

where $\quad u(x,0)=6e^{-3x}$ *(U.P. II Semester summer 2006, A.M.I.E.T.E., Summer 2002)*

Solution. $\quad \frac{\partial u}{\partial x}=2\frac{\partial u}{\partial t}+u \quad ...(1)$

Let $\quad u=X(x).T(t) \quad ...(2)$

where X is a function of x only and T is a function of t only.

Putting the value of u in (1), we get

$$\frac{\partial(X.T)}{\partial x}=2\frac{\partial}{\partial t}(X.T)+X.T$$

$$T\frac{dX}{dx}=2X\frac{dT}{dt}+X.T$$

On separating the variables, we get

$$\frac{1}{X}\frac{dX}{dx}=\frac{2}{T}\frac{dT}{dt}+1=C$$ [On dividing by XT]

$$\frac{1}{X}\frac{dX}{dx}=C \qquad\qquad \frac{2}{T}\frac{dT}{dt}+1=C$$

$$\Rightarrow \quad \frac{dX}{dx}=CX \qquad\qquad \Rightarrow \quad \frac{dT}{dt}+\frac{T}{2}=\frac{CT}{2}$$

$$\Rightarrow \quad DX-CX=0 \qquad\qquad \Rightarrow \quad DT-\left(\frac{C}{2}-\frac{1}{2}\right)T=0$$

$$\Rightarrow \quad (D-C)X=0 \qquad\qquad \text{A.E. is } m-\left(\frac{C}{2}-\frac{1}{2}\right)=0 \Rightarrow m=\frac{C}{2}-\frac{1}{2}$$

A.E. is $m-C=0 \Rightarrow m=C$ $\qquad \Rightarrow T=be^{\frac{1}{2}(c-1)t}$

$\Rightarrow \quad X=ae^{cx}$

Putting the values of X and T in (2), we have

$$u=ae^{cx}.be^{\frac{1}{2}(c-1)t}$$

$$\Rightarrow \quad u=abe^{cx+\frac{1}{2}(c-1)t} \qquad ...(3)$$

On putting $t=0$ and $u=6e^{-3x}$ in (3), we get

$$6e^{-3x}=abe^{cx} \qquad \Rightarrow ab=6 \text{ and } c=-3$$

Putting the values of ab and c in (3), we have

$$u=6e^{-3x+\frac{1}{2}(-3-1)t}$$

$$u=6e^{-3x-2t}$$ **Ans.**

which is the required solution.

Example 4. *Use the method of separation of variables to solve the equation:*

$$\frac{\partial^2 v}{\partial x^2}=\frac{\partial v}{\partial t}$$

given that $v=0$ *when* $t\to\infty$, *as well as* $v=0$ *at* $x=0$ *and* $x=1$.

Solution. $$\frac{\partial^2 v}{\partial x^2}=\frac{\partial v}{\partial t} \qquad ...(1)$$

Let us assume that $v=XT$ where X is a function of x only and T that of t only.

$$\frac{\partial v}{\partial t}=X\frac{dT}{dt} \quad \text{and} \quad \frac{\partial^2 v}{\partial x^2}=T\frac{d^2X}{dx^2}$$

Substituting these values in (1), we get

$$X\frac{dT}{dt}=T\frac{d^2X}{dx^2} \quad \Rightarrow \quad \frac{1}{T}\frac{dT}{dt}=\frac{1}{X}\frac{d^2X}{dx^2} \qquad ...(2)$$

Let each side of (2) be equal to a constant $(-p^2)$

$$\frac{1}{T}\frac{dT}{dt}=-p^2 \qquad\qquad \frac{1}{X}\frac{d^2X}{dx^2}=-p^2$$

$$\frac{dT}{dt}+p^2T=0 \qquad\qquad \Rightarrow \quad \frac{d^2X}{dx^2}+p^2X=0$$

$DT + p^2T = 0$	$D^2X + p^2X = 0$
$(D + p^2)T = 0$	$(D^2 + p^2)X = 0$
A.E. is $m + p^2 = 0$	A.E. is $m^2 + p^2 = 0$
$\Rightarrow \quad m = -p^2$	$m^2 = -p^2 \Rightarrow m = \pm ip$
$T = C_1e^{-p^2t}$	$X = C_2 \cos px + C_3 \sin px$

$$\therefore \quad v = C_1e^{-p^2t}(C_2 \cos px + C_3 \sin px) \quad \text{... (3)}$$

On putting $x = 0, v = 0$ in (3), we get

$$0 = C_1e^{-p^2t}C_2 \quad \Rightarrow \quad C_2 = 0, \text{ since } C_1 \neq 0$$

On putting the value of C_2 in (3), we get

$$v = C_1e^{-p^2t}C_3 \sin px \quad \text{... (4)}$$

Again putting $x = l, v = 0$ in (4), we get

$$0 = C_1e^{-p^2t}.C_3 \sin pl$$

Since C_3 cannot be zero.

$$\therefore \quad \sin pl = 0 = \sin n\pi \Rightarrow pl = n\pi \Rightarrow p = \frac{n\pi}{l}, \text{ where } n \text{ is any integer.}$$

On putting the value of p in (4), it becomes

$$v = C_1C_3e^{-\frac{n^2\pi^2t}{l^2}} \sin \frac{n\pi x}{l}$$

Hence $$v = b_ne^{\frac{-n^2\pi^2t}{l^2}} \sin \frac{n\pi x}{l} \quad \text{where } b_n = C_1C_3$$

This equation satisfies the given condition for all integral values of n. Hence taking $n = 1, 2, 3, ...$ *the most general solution is*

$$v = \sum_{n=1}^{\infty} b_n\, e^{\frac{-n^2\pi^2t}{l^2}} \sin \frac{n\pi x}{l} \quad \textbf{Ans.}$$

Example 5. *Solve the following equation by the method of separation of variables*

$$\frac{\partial^2 u}{\partial x\, \partial t} = e^{-t} \cos x$$

given that $u = 0$ *when* $t = 0$ *and* $\frac{\partial u}{\partial t} = 0$ *when* $x = 0$. [*U.P., II Semester, (SUM) 2008*]

Solution. Let $$u = XT \quad \text{...(1)}$$

where X is a function of x only and T is a function of t only.

Then, $$\frac{\partial u}{\partial t} = \frac{\partial}{\partial t}(XT) = X\frac{dT}{dt}$$

$$\therefore \quad \frac{\partial^2 u}{\partial x \partial t} = \frac{\partial}{\partial x}\left(X\frac{dT}{dt}\right) = \frac{dT}{dt}.\frac{dX}{dx} \quad \text{...(2)}$$

Substituting the value $\frac{\partial^2 u}{\partial x \partial t}$ from (2) in the given equation, we get

$$\frac{dT}{dt}\frac{dX}{dx} = e^{-t} \cos x$$

Separating the variables, we get

$$e^t \frac{dT}{dt} = \frac{\cos x}{\left(\frac{dX}{dx}\right)} = -p^2 \text{ (say)} \quad \text{...(3)}$$

Now, $e^t \dfrac{dT}{dt} = -p^2$

$\Rightarrow$ $dT = [-p^2e^{-t}]\, dt$

On integration, we get

$$T = p^2e^{-t} + c_1 \quad \text{...(4)}$$

Also, $\dfrac{dX}{dx} = -\dfrac{1}{p^2}\cos x$

$dX = -\dfrac{1}{p^2}\cos x\, dx$

On integration, we get

$$X = -\frac{1}{p^2}\sin x + c_2 \quad \text{...(5)}$$

Putting the values of X and T from (4) and (5) in (1), we get

$$u = XT = \left(-\frac{1}{p^2}\sin x + c_2\right)(p^2e^{-t} + c_1) \quad \text{...(6)}$$

On putting $u = 0$ and $t = 0$ in (6), we get

$$0 = \left(-\frac{1}{p^2}\sin x + c_2\right)(p^2 + c_1)$$

$$\Rightarrow \quad p^2 + c_1 = 0 \Rightarrow c_1 = -p^2$$

Differentiating (6) w.r.t. "t", we get

$$\frac{\partial u}{\partial t} = \left(-\frac{1}{p^2}\sin x + c_2\right)(-p^2e^{-t}) \quad \text{...(7)}$$

Putting $\dfrac{\partial u}{\partial t} = 0$ when $x = 0$ in (7), we get

$$0 = c_2(-p^2e^{-t})$$

$$\Rightarrow \quad c_2 = 0$$

Substituting the values of $c_1 = -p^2$ and $c_2 = 0$ in (6), we get

$$u = -\frac{1}{p^2}\sin x\,(p^2e^{-t} - p^2)$$

$$= (1 - e^{-t})\sin x$$

Ans.

Example 6. *Solve the P.D.E. by separation of variables method,*

$$u_{xx} = u_y + 2u,\ u(0, y) = 0$$

$$\frac{\partial}{\partial x} u(0, y) = 1 + e^{-3y}.$$

(U.P. II Semester, 2010, 2009)

Solution. Let $u = XY$ ...(1)

where X is a function of x only and Y is a function of y only.

On differentiating, we get

$$\frac{\partial u}{\partial y} = \frac{\partial}{\partial y}(XY) = X\frac{dY}{dy} = XY' \text{ and}$$

$$\frac{\partial u}{\partial x} = \frac{\partial}{\partial x}(XY) = Y\frac{dX}{dx} \text{ and } \frac{\partial^2 u}{\partial x^2} = \frac{\partial}{\partial x}\left[Y\frac{dX}{dx}\right] = Y\frac{d^2X}{dx^2}$$

Putting the values of u_{xx} and u_y and u in the given equation, we get

$$YX'' = XY' + 2XY$$

On separating the variables, we get

$$\frac{X''}{X} = \frac{Y' + 2Y}{Y}$$

$$\Rightarrow \quad \frac{X''}{X} = \frac{Y'}{Y} + 2 = k \text{ (say)} \qquad ...(2)$$

$$\frac{X''}{X} = k \qquad\qquad \frac{Y'}{Y} + 2 = k$$

$$\Rightarrow \quad X'' - kX = 0 \qquad\qquad \Rightarrow \quad \frac{Y'}{Y} = k - 2$$

A.E. is $m^2 - k = 0 \Rightarrow m = \pm\sqrt{k}$ $\qquad \Rightarrow \quad \frac{dY}{Y} = (k-2)\,dy$

$\therefore \quad \Rightarrow X = C_1 e^{\sqrt{k}x} + C_2 e^{-\sqrt{k}x}$ $\qquad \Rightarrow \quad \log Y = (k-2)\,y + \log C_3$

$\therefore \quad X = C_1 e^{\sqrt{k}x} + C_2 e^{-\sqrt{k}x}$ $\qquad \Rightarrow \quad Y = C_3 e^{(k-2)y}$

On putting the values of X and Y in (1), we get

$$u = (C_1 e^{\sqrt{k}x} + C_2 e^{-\sqrt{k}x})\, C_3 e^{(k-2)y} \qquad ...(3)$$

On putting $x = 0$ and $u = 0$ in (3), we get

$$0 = (C_1 + C_2)\, C_3 e^{(k-2)y} \qquad (C_3 e^{(k-2)y} \neq 0) \qquad ...(4)$$

$$\Rightarrow \quad C_1 + C_2 = 0 \Rightarrow C_2 = -C_1$$

On putting $C_2 = -C_1$ in (4), we get

$$u = \Sigma C_1 C_3 (e^{\sqrt{k}x} - e^{-\sqrt{k}x})\, e^{(k-2)y} \qquad ...(5)$$

On differentiating (5), w.r.t. x, we get

$$\frac{\partial u}{\partial x} = \Sigma C_1 C_3 \sqrt{k}\,(e^{\sqrt{k}x} + e^{-\sqrt{k}x})\, e^{(k-2)y} \qquad ...(6)$$

On putting $x = 0$ and $\frac{\partial u}{\partial x} = 1 + e^{-3y}$ in (6), we get

$$1 + e^{-3y} = \Sigma C_1 C_3 \sqrt{k}\,(2)\, e^{(k-2)y} = \sum_{n=1}^{\infty} b_n e^{(k-2)y}$$

$$1 + e^{-3y} = \Sigma C_1 C_3 \sqrt{k}\,(2)\, e^{(k-2)y} = b_1 e^{(k-2)y} + b_2 e^{(k-2)y} + b_3 e^{(k-2)y} + ...$$

Comparing the coefficients, we get

$$b_1 = 1,\ k - 2 = 0 \Rightarrow k = 2$$

Again comparing, we get

$$2C_1 C_2 \sqrt{k} = 1 \Rightarrow C_1 C_3 = \frac{1}{2\sqrt{k}} = \frac{1}{2\sqrt{2}} \qquad (\because k = 2)$$

Again comparing b_3 on both the sides, we get

$$b_3 = -1,\ k - 2 = -3 \Rightarrow k = -1$$

Again comparing b_3, we get

$$2C_1 C_3 \sqrt{k} = 1, \Rightarrow C_1 C_3 = \frac{1}{2\sqrt{k}} = \frac{1}{2\sqrt{-1}} = \frac{1}{2i} \qquad (\because k = -1)$$

Hence, $u(x,y) = \frac{1}{2\sqrt{2}}(e^{\sqrt{2}x} - e^{-\sqrt{2}x}) + \frac{1}{2i}(e^{ix} - e^{-ix})e^{-3y}$

$$u(x,y) = \frac{1}{\sqrt{2}} \sinh \sqrt{2}\, x + e^{-3y} \sin x. \qquad \textbf{Ans.}$$

EXERCISE 11.1

Using the method of separation of variables, find the solution of the following equations

1. $2x\frac{\partial z}{\partial x} - 3y\frac{\partial z}{\partial y} = 0$ **Ans.** $z = cx^{\frac{k}{2}} y^{\frac{k}{3}}$

2. $\frac{\partial u}{\partial x} + u = \frac{\partial u}{\partial t}$ if $u = 4e^{-3x}$ when $t = 0$ **Ans.** $u = 4e^{-3x-2t}$

3. $4\frac{\partial u}{\partial x} + \frac{\partial u}{\partial y} = 3u$ and $u = e^{-5y}$ when $x = 0$ **Ans.** $u = e^{2x-5y}$

4. $4\frac{\partial u}{\partial t} + \frac{\partial u}{\partial x} = 3u,\ u = 3e^{-x} - e^{-5x}$ at $t = 0$ *(A.M.I.E.T.E., Winter 2000)* **Ans.** $u = 3e^{t-x} - e^{2t-5x}$

5. $3\frac{\partial u}{\partial x} + 2\frac{\partial u}{\partial y} = 0;\ u(x, 0) = 4e^{-x}$ *(A.M.I.E.T.E., Summer 2000)* **Ans.** $u = 4e^{-x+3/2y}$

6. $\frac{\partial u}{\partial x} + \frac{\partial u}{\partial y} = 2(x+y)u$ **Ans.** $u = ce^{x^2+y^2+k(x-y)}$

7. $\frac{\partial u}{\partial t} = 4\frac{\partial^2 u}{\partial x^2}$ if $u(x, 0) = 4x - \frac{1}{2}x^2$ **Ans.** $u = \left(4x - \frac{x^2}{2}\right)e^{-p^2 t}$

8. $\frac{\partial^2 u}{\partial x^2} = 2\frac{\partial u}{\partial t}$ if $u(x,0) = x(4-x)$ **Ans.** $u = x(4-x)e^{\frac{p^2 t}{2}}$

9. $\frac{\partial u}{\partial t} = \frac{\partial^2 u}{\partial x^2}$ if $u(x,0) = 2x$ when $0 \le x \le \frac{l}{2}$

   $= 2(l-x)$ when $\frac{l}{2} \le x \le l$

   Ans. $u = 2xe^{-h^2 t}$ for $0 \le x \le \frac{l}{2}$, $u = 2(l-x)e^{-h^2 t}$ for $\frac{l}{2} \le x \le l$.

10. $\frac{\partial u}{\partial t} = \frac{\partial^2 u}{\partial x^2}$ if $u(x, 0) = \sin \pi x$ **Ans.** $u = \sin \pi x . e^{-pl}$

11. $\frac{\partial^2 u}{\partial x^2} = \frac{\partial u}{\partial t}$ if $u(x,0) = x^2(25 - x^2)$ **Ans.** $u = x^2(25 - x^2)e^{-p^2 t}$

12. $\frac{\partial^2 z}{\partial x^2} - 2\frac{\partial z}{\partial x} - \frac{\partial z}{\partial y} = 0$ **Ans.** $z = c_1 e^{[1+\sqrt{(1+p)}]x + p^2 y} + c_2 e^{[1-\sqrt{1+P}]x + p^2 y}$

13. $\frac{\partial^2 u}{\partial t^2} = \frac{\partial^2 u}{\partial x^2}$ If $u(x,0) = \frac{1}{2}x(1-x)$ **Ans.** $u = \frac{x}{2}(1-x)\cos pt + c_2 \sin pt(c_3 \cos px + c_4 \sin px)$

14. $16\frac{\partial^2 u}{\partial x^2} = \frac{\partial^2 u}{\partial t^2}$ if $u(x,0) = x^2(5-x)$ **Ans.** $u = x^2(5-x)\cos pt + c_4 \sin pt\left(c_1 \cos\frac{px}{4} + c_2 \sin\frac{px}{4}\right)$

15. $\frac{\partial^2 y}{\partial x^2} = \frac{\partial u}{\partial y} + 2u$ if $u = 0$ and $\frac{\partial u}{\partial x} = 1 + e^{-3y}$ when $x = 0$. **Ans.** $u = \frac{1}{2\sqrt{2}}(e^{\sqrt{2}x} - e^{-\sqrt{2}x}) + \frac{1}{2i}(e^{ix} - e^{-ix})e^{-3y}$

16. $x\frac{\partial u}{\partial x} + y\frac{\partial u}{\partial y} = 0$ **Ans.** $u = Ae^{1/2(x^2-y^2)k}$

17. $x^2\frac{\partial^2 u}{\partial x\partial y} + 3y^2 + u = 0$

18. $\frac{\partial u}{\partial x} = \frac{\partial u}{\partial y}$ **Ans.** $u = Ae^{x^2 . y^k}$

19. $\frac{\partial u}{\partial x} = 2\frac{\partial u}{\partial y} + u,\ u(x, 0) = 4e^{-3x}$ (*A.M.I.E.T.E., Summer 2001*) **Ans.** $u = 4e^{-(3x+2y)}$

20. $2\frac{\partial u}{\partial x} + 3\frac{\partial u}{\partial y} + 5u = 0,\ u(0, y) = 2e^{-y}$ **Ans.** $u = 2e^{-x-y}$

21. $\frac{\partial u}{\partial x} + \frac{\partial u}{\partial y} = 3u$ and $u = e^{-5y}$ when $x = 0$ **Ans.** $u = e^{8x-5y}$

22. $\frac{\partial u}{\partial x} - 2\frac{\partial u}{\partial y} = u$, given that $u(x,0) = 3e^{-5x} + 2e^{-3x}$ (*A.M.I.E.T.E., Summer 2001*)

Ans. $u = 3e^{-5x-3y} + 2e^{-3x-2y}$

11.3 EQUATION OF VIBRATING STRING (*U.P. II Semester, June 2007*)

Let us consider small transverse vibrations of an elastic string of length l, which is stretched and then fixed at its two ends. Now we will study the transverse vibration of the string when no external forces act on it. Take an end of the string as the origin and the string in the equilibrium position as the x-axis and the line through the origin and perpendicular to the x-axis as the y-axis. We make the following assumptions:

1. The motion takes place entirely in one plane. This plane is chosen as the *xy plane.*
2. In this plane, each particle of the string moves in a direction perpendicular to the equilibrium position of the string.
3. The tension T caused by stretching the string before fixing it at the end points is constant at all times at all points of the deflected string.
4. The tension T is very large compared with the weight of the string and hence the gravitational force may be neglected.
5. The effect of friction is negligible.
6. The string is perfectly flexible. It can transmit only tension but not bending or shearing forces.
7. The slope of the deflection curve is small at all points and at all times.

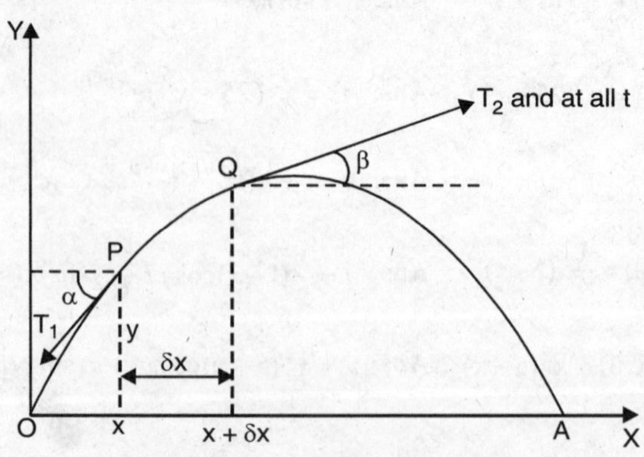

When the string is in motion in the xy-plane, the displacement y of any point of the string is a function of x and time t. Let $P(x, y)$ and $Q(x+\delta x, y+\delta y)$ be two neighbouring points on the string. Let α and β be the inclinations made by the tangents at P and Q respectively with the x-axis. The tension T_1 at P and tension T_2 at Q are tangential. Let m be the mass per unit length of the string which is homogeneous. Consider the motion of the infinitesimal element PQ of the string. The vertical component of the force to which this element is subjected to is

$$m\delta s \frac{\partial^2 y}{\partial t^2} = T_2 \sin\beta - T_1 \sin\alpha \quad \text{...(1)}$$

Since, there is no motion in the horizontal direction,

$$T_1 \cos\alpha = T_2 \cos\beta = T \text{ (constant)} \quad \text{...(2)}$$

Dividing (1) by T, we get

$$\frac{m\delta s \frac{\partial^2 y}{\partial t^2}}{T} = \frac{T_2 \sin\beta}{T} - \frac{T_1 \sin\alpha}{T}$$

$$\frac{m\delta s}{T}\frac{\partial^2 y}{\partial t^2} = \frac{T_2 \sin\beta}{T_2 \cos\beta} - \frac{T_1 \sin\alpha}{T_1 \cos\alpha} \qquad \begin{bmatrix} T = T_1 \cos\alpha \\ T = T_2 \cos\beta \end{bmatrix}$$

$$\Rightarrow \quad \frac{m\delta s}{T}\frac{\partial^2 y}{\partial t^2} = \tan\beta - \tan\alpha$$

$$\Rightarrow \quad \frac{\partial^2 y}{\partial t^2} = \frac{T}{m\delta s}(\tan\beta - \tan\alpha)$$

Slope $\tan\alpha = \left(\frac{\partial y}{\partial x}\right)_x$ at P and slope $\tan\beta = \left(\frac{\partial y}{\partial x}\right)_{x+\delta x}$ at Q.

$$\frac{\partial^2 y}{\partial t^2} = \frac{T}{m\delta x}\left[\left(\frac{\partial y}{\partial x}\right)_{x+\delta x} - \left(\frac{\partial y}{\partial x}\right)_x\right]$$

$$\Rightarrow \quad \frac{\partial^2 y}{\partial t^2} = \frac{T}{m}\left[\frac{\left(\frac{\partial y}{\partial x}\right)_{x+\delta x} - \left(\frac{\partial y}{\partial x}\right)_x}{\delta x}\right]$$

On taking limit as $\delta x \to 0$, we have

$$\frac{\partial^2 y}{\partial t^2} = \frac{T}{m}\frac{\partial^2 y}{\partial x^2},$$

$$\boxed{\frac{\partial^2 y}{\partial t^2} = a^2 \frac{\partial^2 y}{\partial x^2}} \quad \left(\text{take } a^2 = \frac{T}{m}\right)$$

Note. The partial differential equation is knwon as one-dimensional wave equation, because the motion is only in vertical direction (Transverse vibration) not in horizontal direction.

Boundary conditions.

At O, $x = 0$ and $y = 0$, $\frac{\partial y}{\partial t} = 0$ as $t = 0$

At A, $x = l$ and $y = 0$,

Example 7. *Obtain the solution of the wave equation*

$$\frac{\partial^2 y}{\partial t^2} = c^2 \frac{\partial^2 y}{\partial x^2}$$

using the method of separation of variables.

Solution. $\dfrac{\partial^2 y}{\partial t^2} = c^2 \dfrac{\partial^2 y}{\partial x^2}$

Let $y = XT$ where X is a function of x only and T is a function of t only.

$$\frac{\partial y}{\partial t} = X\frac{dT}{dt} \text{ and } \frac{\partial y}{\partial x} = T\frac{dX}{dx}$$

Since T and X are functions of a single variable only.

$$\frac{\partial^2 y}{\partial t^2} = X\frac{d^2T}{dt^2} \text{ and } \frac{\partial^2 y}{\partial x^2} = T\frac{d^2X}{dx^2}$$

Substituting these values in the given equation, we get

$$X\frac{d^2T}{dt^2} = c^2 T\frac{d^2X}{dx^2}$$

By separating the variables, we get

$$\frac{1}{c^2 T}\frac{d^2T}{dt^2} = \frac{1}{X}\frac{d^2X}{dx^2} = k \text{ (say).}$$

(Each side is constant, since the variables x and y are independent).

$$\therefore \quad \frac{d^2T}{dt^2} - kc^2T = 0 \text{ and } \frac{d^2X}{dx^2} - kX = 0$$

Auxiliary equations are

$m^2 - kc^2 = 0 \Rightarrow m = \pm c\sqrt{k}$ | and $m^2 - k = 0$ or $m = \pm\sqrt{k}$

Case 1. If $k > 0$.

$T = C_1 e^{c\sqrt{k}t} + C_2 e^{-c\sqrt{k}t}$, | $X = C_3 e^{\sqrt{k}x} + C_4 e^{-\sqrt{k}x}$

Case 2. If $k < 0$.

$T = C_5 \cos c\sqrt{k}t + C_6 \sin c\sqrt{k}t$, | $X = C_7 \cos\sqrt{k}x + C_8 \sin\sqrt{k}x$

Case 3. If $k = 0$

$T = C_9 t + C_{10}$, | $X = C_{11}x + C_{12}$

These are the three cases depending upon the particular problems. Here we are dealing with wave motion $(k < 0)$.

$$y = TX$$

$$y = (C_5 \cos c\sqrt{k}t + C_6 \sin c\sqrt{k}t) \times (C_7 \cos\sqrt{k}x + C_8 \sin\sqrt{k}x) \qquad \textbf{Ans.}$$

Example 8. *Find the solution of the wave equation*

$$\frac{\partial^2 y}{\partial t^2} = c^2 \frac{\partial^2 y}{\partial x^2}$$

such that $y = P_0 \cos pt$, *(*P_0 *is a constant) when* $x = l$ *and* $y = 0$ *when* $x = 0$.

Solution. $$\frac{\partial^2 y}{\partial t^2} = c^2 \frac{\partial^2 y}{\partial x^2} \qquad ...(1)$$

Its solution is as given in example 7 on previous page.

$$y = (c_1 \cos c\sqrt{k}t + c_2 \sin c\sqrt{k}t)(c_3 \cos\sqrt{k}x + c_4 \sin\sqrt{k}x) \qquad ...(2)$$

On putting $y = 0$ and $x = 0$ in (2), we get

$$0 = (c_1 \cos c\sqrt{k}t + c_2 \sin c\sqrt{k}t)c_3 \quad \Rightarrow \quad c_3 = 0$$

On putting $c_3 = 0$ in (2), we get

$$y = (c_1 \cos c\sqrt{k}t + c_2 \sin c\sqrt{k}t)c_4 \sin\sqrt{k}x$$

$$y = c_1c_4 \cos c\sqrt{k}t \sin\sqrt{k}x + c_2c_4 \sin c\sqrt{k}t \sin\sqrt{k}x \qquad ...(3)$$

On putting $y = P_0 \cos pt$ and $x = l$ in (3), we get

$$P_0 \cos pt = c_1c_4 \cos c\sqrt{k}t \sin\sqrt{k}l + c_2c_4 \sin c\sqrt{k}\,t \sin\sqrt{k}l$$

Equating the coefficients of sin and cos on both sides, we have

$$P_0 = c_1c_4 \sin\sqrt{k}l, \qquad \Rightarrow \quad c_1c_4 = \frac{P_0}{\sin\sqrt{k}\,l}$$

and $$0 = c_2c_4 \sin\sqrt{k}\,l \qquad \Rightarrow \quad c_2 = 0$$

And $$p = c\sqrt{k} \qquad \Rightarrow \quad \frac{p}{c} = \sqrt{k}$$

On putting these values in (3), we get

$$y = \frac{P_0}{\sin\sqrt{kl}} \cos pt \sin\frac{p}{c}x$$

$$y = \frac{P_0}{\sin\frac{P_0}{c}l} \cos p = t \sin\frac{p}{c}x$$

Ans.

Example 9. *Solve completely the equation* $\frac{\partial^2 y}{\partial t^2} = c^2 \frac{\partial^2 y}{\partial x^2}$, *representing the vibrations of a string of length l, fixed at both ends, given that* $y(0, t) = 0$, $y(l, t) = 0$; $y(x, 0) = f(x)$ *and* $\frac{\partial}{\partial t}y(x,0) = 0, 0 < x < l.$ *(U.P. II Semester summer 2005)*

Solution. Here,

$$\frac{\partial^2 y}{\partial t^2} = c^2 \frac{\partial^2 y}{\partial x^2} \qquad ...(1)$$

Let $$y = X(x)\, T(t) \qquad ...(2)$$

$$\frac{\partial^2 y}{\partial t^2} = X\frac{d^2T}{dt^2}$$

$$\frac{\partial^2 y}{\partial x^2} = T\frac{d^2X}{dx^2}$$

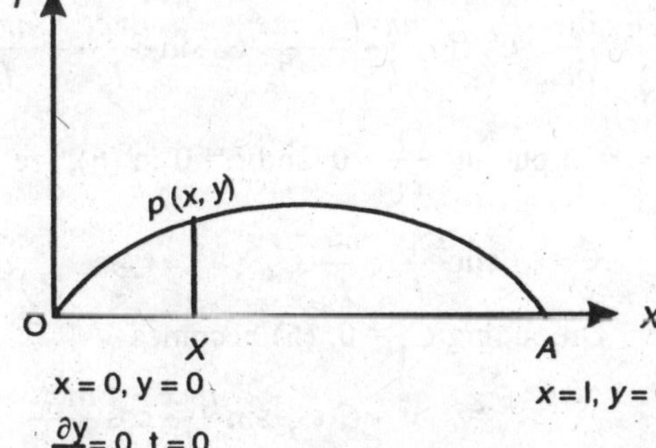

Equation (1) becomes

$$X\frac{d^2T}{dt^2}=c^2T\frac{d^2X}{dx^2}$$

Separating the variables,

$$\frac{1}{X}\frac{d^2X}{dx^2}=\frac{1}{c^2T}\frac{d^2T}{dt^2}=-p^2\,(\text{let})$$

If $\frac{1}{X}\frac{d^2X}{dx^2}=-p^2$

$\Rightarrow \quad \frac{d^2X}{dx^2}=-Xp^2$

$\Rightarrow \quad \frac{d^2X}{dx^2}+Xp^2=0$

$\Rightarrow \quad (D^2+p^2)X=0$

A.E. is $m^2+p^2=0 \quad \Rightarrow \quad m=\pm pi$

$X=(C_1\cos px+C_2\sin px)$

If $\frac{1}{c^2T}\frac{\partial^2T}{\partial t^2}=-p^2$

$\Rightarrow \quad (D^2+p^2c^2)\,T=0$

A.E. is $m^2+p^2c^2=0$

$\Rightarrow \quad m^2=-p^2c^2 \quad \Rightarrow \quad m=\pm pCi$

$T=(C_3\cos pct+C_4\sin pct)$

Putting the values of X and T in (2), we get

$$y=(C_1\cos px+C_2\sin px)(C_3\cos pct+C_4\sin pct) \quad ..(3)$$

Now applying the boundary condition

$x=0$ and $y=0$

Putting these values in (3), we get

$0=C_1\,(C_3\cos pct+C_4\sin pct) \quad \Rightarrow \quad C_1=0$

Equation (3) becomes, $\quad y=C_2\sin px\,(C_3\cos pct+C_4\sin pct) \quad ..(4)$

Putting $x=l$ and $y=0$ in (4), we get

$$0=C_2\sin pl\,(C_3\cos pct+C_4\sin pct) \Rightarrow \quad \sin pl=0=\sin n\pi$$

$$\Rightarrow \quad pl=n\pi \quad \Rightarrow \quad p=\frac{n\pi}{l}$$

On putting $p=\frac{n\pi}{l}$, (4) becomes

$$y=C_2\sin\frac{n\pi x}{l}\left(C_3\cos\frac{n\pi ct}{l}+C_4\sin\frac{n\pi ct}{l}\right) \quad ...(5)$$

On differentiating (5) w.r.t. t, we get

$$\frac{\partial y}{\partial t}=C_2\sin\frac{n\pi x}{l}\left(-\frac{n\pi c}{l}C_3\sin\frac{n\pi ct}{l}+\frac{n\pi c}{l}C_4\cos\frac{n\pi c}{l}t\right) \quad ...(6)$$

On putting $\frac{\partial y}{\partial t}=0$ and $t=0$ in (6), we get

$$0=C_2\sin\frac{n\pi x}{l}.\frac{n\pi c}{l}.C_4 \quad \Rightarrow \quad C_4=0$$

On putting $C_4=0$, (5) becomes

$$y=C_2C_3\sin\frac{n\pi x}{l}\cos\frac{n\pi ct}{l} \quad [\text{let } C_2C_3=b_n] \quad ...(7)$$

Now applying $y = f(x)$ and $t = 0$, (7) becomes

$$f(x) = b_n \sin \frac{n\pi x}{l}$$

C_2C_3 can be calculated using Fourier sine series as

$$b_n = \frac{2}{l}\int_0^l f(x)\sin\frac{n\pi x}{l}\,dx \qquad ...(8)$$

Thus, required solution for the given equation is

$$y = b_n \sin\frac{n\pi x}{l}\cos\frac{n\pi ct}{l}$$

where b_n is given by equation (8). **Ans.**

Example 10. *A tightly stretched string with fixed end points x =0 and x = π is initially at rest in its equilibrium position. If it is set vibrating by giving to each of its points an initial velocity.*

$$\left(\frac{\partial y}{\partial t}\right)_{t=0} = 0.03 \sin x - 0.04 \sin 3x$$

then find the displacement y (x,t) at any point of string at any time t.

Solution. Here we have equation for vibration of a string

$$\frac{\partial^2 y}{\partial t^2} = c^2 \frac{\partial^2 y}{\partial x^2}$$

Its solution is

$$y(x,t) = (c_1 \cos pct + c_2 \sin pct)(c_3 \cos px + c_4 \sin px) \qquad ...(1)$$

[*See Example 7 on page 430*]

Putting $x = 0$ and $y = 0$ in (1), we get

$$0 = (c_1 \cos pct + c_2 \sin pct)(c_3)$$

$$\Rightarrow \qquad c_3 = 0$$

Putting $c_3 = 0$ in (1), we have

$$y = (c_1 \cos pct + c_2 \sin pct)(c_4 \sin px) \qquad ...(2)$$

Putting $x = \pi$ and $y = 0$ in (2), we get

$$0 = (c_1 \cos pct + c_2 \sin pct)\, c_4 \sin p\pi$$

$$\Rightarrow \qquad 0 = \sin p\pi$$

and $\sin n\pi = \sin p\pi \;\Rightarrow\; p = n$

Putting $p = n$ in (2), we get

$$y = (c_1 \cos nct + c_2 \sin nct)(c_4 \sin nx) \qquad ...(3)$$

Putting $t = 0$, $y = 0$ in (3), we get

$$0 = (c_1 + 0)\, c_4 \sin nx \quad (\sin nx = 0 \text{ since } \sin nx \text{ is a part of } X)$$

$$\Rightarrow \qquad c_1 = 0$$

On putting $c_1 = 0$ in (3), we get

$$y = (c_2 \sin nct)(c_4 \sin nx)$$

General equation is $(c_2\, c_4 = b_n)$

$$y = \sum_{n=1}^{\infty} b_n \sin nct \sin nx \qquad ...(4)$$

Differentiating (4) w.r.t. 't', we get

$$\frac{\partial y}{\partial t} = \sum_{n=1}^{\infty} b_n nc \cos nct \sin nx \qquad ...(5)$$

On putting $\frac{\partial y}{\partial t} = 0.03 \sin x - 0.04 \sin 3x$ and $t = 0$ in (5) we get

$$0.03 \sin x - 0.04 \sin 3x = \sum_{n=1}^{\infty} b_n nc \sin nx$$

$\Rightarrow$ $0.03 \sin x - 0.04 \sin 3x = cb_1 \sin x + 2cb_2 \sin 2x + 3c_3b_3 \sin 3x + ...$

Comparing the coefficient, we get

$$0.03 = cb_1 \Rightarrow b_1 = \frac{0.03}{c}$$

$$0 = 2cb_2 \Rightarrow b_2 = 0$$

$$-0.04 = 3c\, b_3 \Rightarrow b_3 = \frac{-0.04}{3c} = \frac{-0.0133}{c}$$

Putting the values of b_1, b_2, b_3 ... in (4), we get

$$y = \frac{0.03}{c} \sin ct \sin x - \frac{0.0133}{c} \sin 3ct \sin 3x$$

Hence $$y = \frac{1}{c}[0.03 \sin ct \sin x - 0.0133 \sin 3x]$$ **Ans.**

Example 11. *A string is stretched and fastened to two points l apart. Motion is started by displacing the string in the form $y = a \sin \frac{\pi x}{l}$ from which it is released at a time $t = 0$. Show that the displacement of any point at a distance x from one end at time t is given by*

$$y(x, t) = a \sin\left(\frac{\pi x}{l}\right) \cos\left(\frac{\pi ct}{l}\right)$$

(*Uttarakhand II Sem. 2012, A.M.I.E.T.E., Winter 2003, U.P., II Semester, 2004, 2009*)

Solution. Consider an elastic string tightly stretched between two points O and A. Let O be the origin and OA as x-axis. On giving a small displacement to the string, perpendicular to its length (parallel to the y-axis). Let y be the displacement at the point P (x, y) at any time. The wave equation.

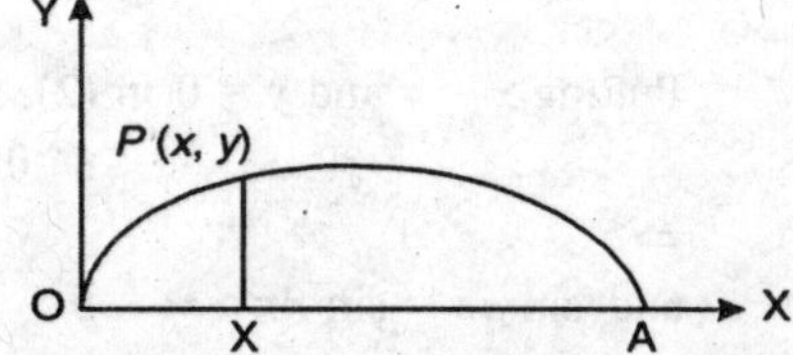

$$\frac{\partial^2 y}{\partial t^2} = c^2 \frac{\partial^2 y}{\partial x^2}$$

where C is a constant. The vibration of the string is given by:

$$\frac{\partial^2 y}{\partial t^2} = c^2 \frac{\partial^2 y}{\partial x^2} \qquad ... (1)$$

As the end points of the string are fixed, for all time,

$$y(0, t) = 0 \qquad ... (2)$$

and $$y(l, t) = 0 \qquad ... (3)$$

Since the initial transverse velocity of any point of the string is zero, therefore,

$$y(0, t) = 0 \quad ... (2)$$

$$\left(\frac{\partial y}{\partial t}\right)_{t=0} = 0 \qquad \text{... (4)}$$

Also
$$y(x,0) = a\sin\frac{\pi x}{l} \qquad \text{... (5)}$$

Now we have to solve (1), subject to the above boundary conditions. Since the vibration of the string is periodic, therefore, the solution of (1) is of the form

$$y(x, t) = (C_1 \cos px + C_2 \sin px)(C_3 \cos c\,pt + C_4 \sin c\,pt) \qquad \text{... (6)}$$

On putting $x = 0$ and $y = 0$ in (6), we get

$$0 = C_1 (C_3 \cos c\,pt + C_4 \sin c\,pt) \Rightarrow C_1 = 0$$

On putting $C_1 = 0$ in (6), we get

$$y(x, t) = C_2 \sin px (C_3 \cos c\,pt + C_4 \sin c\,pt) \qquad \text{... (7)}$$

On differentiating (7) w.r.t. t, we get

$$\frac{\partial y}{\partial t} = C_2 \sin px[C_3(-cp \sin c\,pt) + C_4(cp \cos c\,pt)] \qquad \text{...(8)}$$

On putting $\frac{dy}{dt} = 0$ and $t = 0$ in (8), we get

$$0 = C_2 \sin px(C_4\, cp)$$

$\Rightarrow \quad C_2 C_4\, cp = 0$

If $C_2 = 0$, (7) will lead to the trivial solution $y(x,t) = 0$.

Thus, the only possibility is that $C_4 = 0$

On putting $C_4 = 0$ in (7), we get

$$y(x, t) = C_2 C_3 \sin px \cos c\,pt \qquad \text{...(9)}$$

On putting $x = l$ and $y = 0$ in (9), we get

$$0 = C_2\, C_3 \sin pl \cos cpt, \text{ for all } t.$$

Since C_2 and $C_3 \neq 0$, we have $\sin pl = 0$ $\therefore$ $pl = n\pi$ $\Rightarrow$ $p = \frac{n\pi}{l}$, where n is an integer.

On putting the value of p in (9), we get

$$y(x, t) = C_2 C_3 \sin\frac{n\pi x}{l} \cos\frac{n\pi\, ct}{l} \qquad \text{... (10)}$$

On putting $t = 0$ and $y = a\sin\frac{\pi x}{l}$ in (10), we get

$$a\sin\frac{\pi x}{l} = C_2 C_3 \sin\frac{n\pi x}{l}$$

which will be satisfied by taking $C_2\, C_3 = a$ and $n = 1$

Hence the required solution is

$$y(x, t) = a\sin\frac{\pi x}{l}\cos\frac{\pi ct}{l}$$

Proved.

Example 12. *The vibrations of an elastic string is governed by the partial differential equation*

$$\frac{\partial^2 u}{\partial t^2} = \frac{\partial^2 u}{\partial x^2}$$

The length of the string is π *and the ends are fixed. The initial velocity is zero and the initial deflection is* $u(x, 0) = 2\ (\sin x + \sin 3x)$. *Find the deflection* $u(x, t)$ *of the vibrating string for* $t > 0$.

Solution. $$\frac{\partial^2 u}{\partial t^2} = \frac{\partial^2 u}{\partial x^2}$$

Solution of the given differential equation is

$$\Rightarrow \qquad u = (c_1 \cos pt + c_2 \sin pt)(c_3 \cos px + c_4 \sin px) \qquad \text{... (1)}$$

On putting $x = 0$, $u = 0$ in (1), we get

$$0 = (c_1 \cos pt + c_2 \sin pt)c_3 \qquad \Rightarrow c_3 = 0$$

On putting $c_3 = 0$ in (1), it reduces

$$u = (c_1 \cos pt + c_2 \sin pt)c_4 \sin px \qquad \text{... (2)}$$

On putting $x = \pi$ and $u = 0$ in (2), we have

$$0 = (c_1 \cos pt + c_2 \sin pt)\, c_4 \sin p\pi$$

$$\sin p\pi = 0 = \sin n\pi \qquad n = 1, 2, 3, 4$$

$$\therefore \qquad p\pi = n\pi \quad \text{or} \quad p = n$$

On substituting the value of p in (2), we get

$$u = (c_1 \cos nt + c_2 \sin nt)\, c_4 \sin nx \qquad \text{... (3)}$$

On differentiating (3) w.r.t. "t" we get

$$\frac{du}{dt} = (-\, c_1\, n \sin nt + c_2\, n \cos nt)\; c_4 \sin nx \qquad \text{... (4)}$$

On putting $\frac{du}{dt} = 0$, $t = 0$ in (4) we have

$$0 = (c_2 n)\ (c_4 \sin nx) \;\Rightarrow\; c_2 = 0$$

On putting $c_2 = 0$, (3) becomes

$$u = (c_1 \cos nt)(c_4 \sin nx)$$

$$u = c_1 c_4 \cos nt \sin nx$$

$$u = \Sigma\, b_n \cos nt \sin nx \qquad \text{... (5)}$$

On putting $u = 2(\sin x + \sin 3x)$ and $t = 0$ in (5), we have

$$2\,(\sin x + \sin 3x) = b_1 \sin x + b_2 \sin ax + b_3 \sin 3x + \ldots.$$

Comparing the coefficient of sin, we get

$$b_1 = 2,\ b_2 = 0,\ b_3 = 2,\ b_4 = b_5 = = \ldots. = 0$$

On substituting the values of b's in (5), we get

$$u = 2\ [\ \cos t \sin x + \cos 3t \sin 3x]$$ **Ans.**

Example 13. *A tightly stretched string with fixed end points $x = 0$ and $x = l$ is initially in a position given by* $y = y_0 \sin^3\left(\frac{\pi x}{l}\right)$. *If it is released from rest from this position, find the displacement y (x, t).* *(Uttarakhand, IInd Semester, Summer 2010)*

Solution. Let the equation to the vibrating string be

$$\frac{\partial^2 y}{\partial t^2} = c^2 \frac{\partial^2 y}{\partial x^2} \qquad \text{... (1)}$$

Here the initial conditions are

$$y(0, t) = 0, \ y(l, t) = 0$$

$$\frac{\partial y}{\partial t} = 0 \ \text{ at } t = 0, \ y(x, 0) = y_0 \sin^3 \frac{\pi x}{l} = \frac{y_0}{4}\left[3 \sin \frac{\pi x}{l} - \sin \frac{3\pi x}{l}\right]$$

The solution of (1) is of the form

$$y = (c_1 \cos px + c_2 \sin px)(c_3 \cos pct + c_4 \sin pct) \qquad \text{... (2)}$$

On putting $x = 0$ and $y = 0$ in (2), we get

$$0 = c_1 (c_3 \cos pct + c_4 \sin pct) \qquad \Rightarrow c_1 = 0$$

On putting $c_1 = 0$ in (2), we get

$$y = c_2 \sin px(c_3 \cos pct + c_4 \sin pct) \qquad \text{... (3)}$$

On putting $x = l$ and $y = 0$ in (3), we get

$$0 = c_2 \sin pl(c_3 \cos pct + c_4 \sin pct)$$

$\therefore \ \sin pl = 0 = \sin n\pi$ or $pl = n\pi$, or $p = \frac{n\pi}{l}$ where $n = 0, 1, 2, 3.....$

On putting the value of p in (3), we get

$$y = c_2 \sin \frac{n\pi x}{l}\left(c_3 \cos \frac{n\pi ct}{l} + c_4 \sin \frac{n\pi ct}{l}\right) \qquad \text{... (4)}$$

On differentiating (4), w.r.t. t, we get

$$\frac{\partial y}{\partial t} = c_2 \sin \frac{n\pi x}{l}\left(-\frac{n\pi c}{l} c_3 \sin \frac{n\pi c}{l} t + c_4 \frac{n\pi c}{l} \cos \frac{n\pi ct}{l}\right)$$

On putting $\frac{\partial y}{\partial t} = 0$ and $t = 0$ in (4), we have

$$0 = c_2 \sin \frac{n\pi x}{l} c_4 \frac{n\pi c}{l} \qquad \Rightarrow c_4 = 0$$

On putting $c_4 = 0$ in (4), we get

$$y = c_2 c_3 \sin \frac{n\pi x}{l} \cos \frac{n\pi ct}{l}$$

$$y = b_n \sin \frac{n\pi x}{l} \cos \frac{n\pi c}{l} t \qquad (b_n = c_2 c_3)$$

General solution is

$$y = \sum_{n=1}^{\infty} b_n \sin \frac{n\pi x}{l} \cos \frac{n\pi c}{l} t \qquad \text{... (5)}$$

On putting $t = 0$ and $y = \frac{y_0}{4}\left[3 \sin \frac{\pi x}{l} - \sin \frac{3\pi x}{l}\right]$ in (5), we have

$$\frac{y_0}{4}\left(3\sin\frac{\pi x}{l}-\sin\frac{3\pi x}{l}\right)=\sum_{n=1}^{\infty} b_n \sin\frac{n\pi x}{l}$$

$$\Rightarrow \quad \frac{y_0}{4}\left(3\sin\frac{\pi x}{l}-\sin\frac{3\pi x}{l}\right)=b_1\sin\frac{\pi x}{l}+b_2\sin\frac{2\pi x}{l}+b_3\sin\frac{3\pi x}{l}+\ldots$$

On equating the coefficients of sin on both the sides, we get

$$b_1=\frac{3y_0}{4},\quad b_3=-\frac{y_0}{4}$$

and all others b's are zero.
Hence (5) becomes

$$y=\frac{y_0}{4}\left(3\sin\frac{\pi x}{l}\cos\frac{c\pi t}{l}-\sin\frac{3\pi x}{l}\cos\frac{3c\pi t}{l}\right) \quad \textbf{Ans.}$$

Example 14. *If a string of length l is initially at rest in equilibrium position and each of its points is given the velocity*

$$\left(\frac{\partial y}{\partial t}\right)_{t=0}=b\sin^3\frac{\pi x}{l},\ \textit{find the displacement } y(x,t).$$

(*U.P. II Semester summer 2001, 2006*)

Solution. The equation for the vibrations of the string is

$$\frac{\partial^2 y}{\partial t^2}=c^2\frac{\partial^2 y}{\partial x^2} \quad \ldots(1)$$

The solution of equation (1) is

$$y(x,t)=(c_1\cos cpt+c_2\sin cpt)(c_3\cos px+c_4\sin px) \quad \ldots(2)$$

Boundary conditions are,

$$y(0,t)=0 \quad \ldots(3)$$

$$y(l,t)=0 \quad \ldots(4)$$

$$y(x,0)=0 \quad \ldots(5)$$

$$\left(\frac{\partial y}{\partial t}\right)=b\sin^3\frac{\pi x}{l}\ \text{at } t=0 \quad \ldots(6)$$

Putting $x = 0$ and $y = 0$ in (2), we get

$$0=(c_1\cos cpt+c_2\sin cpt)c_3 \Rightarrow c_3=0$$

Putting the value of c_3 in (2), we get

$$y=(c_1\cos cpt+c_2\sin cpt)c_4\sin px \quad \ldots(7)$$

Putting $x = l$ and $y = 0$ in (7), we get

$$0=(c_1\cos cpt+c_2\sin cpt)c_4\sin pl$$

$$\Rightarrow \sin pl=0=\sin n\pi\ (n\in I) \quad \Rightarrow \quad p=\frac{n\pi}{l}$$

Putting the value of p in (7), we get

$$y=\left(c_1\cos\frac{n\pi ct}{l}+c_2\sin\frac{n\pi ct}{l}\right)c_4\sin\frac{n\pi x}{l} \quad \ldots(8)$$

Putting $t = 0$ and $y = 0$ in (8), we get

$$0=c_1c_4\sin\frac{n\pi x}{l} \Rightarrow c_1=0$$

Putting the value of c_1 in (8), we get

$$y = c_2 c_4 \sin\frac{n\pi ct}{l}\sin\frac{n\pi x}{l}$$

$$\Rightarrow y = b_n \sin\frac{n\pi ct}{l}\sin\frac{n\pi x}{l} \qquad [\text{where } c_2c_4 = b_n]$$

The general solution is $y = \sum_1^\infty b_n \sin\frac{n\pi ct}{l}\sin\frac{n\pi x}{l}$ ...(9)

On differentiating (9) w.r.t. t, we get

$$\frac{\partial y}{\partial t} = \sum_1^\infty b_n.\frac{n\pi c}{l}\cos\frac{n\pi ct}{l}\sin\frac{n\pi x}{l} \qquad ..(10)$$

On putting the values of $\frac{\partial y}{\partial t} = b\sin^3\frac{\pi x}{l}$ and $t = 0$ in (10), we get

$$b\sin^3\frac{\pi x}{i} = \sum_1^\infty b_n.\frac{n\pi c}{l}\sin\frac{n\pi x}{l}$$

$$\Rightarrow \frac{b}{4}\left[3\sin\frac{\pi x}{l} - \sin\frac{3\pi x}{l}\right] = b_1\frac{\pi c}{l}\sin\frac{\pi x}{l} + \frac{2b_2\pi c}{l}\sin\frac{2\pi x}{l} + 3b_3\frac{\pi c}{l}\sin\frac{3\pi c}{l} + \ldots$$

Equating the coefficients of $\sin\frac{\pi x}{l}, \sin\frac{2\pi x}{l}. \sin\frac{3\pi x}{l}$, we get

$$\frac{3b}{4} = b_1\frac{\pi c}{l} \quad \Rightarrow \quad b_1 = \frac{3bl}{4\pi c}$$

$$b_2 = 0 \text{ and } \frac{3b_3\pi c}{l} = -\frac{b}{4} \quad \Rightarrow \quad b_3 = -\frac{bl}{12\pi c}$$

Also, $b_4 = 0 = b_5 = \ldots$ *etc.*,

Putting the values of b's in (9), we get

$$y(x,t) = \frac{3bl}{4\pi c}\sin\frac{\pi ct}{l}\sin\frac{\pi x}{l} - \frac{bl}{12\pi c}\sin\frac{3\pi ct}{l}\sin\frac{3\pi x}{l}$$

$$= \frac{bl}{12\pi c}\left[9\sin\frac{\pi x}{l}\sin\frac{\pi ct}{l} - \sin\frac{3\pi x}{l}\sin\frac{3\pi ct}{l}\right]. \qquad \textbf{Ans.}$$

Example 15. *Solve the wave equation*

$$\frac{\partial^2 u}{\partial t^2} = a^2\frac{\partial^2 u}{\partial x^2}$$

under the conditions :

$$u = 0 \text{ when } x = 0 \text{ and } x = \pi$$

$$\frac{\partial u}{\partial t} = 0 \text{ when } t = 0 \text{ and } u(x, 0) = x,\ 0 < x < \pi.$$

Solution. The solution is of the form

$$u(x, t) = (c_1 \cos px + c_2 \sin px)(c_3 \cos apt + c_4 \sin apt) \qquad \ldots (1)$$

On putting $u = 0$ and $x = 0$ in (1), we get

$$0 = c_1(c_3 \cos apt + c_4 \sin a\,pt) \qquad \Rightarrow c_1 = 0$$

On putting $c_1 = 0$ in (1), we get

$$u(x,t) = c_2 \sin px(c_3 \cos apt + c_4 \sin apt) \qquad \ldots(2)$$

On putting $x = \pi$ and $u = 0$ in (2), we get

$$0 = c_2 \sin p\pi(c_3 \cos apt + c_4 \sin apt) \Rightarrow \sin p\pi = 0 = \sin n\pi \text{ or } p = n$$

On putting $p = n$ in (2), we get

$$u(x, t) = c_2 \sin nx(c_3 \cos ant + c_4 \sin ant)$$

$$u(x, t) = \sin nx(b_1 \cos ant + b_2 \sin ant) \quad [c_2 c_3 = b_1 \text{ and } c_2 c_4 = b_2] \quad \dots (3)$$

On differentiating w.r.t. to t, we get

$$\frac{\partial u}{\partial t} = \sin nx(-ab_1 n \sin ant + ab_2 n \cos ant) \quad \dots (4)$$

On putting $\frac{\partial u}{\partial t} = 0$ and $t = 0$ in (4), we have

$$0 = \sin nx(ab_2 n) \Rightarrow b_2 = 0$$

On putting $b_2 = 0$ in (3), we get

$$u(x, t) = \sin nx(b_1 \cos ant) \quad \dots (5)$$

General solution is

$$u(x, t) = \sum_{n=1}^{\infty} b_n \sin nx \cos ant \quad \dots (6)$$

On putting $u = x$ and $t = 0$ in (6), we get

$$x = \sum_{n=1}^{\infty} b_n \sin nx \quad \text{where } b_n = \frac{2}{\pi}\int_0^{\pi} x \sin nx \, dx$$

$$= \frac{2}{\pi}\left[x\left(-\frac{\cos nx}{n}\right) - 1\left(-\frac{\sin nx}{n^2}\right)\right]_0^{\pi}$$

$$= \frac{2}{\pi}\left[-\frac{\pi}{n}\cos n\pi\right] = -\frac{2}{n}(-1)^n$$

On putting the value of b_n in (6), we get

$$u(x, t) = -2\sum_{n=1}^{\infty} \frac{(-1)^n}{n} \sin nx \cos nat \quad \textbf{Ans.}$$

Example 16. *A string is stretched and fastened to two points l apart. Motion is started by displacing the string into the form $y = k(lx - x^2)$ from which it is released at time $t = 0$. Find the displacement of any point on the string at a distance of x from one end at time t.*

(U.P., III Semester, Summer 2002; A.M.I.E.T.E., Summer 2000)

Solution. The vibration of the string is given by

$$\frac{\partial^2 y}{\partial t^2} = c^2 \frac{\partial^2 y}{\partial x^2} \quad \dots (1)$$

As the end points of the string are fixed for all time,

$$y(0, t) = 0 \quad \dots (2)$$

and

$$y(l, t) = 0 \quad \dots (3)$$

since the initial transverse velocity of any point of the string is zero, therefore,

$$\left(\frac{\partial y}{\partial t}\right)_{t=0} = 0 \quad \dots (4)$$

and $$y(x, 0) = k(lx - x^2) \qquad \ldots (5)$$

Solution of (1) is $$y = (c_1 \cos px + c_2 \sin px)(c_3 \cos c\, pt + c_4 \sin cpt) \qquad \ldots (6)$$

On putting $x = 0$ and $y = 0$ in (6), we get

$$0 = c_1 (c_3 \cos c\, pt + c_4 \sin cpt) \quad \Rightarrow \quad c_1 = 0$$

On putting $c_1 = 0$ in (6), we get

$$y = c_2 \sin px(c_3 \cos cpt + c_4 \sin cpt) \qquad \ldots (7)$$

On differentiating (7) w.r.t. t, we get

$$\frac{\partial y}{\partial t} = c_2 \sin px(-c_3 cp \sin cpt + c_4 cp \cos cpt) \qquad \ldots (8)$$

On putting $\left(\frac{\partial y}{\partial t}\right) = 0$ and $t = 0$ in (8), we get

$$0 = c_2 \sin px(c_4 cp) \quad \Rightarrow \quad \therefore\ c_4 = 0 \text{ since } c_2 \neq 0$$

On putting $c_4 = 0$ in (7), we get

$$y = c_2 \sin px(c_3 \cos cpt)$$

$$y = c_2 c_3 \sin px \cos cpt \qquad \ldots (9)$$

On putting $x = l$ and $y = 0$ in equation (9), we get

$$0 = c_2 c_3 \sin pl \cos cpt \quad \text{or} \quad 0 = \sin pl$$

$$\Rightarrow \quad \sin n\pi = \sin pl \quad \text{or} \quad pl = n\pi, \quad p = \frac{n\pi}{l} \qquad \text{where } n = 1, 2, 3\ldots\ldots$$

On putting $p = \frac{n\pi}{l}$, equation (9) becomes

$$\Rightarrow \quad y = b_n \sin\frac{n\pi x}{l} \cos\frac{n\pi c}{l} t \qquad [c_2 c_3 = b_n]$$

We can have any number of solutions by taking different integral values of n and the complete solution will be the sum of these solutions. Thus,

$$y(x, t) = \sum_{n=1}^{\infty} b_n \sin\frac{n\pi x}{l} \cos\frac{n\pi c}{l} t \qquad \ldots (10)$$

On putting $t = 0$ and $y = k(x - x^2)$ in (10), we get

$$k\,(l\,x - x^2) = \sum_{n=1}^{\infty} b_n \sin\frac{n\pi x}{l} \qquad \ldots (11)$$

Now it is clear that (11) represents the expansion of $f(x)$ in the form of a Fourier sine series and consequently

$$b_n = \frac{2}{l} \int_0^l f(x) \sin\frac{n\pi x}{l} dx \qquad \ldots (12)$$

$$b_n = \frac{2}{l} \int_0^l k\,(lx - x^2) \sin\frac{n\pi x}{l} dx$$

$$= \frac{2k}{l}\left[(lx - x^2)\left(-\cos\frac{n\pi x}{l}\right)\frac{l}{n\pi} - (l - 2x)\left(-\sin\frac{n\pi x}{l}\right)\frac{l^2}{n^2\pi^2} + (-2)\left(\cos\frac{n\pi x}{l}\right)\frac{l^3}{n^3\pi^3}\right]_0^l$$

$$= \frac{2k}{l}\left[(-1)^{n+1}\frac{2l^3}{n^3\pi^3}+\frac{2l^3}{n^3\pi^3}\right]=\frac{8l^2k}{n^3\pi^3} \text{ when } n \text{ is odd.}$$

$= 0,$ when n is even

Putting the value of b_n in (10), we get

$$y=\sum_{n=1}^{\infty}\frac{8l^2k}{n^3\pi^3}\sin\frac{n\pi x}{l}\cos\frac{n\pi c}{l}t, \text{ when } n \text{ is odd.} \qquad \textbf{Ans.}$$

Example 17. *A string is stretched between the fixed points (0, 0) and (l, 0) and released at rest from the initial deflection given by*

$$f(x)=\begin{cases}\dfrac{2kx}{l} & \text{when } 0<x<\dfrac{l}{2}\\[2mm] \dfrac{2k}{l}(l-x) & \text{when } \dfrac{l}{2}<x<l\end{cases}$$

Find the deflection of the string at any time t.

Solution. As we have done in the example 16 (see equation 10)

$$y(x,t)=\sum_{n=1}^{\infty}b_n\cos\frac{n\pi ct}{l}\sin\frac{n\pi x}{l} \qquad \dots (1)$$

$$b_n=\frac{2}{l}\int_0^l f(x)\sin\frac{n\pi x}{l}dx$$

$$b_n=\frac{2}{l}\left[\int_0^{l/2}\frac{2kx}{l}\sin\frac{n\pi x}{l}dx+\int_{\frac{l}{2}}^{l}\frac{2k}{l}(l-x)\sin\frac{n\pi x}{l}dx\right]$$

$$=\frac{4k}{l^2}\left[x\left(-\frac{l}{n\pi}\cos\frac{n\pi x}{l}\right)-1\left(-\frac{l^2}{n\pi^2}\sin\frac{n\pi x}{l}\right)\right]_0^{\frac{l}{2}}$$

$$+\frac{4k}{l^2}\left[(l-x)\left(-\frac{l}{n\pi}\cos\frac{n\pi x}{l}\right)-(-1)\left(\frac{-l^2}{n^2\pi^2}\sin\frac{n\pi x}{l}\right)\right]_{\frac{l}{2}}^{l}$$

$$=\frac{4k}{l^2}\left[-\frac{l}{2}\frac{l}{n\pi}\cos\frac{n\pi}{2}+\frac{l^2}{n^2\pi^2}\sin\frac{n\pi}{2}\right]+\frac{4k}{l^2}\left[\frac{l}{2}\frac{l}{n\pi}\cos\frac{n\pi}{2}+\frac{l^2}{n^2\pi^2}\sin\frac{n\pi}{2}\right]$$

$$=\frac{8k}{l^2}\frac{l^2}{n^2\pi^2}\sin\frac{n\pi}{2}=\frac{8k}{n^2\pi^2}\sin\frac{n\pi}{2}$$

On putting the value of b_n in (1) we get

$$y=\sum_{n=1}^{\infty}\frac{8k}{n^2\pi^2}\sin\frac{n\pi}{2}\cos\frac{n\pi ct}{l}\sin\frac{n\pi x}{l} \qquad \textbf{Ans.}$$

Example 18. *A taut string of length 2l is fastened at both ends. The mid point of the string is taken to a height b and then released from the rest in that position. Find the displacement of the string.*

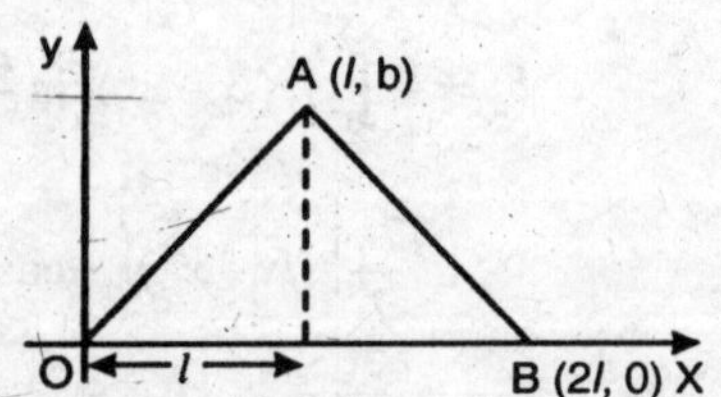

Solution. Taking an end as origin, the boundary conditions are

$$y(0, t) = 0, \qquad t \ge 0 \qquad \text{... (1)}$$

$$y(2l, t) = 0, \qquad t \ge 0 \qquad \text{... (2)}$$

$$\left(\frac{\partial y}{\partial t}\right)_{t=0} = 0,$$

OA passes through (0, 0) and (l, b)

Equation of OA is $y(x, 0) = \frac{b}{l}x,\ 0 \le x \le l, \quad \Rightarrow \quad y = \frac{bx}{l}$

Equation of AB is $y = -\frac{b}{l}(x - 2l),\ l \le x \le 2l$ ...(3) $\left[y - y_1 = \frac{y_2 - y_1}{x_2 - x_1}(x - x_1)\right]$

$\left(\text{Equation of } AB,\ y - 0 = \frac{b-0}{l-2l}(x - 2l)\right)$

Equation of the vibrating string is $\frac{\partial^2 y}{\partial t^2} = c^2 \frac{\partial^2 y}{\partial x^2}$ ... (4)

Starting with the solution of (4), we get

$$y(x, t) = (c_1 \cos px + c_2 \sin px)(c_3 \cos pct + c_4 \sin pct) \qquad \text{... (5)}$$

On putting $x = 0$, $y = 0$ in (5), we get

$$0 = c_1 (c_3 \cos pct + c_4 \sin pct) \Rightarrow c_1 = 0$$

On putting $c_1 = 0$ in (5), we get

$$y(x, t) = c_2 \sin px (c_3 \cos pct + c_4 \sin pct) \qquad \text{... (6)}$$

On putting $y = 0$, $x = 2l$ in (6), we get

$$0 = c_2 \sin p(2l)\ (c_3 \cos pct + c_4 \sin pct) \Rightarrow \sin 2pl = 0 = \sin n\pi \Rightarrow 2pl = n\pi,\ p = \frac{n\pi}{2l}$$

Substituting the value of p in (6), we have

$$y(x, t) = c_2 \sin\frac{n\pi}{2l}x\left(c_3 \cos\frac{n\pi ct}{2l} + c_4 \sin\frac{n\pi ct}{2l}\right) \qquad \text{... (7)}$$

On differentiating (7) w.r.t., t, we get

$$\frac{\partial y}{\partial t} = c_2 \sin\frac{n\pi x}{2l}\left[-c_3 \frac{n\pi c}{2l}\sin\frac{n\pi ct}{2l} + c_4 \frac{n\pi c}{2l}\cos\frac{n\pi ct}{2l}\right] \qquad \text{... (8)}$$

Putting $\left(\frac{\partial y}{\partial t}\right) = 0$, $t = 0$ in (8), we get

$$0 = c_2 \sin\frac{n\pi x}{2l}\left[0 + c_4 \frac{n\pi c}{2l}\right] \Rightarrow c_4 = 0$$

Putting $c_4 = 0$ in (7), we have

$$y(x, t) = \left(c_2 \sin\frac{n\pi x}{2l}\right)\left(c_3 \cos\frac{n\pi ct}{2l}\right)$$

General solution is

$$y(x,t) = \sum_{n=1}^{\infty} b_n \sin\frac{n\pi x}{2l}\cos\frac{n\pi ct}{2l} \qquad (c_2 c_3 = b_n) \qquad \text{... (9)}$$

$$\Rightarrow \quad f(x) = \begin{cases} \dfrac{bx}{l}, & 0 \le x \le l \\[2mm] \dfrac{-b}{l}(x-2l), & l \le x \le 2l \end{cases}$$

$$b_n = \frac{2}{2l}\int_0^{2l} f(x)\sin\frac{n\pi x}{2l}\,dx$$

$$= \frac{1}{l}\int_0^{l} f(x)\sin\frac{n\pi x}{2l}\,dx + \frac{1}{l}\int_l^{2l} f(x)\sin\frac{n\pi x}{2l}\,dx$$

$$= \frac{1}{l}\int_0^{l} \frac{bx}{l}\sin\frac{n\pi x}{2l}\,dx + \frac{1}{l}\int_l^{2l} \frac{-b}{l}(x-2l)\sin\frac{n\pi x}{2l}\,dx$$

$$= \frac{1}{l}\left[\frac{bx}{l}\frac{2l}{n\pi}\left(-\cos\frac{n\pi x}{2l}\right) - \frac{b}{l}\left(-\frac{4l^2}{n^2\pi^2}\sin\frac{n\pi x}{2l}\right)\right]_0^l$$

$$+ \frac{1}{l}\left[\frac{-b}{l}(x-2l)\left(-\frac{2l}{n\pi}\cos\frac{n\pi x}{2l}\right) - \left(-\frac{b}{l}\right) - \frac{4l^2}{n^2\pi^2}\sin\frac{n\pi x}{2l}\right]_l^{2l}$$

$$= \frac{b}{l^2}\left[\frac{-2l^2}{n\pi}\cos\frac{n\pi}{2} + \frac{4l^2}{n^2\pi^2}\sin\frac{n\pi}{2} + \frac{2l^2}{n\pi}\cos\frac{n\pi}{2} + \frac{4l^2}{n^2\pi^2}\sin\frac{n\pi}{2}\right]$$

$$= \frac{b}{l^2}\left[\frac{8l^2}{n^2\pi^2}\sin\frac{n\pi}{2}\right] = \frac{8b}{n^2\pi^2}\sin\frac{n\pi}{2}$$

$= 0$ for n even

$= \dfrac{8b}{n^2\pi^2}\sin\dfrac{n\pi}{2}$ for n odd.

Substituting the value of b_n in (9), we get

$$y(x,t) = \frac{8b}{\pi^2}\sum_{n=1}^{\infty}\frac{1}{(2n-1)^2}\sin(2n-1)\frac{\pi}{2}\sin\frac{(2n-1)\pi x}{2l}\cos\frac{(2n-1)\pi ct}{2l} \qquad \textbf{Ans.}$$

Example 19. *The points of trisection of a string are pulled aside through the same distance on opposite sides of the position of equilibrium and the string is released from rest. Derive an expression for the displacement of the string at subsequent time and show that the mid-point of the string always remains at rest.*

Solution. Let the string OA (l) be trisected at B and C. Initially the string is held in the form $OB'C'A$, where $BB' = CC' = a$

The equation of vibrating string is

$$\frac{\partial^2 y}{\partial t^2} = c^2\frac{\partial^2 y}{\partial x^2} \quad \text{...(1)}$$

Boundary conditions are

$$y(0,t) = 0,\; y(l,t) = 0,\; \left(\frac{\partial y}{\partial t}\right)_{t=0} = 0$$

Fourth condition is that at $t = 0$, the string rests in the form $OB'C'A$.

Equation of OB' is $y = \dfrac{a}{l/3}x \Rightarrow y = \dfrac{3a}{l}x$ $\qquad (y = mx)$

Equation of $B'C'$ is $y-a=\frac{a+a}{\frac{l}{3}-\frac{2l}{3}}\left(x-\frac{l}{3}\right)$ $\left[y-y_1=\frac{y_2-y_1}{x_2-x_1}(x-x_1)\right]$

i.e., $y-a=\frac{2a}{\frac{-l}{3}}\left(x-\frac{l}{3}\right) \Rightarrow y-a=-\frac{6a}{l}\left(x-\frac{l}{3}\right)$

$\Rightarrow y=a-\frac{6ax}{l}+2a \Rightarrow y=3a-\frac{6ax}{l}$

$\Rightarrow \qquad y=\frac{3a}{l}(l-2x)$

Equation of $C'A$ is $y-0=\frac{-a-0}{\frac{2l}{3}-l}(x-l) \qquad \Rightarrow \quad y=\frac{3a}{l}(x-l)$

Hence fourth condition is

$$f(x)=\begin{cases}\frac{3a}{l}x, & 0\le x\le\frac{l}{3}\\ \frac{3a}{l}(l-2x), & \frac{l}{3}\le x\le\frac{2l}{3}\\ \frac{3a}{l}(x-l), & \frac{2l}{3}\le x\le l\end{cases}$$

General solution of (1) is

$$y(x,t)=\sum_{n=1}^{\infty} b_n\sin\frac{n\pi x}{l}\cos\frac{n\pi ct}{l} \qquad \text{... (2)}$$

On putting $t=0$ in (2), we get

$$y(x,0)=\sum_{n=1}^{\infty} b_n\sin\frac{n\pi x}{l}$$

We have to find the value of b_n by Fourier half range formula.

$$b_n=\frac{2}{l}\int_0^l f(x)\sin\frac{n\pi x}{l}\,dx$$

$$=\frac{2}{l}\left[\int_0^{\frac{l}{3}}\frac{3ax}{l}\sin\frac{n\pi x}{l}dx+\int_{\frac{l}{3}}^{\frac{2l}{3}}\frac{3a}{l}(l-2x)\sin\frac{n\pi x}{l}dx+\int_{\frac{2l}{3}}^{l}\frac{3a}{l}(x-l)\sin\frac{n\pi x}{l}dx\right]$$

$$=\frac{6a}{l^2}\left[x\left(-\frac{l}{n\pi}\cos\frac{n\pi x}{l}\right)-(1)\left(-\frac{l^2}{n^2\pi^2}\sin\frac{n\pi x}{l}\right)\right]_0^{\frac{l}{3}}+\frac{6a}{l^2}\left[(l-2x)\left(\frac{-l}{n\pi}\cos\frac{n\pi x}{l}\right)\right.$$

$$\left.-(-2)\left(\frac{-l^2}{n^2\pi^2}\sin\frac{n\pi x}{l}\right)\right]_{\frac{l}{3}}^{\frac{2l}{3}}+\frac{6a}{l^2}\left[(x-l)\left(\frac{-l}{n\pi}\cos\frac{n\pi x}{l}\right)-(1)\left(\frac{-l^2}{n^2\pi^2}\sin\frac{n\pi x}{l}\right)\right]_{\frac{2l}{3}}^{l}$$

$$=\frac{6a}{l^2}\left[\left(-\frac{l^2}{3n\pi}\cos\frac{n\pi}{3}+\frac{l^2}{n^2\pi^2}\sin\frac{n\pi}{3}\right)+\frac{l^2}{3n\pi}\cos\frac{2n\pi}{3}-\frac{2l^2}{n^2\pi^2}\sin\frac{2n\pi}{3}\right.$$

$$\left.+\frac{l^2}{3n\pi}\cos\frac{n\pi}{3}+\frac{2l^2}{n^2\pi^2}\sin\frac{n\pi}{3}-\left(\frac{l^2}{3n\pi}\cos\frac{2n\pi}{3}+\frac{l^2}{n^2\pi^2}\sin\frac{2n\pi}{3}\right)\right]$$

$$= \frac{6a}{l^2}\cdot\frac{3l^2}{n^2\pi^2}\left(\sin\frac{n\pi}{3} - \sin\frac{2n\pi}{3}\right)$$

$$= \frac{18a}{n^2\pi^2}\sin\frac{n\pi}{3}[1+(-1)^n] \qquad \left[\sin\frac{2n\pi}{3} = \sin\left(n\pi - \frac{n\pi}{3}\right) = -(-1)^n\sin\frac{n\pi}{3}\right]$$

$$= 0 \text{ when } n \text{ is odd}$$

$$= \frac{36a}{n^2\pi^2}\sin\frac{n\pi}{3} \text{ when } n \text{ is even}$$

On substituting the value of b_n in (2) we get

$$y(x,t) = \sum_{n=2}^{\infty}\frac{36a}{n^2\pi^2}\sin\frac{n\pi}{3}\sin\frac{n\pi x}{l}\cos\frac{n\pi ct}{l} \qquad \text{...(3) } (n \text{ is even}) \textbf{ Ans.}$$

Putting $x = \frac{l}{2}$ in equation (3), we get

$$y\left(\frac{l}{2}, l\right) = \frac{36a}{n^2\pi^2}\sin\frac{n\pi}{3}\sin\frac{n\pi}{2}\cos\frac{n\pi ct}{l} \qquad \left[\sin\frac{n\pi}{2} = 0 \text{ as } n \text{ is even}\right]$$

Hence, mid-point of the string is always at rest. **Proved.**

11.4 SOLUTION OF WAVE EQUATION BY D'ALEMBERT'S METHOD

$$\frac{\partial^2 y}{\partial t^2} = c^2\frac{\partial^2 y}{\partial x^2} \qquad \text{...(1)}$$

Let us introduce the two new independent variables $u = x + ct$, $v = x - ct$

So that y becomes a function of u and v. Then,

$$\frac{\partial y}{\partial x} = \frac{\partial y}{\partial u}\frac{\partial u}{\partial x} + \frac{\partial y}{\partial v}\frac{\partial v}{\partial x} = \frac{\partial y}{\partial u}(1) + \frac{\partial y}{\partial v}(1) = \frac{\partial y}{\partial u} + \frac{\partial y}{\partial v}$$

$$\frac{\partial}{\partial x} \equiv \frac{\partial}{\partial u} + \frac{\partial}{\partial v}$$

$$\frac{\partial^2 y}{\partial x^2} = \frac{\partial}{\partial x}\left(\frac{\partial y}{\partial x}\right) = \left(\frac{\partial}{\partial u} + \frac{\partial}{\partial v}\right)\left(\frac{\partial y}{\partial u} + \frac{\partial y}{\partial v}\right)$$

$$= \frac{\partial}{\partial u}\left(\frac{\partial y}{\partial u} + \frac{\partial y}{\partial v}\right) + \frac{\partial}{\partial v}\left(\frac{\partial y}{\partial u} + \frac{\partial y}{\partial v}\right) = \frac{\partial^2 y}{\partial u^2} + 2\frac{\partial^2 y}{\partial u\partial v} + \frac{\partial^2 y}{\partial v^2} \qquad \text{... (2)}$$

$$\frac{\partial y}{\partial t} = \frac{\partial y}{\partial u}\frac{\partial u}{\partial t} + \frac{\partial y}{\partial v}\frac{\partial v}{\partial t} = \frac{\partial y}{\partial u}c + \frac{\partial y}{\partial v}(-c) = c\left(\frac{\partial y}{\partial u} - \frac{\partial y}{\partial v}\right) \qquad \left[\because \frac{\partial u}{\partial t} = c,\ \frac{\partial v}{\partial t} = -c\right]$$

$$\frac{\partial}{\partial t} \equiv c\left(\frac{\partial}{\partial u} - \frac{\partial}{\partial v}\right)$$

$$\frac{\partial^2 y}{\partial t^2} = \frac{\partial}{\partial t}\left(\frac{\partial y}{\partial t}\right) = c\left(\frac{\partial}{\partial u} - \frac{\partial}{\partial v}\right)c\left(\frac{\partial y}{\partial u} - \frac{\partial y}{\partial v}\right)$$

$$= c^2\left(\frac{\partial^2 y}{\partial u^2} - 2\frac{\partial^2 y}{\partial u\partial v} + \frac{\partial^2 y}{\partial v^2}\right) \qquad \text{... (3)}$$

Substituting the values of $\frac{\partial^2 y}{\partial x^2}$ and $\frac{\partial^2 y}{\partial t^2}$ from (2) and (3) in (1), we get

$$c^2\left(\frac{\partial^2 y}{\partial u^2}-2\frac{\partial^2 y}{\partial u \partial y}+\frac{\partial^2 y}{\partial v^2}\right)=c^2\left(\frac{\partial^2 y}{\partial u^2}+2\frac{\partial^2 y}{\partial u \partial v}+\frac{\partial^2 y}{\partial v^2}\right) \Rightarrow \frac{\partial^2 y}{\partial u \partial v}=0 \quad \text{... (4)}$$

Integrating (4) *w.r.t.* v, we get $\frac{\partial y}{\partial u}=f(u)$ ... (5)

where $f(u)$ is constant in respect of v. Again integrating (5) *w.r.t.* 'u' we get

$$y=\int f(u)\,du+\psi(v)$$

where $\psi(v)$ is constant in respect of u

$$y=\phi(u)+\psi(v), \text{ where } \phi(u)=\int f(u)\,du$$

$$\Rightarrow \quad y(x,t)=\phi(x+ct)+\psi(x-ct) \quad \text{... (6)}$$

This is D'Alemberts solution of wave equation (1)

To determine ϕ, ψ, let us apply initial conditions, $y(x,0)=f(x)$ and $\frac{\partial y}{\partial t}=0$ when $t=0$.

Differentiating (6) *w.r.t.* "t", we get

$$\frac{\partial y}{\partial t}=c\phi'(x+ct)-c\psi'(x-ct) \quad \text{... (7)}$$

Putting $\frac{\partial y}{\partial t}=0$, and $t=0$ in (7) we get $0=c\phi'(x)-c\psi'(x)$

$$\Rightarrow \quad \phi'(x)=\psi'(x) \text{ or } \phi(x)=\psi(x)+b \quad \text{... (8)}$$

Again substituting $y=f(x)$ and $t=0$ in (6) we get

$$f(x)=\phi(x)+\psi(x) \text{ or } f(x)=[\psi(x)+b]+\psi(x) \quad \text{[Using (8)]}$$

$$\Rightarrow \quad f(x)=2\psi(x)+b$$

So that $\psi(x)=\frac{1}{2}[f(x)-b]$ and $\phi(x)=\frac{1}{2}[f(x)+b]$

On putting the values of $\phi(x+ct)$ and $\psi(x-ct)$ in (6), we get

$$y(x,t)=\frac{1}{2}[f(x+ct)+b]+\frac{1}{2}[f(x-ct)-b]$$

$$\Rightarrow \quad y(x,t)=\frac{1}{2}[f(x+ct)+f(x-ct)] \quad \textbf{Ans.}$$

EXERCISE 11.2

1. Find the solution of the equation of a vibrating string of length l satisfying the initial conditions:

$y = f(x)$ when $t = 0$, and $\frac{\partial y}{\partial t} = g(x)$ when $t = 0$.

It is assumed that the equation of a vibrating string is $\frac{\partial^2 y}{\partial t^2} = a^2 \frac{\partial^2 y}{\partial x^2}$.

Ans. $y(x,t) = \sum_{n=1}^{\infty} \left(b_n \cos \frac{n\pi at}{l} + C_n \sin \frac{n\pi at}{l} \right) \sin \frac{n\pi}{l} x$

where $b_n = \frac{2}{l} \int_0^l f(x) \sin \frac{n\pi x}{l} dx$, $c_n = \frac{2}{cn\pi} \int_0^l g(x) \sin \frac{n\pi x}{l} dx$.

2. A tightly stretched string with fixed end points $x = 0$ and $x = l$ is initially at rest in its equilibrium position. If it is set vibrating by giving to each of its points a velocity $\lambda x(l-x)$, find the displacement of the string at any distance x from one end at any time t.

Ans. $y = \frac{8\lambda l^3}{c\pi^4} \sum_{(n=1)}^{n} \frac{1}{(2n-1)^4} \sin \frac{(2n-1)\pi x}{l} \sin \frac{(2n-1)\pi ct}{l}$

3. Show that the solution of the equation of a vibrating string of length l, satisfying the conditions $y(0, t) = y(l, t) = 0$ and $y = f(x), \frac{\partial y}{\partial t} = g(u)$ at $t = 0$ is $y(x,t) = \sum_{n=1}^{\infty} b_n \sin \frac{n\pi x}{l} \cos \frac{n\pi ct}{l}$.

4. A tightly stretched string of length l fastened at both ends, is disturbed from the position of equilibrium imparting to each of its points an initial velocity of magnitude $f(x)$. Show that the solution of the problem.

$$u(x,t) = \frac{2}{l} \sum_{n=1}^{\infty} \left[\int_0^l f(x) \sin \frac{n\pi}{l} x \, dx \right] \sin \frac{n\pi x}{l} \cos \frac{n\pi at}{l}$$

5. Find the solution of the equation $\frac{\partial^2 y}{\partial x^2} = \frac{1}{c^2} \frac{\partial^2 y}{\partial t^2}$ subject to the boundary conditions

$y(0, t) = 0, y(l, t) = 0, y(x, 0) = \phi(x), \frac{\partial y}{\partial t}(x, 0) = \psi(x)$ **Ans.** $y = \phi(x).\cos \frac{n\pi ct}{l} + \frac{l\psi(x)}{n\pi c} . \frac{\sin \frac{n\pi ct}{l}}{\sin \frac{n\pi x}{l}}$

6. The vibration of an elastic string of length l are governed by the one-dimensional wave equation $\frac{\partial^2 u}{\partial t^2} = c^2 \frac{\partial^2 u}{\partial x^2}$. The string is fixed at the ends.

$u(0,t) = 0 = u(l,t)$ for all t. The initial deflection is $u(x,0) = \begin{cases} x, & 0 < x < \frac{l}{2} \\ l - x, & \frac{l}{2} \le x \le l \end{cases}$

and the initial velocity is zero. Find the deflection of the string at any instant of time.

(A.M.I.E.T.E., Summer 2001)

Ans. $\frac{4l}{\pi^2} \sum_{n=1}^{\infty} \frac{\sin \frac{n\pi}{2}}{n^2} \sin \frac{n\pi x}{l} \cos \frac{n\pi ct}{l}$

7. A tightly stretched violin string of length l is fixed at both ends and is plucked at $x = \frac{l}{3}$ and assumes initially the shape of a triangle of height a. Find the displacement y at any distance x and any time t after the string is released from rest.

11.5 ONE DIMENSIONAL HEAT FLOW

In this article, we shall consider the flow of heat and the accompanying variation of temperature with position and with time in conducting solids the following empirical laws are taken as the basis of investigation.

1. Heat flows from a higher to lower temperature.
2. The amount of heat required to produce a given temperature change in a body is proportional to the mass of the body and to the temperature change. This constant of proportionality is known as the specific heat (c) of the conducting material.
3. The rate at which heat flows through an area is proportional to the area and to the temperature gradient normal to the area. This constant of proportionality is known as the thermal conductivity (k) of the material.

Consider a bar or rod of homogeneous material of density $\rho(\text{gr/cm}^3)$ and having a constant cross-sectional area A (cm^2). We suppose that the sides of the bar are insulated and the loss of heat from the sides by conduction or radiations is negligible.

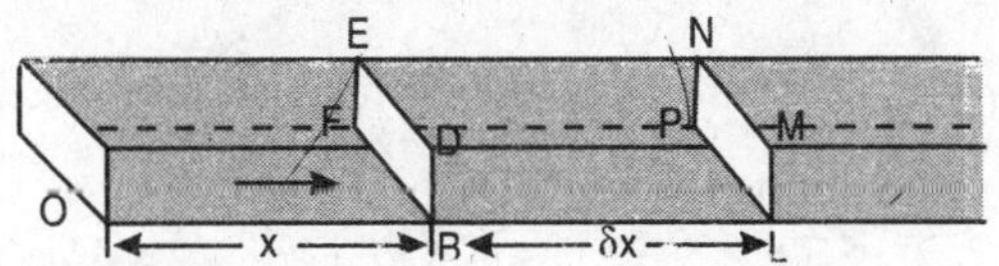

Take an end of the bar as the origin and the direction of heat flow as the positive x-axis.

Let c be the specific heat and k the thermal conductivity of the material.

Consider an element between two parallel sections $BDEF$ and $LMNP$ at distances x and $x + \delta x$ from the origin O, the sections being perpendicular to the x-axis. The mass of the element $= A\,\rho\,\delta\,x$.

Let $u\,(x, t)$ be the temperature at a distance x at time t. By the second law stated above, the rate of increase of heat in the element $= A\,\rho\,\delta\,x\,c\dfrac{\partial u}{\partial t}$. If R_1 and R_2 are respectively the rates (cal/sec) of inflow and outflow, for the sections $x = x$ and $x = x + \delta x$, then

$$R_1 = -kA\left(\frac{\partial u}{\partial x}\right)_x$$

and $R_2 = -kA\left(\dfrac{\partial u}{\partial x}\right)_{x+\delta x}$ the negative sign being due to the fact that heat flows from higher to lower temperature. $\left(i.e., \dfrac{\partial u}{\partial x}\text{ is negative}\right)$

Equating the rates of increase of heat from the two empirical laws.

$$A\rho c\delta x.\frac{\partial u}{\partial t} = R_1 - R_2$$

$$= kA\left[\left(\frac{\partial u}{\partial x}\right)_{x+\delta x} - \left(\frac{\partial u}{\partial x}\right)_x\right]$$

$$\therefore \qquad \frac{\partial u}{\partial t} = \frac{k}{\rho c}\left[\frac{\left(\frac{\partial u}{\partial x}\right)_{x+\delta x} - \left(\frac{\partial u}{\partial x}\right)_x}{\delta x}\right]$$

Taking the limit as $\delta x \to 0$ *i.e.* when $x + \delta x \to x$.

$$\frac{\partial u}{\partial t} = \frac{k}{\rho c} \underset{\delta x \to 0}{\text{Lt}} \left[\frac{\left(\frac{\partial u}{\partial x}\right)_{x+\delta x} - \left(\frac{\partial u}{\partial x}\right)_x}{\delta x}\right]$$

$$= \frac{k}{\rho c}\frac{\partial}{\partial x}\left(\frac{\partial u}{\partial x}\right)$$

i.e.,

$$\frac{\partial u}{\partial t} = \frac{k}{\rho c}\frac{\partial^2 u}{\partial x^2}$$

$\frac{k}{pc}$ is called the diffusivity (cm^2/sec.) of the substance. If we denote it by c^2, the above equation takes the form

$$\frac{\partial u}{\partial t} = c^2 \frac{\partial^2 u}{\partial x^2}$$

Example 20. *By the method of separation of variables solve the P.D.E.*

$$\frac{\partial u}{\partial t} = c^2 \frac{\partial^2 u}{\partial x^2}$$

Solution. Here, we have

$$\frac{\partial u}{\partial t} = c^2 \frac{\partial^2 u}{\partial x^2} \qquad \text{...(1)}$$

Let

$$u = XT \qquad \text{... (2)}$$

where X is the function of x only and T is the function of t only.

Differentiating (2) partially w.r.t. t, we get

$$\frac{\partial u}{\partial t} = \frac{\partial}{\partial t}(XT) = X\frac{\partial T}{\partial t}$$

$$\frac{\partial^2 u}{\partial x^2} = \frac{\partial^2}{\partial x^2}(XT) = \frac{\partial}{\partial x}\left(\frac{\partial XT}{\partial x}\right) = \frac{\partial}{\partial x}\left(T\frac{\partial X}{\partial x}\right) = T\frac{\partial^2 X}{\partial x^2}$$

Putting the values of $\frac{\partial u}{\partial t}$ and $\frac{\partial^2 u}{\partial x^2}$ in (1), we get

$$X\frac{\partial T}{\partial t} = c^2 T\frac{\partial^2 X}{\partial x^2}$$

Separating the variables, we get

$$\frac{1}{c^2}\frac{\frac{\partial T}{\partial t}}{T} = \frac{\frac{\partial^2 X}{\partial x^2}}{X}$$

L.H.S. is constant for x so we take $\frac{1}{c^2}\frac{\frac{\partial T}{\partial t}}{T} = -p^2$

R.H.S is constant for t so we take $\frac{\frac{\partial^2 X}{\partial x^2}}{X} = p^2$

$$\therefore \quad \frac{1}{c^2}\frac{\frac{\partial T}{\partial t}}{T} = \frac{\frac{\partial^2 X}{\partial x^2}}{X} = -p^2$$

$$\frac{1}{c^2}\frac{\frac{\partial T}{\partial t}}{T} = -p^2$$

$$\frac{1}{T}\frac{\partial T}{\partial t} = -p^2c^2$$

A.E. is $m = -p^2c^2$

$$\Rightarrow \quad T = c_1\, e^{-c^2p^2t}$$

$$\frac{\frac{\partial^2 X}{\partial x^2}}{X} = -p^2$$

$$D^2X = -p^2X$$

A.E. is $m^2 = -p^2$

$$\Rightarrow \quad m = ip$$

$$X = c_2 \cos px + c_3 \sin px$$

Putting the values of T and X in (1), we get

$$u = c_1\, e^{-c^2p^2t}\,(c_2 \cos px + c_3 \sin px) \qquad \textbf{Ans.}$$

Example 21. *Find the temperature in a bar of length 2 whose ends kept at zero and lateral surface insulated if the initial temperature is sin* $\frac{\pi x}{2} + 3\sin\frac{5\pi x}{2}$.

[*U.P. II Sem., 2007; U.P. II Semester, 2009*]

Solution. Here, we have

One-dimensional heat flow equation is

$$\frac{\partial u}{\partial t} = c^2\frac{\partial^2 u}{\partial x^2}$$

Its solution is

$$u(x, t) = (c_1 \cos px + c_2 \sin px)c_3 e^{-c^2p^2t} \qquad ...(1)$$

Putting $x = 0$, $u = 0$ in (1), we get

$$0 = c_1c_3e^{-c^2p^2t} \qquad \text{(given)}$$

$$\Rightarrow \quad c_1 = 0$$

Putting $c_1 = 0$, in (1), we get

$$u = c_2c_3 \sin px\, e^{-c^2p^2t} \qquad ...(2)$$

Putting $x = 2$, $u = 0$ in (2), we get (given)

$$0 = c_2c_3 \sin 2p\, e^{-c^2p^2t}$$

$$\Rightarrow \qquad \sin 2p = 0 = \sin n\pi$$

$$\Rightarrow \qquad p = \frac{n\pi}{2}, n \in \mathrm{I}$$

Putting the value of P in (2), we get

$$u = b_n \sin \frac{n\pi x}{2} e^{-\frac{n^2\pi^2c^2t}{4}} \qquad (c_2c_3 = b_n)$$

$$u = \sum_{n=1}^{\infty} b_n \sin \frac{n\pi x}{2} e^{-\frac{n^2\pi^2c^2t}{4}} \qquad \text{... (3) (General solution)}$$

Putting $t = 0$ and $u = \sin \frac{\pi x}{2} + 3 \sin \frac{5\pi x}{2}$ in (3), we get

$$\sin\left(\frac{\pi x}{2}\right) + 3 \sin\left(\frac{5\pi x}{2}\right) = \sum_{n=1}^{\infty} b_n \sin \frac{n\pi x}{2}$$

$$= b_1 \sin\left(\frac{\pi x}{2}\right) + b_2 \sin\left(\frac{2\pi x}{2}\right) + \ldots + b_5 \sin\left(\frac{5\pi x}{2}\right) + \ldots \qquad \text{... (4)}$$

On equating the coefficients of both sides, we get

$$b_1 = 1 \text{ and } b_5 = 3$$

On putting the values of b_1 and b_5 in (4), we get

$$u = \sin\left(\frac{\pi x}{2}\right) e^{-\pi^2c^2t/4} + 3 \sin\left(\frac{5\pi x}{2}\right) e^{-25\pi^2c^2t/4} \qquad \textbf{Ans.}$$

Example 22. *Find the solution of*

$$\frac{\partial^2 u}{\partial x^2} = h^2 \frac{\partial u}{\partial t}$$

for which u (0, t) = u (l, t) = 0, u (x, 0) = $\sin \frac{\pi x}{l}$ *by method of variables separable.*

Solution.
$$\frac{\partial^2 u}{\partial x^2} = h^2 \frac{\partial u}{\partial t} \qquad \text{... (1)}$$

In example 21 the given equation was

$$\frac{\partial^2 u}{\partial x^2} = \frac{1}{c^2} \frac{\partial u}{\partial t} \qquad \text{... (2)}$$

On comparing (1) and (2) we get $h^2 = \frac{1}{c^2}$

Thus solution of (1) is $u = (c_2 \cos px + c_3 \sin px) c_1 e^{-\frac{p^2t}{h^2}}$ ... (3)

On putting $x = 0, u = 0$ in (3), we get

$$0 = c_1c_2 e^{-\frac{p^2t^2}{h^2}} \qquad c_1 \neq 0, \therefore c_2 = 0$$

(3) is reduced to

$$u = c_3 \sin px \, c_1 e^{-\frac{p^2t}{h^2}} \qquad \text{... (4)}$$

On putting $x = l$ and $u = 0$ in (4) we get

$$0 = c_3\, c_1 \sin pl\, e^{-\frac{p^2 t}{h^2}}$$

$\therefore$ $\sin pl = 0 = \sin n\pi$ or $p = \frac{n\pi}{l}$

Now (4) is reduced to

$$u = c_3\, c_1 \sin \frac{n\pi x}{l} e^{-\frac{n^2\pi^2 t}{h^2 l^2}} \quad \text{... (5)}$$

On putting $t = 0,\ u = \sin \frac{\pi x}{l}$ in (5) we get

$$\sin \frac{\pi x}{l} = c_3\, c_1 \sin \frac{n\pi x}{l}$$

This equation will be satisfied if

$$n = 1 \text{ and } c_3\, c_1 = 1$$

On putting the values of $c_3\, c_1$ and n in (5), we have

$$u = \sin \frac{\pi x}{l} e^{-\frac{\pi^2 t}{h^2 l^2}} \quad \textbf{Ans.}$$

Example 23. *Solve the equation* $\frac{\partial u}{\partial t} = \frac{\partial^2 u}{\partial x^2}$ *with boundary condition* $u(x,\ 0) = 3\sin n\pi x$, $u(0,\ t),\ u(l,\ t) = 0$ *where* $0 < x < l$. (*Q.Bank U.P. 2002*)

Solution. Here, we have

$$\frac{\partial u}{\partial t} = \frac{\partial^2 u}{\partial x^2}$$

Its solution is

$$u = e^{-p^2 t}(c_2 \cos px + c_3 \sin px) \quad \text{... (1)}$$

Putting $x = 0$, and $u = 0$ in (1), we get (given)

$$0 = c_1 c_2 e^{-p^2 t}$$

$\Rightarrow$ $c_2 = 0$

Putting the value of c_2 in (1), we get

$$u = c_1 c_3 e^{-p^2 t} \sin px \quad \text{... (2)}$$

Putting $x = l$ and $u = 0$ in (2), we get (given)

$$0 = c_1 c_3 e^{-p^2 t} \sin pl$$

$$\Rightarrow \sin pl = 0 \Rightarrow pl = n\pi \Rightarrow p = \frac{n\pi}{l}$$

Putting the value of p in (2), we get

$$u(x,\ t) = c_1 c_3 e^{-\frac{n^2\pi^2}{l} t} \sin \frac{n\pi x}{l}$$

$$= b_n\, e^{-\frac{n^2\pi^2}{l^2} t} \sin \frac{n\pi x}{l} \quad (b_n = c_1 c_3)$$

$$u = \sum_{n=1}^{\infty} b_n e^{-(n^2\pi^2 t/l^2)} \sin \frac{n\pi x}{l} \qquad \text{... (3) (general solution)}$$

On putting $t = 0$ and $u = 3 \sin n\pi x$ in (3), we get

$$3 \sin n\pi x = \sum_{n=1}^{\infty} b_n \sin \frac{n\pi x}{l}. \qquad \text{... (4)}$$

Equating the coefficients, we get

$$b_n = 3, l = 1.$$

On putting the values of b_n and l in (3), we get

$$u = 3\sum_{n=1}^{\infty} b_n e^{-n^2\pi^2 t} \sin n\pi x$$

Ans.

Example 24. *A rod of length l with insulated sides is initially at a uniform temperature u_0. Its ends are suddenly cooled to 0°C and are kept at that temperature. Prove that the temperature function $u(x,t)$ is given by*

$$u(x, t) = \sum_{n=1}^{\infty} b_n \sin \frac{n\pi x}{l} . e^{-\frac{c^2\pi^2 n^2 t}{l^2}}$$

where b_n is determined from the equation.

$$b_n = \frac{2}{l}\int_0^l u_0(x) \sin \frac{n\pi x}{l} dx$$

Solution. Let heat flow along a bar of uniform cross-section, in the direction perpendicular to the cross-section. Take one end of the bar as origin and the direction of heat flow is along x-axis.

Let the temperature of the bar at any time t at a point x distance from the origin be $u(x, t)$.

Then the equation of one dimensional heat flow is $\frac{\partial u}{\partial t} = c^2 \frac{\partial^2 u}{\partial x^2}$. ...(1)

Let us assume that $u = XT$, where X is a function of x alone and T that of t alone.

$$\therefore \quad \frac{\partial u}{\partial t} = X\frac{dT}{dt} \quad \text{and} \quad \frac{\partial^2 u}{\partial x^2} = T\frac{d^2 X}{dx^2}$$

Substituting these values in (1), we get $X\frac{dT}{dt} = c^2 T\frac{d^2 X}{dx^2}$

$$\Rightarrow \quad \frac{1}{c^2 T}\frac{dT}{dt} = \frac{1}{X}\frac{d^2 X}{dx^2} = -p^2 \text{ (constant)} \qquad \text{...(2)}$$

$\frac{1}{c^2 T}\frac{dT}{dt} = -p^2$	$\frac{1}{X}\frac{d^2 X}{dx^2} = -p^2$
$\frac{dT}{dt} + p^2c^2 T = 0$	$\frac{d^2 X}{dx^2} + p^2 X = 0$
$\Rightarrow \quad DT + p^2c^2 T = 0$	$\Rightarrow \quad D^2X + p^2X = 0$
$\Rightarrow \quad (D + p^2c^2)T = 0$	A.E. is $(m^2 + p^2)X = 0$
A.E. is $m + p^2c^2 = 0$	$\Rightarrow \quad m^2 + p^2 = 0$
$\Rightarrow \quad m = -p^2c^2$	$\Rightarrow \quad m = \pm ip$
$\Rightarrow \quad T = c_1 e^{-p^2c^2t}$	$\Rightarrow \quad X = c_2 \cos px + c_3 \sin px$

$\therefore\ u = XT \quad \Rightarrow \qquad u = c_1 e^{-p^2c^2t}(c_2 \cos px + c_3 \sin px)$ ... (3)

Putting $x = 0$, $u = 0$ in (3), we get

$$0 = c_1 e^{-p^2c^2t}(c_2) \Rightarrow c_2 = 0 \qquad [\text{since } c_1 \neq 0]$$

(3) becomes $\qquad u = c_1 e^{-p^2c^2t} c_3 \sin px$ ... (4)

Again putting $x = l$, $u = 0$ in (4), we get

$$0 = c_1 e^{-p^2c^2t} c_3 \sin pl \Rightarrow \sin pl = 0 = \sin n\pi \quad \Rightarrow \quad pl = n\pi$$

$$\Rightarrow \qquad p = \frac{n\pi}{l},\ n \text{ is any integer}$$

Hence (4) becomes $\qquad u = c_1 c_3 e^{\frac{-n^2c^2\pi^2}{l^2}t} \sin\frac{n\pi x}{l} = b_n e^{-\frac{n^2c^2\pi^2t}{l^2}} \sin\frac{n\pi}{l}x, \qquad b_n = c_1c_3$

This equation satisfies the given conditions for all integral values of n. Hence taking $n = 1, 2, 3, \ldots$, *the most general solution is*

$$u - \sum_{n=1}^{\infty} b_n e^{\frac{-n^2c^2\pi^2t}{l^2}} \sin\frac{n\pi}{l}x$$

Proved.

where, $\quad b_n = \frac{2}{l}\int_0^l u_0 \sin\frac{n\pi x}{l}\,dx = \frac{2}{l}u_0\left[-\frac{l}{n\pi}\cos\frac{n\pi x}{l}\right]_0^l$

$$= \frac{2}{l}u_0\left[-\frac{l}{n\pi}(\cos n\pi - \cos 0)\right] = -\frac{2u_0}{n\pi}\left[(-1)^n - 1\right]$$

$$\Rightarrow \qquad b_n = \begin{cases} 0, & \text{if } n \text{ is even} \\ \frac{4u_0}{n\pi}, & \text{if } n \text{ is odd} \end{cases}$$

Hence the temperature function

$$u(x, t) = \frac{4u_0}{\pi}\sum_{n=1,3,5\ldots}^{\infty} \frac{1}{n}\sin\frac{n\pi x}{l} e^{-\frac{c^2n^2\pi^2t}{l^2}}$$

$$= \frac{4u_0}{\pi}\sum_{n=1}^{\infty}\frac{1}{2n-1}\sin\frac{(2n-1)\pi x}{l} e^{-\frac{c^2(2n-1)^2\pi^2t}{l^2}}$$

Example 25. *Determine the solution of one dimensional heat equation*

$$\frac{\partial u}{\partial t} = c^2\frac{\partial^2 u}{\partial x^2}$$

Subject to the boundary conditions u (0, t) = 0, u (l, t) = 0 (t > 0) and the initial condition u(x, 0) = x, l being the length of the bar. *(U.P. II Semester Summer 2006)*

Solution. $\qquad \frac{\partial u}{\partial t} = c^2\frac{\partial^2 u}{\partial x^2}$ .. (1)

Boundary conditions are

$$u\,(0, t) = 0$$

$$u(l, t) = 0 \qquad (t > 0)$$
$$u(x, 0) = x$$

On solving (1), we get

$$u = c_1 e^{-p^2 c^2 t}(c_2 \cos px + c_3 \sin px) \qquad ...(2)$$

Putting $x = 0$ and $u = 0$ in (2), we get

$$0 = c_1 e^{-p^2 c^2 t}(c_2) \quad \Rightarrow \quad c_2 = 0$$

Putting $c_2 = 0$ in (2), we get

$$u = c_1 e^{-p^2 c^2 t} c_3 \sin px \qquad ...(3)$$

Again putting $x = l$, $u = 0$ in (3), we get

$$0 = c_1 e^{-p^2 c^2 t} c_3 \sin pl$$

$$\Rightarrow \quad \sin pl = 0 = \sin n\pi \quad \Rightarrow \quad pl = n\pi$$

$$\Rightarrow \quad p = \frac{n\pi}{l}, \text{ } n \text{ is any integer}$$

Hence, (3) becomes $u = c_1 c_3 e^{-\frac{n^2 c^2 \pi^2}{l^2} t} \sin \frac{n\pi x}{l} = b_n e^{-\frac{n^2 c^2 \pi^2 t}{l^2}} \sin \frac{n\pi}{l} x$ ...(4)

On putting $t = 0$ and $u = x$ in (4), we get

$$x = b_n \sin \frac{n\pi}{l} x$$

General solution is

$$x = \sum_{n=1}^{\infty} b_n \sin \frac{n\pi}{l} x$$

Now, $\quad b_n = \frac{2}{l} \int_0^l x \sin \frac{n\pi x}{l} dx$

$$= \frac{2}{l}\left[x \cdot \frac{l}{n\pi}\left(-\cos \frac{n\pi x}{l}\right) - (1)\left(\frac{-l^2}{n^2\pi^2} \sin \frac{n\pi x}{l}\right)\right]_0^l$$

$$= \frac{2}{l}\left[\left(l \cdot \frac{l}{n\pi}(-\cos n\pi) + \frac{l^2}{n^2\pi^2} \sin n\pi\right) - 0\right]$$

$$= \frac{2}{l}\left[-\frac{l^2}{n\pi}(-1)^n\right] = (-1)^{n+1} \frac{2l}{n\pi}$$

Putting the value of b_n in (4), we get

$$u = \frac{2l}{\pi} \sum_{n=1}^{\infty} \frac{(-1)^{n+1}}{n} \sin \frac{n\pi x}{l} e^{-\frac{n^2 c^2 \pi^2 t}{l^2}} \qquad \textbf{Ans.}$$

Example 26. *An insulated rod of length l has its ends A and B maintained at 0°C and 100°C respectively until steady state conditions prevail. If B is suddenly reduced to 0°C and maintained at 0°C, find the temperature at a distance x from A at time t.*

(U.P. II Semester Summer, 2004, 2005, AMIE, Summer 2004)

Solution. The initial temperature of the rod can be written as :

$$u(x,t) = 0 + \frac{100}{l}x = \frac{100}{l}x$$

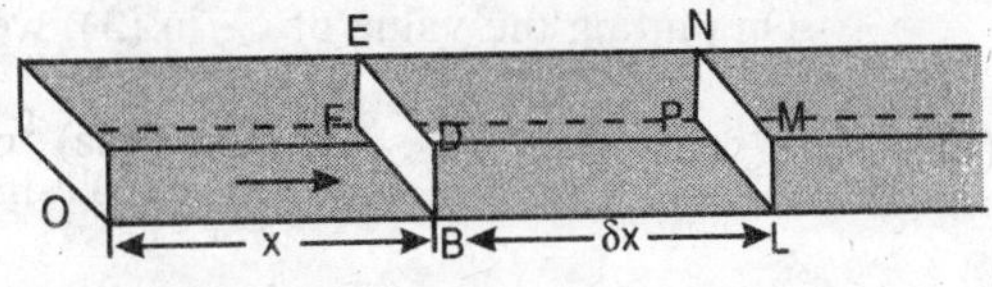

While in steady state, the temperature distribution can be written as

$$u(x,t) = 0 + \frac{0}{l}x = 0$$

To find u in the intermediate period, calculating time from the instant when the end temperature were changed.

$$u = u_1(x) + u_2(x)$$

where $u_2(x)$ is temperature after a sufficient long time and $u_1(x,t)$ is the transient temperature distribution tending to zero as $t \to \infty$. Hence $u_2(x) = 0$.

Also $u_1(x,t)$ satisfies one dimensional heat flow

$$C^2 \frac{\partial^2 u}{\partial x^2} = \frac{\partial u}{\partial t}$$

Thus $$u = (C_1 \cos px + C_2 \sin px)e^{-c^2 p^2 t} \quad \text{... (1)}$$

On putting $x = 0$, $u = 0$ in (1), we get

$$0 = C_1 e^{-p^2 c^2 t} \quad \Rightarrow \quad C_1 = 0$$

On putting $C_1 = 0$ in (1), we get

$$u = C_2 \sin px\, e^{-c^2 p^2 t} \quad \text{... (2)}$$

On putting $x = l$, $u = 0$ in (2), we get

$$0 = C_2 \sin pl e^{-p^2 c^2 t} \quad \Rightarrow \quad \sin pl = 0 = \sin n\pi$$

$$\Rightarrow \quad pl = n\pi \quad \Rightarrow \quad p = \frac{n\pi}{l}$$

On putting the value of p in (2), we get

$$u = C_2 \sin \frac{n\pi x}{l} e^{-\frac{n^2\pi^2c^2}{l^2}t} \quad \text{... (3)}$$

On putting $t = 0$, $u = \frac{100}{l}x$ in (3), we get

$$\frac{100}{l}x = C_2 \sin \frac{n\pi x}{l}$$

$$C_2 = \frac{2}{l}\int_0^l \frac{100}{l} x \cdot \sin \frac{n\pi x}{l}\, dx$$

$$C_2 = \frac{200}{l^2}\int_0^l x \sin \frac{n\pi x}{l}\, dx$$

$$C_2 = \frac{200}{l^2}\left[-\frac{xl}{n\pi}\cos\frac{n\pi x}{l} - (-1)\frac{l^2}{n^2\pi^2}\sin\frac{n\pi x}{l}\right]_0^l$$

$$C_2 = \frac{200}{l^2}\left[\frac{-l^2}{n\pi}\cos n\pi\right]$$

$$\Rightarrow \qquad C_2 = -\frac{200}{n\pi}(-1)^n$$

On putting the value of C_2 in (3), we get

$$u = -\frac{200}{n\pi}(-1)^n \sin\frac{n\pi x}{l} e^{-\frac{n^2\pi^2c^2}{l^2}t}$$

$$\Rightarrow \qquad u = (-1)^{n+1}\cdot\frac{200}{n\pi}\sin\frac{n\pi x}{l} e^{-\frac{n^2\pi^2c^2}{l^2}t} \qquad \textbf{Ans.}$$

Example 27. *The ends A and B of a rod 20 cm long have the temperatures at 30° C and at 80°C until steady state prevails. The temperature of the ends are changed to 40°C and 60°C respectively. Find the temperature distribution in the rod at time t.*

Solution. The initial temperature distribution in the rod is

$$u_1(x,t) = 30 + \frac{80-30}{20}x = 30 + \frac{50}{20}x \;\; i.e.,\; u_1(x,t) = 30 + \frac{5}{2}x$$

and the final distribution (*i.e.* in steady state) is

$$u_2(x) = 40 + \frac{60-40}{2}x = 40 + \frac{20}{20}x = 40 + x$$

To get u in the intermediate period, reckoning time from the instant when the end temperature were changed, we assumed

$$u = u_1(x,t) + u_2(x)$$

where $u_2(x)$ is the steady state temperature distribution in the rod (*i.e.*, temperature after a sufficiently long time) and $u_1(x,t)$ is the transient temperature distribution which tends to zero as t increases.

Thus $\qquad u_2(x) = 40 + x$

Now $u_1(x, t)$ satisfies the one-dimensional heat-flow equation

$$c^2\frac{\partial^2 u}{\partial x^2} = \frac{\partial u}{\partial t}$$

Hence its solution is $(a_k \cos kx + b_k \sin kx)\, e^{-c^2k^2t}$

Hence u is of the form

$$u = 40 + x + \Sigma(a_k \cos kx + b_k \sin kx)e^{-c^2k^2t}$$

Since $\qquad u = 40°$ when $x = 0$ and $u = 60°$ when $x = 20$, we get

$$a_k = 0, \text{ and } k = \frac{n\pi}{20}$$

Hence $$u = 40 + x + \sum_{n=1}^{\infty} b_n \sin\frac{n\pi x}{20} e^{-c^2\left(\frac{n\pi}{20}\right)^2 t} \qquad ...(1)$$

Using the initial conditions *i.e.*,

$$u = 30 + \frac{5}{2}x \text{ when } t = 0, \text{ we get}$$

$$30+\frac{5}{2}x = 40 + x + \Sigma b_n \sin\frac{n\pi}{20}x \Rightarrow \frac{3}{2}x - 10 = \Sigma b_n \sin\frac{n\pi x}{20}$$

Hence $$b_n = \frac{2}{20}\int_0^{20}\left(\frac{3}{2}x-10\right)\sin\frac{n\pi x}{20}\,dx$$

$$\Rightarrow \quad = \frac{1}{10}\left[\left(\frac{3x}{2}-10\right)\left(-\frac{20}{n\pi}\cos\frac{n\pi x}{20}\right)-\frac{3}{2}\left(-\frac{400}{n^2\pi^2}\sin\frac{n\pi x}{20}\right)\right]_0^{20}$$

$$= \frac{1}{10}\left[-20\left(\frac{20}{n\pi}\right)(-1)^n - (-10)\left(\frac{-20}{n\pi}\right)\right] = -\frac{20}{n\pi}[2(-1)^n+1]$$

Putting this value of b_n in (1), we get

$$\therefore \quad u = 40 + x - \frac{20}{\pi}\sum\left[\left(\frac{2(-1)^{n+1}}{n}\right)\sin\frac{n\pi x}{20}.e^{-\left(\frac{cn\pi}{20}\right)^2 t}\right] \quad \textbf{Ans.}$$

EXERCISE 11.3

1. Solve the following boundary value problem which arises in the heat conduction in a rod :

$$\frac{\partial u}{\partial t} = c^2\frac{\partial^2 u}{\partial x^2};\ u(0,t) = u(l,t) = 0,\ u(x,0) = 100\frac{x}{l}$$

Ans. $u(x,t) = \frac{200}{\pi}\sum_{n=1}^{\infty}\frac{(-1)^{n+1}}{n}\sin\frac{n\pi x}{l}e^{-\frac{c^2n^2\pi^2}{l^2}t}$

2. Solve the non-homogeneous heat conduction equation $u = a^2 u_{xx} + \sin 3\pi x$ subject to the following conditions :

$u(x,0) = \sin 2\pi x, \qquad u(0,t) = u(l,t) = 0.$

3. Solve $\frac{\partial u}{\partial t} = a^2\frac{\partial^2 u}{\partial x^2}$, given that

(*i*) $u = 0$ when $x = 0$ and $x = l$ for all t

(*ii*) $u = 3\sin\frac{\pi x}{l}$, when $t = 0$ for all x, $0 < x < l$. **Ans.** $u = 3\sin\frac{\pi x}{l}e^{-\frac{a^2\pi^2 t}{l^2}}$

4. Find by the method of separation of variables the solution of $U(x,t)$ of the boundary value problem

$$\frac{\partial U}{\partial t} = 3\frac{\partial^2 U}{\partial x^2}, \qquad t > 0, \quad 0 < \alpha < 2$$

$U(0,t) = 0,\ U(2,t) = 0, \qquad t = 0$

$U(x,0) = x, \qquad 0 < x < 2.$ **Ans.** $U = \sum_{n=1}^{\infty}\frac{\sin\frac{n\pi x}{2}}{\sin\frac{n\pi}{2}}e^{-\frac{3n^2\pi^2 t}{4}}$

5. The equation $\frac{\partial u}{\partial t} = \mu \frac{\partial^2 u}{\partial x^2}$ refers to the conduction of heat along a bar, without radiation, show that if $u = Ae^{-gx}\sin(nt - gx)$, where A, g and n are positive constants, then $g = \sqrt{\frac{n}{2\mu}}$

6. Solve $\frac{\partial u}{\partial t} = a^2 \frac{\partial^2 u}{\partial x^2}$ under the conditions

$u'(0,t) = 0 \quad t > 0, \quad u'(\pi, t) = 0, \quad u(x,0) = x^2, 0 < x < \pi$

Ans. $u(x,t) = \frac{\pi^2}{3} + 4\sum_{n=1}^{\infty} \frac{(-1)^n}{n^2} \cos nx e^{-a^2 n^2 t}$

7. A uniform rod of length 'a' whose surface is thermally insulated is initially at temperature $\theta = \theta_0$. At time $t = 0$, one end is suddenly cooled to $\theta = 0$ and subsequently maintained at this temperature, the other end remains thermally insulated. Find the temperature $\theta(x,t)$.

Ans. $\theta(x,t) = \frac{4\theta_0}{\pi} \sum_{n=1}^{\infty} \frac{\sin\frac{(2n+1)\pi x}{2a}}{2n+1} e^{-\frac{(2n+1)^2 \pi^2 c^2 t}{4a^2}}$

8. Solve $\frac{\partial U}{\partial t} = a^2 \frac{\partial^2 U}{\partial x^2}$ under the conditions

(*i*) $U \neq \infty$ if $t \to \infty$; (*ii*) $U(0,t) = U(\pi,t) = 0$; (*iii*) $U(x, 0) = \pi x - x^2$

Ans. $u = \frac{8}{\pi} \sum_{n=1}^{\infty} \frac{\sin(2n-1)x}{(2n-1)^3} e^{-a^2 (2n-1)^2 t}$

9. The temperature distribution in a bar of length π, which is prefectly insulated at the ends $x = 0$ and $x = \pi$ is governed by the partial differential equation $\frac{\partial u}{\partial t} = \frac{\partial^2 u}{\partial x^2}$. Assuming the initial temperature as $u(x, 0) = f(x) = \cos 2x$, find the temperature distribution at any instant of time.

(Uttarakhand, IInd Semester, June 2010)

Ans. $u = e^{-4t} \cos 2x$

10. The heat flow in a bar of length 10 cm of homogeneous material is governed by the partial differential equation $u_t = c^2 u_{xx}$. The ends of the bar are kept at temperature 0°C, and the initial temperature is $f(x) = x(10 - x)$. Find the temperature in the bar at any instant of time.

11. Find the temperature $u(x, t)$ in a bar of length π which is perfectly insulated everywhere including the ends $x = 0$ and $x = \pi$. This leads to the conditions $\frac{\partial u}{\partial x}(0, t) = 0, \frac{\partial u}{\partial x}(\pi, t) = 0$. Further the initial conditions are as given below:

$$u(x, 0) = \begin{cases} x, & 0 < x < \pi/2 \\ \pi x, & \pi/2 \leq x < \pi \end{cases}$$

Find the solution by the separation of variables.

11.6 TWO DIMENSIONAL HEAT FLOW

When the heat-flow is along curves instead of along straight lines, all the curves lying in parallel planes, then the flow is called two-dimensional. Let us consider now the flow of heat in a metal-plate in the *xoy* plane. Let the plate be of uniform thickness *h*, density ρ, thermal conductivity *k* and the specific heat *C*. Since the flow is two dimensional, the temperature at any point of the plate is independent of the *z*-coordinate. The heat-flow lies in the *xoy* plane and is zero along the direction normal to the *xoy* plane.

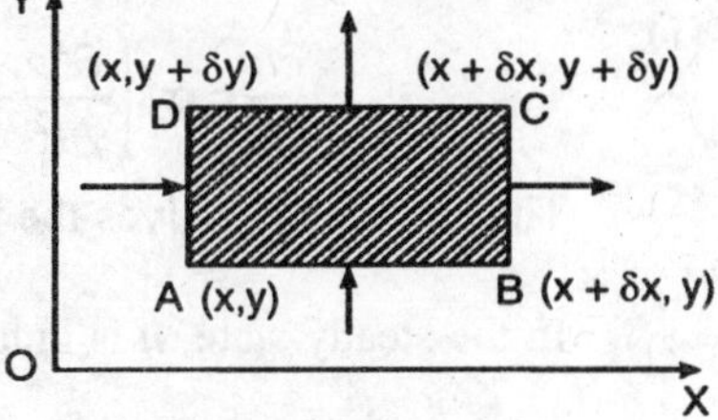

Now, consider a rectangular element *ABCD* of the plate with sides δ*x* and δ*y*, the edges being parallel to the coordinates axes, as shown in the figure, Then the quantity of heat *entering* the element *ABCD* per sec. through the surface *AB* is

$$= -k\left(\frac{\partial u}{\partial y}\right)_y \delta x.h$$

Similarly the quantity of heat entering the element *ABCD* per sec. through the surface *AD* is

$$= -k\left(\frac{\partial u}{\partial x}\right)_x \delta y.h$$

The amount of heat which flows out through the surfaces *BC* and *CD* are

$$-k\left(\frac{\partial u}{\partial x}\right)_{x+\delta x}.\delta y.h \text{ and } -k\left(\frac{\partial u}{\partial y}\right)_{y+\delta y}.\delta x.h \text{ respectively.}$$

Therefore the total gain of heat by the rectangular element *ABCD* per sec.

= inflow – outflow.

$$= kh\left[\left\{\left(\frac{\partial u}{\partial x}\right)_{x+\delta x} - \left(\frac{\partial u}{\partial x}\right)_x\right\}\delta y + \left\{\left(\frac{\partial u}{\partial y}\right)_{y+\delta y} - \left(\frac{\partial u}{\partial y}\right)_y\right\}\delta x\right]$$

$$= kh\delta x.\delta y\left[\frac{\left(\frac{\partial u}{\partial x}\right)_{x+\delta x} - \left(\frac{\partial u}{\partial x}\right)_x}{\delta x} + \frac{\left(\frac{\partial u}{\partial y}\right)_{y+\delta y} - \left(\frac{\partial u}{\partial y}\right)_y}{\delta y}\right] \quad ...(1)$$

The rate of gain of heat by the element *ABCD* is also given by

$$\rho\delta x.\delta y.h.c.\frac{\partial u}{\partial t} \quad ...(2)$$

Equating the two-expressions for gain of heat per sec. from (1) and (2), we have

$$\rho\delta x.\delta y.h.c.\frac{\partial u}{\partial t} = hk\delta x\delta y\left[\frac{\left(\frac{\partial u}{\partial x}\right)_{x+\delta x} - \left(\frac{\partial u}{\partial x}\right)_x}{\delta x} + \frac{\left(\frac{\partial u}{\partial y}\right)_{y+\delta y} - \left(\frac{\partial u}{\partial y}\right)_y}{\delta y}\right]$$

i.e.,

$$\frac{\partial u}{\partial t} = \frac{k}{\rho c}\left[\frac{\left(\frac{\partial u}{\partial x}\right)_{x+\delta x} - \left(\frac{\partial u}{\partial x}\right)_x}{\delta x} + \frac{\left(\frac{\partial u}{\partial y}\right)_{y+\delta y} - \left(\frac{\partial u}{\partial y}\right)_y}{\delta y}\right]$$

Taking the limit as $\delta x \to 0, \delta y \to 0$, the above equation reduces to

$$\frac{\partial u}{\partial t} = \frac{k}{\rho c}\left(\frac{\partial^2 u}{\partial x^2} + \frac{\partial^2 u}{\partial y^2}\right)$$

Putting $\alpha^2 = \dfrac{k}{\rho c}$ as before, the equation becomes,

$$\frac{\partial u}{\partial t} = \alpha^2\left(\frac{\partial^2 u}{\partial x^2} + \frac{\partial^2 u}{\partial y^2}\right) \quad \text{... (3)}$$

The equation (3) gives the temperature distribution of the plate in the transit state.

In the steady-state, u is independent of t, so that $\dfrac{\partial u}{\partial t} = 0$. Hence the temperature distribution of the plate in the steady-state is $\dfrac{\partial^2 u}{\partial x^2} + \dfrac{\partial^2 u}{\partial y^2} = 0.$

Example 28. *Solve* $\dfrac{\partial^2 u}{\partial x^2} + \dfrac{\partial^2 u}{\partial y^2} = 0$ *which satisfies the conditions*

$$u(0, y) = u(l, y) = u(x, 0) = 0$$

and $u(x, a) = \sin \dfrac{n\pi x}{l}$ *(U.P., II Semester, 2004)*

Solution. Consider the heat flow in a metal plate of uniform thickness, in the directions parallel to length and breadth of the plate. There is no heat flow along the normal to the plane of the rectangle.

Let $u(x, y)$ be the temperature at any point (x, y) of the plate at time t is given by

$$\frac{\partial u}{\partial t} = c^2\left(\frac{\partial^2 u}{\partial x^2} + \frac{\partial^2 u}{\partial y^2}\right) \quad \text{... (1)}$$

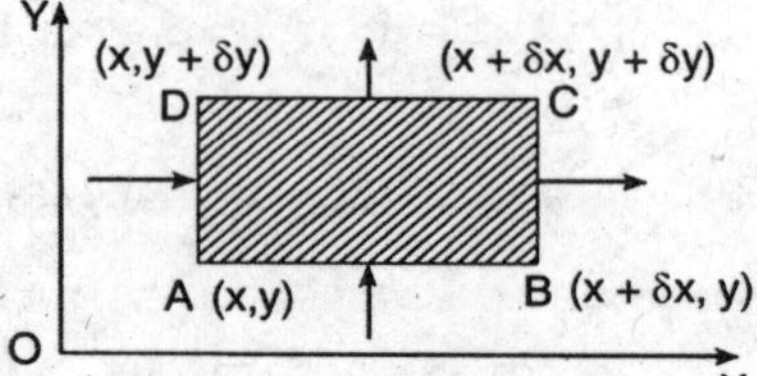

In the steady state, u does not change with t.

$$\therefore \quad \frac{\partial u}{\partial t} = 0$$

(1) becomes $\dfrac{\partial^2 u}{\partial x^2} + \dfrac{\partial^2 u}{\partial y^2} = 0$

This is called Laplace equation in two dimensions.

Let $u = X(x).Y(y)$ ...(1)

Putting the values of $\dfrac{\partial^2 u}{\partial x^2}$ and $\dfrac{\partial^2 u}{\partial y^2}$ in (1), we have

$$X''Y + XY'' = 0 \quad \text{...(2)}$$

$$\Rightarrow \quad \frac{X''}{X} = -\frac{Y''}{Y} = -p^2 \text{ (say)}$$

$D^2X = -p^2X$	$D^2Y = p^2Y$
$\Rightarrow \quad D^2X + p^2X = 0$	$\Rightarrow \quad D^2Y - p^2Y = 0$
$\Rightarrow \quad (D^2 + p^2)X = 0$	$\Rightarrow \quad (D^2 - p^2)Y = 0$
A.E. is $\quad m^2 + p^2 = 0$	A.E. is $m^2 - p^2 = 0$
$\Rightarrow \quad m^2 = -p^2$	$\Rightarrow \quad m^2 = p^2$
$\Rightarrow \quad m = \pm ip$	$\Rightarrow \quad m = \pm p$
$X = c_1 \cos px + c_2 \sin px$	$y = c_3e^{py} + c_4e^{-py}$

Putting the values of X and Y in (1), we have

$$u = (c_1 \cos px + c_2 \sin px)(c_3e^{py} + c_4e^{-py}) \qquad \text{... (3)}$$

Putting $x = 0$, $u = 0$ in (3), we have

$$0 = c_1(c_3e^{py} + c_4e^{-py})$$

$$\therefore \qquad c_1 = 0$$

(3) is reduced to $u = c_2 \sin px(c_3e^{py} + c_4e^{-py})$ ... (4)

On putting $x = l$, $u = 0$ in (4), we have

$$0 = c_2 \sin pl(c_3e^{py} + c_4e^{-py})$$

$$c_2 \neq 0 \qquad \therefore \sin pl = 0 = \sin n\pi \quad \Rightarrow \quad pl = n\pi \quad \Rightarrow \quad p = \frac{n\pi}{l}$$

Now (4) becomes $\quad u = c_2 \sin\frac{n\pi x}{l}\left(c_3e^{\frac{n\pi y}{l}} + c_4e^{\frac{-n\pi y}{l}}\right)$ ... (5)

On putting $u = 0$ and $y = 0$ in (5), we have

$$0 = c_2 \sin\frac{n\pi x}{l}(c_3 + c_4)$$

$$\Rightarrow \qquad c_3 + c_4 = 0 \qquad \Rightarrow \qquad c_3 = -c_4$$

(5) becomes $u = c_2c_3 \sin\frac{n\pi x}{l}\left(e^{\frac{n\pi y}{l}} - e^{\frac{-n\pi y}{l}}\right)$ ... (6)

On putting $y = a$ and $u = \sin\frac{n\pi x}{l}$ in (6), we have

$$\sin\frac{n\pi x}{l} = c_2c_3 \sin\frac{n\pi x}{l}\left(e^{\frac{n\pi a}{l}} - e^{\frac{-n\pi a}{l}}\right) \quad i.e.\ c_2c_3 = \frac{1}{e^{\frac{n\pi a}{l}} - e^{\frac{-n\pi a}{l}}}$$

Putting this value in (6), we have

$$u = \sin\frac{n\pi x}{l}\,\frac{e^{\frac{n\pi y}{l}} - e^{\frac{-n\pi y}{l}}}{e^{\frac{n\pi a}{l}} - e^{\frac{-n\pi a}{l}}} \quad \text{or} \quad u = \sin\frac{n\pi x}{l}\,\frac{\sinh\frac{n\pi y}{l}}{\sinh\frac{n\pi a}{l}}$$

Ans.

Example 29. *A rectangular plate with insulated surfaces is 10 cm wide and so long compared to its width that it may be considered infinite in length without introducing an appreciable error. If the temperature along the short edge y = 0 is given by*

$$u(x, 0) = 20x, \qquad 0 < x \le 5$$
$$= 20(10 - x), \qquad 5 < x < 10$$

while the two long edges x = 0 and x = 10 as well as the other short edges are kept at 0°C. Find the steady state temperature at any point (x, y) of the plate.

Solution. In the steady state, the temperature $u(x, y)$ at any point $p(x, y)$ satisfy the equation

$$\frac{\partial^2 u}{\partial x^2} + \frac{\partial^2 u}{\partial y^2} = 0 \qquad \text{... (1)}$$

P′ (x, y)

0°C 0°C

x = 0 x = 10

The boundary conditions are

$$u(0, y) = 0 \text{ for all values of } y \qquad \text{... (2)}$$
$$u(10, y) = 0 \text{ for all values of } y \qquad \text{... (3)}$$
$$u(x, \infty) = 0 \text{ for all values of } x \qquad \text{... (4)}$$
$$u(x, 0) = 20x, \quad 0 < x \le 5$$
$$= 20(10 - x) \quad 5 < x < 10 \qquad \text{... (5)}$$

Now three possible solutions of (1) are

$$u = (C_1 e^{px} + C_2 e^{-px})(C_3 \cos py + C_4 \sin py) \qquad \text{... (6)}$$
$$u = (C_5 \cos px + C_6 \sin px)(C_7 e^{py} + C_8 e^{-py}) \qquad \text{... (7)}$$
$$u = (C_9 x + C_{10})(C_{11} y + C_{12}) \qquad \text{... (8)}$$

Of these, we have to choose that solution which is consistent with the physical nature of the problem. The solution (6) and (8) cannot satisfy the condition (2), (3) and (4). Thus, only possible solution is (7) *i.e.,* of the form.

$$u(x, y) = (C_1 \cos px + C_2 \sin px)(C_3 e^{py} + C_4 e^{-py}) \qquad \text{... (9)}$$

By (2), $u(0, y) = C_1 (C_3 e^{py} + C_4 e^{-py}) = 0$ for all values of y

$\therefore \quad C_1 = 0$

$\therefore$ (9) reduces to $u(x, y) = C_2 \sin px (C_3 e^{py} + C_4 e^{-py})$ ... (10)

By (3), $u(10, y) = C_2 \sin 10p\,(C_3 e^{py} + C_4 e^{-py}) = 0,\ C_2 \neq 0$

$\therefore \quad \sin 10p = 0 = \sin n\pi \ \Rightarrow\ 10p = n\pi \ \Rightarrow\ p = \dfrac{n\pi}{10}$

Also to satisfy the condition (4) *i.e.* $u = 0$ as $y \to \infty$

$C_3 = 0$

Hence (10) takes the form $u(x, y) = C_2 C_4 \sin px\, e^{-py}$

or $\quad u(x, y) = b_n \sin px . e^{-py}$ where $b_n = C_2 C_4$

$\therefore$ The most general solution that satisfies (2), (3) and (4) is of the form

$$u(x, y) = \sum_{n=1}^{\infty} b_n \sin px\, e^{-py} \qquad \text{... (11)}$$

Putting $y = 0$, $u(x,0) = \sum_{n=1}^{\infty} b_n \sin px$, where $p = \frac{n\pi}{10}$

This requires the expansion of u in Fourier series in the interval $x = 0$ and $x = 5$ and from $x = 5$ to $x = 10$.

$$b_n = \frac{2}{10}\int_0^5 20x \sin px\,dx + \frac{2}{10}\int_5^{10} 20(10-x)\sin px\,dx$$

$$b_n = 4\int_0^5 x \sin px\,dx + 4\int_5^{10} (10-x)\sin px\,dx$$

$$= 4\left[x\left(\frac{-\cos px}{p}\right) - (1)\left(\frac{-\sin px}{p^2}\right)\right]_0^5 + 4\left[(10-x)\left(\frac{-\cos px}{p}\right) - (-1)\left(\frac{-\sin px}{p^2}\right)\right]_5^{10}$$

$$= 4\left[\frac{-5\cos 5p}{p} + \frac{\sin 5p}{p^2}\right] + 4\left[0 - \frac{\sin 10p}{p^2} + \frac{5\cos 5p}{p} + \frac{\sin 5p}{p^2}\right]$$

$$= 4\left[\frac{2\sin 5p}{p^2} - \frac{\sin 10p}{p^2}\right] \qquad \left(p = \frac{n\pi}{10}\right)$$

$$= 4\left[\frac{2\sin 5.\frac{n\pi}{10}}{\frac{n^2\pi^2}{100}} - \frac{\sin 10.\frac{n\pi}{10}}{\frac{n^2\pi^2}{100}}\right] = \frac{800}{n^2\pi^2}\sin\frac{n\pi}{2} - \frac{400}{n^2\pi^2}\sin n\pi$$

$$= \frac{800}{n^2\pi^2}\sin\frac{n\pi}{2} = 0, \text{ if } n \text{ is even.}$$

$$= \pm\frac{800}{n^2\pi^2}, \text{ if } n \text{ is odd.}$$

$$\Rightarrow \quad b_n = \frac{(-1)^{n+1}800}{(2n-1)^2\pi^2}$$

On putting the value of b_n in (11) the temperature at any point (x, y) is given by

$$u(x, y) = \frac{800}{\pi^2}\sum_{n=1}^{\infty}\frac{(-1)^{n+1}}{(2n-1)^2}\sin\frac{(2n-1)\pi x}{10}\, e^{-\frac{(2n-1)\pi y}{10}} \qquad \textbf{Ans.}$$

Example 30. *A thin rectangular plate whose surface is impervious to heat flow has t = 0 an arbitrary distribution of temperature f (x, y). Its four edges x = 0, x = a, y = 0, y = b are kept at zero temperature. Determine the temperature at a point of the plate as t increases.*

(U.P. III Semester, summer 2002)

Solution. The partial differential equation of two dimensional heat conduction problem is

$$\frac{\partial^2 u}{\partial x^2} + \frac{\partial^2 u}{\partial y^2} = \frac{1}{c^2}\frac{\partial u}{\partial t} \qquad \text{... (1)}$$

The boundary conditions are

$$u(0, y) = u(a, y) = u(x, 0) = u(x, b)$$

and the initial condition is

$$u(x, y) = f(x, y)$$

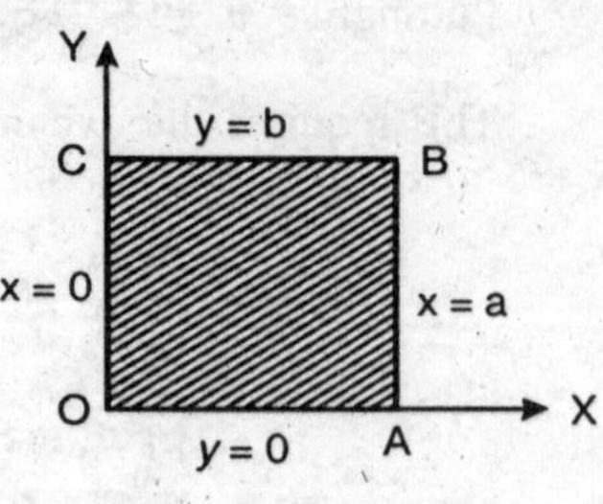

Let the solution be $u = X.Y.T.$

On putting the values of the derivatives in (1), we get

$$YT\frac{d^2X}{dx^2} + XT\frac{d^2Y}{dy^2} = \frac{XY}{c^2}\frac{dT}{dt}$$

Separating the variables, we get, (dividing by XYT)

$$\frac{1}{X}\frac{d^2X}{dx^2} + \frac{1}{Y}\frac{d^2Y}{dy^2} = \frac{1}{c^2T}\frac{dT}{dt}$$

Since X is function of independent variable x alone, Y of y alone and T of t alone, there are three possibilities.

1. $\frac{1}{X}\frac{d^2X}{dx^2} = 0, \quad \frac{1}{Y}\frac{d^2Y}{dy^2} = 0, \quad \frac{1}{c^2T}\frac{dT}{dt} = 0$

2. $\frac{1}{X}\frac{d^2X}{dx^2} = K_1^2, \quad \frac{1}{Y}\frac{d^2Y}{dy^2} = K_2^2, \quad \frac{1}{c^2T}\frac{dT}{dt} = k^2$

3. $\frac{1}{X}\frac{d^2X}{dx^2} = -K_1^2, \quad \frac{1}{Y}\frac{d^2Y}{dy^2} = -K_2^2, \quad \frac{1}{c^2T}\frac{dT}{dt} = -K^2$

$$K^2 = K_1^2 + K_2^2 \qquad \text{... (A)}$$

Out of these three solutions, we have to choose that solution which is consistent with the physical nature of the problem. Accordingly third solution is accepted here.

Then

$$Y = A\cos K_1x + B\sin K_1x$$

$$T = Ee^{-C^2kt}$$

$$u = XYT$$

$$u = (A\cos K_1x + B\sin K_1x)(C\cos K_2y + D\sin K_2y).Ee^{-C^2kt} \qquad \text{... (1)}$$

Now we apply boundary conditions

On putting $u = 0$ and $x = 0$ in (1), we get

$$0 = A(C\cos K_2y + D\sin K_2y)\,Ee^{-C^2kt} \Rightarrow A = 0$$

On putting the value of A in (1), it reduces to

$$u = (B\sin K_1x)(C\cos K_2y + D\sin K_2y).Ee^{-C^2kt}$$

$$\Rightarrow \quad u = (B_1\sin K_1x)(C\cos K_2y + D\sin K_2y).e^{-C^2kt} \qquad \text{... (2)}$$

$(B_1 = A.E.)$

On putting $u = 0$ and $x = a$ in (2), we get

$$0 = (B_1\sin K_1a)(C\cos K_2y + D\sin K_2y).e^{-C^2kt}$$

$$\Rightarrow \quad 0 = B_1\sin K_1a \Rightarrow \sin K_1a = 0 = \sin n\pi \quad \Rightarrow \quad K_1a = n\pi \Rightarrow K_1 = \frac{n\pi}{a}$$

On putting the value of K_1 in (2), we have

$$u = \left(B_1 \sin \frac{n\pi}{a} x\right)(C \cos K_2 y + D \sin K_2 y).e^{-C^2 kt} \qquad \text{... (3)}$$

On putting $u = 0, y = 0$ in (3), we obtain

$$0 = \left(B_1 \sin \frac{n\pi}{a} x\right)(C) e^{-C^2 kt} \qquad \Rightarrow C = 0$$

On substituting the value of C in (3), we have

$$u = \left(B_1 \sin \frac{n\pi x}{a}\right)(D \sin K_2 y) e^{-C^2 kt} \qquad \text{... (4)}$$

On substituting $u = 0$ and $y = b$ in (4), we have

$$0 = \left(B_1 \sin \frac{n\pi x}{a}\right)(D \sin K_2 b).e^{-C^2 kt} \quad \Rightarrow \sin K_2 b = 0 = \sin m\pi \ \Rightarrow \ K_2 = \frac{m\pi}{b}$$

On putting the value of K_2 in (4), we have

$$u = \left(B_1 \sin \frac{n\pi x}{a}\right)\left(D \sin \frac{m\pi y}{b}\right).e^{-C^2 kt} \qquad \text{... (5)}$$

$$u = A_{mn} \sin \frac{n\pi x}{a} \sin \frac{m\pi y}{b}.e^{-C^2 kt}, \qquad (B.D = A_{mn})$$

But from (A), $\quad K^2 = K_1^2 + K_2^2 = \pi^2 \left(\frac{n^2}{a^2} + \frac{m^2}{b^2}\right)$

$$\Rightarrow \qquad K_{mn}^2 = \pi^2 \left(\frac{n^2}{a^2} + \frac{m^2}{b^2}\right)$$

By using K_{mn}, (5) becomes

$$u = \sum_{n=1}^{\infty} \sum_{m=1}^{\infty} A_{mn} \sin \frac{n\pi x}{a} \sin \frac{m\pi y}{b} e^{-C^2 k_{mn} t}$$

On applying the initial condition $u = f(x, y),\ t = 0$

$$f(x, y) = \sum_{n=1}^{\infty} \sum_{m=1}^{\infty} A_{mn} \sin \frac{n\pi x}{a} \sin \frac{m\pi y}{b}$$

Which is the double Fourier sine series of $f(x, y)$

Where $\quad A_{mn} = \frac{2}{a} \frac{2}{b} \int_{x=0}^{a} \int_{y=0}^{b} \sin \frac{n\pi x}{a} \sin \frac{m\pi y}{b} . f(x, y)\, dx\, dy$ **Ans.**

Example 31. *A rectangular plate with insulated surface is 8 cm wide and so long compared to its width that it may be considered infinite in length without introducing an appreciable. If the tempearature along one short edge y = 0 is given by*

$$u(x, 0) = 100 \sin \frac{\pi x}{8},\ 0 < x < 8$$

While the two long edges x = 0 and x = 8 as well as the other short edge are kept at 0°C, show that the steady state temperature at any plate is given by

$$u(x, y) = 100 e^{-\frac{\pi y}{8}} \sin \frac{\pi x}{8}.$$

Soution. Here, we have

Two dimensional heat flow equation in steady state is given by

$$\frac{\partial^2 u}{\partial x^2} + \frac{\partial^2 u}{\partial y^2} = 0$$

Its solution is $u(x, y) = (c_1 \cos px + c_2 \sin px)(c_3 e^{py} + c_4 e^{-py})$ ...(1)

Putting $x = 0$ and $u = 0$ in (1), we get (given)

$$0 = c_1 (c_3 e^{py} + c_4 e^{-py})$$

$$\Rightarrow \qquad c_1 = 0$$

Putting the value of c_1 in (1), we get

$$u = c_2 \sin px (c_3 e^{py} + c_4 e^{-py}) \qquad ...(2)$$

Again putting $x = 8$ and $u = 0$ in (2), we get (given)

$$0 = c_2 \sin 8p (c_3 e^{py} + c_4 e^{-py})$$

$$\Rightarrow \qquad \sin 8p = 0 = \sin n\pi \Rightarrow p = \frac{n\pi}{8} \; (n \in I)$$

Putting the value of p in (2), we get

$$u(x, y) = c_2 \sin \frac{n\pi x}{8} \left(c_3 e^{\frac{n\pi y}{8}} + c_4 e^{-\frac{n\pi y}{8}}\right) \qquad ... (3)$$

Putting $y = \infty$ and u = 0 in (3), we get (given)

$$0 = c_2 \sin \frac{n\pi x}{8} c_3 e^{\frac{n\pi y}{8}}$$

$$\Rightarrow c_3 = 0$$

Putting $c_3 = 0$ in (3), we get

Y
0°C
P(x, y)
0°C
A
x = 0
8 cm
B
x = 8
X

$$u = c_2 c_4 \sin \frac{n\pi x}{8} e^{-\frac{n\pi y}{8}}$$

$$u = b_n \sin \frac{n\pi x}{8} e^{-\frac{n\pi y}{8}} \qquad ... (4)$$

Putting $y = 0$ and $u = 100 \sin \frac{\pi x}{8}$ in (4), we get (given)

$$100 \sin \frac{\pi x}{8} = b_n \sin \frac{n\pi x}{8}$$

$$\Rightarrow b_n = 100,$$

Putting the value of b_n in (4), we get

$$u = 100 \sin\left(\frac{\pi x}{8}\right) e^{-\frac{\pi y}{8}}$$ **Proved.**

Example 32. *Solve* $\frac{\partial^2 u}{\partial x^2} + \frac{\partial^2 u}{\partial y^2} = 0$, $0 < x < \pi$, $0 < y < \pi$, *which satisfies the conditions :*

$u(0, y) = u(\pi, y) = u(x, \pi) = 0$ *and* $u(x, 0) = \sin^2 x$. *(Q.Bank U.P. II Semester 2002)*

Solution. Here, we have

$$\frac{\partial^2 u}{\partial x^2} + \frac{\partial^2 u}{\partial y^2} = 0$$

Its solution is

$$u(x, y) = (c_1 \cos px + c_2 \sin px)(c_3 e^{py} + c_4 e^{-py}) \quad \text{... (1)}$$

Putting $x = 0$ and $u = 0$ in (1), we get (given)

$$0 = c_1(c_3 e^{py} + c_4 e^{-py})$$

$$\Rightarrow c_1 = 0$$

Putting the value of c_1 in (1), we get

$$u(x, y) = c_2 \sin px(c_3 e^{py} + c_4 e^{-py}) \quad \text{... (2)}$$

Putting $x = \pi$ and $u = 0$ in (2), we get (given)

$$0 = c_2 \sin p\pi(c_3 e^{py} + c_4 e^{-py})$$

$$\Rightarrow \quad \sin p\pi = 0 = \sin n\pi$$

$$\Rightarrow \quad p = n$$

Putting $p = n$ in (2), we get

$$u = c_2 \sin nx(c_3 e^{ny} + c_4 e^{-ny})$$

$$= \sin nx\,(Ae^{ny} + Be^{-ny}) \quad \text{... (3) } (c_2 c_3 = A \text{ and } c_2 c_4 = B.)$$

Putting $u = 0$ and $y = \pi$ in (3), we get (given)

$$0 = \sin nx\,(Ae^{n\pi} + Be^{-n\pi})$$

$$\Rightarrow \quad 0 = Ae^{n\pi} + Be^{-n\pi}$$

$$\Rightarrow \quad Ae^{n\pi} = -\,Be^{-n\pi} = -\frac{1}{2}b \Rightarrow A = -\frac{1}{2}be^{-n\pi} \text{ and } B = \frac{1}{2}be^{n\pi}$$

On putting the values of A and B in (3), we get

$$u = \sin nx\left[-\frac{1}{2}be^{-n\pi}e^{ny} + \frac{1}{2}be^{n\pi}e^{-ny}\right]$$

$$= \frac{1}{2}b\,[e^{n(\pi - y)} - e^{-n(\pi - y)}]\sin nx$$

$$= b \sinh n\,(\pi - y) \sin nx.$$

$$u = \sum_{n=1}^{\infty} b_n \sinh n(\pi - y) \sin nx \quad \text{... (4) (General solution)}$$

Putting $y = 0$ and $u = \sin^2 x$ in (4), we get (given)

$$\sin^2 x = \sum_{n=1}^{\infty} b_n \sinh n\pi \sin nx$$

Here

$$b_n \sinh n\pi = \frac{2}{\pi}\int_0^{\pi} \sin^2 x \sin nx\, dx$$

$$= \frac{1}{\pi}\int_0^{\pi} (1 - \cos 2x) \sin nx\, dx$$

$$= \frac{1}{\pi}\int_0^{\pi}\left[\sin nx - \frac{1}{2}\{\sin (n+2)x + \sin(n-2)x\}\right] dx$$

$$= \frac{1}{\pi}\left[\frac{-\cos nx}{n} + \frac{\cos (n+2)x}{2(n+2)} + \frac{\cos (n-2)x}{2(n-2)}\right]_0^{\pi}$$

$$= \frac{1}{\pi}\left[\frac{-\cos n\pi}{n} + \frac{\cos (n+2)\pi}{2(n+2)} + \frac{\cos (n-2)\pi}{2(n-2)} + \frac{1}{n} - \frac{1}{2(n+2)} - \frac{1}{2(n-2)}\right]$$

$$= \frac{1}{\pi}\left[\left(\frac{-(-1)^n}{n} + \frac{(-1)^{n+2}}{2(n+2)} + \frac{(-1)^{n-2}}{2(n-2)}\right) + \left(\frac{1}{n} - \frac{1}{2(n+2)} - \frac{1}{2(n-2)}\right)\right]$$

$$= \frac{1}{2\pi}\left[\left(\frac{1}{n+2} + \frac{1}{n-2} - \frac{2}{n}\right)\{(-1)^n - 1\}\right], \text{ where } n \neq 2$$

$$b_n \sinh n\pi = \frac{-8}{\pi n(n^2-4)}, \text{ when } n \text{ is odd} \quad \ldots (5)$$

$$b_n \sinh \pi = 0, \text{ when } n \text{ is even and } \neq 2$$

Now, we have to find out $b_2 \sinh 2\pi$.

$$b_2 \sinh 2\pi = \frac{2}{\pi}\int_0^\pi \sin^2 x \sin 2x\, dx$$

$$= \frac{1}{\pi}\int_0^\pi (1-\cos 2x)\sin 2x\, dx = \frac{1}{\pi}\int_0^\pi \left(\sin 2x - \frac{1}{2}\sin 4x\right) dx$$

$$= \left(\frac{-\cos 2x}{2} + \frac{1}{8}\cos 4x\right)_0^\pi = 0$$

$$\Rightarrow \qquad b_2 = 0$$

On putting the value of b_n from (5) in (4), we get

$$u = \frac{-8}{\pi}\sum_{n=1,3,5,\ldots}^{\infty} \frac{\sin nx \sinh n(\pi - y)}{n(n^2-4)\sinh n\pi} \qquad \textbf{Ans.}$$

EXERCISE 11.4

1. Find the method of separation of variables, a particular solution of the equation

$$\frac{\partial^2 u}{\partial x^2} + \frac{\partial^2 u}{\partial y^2} = 0$$

that tends to zero as x tends to infinity and is equal to 2 cos y when $x = 0$. **Ans.** $u = 2e^{-x}\cos y$

2. Solve the equation: $u_{xx} + u_{yy} = 0$

$$u(0,y) = u(\pi,y) = 0 \text{ for all } y,$$
$$u(x,0) = k, \qquad 0 < x < \pi$$
$$\lim_{y\to\infty} u(x,y) = 0, \quad 0 < x < \pi$$

Ans. $u(x,y) = \sum_{n=1}^{\infty} b_n \sin nxe^{-ny}, k = \sum_{n=1}^{\infty} b_n \sin nx$

3. An infinitely long uniform plate is bounded by two parallel edges and an end at right angles to them. The breadth is π. This end is maintained at a temperature u_0 at all points and other edges are at zero temperature. Determine the temperature at any point of the plate in the steady state.

Ans. $u(x,y) = \frac{4u_0}{\pi}\left[e^{-y}\sin x + \frac{1}{3}e^{-3y}\sin 3x + \frac{1}{5}e^{-5y}\sin 5x + \ldots\right]$

4. Solve $\frac{\partial^2 V}{\partial x^2} + \frac{\partial^2 V}{\partial y^2} = 0$, given that (*i*) $V = 0$ when $x = 0$ and $x = c$; (*ii*) $V \to 0$ as $y \to \infty$; (*iii*) $V = V_0$ when $y = 0$.

Ans. $V(x,y) = \sum_{n=1}^{\infty} b_n \sin\frac{n\pi x}{c} e^{-\frac{n\pi y}{c}}, V_0 = \sum_{n=1}^{\infty} b_n \sin\frac{n\pi x}{c}$

5. The steady state temperature distribution in a thin plate bounded by the lines $x = 0$, $x = a$, $y = 0$ and $y = \infty$, is governed by the partial differential equation

$$\frac{\partial^2 u}{\partial x^2} + \frac{\partial^2 u}{\partial y^2} = 0.$$

Obtain the steady state temperature distribution under the conditions

$u(0, y) = 0,$ $u(a, y) = 0,$ $u(x, \infty) = 0$

$u(x, 0) = x,$ $0 \le x \le a/2$

$= a - x,$ $a/2 \le x \le a$

6. An infinitely long plane uniform plate is bounded by two parallel edges and an end at right angles to them. The breadth is π. This end is maintained at temperature u_0 at all points and the other edges are at zero temperature. Determine the temperature at any point of the plate in the steady state.

Ans. $u(x,y) = \frac{4u_0}{\pi} \sum_{n=1}^{\infty} \frac{1}{n} \sin x e^{-ny}$

7. A thin rectangular plate whose surface is imperious to heat flow has at $t = 0$ an arbitrary distribution of temperature $f(x, y)$. Its four edges $x = 0$, $x = a$, $y = 0$, $y = b$ are kept at zero temperature. Determine the temperature at a point of a plate as t increases. Discuss the problem when $f(x, y) = \beta \sin(\pi x/a) \sin(\pi y/b)$.

Ans. $u(x,y,t) - \sum_{m=1}^{\infty} \sum_{n=1}^{\infty} A_{mn} \sin\frac{n\pi x}{a} \sin\frac{m\pi x}{b} e^{-c^2 k_{mn}^2 t}$

$$u(x,y,0) = f(x,y) = \sum_{m=1}^{\infty} \sum_{n=1}^{\infty} A_{mn} \sin\frac{n\pi x}{a} \sin\frac{m\pi y}{b}$$

$$u(x,y,t) = 0, \text{ when } f(x,y) = \beta \sin\left(\frac{\pi x}{a}\right) \sin\left(\frac{\pi y}{b}\right)$$

8. The edges of a thin plate of side π cm are kept at temperature zero and faces are perfectly insulated. The initial temperature is $u(x, y, 0) = f(x, y) = xy(\pi - x)(\pi - y)$. Determine the temperature in the plate at time t.

Ans. $u(x, y, t) = \sum_{p=1}^{\infty} \sum_{q=1}^{\infty} A_{pq} \sin(2q-1)x \sin(2p-1)y e^{-\lambda_{pq}^2 t}$

where, $\lambda^2_{pq} = c^2[(2p-1)^2 + (2q-1)^2]$

And $A_{pq} = \frac{64}{\pi^2 (2p-1)^3 (2q-1)^3}$.

9. Find the deflection $u(x, y, t)$ of the square membrane with $a = b = c = 1$, if the initial velocity is zero and the initial deflection $f(x, y) = A \sin \pi x \sin 2\pi y$.

11.7 EQUATION OF HEAT FLOW IN TWO DIMENSIONS IN POLAR COORDINATES

Equation of heat flow in two dimension in polar coordinates.

Consider a sheet of conducting material of uniform density ρ, uniform thickness h, thermal conductivity k and specific heat c. Let O, the pole and OX, the initial line, be taken on the plane of sheet. As we are dealing with two dimensional heat-flow, the temperature function $u(r,\theta,\tau)$ at point (r,θ) of the plate is a function of r, θ and time t.

Consider an element of the sheet included between the circles $r = r$, $r = r + \delta r$ and the straight lines $\theta = \theta$ and $\theta = \theta + \delta\theta$ through the pole. Heat-flow directions are assumed to be positive in the positive directions associated with r and θ. The mass of the element $ABCD = \rho h(r\delta r\rho\theta)$

C
θ = 0
D
B
A
δθ
O
X
r = r

Let δu denote the temperature rise in the element during a short interval of time δt succeeding the time t.

$\therefore$ Rate of increase of heat content in the element

$$= \underset{\delta t\to 0}{\text{Lt}}\ \rho h(r\delta r\delta\theta)c\frac{\delta u}{\delta t}$$

$$= \rho hcr\delta r\delta\theta\frac{\partial u}{\partial t} \qquad \text{... (1)}$$

If R_1, R_2, R_3, R_4 are the rootes of flow of heat across the sides of the element through the edges AB, CD, AD and BC respectively, then

$$R_1 = -k\left(\frac{1}{r}\frac{\partial u}{\partial\theta}\right)_\theta h\delta r$$

$$R_2 = -k\left(\frac{1}{r}\frac{\partial u}{\partial\theta}\right)_{\theta+\delta\theta} h\delta r$$

$$R_3 = -k\left(\frac{\partial u}{\partial r}\right)_r hr\delta\theta$$

$$R_4 = -k\left(\frac{\partial u}{\partial r}\right)_{r+\delta r} h(r+\delta r)\delta\theta$$

The rate of increase of heat in the element $= R_1 - R_2 + R_3 - R_4$ which is equal to the expression in (1)

$$\therefore\ \rho hcr\,\delta r\,\delta\theta\frac{\partial u}{\partial t} = kh\left[\left\{\left(\frac{1}{r}\frac{\partial u}{\partial\theta}\right)_{\theta+\delta\theta} - \left(\frac{1}{r}\frac{\partial u}{\partial\theta}\right)_\theta\right\}\delta r + \left\{\left(\frac{\partial u}{\partial r}\right)_{r+\delta r} - \left(\frac{\partial u}{\partial r}\right)_r\right\}r\delta\theta + \delta r\delta\theta\left(\frac{\partial u}{\partial r}\right)_{r+\delta r}\right]$$

Dividing by $\delta r.\delta\theta.hr\rho c$,

$$\frac{\partial u}{\partial t} = \frac{k}{\rho c}\left[\frac{1}{r^2}\frac{\left\{\left(\frac{\partial u}{\partial\theta}\right)_{\theta+\delta\theta} - \left(\frac{\partial u}{\partial\theta}\right)_\theta\right\}}{\delta\theta} + \frac{\left(\frac{\partial u}{\partial r}\right)_{r+\delta r} - \left(\frac{\partial u}{\partial r}\right)_r}{\delta r} + \frac{1}{r}\left(\frac{\partial u}{\partial r}r + \delta r\right)\right]$$

Taking the limit as $\delta\theta, \delta r \to 0$,

$$\frac{\partial u}{\partial t} = \frac{k}{\rho c}\left[\frac{1}{r^2}\frac{\partial^2 u}{\partial^2\theta} + \frac{\partial^2 u}{\partial r^2} + \frac{1}{r}\frac{\partial u}{\partial r}\right]$$

Therefore the equation of heat-flow in polar coordinates is

$$\frac{\partial^2 u}{\partial r^2}+\frac{1}{r}\frac{\partial u}{\partial r}+\frac{1}{r^2}\frac{\partial^2 u}{\partial \theta^2}=\frac{1}{\alpha^2}\frac{\partial u}{\partial t} \qquad \text{where } \alpha^2=\frac{k}{\rho c}$$

In steady-state, $\frac{\partial u}{\partial t}=0$. Hence the equation of heat-flow in steady-state is

$$\nabla^2 u \equiv \frac{\partial^2 u}{\partial r^2}+\frac{1}{r}\frac{\partial u}{\partial r}+\frac{1}{r^2}\frac{\partial^2 u}{\partial \theta^2}=0$$

Example 33. *Solve* $\frac{\partial^2 u}{\partial r^2}+\frac{1}{r}\frac{\partial u}{\partial r}+\frac{1}{r^2}\frac{\partial^2 u}{\partial \theta^2}=0$

by the method of separation of variables.

Solution. $\frac{\partial^2 u}{\partial r^2}+\frac{1}{r}\frac{\partial u}{\partial r}+\frac{1}{r^2}\frac{\partial^2 u}{\partial \theta^2}=0$ or $r^2\frac{\partial^2 u}{\partial r^2}+r\frac{\partial u}{\partial r}+\frac{\partial^2 u}{\partial \theta^2}=0$ ... (1)

Let $u = R(r).T(\theta)$

$$\frac{\partial u}{\partial r}=\frac{dR}{dr}.T(\theta) \text{ and } \frac{\partial^2 u}{\partial r^2}=\frac{d^2R}{dr^2}.T(\theta)$$

$$\frac{\partial u}{\partial \theta}=R(r).\frac{dT}{d\theta} \text{ and } \frac{\partial^2 u}{\partial \theta^2}=R(r).\frac{d^2T}{d\theta^2}$$

Putting the values of $\frac{\partial^2 u}{\partial r^2}, \frac{\partial u}{\partial r}$ and $\frac{\partial^2 u}{\partial \theta^2}$ in (1), we get

$$r^2.\frac{d^2R}{dr^2}.T(\theta)+r\frac{dR}{dr}.T(\theta)+R(r)\frac{d^2T}{d\theta^2}=0$$

$$\left(r^2\frac{d^2R}{dr^2}+r.\frac{dR}{dr}\right).T+R\frac{d^2T}{d\theta^2}=0$$

$$\frac{r^2\frac{d^2R}{dr^2}+r.\frac{dR}{dr}}{R}=-\frac{1}{T}\frac{d^2T}{d\theta^2}=h \text{ (say)} \qquad \text{... (2)}$$

$$r^2\frac{d^2R}{dr^2}+r.\frac{dR}{dr}-hR=0 \qquad\qquad \frac{d^2T}{d\theta^2}+hT=0$$

Put $r=e^z$ $\qquad\qquad (D^2+h)T=0$

$[D(D-1)+D-h]R=0$ $\qquad$ **A.E. is $m^2+h=0$.... (4) $\Rightarrow m=\pm i\sqrt{h}$**

A.E. is $m^2-h=0$... (3) $\Rightarrow m=\pm\sqrt{h}$. $\qquad$ $T=c_3\cos(\sqrt{h}\theta)+c_4\sin(\sqrt{h}\theta)$

$$R=c_1e^{\sqrt{h}z}+c_2e^{-\sqrt{h}z}$$

$$R=c_1r^{\sqrt{h}}+c_2r^{-\sqrt{h}}$$

$$u = (c_1 r^{\sqrt{h}} + c_2 r^{-\sqrt{h}})[c_3 \cos(\sqrt{h}\theta) + c_4 \sin(\sqrt{h}\theta)] \quad \ldots (5)$$

Case 1. If $h = k^2$

(5) becomes $u = (c_1 r^k + c_2 r^{-k})\,[c_3 \cos(k\theta) + c_4 \sin(k\theta)]$ ... (6)

Case 2. If $h = 0$

On putting $h = 0$ in (3), we get

$D^2 = 0 \quad \Rightarrow \quad D = 0, 0$

$R = (c_5 + z c_6) \quad \Rightarrow \quad R = [c_5 + (\log r)\, c_6]$

Again on putting $h = 0$ in (4), we have

$D^2 = 0 \quad \Rightarrow \quad D = 0, 0$

$T = c_7 + \theta\, c_8$

But $u = R.T.$

$= (c_5 + (\log r)\, c_6)\ (c_7 + \theta\, c_8)$

Case 3. If $h = -p^2$

On putting $h = -p^2$ in (3), we get

$D^2 + p^2 = 0 \quad \Rightarrow \quad D = \pm\, i\, p$

$R = c_9 \cos pz + c_{10} \sin pz$

$= c_9 \cos(p \log r) + c_{10} \sin(p \log r)$

Again on putting $h = -p^2$ in (4), we get

$D^2 - p^2 = 0 \quad \Rightarrow \quad D = \pm\, p$

$T = c_{11}\, e^{p\theta} + c_{12}\, e^{-p\theta}$

$u = RT$

$= [c_9 \cos(p \log r) + c_{10} \sin(p \log r)]\,.\,(c_{11}\, e^{p\theta} + c_{12}\, e^{-p\theta})$

Then there are three possible solutions

$$u = (c_1 r^k + c_2 r^{-k})[c_3 \cos(k\theta) + c_4 \sin(k\theta)]$$

$$u = (c_5 + c_6 \log r)(c_7 + c_8\theta)$$

$$u = [c_9 \cos(p \log r) + c_{10} \sin(p \log r)][c_{11} e^{p\theta} + c_{12} e^{-p\theta}]$$ **Ans.**

Example 34. *The diameter of a semicircular plate of radius a is kept at 0°C and the temperature at the semicircular boundary is T°C. Find the steady state temperature in the plate.*

Solution. Let the centre O of the semicircular plate be the pole and the bounding diameter be as the initial line. Let $u(r,\theta)$ be the steady state temperature at any point $P(r,\theta)$ and u satisfies the equation

$$r^2 \frac{\partial^2 u}{\partial r^2} + r \frac{\partial u}{\partial r} + \frac{\partial^2 u}{\partial \theta^2} = 0 \quad \ldots (1)$$

The boundary conditions are

(*i*) $u(r, 0) = 0 \qquad 0 \le r \le a$

(*ii*) $u(r, \pi) = 0 \qquad 0 \le r \le a$

(*iii*) $u(a, \theta) = T.$

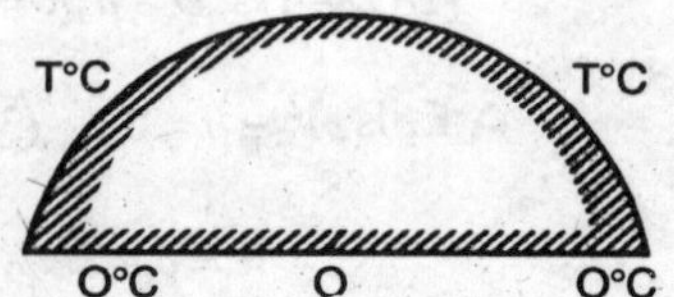

From conditions (*ii*) and (*iii*), we have $u \to 0$ as $r \to 0$. Hence the appropriate solution of (1) is as solved in example 33.

$$u = (c_1 r^p + c_2 r^{-p})(c_3 \cos p\theta + c_4 \sin p\theta) \quad ...(2)$$

Putting $u\,(r, 0) = 0$ in (2), we get

$$0 = (c_1 r^p + c_2 r^{-p})\, c_3 \Rightarrow c_3 = 0$$

On putting $c_3 = 0$ in (2), we get

$$u = (c_1 r^p + c_2 r^{-p}) c_4 \sin p\theta \quad ...(3)$$

Putting $u(r, \pi) = 0$ in (3), we get

$$0 = (c_1 r^p + c_2 r^{-p}) c_4 \sin p\pi \Rightarrow \sin p\pi = 0 = \sin n\pi \quad \Rightarrow \quad p\pi = n\pi \Rightarrow p = n$$

On putting $p = n$, (3) becomes,

$$u = (c_1 r^n + c_2 r^{-n}) c_4 \sin n\theta \quad ...(4)$$

Since, $u = 0$ when $r = 0$

$$0 = c_2$$

(4) becomes, $u = c_1 c_4 r^n \sin n\theta$

The most general solution of (1) is

$$u(r,\theta) = \sum_{n=1}^{\infty} b_n r^n \sin n\theta \qquad (c_1\, c_2 - b_n) \quad ...(5)$$

Putting $r = a$ and $u = T$ in (5), we have

$$T = \sum_{n=1}^{\infty} b_n\, a^n \sin n\theta$$

By Fourier half range series, we get

$$b_n a^n = \frac{2}{\pi}\int_0^{\pi} T \sin n\theta\, d\theta = \frac{2}{\pi} T \left(\frac{-\cos n\theta}{n}\right)_0^{\pi} = \frac{2T}{n\pi}[-(-1)^n + 1]$$

$$b_n a^n = 0 \qquad \text{when } n \text{ is even,}$$

$$b_n a^n = \frac{4T}{n\pi}, \qquad \text{when } n \text{ is odd.}$$

$$\Rightarrow \qquad b_n = \frac{4T}{n\pi\, a^n}$$

Hence, (5) becomes

$$u(r, \theta) = \frac{4T}{\pi}\left[\frac{r/a}{1}\sin\theta + \frac{(r/a)^3}{3}\sin 3\theta + \frac{(r/a)^5}{5}\sin 5\theta + ...\right] \quad \textbf{Ans.}$$

EXERCISE 11.5

1. Solve the steady-state temperature equation

$$\frac{\partial^2 T}{\partial r^2} + \frac{1}{r}\frac{\partial T}{\partial r} + \frac{1}{r^2}\frac{\partial^2 T}{\partial \theta^2} = 0; \quad 10 \le r \le 20, \quad 0 \le \theta \le 2\pi$$

subject to the following conditions:

$T(10,\theta) = 15\cos\theta$ and $T(20,\theta) = 30\sin\theta$ **Ans.** $T(r,\theta) = \frac{4T}{\pi}\left[\frac{r}{a}\sin\theta + \frac{1}{3}\left(\frac{r}{a}\right)^3 \sin 3\theta + ...\right]$

2. A semi-circular plate of radius a has its circumference kept at temperature $u(a,\theta) = k\theta(\pi - \theta)$ while the boundary diameter is kept at zero temperature. Find the steady state temperature distribution $u(r,\theta)$ of the plate assuming the lateral surfaces of the plate to be insulated.

Ans. $u(r,\theta) = \dfrac{8k}{\pi}\sum_{1}^{\infty}\left(\dfrac{r}{a}\right)^{2n-1}\dfrac{\sin(2n-1)\theta}{(2n-1)^3}$

3. Find the steady state temperature in a circular plate of radius a which has one half of its circumference at 0°C and the other half at 60°C.

Ans. $u(r,\theta) = 50 - \dfrac{200}{n}\sum_{n=1}^{\infty}\dfrac{1}{2n-1}\left(\dfrac{r}{a}\right)^{2n-1}\sin(2n-1)\theta.$

11.8 TRANSMISSION LINE EQUATIONS

$$\frac{\partial^2 V}{\partial x^2} = RC\frac{\partial V}{\partial t}$$

$$\frac{\partial^2 i}{\partial x^2} = RC\frac{\partial i}{\partial t}$$ are called telegraph equations,

where V = potential, i = current, C = capacitance, L = inductance

$$\frac{\partial^2 V}{\partial x^2} = LC\frac{\partial^2 V}{\partial t^2}$$

$$\frac{\partial^2 i}{\partial x^2} = LC\frac{\partial^2 i}{\partial t^2}$$

are called radio equations.

Example 35. *Find the current i and voltage v in a transmission line of length l, t seconds after the ends are suddenly grounded, given that* $i(x, 0) = i_0$, $v(x, 0) = v_0 \sin\left(\dfrac{\pi x}{l}\right)$ *and that R and G are negligible.*

Solution. $$\frac{\partial^2 v}{\partial x^2} = LC\frac{\partial^2 v}{\partial t^2}$$

Let $v = XT$ where X and T are the functions of x and t respectively.

$$\frac{\partial^2 v}{\partial x^2} = T\frac{d^2 X}{dx^2} \quad \text{and} \quad \frac{\partial^2 v}{\partial t^2} = X\frac{d^2 T}{dt^2}$$

$$T\frac{d^2 X}{dx^2} = LCX\frac{d^2 T}{dt^2}$$

$$\frac{\frac{d^2 X}{dx^2}}{X} = LC\frac{\frac{d^2 T}{dt^2}}{T} = -p^2 \text{ say}$$

Since the initial conditions suggest the values of v and i are periodic functions,

$\therefore$ $$X = c_1 \cos px + c_2 \sin px$$

$$T = c_3 \cos\frac{pt}{\sqrt{LC}} + c_4 \sin\frac{pt}{\sqrt{LC}}$$

$$\Rightarrow \qquad v = (c_1 \cos px + c_2 \sin px)\left(c_3 \cos\frac{pt}{\sqrt{LC}} + c_4 \sin\frac{pt}{\sqrt{LC}}\right) \qquad ...(1)$$

When $t = 0,\ v = v_0 \sin\dfrac{\pi x}{l}$

$$v_0 \sin\frac{\pi x}{l} = (c_1 \cos px + c_2 \sin px)\, c_3 \qquad ...(2)$$

On equating the coefficients, we get

$$c_1 c_3 = 0 \Rightarrow c_1 = 0 \quad \text{and} \quad c_2 c_3 = v_0,\ p = \frac{\pi}{l}$$

(1) becomes

$$v = \sin\frac{\pi x}{l}\left[v_0 \cos\frac{\pi t}{l\sqrt{LC}} + c_2 c_4 \sin\frac{\pi t}{l\sqrt{LC}}\right] \qquad ...(3)$$

Now when $t = 0,\ i = i_0$ (constant)

Hence, $\dfrac{\partial i}{\partial x} = 0$

$$\frac{\partial i}{\partial x} = \frac{-c\partial v}{\partial t} \qquad \therefore \quad \frac{\partial v}{\partial t} = 0 \text{ when } t = 0$$

Now

$$\frac{\partial v}{\partial t} = \sin\frac{\pi x}{l}\left(\frac{\pi}{l\sqrt{LC}}\right)\left[-v_0 \sin\frac{\pi t}{l\sqrt{LC}} + c_2 c_4 \cos\frac{\pi t}{l\sqrt{LC}}\right] \qquad ...(4)$$

On putting $\dfrac{\partial v}{\partial t} = 0$ and $t = 0$ in (4), we get $\quad c_2\, c_4 = 0 \Rightarrow c_4 = 0$

(3) is reduced to $\quad v = v_0 \sin\dfrac{\pi x}{l} \cos\dfrac{\pi t}{l\sqrt{LC}}$

$$\frac{\partial v}{\partial x} = \frac{\pi}{l} v_0 \cos\frac{\pi x}{l} \cos\frac{\pi t}{l\sqrt{LC}} = -L\frac{\partial i}{\partial t} \qquad ...(5)$$

$$\frac{\partial v}{\partial t} = -\frac{v_0 \pi}{l\sqrt{LC}} \sin\frac{\pi x}{l} \sin\frac{\pi t}{l\sqrt{LC}} = \frac{-1}{C}\frac{\partial t}{\partial x} \qquad ...(6)$$

Integrating (5) and (6), we get

$$i = -v_0\sqrt{\frac{C}{L}} \cos\frac{\pi x}{l} \sin\frac{\pi t}{l\sqrt{LC}} + f(x)$$

$$i = -v_0\sqrt{\frac{C}{L}} \cos\frac{\pi x}{l} \sin\frac{\pi t}{l\sqrt{LC}} + F(t)$$

$\therefore \quad f(x)$ and $F(t)$ must be constant only, since $i = i_0$ when $t = 0$

$\therefore \qquad$ constant $= i_0 = f(x)$

Hence,

$$i = i_0 - v_0\sqrt{\frac{C}{L}} \cos\frac{\pi x}{l} \sin\frac{\pi t}{l\sqrt{LC}}. \qquad \textbf{Ans.}$$

EXERCISE 11.6

1. A transmission line 1,000 miles long is initially under steady state condition with potential 1,300 volts at the sending end ($x = 0$) and 1,200 volts at the receiving end ($x = 1000$). The terminal end of the line is suddenly grounded but the potential at the source is kept at 1,300 volts. Assuming the inductance and leakage to be negligible, find the potential $v(x, t)$, if it satisfies the equation $v_t = (1/RC)v_{xx}$.

Ans. $v(x,y) = \sum_1^{\infty} b_n \sin nx\, e^{-ny}$ and $k = \sum_1^{\infty} b_n \sin nx$

2. Obtain a solution of the telegraph equation

$$\frac{\partial^2 e}{\partial x^2} = RC\frac{\partial e}{\partial t}$$

suitable for the case when e decays with the time and when there is a steady fall of potential from e_0 to 0 along the line of length l initially and the sending end is suddenly earthed.

Ans. $e(x,t) = \frac{2e_0}{\pi}\sum_{n=1}^{\infty}\frac{1}{n}\sin\frac{n\pi x}{l}e^{\frac{-n^2\pi^2 t}{CRl^2}}$

11.9 LAPLACE EQUATION

Laplace equation is used to solve engineering problem and its theory is called potential theory. Its solutions are known as harmonic functions.

Example 36. *Solve the Laplace equation* $\frac{\partial^2 u}{\partial x^2} + \frac{\partial^2 u}{\partial y^2} = 0$ *in a rectangle in the xy-plane with* $u(x, 0) = 0$, $u(x, b) = 0$, $u(0, y)$ *and* $u(a, y) = f(y)$ *parallel to y-axis.* [U.P.T.U. 2010, 2008]

Solution. Here, the Laplace equation is

$$\frac{\partial^2 u}{\partial x^2} + \frac{\partial^2 u}{\partial y^2} = 0 \qquad \text{...(1)}$$

Let $u = XY$

Where X is a function of x only Y is a function of y only.

$$\frac{\partial u}{\partial x} = \frac{\partial (XY)}{\partial x} = Y\frac{\partial X}{\partial x}$$

$$\frac{\partial^2 u}{\partial x^2} = \frac{\partial}{\partial x}\left(\frac{\partial u}{\partial x}\right) = \frac{\partial}{\partial x}\left(Y\frac{\partial X}{\partial x}\right) = Y\frac{\partial^2 X}{\partial x^2}$$

$$\frac{\partial u}{\partial y} = \frac{\partial (XY)}{\partial y} = X\frac{\partial Y}{\partial y}$$

$$\frac{\partial^2 u}{\partial y^2} = \frac{\partial}{\partial y}\left(\frac{\partial}{\partial y}(XY)\right) = \frac{\partial}{\partial y}\left(X\frac{\partial Y}{\partial y}\right) = X\frac{\partial^2 Y}{\partial^2 y}$$

Putting the values of $\frac{\partial^2 u}{\partial x^2}$ and $\frac{\partial^2 u}{\partial v^2}$ in (1), we get

$$Y\frac{\partial^2 u}{\partial x^2} + X\frac{\partial^2 u}{\partial y^2} = 0 \text{ or } YX'' + XY'' = 0$$

On dividing by XY and separating the variables, we get

$$\frac{Y''}{Y} = -\frac{X''}{X}$$

Case I. If $$\frac{Y''}{Y} = -\frac{X''}{X} = p^2 \text{(say)}$$

$$\frac{Y''}{Y} = p^2$$

$$Y'' - p^2 Y = 0$$

$A. E.$ is $m^2 - p^2 = 0 \Rightarrow m = \pm p$

$$Y = c_1 e^{py} + c_2 e^{-py} \quad ...(2)$$

$$-\frac{X''}{X} = p^2$$

$$\Rightarrow \quad X'' + p^2 X = 0$$

Auxiliary equation is

$$m^2 + p^2 = 0 \Rightarrow m = \pm ip$$

$$X = c_3 \cos px + c_4 \sin px \quad ...(3)$$

Putting $y = 0$ and $Y = 0$ in (2), we get

$$c_1 + c_2 = 0 \Rightarrow c_2 = -c_1$$

On putting $y = b$ and Y = 0, we get

$$c_1 e^{pb} + c_2 e^{-pb} = 0$$

$$c_1(e^{pb} - e^{-pb}) = 0$$

$$c_1 = 0 \qquad [c_1 = c_2 = 0]$$

$$\Rightarrow \quad Y = 0$$

But $$u = XY \qquad [Y = 0]$$

$$= X(0) = 0, \text{ which is not possible}$$

Case II. If $$\frac{Y''}{Y} = -\frac{X''}{X} = 0 \text{ (say)}$$

$$\frac{Y''}{Y} = 0$$

$$Y'' = 0 \Rightarrow Y = c_5 + c_6 y \quad ...(4)$$

$$-\frac{X''}{X} = 0$$

$$X'' = 0 \Rightarrow X = c_7 + c_8 x$$

Putting $y = 0$, $Y = 0$ in (4), we get

$$0 = c_5$$

Putting $y = b$, $Y = 0$ in (4), we get

$$0 = c_6 b \Rightarrow c_6 = 0$$

$$\Rightarrow \quad Y = 0$$

$$u = XY \qquad [Y = 0]$$

$$= X(0) = 0, \text{ which is not possible.}$$

Case III. If $$\frac{Y''}{Y} = -\frac{X''}{X} = -p^2$$

(*i*) $$\frac{Y''}{Y} = -p^2$$

$$\Rightarrow \quad Y'' + p^2 Y = 0$$

$A.E.$ is $m^2 + p^2 = 0 \Rightarrow m = \pm pi$

$$Y = c_9 \cos py + c_{10} \sin py \quad ...(5)$$

On putting $y = 0$, $Y = 0$ in (5), we get

$$\Rightarrow \quad c_9 = 0$$

$$-\frac{X''}{X} = -p^2$$

$$\Rightarrow \quad X'' - p^2 X = 0$$

A.E. is $m^2 - p^2 = 0 \Rightarrow m = \pm p$

$$X = c_{11} e^{px} + c_{12} e^{-px}$$

Putting $y = b$ and $Y = 0$ in (5), we get

$$0 = c_9 \cos pb + c_{10} \sin pb.$$

$$c_{10} \sin pb = 0 \qquad (\because c_9 = 0)$$

$$\Rightarrow \quad \sin pb = 0$$

$$\text{and } \sin n\pi = 0$$

$$\Rightarrow \quad pb = n\pi \Rightarrow p = \frac{n\pi}{b}$$

Putting $c_9 = 0,\ p = \frac{n\pi}{b}$ in (5), we get

$$Y = c_{10} \sin \frac{n\pi y}{b}$$

$$u = XY$$

$$u = c_{10} \sin \frac{n\pi y}{b}\left(c_{11}e^{\frac{n\pi x}{b}} + c_{12}e^{-\frac{n\pi x}{b}}\right) \qquad \text{... (6)}$$

Putting $x = 0$ and $u = 0$ in (6), we get

$$0 = c_{10} \sin \frac{n\pi y}{b}(c_{11} + c_{12})$$

$$c_{11} + c_{12} = 0 \Rightarrow c_{12} = -c_{11}$$

Putting the value of $c_{12} = -c_{11}$ in (6), we get

$$u = c_{10}c_{11} \sin \frac{n\pi y}{b}\left(e^{\frac{n\pi x}{b}} - e^{-\frac{n\pi x}{b}}\right)$$

$$u = 2c_{10}c_{11} \sin \frac{n\pi y}{b} \sinh \frac{n\pi x}{b}$$

$$u = b_n \sin \frac{n\pi y}{b} \sinh \frac{n\pi x}{b} \qquad (b_n = 2\, c_{10}\, c_{11})$$

$$u = \sum_{n=1}^{\infty} b_n \sin \frac{n\pi y}{b} \sinh \frac{n\pi x}{b} \qquad \text{... (7) (General solution)}$$

Putting $x = a$ and $u = f(y)$ in (7), we get

$$u = f(y) = \sum_{n=1}^{\infty} b_n \sin \frac{n\pi y}{b} \sinh \frac{n\pi x}{b}$$ **Ans.**

We also know that

$$\left(\sin \frac{n\pi a}{b}\right) b_n = \frac{2}{b}\int_0^b f(y) \sin \frac{n\pi y}{b}\, dy$$

$$b_n = \frac{2}{b \sinh\left(\frac{n\pi a}{b}\right)} \int_0^b f(y) \sin \frac{n\pi y}{b}\, dy.$$

Example 37. *Find the deflection $u(x, y, t)$ of the square membrane with $a = b = c = 1$, if the initial velocity is zero and the initial deflection $f(x, y) = A \sin \pi x \sin 2\pi y$.*

(Q.Bank U.P.T.U. 2001)

Solution. Here, the equation of the vibrations of the square membrane

$$\frac{\partial^2 u}{\partial t^2} = c^2\left(\frac{\partial^2 u}{\partial x^2}+\frac{\partial^2 u}{\partial y^2}\right) \qquad ...(1)$$

Let $u = XYT$ ... (2)

Where X is function of x only, Y is function of y only, and T is function of t only.

$$\frac{\partial^2 u}{\partial t^2} = \frac{\partial^2}{\partial t^2}(XYT) = XY\frac{d^2T}{dt^2}$$

$$\frac{\partial^2 u}{\partial x^2} = \frac{\partial^2}{\partial x^2}(XYT) = YT\frac{d^2X}{dx^2}$$

$$\frac{\partial^2 u}{\partial y^2} = \frac{\partial^2}{\partial y^2}(XYT) = XT\frac{d^2Y}{dy^2}$$

On putting the values of $\frac{\partial^2 u}{\partial t^2}$, $\frac{\partial^2 u}{\partial x^2}$ and $\frac{\partial^2 u}{\partial y^2}$ in (1), we get

$$XY\frac{d^2T}{dy^2} = c^2\left(YT\frac{d^2X}{dx^2}+XT\frac{d^2Y}{dy^2}\right)$$

On dividing by $c^2\,XYT$, we get

$$\frac{\frac{d^2T}{dt^2}}{c^2T} = \frac{\frac{d^2X}{dx^2}}{X}+\frac{\frac{d^2Y}{dy^2}}{Y}$$

This will be true only when each member is a constant. Choosing suitably, we have

$$\frac{\frac{d^2X}{dx^2}}{X} = -k^2, \quad \frac{\frac{d^2Y}{dy^2}}{Y} = -l^2 \quad \Rightarrow \quad \frac{\frac{d^2T}{dt^2}}{c^2T} = -(k^2+l^2)$$

$\frac{d^2X}{dx^2}+k^2X=0$	$\frac{d^2Y}{dy^2}+l^2Y=0$	$\frac{d^2T}{dt^2}+(k^2+l^2)c^2T=0$
$A.E.\ is\ m^2+k^2=0$	$A.E.\ is\ m^2+l^2=0$	$A.E.\ is\ m^2+(k^2+l^2)c^2=0$
$\Rightarrow m=ik, m=-ik$	$\Rightarrow m=il, m=-il$	$m=\sqrt{k^2+l^2}\,c,\ m=-i\sqrt{k^2+l^2}\,c$

$X = c_1\cos kx + c_2\sin kx \quad Y = c_3\cos ly + c_4\sin ly \quad T = c_5\cos\sqrt{k^2+l^2}\,ct + c_6\sin\sqrt{k^2+l^2}\,ct.$

Putting the values of X; Y and T in (2), we get (given)

$$u = (c_1\cos kx + c_2\sin kx)(c_3\cos ly + c_4\sin l\,y)[c_5\cos\sqrt{k^2+l^2}\ ct + c_6\sin\sqrt{k^2+l^2}\ ct] \quad ...(3)$$

On putting $x = 0$ and $u = 0$ in (3), we get (given)

$$0 = c_1(c_3\cos ly + c_4\sin ly)[c_5\cos\sqrt{k^2+l^2}\ ct + c_6\sin\sqrt{k^2+l^2}\ ct] \quad ...(4)$$

$c_1 = 0$

On putting $c_1 = 0$ in (3), we get (given)

$$u = c_2\sin kx(c_3\cos ly + c_4\sin ly)(c_5\cos\sqrt{k^2+l^2}\ ct + c_6\sin\sqrt{k^2+l^2}\ ct) \quad ...(5)$$

Putting $x = 1$ and $u = 0$ in (5), we get (given)

$$0 = c_2 \sin k(c_3 \cos ly + c_4 \sin ly)(c_5 \cos\sqrt{k^2+l^2}\, ct + c_6 \sin\sqrt{k^2+l^2}\, ct) \quad \text{... (6)}$$

$\Rightarrow \sin k = 0 = \sin m\pi \quad \Rightarrow k = m\pi$

Putting the value k in (5), we get

$$u = c_2 \sin m\pi x(c_3 \cos ly + c_4 \sin ly)(c_5 \cos\sqrt{m^2\pi^2+l^2}\, ct + c_6 \sin\sqrt{m^2\pi^2+l^2}\, ct) \quad \text{... (7)}$$

Putting $y = 0$, $u = 0$ in (7), we get

$$0 = c_2 \sin m\pi x . c_3 (c_5 \cos\sqrt{m^2\pi^2+l^2}\, ct + c_6 \sin\sqrt{m^2\pi^2+l^2}\, ct) \quad \text{... (8)}$$

$\Rightarrow c_3 = 0$

Putting the value of c_3 in (7), we get

$$u = c_2 c_4 \sin m\pi x \sin ly\, (c_5 \cos\sqrt{m^2\pi^2+l^2}\, ct + c_6 \sin\sqrt{m^2\pi^2+l^2}\, ct) \quad \text{... (9)}$$

Putting $y = 1$ and $u = 0$ in (9), we get (given)

$$0 = c_2 c_4 \sin m\pi x \sin l\, (c_5 \cos\sqrt{m^2\pi^2+l^2}\, ct + c_6 \sin\sqrt{m^2\pi^2+l^2}\, ct) \quad \text{... (6)}$$

$\Rightarrow \sin l = 0 = \sin n\pi \quad \Rightarrow \quad l = n\pi$

Putting the value $l = n\pi$ in (9), we get

$$u = c_2 c_4 \sin m\pi x \sin n\pi y\, (c_5 \cos\sqrt{m^2\pi^2+n^2\pi^2}\, ct + c_6 \sin\sqrt{m^2\pi^2+n^2\pi^2}\, ct) \quad \text{... (10)}$$

Putting $c_2 c_4 c_5 = A_{mn}$, $c_2 c_4 c_6 = B_{mn}$ and $p = \pi c\sqrt{m^2+n^2}$ in (10), we get the general solution as

$$u(x, y, t) = \sum_{m=1}^{\infty}\sum_{n=1}^{\infty} \sin m\pi x \sin n\pi y\, (A_{mn} \cos pt + B_{mn} \sin pt) \quad \text{... (11)}$$

Differentiating (11) w.r.t. 't', we get

$$\frac{du}{dt} = \sum_{m=1}^{\infty}\sum_{n=1}^{\infty} \sin m\pi x \sin n\pi y(-pA_{mn} \sin pt + PB_{mn} \cos pt)$$

Applying the condition $\frac{\partial u}{\partial t} = 0$, when $t = 0$

$$0 = \sum_{m=1}^{\infty}\sum_{n=1}^{\infty} \sin m\pi x \sin \pi y(PB_{mn}) \Rightarrow B_{mn} = 0 \quad \text{... (12)}$$

Putting $B_{mn} = 0$ in (11), we get

$$u(x,y,t) = \sum_{m=1}^{\infty}\sum_{n=1}^{\infty} \sin m\pi x \sin n y(A_{mn} \cos pt) \quad \text{... (13)}$$

Using the condition $u = f(x,y) = A \sin \pi x \sin 2\pi y$, when $t = 0$

On putting the value of u and t in (13), we get

$$A \sin \pi x \sin 2\pi y = \sum_{m=1}^{\infty}\sum_{n=1}^{\infty} \sin m\pi n \sin n\pi\, y(A_{mn})$$

This is a double Fourier series.

$$A_{mn} = \frac{2}{1}, \frac{2}{1} \int_0^1 \int_0^1 A \sin \pi x \sin 2\pi y \sin m\pi x \sin n\pi y\, dx\, dy$$

Clearly $A_{m_1} = A_{m_3} = A_{m_4} = \text{......} = 0$

But $A_{m2} = 4A \int_0^1 \int_0^1 \sin \pi x \sin m \pi x \sin^2 2 \pi y \, dx \, dy$

$$= 2A \int_0^1 \int_0^1 \sin \pi x \sin m \pi x \, dx \int_0^1 (1 - \cos 4 \pi y) \, dy$$

$$= 2A \int_0^1 \sin \pi x \sin m \pi x \, dx \left(y - \sin \frac{4\pi y}{4\pi} \right)_0^1$$

$$= 2A \int_0^1 \sin \pi x \sin m \pi x \, dx \left(1 - \sin \frac{4\pi}{4\pi} \right)$$

$$= 2A \int_0^1 \sin \pi x \sin m \pi x \, dx \qquad (\sin 4\pi = 0)$$

Again clearly, $A_{22} = A_{32} = A_{42} = \ldots\ldots = 0$

But $A_{12} = 2A \int_0^1 \sin^2 \pi x \, dx = A \int_0^1 (1 - \cos 2 \pi x) \, dx$

$$\Rightarrow A_{12} = A \left(x - \frac{\sin 2\pi x}{2\pi} \right)_0^1 = A \left(1 - \frac{\sin 2\pi}{2\pi} \right) = A \qquad \ldots(14) \quad (\sin \pi 2 = 0)$$

Form (11), (12) and (14), we get,

$$u(x, y, t) = A \sin \pi x \sin 2 \pi y \cos pt \qquad \ldots(15)$$

Where $p = \pi c \sqrt{m^2 + n^2}$

Putting $c = 1$ (givne n), $m = 1$, $n = 2$ in p, we get

$$p = \pi (11) \sqrt{1^2 + (2)^2} = \pi\sqrt{5}$$

Puting $p = \pi \sqrt{5}$ in (15), we get

$u(x,y,t) = A \cos \pi \sqrt{5} \, t \sin \pi x \sin 2 \pi y.$ **Ans.**

OBJECTIVE TYPE QUESTIONS

Choose the correct alternative

1. The general solution the wave equations

$$\frac{\partial^2 y}{\partial t^2} = c^2 \frac{\partial^2 y}{dx^2} \text{ is}$$

(a) $y = (c_5 \cos c \sqrt{k}\, t + c_6 \sin c \sqrt{k}\, t)(c_7 \cos \sqrt{k}\, x + c_8 \sin \sqrt{k}\, x)$

(b) $y = (c_5 \cos c \sqrt{k}\, t + \sin c \sqrt{k}\, t)(c_6 \cos \sqrt{k}\, x + c_7 \sin \sqrt{k}\, x)$

(c) $y = (c_5 \cos c \sqrt{k}\, t + c_6 \sin c \sqrt{k}\, t)(c_7 \cos \sqrt{k}\, x + \sin \sqrt{k}\, x)$

(d) $y = (c_5 \cos c \sqrt{k}\, t + c_6 \sin c \sqrt{k}\, t)(\cos \sqrt{k}\, x + \sin \sqrt{k}\, x)$ **Ans.** (*a*)

2. The general solution of the *P.D.E.*

$$\frac{\partial u}{\partial x} = a^2 \frac{\partial^2 u}{\partial x^2} \text{ is}$$

(a) $u = \sum_{n=1}^{\infty} b_n (c_1 \cos pt + c_2 \sin pt) \, e^{p^2 c^2 x}$

(b) $u = \sum_{n=1}^{\infty} b_n (c_1 \cos pt + c_2 \sin pt)\, e^{-p^2 c^2 x}$

(c) $u = \sum_{n=1}^{\infty} b_n (c_1 \cos px + c_2 \sin px)\, e^{-p^2 c^2 t}$

(d) $u = \sum_{n=1}^{\infty} b_n (c_1 \cos px + c_2 \sin px)\, e^{p^2 c^2 t}$ **Ans.** (c)

3. The general solution of two dimensional heat flow

$$\frac{\partial^2 u}{\partial x^2} + \frac{\partial^2 u}{\partial t^2} = 0 \text{ is}$$

(a) $u = (c_1 \cos px + c_2 \sin px)(e^{py} + e^{-py})$

(b) $u = (c_1 \cosh px + c_2 \sinh px)(c_3 e^{py} + c_4 e^{-py})$

(c) $u = (c_1 \cos py + c_2 \sin py)(c_3 e^{px} + c_4 e^{-px})$

(d) $u = (c_1 \cos px + c_2 \sin px)(c_3 e^{py} + c_4 e^{-py})$ **Ans.** (d)

4. The formula of b_n is the equation

$$u = \sum_{n=1}^{\infty} b_n \sin\left(\frac{n\pi x}{l}\right) \text{ is}$$

(a) $\frac{2}{l}\int_0^l u_0 \sin\left(\frac{n\pi x}{l}\right) dx$

(b) $\frac{1}{l}\int_0^{2l} u_0 \sin\left(\frac{n\pi x}{l}\right) dx$

(c) $\frac{1}{l}\int_0^l u_0 \sin\left(\frac{n\pi x}{l}\right) dx$

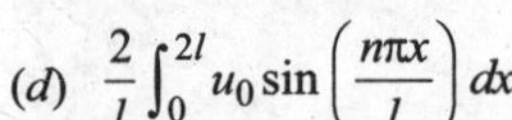

(d) $\frac{2}{l}\int_0^{2l} u_0 \sin\left(\frac{n\pi x}{l}\right) dx$ **Ans.** (a)

5. In the figure a string *OAB* is stretched to a height *b*, the equation of the *OA* portion of the string is:

Y, A (l, b), b, O, l, B $(2l, 0)$, X

(a) $x = -\frac{by}{l}$

(b) $y = \frac{by}{l}$

(c) $y = -\frac{by}{l}$

(d) $x = \frac{b}{l} y$ **Ans.** (b)

6. In the given figure write down OB portion of the string :

(a) $y = -\frac{b}{l}(x + l)$

(b) $y = -\frac{b}{l}(x - l)$

(c) $y = -\frac{b}{l}(x - 2l)$

(d) $y = \frac{b}{l}(x - 2l)$ **Ans.** (c)

7. If $u = x^2 + t^2$ is a solution of $c^2 \dfrac{\partial^2 u}{\partial x^2} = \dfrac{\partial^2 u}{\partial t^2}$, then the value of C is

(*a*) 1. (*b*) 2 (*c*) – 2 (*d*) – 1/2 **Ans.** (*a*)

8. Laplace's equation in polar coordinates is

(*a*) $\dfrac{\partial^2 u}{\partial r^2} + \dfrac{\partial u}{\partial r} + \dfrac{\partial u}{\partial \theta} = 0$ (*b*) $r^2 \dfrac{\partial^2 u}{\partial r^2} + r\dfrac{\partial u}{\partial r} + \dfrac{\partial^2 u}{\partial \theta^2} = 0$

(*c*) $\dfrac{\partial^2 u}{\partial r^2} + \dfrac{1}{r}\dfrac{\partial u}{\partial r} = 0$ (*d*) $\dfrac{\partial^2 u}{\partial r^2} + \dfrac{1}{r}\dfrac{\partial u}{\partial r} + \dfrac{\partial^2 u}{\partial \theta^2} = 0$ **Ans.** (*b*)

9. In one dimensional heat flow, the condition on temperature is

(*a*) temperature always increases

(*b*) temperature decreases as time increases

(*c*) temperature always decreases

(*d*) temperature remains always non zero at all times.

10. If the ends $x = 0$ and $x = L$ are insulated in one dimensional heat flow problems, then the boundary conditions are

(*a*) $\dfrac{\partial u(o,t)}{\partial x} = 0,\ \dfrac{\partial u(L,t)}{\partial x} = 1$ at $t = 0$.

(*b*) $\dfrac{\partial u(o,t)}{\partial x} = 1,\ \dfrac{\partial u(L,t)}{\partial x} = 1$ at $t = 0$.

(*c*) $\dfrac{\partial u(o,t)}{\partial x} = 0,\ \dfrac{\partial u(L,t)}{\partial x} = 0$ for all t.

(*d*) $\dfrac{\partial u(o,t)}{\partial x} = 0,\ \dfrac{\partial u(L,t)}{\partial x} = 1$ for all t. **Ans.** (*c*)

Indicate True or False for the following :

11. The small transverse vibrations of a string are governed by one dimensional heat equation $y_t = a^2 y_{xx}$. (True/False) (*U.P. II Semester, 2009*) **Ans.** False

12. Two dimensional steady state heat flow is given by Laplace's equation $u_t = a^2(u_{xx} + u_{tt})$. (True/False) (*U.P. II Semester, 2009*) **Ans.** False

13. $\dfrac{\partial^2 y}{\partial t^2} = c^2 \dfrac{\partial^2 y}{\partial x^2}$ is a two-dimensional wave equation **Ans.** True

14. Radio equations are $V_{xx} = LCV_n$ and $I_{xx} = LCI_{tt}$ **Ans.** True

15. The small transverse vibrations of a string are $y_t^2 = a^2 y_{xx}$. **Ans.** False

Fill in the Blanks

16. The general solution of the equation $\dfrac{\partial^2 z}{\partial x^2} - \dfrac{\partial^2 z}{\partial t^2} = 0$ is ......

Ans. $(c_1 \cos px + c_2 \sin px)(c_3 \cos pt + c_4 \sin pt)$.

17. The general solution of the equation $\dfrac{\partial^2 z}{\partial x \partial y} = 0$ is ...... **Ans.** $f_1(x) + f_2(y)$.

18. The solution of $z(x, y)$ of the equation $\dfrac{\partial z}{\partial x} - y\dfrac{\partial z}{\partial y} = 0$ is ...... **Ans.** $f(x + \log y, z) = 0$.

19. The solution of $3x\dfrac{\partial z}{\partial x} - 5y\dfrac{\partial z}{\partial y} = 0$ is ...... **Ans.** $f(x^5 y^3, z)$

20. The solution of $\frac{\partial^2 z}{\partial x^2} = \sin(xy)$ is ...... **Ans.** $\frac{1}{y^2}\sin(xy) + x\, f_1(y) + f_2(y)$

21. The solution of $\frac{\partial u}{\partial t} = 2\frac{\partial^2 u}{\partial x^2}$ if $u(0,t) = u(3,t) = 0$ and $u(x,0) = 5\sin 4\pi x - 8\pi x$ is ......

Ans. $(5\sin 4\pi x e^{-32x^2 t} - 3\sin 8\pi x e^{-12n^2 t})$

The solution to the P.D.E.

22. $3u_x + 2u_y = 0$ is .... where $u_x = \frac{\partial u}{\partial x}, u_y = \frac{\partial u}{\partial y}$ *(U.P. II Semester 2009)* **Ans.** $u(x, y) = ce^{\frac{k}{6}(2x-3y)}$

23. $\frac{\partial^2 u}{\partial x^2} + \frac{\partial^2 u}{\partial y^2} = 0$ is called ...... **Ans.** Laplace equation.

24. On solving $\frac{\partial u}{\partial x} + 2\frac{\partial u}{\partial y} = 0$ by the method of separating of variable we suppose $u =$ ......

Ans. $u = XY$

25. In the separating of variable, we assume $u = XY$, then X is a function of .... and Y is a function of ......

Ans. x only, y only.

26. Transverse vibration in one dimensional wave equation, the motion in horizontal direction is ......

Ans. zero.

27. The Laplace equation in two dimension is ...... **Ans.** $\frac{\partial^2 u}{\partial x^2} + \frac{\partial^2 u}{\partial y^2} = 0$

28. D'Alembert's solution of the wave equation is ..... **Ans.** $y(x, t) = \phi(x + ct) + \psi(x - ct)$

Match the following

29. (*i*) $\frac{\partial^2 y}{\partial t^2} = c^2 \frac{\partial^2 y}{\partial x^2}$ (*a*) $\sum_{n=1}^{\infty} b_n(c_1 \cos px + c_2 \sin px)e^{-1p^2c^2t}$

(*ii*) $\frac{\partial^2 u}{\partial t^2} = a^2 \frac{\partial^2 u}{\partial x^2}$ (*b*) $Ae^{x\left(\frac{x}{3} - \frac{y}{2}\right)}$

(*iii*) $3\frac{\partial u}{\partial x} + \frac{\partial u}{\partial y} = 0$ (*c*) $(c_5 \cos c\sqrt{k}\, t + c_6 \sin c\sqrt{k}\, t)(c_7 \cos \sqrt{k}\, t + c_8 \sin \sqrt{k}\, x)$

(*iv*) $\frac{\partial^2 u}{\partial x^2} + \frac{\partial^2 u}{\partial t^2} = 0$ (*d*) $(c_1 \cos px + c_2 \sin px)(c_3 e^{py} + c_4 e^{-py})$

Ans. (*i*) (*c*), (*ii*) (*a*), (*iii*) (*b*), (*iv*) (*d*).

Match the following equations

30. (*i*) One dimensional heat flow is (*a*) $\frac{\partial^2 y}{\partial t^2} = c^2 \frac{\partial^2 y}{\partial x^2}$

(*ii*) The transverse vibration of a string is (*b*) $\frac{\partial^2 u}{\partial x^2} + \frac{\partial^2 u}{\partial y^2} = 0$

(*iii*) Two dimensional heat flow is (*c*) $\frac{\partial^2 u}{\partial r^2} + \frac{1}{r}\frac{\partial u}{\partial r} + \frac{1}{r^2}\frac{\partial^2 u}{\partial \theta^2} = 0$

(*iv*) Two dimensional heat flow in polar form is (*d*) $\frac{\partial u}{\partial t} = c^2 \frac{\partial^2 u}{\partial x^2} = 0$

Ans. (*i*) (*d*), (*ii*) (*a*), (*iii*) (*b*), (*iv*) (*c*).

UTTARAKHAND TECHNICAL UNIVERSITY

JUNE 2010

B.TECH.

(IInd Semester)

MATHEMATICS II

Time 3 Hours Total Marks : 100

Note : Attempt all question. All question carry equal marks.

1. Attempt any four of the following : **$5 \times 4 = 20$**

(*a*) Solve $(D^4 - 1)\,y = e^x \cos x$. (*See Solved Example 25 on page 56*)

(*b*) Solve $x^2 d^2y/dx^2 - x\,dy/dx + y = \log x$ (*See Solved Example 45 on page 70*)

(*c*) Find the value of λ for which the differential equation :

$(xy^2 + \lambda x^3 y2)\,dx + (x^3 y + yx)\,x dy = 0$, is exact. Hence solve.

(*See Solved Example 28 on page 20*)

(*d*) Solve by the method of variation of parameter

$$\frac{d^2y}{dx^2} + y = \tan x.$$ (*See Solved Example 32 on page 135*)

(*e*) Solve. $(D^2 - 2D + 1)\,y = x \sin x$ (*See Solved Example 38 on page 66*)

(*f*) Solve the following simultaneous equation

$$\frac{d^2y}{dt^2} - 3x - 4y = 0, \quad \frac{d^2y}{dt^2} + x + y - 0$$ (*See Solved Example 14 on page 100*)

2. Attempt any four of the following : **$5 \times 4 = 20$**

(*a*) Find Laplace transform of

$$f(t) = \frac{\cos at - \cos bt}{t}, t > 0$$ (*See Solved Example 25 on page 172*)

(*b*) Find $L^{-1}\left\{\log\left(\frac{S+1}{S-1}\right)\right\}$ (*See Solved Example 13 on page 207*)

(*c*) Find $L\,\{t^2 e^{2t} \sin t\}$ (*See Solved Example 22 on page 170*)

(*d*) Find by using convolution theorem

$$L^{-1}\left[\frac{S^2}{(S^2 + a^2)(S^2 + b^2)}\right]$$ (*See Solved Example 36 on page 218*)

(*e*) Solve by using Laplace trasnform

$$\frac{d^2y}{dt^2} + 9y = \cos 2t,\ y(0) = 1,\ y(\pi/2) = -1$$ (*See Solved Example 48 on page 225*)

(*f*) Evaluate by Laplace transform $\int_0^\infty te^{-3t} \sin t\,dt$ (*See Solved Example 15 on page 166*)

3. Attempt any two of the following : **$10 \times 2 = 20$**

(*a*) Find the Fourier series of the function $f(x)$, defined as $f(x) = \sin x, \ -\pi < x < \pi$

Hence show that

$$\frac{\pi}{2} = 1 + \frac{2}{1.3} - \frac{2}{3.5} + \ldots\ldots$$

(*b*) Find the Haff-range cosine series of the following function

$f(x) = |\cos x|, 0 < x < \pi$

Draw its diagram also. *(See Solved Example 15 on page 339)*

(*c*) Solve the partial differential equation

$(D^2 + DD' - 6D'^2) \ z = y \cos x$ *(See Solved Example 27 on page 401)*

4. Attempt any two of the following : **$10 \times 2 = 20$**

(*a*) A tightly stretched string with fixed end points $x = 0$ and $x = l$ is initially in a position given by $y(x, 0) = A \sin^3 \frac{\pi x}{\ell}$. It is released from rest from this position. Find the displacement $y(x, t)$.

(See Solved Example 13 on page 437)

(*b*) Solve by the method of separation of variable

$$\frac{\partial u}{\partial x} + \frac{\partial u}{\partial y} = 3u; \text{given } u(0, y) = 4e^{-y} - e^{-5y}$$

(*c*) The temperature distribution in a bar of length π, which is perfectly insulated at ends $x = 0$ and $x = \pi$ is governed by the partial differential equation

$$\frac{\partial u}{\partial t} = \frac{\partial^2 u}{\partial s^2}$$ *(See Q. 9 on page 460)*

Assuming the initial temperature distribution as $u(x, 0) = f(x) = \cos 2x$. Find the temperature distribution at any instant of time.

5. Attempt any four of the following : **$5 \times 4 = 20$**

(*a*) Show that the G.P. $\sum_{n=0} r^u, (r > 0)$ is convergent when $r < 1$ and divergent when $r \geq 1$.

(See Art. 7.10 on page 265)

(*b*) Test the convergence of $\sum_{n=1}^{\infty} \cos(1/n)$ *(See Solved Example 8 on page 268)*

(*c*) Test whether the following series is convergent or divergnet

$$\sum_{n=1}^{\infty} \frac{1}{n} \sin\left(\frac{1}{n}\right)$$ *(See Solved Example 9 on page 268)*

(*d*) By D'Alembert's ratio test, discuss the convergence of the series,

$$\frac{1^2 2^2}{1!} + \frac{2^2 3^2}{2!} + \frac{3^2 4^2}{3!} + \ldots\ldots.$$ *(See Solved Example 24 on page 280)*

(*e*) Discuss the convergence of the series,

$$\frac{1}{2} + \frac{1 \cdot 3}{2 \cdot 4} + \frac{1 \cdot 3 \cdot 5}{2 \cdot 4 \cdot 6} + \ldots\ldots$$ *(See Solved Example 27 on page 282)*

(*f*) By Cauchy's root test, discuss the convergence of the series,

$$\sum \left(\frac{n}{n+1}\right)^{n^2}$$ *(See Solved Example 41 on page 293)*

UTTARAKHAND TECHNICAL UNIVERSITY

JAN. 2011

B.TECH.

(II Semester)

MATHEMATICS II

Time 3 Hours   Total Marks : 100

Note : Attempt all question. All question carry equal marks.

1. Attempt any **four** of the following : **5 × 4 = 20**

(*i*) *Solve:* $x\,dx + y\,dy = \dfrac{a^2\,(x\,dy - y\,dx)}{x^2 + y^2}$ *(See Solved Example 31 on page 21)*

(*ii*) Solve: $\dfrac{dz}{dx} + \dfrac{z}{x}\log x = \dfrac{z}{x^2}(\log z)^2$ *(See Q. 24 on page 19)*

(*iii*) $\dfrac{d^2y}{dx^2} + a^2y = \sec ax$ *(See Q. 1 on page 67)*

(*iv*) Solve the simultaneous equations

$$\frac{dx}{dt} + \frac{dy}{dt} - 2y = 2\cos t - 7\sin t$$

$$\frac{dx}{dt} - \frac{dy}{dt} + 2x = 4\cos t - 3\sin t$$

(See Solved Example 10 on page 96)

(*v*) Solve: $\dfrac{d^2y}{dx^2} + \dfrac{2}{x}\dfrac{dy}{dx} + \dfrac{a^2}{x^4}y = 0$ *(See Q. 1 on page 134)*

(*vi*) Apply the method of variation of parameters to

$\dfrac{d^2y}{dx^2} + y = \tan x$ *(See Solved Example 31 on page 135)*

2. Attempt any **two** of the following : **5 × 4 = 20**

(*i*) (*a*) If f (t) is continous and of exponential order a as $t \to \infty$, then prove that Laplace trans form $f(t)$ exists for $s > a$. *(See Art. 5.7 on page 163)*

(*b*) Find the Laplace transform of the following periodic function

$f(t) = \dfrac{1}{T}$ for $0 < t < T$ *(See Similar Example 44 on page 183)*

(*ii*) If $\bar{f}_1(s)$ and $\bar{f}_2(s)$ are the Laplace transforms of $f_1(t)$ and $f_2(t)$ respectively, then prove that

$\bar{f}_1(s) \,.\, \bar{f}_2(s) = L\left\{\int_0^t f_1(u) \,.\, f_2(t-u)\,du\right\}$ *(See Art. 5.23 on page 187)*

Hence evaluate $L^{-1}\left\{\dfrac{s^2}{(s^2+a^2)(s^2+b^2)}\right\}$ (*See Solved Example 32 on page 215*)

(*iii*) Solve the following simultaneous differential equations by using Laplace transformation: $Dx - y = e^t$, $Dy + x = \sin t$, $x_0 = 1$, $y_0 = 0$. (*See Solved Example 71 on page 248*)

3. Attempt any **two** of the following : **10 × 2 = 20**

(*i*) (*a*) Test the convergence or divergence of the following series:

$$\sum_{n=2}^{\infty} \frac{1}{n(\log n)^k} \ (p > 0)$$ (*See Q. 6 on page 295*)

(*b*) $\displaystyle\sum_{n=1}^{\infty} \frac{1}{\sqrt{n} + \sqrt{(n+1)}}$ (*See Q. 11 on page 277*)

(*ii*) State the prove the Cauchy's root test. Hence test for convergence the following series: (*See Art. 7.20 on page 292*)

$$\sum_{n=1}^{\infty} \left(1 + \frac{1}{\sqrt{n}}\right)^{-n^{3/2}}$$ (*See Q. 3 on page 295*)

(*iii*) (*a*) Show that the following series converges uniformly in any interval:

$$\sum_{n=1}^{\infty} \frac{\cos nx}{n^p}$$

(*b*) Prove that $\displaystyle\int_0^1 \left(\sum_{n=1}^{\infty} \frac{x^n}{n^2}\right) dx = \sum \frac{1}{n^2(n+1)}$

4. Attempt any two of the following : **10 × 2 = 20**

(*i*) Find the Fourer series for the function f (x) = x + x² in the interval $-\pi < x < \pi$.

Hence show $\dfrac{\pi^2}{6} = 1 + \dfrac{1}{2^2} + \dfrac{1}{3^2} + \dfrac{1}{4^2} + \ldots$ (*See Solved Example 2 on page 322*)

(*ii*) Obtain Fourier series for the function $f(x)$ given by $f(x) = 1 + 2x/\pi, \ -\pi \le x \le 0$

$= 1 - 2x/\pi, \ 0 \le x \le \pi$

(*See Similar Example 21 on page 346*)

(*iii*) Express $f(x) = x$ as a half-range sine series in $0 < x < 2$.

(*iv*) Eliminate the arbitary function f and g from the following equation :

$v = f(x + iy) + g(x - iy)$, where $i = \sqrt{(-1)}$

(*v*) Find the particular integral of

$(D - D' - 1)(D - D' - 2)\, z = \sin(2x + 3y)$

where $D \equiv \dfrac{\partial}{\partial x}$, $D' \equiv \dfrac{\partial}{\partial y}$. (*Similar Similar Example 7 on page 386*)

$$= 4.\frac{1}{(2)^2+4(2)-5}e^{2t}+3.\frac{1}{-5\left(1-\frac{D^2}{5}-\frac{4}{5}D\right)}t$$

$$= \frac{4}{7}e^{2t}-\frac{3}{5}\left(1-\frac{4}{5}D-\frac{D^2}{5}\right)^{-1}t$$

$$= \frac{4}{7}e^{2t}-\frac{3}{5}\left(1+\frac{4}{5}D+\frac{D^2}{5}\right)t$$

$$= \frac{4}{7}e^{2t}-\frac{3}{5}t-\frac{12}{25}$$

$$y = C_1e^t+C_2e^{-5t}+\frac{4}{7}e^{2t}-\frac{3}{5}t-\frac{12}{25}$$

Putting the value of y in (2), we get

$$3\,x = \frac{d}{dt}\left(C_1e^t+C_2e^{-5t}+\frac{4}{7}e^{2t}-\frac{3}{5}t-\frac{12}{25}\right)+2\left(C_1e^t+C_2e^{-5t}+\frac{4}{7}e^{2t}-\frac{3}{5}t-\frac{12}{25}\right)-e^{2t}$$

$$= \left(C_1e^t-5C_2e^{-5t}+\frac{4}{7}(2e^{2t})-\frac{3}{5}\right)+2C_1e^t+2C_2e^{-5t}+\frac{8}{7}e^{2t}-\frac{6}{5}t-\frac{24}{25}-e^{2t}$$

$$= 3C_1e^t-3C_2e^{-5t}+\frac{9}{7}e^{2t}-\frac{6}{5}t-\frac{39}{25}$$

$$x = C_1e^t-C_2e^{-5t}+\frac{3}{7}e^{2t}-\frac{2}{5}t-\frac{13}{25}$$

and $y = C_1e^t+C_2e^{-5t}+\frac{4}{7}e^{2t}-\frac{3}{5}t-\frac{12}{25}$ **Ans.**

(*e*) Solve :

$$\frac{d^2y}{dx^2}+\cot x\frac{dy}{dx}+4y\,\text{cosec}^2x = 0$$

by changing the independent variable. (*See Solved Example 23 on page 125*)

(*f*) Apply the method of variation of parameters to solve

$$\frac{d^2y}{dx^2}+n^2 = \sec nx$$

Solution. We have,

$$\frac{d^2y}{dx^2}+n^2y = \sec nx \qquad ...(1)$$

$$(D^2+n^2)\,y = \sec nx$$

A.E. is $m^2+n^2=0 \Rightarrow m^2=-n^2 \Rightarrow m=\pm ni$

Complementary function of (1) is

C.F. = A cos nx + B sin nx

Here $y_1 = \cos nx$, $y_2 = \sin nx$

P.I. = $uy_1 + vy_2$ ...(2) $\left[\begin{array}{l} y_1y'_2-y'_1y_2 = n\cos nx\cos nx-(-n\sin nx)\sin nx \\ \qquad = n(\cos^2 nx+\sin^2 nx) = n\end{array}\right]$

where $$u = \int \frac{-y_2 \sec nx}{y_1 y_2' - y_1' y_2} dx$$

On putting the values of y_2 and $y_1 y_2' - y_1' y_2$, we get

$$u = \int \frac{-\sin nx . \sec nx}{n} dx = -\frac{1}{n} \int \tan nx \, dx = \frac{1}{n^2} \log \cos nx$$

and $$v = \int \frac{y_1 \sec nx}{y_1 y_2' - y_1' y_2} dx$$

On putting the values of y_1 and $y_1 y_2' - y_1' y_2$, we get

$$v = \int \frac{\cos nx . \sec nx}{n} dx = \frac{1}{n} \int dx = \frac{x}{n}$$

Putting the values of u and v, in (2), we get

$$\text{P.I.} = \left(\frac{1}{n^2} \log \cos nx\right) \cos nx + \left(\frac{x}{n}\right) \sin nx$$

Complete solution is

$$y = \text{C.F.} + \text{P.I.}$$

$$y = A \cos nx + B \sin nx + \frac{1}{n^2} \cos nx \log \cos nx + \frac{x}{n} \sin nx$$ **Ans.**

Q. 2. Attempt any two parts of the following :

(a) Find Laplace Transform of

(*i*) $te^{at} \sin at$ *(See Q 8 on page 170)*

(*ii*) $(\cos at - \cos bt)/t$ *(See Solved Example 25 on page 172)*

Q. 2. (b) (*i*) Find $L^{-1}\left[\log \frac{s+1}{s-1}\right]$ *(See Solved Example 13 on page 207)*

(*ii*) Find $L^{-1}\left[\frac{1}{s(s+a)^3}\right]$

Solution. (*ii*) Since $L^{-1}\left[\frac{1}{(s+a)^3}\right] = \frac{e^{-at} t^2}{2!}$ (By I shifting property)

$$\therefore L^{-1}\left[\frac{1}{S(s+a)^3}\right] = \int_o^t L^{-1}\left(\frac{1}{(s+a)^3}\right) dt$$

$$= \int_o^t \left(\frac{e^{-at} t^2}{2!}\right) dt = \frac{1}{2!} \int_o^t t^2 e^{-at} dt$$

$$= \frac{1}{2!}\left[t^2 \frac{e^{-at}}{-a} - \int 2t \frac{e^{-at}}{-a} dt\right]_o^t$$

$$= \frac{1}{2!}\left[\frac{t^2}{-a}e^{-at} + \frac{2}{a}\{t\frac{e^{-at}}{-a} - \int 1\frac{e^{-at}}{-a}dt\}\right]_o^t$$

$$= \frac{1}{2!}\left[-\frac{t^2}{a}e^{-at} - \frac{2}{a^2}te^{-at} - \frac{2}{a^3}e^{-at}\right]_o^t$$

$$= -\frac{1}{2!}\left[\frac{t^2}{a}e^{-at} + \frac{2}{a^2}te^{-at} + \frac{2}{a^3}e^{-at} - \frac{2}{a^3}\right]$$

$$= \frac{1}{a^3} - e^{-at}\left[\frac{t^2}{2a} + \frac{t}{a^2} + \frac{1}{a^3}\right]$$ **Ans.**

Q. 2. (c) Using Laplace Transform solve the following equations :

$$\frac{d^2x}{dt^2} + x = t\cos 2t$$

Given $x(0) = x'(0) = 0$

(See Solved Example 55 on page 230)

Q. 3. Attempt any two parts of the following :

(*a*) (*i*) Test for convergence the series

$$\frac{1}{1.2.3} + \frac{3}{2.3.4} + \frac{5}{3.4.5} + - - - - - \infty$$

(See Solved Example 16 on page 274)

(*ii*) Test for convergence the series

$$\frac{1}{2\sqrt{1}} + \frac{x^2}{3\sqrt{2}} + \frac{x^4}{4\sqrt{3}} + \frac{x^6}{5\sqrt{4}} - - - - - - \infty$$

(See Q 25 on page 313)

Q. 3(b) Discuss the convergence of the series

$$\sum_{n=2}^{\infty}\frac{1}{n(\log n)^p}; p > 0, \text{ by Integral Test.}$$

Solution. We have,

$$u_n = \frac{1}{n(\log n)^p} = f(n)$$

$$f(x) = \frac{1}{x(\log x)^p}$$

For $x \geq 2$, $p > 0$, $f(x)$ is positive and monotonic decreasing.

By cauchy Integral Test,

$\sum_{n=2}^{\infty} u_n$ and $\int_2^{\infty} f(x)\,dx$ converge or diverge together.

Case I. When $p \neq 1$

$$\int_2^\infty f(x)dx = \int_2^\infty (\log x)^{-p} \cdot \frac{1}{x} dx = \left[\frac{(\log x)^{-p+1}}{-p+1}\right]_2^\infty$$

Sub case 1. When $p > 1$, $p - 1$ is + ve, so that

$$\int_2^\infty f(x)dx = -\frac{1}{p-1}\left[\frac{1}{(\log x)^{p-1}}\right]_2^\infty$$

$$= -\frac{1}{p-1}\left[0 - \frac{1}{(\log 2)^{p-1}}\right] = \frac{1}{(p-1)(\log 2)^{p-1}} = \text{finite}$$

$$\Rightarrow \int_2^\infty f(x)\, dx \text{ converges } \Rightarrow \sum_{n=2}^\infty u_n \text{ converges}$$

Sub-case 2. When $p < 1$, $1 - p$ is positive, so that

$$\int_2^\infty f(x)dx = \frac{1}{1-p}\left[(\log x)^{1-p}\right]_2^\infty = \frac{1}{1-p}\left[\infty - (\log 2)^{1-p}\right] = \infty$$

$$\Rightarrow \quad \int_2^\infty f(x)dx \text{ diverges } \Rightarrow \sum_{n=2}^\infty u_n \text{ diverges}$$

Case II. When $p = 1$, $f(x) = \dfrac{1}{x \log x}$

$$\int_2^\infty f(x)dx = \int_2^\infty \frac{dx}{x \log x} = \int_2^\infty \frac{\frac{1}{x}}{\log x} dx$$

$$= [\log \log x]_2^\infty = \infty - \log \log 2 = \infty$$

$$\Rightarrow \quad \int_2^\infty f(x)dx \text{ diverges } \Rightarrow \sum_{n=2}^\infty u_n \text{ diverges.}$$

Hence, $\sum u_n$ converges if $p > 1$ and diverges if $0 < p \leq 1$. **Ans.**

Q.3. (c). Show that the Geometric series $\sum_{n=0}^\infty r^n$, $(r > 0)$ is convergent when $r < 1$ and divergent when $r \geq 1$

(See Art 7.10 on page 265)

Q.4. Attempt any two parts of the following :

(a) Find the Fourier series expansion for $f(x)$, if

$$f(x) = \begin{cases} -\pi, & -\pi < x < 0 \\ x, & 0 < x < \pi \end{cases}$$

Deduce that $\frac{1}{1^2}+\frac{1}{3^2}+\frac{1}{5^2}+---=\frac{\pi^2}{8}$ (*See Solved Example 7 on page 329*)

Q.4. (*b*) Expand $f(x) = \begin{cases} \frac{1}{4}x, & 0 < x < \frac{1}{2} \\ x - 3/4, & \frac{1}{2} < x < 1 \end{cases}$

as the Fourier sine series.

Q. 4. (*c*) Solve $(D^3 - 4D^2D' + 4DD'^2)z = 6\sin(3x + 2y)$

(*See Solved Example 9 on page 387*)

Q. 5. Attempt any two parts of the following : **10×2 = 20**

(a) A tightly stretched string with fixed end points $x = 0$ and $x = 1$ is initially in a position given by $y(x, 0) = y_0 \sin(\pi x / l)$. If it is released from rest from this position, find the displacement y at any distance x from one end at any time t.

(*See Solved Example 11 on page 434*)

Q. 5. (b) A homogeneous rod of conducting material of length 'I' has its ends kept at zero temperature. The temperature at the centre is T and falls uniformly to zero at the two ends. Find the temperature distribution.

Q. 5(c) Solve $\frac{\partial u}{\partial x}+\frac{\partial u}{\partial y} = 3u$ given $u(0,y) = 4e^{-y} - e^{-5y}$, by the method of separation of variables.

Solution. Here, we have

$$\frac{\partial u}{\partial x}+\frac{\partial u}{\partial y} = 3u \qquad ...(1)$$

Let $$u = X(x)\,Y(y) \qquad ...(2)$$

where X is a function of x only and Y is a function of y only. On differentiating (2) partially w.r.t. 'x', we get

$$\frac{\partial u}{\partial x} = \frac{\partial X}{\partial x}.Y \qquad ...(3)$$

On differentiating (2) partially w.r.t. 'y', we get

$$\frac{\partial u}{\partial y} = X\frac{\partial Y}{\partial y} \qquad ...(4)$$

Putting the values of $\frac{\partial u}{\partial x}$ and $\frac{\partial u}{\partial y}$ from (3) and (4) in (1), we get

$$\frac{\partial X}{\partial x}.Y + X\frac{\partial Y}{\partial y} = 3\,XY$$

On dividing by $X\,Y$, we get

$$\frac{1}{X}.\frac{\partial X}{\partial x}+\frac{1}{Y}.\frac{\partial Y}{\partial y} = 3$$

On separating the variables, we get

$$\frac{1}{x}\frac{\partial X}{\partial x} = 3 - \frac{1}{Y}\frac{\partial Y}{\partial y} = k \text{ (say)}$$

$$\frac{1}{x}\frac{\partial X}{\partial x} = k$$

$$\Rightarrow \quad \frac{\partial X}{\partial x} = kX$$

$$\Rightarrow \quad DX - kx = 0$$

A.E. is $m - k = 0 \Rightarrow m = k$

$$\Rightarrow \quad X = ae^{kx}$$

$$\Rightarrow \quad 3 - \frac{1}{Y}\frac{\partial Y}{\partial y} = k$$

$$\Rightarrow \quad \frac{\partial Y}{\partial y} = (3-k)\,Y$$

$$\Rightarrow \quad DY - (3-k)Y = 0$$

A.E. is $m - (3 - k) = 0 \Rightarrow m = 3 - k$

$$\Rightarrow \quad Y = be^{(3-k)\,y}$$

Putting the values of X and Y in (2), we get

$$u = ae^{kx}\, be^{(3-k)\,y}$$

$$u = ab\, e^{kx+(3-k)y} \qquad ...(5)$$

Putting $x = 0$, $u = 4e^{-y} - e^{-5y}$ in (5), we get

$$4e^{-y} - e^{-5y} = ab\, e^{0+(3-k)y}$$

$$\Rightarrow \quad 4e^{-y} - e^{-5y} = ab\, e^{3y-ky}$$

$$\Rightarrow \quad 4e^{-y} - e^{-5y} = ab\, e^{3y} \,.\, e^{-ky}$$

$$\Rightarrow \quad \frac{4e^{-y} - e^{-5y}}{e^{3y}} = ab\, e^{-ky}$$

$$\Rightarrow \quad 4e^{-4y} - e^{-8y} = ab\, e^{-ky}$$

$$\Rightarrow \quad e^{-4y}(4 - e^{-4y}) = ab\, e^{-ky}$$

Comparing the like terms, we have

$$k = 4,\ 4 - e^{-4y} = ab$$

Putting these values in (5), we get

$$u = (4 - e^{-4y})\, e^{4x+(3-4)y}$$

$$u = (4 - e^{-4y})\, e^{4x-y}$$

Ans.

UTTARAKHAND TECHNICAL UNIVERSITY

2015

B.TECH. (SEM. II)

(IInd Semester)

(All Branches)

MATHEMATICS-II

Time: 3 hours *Total Marks : 100*

Note:

(*i*) This question paper contains *five* questions with alternative choice.
(*ii*) All questions are compulsory.
(*iii*) Each question carries four Parts (*a*) and (*b*) or Parts (*c*) and (*d*) of each question.
(*iv*) Each part carries ten marks. Total marks assigned to each question are **twenty**.

1. (*a*) Solve the differential equation:

$$\frac{d^2y}{dx^2} - 4\frac{dy}{dx} + 4y = 8x^2e^{2x}\sin 2x$$ (*See Solved Example 35 on Page 62*)

(*b*) Solve the homogeneous differential equation:

$$x^2\frac{d^2y}{dx^2} - x\frac{dy}{dx} - 3y = x^2\log x$$ (*See Solved Example 49 on Page 73*)

Or

(*c*) Solve the differential equation:

$$\frac{d^4y}{dx^4} - y = x\sin x$$

Solution. $\frac{d^4y}{dx^4} - y = x\sin x$

$(D^4 - 1)y = x\sin x$

A.E. is $m^4 - 1 = 0$

$\Rightarrow (m-1)(m+1)(m^2+1) = 0$

$\Rightarrow (m-1)(m+1)(m+i)(m-i) = 0 \quad \Rightarrow \quad m = -1, 1, i, -i$

C.F. $= C_1e^{-x} + C_2e^{x} + (C_3\cos x + C_4\sin x)$

$$\text{P.I} = \frac{1}{D^4-1}x\sin x = \text{Imaginary part of } \frac{1}{D^4-1}x(\cos + i\sin x) = \text{I.P. } \frac{1}{D^4-1}xe^{ix}$$

$$= \text{I.P. } e^{ix}\frac{1}{(D+i)^4-1}x = \text{I.P. } e^{ix}\frac{1}{D^4+4iD^3-6D^2-4iD+1-1}x$$

$$= \text{I.P. } e^{ix}\frac{1}{D^4+4iD^3-6D^2-4iD}x = \text{I.P. } \frac{e^{ix}}{-4iD}\left[1-\frac{3}{2}iD-D^2+\frac{i}{4}D^3\right]^{-1}x$$

$$= \text{I.P.}\left(\frac{ie^{ix}}{4D}\right)\left[1+\frac{3}{2}iD\right]x$$

$$= \text{I.P.}\ \frac{ie^{ix}}{4D}\left(x+\frac{3i}{2}\right) = \text{I.P}\ \frac{ie^{ix}}{4}\int\left(x+\frac{3}{2}i\right)dx$$

$$= \text{I.P}\ \frac{ie^{ix}}{4}\int\left(\frac{x^2}{2}+\frac{3}{2}ix\right) = \text{I.P}\ \frac{(i\cos x-\sin x)}{4}\left(\frac{x^2}{2}+\frac{3i}{2}x\right)$$

$$= \frac{x^2}{2}\cos x - \frac{3}{2}x\sin x$$

Complete solution = C.F. + P.I.

$$y = C_1e^{-x} + C_2e^{x} + (C_3\cos x + C_4\sin x) + \frac{x^2}{2}\cos x - \frac{3}{2}x\sin x \quad \textbf{Ans.}$$

(*d*) Solve by method of variation of parameters:

$\frac{d^2y}{dx^3} - 2\frac{dy}{dx} = e^x \sin x$ (*See Solved Example 39 on Page 141*)

2. (*a*) Obtain the Laplace Transform of $t^2e^{2t}\sin t$. (*See Solved Example 22 on Page 170*)

2. (*b*) (*i*) Express the following function in terms of unit step function:

$$f(t) = \begin{cases} t-1, & 1<t<2 \\ 3-t, & 2\le t<3 \end{cases}$$

and find its Laplace transform. (*See Solved Example 33 on Page 177*)

2. (*c*) (*ii*) Evaluate $\int_0^\infty \frac{e^{-at}-e^{-bt}}{t}dt$ (*See Solved Example 53 on Page 190*)

Or

(*c*) State convolution theorem and hence find:

$L^{-1}\left\{\frac{s^2}{(s^2+a^2)(s^2+b^2)}\right\}\ a\neq b$ (*See Solved Example 36 on Page 218*)

(*d*) Solve, by method of Laplace transform, the differential equation:

$\frac{d^2y}{dt^2} + n^2y = a\sin(nt+\alpha)$ given $y(0) = y'(0) = 0$.

(*See Solved Example 58 on Page 233*)

3. (*a*) Given that $f(x) = x + x^2$, for $-\pi < x < \pi$.
find the Fourier expression of $f(x)$. Deduce that:

$\frac{\pi^2}{6} = 1 + \frac{1}{2^2} + \frac{1}{3^2} + \frac{1}{4^2}$ (*See Solved Example .. on Page 322*)

3. (*b*) Find the Fourier series of the function defined as

$$f(x) = \begin{cases} x+\pi, & \text{for } 0\le x\le\pi \\ -x-\pi, & \text{for } -\pi\le x<0 \end{cases}$$

and $f(x+2\pi) = f(x)$. (*See Solved Example 10 on Page 332*)

3. (*c*) Find the Fourier Cosine series expansion of the periodic function defined by:

$$f(x) = \sin\left(\frac{\pi x}{l}\right). \qquad 0 < x < l$$

(*See Solved Example 24 on Page 348*)

3. (*d*) Determine the first four terms of the Fourier series for the following function:

$$f(x) = \begin{cases} \dfrac{2k}{\pi}x, & \text{for } 0 \le x \le \pi/2 \\ -\dfrac{2k}{\pi}x + 2k, & \text{for } \dfrac{\pi}{2} \le x \le \dfrac{3\pi}{2} \\ \dfrac{2k}{\pi}x - 4k, & \text{for } \dfrac{3\pi}{2} \le x \le 2\pi \end{cases}$$

Solution. The function is odd.

$$a_0 = 0 = a_n$$

$$b_n = \frac{1}{\pi}\int_0^{2\pi} f(x)\sin nx\,dx$$

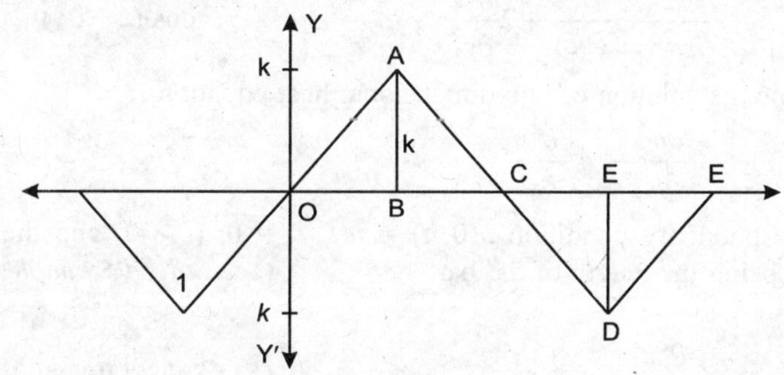

$$= \frac{1}{\pi}\int_0^{\frac{\pi}{2}} \frac{2k}{\pi}(x)\sin nx\,dx + \frac{1}{\pi}\int_{\frac{\pi}{2}}^{\frac{3\pi}{2}}\left[\frac{-2k}{\pi}x + 2k\right]\sin nx\,dx \;+\; \frac{1}{\pi}\int_{\frac{3\pi}{2}}^{2\pi}\left(\frac{2kx}{\pi} - 4k\right)\sin nx\,dx$$

$$= \frac{2k}{\pi^2}\left[-(x)\frac{\cos nx}{n} + \frac{\sin nx}{n^2}\right]_0^{\frac{\pi}{2}} + \frac{1}{\pi}\left[-\left(-\frac{2k}{\pi}x + 2k\right)\frac{\cos nx}{n} + \left(-\frac{2k}{\pi}\sin\frac{nx}{n^2}\right)\right]_{\frac{\pi}{2}}^{\frac{3\pi}{2}}$$

$$+ \frac{1}{\pi}\left[-\left(\frac{2k}{\pi}(x) - 4k\right)\frac{\cos nx}{n} + \frac{2k}{\pi}\frac{\sin nx}{n^2}\right]_{\frac{3\pi}{2}}^{\pi}$$

$$= \frac{2k}{\pi^2}\left[-\frac{\pi}{2}\frac{\cos\frac{n\pi}{2}}{n} + \frac{\sin n\frac{\pi}{2}}{n^2}\right] + \frac{1}{\pi}\left[-(-3k+2k)\frac{\cos\frac{3n\pi}{2}}{n} - \frac{2k}{\pi}\frac{\sin 3n\frac{\pi}{2}}{n^2} + (-k+2k)\frac{\cos\frac{n\pi}{2}}{n} + \frac{2k}{\pi}\frac{\sin\frac{n\pi}{2}}{n^2}\right]$$

$$\frac{1}{\pi}\left[(2k-4k)\cos\frac{n\pi}{n} + \frac{2k}{\pi}\frac{\sin n\pi}{n^2}\right] + (3k-4k)\cos\frac{3n\frac{\pi}{2}}{n^2} - \frac{2k}{\pi}\frac{\sin\frac{3n\pi}{2}}{n^2}\Bigg]$$

$$\frac{1}{k}\left[-(2k-4k)\frac{\cos n\pi}{n}+\frac{2k}{\pi}\frac{\sin n\pi}{n^2}\right]$$

$$=\frac{k}{\pi}\left[-\frac{1}{n}\cos\frac{n\pi}{2}+\frac{2}{n^2\pi}\sin\frac{n\pi}{2}+\frac{1}{n}\cos\frac{3n\pi}{2}-\frac{2}{n^2\pi}\sin\frac{3n\pi}{2}+\frac{1}{n}\cos\frac{n\pi}{2}\right.$$

$$\left.+\frac{2}{n^2\pi}\sin\frac{n\pi}{2}-\frac{1}{n}\cos\frac{3n\pi}{2}-\frac{2}{n^2\pi}\sin\frac{3n\pi}{2}\right]$$

$$=\frac{k}{\pi}\left[\frac{4}{n^2\pi}\sin\frac{n\pi}{2}-\frac{4}{n^2\pi}\sin\frac{3n\pi}{2}\right]=\frac{4k}{n^2\pi^n}\left(\sin\frac{n\pi}{2}-\sin\frac{3n\pi}{2}\right)=\frac{8}{n^2\pi^2}\sin\frac{n\pi}{2}$$

Hence Fourier expansion $y = f(x)$ is given by

$$f(x)=\frac{8k}{\pi^2}\left[\frac{1}{1^2}\sin x-\frac{1}{3^2}\sin 3x+\frac{1}{5^2}\sin 5x-\frac{1}{7^2}\sin 7x+\ldots\right]$$ **Ans.**

4. (*a*) Solve the partial different equation: (*See Solved Example 12 on Page 389*)

$$\frac{\partial^2 z}{\partial x^2}-3\frac{\partial^2 z}{\partial x\partial y}+2\frac{\partial^2 z}{\partial y^2}=e^{2x-y}+e^{x+y}+\cos(x+2y)$$

(*b*) Determine the solution of one-dimensional heat equation:

$$\frac{\partial u}{\partial t}=c^2\frac{\partial^2 u}{\partial x^2}$$

Subject to the boundary condition $u(0, t) = u(l, t) = 0$, $(t > 0)$ and the initial condition $u(x, 0) = x$, l being the length of the bar. (*See Solved Example 25 on Page 455*)

(*c*) Solve:

$$\frac{\partial^2 z}{\partial x^2}+\frac{\partial^2 z}{\partial x\partial y}-6\frac{\partial^2 z}{\partial y^2}-y\cos x.$$ (*See Solved Example 27 on Page 401*)

(*d*) Solve $\frac{\partial^2 u}{\partial x^2}+\frac{\partial^2 u}{\partial y^2}=0$ which satisfies the conditons :

$$u(0, y) = u(l, y) = u(x, 0) = 0 \text{ and } u(x, a) = \sin\frac{n\pi}{l}.$$

(*See Solved Example 28 on Page 462*)

5. (*a*) (*i*) Show that $J_{-n}(x) = (-1)^n Jn(x)$

Solution. We know that $J_{-n}(x)=\sum_{r=0}^{\infty}(-1)^r\frac{1}{r!\,\overline{|(r-n+1)}}\left(\frac{x}{2}\right)^{-n+2r}$

$$\Rightarrow \quad =\sum_{r=0}^{r=n-1}\frac{(-1)^r\left(\frac{x}{2}\right)^{-n+2r}}{r!\,\overline{|(-n+r+1)}}+\sum_{r=n}^{\infty}\frac{(-1)^r\left(\frac{x}{2}\right)^{-n+2r}}{r!\,\overline{|(-n+r+1)}}$$

$$\Rightarrow \quad =0+\sum_{r=n}^{\infty}\frac{(-1)^r\left(\frac{x}{2}\right)^{-n+2r}}{r!\,\overline{|(-n+r+1)}}$$ (since – ve integer = ∞)

On putting $r = n + k$, we get

$$J_{-n}(x) = \sum_{k=0}^{\infty} \frac{(-1)^{n+k}\left(\frac{x}{2}\right)^{n+2k}}{(n+k)!\lceil(k+1)} = (-1)^n \sum_{k=0}^{\infty} \frac{(-1)^k \left(\frac{x}{2}\right)^{n+2k}}{(n+k)!k!}$$

$= (-1)^n J_n(x)$ **Proved**

(*ii*) Prove that:

$J_n(-x) = (-1)^n J_n(x)$

(*b*) Prove that:

$$\int_{-1}^{1} [P_n(x)]^2 dx = \frac{2}{2n+1}, \text{ if } m = n.$$

Solution. We know that $(1 - 2xz + z^2)^{-1/2} = \Sigma z^n P_n(x)$

Squaring both sides, we get

$$(1 - 2xz = z^2)^{-1} = \Sigma z^{2n} P_n^2(x) + 2\Sigma z^{m+n} P_m(x) \cdot P_n(x)$$

Integrating both sides between –1 and + 1, we have

$$\int_{-1}^{+1} \Sigma z^{2n} \cdot P_n^2(x)\, dx + \int_{-1}^{+1} 2\Sigma z^{m+n} \cdot P_m(x)\, dx = \int_{-1}^{+1} (1 - 2xz + z^2)^{-1} dx$$

$$\int_{-1}^{+1} \Sigma z^{2n} P_n^2(x)\, dx + 0 = \int_{-1}^{+1} \frac{1}{1 - 2xz + z^2} dx$$

$$\Rightarrow \quad \Sigma z^{2n} \int_{-1}^{+1} P_n^2(x)\, dx = -\frac{1}{2z}\{\log(1 - 2xz + z^2)\}_{-1}^{+1}$$

$$= -\frac{1}{2z}\log\frac{1 - 2z + z^2}{1 + 2z + z^2} = -\frac{1}{2z}\log\left(\frac{1-z}{1+z}\right)^2$$

$$= \frac{1}{z}\log\frac{1+z}{1-z} = \frac{1}{z}[\log(1+z) - \log(1-z)]$$

$$= \frac{1}{z}\left[\left(z - \frac{z^2}{2} + \frac{z^3}{3} - \frac{z^4}{4} + \ldots + \frac{z^{2n+1}}{2n+1} + \ldots 0\right) - \left(-z - \frac{z^2}{2} - \frac{z^3}{3} - \frac{z^4}{4} - \ldots - \frac{z^{2n+1}}{2n+1} - \ldots\right)\right]$$

$$\Sigma z^{2n} \int_{-1}^{+1} P_n^2(x)\, dx = \frac{2}{z}\left[z + \frac{z^3}{3} + \frac{z^5}{5} + \ldots + \frac{z^{2n+1}}{2n+1} + \ldots\right] = 2\left[1 + \frac{z^2}{3} + \frac{z^4}{5} + \ldots + \frac{z^{2n}}{2n+1} + \ldots\right]$$

Equating the coefficient of z^{2n} on both sides, we have

$$\int_{-1}^{1+} [P_n(x)]^2 dx = \frac{2}{2n+1}.$$ **Proved**

(*c*) Solve and prove Rodrigue formula.

Prove that:

$nP_n = (2n - 1)\, x\, P_{n-1} - (n - 1)\, P_{n-2}$

Solution. We know that, $(1 - 2xz + z^2)^{-\frac{1}{2}} = \Sigma z^n P_n(x)$

Differentiating w.r.t, 'z', we get

$$-\frac{1}{2}(1 - 2xz + z^2)^{-3/2}(-2x + 2z) = \Sigma n z^{n-1} P_n(x)$$

Mutiplying both sides by $(1 - 2xz + z^2)$, we get

$$(1 - 2xz + z^2)^{-1/2}\,(x - z) = (1 - 2xz + z^2)\,\Sigma nz^{n-1}\,P_n(x)$$

$$(x - z)\,\Sigma z^n P_n(x) = (1 - 2xz + z^2)\Sigma nz^{n-1}\,P_n(x) \qquad \ldots (1)$$

Equating the coefficients of z^n on both sides of (1), we get

$$xP_n(x) - P_{n-1}(x) = (n + 1)P_{n+1}\,(x) - 2x\,(n)P_n(x) + (n - 1)P_{n-1}(x)$$

$$(n + 1)\,P_{n+1}\,(x) = (2n + 1)x\,P_n - nP_{n-1}$$

Replacing n by $n - 1$, we get

$$nP_n\,(x) = (2n - 1)\,P_{n-1}\,(x) - (n - 1)\,P_{n-2}(x)$$ **Proved**

$$= \frac{1}{\pi}\left[-2\pi + (\sin x)_0^{2\pi}\right] = -2$$

$$a_n = \frac{1}{\pi}\int_0^{2\pi} f(x)\cos nx\, dx$$

$$= \frac{1}{\pi}\int_0^{2\pi} f(x)\cos nx\, dx = \frac{1}{\pi}\int_0^{2\pi} x\sin x\cos nx\, dx$$

$$= \frac{1}{2\pi}\int_0^{2\pi} x\sin(n+1)x dx - \frac{1}{2\pi}\int_0^{2\pi} x\sin(n-1)x\, dx$$

$$= \frac{1}{2\pi}\left[\left\{\frac{x(-\cos(n+1)x)}{n+1}\right\}_0^{2\pi} - \left\{\frac{1(-\sin(n+1)x)}{(n+1)^2}\right\}_0^{2\pi}\right] - \frac{1}{2}\pi\left[\left\{\frac{x(-\cos(n-1)x)}{n-1}\right\}\right]_0^{2\pi} - \left[\left\{\frac{1(-\sin(n-1)x}{(n-1)^2}\right\}_0^{2\pi}\right]$$

$$= -\frac{2\pi}{2\pi}\frac{\cos(n+1)2\pi}{n+1} + 0 + \frac{2\pi}{2\pi}\frac{\cos(n-1)2\pi}{n-1} - 0 = -\frac{1}{n+1} + \frac{1}{n-1} = \frac{2}{n^2-1}, n \neq 1$$

$$a_1 = \frac{1}{\pi}\int_0^{2\pi} x\sin x\cos x\, dx = \frac{1}{2\pi}\int_0^{2\pi} x\sin 2x\, dx$$

$$= \frac{1}{2\pi}\left[\left\{\left(\frac{-\cos 2x}{2}\right)\right\}_0^{2\pi} - \left\{\frac{1(-\sin 2x)}{4}\right\}_0^{2\pi}\right] = -\frac{2\pi}{2\pi}\frac{1}{2} - 0 = -\frac{1}{2}$$

$$b_n = \frac{1}{\pi}\int_0^{2\pi} x\sin x\sin nx\, dx = \frac{1}{2\pi}\int_0^{2\pi} x\{\cos(n-1)x - \cos(n+1)x\}\, dx$$

$$= \frac{1}{2\pi}\left[\left\{\frac{x\sin(n-1)x}{n-1} - 1\left\{\frac{-\cos(n-1x)}{(n-1)^2}\right\}_0^{2\pi} - \left\{\frac{x\sin(n+1)x}{n+1}\right\}_0^{2\pi}\right\} + \left\{\frac{1(-\cos(n+1)x)}{(n+1)^2}\right\}_0^{2\pi}\right]$$

$$= \frac{1}{2\pi}\left[2\pi\frac{\sin(n-1)2\pi}{n-1} + \frac{\cos(n-1)2\pi}{(n-1)^2} - \frac{1}{(n-1)^2} - 2\pi\frac{\sin(n-1)2\pi}{n-1}\frac{1}{(n-1)^2} - \frac{\cos(n+1)2\pi}{(n+1)^2} + \frac{1}{(n+1)^2}\right]$$

$$= \frac{1}{2\pi}\left[0 + \frac{1}{(n-1)^2} - \frac{1}{(n-1)^2} + 0 + \frac{1}{(n+1)^2} - \frac{1}{(n+1)^2}\right] = 0 \;\; n \neq 1$$

$$b_1 = \frac{1}{\pi}\int_0^{2\pi} x\sin x\sin x\, dx = \frac{1}{2}\pi\int_0^{2\pi} x(1-\cos 2x)\, dx = \frac{1}{2\pi}\int_0^{2\pi} x dx - \frac{1}{2\pi}\int_0^{2\pi} x\cos 2x\, dx$$

$$= \frac{1}{2\pi}\left[\frac{x^2}{2}\right]_0^{2\pi} - \frac{1}{2\pi}\left[\left(x\frac{\sin 2x}{2}\right)_0^{2\pi} - 1\cdot\left(\frac{-\cos 2x}{4}\right)\right]_0^{2\pi} = \pi - 0 + \frac{1}{4} - \frac{1}{4} = \pi$$

Hence the Fourier expansionof f is

$$\frac{a_0}{2} + a_1\cos x + a_2\cos 2x + a_3\cos 3x + a_4\cos 4x + \ldots$$

$$+ b_1\sin x + b_2\sin 2x + b_3\sin 3x + b_4\sin 4x + \ldots$$

$$x\sin x = -1 - \frac{1}{2}\cos x + 2\left[\frac{\cos 2x}{2^2-1} + \frac{\cos 3x}{3^2-1} + \frac{\cos 4x}{4^2-1} + \ldots\right] + \pi\sin x \quad \ldots (2)$$

Put $x = \frac{\pi}{2}$ in (2), we get

$$\frac{\pi}{2}\sin\frac{\pi}{2} = -1 - \frac{1}{2}\cos\frac{\pi}{2} + 2\left[\frac{\cos\pi}{2^2-1} + \frac{\cos\frac{3\pi}{2}}{3^2-1} + \frac{\cos 2\pi}{4^2-1} + \ldots\pi\sin\frac{\pi}{3}\right.$$

$$\Rightarrow \quad \frac{\pi}{2} = -1 + 2\left[-\frac{1}{z^2-1} + \frac{1}{4^2-1} + \ldots\right] + \pi$$

$$\Rightarrow \quad \frac{\pi}{4} = -\frac{1}{2} - \frac{1}{2^2-1} + \frac{1}{4^2-1} + \ldots + \frac{\pi}{2}$$

$$\Rightarrow \quad \frac{\pi}{4} - \frac{\pi}{2} = -\frac{1}{2} - \frac{1}{1.3} + \frac{1}{3.5} - \ldots$$

$$\Rightarrow \quad -\frac{\pi}{4} = -\frac{1}{2} - \frac{1}{1.3} + \frac{1}{3.5}$$

$$\Rightarrow \quad \frac{\pi}{4} = \frac{1}{2} + \frac{1}{1.3} - \frac{1}{3.5} \ldots$$

Ans.

5. Attempt any two of the following: (10 × 2 = 20)

(*a*) Using the method of separation of variable, solve $\frac{\partial u}{\partial x} = 2\frac{\partial u}{\partial t} + u$. Where $u(x, t) = 6e^{-3x}$, when $t = 0$. *(See Solved Example 3 on Page 422)*.

(*b*) If a string of length '*l*' is initially at rest in equilibrium position and each of its point is given the velocity $\left(\frac{\partial u}{\partial t}\right)_{t=0} = b\sin^3\left(\frac{\pi x}{l}\right)$, find the displacement $u(x, t)$.

(See Solved Example 14 on Page 438)

(c) Solve completely the equation $\frac{\partial^2 u}{\partial t^2} = c^2\frac{\partial^2 u}{\partial x^2}$ representing the vibrations of a string of length '*l*', fixed at both ends, given that $u(0, t) = 0$, $u(1, t) = 0$, $u(x, 0) = f(x)$ and $\left(\frac{\partial u}{\partial t}\right)_{i=0} = 0 ; 0 < x < 1$.

(See Solved Example 9 on Page 431)

INDEX

M

O

P

Q

R

S

T

U

V